Y0-BZW-825

IES LIGHTING HANDBOOK

1984 Reference Volume

"Its object shall be . . . the advancement of the theory and practice of illuminating engineering and the dissemination of knowledge relating thereto."

IES LIGHTING HANDBOOK

1984 **Reference Volume**

"Its object shall be . . . the advancement of the theory and practice of illuminating engineering and the dissemination of knowledge relating thereto."

IES LIGHTING HANDBOOK

1984 **Reference Volume**

JOHN E. KAUFMAN, PE, FIES
Editor

JACK F. CHRISTENSEN
Associate Editor

Published by
ILLUMINATING ENGINEERING SOCIETY OF NORTH AMERICA
345 East 47th Street, New York, N.Y. 10017

ISBN 0-87995-015-3
Library of Congress Catalog Card Number: 84-81835

Composed and Printed by WAVERLY PRESS, INC., Baltimore, Maryland

PRINTED IN THE UNITED STATES OF AMERICA

Preface

Since its first publication in 1949, the *IES Lighting Handbook* has been completely revised five times, the first in 1952, the second in 1959, the third in 1966, the fourth in 1972, and the fifth in 1981. Until 1981 each Handbook was identified by an edition number—the first through fifth editions. In 1981 the Handbook was split into two volumes—a Reference Volume and an Application Volume—each to be updated more frequently than the single volumes of the past. At that time each volume was identified by its publication date (1981) so that each might be revised independently of the other.

This *IES Lighting Handbook—1984 Reference Volume* is a revision of the 1981 Reference Volume. A revision of the Application Volume should follow later. Thus, the Index at the end of this book contains the locations of subject matter in both this book and the *IES Lighting Handbook—1981 Application Volume* using an *R* and an *A* respectively, before page numbers.

All Handbook material has been reviewed for content and accuracy either by the Society's technical committees, under the counsel of the Handbook Editor as Technical Director, or by individuals knowledgeable in the subjects, and for that we wish to acknowledge sincere appreciation to the Society's technical committees and to the individuals listed on the contributors' page.

Major changes in content reflect changes in technology and the needs of the individual users as a result of a user survey. These changes, not necessarily in order of extent or importance in change, include:

1. Over 85 new or revised lighting terms and definitions.
2. The use of candelas per square foot in place of the deprecated term footlambert, as a unit of luminance.
3. New calculation procedures for daylighting, indirect electric lighting and lighting for spaces containing partitions and coves or valances.
4. An expanded use of algorithms to supplement or replace tabular material for computer applications.
5. Recalculated coefficients of utilization and exitance coefficients, based on a new IES algorithm.
6. A criteria rating technique for evaluating designs (where objective evaluation methods exist).
7. Addition of data on newly available light sources.
8. The avoidance of the use of the term "light control" by referring to "optical control of light" to avoid confusion with the now popular term "lighting controls", when referring to switching or dimming devices. Lighting controls are to be covered in the next Application Volume as a part of energy management.
9. Removal of Illuminance Selection (Appendix A), which can be found in the latest Application Volume.

Appreciation is also given to the Handbook Committee (listed below), which developed policy and plans for operation, and to the officers of the Illuminating Engineering Society during the Society's years 1982–1984 under whose administration and stimulation these volumes were prepared.

Handbook Committee

J. H. Jensen, *Chairman*
L. Erhardt, *Advisory*
C. M. Cling
M. H. Lerner
R. P. Meden
J. J. Neidhart, *Advisory*
N. F. Schnitker
D. H. Shapiro

Editing, coordination, artwork and production of the Handbook were carried out through the Society's Technical Office with Jack F. Christensen as Associate Editor, and with LaRee Di Stasio and Peter Ruffett.

With the continued objective of providing essential information on light and lighting in a simple, condensed style and with revisions and additions of new subjects and material it is evident that errors or omissions will occur. Whenever such errors or omissions are found, your comments will be appreciated.

John E. Kaufman
Editor

Contributors

The numbers in parentheses following each name refer to the Sections where contributions were made. Credits for Illustrations and Tables are found on page C-1.

Contributing IES Technical Committees

Color (5)
Daylighting (7)
Design Practice (9)
Light Control & Equipment Design (2) (6)
Light Sources (2) (8)
Nomenclature (1)
Recommendations on Quality and Quantity of Illumination (3)
Testing Procedures (4)

Contributing Individuals

A. Abramowitz (1)
D. L. DiLaura (9)
C. A. Douglas (3)
R. E. Levin (9)
A. L. Lewis (3)
W. Pierpoint (7)
A. R. Robertson (5)
W. E. Thouret (2) (8)

1984 Reference Volume

Contents

1981 Application Volume

Contents

SECTION 1

Dictionary of Lighting Terminology

As the title implies this first section contains terminology directly related to light and lighting practice. All terms are presented in alphabetical order and are followed by their standard symbols and defining equations where applicable, by their definitions and by other related terms of interest. No attempt has been made to provide information on pronunciations or etymologies. Definitions of electrical terms common to lighting and to other fields are available in *American National Standard Dictionary of Electrical and Electronics Terms* (ANSI/IEEE 100-1984).

Any of the radiometric and photometric quantities that follow may be restricted to a narrow wavelength interval $\Delta\lambda$ by the addition of the word spectral and the specification of the wavelength λ. The corresponding symbols are changed by adding a subscript λ, *i.e.*, Q_λ, for a spectral concentration, or a λ in parentheses, *i.e.*, K (λ) for a function of wavelength. Fig. 1-1 is a tabulated summary of standard units, symbols and defining equations for the fundamental photometric and radiometric quantities. Other symbols, abbreviations and conversion factors are given in Figs. 1-7 to 1-15 at the end of this section.

DEFINITIONS

A

absolute luminance threshold: luminance threshold for a bright object like a disk on a totally dark background.

absorptance, $\alpha = \Phi_a/\Phi_i$**:** the ratio of the flux absorbed by a medium to the incident flux. See *absorption.*

NOTE: The sum of the hemispherical reflectance, the hemispherical transmittance, and the absorptance is one.

NOTE: References are listed at the end of each section.

absorption: a general term for the process by which incident flux is converted to another form of energy, usually and ultimately to heat.

NOTE: All of the incident flux is accounted for by the processes of reflection, transmission and absorption.

accent lighting: directional lighting to emphasize a particular object or to draw attention to a part of the field of view. See *directional lighting.*

accommodation: the process by which the eye changes focus from one distance to another.

adaptation: the process by which the visual system becomes accustomed to more or less light or light of a different color than it was exposed to during an immediately proceding period. It results in a change in the sensitivity of the eye to light. See *scotopic vision, photopic vision, chromatic adaptation.*

NOTE: Adaptation is also used to refer to the final state of the process, as reaching a condition of dark adaptation or light adaptation.

adaptive color shift: the change in the perceived object color caused solely by change of the state of chromatic adaptation.

adverse weather lamp: See *fog lamp.*

aerodrome beacon: an aeronautical beacon used to indicate the location of an aerodrome.

NOTE: An aerodrome is any defined area on land or water—including any buildings, installations, and equipment—intended to be used either wholly or in part for the arrival, departure and movement of aircraft.

aeronautical beacon: an aeronautical ground light visible at all azimuths, either continuously or intermittently, to designate a particular location on the surface of the earth. See *aerodrome beacon, airway beacon, hazard or obstruction beacon, landmark beacon.*

aeronautical ground light: any light specially provided as an aid to air navigation, other than a light displayed on an aircraft. See *aeronautical beacon, angle-of-approach lights, approach lights, approach-light beacon, bar (of lights), boundary lights, circling guidance lights, course light, channel lights, obstruction lights, runway alignment indicator, runway-end identification light, perimeter lights, runway lights, taxichannel lights, taxiway lights.*

aeronautical light: any luminous sign or signal specially provided as an aid to air navigation.

Fig. 1-1. Units, Symbols and Defining Equations for Fundamental Photometric and Radiometric Quantities*

Quantity†	Symbol†	Defining Equation	Unit	Symbol
Radiant energy	$Q, (Q_e)$		erg joule‡ calorie kilowatt-hour	erg J cal kWh
Radiant energy density	$w, (w_e)$	$w = dQ/dV$	joule per cubic meter‡ erg per cubic centimeter	J/m^3 erg/cm^3
Radiant flux	$\Phi, (\Phi_e)$	$\Phi = dQ/dt$	erg per second watt‡	erg/s W
Radiant flux density at a surface Radiant exitance (Radiant emittance§) Irradiance	 $M, (M_e)$ $E, (E_e)$	 $M = d\Phi/dA$ $E = d\Phi/dA$	watt per square centimeter, watt per square meter,‡ etc.	W/cm^2 W/m^2
Radiant intensity	$I, (I_e)$	$I = d\Phi/d\omega$ (ω = solid angle through which flux from point source is radiated)	watt per steradian‡	W/sr
Radiance	$L, (L_e)$	$L = d^2\Phi/(d\omega dA \cos\theta)$ $= dI/(dA \cos\theta)$ (θ = angle between line of sight and normal to surface considered)	watt per steradian and square centimeter watt per steradian and square meter‡	W/sr·cm^2 W/sr·m^2
Emissivity, spectral-total hemispherical	ϵ	$\epsilon = M/M_{blackbody}$ (M and $M_{blackbody}$ are respectively the radiant exitance of the measured specimen and that of a blackbody at the same temperature as the specimen)	one (numeric)	
Emissivity, spectral-total directional	$\epsilon(\theta, \phi, T)$	$\epsilon(\theta, \phi, T) = L(T)/L_{blackbody}(T)$[$L(T)$ and $L_{blackbody}(T)$ are, respectively, the radiance of the measured specimen and that of a blackbody at the same temperature (that of the specimen)].	one (numeric)	
Emissivity, spectral directional	$\epsilon(\theta, \phi, \lambda, T)$	$\epsilon(\lambda, \theta, \phi, T) = L_\lambda(\lambda, \theta, \phi, T)/L_{\lambda,blackbody}(\lambda, T)$ (L_λ and $L_{\lambda blackbody}$ are respectively the spectral radiance of the measured specimen and that of a blackbody at the same temperature of the specimen)	one (numeric)	
Emissivity, spectral hemispherical	$\epsilon(\lambda, T)$	$\epsilon(\lambda, T) = M_\lambda(\lambda, T)/M_{\lambda,blackbody}(\lambda, T)$ are respectively the spectral radiant exitance of the measured specimen and that of a blackbody at the same temperature of the specimen	one (numeric)	

after image: a visual response that occurs after the stimulus causing it has ceased.

aircraft aeronautical light: any aeronautical light specially provided on an aircraft. See *navigation light system, anti-collision light, ice detection light, fuselage lights, landing light, position lights, taxi light.*

airway beacon: an aeronautical beacon used to indicate a point on the airway.

altitude (in daylighting): the angular distance of a heavenly body measured on the great circle that passes perpendicular to the plane of the horizon through the body and through the zenith. It is measured positively from the horizon to the zenith, from 0 to 90 degrees.

ambient lighting: lighting throughout an area that produces general illumination.

anchor light (aircraft): a light designed for use on a seaplane or amphibian to indicate its position when at anchor or moored.

angle-of-approach lights: aeronautical ground lights arranged so as to indicate a desired angle of descent during an approach to an aerodrome runway. (Also called optical glide path lights.)

angle of collimation: the angle subtended by a light source at a point on an irradiated surface.

angstrom, Å: unit of wavelength equal to 10^{-10} (one ten-billionth) meter.

anti-collision light: a flashing aircraft aeronautical light or system of lights designed to provide a red signal throughout 360 degrees of azimuth for the purpose of giving long-range indication of an aircraft's location to pilots of other aircraft.

aperture color: perceived color of the sky or a patch of color seen through an aperture where it cannot be identified as belonging to a specific object.

apostilb (asb): a lambertian unit of luminance equal to $1/\pi$ candela per square meter. The use of this unit is deprecated.

apparent luminous intensity of an extended

Fig. 1-1. *Continued*

Quantity†	Symbol†	Defining Equation	Unit	Symbol
Absorptance	α	$\alpha = \Phi_a/\Phi_i$‖	one (numeric)	
Reflectance	ρ	$\rho = \Phi_r/\Phi_i$‖	one (numeric)	
Transmittance	τ	$\tau = \Phi_t/\Phi_i$‖	one (numeric)	
Luminous efficacy	K	$K = \Phi_v/\Phi_e$	lumen per watt‡	lm/W
Luminous efficiency	V	$V = K/K_{maximum}$ ($K_{maximum}$ = maximum value of $K(\lambda)$ function)	one (numeric)	
Luminous energy (quantity of light)	Q, (Q_v)	$Q_v = \int_{380}^{770} K(\lambda)Q_{e\lambda}d\lambda$	lumen-hour lumen-second‡ (talbot)	lm·h lm·s
Luminous energy density	w, (w_v)	$w = dQ/dV$	lumen-hour per cubic centimeter	$lm \cdot h/cm^3$
Luminous flux	Φ, (Φ_v)	$\Phi = dQ/dt$	lumen‡	lm
Luminous flux density at a surface				
Luminous exitance (Luminous emittance§)	M, (M_v)	$M = d\Phi/dA$	lumen per square foot	lm/ft^2
Illuminance (Illumination§)	E, (E_v)	$E = d\Phi/dA$	footcandle (lumen per square foot) lux (lm/m^2)‡ phot(lm/cm^2)	fc lx ph
Luminous intensity (candlepower)	I, (I_v)	$I = d\Phi/d\omega$ (ω = solid angle through which flux from point source is radiated)	candela‡ (lumen per steradian)	cd
Luminance	L, (L_v)	$L = d^2\Phi/(d\omega dA \cos\theta)$ $= dI/(dA \cos\theta)$ (θ = angle between line of sight and normal to surface considered)	candela per unit area stilb (cd/cm^2) nit (cd/m^2‡) footlambert ($cd/\pi ft^2$)§ lambert ($cd/\pi cm^2$)§ apostilb ($cd/\pi m^2$)§	cd/in^2, etc. sb nt, cd/m^2 fL§ L§ asb§

* The symbols for photometric quantities are the same as those for the corresponding radiometric quantities. When it is necessary to differentiate them the subscripts v and e respectively should be used, *e.g.*, Q_v and Q_e.

† Quantities may be restricted to a narrow wavelength band by adding the word spectral and indicating the wavelength. The corresponding symbols are changed by adding a subscript λ, *e.g.*, Q_λ, for a spectral concentration or a λ in parentheses, *e.g.*, $K(\lambda)$, for a function of wavelength.

‡ International System (SI) unit

§ Use is deprecated

‖ Φ_i = incident flux, Φ_α = absorbed flux, Φ_r = reflected flux, Φ_t = transmitted flux

source at a specific distance: See *equivalent luminous intensity of an extended source.*

approach-light beacon: an aeronautical ground light placed on the extended centerline of the runway at a fixed distance from the runway threshold to provide an early indication of position during an approach to a runway.

NOTE: The runway threshold is the beginning of the runway usable for landing.

approach lights: a configuration of aeronautical ground lights located in extension of a runway or channel before the threshold to provide visual approach and landing guidance to pilots. See *angle-of-approach lights, approach-light beacon, VASIS.*

arc discharge: an electric discharge characterized by high cathode current densities and a low voltage drop at the cathode.

NOTE: The cathode voltage drop is small compared with that in a glow discharge, and secondary emission plays only a small part in electron emission from the cathode.

arc lamp: a discharge lamp in which the light is emitted by an arc discharge or by its electrodes.

NOTE: The electrodes may be either of carbon (operating in air) or of metal.

artificial pupil: a device or arrangement for confining the light passing through the pupil of the eye to an area smaller than the natural pupil.

atmospheric transmissivity: the ratio of the directly transmitted flux incident on a surface after passing through unit thickness of the atmosphere to the flux that would be incident on the same surface if the flux had passed through a vacuum.

average luminance (of a surface): the average luminance of a surface may be expressed in terms of the total luminous flux (lumens) leaving the surface per unit solid angle and unit area.

average luminance (of a luminaire): the luminous intensity at a given angle divided by the projected area of the luminaire at that angle.

azimuth: the angular distance between the vertical plane containing a given line or celestial body and the plane of the meridian.

B

back light: illumination from behind (and usually above) a subject to produce a highlight along its edge and consequent separation between the subject and its background. See *side back light.*

backing lighting: the illumination provided for scenery in off-stage areas visibile to the audience.

backup lamp: a lighting device mounted on the rear of a vehicle for illuminating the region near the rear of the vehicle while moving or about to move in reverse. It normally can be used only while backing up.

bactericidal (germicidal) effectiveness: the capacity of various portions of the ultraviolet spectrum to destroy bacteria, fungi and viruses.

bactericidal (germicidal) efficiency of radiant flux (for a particular wavelength): the ratio of the bactericidal effectiveness at a particular wavelength to that at wavelength 265.0 nanometers, which is rated as unity.

NOTE: Tentative bactericidal efficiency of various wavelengths of radiant flux is given in Fig. 19-12 in 1981 Application Volume.

bactericidal (germicidal) exposure: the product of bactericidal flux density on a surface and time. It usually is measured in bactericidal microwatt-minutes per square centimeter or bactericidal watt-minutes per square foot.

bactericidal (germicidal) flux: radiant flux evaluated according to its capacity to produce bactericidal effects. It usually is measured in microwatts of ultraviolet radiation weighted in accordance with its bactericidal efficiency. Such quantities of bactericidal flux would be in bactericidal microwatts.

NOTE: Ultraviolet radiation of wavelength 253.7 nanometers usually is referred to as "ultraviolet microwatts" or "UV watts." These terms should not be confused with "bactericidal microwatts" because the radiation has not been weighted in accordance with the values given in Fig. 19-12 in 1981 Application Volume.

bactericidal (germicidal) flux density: the bactericidal flux per unit area of the surface being irradiated. It is equal to the quotient of the incident bactericidal flux divided by the area of the surface when the flux is uniformly distributed. It usually is measured in microwatts per square centimeter or watts per square foot of bactericidally weighted ultraviolet radiation (bactericidal microwatts per square centimeter or bactericidal watts per square foot).

bactericidal lamp: an ultraviolet lamp that radiates a significant portion of its radiative power in the UV-C band (100 to 280 nanometers).

baffle: a single opaque or translucent element to shield a source from direct view at certain angles, or to absorb unwanted light.

balcony lights: luminaires mounted on the front edge of the auditorium balcony.

ballast: a device used with an electric-discharge lamp to obtain the necessary circuit conditions (voltage, current and wave form) for starting and operating. See *reference ballast.*

ballast factor: the fractional flux of a lamp(s) operated on a ballast compared to the flux when operated on the reference ballasting specified for rating lamp lumens.

bar (of lights): a group of three or more aeronautical ground lights placed in a line transverse to the axis, or extended axis, of the runway. See *barette.*

bare (exposed) lamp: a light source with no shielding.

barn doors: a set of adjustable flaps, usually two or four (two-way or four-way) that may be attached to the front of a luminaire (usually a Fresnel spotlight) in order to partially control the shape and spread of the light beam.

barrette (in aviation): a short bar in which the lights are closely spaced so that from a distance they appear to be a linear light.

NOTE: Barettes are usually less than 4.6 meters (15 feet) in length.

base light: uniform, diffuse illumination approaching a shadowless condition, which is sufficient for a television picture of technical acceptability, and which may be supplemented by other lighting.

beacon: a light (or mark) used to indicate a geographic location. See *aerodrome beacon, aeronautical beacon, airway beacon, approach-light beacon, hazard or obstruction beacon, identification beacon, landmark beacon.*

beam angle: the included angle between those points on opposite sides of the beam axis at which the luminous intensity from a theatrical luminaire is 50 per cent of maximum. This angle may be determined from a candlepower curve, or may be approximated by use of an incident light meter.

beam axis of a projector: a line midway between two lines that intersect the candlepower distribution curve at points equal to a stated per cent of its maximum (usually 50 per cent).

beam projector: a luminaire with the light source at or near the focus of a paraboloidal reflector producing near parallel rays of light in a beam of small divergence. Some are equipped with spill rings to reduce spill and glare. In most types, the lamp may be moved toward or away from the reflector to vary the beam spread.

beam spread (in any plane): the angle between the two directions in the plane in which the intensity is equal to a stated percentage of the maximum beam intensity. The percentage typically is 10 per cent for floodlights and 50 per cent for photographic lights.

biconical reflectance, $\rho(\omega_i; \omega_r)$: the ratio of reflected flux collected through a conical solid angle to the incident flux limited to a conical solid angle.

NOTE: The directions and extent of each cone must be specified; the solid angle is not restricted to a right-circular cone.

biconical transmittance, $\tau(\omega_i; \omega_t)$: ratio of transmitted flux collected through a conical solid angle to the incident flux limited to a conical solid angle.
NOTE: The directions and extent of each cone must be specified; the solid angle is not restricted to a right-circular cone.

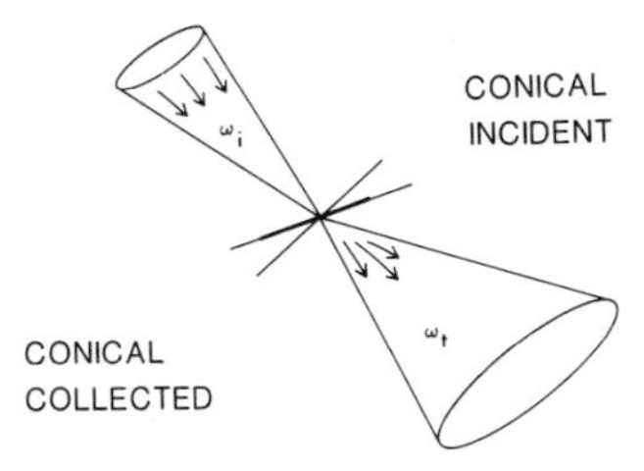

bidirectional reflectance, $\rho(\theta_i, \phi_i; \theta_r, \phi_r)$: ratio of reflected flux collected over an element of solid angle surrounding the given direction to essentially collimated incident flux.
NOTE: The directions of incidence and collections and the size of the solid angle "element" of collection must be specified. In each case of conical incidence or collection, the solid angle is not restricted to a right circular cone, but may be of any cross section, including rectangular, a ring, or a combination of two or more solid angles.

bidirectional reflectance-distribution function (BRDF): the ratio of the differential luminance of a ray $dL_r(\theta_r, \phi_r)$ reflected in a given direction (θ_r, ϕ_r) to the differential luminous flux density $dE_i(\theta_i, \phi_i)$ incident from a given direction of incidence, (θ_i, ϕ_i), which produces it.

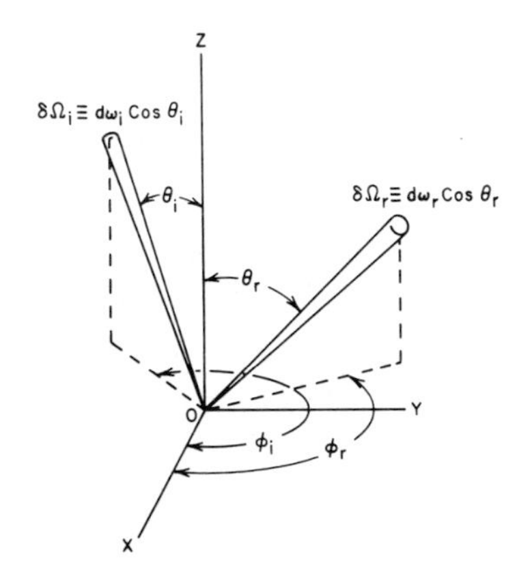

$$f_r(\theta_i, \phi_i; \theta_r, \phi_r) \equiv dL_r(\theta_r, \phi_r)/dE_i(\theta_i, \phi_i)(sr)^{-1}$$

$$= dL_r(\theta_r, \phi_r)/L_i(\theta_i, \phi_i)d\Omega_i$$

where $d\Omega \equiv d\omega cos\theta$
NOTE: This distribution function is the basic parameter for describing (geometrically) the reflecting properties of an opaque surface element (negligible internal scattering). It may have any positive value and will approach infinity in the specular direction for ideally specular reflectors. The spectral and polarization aspects must be defined for complete specification, since the BRDF as given above only defines the geometric aspects.

bidirectional transmittance, $\tau\ (\theta_i, \phi_i; \theta_t, \phi_t)$: ratio of incident flux collected over an element of solid angle surrounding the given direction to essentially collimated incident flux.
NOTE: The direction of incidence, collection and size of the solid angle "element" must be specified.

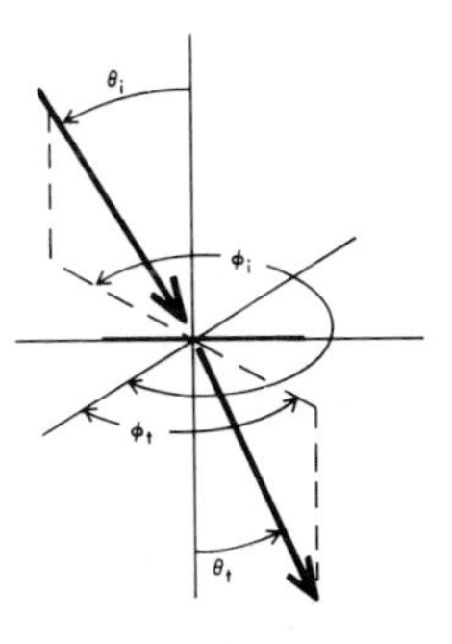

bihemispherical reflectance, $\rho\ (2\pi; 2\pi)$: ratio of reflected flux collected over the entire hemisphere to the incident flux from the entire hemisphere.

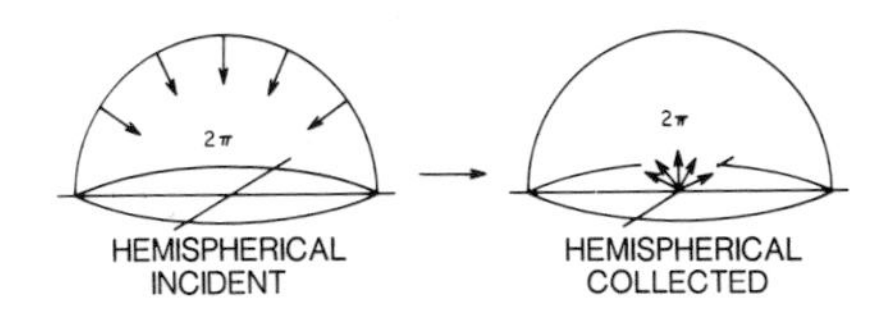

bihemispherical transmittance, $\tau\ (2\ \pi; 2\ \pi)$: ratio of transmitted flux collected over the entire hemisphere to the incident flux from the entire hemisphere.

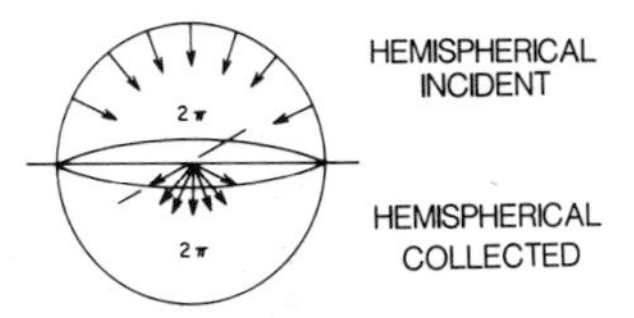

binocular portion of the visual field: that portion of space where the fields of the two eyes overlap.

blackbody: a temperature radiator of uniform temperature whose radiant exitance in all parts of the spectrum is the maximum obtainable from any temperature radiator at the same temperature.

Such a radiator is called a blackbody because it will absorb all the radiant energy that falls upon it. All other temperature radiators may be classed as nonblackbodies. They radiate less in some or all wavelength intervals than a blackbody of the same size and the same temperature.

NOTE: the blackbody is practically realized in the form of a cavity with opaque walls at a uniform temperature and with a small opening for observation purposes. It also is called a full radiator, standard radiator, complete radiator or ideal radiator.

blackbody (Planckian) locus: the locus of points on a chromaticity diagram representing the chromaticities of blackbodies having various (color) temperatures.

"black light:" the popular term for ultraviolet energy near the visible spectrum.

NOTE: For engineering purposes the wavelength range 320–400 nanometers has been found useful for rating lamps and their effectiveness upon fluorescent materials (excluding phosphors used in fluorescent lamps). By confining "black light" applications to this region, germicidal and erythemal effects are, for practical purposes, eliminated.

"black light" flux: radiant flux within the wavelength range 320 to 400 nanometers. It is usually measured in milliwatts. See *fluoren.*

NOTE: Because of the variability of the spectral sensitivity of materials irradiated by "black light" in practice, no attempt is made to evaluate "black light" flux according to its capacity to produce effects.

"black light" flux density: "black light" flux per unit area of the surface being irradiated. It is equal to the incident "black light" flux divided by the area of the surface when the flux is uniformly distributed. It usually is measured in milliwatts per square foot of "black light" flux.

"black light" lamp: An ultraviolet lamp that emits a significant portion of its radiative power in the UV-A band (315–400 nanometers).

blending lighting: general illumination used to provide smooth transitions between the specific lighting areas on a stage.

blinding glare: glare which is so intense that for an appreciable length of time after it has been removed, no object can be seen.

borderlight: a long continuous striplight hanging horizontally above the stage and aimed down to provide general diffuse illumination and/or to light the cyclorama or a drop, usually wired in three or four color circuits.

borderline between comfort and discomfort (BCD): the average luminance of a source in a field of view which produces a sensation between comfort and discomfort.

boundary lights: aeronautical ground lights delimiting the boundary of a land aerodrome without runways. See *range lights.*

bowl: an open top diffusing glass or plastic enclosure used to shield a light source from direct view and to redirect or scatter the light.

bracket (mast arm): an attachment to a lamp post or pole from which a luminaire is suspended.

brightness: See *subjective brightness, luminance, veiling luminance, brightness of a perceived light-source color.*

brightness contrast threshold: when two patches of color are separated by a brightness contrast border as in the case of a bipartite photometric field or in the case of a disk shaped object surrounded by its background, the border between the two patches is a brightness contrast border. The contrast which is just detectable is known as the brightness contrast threshold.

brightness of perceived light-source color: the attribute in accordance with which the source seems to emit more or less luminous flux per unit area.

bulb: a source of electrically powered light. This term is used to distinguish between an assembled unit consisting of a light source in a housing called a *lamp* and the internal source. See *lamp.*

C

candela, cd: the SI unit of luminous intensity. One candela is one lumen per steradian. Formerly, candle. See Fig. 1–2.

NOTE: The fundamental luminous intensity definition in the SI is the candela in terms of monochromatic radiation at 540×10^{12} hertz (approximately 555 nm). One candela is defined as the luminous intensity, in a given direction, of a source that emits monochromatic radiation of frequency 540×10^{12} hertz and of which the radiant intensity in that direction is 1/683 watt per steradian.

From 1909 until 1948, the unit of luminous intensity in the United States, as well as in France and Great Britain, was the "international candle" which was maintained by a group of carbon-filament vacuum lamps. From 1948 to 1979 the unit of luminous intensity was defined in terms of the luminance of a blackbody at the freezing point of platinum. Since 1948, the internationally accepted name for the unit of luminous intensity is *candela.* The difference between the candela and the old international candle is so small that only measurements of high accuracy are affected.

candlepower, $I = d\Phi/d\omega$; cp.: luminous intensity expressed in candelas.

candlepower (intensity) distribution curve: a curve, generally polar, representing the variation of luminous intensity of a lamp or luminaire in a plane through the light center.

NOTE: A vertical candlepower distribution curve is obtained by taking measurements at various angles of elevation in a vertical plane through the light center; unless the plane is specified, the vertical curve is assumed to represent an average such as would be obtained by rotating the lamp or luminaire about its vertical axis. A horizontal candlepower distribution curve represents measurements made at various angles of azimuth in a horizontal plane through the light center.

carbon-arc lamp: an electric-discharge lamp employing an arc discharge between carbon electrodes.

One or more of these electrodes may have cores of special chemicals that contribute importantly to the radiation.

cavity ratio, CR: a number indicating cavity proportions calculated from length, width and height. See *ceiling cavity ratio, floor cavity ratio and room cavity ratio.*

ceiling area lighting: a general lighting system in which the entire ceiling is, in effect, one large luminaire.

NOTE: Ceiling area lighting includes *luminous ceilings* and *louvered ceilings.*

ceiling cavity: the cavity formed by the ceiling, the plane of the luminaires, and the wall surfaces between these two planes.

ceiling cavity ratio, CCR: a number indicating ceiling cavity proportions calculated from length, width and height. See *Section 9.*

ceiling projector: a device designed to produce a well-defined illuminated spot on the lower portion of a cloud for the purpose of providing a reference mark for the determination of the height of that part of the cloud.

ceiling ratio: the ratio of the luminous flux reaching the ceiling directly to the upward component from the luminaire.

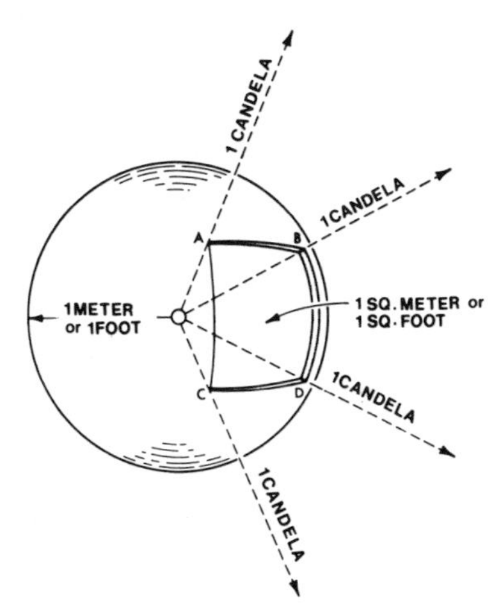

Fig. 1-2. Relationship between candelas, lumens, lux and footcandles.

A uniform point source (luminous intensity or candlepower = one candela) is shown at the center of a sphere of one meter or one foot radius. It is assumed that the sphere surface has zero reflectance.

The illuminance at any point on the sphere is one lux (one lumen per square meter) when the radius is one meter, or one footcandle (one lumen per square foot) when the radius is one foot.

The solid angle subtended by the area, *A, B, C, D* is one steradian. The flux density is therefore one lumen per steradian, which corresponds to a luminous intensity of one candela, as originally assumed.

The sphere has a total area of 12.57 (4π) square meters or square feet, and there is a luminous flux of one lumen falling on each square meter or square foot. Thus the source provides a total of 12.57 lumens.

central (foveal) vision: the seeing of objects in the central or foveal part of the visual field, approximately two degrees in diameter. It permits seeing much finer detail than does peripheral vision.

central visual field: that region of the visual field corresponding to the foveal portion of the retina.

channel: an enclosure containing the ballast, starter, lamp holders and wiring for a fluorescent lamp, or a similar enclosure on which filament lamps (usually tubular) are mounted.

channel lights: aeronautical ground lights arranged along the sides of a channel of a water aerodrome. See *taxi-channel lights.*

characteristic curve: a curve which expresses the relationship between two variable properties of a light source, such as candlepower and voltage, flux and voltage, etc.

chromatic adaptation: the process by which the chromatic properties of the visual system are modified by the observation of stimuli of various chromaticities and luminances. See *state of chromatic adaptation.*

chromatic contrast thresholds (color contrast thresholds): two patches of color juxtaposed and separated only by a color contrast border cannot be perceived as different in chromaticness or separated by a contrast border if the difference in chromaticity on the two sides of the border is reduced below the threshold of visibility. A contrast border can involve both differences in luminance and chromaticity on the two sides of the border.

chromaticity coordinates (of a color), *x, y, z*: the ratios of each of the tristimulus values of the light to the sum of the three tristimulus values.

chromaticity diagram: a plane diagram formed by plotting one of the three chromaticity coordinates against another.

chromaticity difference thresholds (color difference thresholds): two patches of color of the same luminance may have a threshold difference in chromaticity which makes them perceptibly different in chromaticness. The difference may be a difference in hue or saturation or a combination of the two. The threshold can also involve a combination of differences in both luminance and chromaticity.

chromaticity of a color: consists of the dominant or complementary wavelength and purity aspects of the color taken together, or of the aspects specified by the chromaticity coordinates of the color taken together.

chromaticness: the attribute of a visual sensation according to which the (perceived) color of an area appears to be more or less chromatic.

CIE standard chromaticity diagram: a diagram in which the *x* and *y* chromaticity coordinates are plotted in rectangular coordinates.

circling guidance lights: aeronautical ground lights provided to supply additional guidance during a circling approach when the circling guidance furnished by the approach and runway lights is inadequate.

clear sky: a sky that has less than 30 per cent cloud cover.

clearance lamp: a lighting device mounted on a vehicle for the purpose of indicating the overall width and height of the vehicle.

clerestory: that part of a building rising clear of the roofs or other parts and whose walls contain windows for lighting the interior.

cloudy sky: a sky having more than 70 per cent cloud cover.

coefficient of attenuation, μ: the decrement in flux per unit distance in a given direction within a medium and defined by the relation: $\Phi_x = \Phi_o e^{-\mu x}$ where Φ_x is the flux at any distance x from a reference point having flux Φ_o.

coefficient of beam utilization, CBU: the ratio of the luminous flux (lumens) reaching a specified area directly from a floodlight or projector to the total beam luminous flux (lumens).

coefficient of utilization, CU: the ratio of the luminous flux (lumens) from a luminaire calculated as received on the work-plane to the luminous flux emitted by the luminaire's lamps alone.

coffer: a recessed panel or dome in the ceiling.

cold-cathode lamp: an electric-discharge lamp whose mode of operation is that of a glow discharge, and having electrodes so spaced that most of the light comes from the positive column between them.

color: the characteristic of light by which a human observer may distinguish between two structure-free patches of light of the same size and shape. See *light-source color* and *object color.*

color comparison, or color grading (CIE, object color inspection): the judgment of equality, or of the amount and character of difference, of the color of two objects viewed under identical illumination.

color contrast thresholds: See *chromaticity difference thresholds.*

color correction (of a photograph or printed picture): the adjustment of a color reproduction process to improve the perceived-color conformity of the reproduction to the original.

color discrimination: the perception of differences between two or more colors.

color matching: action of making a color appear the same as a given color.

color-matching functions (spectral tristimulus values), $\bar{x}(\lambda) = X_\lambda/\Phi_{e\lambda}; \bar{y}(\lambda) = Y_\lambda/\Phi_{e\lambda}; \bar{z}(\lambda) = Z_\lambda/\Phi_{e\lambda}$: the tristimulus values per unit wavelength interval and unit spectral radiant flux.

NOTE: Color-matching functions have been adopted by the International Commission on Illumination. They are tabulated as functions of wavelength throughout the spectrum and are the basis for the evaluation of radiant energy as light and color. The standard values adopted by the CIE in 1931 are given in Fig. 5-1. The $\bar{y}$ values are identical with the values of spectral luminous efficiency for photopic vision.

The $\bar{x}$, $\bar{y}$, and $\bar{z}$ values for the 1931 Standard Observer are based on a two-degree bipartite field, and are recommended for predicting matches for stimuli subtending between one and four degrees. Supplementary data based on a 10-degree bipartite field were adopted in 1964 for use for angular subtenses greater than four degrees.

Tristimulus computational data for CIE standard sources *A* and *C* are given in Fig. 5-3.

color preference index, CPI: See *flattery index.*

color rendering: general expression for the effect of a light source on the color appearance of objects in conscious or subconscious comparison with their color appearance under a reference light source.

color rendering improvement (of a light source): the adjustment of spectral composition to improve color rendering.

color rendering index (of a light source) (CRI): measure of the degree of color shift objects undergo when illuminated by the light source as compared with the color of those same objects when illuminated by a reference source of comparable color temperature.

color temperature of a light source: the absolute temperature of a blackbody radiator having a chromaticity equal to that of the light source. See also *correlated color temperature* and *distribution temperature.*

colorfulness: See *chromaticness.*

colorimetric purity (of a light), p_c: the ratio L_1/L_2 where L_1 is the luminance of the single frequency component that must be mixed with a reference standard to match the color of the light and L_2 is the luminance of the light. See *excitation purity.*

colorimetric shift: the change of chromaticity and luminance factor of an object color due to change of the light source. See *adaptive color shift* and *resultant color shift.*

colorimetry: the measurement of color.

compact-arc lamp: See *short-arc lamp.*

compact source iodide (CSI): an ac arc source utilizing a mercury vapor arc with metal halide additives to produce illumination in the 4000- to 6000-K range. Requires a ballast and ignition system for operation.

comparison lamp: a light source having a constant but not necessarily known, luminous intensity with which standard and test lamps are compared successively.

complementary wavelength (of a light), λ_c: the wavelength of radiant energy of a single frequency that, when combined in suitable proportion with the light, matches the color of a reference standard. See *dominant wavelength.*

complete diffusion: diffusion in which the diffusing medium redirects the flux incident by scattering so that none is in an image-forming state.

cone: a retinal receptor which dominates the retinal response when the luminance level is high and provides the basis for the perception of color.

configuration factor, C_{2-1}: the ratio of illuminance on a surface at point 2 (due to flux directly received from lambertian surface 1) to the exitance of surface 1. It is used in *flux transfer theory.* $C_{2-1} = (E_2)/(M_1)$

conical-directional reflectance, $\rho\ (\omega_i;\ \theta_r,\ \phi_r)$: ratio of reflected flux collected over an element of solid angle surrounding the given direction to the incident flux limited to a conical solid angle.

NOTE: The direction and extent of the cone must be specified and the direction of collection and size of the solid angle "element" must be specified.

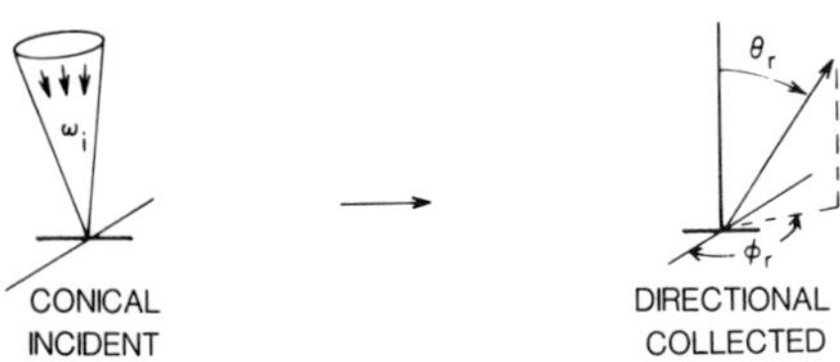

conical-directional transmittance, τ $(\omega_i;\theta_t,\phi_t)$: ratio of transmitted flux collected over an element of solid angle surrounding the direction to the incident flux limited to a conical solid angle.

NOTE: The direction and extent of the cone must be specified and the direction of collection and size of the solid angle "element" must be specified.

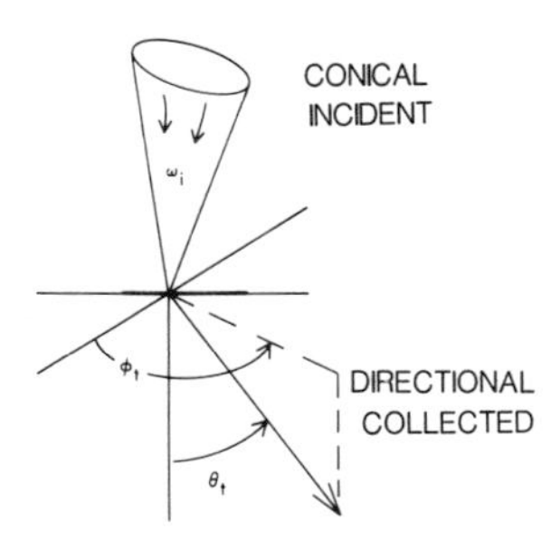

conical-hemispherical reflectance, ρ $(\omega_i;\ 2\pi)$: ratio of reflected flux collected over the entire hemisphere to the incident flux limited to a conical solid angle.

NOTE: The direction and extent of the cone must be specified.

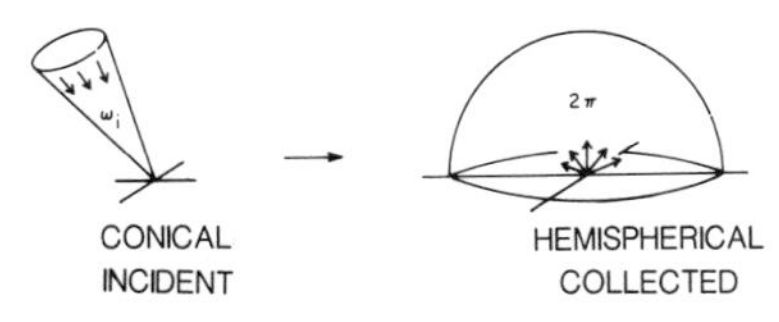

conical-hemispherical transmittance, τ $(\omega_t;\ 2\ \pi)$: ratio of transmitted flux collected over the entire hemisphere to the incident flux limited to a conical solid angle.

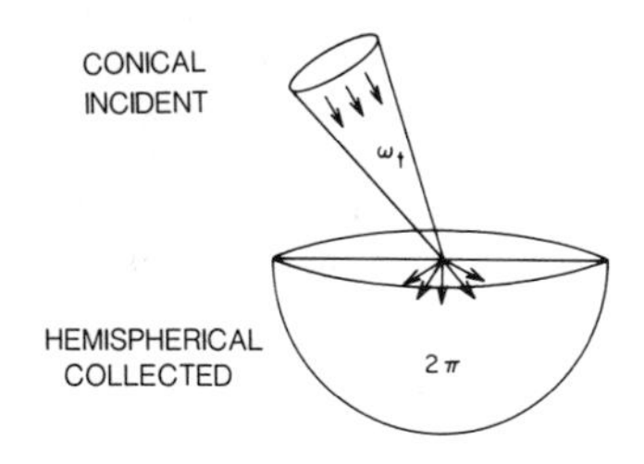

NOTE: The direction and extent of the cone must be specified.

conspicuity: the capacity of a signal to stand out in relation to its background so as to be readily discovered by the eye.

contrast: See *luminance contrast.*

contrast rendition factor, CRF: the ratio of visual task contrast with a given lighting environment to the contrast with sphere illumination. Also known as *contrast rendering factor.*

contrast sensitivity: the ability to detect the presence of luminance differences. Quantitatively, it is equal to the reciprocal of the contrast threshold.

cornice lighting: lighting comprising sources shielded by a panel parallel to the wall and attached to the ceiling, and distributing light over the wall.

correlated color temperature (of a light source): the absolute temperature of a blackbody whose chromaticity most nearly resembles that of the light source.

cosine law: the law that the illuminance on any surface varies as the cosine of the angle of incidence. The angle of incidence θ is the angle between the normal to the surface and the direction of the incident light. The inverse-square law and the cosine law can be combined as $E = (I \cos \theta)/d^2$. See *cosine-cubed law* and *inverse-square law.*

cosine-cubed law: an extension of the cosine law in which the distance d between the source and surface is replaced by $h/\cos \theta$, where h is the perpendicular distance of the source from the plane in which the point is located. It is expressed by $E = (I \cos^3 \theta)/h^2$.

counter-key light: illumination on a subject from a direction that is opposite to that of the key light.

country beam: See *upper (driving) beams.*

course light: an aeronautical ground light, supplementing an airway beacon, for indicating the direction of the airway and to identify by a coded signal the location of the airway beacon with which it is associated.

cove lighting: lighting comprising sources shielded by a ledge or horizontal recess, and distributing light over the ceiling and upper wall.

critical fusion frequency, cff: See *flicker fusion frequency.*

critical flicker frequency, cff: See *flicker fusion frequency.*

cross light: equal illuminance in front of the subject from two directions at substantially equal and opposite angles with the optical axis of the camera and a horizontal plane.

cucoloris: an opaque cutout panel mounted between a light source (sun, arc, etc.) and a target surface in order to project a shadow pattern (clouds or leaves are typical) upon scenery, cyclorama or acting area.

cut-off angle (of a luminaire): the angle, measured up from nadir, between the vertical axis and the first line of sight at which the bare source is not visible.

D

dark adaptation: the process by which the retina becomes adapted to a luminance less than about 0.034 candela per square meter.

daylight availability: the amount of light from the sun and the sky at a specific location, time, day and sky condition.

daylight factor: a measure of daylight illuminance at a point on a given plane expressed as a ratio of the illuminance on the given plane at that point to the simultaneous exterior illuminance on a horizontal plane from the whole of an unobstructed sky of assumed or known luminance distribution. Direct sunlight is excluded from both interior and exterior values of illuminance.

daylight lamp: a lamp producing a spectral distribution approximating that of a specified daylight.

densitometer: a photometer for measuring the optical density (common logarithm of the reciprocal of the transmittance) of materials.

diffuse reflectance: the ratio of the flux leaving a surface or medium by diffuse reflection to the incident flux.

diffuse reflection: the process by which incident flux is re-directed over a range of angles.

diffuse transmission: the process by which the incident flux passing through a surface or medium is scattered.

diffuse transmittance: the ratio of the diffusely transmitted flux leaving a surface or medium to the incident flux.

diffused lighting: lighting provided on the work-plane or on an object, that is not predominantly incident from any particular direction.

diffuser: a device to redirect or scatter the light from a source, primarily by the process of diffuse transmission.

diffusing panel: a translucent material covering the lamps in a luminaire to reduce the luminance by distributing the flux over an extended area.

diffusing surfaces and media: those that redistribute some of the incident flux by scattering in all directions. See *complete diffusion, incomplete diffusion, perfect diffusion, narrow-angle diffusion, wide-angle diffusion.*

digital display: See *numerical display.*

dimmer: a device used to control the intensity of light emitted by a luminaire by controlling the voltage or current available to it.

direct component: that portion of the light from a luminaire which arrives at the work-plane without being reflected by room surfaces. See *indirect component.*

direct glare: glare resulting from high luminances or insufficiently shielded light sources in the field of view. It usually is associated with bright areas, such as luminaires, ceilings and windows which are outside the visual task or region being viewed.

direct-indirect lighting: a variant of general diffuse lighting in which the luminaires emit little or no light at angles near the horizontal.

direct lighting: lighting by luminaires distributing 90 to 100 per cent of the emitted light in the general direction of the surface to be illuminated. The term usually refers to light emitted in a downward direction.

direct ratio: the ratio of the luminous flux reaching the work-plane directly to the downward component from the luminaire.

directional-conical reflectance, ρ (θ_i, ϕ_i; ω_r): ratio of reflected flux collected through a conical solid angle to essentially collimated incident flux.

NOTE: The direction of incidence must be specified, and the direction and extent of the cone must be specified.

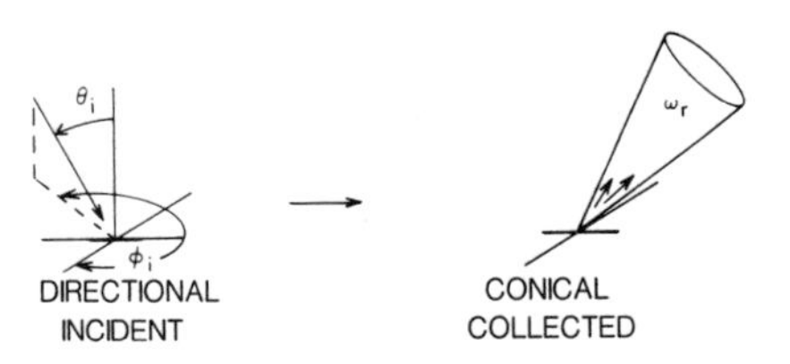

directional-conical transmittance, τ (θ_i, ϕ_i; ω_t): ratio of transmitted flux collected through a conical solid angle to essentially collimated incident flux.

NOTE: The direction of incidence must be specified, and the direction and extent of the cone must be specified.

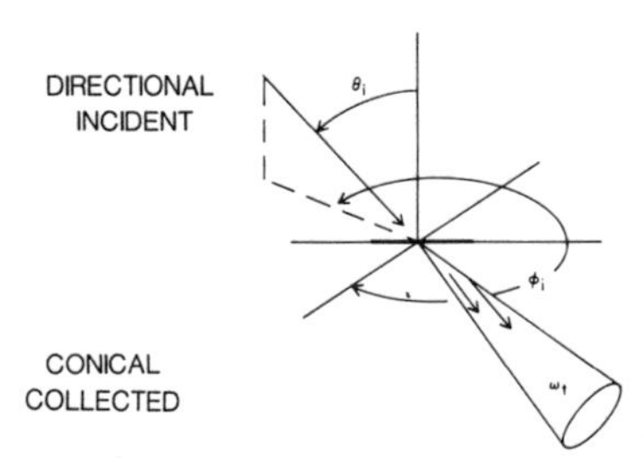

directional-hemispherical reflectance, ρ (θ_i, ϕ_i; 2 π): ratio of reflected flux collected over the entire hemisphere to essentially collimated incident flux.

NOTE: The direction of incidence must be specified.

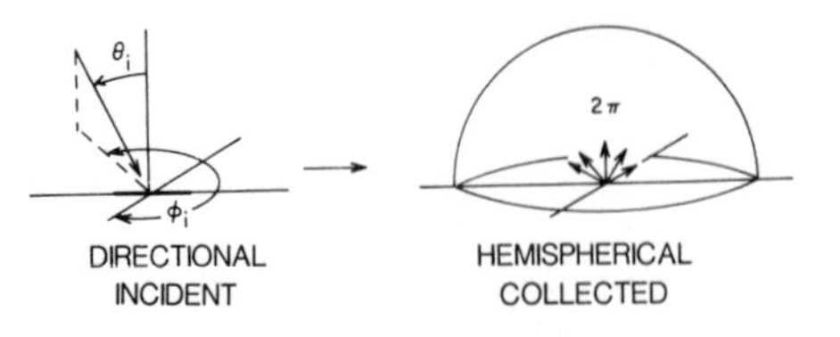

directional-hemispherical transmittance, τ (θ_i, ϕ_i; 2 π): ratio of transmitted flux collected over the entire hemisphere to essentially collimated incident flux.

NOTE: The direction of incidence must be specified.

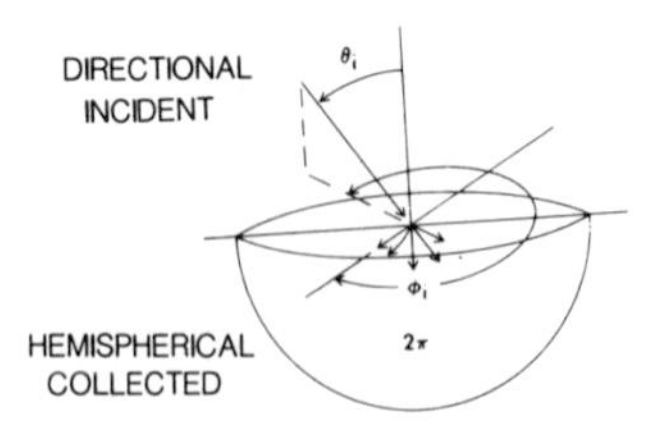

directional lighting: lighting provided on the work-plane or on an object predominantly from a preferred direction. See *accent lighting, key light, cross light.*

disability glare: glare resulting in reduced visual performance and visibility. It often is accompanied by discomfort. See *veiling luminance.*

disability glare factor (DGF): a measure of the visibility of a task in a given lighting installation in comparison with its visibility under reference lighting conditions, expressed in terms of the ratio of luminance contrasts having an equivalent effect upon task visibility. The value of DGF takes account of the equivalent veiling luminance produced in the eye by the pattern of luminances in the task surround.

discomfort glare: glare producing discomfort. It does not necessarily interfere with visual performance or visibility.

discomfort glare factor: the numerical assessment of the capacity of a single source of brightness, such as a luminaire, in a given visual environment for producing discomfort. (This term is obsolete and is retained for reference and literature searches.) See *glare* and *discomfort glare.*

discomfort glare rating (DGR): a numerical assessment of the capacity of a number of sources of luminance, such as luminaires, in a given visual environment for producing discomfort. It is the net effect of the individual values of index of sensation for all luminous areas in the field of view. See *discomfort glare factor.*

distal stimuli: in the physical space in front of the eye one can identify points, lines and surfaces and three dimensional arrays of scattering particles which constitute the distal physical stimuli which form optical images on the retina. Each element of a surface or volume to which an eye is exposed subtends a solid angle at the entrance pupil. Such elements of solid angle make up the field of view and each has a specifiable luminance and chromaticity. Points and lines are specific cases which have to be dealt with in terms of total candlepower and candlepower per unit length.

distribution temperature (of a light source): the absolute temperature of a blackbody whose relative spectral distribution is the same (or nearly so) in the visible region of the spectrum as that of the light source.

dominant wavelength (of a light), λ_d: the wavelength of radiant energy of a single frequency that, when combined in suitable proportion with the radiant energy of a reference standard, matches the color of the light. See *complementary wavelength.*

downlight: a small direct lighting unit which directs the light downward and can be recessed, surface mounted or suspended.

downward component: that portion of the luminous flux from a luminaire emitted at angles below the horizontal. See *upward component.*

driving beam: See *upper beam.*

dual headlighting system: headlighting by means of two double units, one mounted on each side of the front end of a vehicle. Each unit consists of two lamps mounted in a single housing. The upper or outer lamps may have two filaments supplying the lower beam and part of the upper beam, respectively. The lower or inner lamps have one filament providing the primary source of light for the upper beam.

dust-proof luminaire: a luminaire so constructed or protected that dust will not interfere with its successful operation.

dust-tight luminaire: a luminaire so constructed that dust will not enter the enclosing case.

E

effective ceiling cavity reflectance, ρ_{CC}: a number giving the combined reflectance effect of the walls and ceiling of the ceiling cavity. See *ceiling cavity ratio.*

effective floor cavity reflectance, ρ_{FC}: a number giving the combined reflectance effect of the walls and floor of the floor cavity. See *floor cavity ratio.*

efficacy: See *luminous efficacy of a source of light* and *spectral luminous efficacy of radiant flux.*

efficiency: See *luminaire efficiency, luminous efficacy of a source of light* and *spectral luminous efficiency of radiant flux.*

electric discharge: See *arc discharge, gaseous discharge and glow discharge.*

electric-discharge lamp: a lamp in which light (or radiant energy near the visible spectrum) is produced by the passage of an electric current through a vapor or a gas. See *fluorescent lamp, cold-cathode lamp, hot-cathode lamp, carbon-arc lamp, glow lamp, fluorescent lamp, high intensity discharge lamp.*

NOTE: Electric-discharge lamps may be named after the filling gas or vapor that is responsible for the major portion of the radiation; *e.g.* mercury lamps, sodium lamps, neon lamps, argon lamps, etc.

A second method of designating electric-discharge lamps is by psysical dimensions or operating parameters; *e.g.* short-arc lamps, high-pressure lamps, low-pressure lamps, etc.

A third method of designating electric-discharge lamps is by their application; in addition to lamps for illumination there are photochemical lamps, bactericidal lamps, blacklight lamps, sun lamps, etc.

electroluminescence: the emission of light from a phosphor excited by an electromagnetic field.

electromagnetic spectrum: a continuum of electric and magnetic radiation encompassing all wavelengths. See *regions of electromagnetic spectrum.*

elevation: the angle between the axis of a searchlight drum and the horizontal. For angles above the horizontal, elevation is positive, and below the horizontal negative.

ellipsoidal reflector spotlight: a spotlight in which a lamp and an ellipsoidal reflector are mounted in a fixed relationship directing a beam of light into an aperture where it may be shaped by a pattern, iris, shutter system or other insertion. The beam then passes through a single or compound lens system that focuses it as required, producing a sharply defined beam with variable edge definition.

emergency lighting: lighting designed to supply illumination essential to safety of life and property in the event of failure of the normal supply.

emissivity, ϵ: the ratio of the radiance (for directional emissivity) or radiant exitance (for hemispherical emissivity) of an element of surface of a temperature radiator to that of a blackbody at the same temperature.

emittance, ϵ: (1) The ratio of the radiance in a given direction (for directional emittance) or radiant exitance (for hemispherical emittance) of a sample of a thermal radiator to that of a blackbody radiator at the same temperature. (2) See *exitance.* Use of the term with this meaning is deprecated.

enclosed and gasketed: See *vapor-tight.*

equal interval (isophase) light: a rhythmic light in which the light and dark periods are equal.

equipment operating factor: the fractional flux of a high intensity discharge (HID) lamp-ballast-luminaire combination in a given operating position compared to the flux of the lamp-luminaire combination (a) operated in the position for rating lamp lumens and (b) using the reference ballasting specified for rating lamp lumens.

equivalent contrast, $\tilde{C}$: a numerical description of the relative visibility of a task. It is the contrast of the standard visibility reference task giving the same visibility as that of a task whose contrast has been reduced to threshold when the background luminances are the same. See *visual task evaluator.*

equivalent contrast, $\tilde{C}_e$: the actual equivalent contrast in a real luminous environment with non-diffuse illumination. The actual equivalent contrast ($\tilde{C}_e$) may be less than the equivalent contrast due to veiling reflection. $\tilde{C}_e = \tilde{C} \times \mathrm{CRF}$. See *contrast rendition factor.*

equivalent luminous intensity (of an extended source at a specified distance): the intensity of a point source which would produce the same illuminance at that distance. Formerly, *apparent luminous intensity of an extended source.*

equivalent sphere illumination (ESI): the level of sphere illumination which would produce task visibility equivalent to that produced by a specific lighting environment.

erythema: the temporary reddening of the skin produced by exposure to ultraviolet energy.

NOTE: The degree of erythema is used as a guide to dosages applied in ultraviolet therapy.

erythemal effectiveness: the capacity of various portions of the ultraviolet spectrum to produce erythema.

erythemal efficiency of radiant flux (for a particular wavelength): the ratio of the erythemal effectiveness of a particular wavelength to that of wavelength 296.7 nanometers, which is rated as unity.

NOTE: This term formerly was called "relative erythemal factor."

The erythemal efficiency of radiant flux of various wavelengths for producing a minimum perceptible erythema (MPE) is given in Fig. 19-12 in the 1981 Application Volume. These values have been accepted for evaluating the erythemal effectiveness of sun lamps.

erythemal exposure: the product of erythemal flux density on a surface and time. It usually is measured in erythemal microwatt-minutes per square centimeter.

NOTE: For average untanned skin a minimum perceptible erythema requires about 300 microwatt-minutes per square centimeter of radiation at 296.7 nanometers.

erythemal flux: radiant flux evaluated according to its capacity to produce erythema of the untanned human skin. It usually is measured in microwatts of ultraviolet radiation weighted in accordance with its erythemal efficiency. Such quantities of erythemal flux would be in erythemal microwatts. See *erythemal efficiency of radiant flux.*

NOTE: A commonly used practical unit of erythemal flux is the erythemal unit (EU) or E-viton (erytheme) which is equal to the amount of radiant flux that will produce the same erythemal effect as 10 microwatts of radiant flux at wavelength 296.7 nanometers. See also *erythemal unit or E-viton.*

erythemal flux density: the erythemal flux per unit area of the surface being irradiated. It is equal to the quotient of the incident erythemal flux divided by the area of the surface when the flux is uniformly distributed. It usually is measured in microwatts per square centimeter of erythemally weighted ultraviolet radiation (erythemal microwatts per square centimeter). See *Finsen.*

erythemal threshold: See *minimal perceptible erythema.*

erythemal unit, EU: a unit of erythemal flux that is equal to the amount of radiant flux that will produce the same erythemal effect as 10 microwatts of radiant flux at wavelength 296.7 nanometers. Also called E-viton.

E-viton (erytheme): See *erythemal unit.*

exitance: See *luminous exitance* and *radiant exitance.*

exitance coefficient: a coefficient similar to the coefficient of utilization used to determine wall and ceiling exitances. These coefficients are numerically equal to the previously used luminance coefficients.

excitation purity (of a light), p_e: the ratio of the distance on the CIE (x,y) chromaticity diagram between the reference point and the light point to the distance in the same direction between the reference point and the spectrum locus or the purple boundary. See *colorimetric purity.*

explosion-proof luminaire: a luminaire which is completely enclosed and capable of withstanding an explosion of a specific gas or vapor that may occur within it, and preventing the ignition of a specific gas or vapor surrounding the enclosure by sparks, flashes or explosion of the gas or vapor within. It must operate at such an external temperature that a surrounding flammable atmosphere will not be ignited thereby.

eye light: illumination on a person to provide a specular reflection from the eyes (and teeth) without adding a significant increase in light on the subject.

F

far (long wavelength) infrared: region of the electromagnetic spectrum extending from 5,000 to 1,000,000 nm.

far ultraviolet: region of the electromagnetic spectrum extending from 100 to 200 nm.

fay light: a luminaire that uses incandescent parabolic reflector lamps with a dichroic coating to provide "daylight" illumination.

fenestra method: a procedure for predicting the interior illuminance received from daylight through windows.

fenestration: any opening or arrangement of openings (normally filled with media for control) for the admission of daylight.

field angle: the included angle between those points on opposite sides of the beam axis at which the luminous intensity from a theatrical luminaire is 10 per cent of the maximum value. This angle may be determined from a candlepower curve, or may be approximated by use of an incident light meter.

fill light: supplementary illumination to reduce shadow or contrast range.

film (or aperture) color: perceived color of the sky or a patch of color seen through an aperture.

filter: a device for changing, usually by transmission, the magnitude and/or the spectral composition of the flux incident on it. Filters are called *selective* (or *colored*) or *neutral*, according to whether or not they alter the spectral distribution of the incident flux.

filter factor: the transmittance of "black light" by a filter.

NOTE: The relationship among these terms is illustrated by the following formula for determining the luminance of fluorescent materials exposed to "black light":

candelas per square meter

$$= \frac{I}{\pi^*} \frac{\text{fluorens}}{\text{square meter}} \times \text{glow factor} \times \text{filter factor}$$

When integral-filter "black light" lamps are used, the filter factor is dropped from the formula because it already has been applied in assigning fluoren ratings to these lamps.

Finsen: a suggested practical unit of erythemal flux density equal to one E-viton per square centimeter.

fixed light: a light having a constant luminous intensity when observed from a fixed point.

fixture: See *luminaire.*

flashing light: a rhythmic light in which the periods of light are of equal duration and are clearly shorter than the periods of darkness. See *group flashing light, interrupted quick-flashing light, quick-flashing light.*

flashtube: a tube of glass or fused quartz with electrodes at the ends and filled with a gas, usually xenon. It is designed to produce high intensity flashes of light of extremely short duration.

flattery index (of a light source) (R_f): measure appraising a light source for appreciative viewing of colored objects, or for promoting an optimistic viewpoint by flattery (to make more pleasant), or for enhancing the perception of objects in terms of color. Also sometimes called *color preference index (CPI).*

flicker fusion frequency, fff: the frequency of intermittent stimulation of the eye at which flicker disappears. It also is called critical fusion frequency (cff) or critical flicker frequency (cff).

flicker photometer: See *visual photometer.*

floodlight: a projector designed for lighting a scene or object to a luminance considerably greater than its surroundings. It usually is capable of being pointed in any direction and is of weatherproof construction. See *heavy duty floodlight, general purpose floodlight, ground-area open floodlight, ground-area floodlight with reflector insert.*

floodlighting: a system designed for lighting a scene or object to a luminance greater than its surroundings. It may be for utility, advertising or decorative purposes.

floor cavity: the cavity formed by the work-plane, the floor, and the wall surfaces between these two planes.

floor cavity ratio, FCR: a number indicating floor cavity proportions calculated from length, width and height. See *Section 9.*

floor lamp: a portable luminaire on a high stand suitable for standing on the floor. See *torchere.*

fluoren: a unit of "black light" flux equal to one milliwatt of radiant flux in the wavelength range 320 to 400 nanometers.

fluorescence: the emission of light (luminescence) as the result of, and only during, the absorption of radiation of other (mostly shorter) wavelengths.

fluorescent lamp: a low-pressure mercury electric-discharge lamp in which a fluorescing coating (phosphor) transforms some of the ultraviolet energy generated by the discharge into light. See *instant start fluorescent lamp, preheat (switch start) fluorescent lamp, rapid start fluorescent lamp.*

flush mounted or recessed: a luminaire which is mounted above the ceiling (or behind a wall or other surface) with the opening of the luminaire level with the surface.

flux transfer theory: a method of calculating the illuminance in a room by taking into account the interreflection of the light flux from the room sur-

* π is omitted when luminance is in footlamberts and the area is in square feet.

faces based on the average flux transfer between surfaces.

fog (adverse-weather) lamps: units which may be used in lieu of headlamps or in connection with the lower beam headlights to provide road illumination under conditions of rain, snow, dust or fog.

follow spot: any instrument operated so as to follow the movement of an actor. Follow spots are usually high intensity controlled beam luminaires.

footcandle, fc: the unit of illuminance when the foot is taken as the unit of length. It is the illuminance on a surface one square foot in area on which there is a uniformly distributed flux of one lumen, or the illuminance produced on a surface all points of which are at a distance of one foot from a directionally uniform point source of one candela. See Fig. 1-2.

footcandle meter: See *illuminance meter.*

footlambert, fL: a unit of luminance equal to $1/\pi$ candela per square foot, or to the uniform luminance of a perfectly diffusing surface emitting or reflecting light at the rate of one lumen per square foot, or to the average luminance of any surface emitting or reflecting light at that rate. See *units of luminance.* The use of this unit is deprecated.

NOTE: The average luminance of any reflecting surface in footlamberts is, therefore, the product of the illumination in footcandles by the luminous reflectance of the surface.

footlights: a set of striplights at the front edge of the stage platform used to soften face shadows cast by overhead luminaires and to add general toning lighting from below.

form factor, f_{1-2}: the ratio of the flux directly received by surface 2 (and due to lambertian surface 1) to the total flux emitted by surface 1. It is used in *flux transfer theory.*

$$f_{1-2} = (\Phi_{1-2})/(\Phi_1)$$

formation light: a navigation light especially provided to facilitate formation flying.

fovea: a small region at the center of the retina, subtending about two degrees, containing only cones but no rods and forming the site of most distinct vision.

foveal vision: See *central vision.*

Fresnel spotlight: a luminaire containing a lamp and a Fresnel lens (stepped "flat" lens with a textured back) which has variable field and beam angles obtained by changing the spacing between lamp and lens (flooding and spotting). The Fresnel produces a smooth, soft edge, defined beam of light.

fuselage lights: aircraft aeronautical lights, mounted on the top and bottom of the fuselage, used to supplement the navigation light.

G

gas-filled lamp: an incandescent lamp in which the filament operates in a bulb filled with one or more inert gases.

gaseous discharge: the emission of light from gas atoms excited by an electric current.

general color rendering index (R_a): measure of the average shift of 8 standardized colors chosen to be of intermediate saturation and spread throughout the range of hues. If the Color Rendering Index is not qualified as to the color samples used, R_a is assumed.

general diffuse lighting: lighting involving luminaires which distribute 40 to 60 per cent of the emitted light downward and the balance upward, sometimes with a strong component at 90 degrees (horizontal). See *direct-indirect lighting.*

general lighting: lighting designed to provide a substantially uniform level of illumination throughout an area, exclusive of any provision for special local requirements. See *direct lighting, semi-direct lighting, general diffuse lighting, direct-indirect lighting, semi-indirect lighting, indirect lighting, ceiling area lighting, localized general lighting.*

general purpose floodlight (GP): a weatherproof unit so constructed that the housing forms the reflecting surface. The assembly is enclosed by a cover glass.

germicidal effectiveness: See *bactericidal effectiveness.*

germicidal efficiency of radiant flux: See *bactericidal efficiency of radiant flux.*

germicidal exposure: See *bactericidal exposure.*

germicidal flux and flux density: See *bactericidal flux and flux density.*

germicidal lamp: a low pressure mercury lamp in which the envelope has high transmittance for 254-nanometer radiation. See *bactericidal lamp.*

glare: the sensation produced by luminance within the visual field that is sufficiently greater than the luminance to which the eyes are adapted to cause annoyance, discomfort, or loss in visual performance and visibility. See *blinding glare, direct glare, disability glare, discomfort glare.*

NOTE: The magnitude of the sensation of glare depends upon such factors as the size, position and luminance of a source, the number of sources and the luminance to which the eyes are adapted.

globe: a transparent or diffusing enclosure intended to protect a lamp, to diffuse and redirect its light, or to change the color of the light.

glossmeter: an instrument for measuring gloss as a function of the directionally selective reflecting properties of a material in angles near to and including the direction giving specular reflection.

glow discharge: an electric discharge characterized by a low, approximately constant, current density at the cathode, low cathode temperature, and a high, approximately constant, voltage drop.

glow factor: a measure of the visible light response of a fluorescent material to "black light." It is equal to π* times the luminance in candelas per square meter produced on the material divided by the in-

*π is omitted when the luminance is in footlamberts and flux density is in milliwatts per square foot.

cident "black light" flux density in milliwatts per square meter. It may be measured in lumens per milliwatt.

glow lamp: an electric-discharge lamp whose mode of operation is that of a glow discharge, and in which light is generated in the space close to the electrodes.

goniophotometer: a photometer for measuring the directional light distribution characteristics of sources, luminaires, media and surfaces.

graybody: a temperature radiator whose spectral emissivity is less than unity and the same at all wavelengths.

ground-area open floodlight (O): a unit providing a weatherproof enclosure for the lamp socket and housing. No cover glass is required.

ground-area open floodlight with reflector insert (OI): a weatherproof unit so constructed that the housing forms only part of the reflecting surface. An auxiliary reflector is used to modify the distribution of light. No cover glass is required.

ground light: visible radiation from the sun and sky reflected by surfaces below the plane of the horizon.

group flashing light: a flashing light in which the flashes are combined in groups, each including the same number of flashes, and in which the groups are repeated at regular intervals. The duration of each flash is clearly less than the duration of the dark periods between flashes, and the duration of the dark periods between flashes is clearly less than the duration of the dark periods between groups.

H

hard light: light that causes an object to cast a sharply defined shadow.

hazard or obstruction beacon: an aeronautical beacon used to designate a danger to air navigation.

hazardous location: an area where ignitable vapors or dust may cause a fire or explosion created by energy emitted from lighting or other electrical equipment or by electrostatic generation.

headlamp: a major lighting device mounted on a vehicle and used to provide illumination ahead of it. Also called headlight. See *multiple-beam headlamp* and *sealed beam headlamp.*

headlight: an alternate term for *headlamp.*

heavy duty floodlight (HD): a weatherproof unit having a substantially constructed metal housing into which is placed a separate and removable reflector. A weatherproof hinged door with cover glass encloses the assembly but provides an unobstructed light opening at least equal to the effective diameter of the reflector.

hemispherical-conical reflectance, ρ $(2\pi;\ \omega_r)$: ratio of reflected flux collected over a conical solid angle to the incident flux from the entire hemisphere.

NOTE: The direction and extent of the cone must be specified.

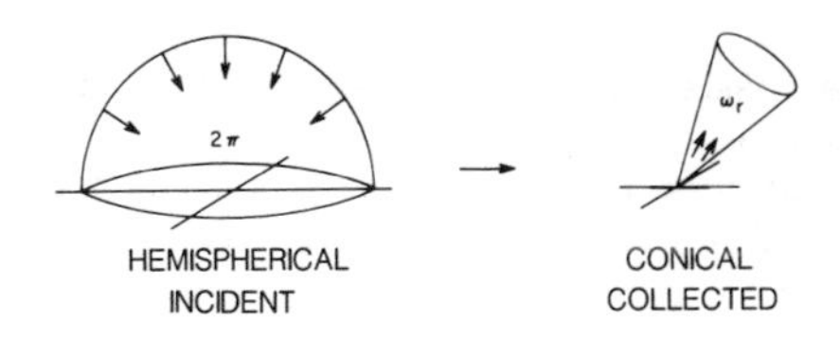

hemispherical-conical transmittance, τ $(2\pi;\ \omega_t)$: ratio of transmitted flux collected over a conical solid angle to the incident flux from the entire hemisphere.

NOTE: The direction and extent of the cone must be specified.

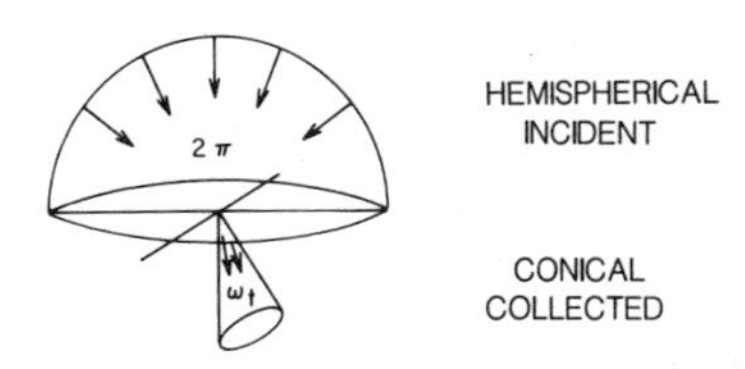

hemispherical-directional reflectance, ρ $(2\pi;\ \theta_r,\ \phi_r)$: ratio of reflected flux collected over an element of solid angle surrounding the given direction to the incident flux from the entire hemisphere.

NOTE: The direction of collection and the size of the solid angle "element" must be specified.

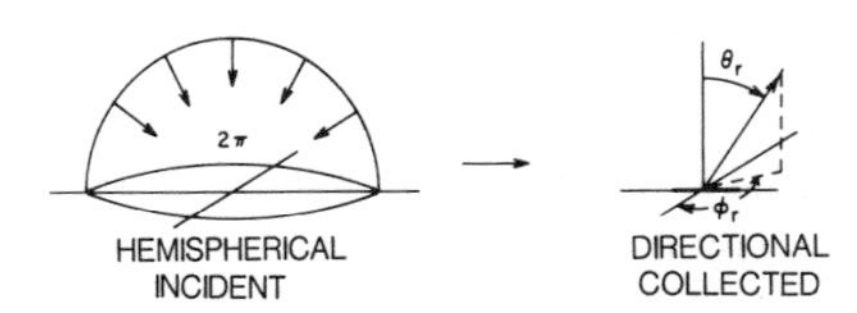

hemispherical-directional transmittance, $\tau(2\pi;\ \theta_t,\ \phi_t)$: ratio of transmitted flux collected over an element of solid angle surrounding the given direction to the incident flux from the entire hemisphere.

NOTE: The direction of collection and size of the solid angle "element" must be specified.

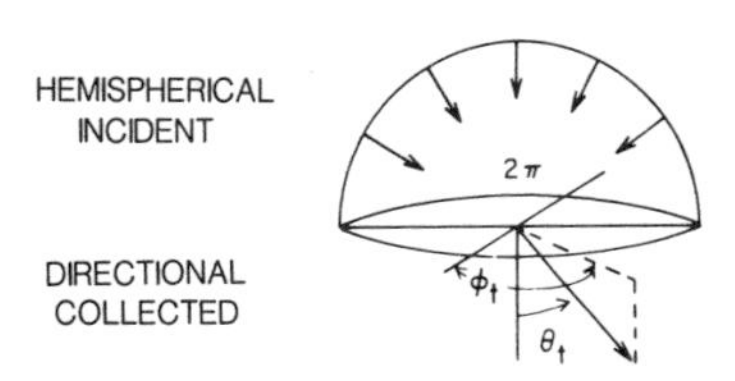

hemispherical reflectance: the ratio of all of the flux leaving a surface or medium by reflection to the incident flux. See *hemispherical transmittance.*

NOTE: If reflectance is not preceded by an adjective descriptive of the angles of view, hemispherical reflectance is implied.

hemispherical transmittance: the ratio of the transmitted flux leaving a surface or medium to the incident flux.

NOTE: If transmittance is not preceded by an adjective, descriptive of the angles of view, hemispherical reflectance is implied.

high intensity discharge (HID) lamp: an electric discharge lamp in which the light producing arc is stabilized by wall temperature, and the arc tube has a bulb wall loading in excess of three watts per square centimeter. HID lamps include groups of lamps known as mercury, metal halide, and high pressure sodium.

high-key lighting: a type of lighting that, applied to a scene, results in a picture having gradations falling primarily between gray and white; dark grays or blacks are present, but in very limited areas. See *low-key lighting.*

high mast lighting: illumination of a large area by means of a group of luminaires which are designed to be mounted in fixed orientation at the top of a high mast, generally 20 meters (65 feet) or higher.

high pressure sodium (HPS) lamp: high intensity discharge (HID) lamp in which light is produced by radiation from sodium vapor operating at a partial pressure of about 1.33×10^4 Pa (100 torr). Includes clear and diffuse-coated lamps.

horizontal plane of a searchlight: the plane which is perpendicular to the vertical plane through the axis of the searchlight drum and in which the train lies.

hot-cathode lamp: an electric-discharge lamp whose mode of operation is that of an arc discharge. The cathodes may be heated by the discharge or by external means.

house lights: the general lighting system installed in the audience area (house) of a theatre, film or television studio or arena.

hue of a perceived color: the attribute that determines whether the color is red, yellow, green, blue, or the like.

hue of a perceived light-source color: the attribute that determines whether the color is red, yellow, green, blue, or the like. See *hue of a perceived color.*

hydrargyrum, medium-arc-length, iodide (HMI): an ac arc light source utilizing mercury vapor and metallic iodide additives for an approximation of daylight (5600 K) illumination. Requires a ballast and ignition system for operation.

I

ice detection light: an inspection light designed to illuminate the leading edge of an aircraft wing to check for ice formation.

ideal radiator: See *blackbody*

identification beacon: an aeronautical beacon emitting a coded signal by means of which a particular point of reference can be identified.

ignitor: a device, either by itself or in association with other components, that generates voltage pulses to start discharge lamps without preheating of electrodes.

illuminance, $E = d\Phi/dA$: the density of the luminous flux incident on a surface; it is the quotient of the luminous flux by the area of the surface when the latter is uniformly illuminated. See Fig. 1-7 for units and conversion factors.

illuminance (lux or footcandle) meter: an instrument for measuring illuminance on a plane. Instruments which accurately respond to more than one spectral distribution are color corrected, *i.e.*, the spectral response is balanced to $V(\lambda)$ or $V'(\lambda)$. Instruments which accurately respond to more than one spatial distribution of incident flux are cosine corrected, *i.e.*, the response to a source of unit luminous intensity, illuminating the detector from a fixed distance and from different directions decreases as the cosine of the angle between the incident direction and the normal to the detector surface. The instrument is comprised of some form of photodetector, with or without a filter, driving a digital or analog readout through appropriate circuitry.

illumination: the act of illuminating or state of being illuminated. This term has been used for density of luminous flux on a surface (illuminance) and such use is to be deprecated.

incandescence: the self-emission of radiant energy in the visible spectrum due to the thermal excitation of atoms or molecules.

incandescent filament lamp: a lamp in which light is produced by a filament heated to incandescence by an electric current.

NOTE: Normally, the filament is of coiled or coiled-coil (doubly coiled) tungsten wire. However, it may be uncoiled wire, a flat strip, or of material other than tungsten.

incomplete diffusion (partial diffusion): that in which the diffusing medium partially re-directs the incident flux by scattering while the remaining fraction of incident flux is redirected without scattering, *i.e.*, a fraction of the incident flux can remain in an image-forming state.

index of sensation (M) (of a source): a number which expresses the effects of source luminance, *solid angle factor, position index*, and the field luminance on *discomfort glare rating.*

indirect component: the portion of the luminous flux from a luminaire arriving at the work-plane after being reflected by room surfaces. See *direct component.*

indirect lighting: lighting by luminaires distributing 90 to 100 per cent of the emitted light upward.

infrared lamp: a lamp that radiates predominately in the infrared; the visible radiation is not of principal interest.

infrared radiation: for practical purposes any radiant energy within the wavelength range of 770 to 10^6 nanometers. This radiation is arbitrarily divided as follows:

Near (short wavelength) infrared 770–1400 nm
Intermediate infrared 1400–5000 nm
Far (long wavelength) infrared 5000–1,000,000 nm

NOTE: In general, unlike ultraviolet energy, infrared energy is not evaluated on a wavelength basis but rather in terms of all of such energy incident upon a surface.

Examples of these applications are industrial heating, drying, baking and photoreproduction. However, some applications, such as infrared viewing devices, involve detectors sensitive to a restricted range of wavelengths; in such cases the spectral characteristics of the source and receiver are of importance.

initial luminous exitance: the density of luminous flux leaving a surface within an enclosure before interreflections occur.

NOTE: For light sources this is the luminous exitance as defined in *luminous flux density at a surface.* For non-self-luminous surfaces it is the reflected luminous exitance of the flux received directly from sources within the enclosure or from daylight.

instant start fluorescent lamp: a fluorescent lamp designed for starting by a high voltage without preheating of the electrodes.

NOTE: Also known as a cold-start lamp in some countries.

integrating photometer: a photometer that enables total luminous flux to be determined by a single measurement. The usual type is the Ulbricht sphere with associated photometric equipment for measuring the indirect luminance of the inner surface of the sphere. (The measuring device is shielded from the source under measurement.)

intensity: a shortening of the terms *luminous intensity* and *radiant intensity.* Often misused for level of illumination or illuminance.

interflectance: an alternate term for *room utilization factor.*

interflectance method: a lighting design procedure for predetermining the luminances of walls, ceiling and floor and the average illuminance on the work-plane based on integral equations. It takes into account both direct and reflected flux.

interflected component: the portion of the luminous flux from a luminaire arriving at the work-plane after being reflected one or more times from room surfaces, as determined by the *flux transfer theory.*

interflection: the multiple reflection of light by the various room surfaces before it reaches the work-plane or other specified surface of a room.

inter-reflectance: the portion of the luminous flux (lumens) reaching the work-plane that has been reflected one or more times as determined by the flux transfer theory. See Section 9.

interrupted quick-flashing light: a quick flashing light in which the rapid alternations are interrupted by periods of darkness at regular intervals.

inverse-square law: the law stating that the illuminance E at a point on a surface varies directly with the intensity I of a point source, and inversely as the square of the distance d between the source and the point. If the surface at the point is normal to the direction of the incident light, the law is expressed by $E = I/d^2$.

NOTE: For sources of finite size having uniform luminance, this gives results that are accurate within one percent when d is at least five times the maximum dimension of the source as viewed from the point on the surface. Even though practical interior luminaires do not have uniform luminance, this distance, d, is frequently used as the minimum for photometry of such luminaires, when the magnitude of the measurement error is not critical.

iris: an assembly of flat metal leaves arranged to provide an easily adjustable near-circular opening, placed near the focal point of the beam (as in an ellipsoidal reflector spotlight), or in front of the lens to act as a mechanical dimmer as in older types of carbon arc follow spotlights.

irradiance, *E*: the density of radiant flux incident on a surface.

isocandela line: a line plotted on any appropriate set of coordinates to show directions in space, about a source of light, in which the intensity is the same. A series of such curves, usually for equal increments of intensity, is called an isocandela diagram.

isolux (isofootcandle) line: a line plotted on any appropriate set of coordinates to show all the points on a surface where the illuminance is the same. A series of such lines for various illuminance values is called an isolux (isofootcandle) diagram.

K

key light: the apparent principal source of directional illumination falling upon a subject or area.

kicker: a luminaire used to provide an additional highlight or accent on a subject.

klieg light: a high intensity carbon arc spotlight, typically used in motion picture lighting.

L

laboratory reference standards: the highest ranking order of standards at each laboratory.

lambert, L: a lambertian unit of luminance equal to $1/\pi$ candela per square centimeter. The use of this unit is deprecated.

lambertian surface: a surface that emits or reflects light in accordance with Lambert's cosine law. A lambertian surface has the same luminance regardless of viewing angle.

Lambert's cosine law, $I_\theta = I_0 \cos\theta$: the law stating that the luminous intensity in any direction from an element of a perfectly diffusing surface varies as the cosine of the angle between that direction and the perpendicular to the surface element.

lamp: a generic term for a man-made source of light. By extension, the term is also used to denote sources that radiate in regions of the spectrum adjacent to the visible.

NOTE: A lighting unit consisting of a lamp with shade, reflector, enclosing globe, housing, or other accessories is also called a "lamp." In such cases, in order to distinguish between the assembled unit and the light source within it, the latter is often called a "bulb" or "tube," if it is electrically powered. See also *luminaire.*

lamp burnout factor: the fractional loss of task illuminance due to burned out lamps left in place for long periods.

lamp lumen depreciation factor, LLD: the multiplier to be used in illumination calculations to relate the initial rated output of light sources to the anticipated minimum rated output based on the relamping program to be used.

lamp position factor: The fractional flux of a high intensity discharge (HID) lamp at a given operating position compared to the flux when the lamp is operated in the position at which the lamp lumens are rated.

lamp post: a standard support provided with the necessary internal attachments for wiring and the external attachments for the bracket and luminaire.

lamp shielding angle: the angle between the plane of the baffles or louver grid and the plane most nearly horizontal that is tangent to both the lamps and the louver blades. See Fig. 1-3.

NOTE: The lamp shielding angle frequently is larger than the louver shielding angle, but never smaller. See *louver shielding angle.*

landing direction indicator: a device to indicate visually the direction currently designated for landing and take-off.

landing light: an aircraft aeronautical light designed to illuminate a ground area from the aircraft.

landmark beacon: an aeronautical beacon used to indicate the location of a landmark used by pilots as an aid to enroute navigation.

laser: an acronym for *L*ight *A*mplification by *S*timulated *E*mission of *R*adiation. The laser produces a highly monochromatic and coherent (spatial and temporal) beam of radiation. A steady oscillation of nearly a single electromagnetic mode is maintained in a volume of an active material bounded by highly reflecting surfaces, called a resonator. The frequency of oscillation varies according to the material used and by the methods of initially exciting or pumping the material.

lateral width of a light distribution: (in roadway lighting) the lateral angle between the reference line and the width line, measured in the cone of maximum candlepower. This angular width includes the line of maximum candlepower. See *reference line* and *width line.*

lens: a glass or plastic element used in luminaires to change the direction and control the distribution of light rays.

level of illumination: See *illuminance.*

life performance curve: a curve which represents the variation of a particular characteristic of a light source (luminous flux, intensity, etc.) throughout the life of the source.

NOTE: Life performance curves sometimes are called maintenance curves as, for example, lumen maintenance curves.

life test of lamps: a test in which lamps are operated under specified conditions for a specified length of time, for the purpose of obtaining information on lamp life. Measurements of photometric and electrical characteristics may be made at specified intervals of time during this test.

light: radiant energy that is capable of exciting the retina and producing a visual sensation. The visible portion of the electromagnetic spectrum extends from about 380 to 770 nm.

NOTE: The subjective impression produced by stimulating the retina is sometimes designated as light. Visual sensations are sometimes arbitrarily defined as sensations of light, and in line with this concept it is sometimes said that light cannot exist until an eye has been stimulated. Electrical stimulation of the retina or the visual cortex is described as producing flashes of light. In illuminating engineering, however, light is a physical entity—radiant energy weighted by the luminous efficiency function. It is a physical stimulus which can be applied to the retina. (See *spectral luminous efficacy of radiant flux, values of spectral luminous efficiency for photopic vision.*)

light adaptation: the process by which the retina becomes adapted to a luminance greater than about 3.4 candelas per square meter. See also *dark adaptation.*

light center (of a lamp): the center of the smallest sphere that would completely contain the light-emitting element of the lamp.

light center length (of a lamp): the distance from the light center to a specified reference point on the lamp.

light loss factor, LLF: a factor used in calculating illuminance after a given period of time and under given conditions. It takes into account temperature and voltage variations, dirt accumulation on luminaire and room surfaces, lamp depreciation, maintenance procedures and atmosphere conditions. Formerly called *maintenance factor.*

light meter: See *illuminance meter.*

light-source color: the color of the light emitted by the source.

NOTE: The color of a point source may be defined by its luminous intensity and chromaticity coordinates; the color of an extended source may be defined by its luminance and chromaticity coordinates. See *perceived light-source color, color temperature, correlated color temperature.*

lighting effectiveness factor, LEF_V: the ratio of equivalent sphere illumination to ordinary measured or calculated illumination.

lightness (of a perceived patch of surface color): the attribute by which it is perceived to transmit or reflect a greater or lesser fraction of the incident light.

light-watt: radiation weighted by the spectral luminous efficiency for photopic vision.

linear light: a luminous signal having a perceptible physical length.

linear polarization: the process by which the transverse vibrations of light waves are oriented in a specific plane. Polarization may be obtained by using either transmitting or reflecting media.

Linnebach projector: a lensless scenic projector, using a concentrated source in a black box and a slide or cutout between the source and the projection surface.

liquid crystal display, (LED): a display made of material whose reflectance or transmittance changes when an electric field is applied.

local lighting: lighting designed to provide illuminance over a relatively small area or confined space without providing any significant general surrounding lighting.

localized general lighting: lighting that utilizes luminaires above the visual task and contributes also to the illumination of the surround.

long-arc lamp: an arc lamp in which the distance between the electrodes is large.

NOTE: This type of lamp (*e.g.*, xenon) is generally of high pressure. The arc fills the discharge tube and is therefore wall stabilized.

louver: a series of baffles used to shield a source from view at certain angles or to absorb unwanted light. The baffles usually are arranged in a geometric pattern.

louver shielding angle, θ: the angle between the horizontal plane of the baffles or louver grid and the plane at which the louver conceals all objects above. See Fig. 1-4 and *lamp shielding angle.*

NOTE: The planes usually are so chosen that their intersection is parallel with the louvered blade.

louvered ceiling: a ceiling area lighting system comprising a wall-to-wall installation of multicell louvers shielding the light sources mounted above it. See *luminous ceiling.*

lower (passing) beams: one or more beams directed low enough on the left to avoid glare in the eyes of oncoming drivers, and intended for use in congested areas and on highways when meeting other vehicles within a distance of 300 meters (1000 feet). Formerly "traffic beam."

low pressure mercury lamp: a discharge lamp (with or without a phosphor coating) in which the partial pressure of the mercury vapor during operation does not exceed 100 Pa.

low pressure sodium lamp: a discharge lamp in which light is produced by radiation from sodium vapor operating at a partial pressure of 0.1 to 1.5 Pa (approximately 10^{-3} to 10^{-2} torr).

low-key lighting: a type of lighting that, applied to a scene, results in a picture having gradations falling primarily between middle gray and black, with comparatively limited areas of light grays and whites. See *high-key lighting.*

lumen, lm: SI unit of luminous flux. Radiometrically, it is determined from the radiant power. Photometrically, it is the luminous flux emitted within a unit solid angle (one steradian) by a point source having a uniform luminous intensity of one candela.

lumen-hour, lm·h: a unit of quantity of light (luminous energy). It is the quantity of light delivered in one hour by a flux of one lumen.

lumen (or flux) method: a lighting design procedure used for predetermining the relation between the number and types of lamps or luminaires, the room characteristics, and the average illuminance on the work-plane. It takes into account both direct and reflected flux.

lumen-second, lm·s: a unit of quantity of light, the SI unit of luminous energy (also called a talbot). It is the quantity of light delivered in one second by a luminous flux of one lumen.

luminaire: a complete lighting unit consisting of a lamp or lamps together with the parts designed to distribute the light, to position and protect the lamps and to connect the lamps to the power supply.

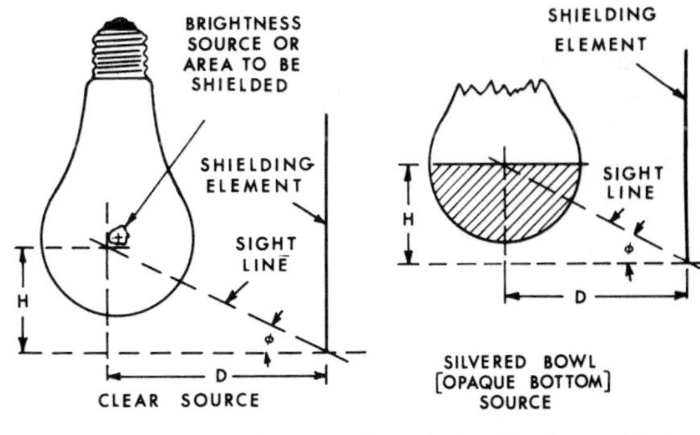

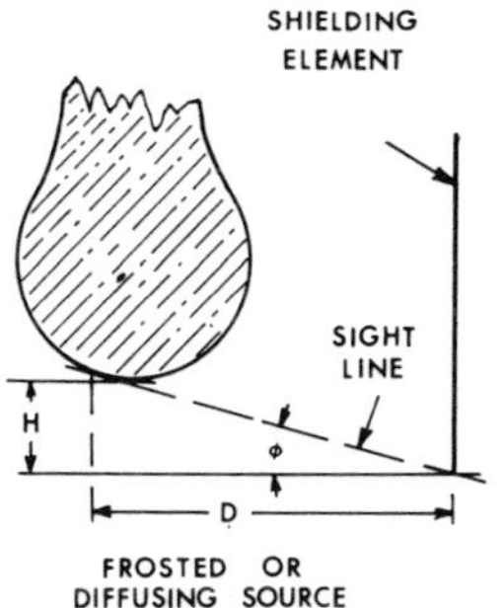

Fig. 1-3. The lamp shielding angle is formed by a sight line tangent to the lowest part of the brightness area to be shielded. *H* is the vertical distance from the brightness source to the bottom of the shielding element. *D* is the horizontal distance from the brightness source to the shielding element. Lamp shielding angle $\phi = \tan^{-1} H/D$.

luminaire ambient temperature factor: the fractional loss of task illuminance due to improper operating temperature of a gas discharge lamp.

luminaire dirt depreciation factor, LDD: the multiplier to be used in illuminance calculations to relate the initial illuminance provided by clean, new luminaires to the reduced illuminance that they will provide due to dirt collection on the luminaires at the time at which it is anticipated that cleaning procedures will be instituted.

luminaire efficiency: the ratio of luminous flux (lumens) emitted by a luminaire to that emitted by the lamp or lamps used therein.

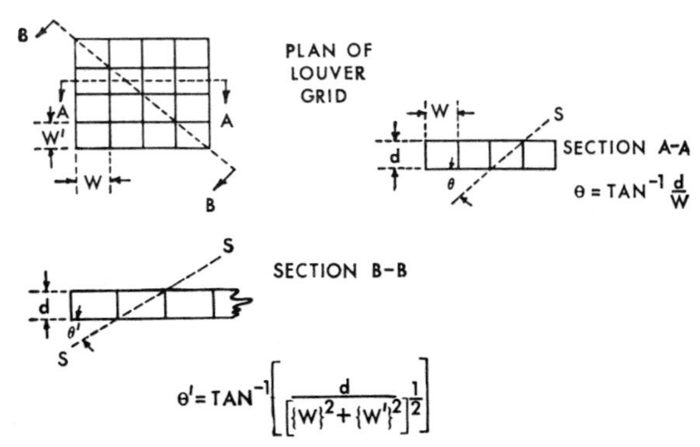

Fig. 1-4. Louver shielding angles θ and θ'.

luminaire surface depreciation factor: the loss of task illuminance due to permanent deterioration of luminaire surfaces.

luminance, $L = d^2\Phi/(d\omega\ dA\ \cos\theta)$ (in a direction and at a point of a real or imaginary surface): the quotient of the luminous flux at an element of

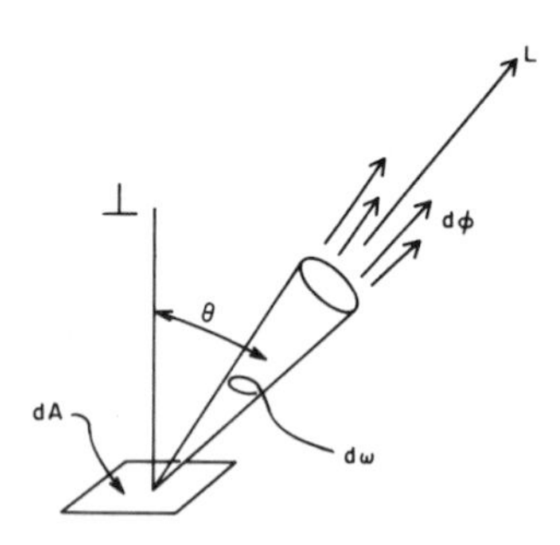

the surface surrounding the point, and propagated in directions defined by an elementary cone containing the given direction, by the product of the solid angle of the cone and the area of the orthogonal projection of the element of the surface on a plane perpendicular to the given direction. The luminous flux may be leaving, passing through, and/or arriving at the surface. Formerly, *photometric brightness.*

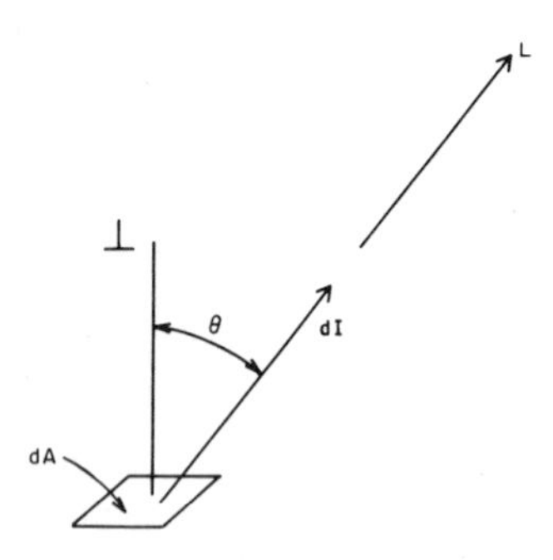

By introducing the concept of luminous intensity, luminance may be expressed as $L = dI/(dA\ \cos\theta)$. Here, luminance at a point of a surface in a direction is interpreted as the quotient of luminous intensity in the given direction produced by an element of the surface surrounding the point by the area of the orthogonal projection of the element of surface on a plane perpendicular to the given direction. (Luminance may be measured at a receiving surface by using $L = dE/(d\omega\ \cos\theta)$. This value may be less than the luminance of the emitting surface due to the attenuation of the transmitting media.)

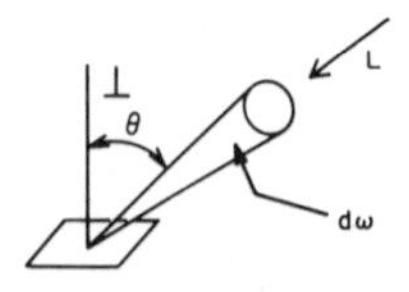

NOTE: In common usage the term "brightness" usually refers to the strength of sensation which results from viewing surfaces or spaces from which light comes to the eye. This sensation is determined in part by the definitely measurable luminance defined above and in part by conditions of observation such as the state of adaptation of the eye.

In much of the literature the term brightness, used alone, refers to both luminance and sensation. The context usually indicates which meaning is intended. Previous usage notwithstanding, neither the term brightness nor the term photometric brightness should be used to denote the concept of luminance.

luminance coefficient: a coefficient similar to the coefficient of utilization used to determine wall and ceiling luminances. An obsolete term, see *exitance coefficient.*

luminance contrast: the relationship between the luminances of an object and its immediate background. It is equal to $(L_1 - L_2)/L_1$, $(L_2 - L_1)/L_1$ or $\Delta L/L_1$, where L_1 and L_2 are the luminances of the background and object, respectively. The form of the equation must be specified. The ratio $\Delta L/L_1$ is known as Weber's fraction.

NOTE: See last paragraph of the note under *luminance.* Because of the relationship among luminance, illuminance and reflectance, contrast often is expressed in terms of reflectance when only reflecting surfaces are involved. Thus, contrast is equal to $(\rho_1 - \rho_2)/\rho_1$ or $(\rho_2 - \rho_1)/\rho_1$ where ρ_1 and ρ_2 are the reflectances of the background and object, respectively. This method of computing contrast holds only for perfectly diffusing surfaces; for other surfaces it is only an approximation unless the angles of incidence and view are taken into consideration. (See *reflectance.*)

luminance difference: the difference in luminance between two areas. It usually is applied to contiguous areas, such as the detail of a visual task and its immediate background, in which case it is quantitatively equal to the numerator in the formula for luminance contrast.

NOTE: See last paragraph of the note under *luminance.*

luminance factor, β: the ratio of the luminance of a surface or medium under specified conditions of incidence, observation, and light source, to the luminance of a completely reflecting or transmitting, perfectly diffusing surface or medium under the same conditions.

NOTE: Reflectance or transmittance cannot exceed unity, but luminance factor may have any value from zero to values approaching infinity.

luminance factor of room surfaces: factors by which the average work-plane illuminance is multiplied to obtain the average luminances of walls, ceilings and floors.

luminance ratio: the ratio between the luminances of any two areas in the visual field.

luminance threshold: the minimum perceptible difference in luminance for a given state of adaptation of the eye.

luminescence: any emission of light not ascribable directly to incandescence. See *electroluminescence, fluorescence, phosphorescence.*

luminosity factor: previously used term for *spectral luminous efficacy of radiant flux.*

luminous ceiling: a ceiling area lighting system comprising a continuous surface of transmitting material of a diffusing or light-controlling character

with light sources mounted above it. See *louvered ceiling.*

luminous density, $w = dQ/dV$: quantity of light (luminous energy) per unit volume.

luminous efficacy of radiant flux: the quotient of the total luminous flux by the total radiant flux. It is expressed in lumens per watt.

luminous efficacy of a source of light: the quotient of the total luminous flux emitted by the total lamp power input. It is expressed in lumens per watt.

NOTE: The term luminous efficiency has in the past been extensively used for this concept.

luminous efficiency: See *spectral luminous efficiency of radiant flux.*

luminous energy: See *quantity of light.*

luminous exitance, $M = d\Phi/dA$: the density of luminous flux leaving a surface at a point. Formerly luminous emittance.

NOTE: This is the total luminous flux emitted, reflected, and transmitted from the surface into a complete hemisphere.

luminous flux, Φ: the time rate of flow of light.

luminous flux density at a surface, $d\Phi/dA$: the luminous flux per unit area at a point on a surface.

NOTE: This need not be a physical surface; it may equally well be a mathematical plane. Also see *illuminance* and *luminous exitance.*

luminous intensity, $I = d\Phi/d\omega$ (of a point source of light in a given direction): the luminous flux per unit solid angle in the direction in question. Hence, it is the luminous flux on a small surface centered on and normal to that direction divided by the solid angle (in steradians) which the surface subtends at the source. Luminous intensity may be expressed in candelas or in lumens per steradian (lm/sr).

NOTE: Mathematically a solid angle must have a point as its apex; the definition of luminous intensity, therefore, applies strictly only to a point source. In practice, however, light emanating from a source whose dimensions are negligible in comparison with the distance from which it is observed may be considered as coming from a point. Specifically, this implies that with change of distance (1) the variation in solid angle subtended by the source at the receiving point approaches $1/(\text{distance})^2$ and that (2) the average luminance of the projected source area as seen from the receiving point does not vary appreciably.

luminous intensity distribution curve: See *candlepower distribution curve.*

luminous reflectance: any reflectance (regardless of beam geometry) in which both the incident and reflected flux are weighted by the spectral luminous efficiency of radiant flux, $V(\lambda)$; *i.e.*, where they are expressed as luminous flux. Thus, luminous reflectance is not a unique property of a reflecting surface but depends also on the spectral distribution of the incident radiant flux.

NOTE: Unless otherwise qualified, the term luminous reflectance is meant by the term reflectance.

luminous transmittance: any transmittance (regardless of beam geometry) in which both the incident and transmitted flux are weighted by the spectral luminous efficiency of radiant flux, $V(\lambda)$; *i.e.*, where they are expressed as luminous flux. Thus, luminous transmittance is not a unique property of a transmitting element but depends also on the spectral distribution of the incident radiant flux.

NOTE: Unless otherwise qualified, the term luminous transmittance is meant by the term transmittance.

lux, lx: the SI unit of illuminance. One lux is one lumen per square meter (lm/m^2). (See Fig. 1-7 for conversion values.)

lux meter: See *illuminance meter.*

M

maintenance factor, MF: a factor formerly used to denote the ratio of the illuminance on a given area after a period of time to the initial illuminance on the same area. See *light loss factor.*

matte surface: one from which the reflection is predominantly diffuse, with or without a negligible specular component. See *diffuse reflection.*

mean horizontal intensity (candlepower): the average intensity (candelas) of a lamp in a plane perpendicular to the axis of the lamp and which passes through the luminous center of the lamp.

mean spherical luminous intensity: average value of the luminous intensity in all directions for a source. Also, the quotient of the total emitted luminous flux of the source divided by 4π.

mean zonal candlepower: the average intensity (candelas) of a symmetrical luminaire or lamp at an angle to the luminaire or lamp axis which is in the middle of the zone under consideration.

mechanical equivalent of light: See *spectral luminous efficacy of radiant flux, $K(\lambda) = \Phi_{v\lambda}/\Phi_{e\lambda}$.*

mercury lamp: a high intensity discharge (HID) lamp in which the major portion of the light is produced by radiation from mercury operating at a partial pressure in excess of 10^5 Pa (approximately 1 atmosphere). Includes clear, phosphor-coated (mercury-fluorescent), and self-ballasted lamps.

mercury-fluorescent lamp: an electric-discharge lamp having a high-pressure mercury arc in an arc tube, and an outer envelope coated with a fluorescing substance (phosphor) which transforms some of the ultraviolet energy generated by the arc into light.

mesopic vision: vision with fully adapted eyes at luminance conditions between those of photopic and scotopic vision, that is, between about 3.4 and 0.034 candelas per square meter.

metal halide lamp: a high intensity discharge (HID) lamp in which the major portion of the light is produced by radiation of metal halides and their products of dissociation-possibly in combination with metallic vapors such as mercury. Includes clear and phosphor coated lamps.

metamers: lights of the same color but of different spectral energy distribution.

middle ultraviolet: a portion of the electromagnetic spectrum in the range of 200–300 nanometers.

minimal perceptible erythema, MPE: the erythemal threshold.

mired: a unit of reciprocal color temperature; microreciprocal degrees; $1/T_k$ times 10^6.

NOTE: The unit of thermodynamic temperature is now denoted by "kelvin" and not "degree-kelvin". Consequently the acronym "mirek" for microreciprocal-kelvin occasionally has been used in the literature for "mired."

mirek: see *mired.*

modeling light: that illumination which reveals the depth, shape and texture of a subject. Key light, cross light, counter-key light, side light, back light, and eye light are types of modeling light.

modulation threshold: in the case of a square wave or sine wave grating, manipulation of luminance differences can be specified in terms of modulation and the threshold may be called the modulation threshold.

$$\text{modulation} = \frac{L_{max} - L_{min}}{L_{max} + L_{min}}$$

Periodic patterns that are not sine wave can be specified in terms of the modulation of the fundamental sine wave component. The number of periods or cycles per degree of visual angle represents the spatial frequency.

monocular visual field: the visual field of a single eye. See *binocular visual field.*

mounting height (roadway): the vertical distance between the roadway surface and the center of the apparent light source of the luminaire.

mounting height above the floor, MH_f: the distance from the floor to the light center of the luminaire or to the plane of the ceiling for recessed equipment.

mounting height above the work-plane, MH_{wp}: the distance from the work-plane to the light center of the luminaire or to the plane of the ceiling for recessed equipment.

multiple-beam headlamp: a headlamp so designed to permit the driver of a vehicle to use any one of two or more distributions of light on the road.

Munsell chroma, C: an index of perceived chroma of the object color defined in terms of the luminance factor (Y) and chromaticity coordinates (x, y) for CIE Standard Illuminant C and the CIE 1931 Standard Observer.

Munsell color system: a system of surface-color specification based on perceptually uniform color scales for the three variables: Munsell hue, Munsell value, and Munsell chroma. For an observer of normal color vision, adapted to daylight, and viewing a specimen when illuminated by daylight and surrounded with a middle gray to white background, the Munsell hue, value and chroma of the color correlate well with the hue, lightness and perceived chroma.

Munsell hue, H: an index of the hue of the perceived object color defined in terms of the luminance factor (Y) and chromaticity coordinates (x, y) for CIE Standard Illuminant C and the CIE 1931 Standard Observer.

Munsell Value, V: an index of the lightness of the perceived object color defined in terms of the luminance factor (Y) for CIE Standard Illuminant C and the CIE 1931 Standard Observer.

NOTE: The exact definition gives Y as a 5th power function of V so that tabular or iterative methods are needed to find V as a function of Y. However, V can be estimated within ±0.1 by $V = 11.6\ (Y/100)^{1/3} - 1.6$ or within ±0.6 by $V = Y^{1/2}$ where Y is the luminance factor expressed in per cent.

N

nanometer, nm: unit of wavelength equal to 10^{-9} meter. See Fig. 1-11.

narrow-angle diffusion: that in which flux is scattered at angles near the direction that the flux would take by regular reflection or transmission. See *wide-angle-diffusion.*

narrow angle luminaire: a luminaire that concentrates the light within a cone of comparatively small solid angle. See *wide angle luminaire.*

national standard of light: a primary standard of light which has been adopted as a national standard. See *primary standard of light.*

navigation lights: an alternate term for *position lights.*

navigation light system: a set of aircraft aeronautical lights provided to indicate the position and direction of motion of an aircraft to pilots of other aircraft or to ground observers.

near infrared: the region of the electromagnetic spectrum from 770 to 1400 nanometers.

near ultraviolet: the region of the electromagnetic spectrum from 300 to 380 nanometers.

night: the hours between the end of evening civil twilight and the beginning of morning civil twilight.

NOTE: Civil twilight ends in the evening when the center of the sun's disk is six degrees below the horizon and begins in the morning when the center of the sun's disk is six degrees below the horizon.

nit, nt: a unit of luminance equal to one candela per square meter.

NOTE: Candela per square meter is the International Standard (SI) unit of luminance.

numerical display (digital display): an electrically operated display of digits. Tungsten filaments, gas discharges, light emitting diodes, liquid crystals, projected numerals, illuminated numbers and other principles of operation may be used.

O

object color: the color of the light reflected or transmitted by the object when illuminated by a standard light source, such as CIE source A, B, C or D_{65}. See *standard source* and *perceived object color.*

obstruction beacon: See *hazard beacon*

obstruction lights: aeronautical ground lights provided to indicate obstructions.

occulting light: a rhythmic light in which the periods of light are clearly longer than the periods of darkness.

orientation: the relation of a building with respect to compass directions.

Ostwald color system: a system of describing colors in terms of color content, white content and black content.

overcast sky: one that has 100 per cent cloud cover; the sun is not visible.

overhang: the distance between a vertical line passing through the luminaire and the curb or edge of the roadway.

ozone-producing radiation: ultraviolet energy shorter than about 220 nanometers that decomposes oxygen O_2 thereby producing ozone O_3. Some ultraviolet sources generate energy at 184.9 nanometers that is particularly effective in producing ozone.

P

PAR lamp: See *pressed reflector lamp.*

parking lamp: a lighting device placed on a vehicle to indicate its presence when parked.

partial diffusion: See *incomplete diffusion.*

partly cloudy sky: one that has 30 to 70 per cent cloud cover.

passing beams: See *lower beams.*

pendent luminaire: See *suspended luminaire.*

perceived light-source color: the color perceived to belong to a light source.

perceived object color: the color perceived to belong to an object, resulting from characteristics of the object, of the incident light, and of the surround, the viewing direction and observer adaptation. See *object color.*

perfect diffusion: that in which flux is uniformly scattered such that the luminance (radiance) is the same in all directions.

perimeter lights: aeronautical ground lights provided to indicate the perimeter of a landing pad for helicopters.

peripheral vision: the seeing of objects displaced from the primary line of sight and outside the central visual field.

peripheral visual field: that portion of the visual field that falls outside the region corresponding to the foveal portion of the retina.

phosphor mercury lamp: See *mercury-fluorescent lamp.*

phosphorescence: the emission of light (luminescence) as the result of the absorption of radiation, and continuing for a noticeable length of time after excitation.

phot, ph: the unit of illuminance when the centimeter is taken as the unit of length; it is equal to one lumen per square centimeter.

photochemical radiation: energy in the ultraviolet, visible and infrared regions capable of producing chemical changes in materials.

NOTE: Examples of photochemical processes are accelerated fading tests, photography, photoreproduction and chemical manufacturing. In many such applications a specific spectral region is of importance.

photoelectric receiver: a device that reacts electrically in a measurable manner in response to incident radiant energy.

photoflash lamp: a lamp in which combustible metal or other solid material is burned in an oxidizing atmosphere to produce light of high intensity and short duration for photographic purposes.

photoflood lamp: an incandescent filament lamp of high color temperature for lighting objects for photography or videography.

photometer: an instrument for measuring photometric quantities such as luminance, luminous intensity, luminous flux and illuminance. See *densitometer, goniophotometer, illuminance meter, integrating photometer, reflectometer, spectrophotometer, transmissometer.*

photometry: the measurement of quantities associated with light.

NOTE: Photometry may be visual in which the eye is used to make a comparison, or physical in which measurements are made by means of physical receptors.

photometric brightness: a term formerly used for *luminance.*

photopic vision: vision mediated essentially or exclusively by the cones. It is generally associated with adaptation to a luminance of at least 3.4 candelas per square meter. See *scotopic vision.*

physical photometer: an instrument containing a physical receptor (photoemissive cell, barrier-layer cell, thermopile, etc.) and associated filters, that is calibrated so as to read photometric quantities directly. See *visual photometer.*

pilot house control: a mechanical means for controlling the elevation and train of a searchlight from a position on the other side of the bulkhead or deck on which it is mounted.

Planck radiation law: an expression representing the spectral radiance of a blackbody as a function of the wavelength and temperature. This law commonly is expressed by the formula

$$L_\lambda = dI_\lambda/dA' = c_{1L}\lambda^{-5}[e^{(c_2/\lambda T)} - 1]^{-1}$$

in which L_λ is the spectral radiance, dI_λ is the spectral radiant intensity, dA' is the projected area ($dA \cos \theta$) of the aperture of the blackbody, e is the base of natural logarithms (2.71828), T is absolute temperature, c_{1L} and c_2 are constants designated as the first and second radiation constants.

NOTE: The designation c_{1L} is used to indicate that the equation in the form given here refers to the radiance L, or to the intensity I per unit projected area A', of the source. Numerical values are commonly given not for c_{1L} but for c_1 which applies to the total flux radiated from a blackbody aperture, that is, in a hemisphere (2π steradians), so that, with the Lambert cosine law taken into account, $c_1 = \pi c_{1L}$. The currently recommended value of c_1 is 3.741832×10^{-16} W·m^2 or 3.741832×10^{-12} W·cm^2. Then c_{1L} is $1.1910621 \times 10^{-16}$ W·m^2·sr^{-1} or $1.1910621 \times 10^{-12}$ W·cm^2·sr^{-1}. If, as is more convenient, wavelengths are expressed in micrometers and area in square centimeters. $c_{1L} = 1.1910621 \times 10^4$ W μm^4·cm^{-2}·sr^{-1}, L_λ being given in W cm^{-2}·sr^{-1}·μm^{-1}. The currently recommended value of c_2 is 1.438786×10^{-2} m·K.

The Planck law in the following form gives the

energy radiated from the blackbody in a given wavelength interval ($\lambda_1 - \lambda_2$):

$$Q = \int_{\lambda_1}^{\lambda_2} Q_\lambda \, d\lambda = Atc_1 \int_{\lambda_1}^{\lambda_2} \lambda^{-5}(e^{(c_2/\lambda T)} - 1)^{-1} \, d\lambda$$

If A is the area of the radiation aperture or surface in square centimeters, t is time in seconds, λ is wavelength in micrometers, and $c_1 = 3.7418 \times 10^4$ $W \cdot \mu m^4 \cdot cm^{-2}$, then Q is the total energy in watt seconds emitted from this area (that is, in the solid angle 2π), in time t, within the wavelength interval ($\lambda_1 - \lambda_2$).

planckian locus: See *blackbody locus.*

point of fixation: a point or object in the visual field at which the eyes look and upon which they are focused.

point of observation: the midpoint of the base line connecting the centers of the entrance pupils of the two eyes. For practical purposes, the center of the pupil of the eye often is taken as the point of observation.

point by point method: a method of lighting calculation now called *point method.*

point method: a lighting design procedure for predetermining the illuminance at various locations in lighting installations, by use of luminaire photometric data.

NOTE: The direct component of illuminance due to the luminaires and the interreflected component of illuminance due to the room surfaces are calculated separately. The sum is the total illuminance at a point.

point source: a source of radiation the dimensions of which are small enough, compared with the distance between the source and the irradiated surface, for them to be neglected in calculations and measurements.

polarization: the process by which the transverse vibrations of light waves are oriented in a specific plane. Polarization may be obtained by using either transmitting or reflecting media.

pole (roadway lighting): a standard support generally used where overhead lighting distribution circuits are employed.

portable lighting: lighting involving lighting equipment designed for manual portability.

portable luminaire: a lighting unit that is not permanently fixed in place. See *table lamp* and *floor lamp.*

portable traffic control light: a signalling light producing a controllable distinctive signal for purposes of directing aircraft operations in the vicinity of an aerodrome.

position index, P: a factor which represents the relative average luminance for a sensation at the borderline between comfort and discomfort (BCD), for a source located anywhere within the visual field.

position lights: aircraft aeronautical lights forming the basic or internationally recognized navigation light system.

NOTE: The system is composed of a red light showing from dead ahead to 110 degrees to the left, a green light showing from dead ahead to 110 degrees to the right, and a white light showing to the rear through 140 degrees. Position lights are also called navigation lights.

prefocus lamp: a lamp in which, during manufacture, the luminous element is accurately adjusted to a specified position with respect to the physical locating element (usually the base).

preheat (switch start) fluorescent lamp: a fluorescent lamp designed for operation in a circuit requiring a manual or automatic starting switch to preheat the electrodes in order to start the arc.

pressed reflector lamp: an incandescent filament or electric discharge lamp of which the outer bulb is formed of two pressed parts that are fused together; namely, a reflectorized bowl and a cover which may be clear or patterned for optical control.

NOTE: Often called a projector or PAR lamp.

primary (light): any one of three lights in terms of which a color is specified by giving the amount of each required to match it by additive combination.

primary line of sight: the line connecting the point of observation and the point of fixation.

primary standard of light: a light source by which the unit of light is established and from which the values of other standards are derived. See *national standard of light.*

NOTE: A satisfactory primary (national) standard must be reproducible from specifications (see *candela*). Primary (national) standards usually are found in national physical laboratories such as the National Bureau of Standards.

projection lamp: a lamp with physical and luminous characteristics suited for projection systems (*e.g.*, motion picture projectors, microfilm viewers, etc.).

projector: a lighting unit that by means of mirrors and lenses, concentrates the light to a limited solid angle so as to obtain a high value of luminous intensity. See *floodlight, searchlight, signalling light.*

protective lighting: a system intended to facilitate the nighttime policing of industrial and other properties.

proximal stimuli: the distribution of illuminance on the retina constitutes the proximal stimulus.

pupil (pupillary aperture): the opening in the iris that admits light into the eye. See *artificial pupil.*

Purkinje phenomenon: the reduction in subjective brightness of a red light relative to that of a blue light when the luminances are reduced in the same proportion without changing the respective spectral distributions. In passing from photopic to scotopic vision, the curve of spectral luminous efficiency changes, the wavelength of maximum efficiency being displaced toward the shorter wavelengths.

purple boundary: the straight line drawn between the ends of the spectrum locus on a chromaticity diagram.

Q

quality of lighting: pertains to the distribution of luminance in a visual environment. The term is used in a positive sense and implies that all luminances contribute favorably to visual performance, visual comfort, ease of seeing, safety, and esthetics for the specific visual tasks involved.

quantity of light (luminous energy, $Q = \int \Phi dt$): the product of the luminous flux by the time it is maintained. It is the time integral of luminous flux (compare *light* and *luminous flux.*)

quartz-iodine lamp: an obsolete term for *tungsten-halogen lamp.*

quick-flashing light: a rhythmic light exhibiting very rapid regular alternations of light and darkness. There is no restriction on the ratio of the durations of the light to the dark periods.

R

radiance, $L = d^2\Phi/(d\omega dA \cos\theta) = dI/(dA \cos\theta)$ (in a direction, at a point of the surface of a source, of a receiver, or of any other real or virtual surface): the quotient of the radiant flux leaving, passing through, or arriving at an element of the surface surrounding the point, and propagated in directions defined by an elementary cone containing the given direction, by the product of the solid angle of the cone and the area of the orthogonal projection of the element of the surface on a plane perpendicular to the given direction.

NOTE: In the defining equation θ is the angle between the normal to the element of the source and the given direction.

radiant energy density, $w = dQ/dV$: radiant energy per unit volume; *e.g.*, joules per cubic meter.

radiant energy, Q: energy traveling in the form of electromagnetic waves. It is measured in units of energy such as joules, ergs, or kilowatt-hours. See *spectral radiant energy.*

radiant exitance, M: the density of radiant flux leaving a surface. It is expressed in watts per unit area of the surface.

radiant flux, $\Phi = dQ/dt$: the time rate of flow of radiant energy. It is expressed preferably in watts, or in joules per second. See *spectral radiant flux.*

radiant flux density at a surface: the quotient of radiant flux of an element of surface to the area of that element; *e.g.*, watts per square meter. When referring to radiant flux emitted from a surface, this has been called *radiant emittance* (deprecated). The preferred term for radiant flux leaving a surface is *radiant exitance.* The radiant exitance per unit wavelength interval is called *spectral radiant exitance.* When referring to radiant flux incident on a surface, it is called *irradiance* (E).

radiant intensity, $I = d\Phi/d\omega$: the radiant flux proceeding from the source per unit solid angle in the direction considered; *e.g.*, watts per steradian. See *spectral radiant intensity.*

radiator: an emitter of radiant energy.

radiometry: the measurement of quantities associated with radiant energy.

range lights: groups of color-coded boundary lights provided to indicate the direction and limits of a preferred landing path normally on an aerodrome without runways but exceptionally on an aerodrome with runways.

rapid start fluorescent lamp: a fluorescent lamp designed for operation with a ballast that provides a low-voltage winding for preheating the electrodes and initiating the arc without a starting switch or the application of high voltage.

reaction time: the interval between the beginning of a stimulus and the beginning of the response of an observer.

recessed luminaire: See *flush mounted luminaire.*

reciprocal color temperature: color temperature T_k expressed on a reciprocal scale ($1/T_k$). An important use stems from the fact that a given small increment in reciprocal color temperature is approximately equally perceptible regardless of color temperature. Also, color temperature conversion filters for sources approximating graybody sources change the reciprocal color temperature by nearly the same amount anywhere on the color temperature scale. See *mired.*

recoverable light loss factors: factors which give the fractional light loss that can be recovered by cleaning or lamp replacement.

redirecting surfaces and media: those surfaces and media that change the direction of the flux without scattering the redirected flux.

reference ballast: a ballast specially constructed to have certain prescribed characteristics for use in testing electric discharge lamps and other ballasts.

reference line (roadway lighting): either of two radial lines where the surface of the cone of maximum candlepower is intersected by a vertical plane parallel to the curb line and passing through the light-center of the luminaire.

reference standard: an alternate term for *secondary standard.*

reflectance of a surface or medium, $\rho = \Phi_r/\Phi_i$: the ratio of the reflected flux to the incident flux. See *biconical reflectance, bidirectional reflectance, bihemispherical reflectance, conical-directional reflectance, conical-hemispherical reflectance, diffuse reflectance, directional hemispherical reflectance, effective ceiling cavity reflectance, effective floor cavity reflectance, reflectance factor, hemispherical reflectance, hemispherical-conical reflectance, hemispherical directional reflectance, luminous reflectance, spectral reflectance, specular reflectance.*

NOTE: Measured values of reflectance depend upon the angles of incidence and view and on the spectral character of the incident flux. Because of this dependence, the angles of incidence and view and the spectral characteristics of the source should be specified. See *reflection.*

reflectance factor, R: ratio of the radiant (or luminous) flux reflected in directions delimited to that reflected in the same directions by a perfect reflecting diffuser identically irradiated (or illuminated).

reflected glare: glare resulting from specular reflections of high luminances in polished or glossy surfaces in the field of view. It usually is associated with reflections from within a visual task or areas in close proximity to the region being viewed. See *veiling reflection.*

reflection: a general term for the process by which

the incident flux leaves a surface or medium from the incident side, without change in frequency.

NOTE: Reflection is usually a combination of regular and diffuse reflection. See *regular (specular) reflection, diffuse reflection and veiling reflection.*

reflectivity: reflectance of a layer of a material of such a thickness that there is no change of reflectance with increase in thickness.

reflectometer: a photometer for measuring reflectance.

NOTE: Reflectometers may be visual or physical instruments.

reflector: a device used to redirect the luminous flux from a source by the process of reflection. See *retroreflector.*

reflector lamp: an incandescent filament or electric discharge lamp in which the outer blown glass bulb is coated with a reflecting material so as to direct the light (*e.g.*, R- or ER-type lamps). The light transmitting region may be clear, frosted, patterned or phosphor coated.

reflex reflector: See *retro-reflector.*

refraction: the process by which the direction of a ray of light changes as it passes obliquely from one medium to another in which its speed is different.

refractor: a device used to redirect the luminous flux from a source, primarily by the process of refraction

regions of electromagnetic spectrum: for convenience of reference the electromagnetic spectrum is arbitrarily divided as follows:

Vacuum ultraviolet	
Extreme ultraviolet	10–100 nm
Far ultraviolet	100–200 nm
Middle ultraviolet	200–300 nm
Near ultraviolet	300–380 nm
Visible	380–770 nm
Near (short wavelength) infrared	770–1400 nm
Intermediate infrared	1400–5000 nm
Far (long wavelength) infrared	5000–1,000,000 nm

NOTE: The spectral limits indicated above have been chosen as a matter of practical convenience. There is a gradual transition from region to region without sharp delineation. Also, the division of the spectrum is not unique. In various fields of science the classifications may differ due to the phenomena of interest.

regressed luminaire: a luminaire mounted above the ceiling with the opening of the luminaire above the ceiling line. See *flush-mounted, surface-mounted, suspended* and *troffer.*

regular (specular) reflectance: the ratio of the flux leaving a surface or medium by regular (specular) reflection to the incident flux. See *regular (specular) reflection.*

regular (specular) reflection: that process by which incident flux is re-directed at the specular angle. See *specular angle.*

regular transmission: that process by which incident flux passes through a surface or medium without scattering. See *regular transmittance.*

regular transmittance: the ratio of the regularly transmitted flux leaving a surface or medium to the incident flux.

relative contrast sensitivity RCS: the relation between the reciprocal of the luminous contrast of a task at visibility threshold and the background luminance expressed as a percentage of the value obtained under a very high level of diffuse task illumination.

relative erythemal factor: See *erythemal efficiency of radiant flux.*

relative luminosity: previously used term for *spectral luminous efficiency of radiant flux.*

relative luminosity factor: previously used term for *spectral luminous efficiency of radiant flux.*

resolving power: the ability of the eye to perceive the individual elements of a grating or any other periodic pattern with parallel elements. It is measured by the number of cycles per degree that can be resolved. The resolution threshold is the period of the pattern that can be just resolved. Visual acuity, in such a case, is the reciprocal of one-half of the period expressed in minutes. The resolution threshold for a pair of points or lines is the distance between them when they can just be distinguished as two, not one, expressed in minutes of arc.

resultant color shift: the difference between the perceived color of an object illuminated by a test source and that of the same object illuminated by the reference source, taking account of the state of chromatic adaptation in each case; *i.e.*, the resultant of colorimetric shift and adaptive color shift.

retina: a membrane lining the more posterior part of the inside of the eye. It comprises photoreceptors (cones and rods) that are sensitive to light and nerve cells that transmit to the optic nerve the responses of the receptor elements.

retro-reflector (reflex reflector): a device designed to reflect light in a direction close to that at which it is incident, whatever the angle of incidence.

rhythmic light: a light, when observed from a fixed point, having a luminous intensity that changes periodically. See *equal interval light, flashing light, group flashing light, interrupted quick-flashing light, quick flashing light, occulting light.*

ribbon filament lamp: an incandescent lamp in which the luminous element is a tungsten ribbon.

NOTE: This type of lamp is often used as a standard in pyrometry and radiometry.

rods: retinal receptors which respond at low levels of luminance even down below the threshold for cones. At these levels there is no basis for perceiving differences in hue and saturation. No rods are found in the center of the fovea.

room cavity: the cavity formed by the plane of the luminaires, the work-plane, and the wall surfaces between these two planes.

room cavity ratio, RCR: a number indicating room cavity proportions calculated from length, width and height. See Section 9.

room index: a letter designation for a range of room ratios.

room ratio: a number indicating room proportions,

calculated from the length, width and ceiling height (or luminaire mounting height) above the work-plane.

room utilization factor (utilance): the ratio of the luminous flux (lumens) received on the work-plane to that emitted by the luminaire.

NOTE: This ratio sometimes is called interflectance. Room utilization factor is based on the flux emitted by a complete luminaire, whereas coefficient of utilization is based on the flux generated by the bare lamps in a luminaire.

room surface dirt depreciation (RSDD): the fractional loss of task illuminance due to dirt on the room surface.

runway alignment indicator: a group of aeronautical ground lights arranged and located to provide early direction and roll guidance on the approach to a runway.

runway centerline lights: lights installed in the surface of the runway along the centerline indicating the location and direction of the runway centerline; of particular value in conditions of very poor visibility.

runway-edge lights: lights installed along the edges of a runway marking its lateral limits and indicating its direction.

runway-end identification light: a pair of flashing aeronautical ground lights symmetrically disposed on each side of the runway at the threshold to provide additional threshold conspicuity.

runway exit lights: lights placed on the surface of a runway to indicate a path to the taxiway centerline.

runway lights: aeronautical ground lights arranged along or on a runway. See *runway centerline lights, runway-edge lights, runway-end identification light, runway-exit lights.*

runway threshold: the beginning of the runway usable for landing.

runway visibility: the meteorological visibility along an identified runway. Where a transmissometer is used for measurement, the instrument is calibrated in terms of a human observer; *e.g.*, the sighting of dark objects against the horizon sky during daylight and the sighting of moderately intense unfocused lights of the order of 25 candelas at night. See *visibility (meteorological).*

runway visual range, RVR: in the United States an instrumentally derived value, based on standard calibrations, representing the horizontal distance a pilot will see down the runway from the approach end; it is based on the sighting of either high intensity runway lights or on the visual contrast of other targets—whichever yields the greater visual range.

S

saturation of a perceived color: the attribute according to which it appears to exhibit more or less chromatic color judged in proportion to its brightness. In a given set of viewing conditions, and at luminance levels that result in photopic vision, a stimulus of a given chromaticity exhibits approximately constant saturation for all luminances.

scoop: a floodlight consisting of a lamp in an ellipsoidal or paraboloidal matte reflector, usually in a fixed relationship, though some types permit adjustment of the beam shape.

scotopic vision: vision mediated essentially or exclusively by the rods. It is generally associated with adaptation to a luminance below about 0.034 candela per square meter. *See photopic vision.*

sealed beam lamp. A pressed glass reflector lamp (PAR) that provides a closely controlled beam of light.

NOTE: This term is generally applied in transportation lighting (automotive headlamps, aircraft landing lights, ect.) to distinguish from similar devices in which the light source is replaceable within the reflector-lens unit.

sealed beam headlamp: an integral optical assembly designed for headlighting purposes, identified by the name "Sealed Beam" branded on the lens.

searchlight: a projector designed to produce an approximately parallel beam of light, and having an optical system with an aperture of 20 centimeters (8 inches) or more.

secondary standard source: a constant and reproducible light source calibrated directly or indirectly by comparison with a primary standard. This order of standard also is designated as a reference standard.

NOTE: National secondary (reference) standards are maintained at national physical laboratories; laboratory secondary (reference) standards are maintained at other photometric laboratories.

self-ballasted lamp: any arc discharge lamp of which the current-limiting device is an integral part.

NOTE: Known as blended lamp in some countries.

semi-direct lighting: lighting by luminaires distributing 60 to 90 per cent of their emitted light downward and the balance upward.

semi-indirect lighting: lighting by luminaires distributing 60 to 90 per cent of their emitted light upward and the balance downward.

service period: the number of hours per day for which daylighting provides a specified illuminance. It often is stated as a monthly average.

set light: the separate illumination of the background or set, other than that provided for principal subjects or areas.

shade: a screen made of opaque or diffusing material designed to prevent a light source from being directly visible at normal angles of view.

shielding angle (of a luminaire): the angle between a horizontal line through the light center and the line of sight at which the bare source first becomes visible. See *cut-off angle (of a luminaire).*

short-arc lamp: an arc lamp in which the distance between the electrodes is small (on the order of 1 to 10 millimeters).

NOTE: This type of lamp (*e.g.*, xenon or mercury) is generally of very high pressure.

side back light: illumination from behind the subject in a direction not parallel to a vertical plane through the optical axis of the camera. See *back light*.

side light: lighting from the side to enhance subject modeling and place the subject in depth; apparently separated from background.

side marker lamp: lights indicating the presence and overall length of a vehicle seen from the side.

signal shutter: a device that modulates a beam of light by mechanical means for the purpose of transmitting intelligence.

signalling light: a projector used for directing light signals toward a designated target zone.

size threshold: the minimum perceptible size of an object. It also is defined as the size that can be detected some specific fraction of the times it is presented to an observer, usually 50 per cent. It usually is measured in minutes of arc. See *visual acuity*.

sky factor: the ratio of the illuminance on a horizontal plane at a given point inside a building due to the light received directly from the sky, to the illuminance due to an unobstructed hemisphere of sky of uniform luminance equal to that of the visible sky.

sky light: visible radiation from the sun redirected by the atmosphere.

sky luminance distribution function: for a specified sky condition, the luminance of each point in the sky relative to the zenith luminance.

soft light: diffuse illumination that produces soft edged poorly defined shadows on the background when an object is placed in its path. Also, a luminaire designed to produce such illumination.

solar efficacy: the ratio of the solar illuminance constant to the solar irradiance constant. The current accepted value is 94.2 lumens per watt.

solar illuminance constant: the solar illuminance at normal incidence on a surface in free space at the earth's mean distance from the sun. The current accepted value is 127.5 klx (11,850 footcandles).

solar irradiance constant: the solar irradiance at normal incidence on a surface in free space at the earth's mean distance from the sun. The current accepted value is 1353 watts per square meter (125.7 watts per square foot).

solar radiation simulator: a device designed to produce a beam of collimated radiation having a spectrum, flux density, and geometric characteristics similar to those of the sun outside the earth's atmosphere.

solid angle, ω: a measure of that portion of space about a point bounded by a conic surface whose vertex is at the point. It can be measured by the ratio of intercepted surface area of a sphere centered on that point to the square of the sphere's radius. It is expressed in steradians.

NOTE: Solid angle is a convenient way of expressing the area of light sources and luminaires for computations of discomfort glare factors. It combines into a single number the projected area A_p of the luminaire and the distance D between the luminaire and the eye. It usually is computed by means of the approximate formula

$$\omega = \frac{A_p}{D^2}$$

in which A_p and D^2 are expressed in the same units. This formula is satisfactory when the distance D is greater than about three times the maximum linear dimension of the projected area of the source. Larger projected areas should be sub-divided into several elements.

solid angle factor, Q: a function of the solid angle (ω) subtended by a source at a viewing location and is given by: $Q = 20.4\,\omega + 1.52\,\omega^{0.2} - 0.075$. See *index of sensation*.

spacing: for roadway lighting the distance between successive lighting units, measured along the center line of the street. For interior applications see Section 9.

spacing-to-mounting height ratio, S/MH_{wp}: the ratio of the distance between luminaire centers to the mounting height above the work-plane.

special color rendering index (R_i): measure of the color shift of various standardized special colors including saturated colors, typical foliage, and Caucasian skin. It also can be defined for other color samples when the spectral reflectance distributions are known.

spectral directional emissivity, $\epsilon\,(2\pi, \lambda, T)$ (of an element of surface of a temperature radiator at any wavelength and in a given direction): the ratio of its spectral radiance at that wavelength and in the given direction to that of a blackbody at the same temperature and wavelength.

$$\epsilon(2\pi, \lambda, T) \equiv L_\lambda\,(\lambda, \theta, \phi, T)/L_{\lambda,\text{blackbody}}(\lambda, T)$$

spectral hemispherical emissivity, $\epsilon(\lambda, T)$ (of an element of surface of a temperature radiator): the ratio of its spectral radiant exitance to that of a blackbody at the same temperature.

$$\epsilon(\lambda, T) = \int \epsilon(\theta, \phi, \lambda, T)\cos\theta \cdot d\omega$$

$$\equiv M_\lambda(\lambda, T)/M_{\lambda,\text{blackbody}}(\lambda, T)$$

NOTE: Hemispherical emissivity is frequently called "total" emissivity. "Total" by itself is ambiguous, and should be avoided since it may also refer to "spectral-total" (all wavelengths) as well as directional-total (all directions).

spectral (spectroscopic) lamp: a discharge lamp that radiates a significant portion of its radiative power in a line spectrum and which, in combination with filters, may be used to obtain monochromatic radiation.

spectral luminous efficacy of radiant flux, $K(\lambda) = \Phi_{v\lambda}/\Phi_{e\lambda}$: the quotient of the luminous flux at a given wavelength by the radiant flux at that wavelength. It is expressed in lumens per watt.

NOTE: This term formerly was called "luminosity factor." The reciprocal of the maximum luminous efficacy of radiant flux is sometimes called "mechanical equivalent of light;" that is, the ratio between radiant and luminous flux at the wavelength of maximum luminous efficacy. The most probable value is 0.00146 watt per lumen, corresponding to 683 lumens per watt as the maximum possible luminous efficacy. For scotopic vision values (Fig. 1-5) the maximum luminous efficacy is 1754 "scotopic" lumens per watt.

spectral luminous efficiency for photopic vision, $V(\lambda)$: See *values of spectral luminous efficiency for photopic vision.*

spectral luminous efficiency for scotopic vision, $V'(\lambda)$: See *values of spectral luminous efficiency for scotopic vision.*

spectral luminous efficiency of radiant flux: the ratio of the luminous efficacy for a given wavelength to the value for the wavelength of maximum luminous efficacy. It is dimensionless.

NOTE: The term *spectral luminous efficiency* replaces the previously used terms *relative luminosity* and *relative luminosity factor.*

spectral radiant energy, $Q_\lambda = dQ/d\lambda$: radiant energy per unit wavelength interval at wavelength λ; *e.g.*, joules/nanometer.

spectral radiant exitance: See *radiant flux density.*

spectral radiant flux, $\Phi_\lambda = d\Phi/d\lambda$: radiant flux per unit wavelength interval at wavelengh λ; *e.g.*, watts per nanometer.

spectral radiant intensity, $I_\lambda = dI/d\lambda$: radiant intensity per unit wavelength interval; *e.g.*, watts per (steradian-nanometer).

spectral reflectance of a surface or medium, $\rho(\lambda) = \Phi_{r\lambda}/\Phi_{i\lambda}$: the ratio of the reflected flux to the incident flux at a particular wavelength, λ, or within a small band of wavelengths, $\Delta\lambda$, about λ.

NOTE: The terms *hemispherical, regular,* or *diffuse reflectance* may each be considered restricted to a specific region of the spectrum and may be so designated by the addition of the adjective "spectral."

Fig. 1-5. Scotopic Spectral Luminous Efficiency Values, $V'(\lambda)$
(Unity at Wavelength of Maximum Luminous Efficacy)

Wavelength λ (nanometers)	Relative Value	Wavelength λ (nanometers)	Relative Value
380	0.000589	590	0.0655
390	.002209	600	.03315
400	.00929	610	.01593
410	.03484	620	.00737
420	.0966	630	.003335
430	.1998	640	.001497
440	.3281	650	.000677
450	.455	660	.0003129
460	.567	670	.0001480
470	.676	680	.0000715
480	.793	690	.00003533
490	.904	700	.00001780
500	.982	710	.00000914
510	.997	720	.00000478
520	.935	730	.000002546
530	.811	740	.000001379
540	.650	750	.000000760
550	.481	760	.000000425
560	.3288	770	.0000002413
570	.2076	780	.0000001390
580	.1212		

spectral-total directional emissivity, $\epsilon\,(\theta,\phi,T)$ (of an element of surface of a temperature radiator in a given direction): the ratio of its radiance to that of a blackbody at the same temperature.

$$\epsilon(\theta,\phi,T) \equiv L\,(\theta,\phi,T)/L_{\text{blackbody}}(T)$$

where θ and ϕ are directional angles and T is temperature.

spectral-total hemispherical emissivity, ϵ (of an element of surface of a temperature radiator): the ratio of its radiant exitance to that of a blackbody at the same temperature.

$$\epsilon = \int \epsilon(\theta,\phi)\cdot\cos\theta\cdot d\omega$$

$$\int\int \epsilon(\lambda,\theta,\phi)\cdot\cos\theta\cdot d\omega\cdot d\lambda \equiv M(T)/M_{\text{blackbody}}\,(T)$$

spectral transmittance of a medium $\tau(\lambda) = \Phi_{t\lambda}/\Phi_{i\lambda}$: the ratio of the transmitted flux to the incident flux at a particular wavelength, λ or within a small band of wavelengths, $\Delta\lambda$, about λ.

NOTE: The terms *hemispherical, regular,* or *diffuse transmittance* may each be considered restricted to a specific region of the spectrum and may be so designated by the addition of the adjective "spectral."

spectral tristimulus values: See *color matching functions.*

spectrophotometer: an instrument for measuring the transmittance and reflectance of surfaces and media as a function of wavelength.

spectroradiometer: an instrument for measuring radiant flux as a function of wavelength.

spectrum locus: the locus of points representing the colors of the visible spectrum in a chromaticity diagram.

specular angle: the angle between the perpendicular to the surface and the reflected ray that is numerically equal to the angle of incidence and that lies in the same plane as the incident ray and the perpendicular but on the opposite side of the perpendicular to the surface.

specular reflectance: see *regular reflectance.*

specular reflection: see *regular reflection.*

specular surface: one from which the reflection is predominantly regular. See *regular (specular) reflection.*

speed of light: the speed of all radiant energy, including light, is 2.997925×10^8 meters per second in vacuum (approximately 186,000 miles per second). In all material media the speed is less and varies with the material's index of refraction, which itself varies with wavelength.

speed of vision: the reciprocal of the duration of the exposure time required for something to be seen.

sphere illumination: the illumination on a task from a source providing equal luminance in all directions about that task, such as an illuminated sphere with the task located at the center.

spherical reduction factor: the ratio of the *mean spherical luminous intensity* to the *mean horizontal intensity.*

spotlight: any of several different types of luminaires with relatively narrow beam angle designed to illuminate a specifically defined area. In motion pictures, generic for Fresnel lens luminaires.

standard illuminant: a hypothetical light source of specified relative spectral power distribution.

NOTE: The CIE has specified spectral power distributions for standard illuminants A, B, and C and several D-illuminants. See Fig. 5-2.

standard source: in colorimetry, a source that has a specified spectral distribution and is used as a standard.

standard source A: a tungsten filament lamp operated at a color temperature of 2856 K, and approximating a blackbody operating at that temperature.

standard source B: an approximation of noon sunlight having a correlated color temperature of approximately 4874 K. It is obtained by a combination of Source A and a special filter consisting of a layer one centimeter thick of each of the following solutions, contained in a double cell constructed of nonselective optical glass:

No. 1	
Copper sulphate ($CuSO_4 \cdot 5H_2O$)	2.452 g
Mannite ($C_6H_8(OH)_6$)	2.452 g
Pyridine (C_5H_5N)	30.0 cm^3
Distilled water to make	1000.0 cm^3
No. 2	
Cobalt-ammonium sulphate ($CoSO_4 \cdot (NH_4)_2SO_4 \cdot 6H_2O$)	21.71 g
Copper sulphate ($CuSO_4 \cdot 5H_2O$)	16.11 g
Sulphuric acid (density 1.835)	10.0 cm^3
Distilled water to make	1000.0 cm^3

standard source C: an approximation of daylight provided by a combination of direct sunlight and clear sky having a correlated color temperature of approximately 6774 K. It is obtained by a combination of Source A plus a cell identical with that used for Source B, except that the solutions are:

No. 1	
Copper sulphate ($CuSO_4 \cdot 5H_2O$)	3.412 g
Mannite ($C_6H_8(OH)_6$)	3.412 g
Pyridine (C_5H_5N)	30.0 cm^3
Distilled water to make	1000.0 cm^3
No. 2	
Cobalt-ammonium sulphate ($CoSO_4 \cdot (NH_4)_2SO_4 \cdot 6H_2O$)	30.580 g
Copper sulphate ($CuSO_4 \cdot 5H_2O$)	22.520 g
Sulphuric acid (density 1.835)	10.0 cm^3
Distilled water to make	1000.0 cm^3

starter: a device used in conjunction with a ballast for the purpose of starting an electric-discharge lamp.

state of chromatic adaptation: the condition of the eye in equilibrium with the average color of the visual field.

Stefan-Boltzmann law: the statement that the radiant exitance of a blackbody is proportional to the fourth power of its absolute temperature; that is,

$$M = \sigma T^4$$

NOTE: The currently recommended value of the Stefan-Boltzmann constant σ is 5.67032×10^{-8} W·m^{-2}K^{-4} or 5.67032×10^{-12} W·cm^{-2}K^{-4}.

steradian, sr (unit solid angle): a solid angle subtending an area on the surface of a sphere equal to the square of the sphere radius.

stilb: a CGS unit of luminance. One stilb equals one candela per square centimeter. The use of this term is deprecated.

Stiles-Crawford effect: the reduced luminous efficiency of rays entering the peripheral portion of the pupil of the eye.

stop lamp: a device giving a steady warning light to the rear of a vehicle or train of vehicles, to indicate the intention of the operator to diminish speed or to stop.

stray light (in the eye): light from a source that is scattered onto parts of the retina lying outside the retinal image of the source.

street lighting luminaire: a complete lighting device consisting of a light source and ballast, where appropriate, together with its direct appurtenances such as globe, reflector, refractor housing, and such support as is integral with the housing. The pole, post or bracket is not considered part of the luminaire.

street lighting unit: the assembly of a pole or lamp post with a bracket and a luminaire.

striplight (theatrical): a compartmentalized luminaire, with each compartment containing a lamp, reflector and color frame holder, wired in rotation in three- or four-circuit and used as borderlights, footlights or cyclorama lighting from above or below.

stroboscopic lamp (strobe light): a flash tube designed for repetitive flashing.

subjective brightness: the subjective attribute of any light sensation giving rise to the percept of luminous magnitude, including the whole scale of qualities of being bright, light, brilliant, dim or dark. See *saturation of a perceived color.*

NOTE: The term brightness often is used when referring to the measurable luminance. While the context usually makes it clear as to which meaning is intended, the preferable term for the photometric quantity is *luminance*, thus reserving *brightness* for the subjective sensation.

sun bearing: the angle measured in the plane of the horizon between a vertical plane at a right angle to the window wall and the position of this plane after it has been rotated to contain the sun.

sun light: direct visible radiation from the sun.

sunlamp: an ultraviolet lamp that radiates a significant portion of its radiative power in the UV-B band (280 to 315 nanometers).

supplementary lighting: lighting used to provide an additional quantity and quality of illumination that cannot readily be obtained by a general lighting system and that supplements the general lighting level, usually for specific work requirements.

supplementary standard illuminant D_{55}: a representation of a phase of daylight at a correlated color temperature of approximately 5500 K.

supplementary standard illuminant D_{75}: a representation of a phase of daylight at a correlated color temperature of approximately 7500 K.

surface mounted luminaire: a luminaire mounted

directly on the ceiling.

suspended (pendant) luminaire: a luminaire hung from a ceiling by supports.

switch start fluorescent lamps: see *preheat fluorescent lamp.*

T

table lamp: a portable luminaire with a short stand suitable for standing on furniture.

tail lamp: a lamp used to designate the rear of a vehicle by a warning light.

talbot, T.: a unit of light; equal to one lumen-second.

tanning lamp: an ultraviolet lamp that radiates a significant portion of its radiative power in the UV-A and/or B band.

task lighting: lighting directed to a specific surface or area that provides illumination for visual tasks.

task-ambient lighting: a combination of task lighting and ambient lighting within an area such that the general level of ambient lighting is lower than and complementary to the task lighting.

taxi-channel lights: aeronautical ground lights arranged along a taxi-channel of a water aerodrome to indicate the route to be followed by taxiing aircraft.

taxi light: an aircraft aeronautical light designed to provide necessary illumination for taxiing.

taxiway lights: aeronautical ground lights provided to indicate the route to be followed by taxiing aircraft. See *taxiway-centerline lights, taxiway-edge lights, taxiway holding-post light.*

taxiway-centerline lights: taxiway lights placed along the centerline of a taxiway except that on curves or corners having fillets, these lights are placed a distance equal to half the normal width of the taxiway from the outside edge of the curve or corner.

taxiway-edge lights: taxiway lights placed along or near the edges of a taxiway.

taxiway holding-post light: a light or group of lights installed at the edge of a taxiway near an entrance to a runway, or to another taxiway, to indicate the position at which the aircraft should stop and obtain clearance to proceed.

temperature radiator: a radiator whose radiant flux density (radiant exitance) is determined by its temperature and the material and character of its surface, and is independent of its previous history. See *blackbody* and *graybody.*

thermopile: a thermal radiation detector consisting of a number of thermocouples interconnected in order to increase the sensitivity to incident radiant flux.

threshold: the value of a variable of a physical stimulus (such as size, luminance, contrast or time) that permits the stimulus to be seen a specific percentage of the time or at a specific accuracy level. In many psychophysical experiments, thresholds are presented in terms of 50 per cent accuracy or accurately 50 per cent of the time. However, the threshold also is expressed as the value of the physical variable that permits the object to be just barely seen. The threshold may be determined by merely detecting the presence of an object or it may be determined by discriminating certain details of the object. See *absolute luminance threshold, brightness contrast threshold, luminance threshold, modulation size threshold.*

threshold lights: runway lights placed to indicate the longitudinal limits of that portion of a runway, channel or landing path usable for landing.

top light: illumination of a subject directly from above employed to outline the upper margin or edge of the subject.

torchere: an indirect floor lamp sending all or nearly all of its light upward.

tormentor light: luminaire mounted directly behind the sides of the stage arch.

total emissivity: See *spectral-total directional emissivity* and *spectral-total hemispherical emissivity.*

touchdown zone lights: barettes of runway lights installed in the surface of the runway between the runway edge lights and the runway centerline lights to provide additional guidance during the touchdown phase of a landing in conditions of very poor visibility.

traffic beam: See *lower (passing) beams.*

train: the angle between the vertical plane through the axis of the searchlight drum and the plane in which this plane lies when the search light is in a position designated as having zero train.

transient adaptation factor, TAF: a factor which reduces the *equivalent contrast* due to readaptation from one luminous background to another.

transmission: a general term for the process by which incident flux leaves a surface or medium on a side other than the incident side, without change in frequency.

NOTE: Transmission through a medium is often a combination of regular and diffuse transmission. See *regular transmission, diffuse transmission,* and *transmittance.*

transmissometer: a photometer for measuring transmittance.

NOTE: Transmissometers may be visual or physical instruments.

transmittance, $\tau = \Phi_t/\Phi_i$**:** the ratio of the transmitted flux to the incident flux.

NOTE: Measured values of transmittance depend upon the angle of incidence, the method of measurement of the transmitted flux, and the spectral character of the incident flux. Because of this dependence complete information on the technique and conditions of measurement should be specified.

It should be noted that transmittance refers to the ratio of flux emerging to flux incident; therefore, reflections at the surface as well as absorption within the material operate to reduce the transmittance.

tristimulus values of a light, X, Y, Z**:** the amounts of each of three primaries required to match the color of the light.

troffer: a recessed lighting unit, usually long and installed with the opening flush with the ceiling. The term is derived from "trough" and "coffer."

troland: a unit of retinal illuminance which is based upon the fact that retinal illuminance is proportional to the product of the luminance of the distal stimulus and the area of entrance pupil. One troland is the retinal illuminance produced when the luminance of the distal stimulus is one candela per square meter and the area of the pupil is one square millimeter.

NOTE: The troland makes no allowance for interocular attenuation or for the *Stiles-Crawford effect.*

tube: See *lamp.*

tungsten-halogen lamp: a gas filled tungsten incandescent lamp containing a certain proportion of halogens in an inert gas whose pressure exceeds three atmospheres.

NOTE: The tungsten-iodine lamp (UK) and quartz-iodine lamp (USA) belong to this category.

turn signal operating unit: that part of a signal system by which the operator of a vehicle indicates the direction a turn will be made, usually by a flashing light.

U

ultraviolet lamp: a lamp which radiates a significant portion of its radiative power in the ultraviolet (UV) part of the spectrum; the visible radiation is not of principal interest.

ultraviolet radiation: for practical purposes any radiant energy within the wavelength range 10 to 380 nanometers. See *regions of electromagnetic spectrum.*

NOTE: On the basis of practical applications and the effect obtained, the ultraviolet region often is divided into the following bands:

Ozone-producing	180–220 nanometers
Bactericidal (germicidal)	220–300 nanometers
Erythemal	280–320 nanometers
"Black light"	320–400 nanometers

There are no sharp demarcations between these bands, the indicated effects usually being produced to a lesser extent by longer and shorter wavelengths. For engineering purposes, the "black light" region extends slightly into the visible portion of the spectrum. Another division of the ultraviolet spectrum often used by photobiologists is given by the CIE:

UV-A	315-400 nanometers
UV-B	280-315 nanometers
UV-C	100–280 nanometers

units of luminance: the luminance of a surface in a specified direction may be expressed in luminous intensity per unit of projected area of surface or in luminous flux per unit of solid angle and per unit of projected surface area.

NOTE: Typical units are the candela per square meter (lumen per steradian and per square meter) and the candela per square foot (lumen per steradian and per square foot).

The luminance of a surface in a specified direction is also expressed (incorrectly) in lambertian units as the number of lumens per unit area that would leave the surface *if the luminance in all directions within the hemisphere on the side of the surface being considered were the same as the luminance in the specified direction.*

NOTE: A typical unit in this system is the footlambert, equal to one lumen per square foot.

This method of specifying luminance is equivalent to stating the number of lumens that would leave the surface *if the surface were replaced by a perfectly diffusing surface with a luminance in all directions within the hemisphere equal to the luminance of the actual surface in the direction specified.* In practice no surface follows exactly the cosine formula of emission or reflection; hence the luminance is not uniform but varies with the angle from which it is viewed. For this reason, this practice is denigrated.

unrecoverable light loss factors: factors which give the fractional light loss that cannot be recovered by cleaning or lamp replacement.

upper (driving) beams: one or more beams intended for distant illumination and for use on the open highway when not meeting other vehicles. Formerly "country beam." See *lower (passing) beams.*

upward component: that portion of the luminous flux from a luminaire emitted at angles above the horizontal. See *downward component.*

utilance: See *room utilization factor.*

V

vacuum lamp: an incandescent lamp in which the filament operates in an evacuated bulb.

valance: a longitudinal shielding member mounted across the top of a window or along a wall and usually parallel to the wall, to conceal light sources giving both upward and downward distributions.

valance lighting: lighting comprising light sources shielded by a panel parallel to the wall at the top of a window.

values of spectral luminous efficiency for photopic vision, $V(\lambda)$: values for spectral luminous efficiency at 10-nanometer intervals (see Fig. 1–6) were provisionally adopted by the CIE in 1924 and were adopted in 1933 by the International Committee on Weights and Measures as a basis for the establishment of photometric standards of types of sources differing from the primary standard in spectral distribution of radiant flux. These values are given in the second column of Fig. 1–6; the intermediate values given in the other columns have been interpolated.

NOTE: These standard values of spectral luminous efficiency were determined by observations with a two-degree photometric field having a moderately high luminance, and photometric evaluations based upon them consequently do not apply exactly to other conditions of observation. Power in watts weighted in accord with these standard values are often referred to as *lightwatts.*

values of spectral luminous efficiency for scotopic vision $V'(\lambda)$: values of spectral luminous efficiency at 10-nanometer intervals (see Fig. 1–5) were provisionally adopted by the CIE in 1951.

Fig. 1-6. Photopic Spectral Luminous Efficiency, $V(\lambda)$
(Unity at Wavelength of Maximum Luminous Efficacy)

Wavelength λ (nanometers)	Standard Values	Values Interpolated at Intervals of One Nanometer								
		1	2	3	4	5	6	7	8	9
380	0.00004	0.000045	0.000049	0.000054	0.000058	0.000064	0.000071	0.000080	0.000090	0.000104
390	.00012	.000138	.000155	.000173	.000193	.000215	.000241	.000272	.000308	.000350
400	.0004	.00045	.00049	.00054	.00059	.00064	.00071	.00080	.00090	.00104
410	.0012	.00138	.00156	.00174	.00195	.00218	.00244	.00274	.00310	.00352
420	.0040	.00455	.00515	.00581	.00651	.00726	.00806	.00889	.00976	.01066
430	.0116	.01257	.01358	.01463	.01571	.01684	.01800	.01920	.02043	.02170
440	.023	.0243	.0257	.0270	.0284	.0298	.0313	.0329	.0345	.0362
450	.038	.0399	.0418	.0438	.0459	.0480	.0502	.0525	.0549	.0574
460	.060	.0627	.0654	.0681	.0709	.0739	.0769	.0802	.0836	.0872
470	.091	.0950	.0992	.1035	.1080	.1126	.1175	.1225	.1278	.1333
480	.139	.1448	.1507	.1567	.1629	.1693	.1761	.1833	.1909	.1991
490	.208	.2173	.2270	.2371	.2476	.2586	.2701	.2823	.2951	.3087
500	.323	.3382	.3544	.3714	.3890	.4073	.4259	.4450	.4642	.4836
510	.503	.5229	.5436	.5648	.5865	.6082	.6299	.6511	.6717	.6914
520	.710	.7277	.7449	.7615	.7776	.7932	.8082	.8225	.8363	.8495
530	.862	.8739	.8851	.8956	.9056	.9149	.9238	.9320	.9398	.9471
540	.954	.9604	.9661	.9713	.9760	.9803	.9840	.9873	.9902	.9928
550	.995	.9969	.9983	.9994	1.0000	1.0002	1.0001	.9995	.9984	.9969
560	.995	.9926	.9898	.9865	.9828	.9786	.9741	.9691	.9638	.9581
570	.952	.9455	.9386	.9312	.9235	.9154	.9069	.8981	.8890	.8796
580	.870	.8600	.8496	.8388	.8277	.8163	.8046	.7928	.7809	.7690
590	.757	.7449	.7327	.7202	.7076	.6949	.6822	.6694	.6565	.6437
600	.631	.6182	.6054	.5926	.5797	.5668	.5539	.5410	.5282	.5156
610	.503	.4905	.4781	.4568	.4535	.4412	.4291	.4170	.4049	.3929
620	.381	.3690	.3570	.3449	.3329	.3210	.3092	.2977	.2864	.2755
630	.265	.2548	.2450	.2354	.2261	.2170	.2082	.1996	.1912	.1830
640	.175	.1672	.1596	.1523	.1452	.1382	.1316	.1251	.1188	.1128
650	.107	.1014	.0961	.0910	.0862	.0816	.0771	.0729	.0688	.0648
660	.061	.0574	.0539	.0506	.0475	.0446	.0418	.0391	.0366	.0343
670	.032	.0299	.0280	.0263	.0247	.0232	.0219	.0206	.0194	.0182
680	.017	.01585	.01477	.01376	.01281	.01192	.01108	.01030	.00956	.00886
690	.0082	.00759	.00705	.00656	.00612	.00572	.00536	.00503	.00471	.00440
700	.0041	.00381	.00355	.00332	.00310	.00291	.00273	.00256	.00241	.00225
710	.0021	.001954	.001821	.001699	.001587	.001483	.001387	.001297	.001212	.001130
720	.00105	.000975	.000907	.000845	.000788	.000736	.000688	.000644	.000601	.000560
730	.00052	.000482	.000447	.000415	.000387	.000360	.000335	.000313	.000291	.000270
740	.00025	.000231	.000214	.000198	.000185	.000172	.000160	.000149	.000139	.000130
750	.00012	.000111	.000103	.000096	.000090	.000084	.000078	.000074	.000069	.000064
760	.00006	.000056	.000052	.000048	.000045	.000042	.000039	.000037	.000035	.000032

NOTE: These values of spectral luminous efficiency were determined by observation by young dark-adapted observers using extra-foveal vision at near-threshold luminance.

vapor-tight luminaire: a luminaire designed and approved for installation in damp or wet locations. It also is described as "enclosed and gasketed."

VASIS (Visual Approach Slope Indicator System): the system of angle-of-approach lights accepted as a standard by the International Civil Aviation Organization, comprising two bars of lights located at each side of the runway near the threshold and showing red or white or a combination of both (pink) to the approaching pilot depending upon his position with respect to the glide path.

veiling luminance: a luminance superimposed on the retinal image which reduces its contrast. It is this veiling effect produced by bright sources or areas in the visual field that results in decreased visual performance and visibility.

veiling reflection: regular reflections superimposed upon diffuse reflections from an object that partially or total obscure the details to be seen by reducing the contrast. This sometimes is called reflected glare.

vertical plane of a searchlight: the plane through the axis of the searchlight drum which contains the elevation angle. See *horizontal plane of a searchlight*

visibility: the quality or state of being perceivable by the eye. In many outdoor applications, visibility is defined in terms of the distance at which an object can be just perceived by the eye. In indoor applications it usually is defined in terms of the contrast or size of a standard test object, observed under standardized view-conditions, having the same threshold as the given object. See *visibility (meteorological)*.

visibility (meteorological): a term that denotes the greatest distance, expressed in kilometers or miles, that selected objects (visibility markers) or lights of moderate intensity (25 candelas) can be seen and identified under specified conditions of observation.

visibility level, VL: a contrast multiplier to be

applied to the visibility reference function to provide the luminance contrast required at different levels of task background luminance to achieve visibility for specified conditions relating to the task and observer.

visibility performance criteria function, VL8: a function representing the luminance contrast required to achieve 99 percent visual certainty for the same task used for the visibility reference function, including the effects of dynamic presentation and uncertainty in task location.

visibility reference function, VL1: a function representing the luminance contrast required at different levels of task background luminance to achieve visibility threshold for the visibility reference task consisting of a 4 minute disk exposed for ⅕ second.

vision: See *central vision, foveal vision, mesopic vision, peripheral vision, photopic vision and scotopic vision.*

visual acuity: a measure of the ability to distinguish fine details. Quantitatively, it is the reciprocal of the minimum angular separation in minutes of two lines of width subtending one minute of arc when the lines are just resolvable as separate. See *size threshold.*

visual angle: the angle subtended by an object or detail at the point of observation. It usually is measured in minutes of arc.

visual approach slope indicator system: see *VASIS*

visual comfort probability, VCP: the rating of a lighting system expressed as a per cent of people who, when viewing from a specified location and in a specified direction, will be expected to find it acceptable in terms of discomfort glare. *Visual comfort probability* is related to *discomfort glare rating (DGR).*

visual field: the locus of objects or points in space that can be perceived when the head and eyes are kept fixed. The field may be monocular or binocular. See *monocular visual field, binocular visual field, central visual field, peripheral visual field.*

visual perception: the interpretation of impressions transmitted from the retina to the brain in terms of information about a physical world displayed before the eye.

NOTE: Visual perception involves any one or more of the following: recognition of the presence of something (object, aperture or medium); identifying it; locating it in space; noting its relation to other things; identifying its movement, color, brightness or form.

visual performance: the quantitative assessment of the performance of a visual task, taking into consideration speed and accuracy.

visual photometer: a photometer in which the equality of brightness of two surfaces is established visually. See *physical photometer.*

NOTE: The two surfaces usually are viewed simultaneously side by side. This method is used in portable visual luminance meters. This is satisfactory when the color difference between the test source and comparison source is small. However, when there is a color difference, a flicker photometer provides more precise measurements. In this type of photometer the two surfaces are viewed alternately at such a rate that the color sensations either nearly or completely blend and the flicker due to brightness difference is balanced by adjusting the comparison source.

visual range (of a light or object): the maximum distance at which a particular light (or object) can be seen and identified.

visual surround: includes all portions of the visual field except the visual task.

visual task: conventionally designates those details and objects that must be seen for the performance of a given activity, and includes the immediate background of the details or objects.

NOTE: The term visual task as used is a misnomer because it refers to the visual display itself and not the task of extracting information from it. The task of extracting information also has to be differentiated from the overall task performed by the observer.

visual task evaluator, VTE: a contrast reducing instrument which permits obtaining a value of luminance contrast, called the *equivalent contrast* $\tilde{C}$ of a standard visibility reference task giving the same visibility as that of a task whose contrast has been reduced to threshold when the background luminances are the same for the task and the reference task. See *equivalent contrast.*

voltage to luminaire factor: the fractional loss of task illuminance due to improper voltage at the luminaire.

W

wavelengths: wavelength is the distance between two successive points of a periodic wave in the direction of propagation, in which the oscillation has the same phase. The three commonly used units are listed in the following table:

Name	Symbol	Value
Micrometer	μm	1 μm = 10^{-6} m
Nanometer	nm	1 nm = 10^{-9} m
Angstrom*	Å	1 Å = 10^{-10} m

Weber's fraction: See *luminous luminance contrast*

wide-angle diffusion: diffusion in which flux is scattered at angles far from the direction that the flux would take by regular reflection or transmission. See *narrow-angle diffusion.*

wide-angle luminaire: a luminaire distributing the light through a comparatively large solid angle. See also *narrow-angle luminaire.*

width line: the radial line (the one that makes the larger angle with the reference line) that passes through the point of one-half maximum candlepower on the lateral candlepower distribution curve plotted on the surface of the cone of maximum candlepower.

* The use of this unit is deprecated.

Wien displacement law: an expression representing, in a functional form, the spectral radiance of a blackbody as a function of the wavelength and the temperature.

$$L_\lambda = dI_\lambda/dA' = C_1\lambda^{-5}f(\lambda T)$$

The two principal corollaries of this law are:

$$\lambda_m T = b$$
$$L_m/T^5 = b'$$

which show how the maximum spectral radiance L_m and the wavelength λ_m at which it occurs are related to the absolute temperature T. See *Wien radiation law*.

NOTE: The currently recommended value of b is 2.898×10^{-3} m·K or 2.898×10^{-1} cm·K. From the Planck radiation law, and with the use of the value of b, c_1, and c_2 as given above, b' is found to be 4.10×10^{-12} W·cm^{-3}·K^{-5}·sr^{-1}.

Wien radiation law: an expression representing approximately the spectral radiance of a blackbody as a function of its wavelength and temperature. It commonly is expressed by the formula

$$L_\lambda = dI_\lambda/dA' = c_{1L}\lambda^{-5}e^{-(c_2/\lambda T)}$$

This formula is accurate to one per cent or better for values of λT less than 3000 micrometer-kelvin.

wing clearance lights: aircraft lights provided at the wing tips to indicate the extent of the wing span when the navigation lights are located an appreciable distance inboard of the wings tips.

work-plane: the plane at which work usually is done, and on which the illuminance is specified and measured. Unless otherwise indicated, this is assumed to be a horizontal plane 0.76 meters (30 inches) above the floor.

working standard: a standardized light source for regular use in photometry.

Z

zonal-cavity inter-reflectance method: a procedure for calculating coefficients of utilization, wall luminance coefficients, and ceiling cavity luminance coefficients taking into consideration the luminaire intensity distribution, room size and shape (cavity ratio concepts), and room reflectances. It is based on *flux transfer theory*.

zonal constant: a factor by which the mean candlepower emitted by a source of light in a given angular zone is multiplied to obtain the lumens in the zone. See Fig. 4-17.

zonal-factor method: a procedure for predetermining, from typical luminaire photometric data in discrete angular zones, the proportion of luminaire output which would be incident initially (without interreflections) on the work-plane, ceiling, walls and floor of a room.

zonal factor interflection method: a former procedure for calculating coefficients of utilization.

Fig. 1-7. Illuminance Conversion Factors.

1 lumen = 1/683 light-watt
1 lumen-hour = 60 lumen-minutes
1 footcandle = 1 lumen/square foot
1 watt-second = 10^7 ergs
1 phot = 1 lumen/square centimeter
1 lux = 1 lumen/square meter = 1 metercandle

Multiply Number of → By / To Obtain Number of ↓	Footcandles	Lux	Phot	Milliphot
Footcandles	1	0.0929	929	0.929
Lux	10.76	1	10,000	10
Phot	0.00108	0.0001	1	0.001
Milliphot	1.076	0.1	1,000	1

Fig. 1-8. Greek Alphabet (Capital and Lower Case).

Capital	Lower Case	Greek Name
Α	α	Alpha
Β	β	Beta
Γ	γ	Gamma
Δ	δ	Delta
Ε	ϵ	Epsilon
Ζ	ζ	Zeta
Η	η	Eta
Θ	θ	Theta
Ι	ι	Iota
Κ	κ	Kappa
Λ	λ	Lambda
Μ	μ	Mu
Ν	ν	Nu
Ξ	ξ	Xi
Ο	o	Omicron
Π	π	Pi
Ρ	ρ	Rho
Σ	σ, ς	Sigma
Τ	τ	Tau
Υ	υ	Upsilon
Φ	φ, ϕ	Phi
Χ	χ	Chi
Ψ	ψ	Psi
Ω	ω	Omega

Fig. 1-9. Unit Prefixes.

Prefix	Symbol	Factor by Which the Unit is Multiplied
exa	E	1,000,000,000,000,000,000 = 10^{18}
peta	P	1,000,000,000,000,000 = 10^{15}
tera	T	1,000,000,000,000 = 10^{12}
giga	G	1,000,000,000 = 10^9
mega	M	1,000,000 = 10^6
kilo	k	1,000 = 10^3
hecto	h	100 = 10^2
deka	da	10 = 10^1
deci	d	0.1 = 10^{-1}
centi	c	0.01 = 10^{-2}
milli	m	0.001 = 10^{-3}
micro	μ	0.000,001 = 10^{-6}
nano	n	0.000,000,001 = 10^{-9}
pico	p	0.000,000,000,001 = 10^{-12}
femto	f	0.000,000,000,000,001 = 10^{-15}
atto	a	0.000,000,000,000,000,001 = 10^{-18}

Fig. 1-10. Partial list of abbreviations and acronyms used in the Handbook.

A

A	ampere
Å	Angstrom unit
ac	alternating current
AIA	American Institute of Architects
ANSI	American National Standards Institute
ASID	American Society of Interior Designers
ASTM	American Society for Testing and Materials
ASHRAE	American Society of Heating, Refrigerating and Air-conditioning Engineers
atm	atmosphere

B

BCD	borderline between comfort and discomfort
BCP	beam candlepower
BL	blacklight
BRDF	bidirectional reflectance-distribution function
Btu	British thermal unit

C

°C	degree Celsius
cal	calorie
CBM	Certified Ballast Manufacturers
CBU	coefficient of beam utilization
CCR	ceiling cavity ratio
cgs	centimeter-gram-second (system)
CIE	Commission Internationale de l'Eclairage (International Commission on Illumination)
cm	centimeter
cos	cosine
cp	candlepower
CPI	color preference index
CRF	contrast rendition factor
CRI	color rendering index
CRT	cathode ray tube
CSA	Canadian Standards Association
CSI	compact source iodide
CU	coefficient of utilization
CW	cool white
CWX	cool white deluxe

D

dB	decibel
dc	direct current
DGF	disability glare factor
DGR	disability glare rating
DIC	direct illumination component

E

emf	electromotive force
ESI	equivalent sphere illumination
EU	erythemal unit

F

°F	degree Fahrenheit
fc	footcandle
FCR	floor cavity ratio
fff	flicker fusion frequency
ft	foot
ft^2	square foot

H

h	hour
HID	high intensity discharge
HMI	hydrargyrum, medium-arc-length, iodide
hp	horse power
HPS	high pressure sodium
Hz	hertz

I

IALD	International Association of Lighting Designers
IDSA	Industrial Designers Society of America
IEEE	Institute of Electrical and Electronics Engineers
IERI	Illuminating Engineering Research Institute
in	inch
in^2	square inch
IR	infrared
ISO	International Organization for Standardization

J

J	joule

K

K	kelvin
kcal	kilocalorie
kg	kilogram
kHz	kilohertz
km	kilometer
km^2	square kilometer
km/s	kilometer per second
kV	kilovolt
kVA	kilovolt-ampere
kVAr	reactive kilovolt-ampere
kW	kilowatt
kWh	kilowatt-hour

L

LBO	lamp burnout
LCD	liquid crystal display
LDD	luminaire dirt depreciation
LED	light emitting diode
LEF_v	lighting effectiveness factor
LLD	lamp lumen depreciation
LLF	light loss factor
lm	lumen
ln	logarithm (natural)
LPS	low pressure sodium
lx	lux

M

m	meter
m^2	square meter
mA	milliampere
max	maximum
MF	maintenance factor
MH	mounting height
MHz	megahertz
min	minimum
min	minute (time)
mm	millimeter
mm^2	square millimeter
mol wt	molecular weight
MPE	minimal perceptible erythema
mph	mile per hour

N

NBS	National Bureau of Standards
NEC	National Electrical Code
NEMA	National Electrical Manufacturers Association
nm	nanometer

Fig. 1-10. *Continued*

O

OSA	Optical Society of America

P

PAR	pressed reflector lamp
pf	power factor

R

R	reflectance factor
rad	radian
RCR	room cavity ratio
RCS	relative contrast sensitivity
rms	root mean square
RSDD	room surface dirt depreciation
RTP	relative task performance
RVP	relative visual performance
RVR	runway visual range

S

s	second (time)
sin	sine
SPD	spectral power distribution
sq	square
sr	steradian

T

TAF	transient adaptation factor
tan	tangent
temp	temperature

U

UV	ultraviolet
UL	Underwriters Laboratories

V

V	volt
VA	volt-ampere
VAr	reactive volt-ampere
VASIS	visual approach slope indicator system
VCP	visual comfort probability
VDU	visual display unit
VHO	very high output (lamp)
VI	visibility index
VL	visibility level
VTE	visual task evaluator
VTP	visual task photometer

W

W	watt
WW	warm white
WWX	warm white deluxe
μA	microampere
μV	microvolt
μW	microwatt
ρ_{CC}	effective ceiling cavity reflectance
ρ_{FC}	effective floor cavity reflectance
′	minute (angular measure)
″	second (angular measure)
°	degree

Fig. 1-11. Conversion Factors for Units of Length

Multiply Number of → / To Obtain Number of ↓ By ↘	Angstroms	Nanometers	Micrometers (Microns)	Millimeters	Centimeters	Meters	Kilometers	Mils	Inches	Feet	Miles
Angstroms	1	10	10^4	10^7	10^8	10^{10}	10^{13}	2.540×10^5	2.540×10^8	3.048×10^9	1.609×10^{13}
Nanometers	10^{-1}	1	10^3	10^6	10^7	10^9	10^{12}	2.540×10^4	2.540×10^7	3.048×10^8	1.609×10^{12}
Micrometers (Microns)	10^{-4}	10^{-3}	1	10^3	10^4	10^6	10^9	2.540×10	2.540×10^4	3.048×10^5	1.609×10^9
Millimeters	10^{-7}	10^{-6}	10^{-3}	1	10	10^3	10^6	2.540×10^{-2}	2.540×10	3.048×10^2	1.609×10^6
Centimeters	10^{-8}	10^{-7}	10^{-4}	10^{-1}	1	10^2	10^5	2.540×10^{-3}	2.540	3.048×10	1.609×10^5
Meters	10^{-10}	10^{-9}	10^{-6}	10^{-3}	10^{-2}	1	10^3	2.540×10^{-5}	2.540×10^{-2}	3.048×10^{-1}	1.609×10^3
Kilometers	10^{-13}	10^{-12}	10^{-9}	10^{-6}	10^{-5}	10^{-3}	1	2.540×10^{-8}	3.048×10^{-5}	3.048×10^{-4}	1.609
Mils	3.937×10^{-6}	3.937×10^{-5}	3.937×10^{-2}	3.937×10	3.937×10^2	3.937×10^4	3.937×10^7	1	10^3	1.2×10^4	6.336×10^7
Inches	3.937×10^{-9}	3.937×10^{-8}	3.937×10^{-5}	3.937×10^{-2}	3.937×10^{-1}	3.937×10	3.937×10^4	10^{-3}	1	12	6.336×10^4
Feet	3.281×10^{-10}	3.281×10^{-9}	3.281×10^{-6}	3.281×10^{-3}	3.281×10^{-2}	3.281	3.281×10^3	8.333×10^{-5}	8.333×10^{-2}	1	5.280×10^3
Miles	6.214×10^{-14}	6.214×10^{-13}	6.214×10^{-10}	6.214×10^{-7}	6.214×10^{-6}	6.214×10^{-4}	6.214×10^{-1}	1.578×10^{-8}	1.578×10^{-5}	1.894×10^{-4}	1

Fig. 1-12. Conversion from Values in SI Units.

m / cm / kcd/m² / cd/m² / lx*	lx* → fc	cd/m² → fL	kcd/m² → cd/in²	cm → in	m → ft
1	.09	.29	.65	.39	3.3
2	.19	.58	1.29	.79	6.6
3	.28	.88	1.94	1.18	9.8
4	.37	1.17	2.58	1.57	13.1
5	.47	1.46	3.23	1.97	16.4
6	.56	1.75	3.87	2.36	19.7
7	.65	2.04	4.52	2.76	23.0
8	.74	2.34	5.16	3.15	26.2
9	.84	2.63	5.81	3.54	29.5
100	9.3	29.2	64.5	39.4	328
110	10.2	32.1	71.0	43.3	361
120	11.1	35.0	77.4	47.2	394
130	12.1	37.9	83.9	51.2	427
140	13.0	40.9	90.3	55.1	459
150	13.9	43.8	96.8	59.1	492
160	14.9	46.7	103.2	63.0	525
170	15.8	49.6	109.7	66.9	558
180	16.7	52.5	116.1	70.9	591
190	17.7	55.5	122.6	74.8	623
200	18.6	58.4	129.0	78.7	656
210	19.5	61.3	135.5	82.7	689
220	20.4	64.2	141.9	86.6	722
230	21.4	67.1	148.4	90.6	755
240	22.3	70.1	154.8	94.5	787
250	23.2	73.0	161.3	98.4	820
260	24.2	75.9	167.7	102.4	853
270	25.1	78.8	174.2	106.3	886
280	26.0	81.7	180.6	110.2	919
290	26.9	84.7	187.1	114.2	951
300	27.9	87.6	193.5	118.1	984
310	28.8	90.5	200.0	122.0	1017
320	29.7	93.4	206.4	126.0	1050
330	30.7	96.3	212.9	130.0	1083
340	31.6	99.2	219.3	133.9	1116
350	32.5	102.2	225.8	137.8	1148
360	33.4	105.8	232.2	141.7	1181
370	34.4	108.0	238.7	145.7	1214
380	35.3	110.9	245.1	149.6	1247
390	36.2	113.8	251.6	153.5	1280
400	37.2	116.8	258.0	157.5	1312
410	38.1	119.7	264.5	161.4	1345
420	39.0	122.6	270.9	165.4	1378
430	39.9	125.5	277.4	169.3	1411
440	40.9	128.4	283.8	173.2	1444
450	41.8	131.4	290.3	177.2	1476
460	42.7	134.3	296.7	181.1	1509
470	43.7	137.2	303.2	185.0	1542
480	44.6	140.1	309.6	189.0	1575
490	45.5	143.0	316.1	192.9	1608

m / cm / kcd/m² / cd/m² / lx*	lx* → fc	cd/m² → fL	kcd/m² → cd/in²	cm → in	m → ft
500	46.5	146.0	322.5	196.9	1641
510	47.4	148.9	329.0	200.8	1673
520	48.3	151.8	335.4	204.7	1706
530	49.2	154.7	341.9	208.7	1739
540	50.2	157.6	348.3	212.6	1772
550	51.1	160.5	354.8	216.5	1805
560	52.0	163.5	361.2	220.5	1837
570	53.0	166.4	367.7	224.4	1870
580	53.9	169.3	374.1	228.3	1903
590	54.8	172.2	380.6	232.3	1936
600	55.7	175.1	387.0	236.2	1969
610	56.7	178.1	393.5	240.2	2001
620	57.6	181.0	399.9	244.1	2034
630	58.5	183.9	406.4	248.0	2067
640	59.5	186.8	412.8	252.0	2100
650	60.4	189.7	419.3	255.9	2133
660	61.3	192.7	425.7	259.8	2165
670	62.2	195.6	432.2	263.8	2198
680	63.2	198.5	438.6	267.7	2231
690	64.1	201.4	445.1	271.7	2264
700	65.0	204.3	451.5	275.6	2297
710	66.0	207.2	458.0	279.5	2330
720	66.9	210.2	464.4	283.5	2362
730	67.8	213.1	470.9	287.4	2395
740	68.7	216.0	477.3	291.3	2428
750	69.7	218.9	483.8	295.3	2461
760	70.6	221.8	490.2	299.2	2494
770	71.5	224.8	496.7	303.1	2526
780	72.5	227.7	503.1	307.1	2559
790	73.4	230.6	509.6	311.0	2592
800	74.3	233.5	516.0	315.0	2625
810	75.2	236.4	522.5	318.9	2658
820	76.2	239.4	528.9	322.8	2690
830	77.1	242.3	535.4	326.8	2723
840	78.0	245.2	541.8	330.7	2756
850	79.0	248.1	548.3	334.6	2789
860	79.9	251.0	554.7	338.6	2822
870	80.8	254.0	561.2	342.5	2854
880	81.8	256.9	567.6	346.5	2887
890	82.7	259.8	574.1	350.4	2920
900	83.6	262.7	580.5	354.3	2953
910	84.5	265.6	587.0	358.3	2986
920	85.5	268.5	593.4	362.2	3019
930	86.4	271.5	600.0	366.1	3051
940	87.3	274.4	606.3	370.1	3084
950	88.3	277.3	612.8	374.0	3117
960	89.2	280.2	619.2	378.0	3150
970	90.1	283.1	625.7	381.9	3183
980	91.0	286.1	632.1	385.8	3215
990	92.0	289.0	638.6	389.8	3248

* Also useful for converting from ft² to m².

Fig. 1-13. Conversion to Values in SI Units

ft / in / cd/in² / fL / fc*	lx	cd/m²	kcd/m²	cm	m
1	10.76	3.4	1.55	2.54	.30
2	21.5	6.9	3.00	5.08	.61
3	32.3	10.3	4.65	7.62	.91
4	43.0	13.7	6.20	10.16	1.22
5	53.8	17.1	7.75	12.70	1.52
6	64.6	20.6	9.30	15.24	1.83
7	75.3	24.0	10.85	17.78	2.13
8	86.1	27.4	12.40	20.32	2.44
9	96.8	30.8	13.95	22.86	2.74
100	1076	343	155.0	254	30.5
110	1184	377	170.5	279	33.5
120	1291	411	186.0	305	36.6
130	1399	445	201.5	330	39.6
140	1506	480	217.0	356	42.7
150	1614	514	232.5	381	45.7
160	1722	548	248.0	406	48.8
170	1829	582	263.5	432	51.8
180	1937	617	279.0	457	54.9
190	2044	651	294.5	483	57.9
200	2152	685	310.0	508	61.0
210	2260	719	325.5	533	64.0
220	2367	754	341.0	559	67.1
230	2475	788	356.5	584	70.1
240	2582	822	372.0	610	73.2
250	2690	857	387.5	635	76.2
260	2798	891	403.0	660	79.2
270	2905	925	418.5	686	82.3
280	3013	959	434.0	711	85.3
290	3120	994	449.5	737	88.4
300	3228	1028	465.0	762	91.4
310	3336	1062	480.5	787	94.5
320	3443	1096	496.0	813	97.5
330	3551	1131	511.5	838	100.6
340	3658	1165	527.0	864	103.6
350	3766	1199	542.5	889	106.7
360	3874	1233	558.0	914	109.7
370	3981	1268	573.5	940	112.8
380	4089	1302	589.0	965	115.8
390	4196	1336	604.5	991	118.9
400	4304	1370	620.0	1016	121.9
410	4412	1405	635.5	1041	125.0
420	4519	1439	651.0	1067	128.0
430	4627	1473	666.5	1092	131.1
440	4734	1507	682.0	1118	134.1
450	4842	1542	697.5	1143	137.2
460	4950	1576	713.0	1168	140.2
470	5057	1610	728.5	1194	143.3
480	5165	1644	744.0	1219	146.3
490	5272	1679	759.5	1245	149.4

ft / in / cd/in² / fL / fc*	lx	cd/m²	kcd/m²	cm	m
500	5380	1713	775.0	1270	152.4
510	5488	1747	790.5	1295	155.4
520	5595	1782	806.0	1321	158.5
530	5703	1816	821.6	1346	161.5
540	5810	1850	837.0	1372	164.6
550	5918	1884	852.5	1397	167.6
560	6026	1919	868.0	1422	170.7
570	6133	1953	883.5	1448	173.7
580	6241	1987	899.0	1473	176.8
590	6348	2021	914.5	1499	179.8
600	6456	2056	930.0	1524	182.9
610	6564	2090	945.5	1549	185.9
620	6671	2124	961.0	1575	189.0
630	6779	2158	976.5	1600	192.0
640	6886	2193	992.0	1626	195.1
650	6994	2227	1007.5	1651	198.1
660	7102	2261	1023.0	1676	201.2
670	7209	2295	1038.5	1702	204.2
680	7317	2330	1054.0	1727	207.3
690	7424	2364	1069.5	1753	210.3
700	7532	2398	1085.0	1778	213.4
710	7640	2432	1100.5	1803	216.4
720	7747	2467	1116.0	1829	219.5
730	7855	2501	1131.5	1854	222.5
740	7962	2535	1147.0	1880	225.6
750	8070	2570	1162.5	1905	228.6
760	8178	2604	1178.0	1930	231.6
770	8285	2638	1193.5	1956	234.7
780	8393	2672	1209.0	1981	237.7
790	8500	2702	1224.5	2007	240.8
800	8608	2741	1240.0	2032	243.8
810	8716	2775	1255.5	2057	246.9
820	8823	2809	1271.0	2083	249.9
830	8931	2844	1286.5	2108	253.0
840	9038	2878	1302.0	2134	256.0
850	9146	2912	1317.5	2159	259.1
860	9254	2946	1333.0	2184	262.1
870	9361	2981	1348.5	2210	265.2
880	9469	3015	1364.0	2235	268.2
890	9576	3049	1379.5	2261	271.3
900	9684	3083	1395.0	2286	274.3
910	9792	3118	1410.5	2311	277.4
920	9899	3152	1426.0	2337	280.4
930	10010	3186	1441.5	2362	283.5
940	10110	3220	1457.0	2388	286.5
950	10220	3255	1472.5	2413	289.6
960	10330	3289	1488.0	2438	292.6
970	10440	3323	1503.5	2464	295.7
980	10540	3357	1519.0	2489	298.7
990	10650	3392	1534.5	2515	301.8

* Also useful for converting from m² to ft²

Fig. 1-14. Luminance Conversion Factors.

1 nit = 1 candela/square meter
1 stilb = 1 candela/square centimeter
1 apostilb (international) = 0.1 millilambert = 1 blondel
1 apostilb (German Hefner) = 0.09 millilambert
1 lambert = 1000 millilamberts

Multiply Number of → / To Obtain Number of ↓ By ↘	Footlambert*	Candela/ square meter	Millilambert*	Candela/ square inch	Candela/ square foot	Stilb
Footlambert*	1	0.2919	0.929	452	3.142	2,919
Candela/square meter	3.426	1	3.183	1,550	10.76	10,000
Millilambert*	1.076	0.3142	1	487	3.382	3,142
Candela/square inch	0.00221	0.000645	0.00205	1	0.00694	6.45
Candela/square foot	0.3183	0.0929	0.2957	144	1	929
Stilb	0.00034	0.0001	0.00032	0.155	0.00108	1

* Deprecated unit of luminance.

Fig. 1-15. Angular Measure, Temperature, Power and Pressure Conversion Equations.

Angle
1 radian = 57.29578 degrees
Temperature
(F to C) C = 5/9 (F − 32)
(C to F) F = 9/5 C + 32
(C to K) K = C + 273
Power
1 kilowatt = 1.341 horsepower
= 56.89 Btu per minute
Pressure
1 atmosphere = 760 millimeters of mercury at 0°C
= 29.92 inches of mercury at 0°C
= 14.7 pounds per square inch
= 101.3 kilopascals

REFERENCES

1. *American National Standard Nomenclature and Definitions for Illuminating Engineering,* ANSI/IES-RP-16, 1980, American National Standard Institute, New York, 1980. Sponsored by the Illuminating Engineering Society.
2. *American National Standard Practice for the Use of Metric (SI) Units in Building Design and Construction,* ANSI/ASTM E621-78, American National Standards Institute, New York, 1978. Sponsored by American Society for Testing and Materials.
3. Barbrow, L. E.: "The Metric System in Illuminating Engineering," *Illum. Eng.,* Vol. 62, p. 638, November, 1967.
4. *Colorimetry, CIE Publication No. 15,* Bureau Central de la Commission Internationale de L'Eclairage, Paris, France, 1971.
5. Commission Internationale de l'Eclairage, *International Lighting Vocabulary,* 3rd Edition, Paris, France, 1970.
6. Committee of Testing Procedures of the IES: "IES General Guide to Photometry," *Illum. Eng.,* Vol. L., p. 201, April, 1955.
7. Committee on Colorimetry of the Optical Society of America: *The Science of Color,* Edward Brothers Inc., New York, 1973.
8. Theatre, Television and Film Lighting Committee of the IES, "A Glossary of Commonly Used Terms in Theatre, Television and Film Lighting," *Light. Des. Appl.,* Vol. 13, p. 43, November 1983.
9. Hollander, A., ed: *Radiation Biology, Volume II, Ultraviolet and Related Radiations,* McGraw-Hill Book Company, New York, 1955.
10. Kaufman, J. E.: "Introducing SI Units," *Illum. Eng.* Vol. 63, p. 537, October, 1968.
11. Koller, L. R.: *Ultraviolet Radiation,* John Wiley & Sons Inc., New York, 1952.
12. Levin, R. E.: "Luminance—A Tutorial Paper," *J. Soc. Motion Pict. Telev. Engineers,* Vol. 77, p. 1005, October, 1968.
13. Nicodemus, F. E., Richmond, J. C., Hsia, J. J., Ginsberg, I. W. and Limperis, T.: *Geometrical Considerations and Nomenclature for Reflectance, NBS Monograph No. 160,* National Bureau of Standards, U.S. Department of Commerce, Washington, D.C., October, 1977.
14. "Nomenclature and Definitions Applicable to Radiometric and Photometric Characteristics of Matter," *ASTM Spec. Tech. Publ. 475,* American Society for Testing and Materials, Philadelphia, 1970.
15. Schapero, M., Cline, D. and Hofstetter, A. W.: *Dictionary of Visual Science,* Chilton Company, Philadelphia and New York, 1960.
16. *Supplement No. 2 to CIE Publication No. 15, Uniform Color Spaces, Color Difference Equations, Psychometric Color Terms,* Bureau Central de la Commission Internationale de l'Eclairage, Paris, France, 1978.
17. Wyszecki, G. and Stiles, W. S.: *Color Science: Concepts and Methods, Quantitative Data and Formulae,* John Wiley & Sons Inc., New York, 1982.

Light and Optics

SECTION 2

LIGHT CONCEPTS

For illuminating engineering purposes, the Illuminating Engineering Society of North America has defined light as *radiant energy that is capable of exciting the human retina and creating a visual sensation.*

From the viewpoint of physics, light is regarded as that portion of the electromagnetic spectrum which lies between the wavelength limits of 380 nanometers and 770 nanometers. Visually there is some individual variation in these limits.

Radiant energy of the proper wavelength makes visible anything from which it is emitted or reflected in sufficient quantity to activate the receptors in the eye.

Radiant energy may be evaluated in a number of different ways. Two are: *radiant flux*, measured in joules per second or in watts, and *luminous flux*, measured in lumens. For further information on these terms see Section 1.

Theories

Several theories describing radiant energy have been advanced.[1] They are briefly discussed below.

Corpuscular Theory. The theory advocated by Newton, based on these premises:
1. That luminous bodies emit radiant energy in particles.
2. That these particles are intermittently ejected in straight lines.
3. That the particles act on the retina of the eye stimulating the optic nerves to produce the sensation of light.

Wave Theory. The theory advocated by Huygens, based on these premises:
1. That light results from the molecular vibration in the luminous material.
2. That vibrations are transmitted through an "ether" as wavelike movements (comparable to ripples in water).
3. That the vibrations thus transmitted act on the retina of the eye stimulating the optic nerves to produce visual sensation.

Electromagnetic Theory.[2] The theory advanced by Maxwell, based on these premises:
1. That luminous bodies emit light in the form of radiant energy.
2. That this radiant energy is propagated in the form of electromagnetic waves.
3. That the electromagnetic waves act upon the retina of the eye thus stimulating the optic nerves to produce the sensation of light.

Quantum Theory. A modern form of the corpuscular theory advanced by Planck, based on these premises:
1. That energy is *emitted* and *absorbed* in discrete quanta (photons).
2. That the magnitude of each quantum is $h\nu$, where $h = 6.626 \times 10^{-34}$ joule second (Planck's constant), and
ν = frequency in hertz.

Unified Theory. The theory proposed by De Broglie and Heisenberg and based on these premises:
1. Every moving element of mass has associated with it a wave whose length is given by $\lambda = h/mv$
where λ = wavelength of the wave motion
h = Planck's constant
m = mass of the particle
v = velocity of the particle
2. It is impossible to simultaneously determine all of the properties that are distinctive of a wave or a corpuscle.

The quantum and electromagnetic wave theories provide an explanation of those characteristics of radiant energy of concern to the illumi-

NOTE: References are listed at the end of each section.

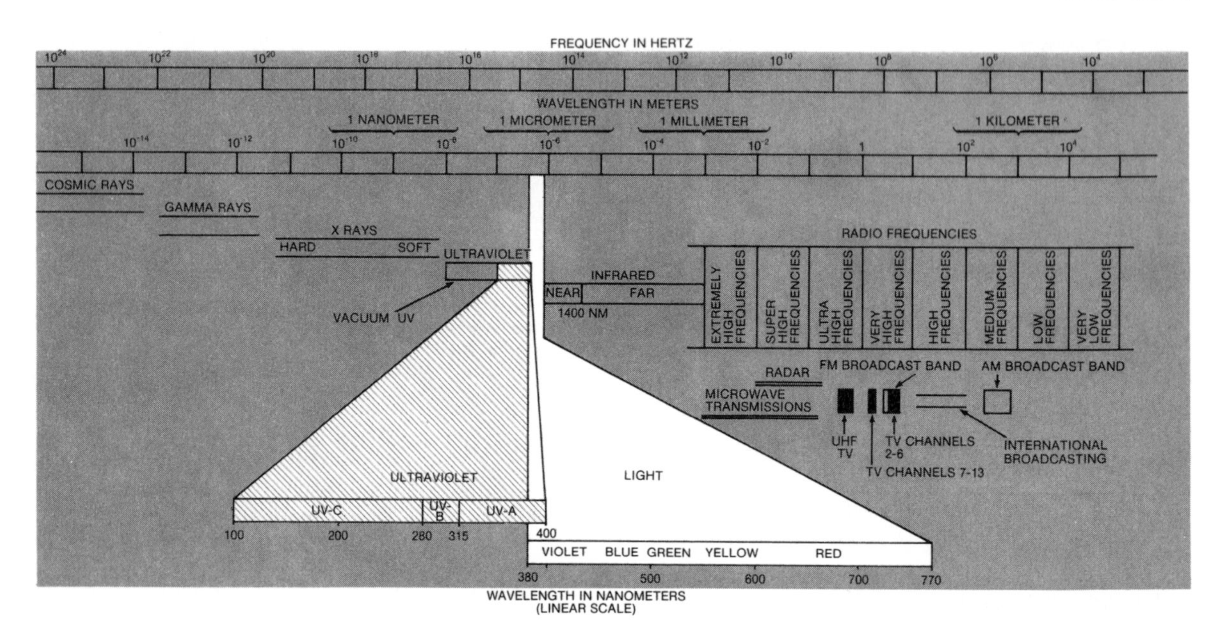

Fig. 2-1. The radiant energy (electromagnetic) spectrum.

nating engineer. Whether light is thought of as wave-like or photon-like in nature, it is radiation that is produced by electronic processes in the most exact sense of the term. It is produced in an incandescent body, a gas discharge or a solid-state device, by excited electrons just having reverted to more stable positions in their respective atoms, releasing energy.

Light and the Energy Spectrum

The wave theory permits a convenient graphical representation of radiant energy in an orderly arrangement according to its wavelength or frequency. This arrangement is called a *spectrum* (see Fig. 2-1). It is useful in indicating the relationship between various radiant energy wavelength regions. Such a graphical representation should not be construed to indicate that each region of the spectrum is divided from the others in any physical way whatsoever. Actually there is a gradual transition from one region to another.

The general limits of the radiant energy spectrum extend over a range of wavelengths varying from 10^{-16} to 10^{5} meters. Radiant energy in the visible spectrum has wavelengths between 380×10^{-9} and 770×10^{-9} meters (380 and 770 nanometers).

The Angstrom unit (Å), the nanometer (nm) and the micrometer (μm), which are respectively 10^{-10}, 10^{-9}, and 10^{-6} meters, are commonly used units of length in the visible spectrum region. The relationship of these units for measuring wavelength is given in Fig. 1-11 in Section 1.

All forms of radiant energy are transmitted at the same speed in vacuum (299,793 kilometers per second or 186,282 miles per second). However, each form differs in wavelengths and thus in frequency. The wavelength and velocity may be altered by the medium through which it passes, but the frequency is fixed independently of the medium. Thus, through the equation

$$v = \frac{\lambda \nu}{n}$$

where v = velocity of waves in the medium, *e.g.*, in meters per second,
n = index of refraction of the medium,
λ = wavelength in a vacuum, *e.g.*, in meters, and
ν = frequency, *e.g.*, in hertz,

it is possible to determine the velocity of radiant energy and also to indicate the relationship between frequency and wavelength.

Fig. 2-2 gives the speed of light in different media for a frequency corresponding to wavelength of 589 nanometers in air.

Fig. 2-2. Speed of Light for a Wavelength of 589 Nanometers (Sodium D-Lines)

Medium	Speed (meters per second)
Vacuum	2.99793×10^8
Air (760 mm at 0°C)	2.99724×10^8
Crown Glass	1.98223×10^8
Water	2.24915×10^8

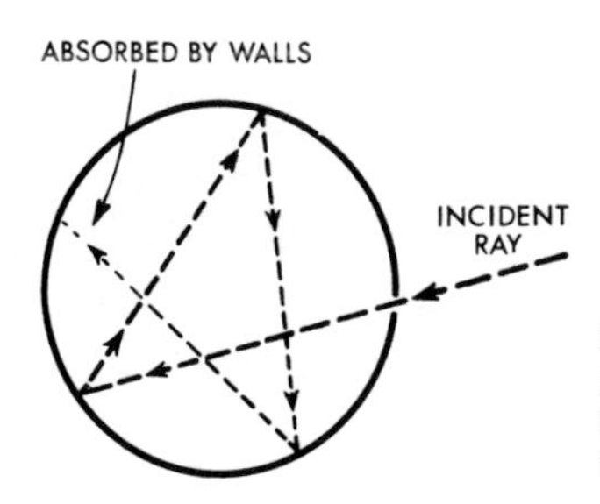

Fig. 2-3. Small aperture in an enclosure exhibits blackbody characteristics.

Blackbody Radiation

The light from practical light sources, particularly that from incandescent lamps, is often described by comparison with that from a blackbody. A blackbody will for equal area radiate more total power and more power at any given wavelength than any other source operating at the same temperature.

For experimental purposes, laboratory radiation sources have been devised which approach very closely a blackbody. Designs of these sources are based on the fact that a hole in the wall of a closed chamber, small in size as compared with the size of the enclosure exhibits blackbody characteristics. This is understood if one considers what happens to a ray of light entering such an enclosure. See Fig. 2-3. At reflection (A, B, C, etc.) some energy is absorbed. In time, all incoming energy will be absorbed by the walls.

From 1948 to 1979 the luminance of a blackbody operated at the temperature of freezing platinum has been used as an international reference standard for maintaining the unit of luminous intensity. (See Section 1, definition of *candela.*)

Planck Radiation Law. Data describing blackbody radiation curves were obtained by Lummer and Pringsheim using a specially constructed and uniformly heated tube as the source. Planck, introducing the concept of discrete quanta of energy, developed an equation depicting these curves. It gives the spectral radiance of a blackbody as a function of the wavelength and temperature. See definition of Planck's Radiation Law in Section 1.

Fig. 2-4 shows the spectral radiance of a blackbody as a function of wavelength for several values of absolute temperature, plotted on a logarithmic scale.

Wien Radiation Law. In the temperature range of incandescent filament lamps (2000 K to 3400 K) and in the visible wavelength region (380 to 770 nm), a simplification of the Planck equation, known as the Wien Radiation Law, gives a good representation of the blackbody distribution of spectral radiance. See Section 1.

Wien Displacement Law. This gives the relation between blackbody distributions for various temperatures. (See line AB, Fig. 2-4 and Section 1.)

Stefan-Boltzmann Law. This law, obtained by integrating Planck's expression for L_λ from zero to infinity, states that the total radiant power per unit area of a blackbody varies as the fourth power of the absolute temperature. See Section 1.

It should be noted that this law applies to the total power, that is, the whole spectrum. It cannot be used to estimate the power in the visible portion of the spectrum alone.

Spectral Emissivity

No known radiator has the same emissive power as a blackbody. The ratio of the output of a radiator at any wavelength to that of a blackbody at the same temperature and the same wavelength is known as the spectral emissivity, $\epsilon(\lambda)$, of the radiator.

Graybody Radiation

When the spectral emissivity is constant for all wavelengths, the radiator is known as a gray-

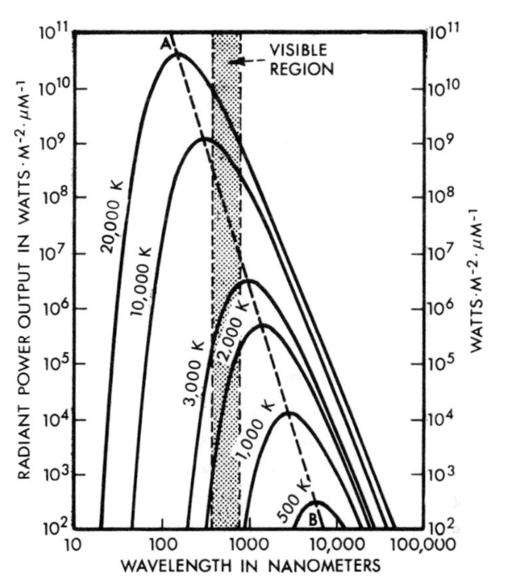

Fig. 2-4. Blackbody radiation curves for operating temperatures between 500 K and 20,000 K showing Wien displacement of peaks. Shaded area is region of visible wavelengths.

body. No known radiator has a constant spectral emissivity for all visible, infrared, and ultraviolet wavelengths, but in the visible region a carbon filament exhibits very nearly uniform emissivity, that is, nearly a graybody.

Selective Radiators

The emissivity of all known material varies with wavelength. Therefore, they are called selective radiators. In Fig. 2-5 the radiation curves for a blackbody, a graybody, and a selective radiator (tungsten), all operating at 3000 K, are plotted on the same logarithmic scale to show the characteristic differences in output.

Radiation Equations. When spectral-total directional emissivity is introduced as a multiplier in the Planck radiation law and the Wien radiation law equations, and spectral-total hemispherical emissivity in the Stefan-Boltzman law equation, those equations are applicable to any incandescent source.

Color Temperature

The radiation characteristics of a blackbody of unknown area may be specified with the aid of the above equations by fixing only two quantities; the magnitude of the radiation at any given wavelength and the absolute temperature. The same type of specification may be used with reasonable accuracy in the visible region of the spectrum for tungsten filaments and other incandescent sources. However, the temperature used in the case of selective radiators is not that of the filament but a value called the color temperature.

The color temperature of a selective radiator is that temperature at which a blackbody must be operated in order that its output be the closest possible approximation to a perfect color match with the output of the selective radiator (see Section 5). While the match is never perfect the small deviations that occur in the case of incandescent filament lamps are not of practical importance.

The apertures between coils of the filaments used in many tungsten lamps act somewhat as a blackbody because of the interreflections which occur at the inner surfaces of the helix formed by the coil. For this reason the distribution from coiled filaments exhibits a combination of the characteristics of the straight filament and of a blackbody operating at the same temperature.

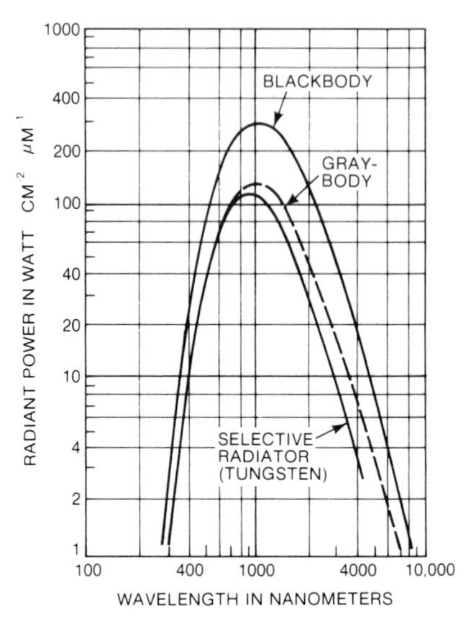

Fig. 2-5. Radiation curves for blackbody, graybody and selective radiators operating at 3000 K.

The application of the color temperature method to deduce the spectral distribution from other than incandescent sources even in the visible region of the spectrum will usually result in appreciable error. Color temperature values associated with light sources other than incandescent are correlated color temperatures and not true color temperatures (see Section 5).

Atomic Structure and Radiation

The atomic theories first proposed by Rutherford and Bohr in 1913 have since been expanded upon and confirmed by an overwhelming amount of experimental evidence. They hypothesize that each atom in reality resembles a minute solar system, such as that shown in Fig. 2-6.

The *atom* consists of a central nucleus possessing a positive charge n about which rotate n negatively charged electrons. In the normal state these electrons remain in particular orbits or energy levels and radiation is not emitted by the atom.

The *orbit* described by a particular electron rotating about the nucleus is determined by the energy of that electron. That is to say, there is a particular energy associated with each orbit. The system of orbits or energy levels is characteristic of each element and remains stable until disturbed by external forces.

The *electrons* of an atom can be divided into two classes. The first includes the inner shell electrons which are not readily removed or ex-

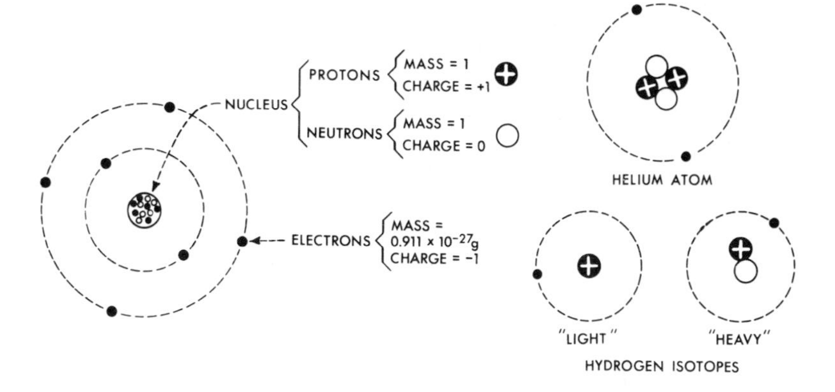

Fig. 2-6. Schematic structure of the atom showing electron orbits around central nucleus. Hydrogen isotopes and helium atom are simplest of all atomic structures.

cited except by high energy radiation. The second class includes the outer shell valence electrons which cause chemical bonding into molecules. Valence electrons are readily excited by ultraviolet or visible radiation or by electron impact and can be removed completely with relative ease. The valence electrons of an atom in a solid when removed from their associated nucleus enter the so-called conduction band and confer on the solid the property of electrical conductivity.

Upon the absorption of sufficient energy by an atom in the gaseous state, the valence electron is pushed to a higher energy level further from the nucleus. Eventually, the electron returns to the normal orbit, or an intermediate one, and in so doing the energy that the atom loses is emitted as a quantum of radiation. The wavelength of the radiation is determined by Planck's formula:

$$E_1 - E_2 = h\nu$$

where E_1 = energy associated with the excited orbit
E_2 = energy associated with the normal orbit
h = Planck's constant
ν = frequency of the emitted radiation.

This formula can be converted to a more usable form:

$$\text{wavelength} = \frac{1239.76}{V_d} \text{ nanometers}$$

where V_d is the potential difference (volts) between two energy levels through which the displaced electron has fallen in one transition.

Luminous Efficiency of Radiant Energy

Many apparent differences in intensity between radiant energy of various wavelengths are in reality differences in ability of various receiving and sensing devices to detect them uniformly.[3]

The reception characteristics of the human eye have been subject to extensive investigations. The results may be summarized as follows:

1. The spectral response characteristic of the human eye varies between individuals, with time, and with the age and the state of health of any individual, to the extent that the selection of any individual to act as a standard observer is not scientifically feasible.

2. However, from the wealth of data available, a luminous efficiency curve has been selected for engineering purposes to represent a typical human observer. This curve may be applied mathematically to the solution of photometric problems. (See also Section 3.)

Recognizing these facts, the Illuminating Engineering Society in 1923 and the International Commission on Illumination (CIE) in 1924 adopted the values of photopic spectral luminous efficiency of Fig. 1-6 from which the spectral luminous efficiency curve of Fig. 3-7 was plotted.

The standard spectral luminous efficiency curve for photopic vision represents a typical characteristic, adopted arbitrarily to give unique solutions to photometric problems, from which the characteristics of any individual may be expected to vary. Goodeve's data (see Reference 14) indicate that most human observers are capable of experiencing a visual sensation upon exposure to radiant energy of wavelengths longer than 770 nanometers, usually termed "infrared," provided the radiant energy reaches the eye at a sufficiently high rate. It also is known that "ultraviolet" radiant energy (wavelength less than 380 nanometers) can be seen if it reaches the retina at even a moderate rate. Most observers, however, yield only a slight response to ultraviolet radiant energy because the lens of the eye absorbs nearly all of it.

Luminous Efficacy of Light Sources

The luminous efficacy of a light source is defined as the ratio of the total luminous flux (lumens) to the total power input (watts or equivalent).

The maximum luminous efficacy of an "ideal" white source, defined as a radiator with constant output over the visible spectrum and no radiation in other parts of the spectrum is approximately 220 lumens per watt.

LIGHT GENERATION

Natural Phenomena

Sunlight. Energy of color temperature about 6500 K is received from the sun at the outside of the earth's atmosphere at an average rate of about 1350 watts per square meter.[4] About 75 per cent of this energy reaches the earth's surface at sea level (equator) on a clear day.

The average luminance of the sun is approximately 1600 megacandelas per square meter viewed from sea level. The illuminance on the earth's surface by the sun may exceed 100 kilolux [10,000 footcandles]; on cloudy days the illuminance drops to less than 10 kilolux [1000 footcandles]. See Section 7.

Sky Light. A considerable amount of light is scattered in all directions by the earth's atmosphere. The investigations of Rayleigh first showed that this was a true scattering effect. On theoretical grounds the scattering should vary inversely with the fourth power of the wavelength when the size of the scattering particles is small compared to the wavelength of light, as in the case of the air molecules themselves. The blue color of a clear sky and the reddish appearance of the rising or setting sun are common examples of this scattering effect. If the scattering particles are of appreciable size (the water droplets in a cloud, for example), scattering is essentially the same for all wavelengths. (Clouds appear white.) Polarization in parts of the sky may amount to 50 per cent.

Moonlight. The moon shines purely by virtue of its ability to reflect sunlight. Since the reflectance of its surface is rather low, its luminance is approximately 2500 candelas per square meter. Illumination of the earth's surface by the moon may be as high as 0.1 lux [0.01 footcandle].

Lightning. Lightning is a meterological phenomenon arising from the accumulation, in the formation of clouds, of tremendous electrical charges, usually positive, which are suddenly released in a spark type of discharge. The lightning spectrum corresponds closely to that of an ordinary spark in air, consisting principally of nitrogen bands, though hydrogen lines may sometimes appear owing to dissociation of water vapor.

Aurora Borealis (Northern Lights) and Aurora Australis (Southern Lights). These hazy horizontal patches or bands of greenish light on which white, pink, or red streamers sometimes are superposed appear 60 to 120 miles above the earth. They are caused by electron streams spiraling into the atmosphere, primarily at polar latitudes. Some of the lines in their spectrum have been identified with transitions of valence electrons from metastable states of oxygen and nitrogen atoms.

Bioluminescence. "Living light" is a form of chemiluminescence (see page 2-13) in which special compounds manufactured by plants and animals are oxidized, producing light. Although it has been proven that oxygen is required to produce bioluminescence, there is no evidence that the light producing compound must be a "living" material. The light producing compound can be dried and stored many years and upon exposure to oxygen emit light.

"Man-Made" Sources

Historically, light sources have been divided into two types—incandescent and luminescent. Fundamentally, the cause of light emission is the same, *i.e.*, electronic transitions from higher to lower energy states. The mode of electron excitation is different, however, as well as the spectral distribution of radiation. Incandescent solid substances basically emit a continuous spectrum while gaseous discharges radiate mainly in discrete spectral lines; however, there is some overlapping. Incandescent rare earth elements can emit lines, whereas high pressure discharges produce a continuous spectrum.

The two classical types, with subdivisions showing associated devices or processes, are listed as follows (see Section 8):

1. Incandescence
   - Filament lamps
   - Pyroluminescence (flames)

Candoluminescence (gas mantle)
Carbon arc radiation
2. Luminescence
Photoluminescence
a) Gaseous discharges
b) Fluorescence
c) Phosphorescence
d) Lasers
Electroluminescence
a) Electroluminescent lamps (ac capacitive)
b) Light emitting diodes
c) Cathodoluminescence (electron exitation)
Miscellaneous luminescence phenomena
a) Galvanoluminescence (chemical)
b) Crystalloluminescence (crystallization)
c) Chemiluminescence (oxidation)
d) Thermoluminescence (heat)
e) Triboluminescence (friction/fracture)
f) Sonoluminescence (ultrasonics)
g) Radioluminescence (α, β, γ, X-rays)

INCANDESCENCE

Incandescent Filament Lamps

All familiar physical objects are simple or complex combinations of chemically identifiable molecules, which in turn are made up of atoms. In solid materials the molecules are packed together and the substances hold their shape over a wide range of physical conditions. In contrast, the molecules of a gas are highly mobile and occupy only a small part of the space filled by the gas.

Molecules of both gases and solids are constantly in motion at temperatures above absolute zero and their movement is a function of temperature. If the solid or gas is hot, the molecules move rapidly; if it is cold, they move more slowly.

At temperatures below about 873 K (600°C) only invisible energy of the longer infrared (heat) wavelengths is emitted by any body; a coal stove or an electric iron, for example. Electronic transitions in atoms and molecules at temperatures of about 600°C result in the release of visible radiation along with the heat.

The incandescence of a lamp filament is caused by the heating action of an electric current. This heating action raises the filament temperature substantially above 600°C, producing light.

Pyroluminescence (Flame Luminescence)

A flame is the most often noted "visible" evidence of combustion. Flame luminescent light may be due to recombination of ions to form molecules, reflection from solid particles in the flame, incandescence of carbon or other solid particles, and any combination of these.

The combustion process is a high temperature energy exchange between highly excited molecules and atoms. The process releases and radiates energy, some of which is in that portion of the electromagnetic spectrum called light. The quality and the amount of light generated depend upon the material undergoing combustion. For example, a flashbulb containing zirconium yields the equivalent of 56 lumens per watt whereas an acetylene flame yields 0.2 lumen per watt.

Candoluminescence (Gas Mantle)

Incandescence is exhibited by heated bodies which give off shorter wavelength radiation than would be expected by radiation laws—the effect due to fluorescence excited by incandescent radiation. Such materials are zinc oxide and rare earth elements (cerium, thorium) used in the Welsbach gas mantle.

Carbon Arc Radiation

A carbon arc source radiates because of incandescence of the electrodes and because of luminescence of vaporized electrode material and other constituents of the surrounding gaseous atmosphere. Considerable spread in the luminance, total radiation, and spectral power distribution may be achieved by varying the electrode materials.

LUMINESCENCE[5-8]

Radiation from luminescent sources results from the excitation of single valence electrons of an atom, either in a gaseous state, where each atom is free from interference from its neighbors, or in a crystalline solid or organic molecule, where the action of its neighbors exerts a marked effect. In the first case line spectra, such as those of mercury or sodium arcs, result. In the second case relatively narrow emission bands, which cover a portion of the spectrum (usually in the visible region) result. Both of these cases contrast with the radiation from incandescent sources, where the irregular excitation at high temperature of the free electrons of innumerable atoms gives rise to all wavelengths of radiation to form a continuous spectrum of radiation as discussed in the section on blackbody radiation.

Photoluminescence

Gaseous Discharge. Radiation, including light, can be produced by gaseous discharges as discussed above and previously under *Atomic Structure.* A typical mechanism for generating light (photons) from a gaseous discharge (such as in a fluorescent lamp) is described below. See Fig. 2-7.

1. A free electron emitted from the cathode collides with one of the two valence electrons of a mercury atom and excites it by imparting to it part of the kinetic energy of the moving electron, thus raising the valence electron from its normal energy level to a higher one.

2. The conduction electron loses speed at the impact and changes direction, but continues along the tube to excite or ionize one or more additional atoms before losing its energy stepwise and completing its path. It generally ends at the wall of the tube where it recombines with an ionized atom. A part of the electron current is collected at the anode.

3. Conduction electrons, either from the cathode or formed by collision processes, gain energy from the electric field thus maintaining the discharge along the length of the tube.

4. After a short delay the valence electron returns to its normal energy level either in a single transition or by a series of steps from one excited level to a lower level. At each of these steps a photon (quantum of radiant energy) is emitted. If the electron returns to its normal energy level in a single transition the emitted radiation is called "resonance" radiation. See Fig. 2-8.

5. In some cases (as in the high pressure sodium lamp) a portion of the resonance radiation is self absorbed by the gas of the discharge before it leaves the discharge envelope. The absorbed energy is then reradiated as a continuum on either side of the resonant wavelength, leaving a depressed or dark region at that point in the spectrum.

Fluorescence. In the fluorescent lamp and in the fluorescent mercury lamp ultraviolet radiation resulting from luminescence of the mercury vapor due to a gas discharge is converted into visible light by a phosphor coating on the inside of the tube or outer jacket. If this emission continues only during the excitation it is called "fluorescence."

Fig. 2-7 shows schematically a greatly magnified section of a part of a fluorescent lamp.

Ultraviolet photons generated in an arc discharge such as in a fluorescent or fluorescent-mercury lamp eventually strike one of the phosphor crystals on the glass surface of the tube or outer envelope.

The phosphor will transmit this energy through the crystal until it reaches an activator

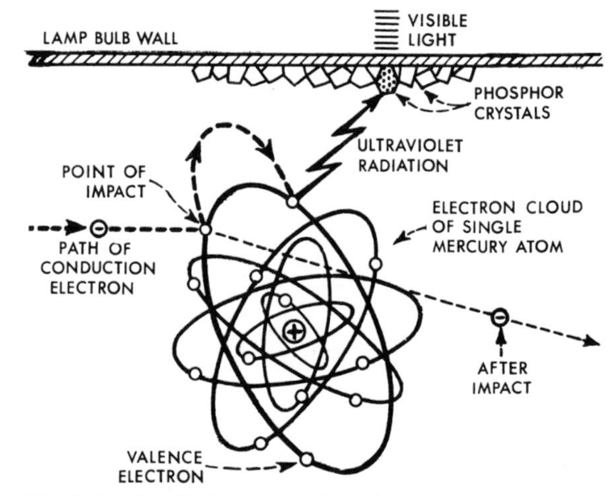

Fig. 2-7. Magnified cross section of fluorescent lamp schematically showing progressive steps in luminescent process which finally result in the release of visible light.

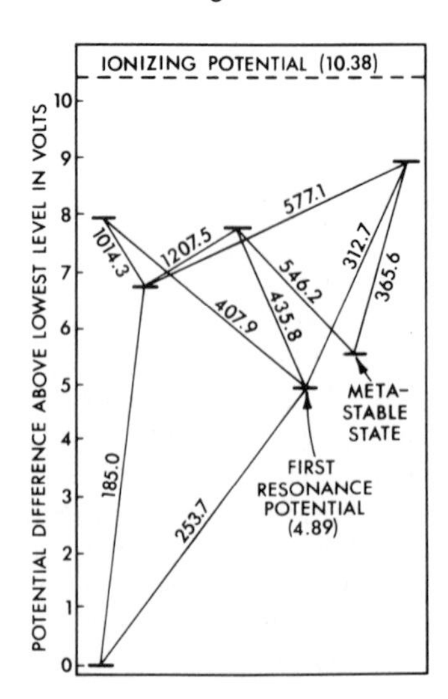

Fig. 2-8. Simplified energy diagram for mercury showing a few of the characteristic spectral lines.

ion. At this point, if the photon has a wavelength within the excitation band of the phosphor, it will be absorbed and converted into a photon of longer wavelength usually in the near ultraviolet or in the visible spectrum (see Fig. 2-9).

The phosphors used in fluorescent lamps are crystalline inorganic compounds, of exceptionally high chemical purity and of controlled composition to which small percentages of intentional impurities (the activators) have been added to convert them into efficient fluorescent materials. By choice of the right combination of activator and inorganic compound the color of emission can be controlled. A typical schematic model for a phosphor is given in Fig. 2-10. In the normal state the electron oscillates about position A on the energy curve, as the lattice expands and contracts due to thermal vibration. For the phosphor to emit light it must first absorb radiation. In the fluorescent lamp this is chiefly that at 253.7 nanometers, while in the mercury lamp it may be ultraviolet of this and longer wavelengths generated in the arc. The absorbed energy transfers the electron to an excited state at position B. After loss of excess energy to the lattice as vibrational (heat) energy the electron again oscillates around a stable position, C, for a very short time after which it returns to position D on the normal energy curve with simultaneous emission of a photon of radiation. Stokes Law, stating that the radiation emitted must be of longer wavelength than that absorbed, is readily explained by this model. It then returns to A with a further loss of energy as heat and is then ready for another cycle of excitation and emission.

Because of the oscillation around both stable positions A and C the excitation and emission processes cover a range of wavelength commonly referred to as bands.

In some phosphors two activators are present. One of these, the primary activator, determines the absorption characteristics and can be used alone as it also gives emission. The other, or secondary activator, does not enter into the absorption mechanism but receives its energy by transfer within the crystal from a neighboring primary activator. The emitted light from the secondary activator is longer in wavelength than that from the primary activator. The relative amount of emission from the two activators is determined by the concentration of the secondary activator. The phosphors now used in most "white" fluorescent lamps are doubly activated calcium halophosphate phosphors.

Fig. 2-11 shows the characteristic colors and uses of phosphors currently employed in manufacture of fluorescent lamps. Fig. 2-12 gives the characteristics of some phosphors useful in fluorescent-mercury lamps.

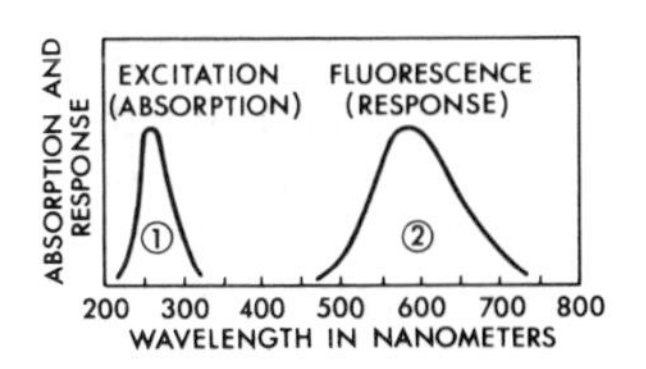

Fig. 2-9. Fluorescence curve of typical phosphor showing initial excitation by ultraviolet rays and subsequent release of visible radiation.

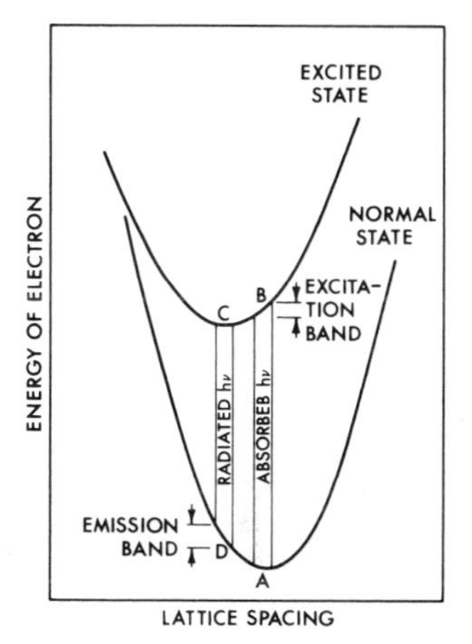

Fig. 2-10. Simplified energy diagram for a typical phosphor.

Impurities other than activators and excessive amounts of activators have a serious deleterious effect on the efficiency of a phosphor.[9]

Phosphorescence.[7,8] In some fluorescent materials there are metastable excited states in which the electron can be trapped, for a time varying from milliseconds to days. After release from these traps they emit light. This phenomenon is called "phosphorescence." The metastable states lie slightly below the usual excited states responsible for fluorescence, and energy usually derived from heat is required to transfer the electron from the metastable state to the emitting state. Since the same emitting state is usually involved, the color of fluorescence and phosphorescence is generally the same for a given phosphor. In doubly activated phosphors the secondary activator has longer phosphorescence than the primary activator so the color changes with time.

Short duration phosphorescence is important in fluorescent lamps in reducing the stroboscopic effect with alternating current operation.

The infrared stimulable phosphors have an unusual type of phosphorescence. After excitation they show phosphorescence, but decay to invisibility in a few seconds. However, they retain a considerable amount of energy trapped in metastable states which can be released as visible light by infrared radiation, when a wavelength

Fig. 2–11. Color Characteristics of Important Fluorescent Lamp Phosphors.

Material	Activator	Peak of Fluorescent Band*	Color of Fluorescence
Calcium phosphate	Thallium	310	Ultraviolet
Barium disilicate	Lead	355	Pale blue
Barium strontium magnesium silicate	Lead	370	Pale blue
Calcium tungstate	Lead	440	Blue
Strontium chloroapotite	Europium	445	Blue
Barium magnesium aluminate	Europium	450	Blue
Strontium pyrophosphate	Tin	470	Blue
Magnesium tungstate	None	480	Blue white
Calcium halophosphate	Antimony	480	Blue white
Barium titanium phosphate	Titanium	490	Blue green
Zinc silicate	Manganese	520	Green
Cerium terbium magnesium aluminate	Terbium	545	Green
Calcium halophosphate	Antimony and manganese	590	White to yellow
Calcium silicate	Lead and manganese	610	Pink
Yttrium oxide	Europium	612	Orange red
Cadmium borate	Manganese	615	Pink
Strontium magnesium phosphate	Tin	620	Pink
Calcium strontium phosphate	Tin	640	Pink
Magnesium fluorogermanate	Manganese	660	Red
Lithium pentaaluminate	Iron	743	Infrared

* Wavelength in nanometers

within the stimulation band is allowed to fall on the phosphor.

Solid Laser.[11–13] A laser is defined by its component letters as a light source in which there is *L*ight *A*mplification by *S*timulated *E*mission of *R*adiation. However, there are other characteristics of laser light which are of major interest to illuminating engineers. In addition to amplifying light, lasers produce intense, highly monochromatic, well collimated, coherent light.

Coherent light consists of radiation whose waves are in phase with regard to time and space. Ordinary light, although it may contain a finite proportion of coherent light, is basically incoherent because the atomic processes that cause light emission occur in a random fashion.

In a laser, however, electronic transitions are triggered (stimulated) by a wave of the same frequency as the emitted light instead of occurring at random. As a consequence, a beam of light is emitted, all of whose waves are in phase and of the same frequency.

A prerequisite to laser action is a pumping process whereby an upper and a lower electron level in the active material undergo a population inversion. The pumping source may be a light as in a ruby laser or electronic excitation as in a gas laser.

The choice of laser materials is quite limited. First, it must be possible to highly populate an upper electronic level; second, there must be a light emitting transition from this upper level with a long lifetime; and third, a lower level must exist which can be depopulated either spontaneously or through pumping.

Laser construction is as important to laser action as is the source material. Since light wavelengths are too short to allow building a resonant cavity, long multi-nodal chambers are made with parallel reflectors at each end to feed back radiation until lasing takes place. The effect is to produce well collimated light that is highly directional.

An example of a laser is the pink ruby device whose electronic transitions are shown in Fig. 2-13 and whose mechanical construction is indicated in Fig. 2-14. This laser is pumped by a flash tube (a) and electrons in the ruby (b) are raised from level E_1 to E_3. The electrons decay rapidly and spontaneously from E_3 to E_2. They

Fig. 2–12. Color Characteristics of Some Phosphors for Mercury and Metal Halide Lamps

Material	Activator	Peak of Fluorescent Band*	Color of Fluorescence
Strontium chloropatite	Europium	445	Blue
Strontium-magnesium phosphate	Tin	610	Orange red
Strontium-zinc phosphate	Tin	610	Orange red
Yttrium-vanadate	Europium	612	Orange red
Yttrium-vanadate phosphate	Europium	612	Orange red
Magnesium fluorogermanate	Manganese	660	Deep red
Magnesium arsenate	Manganese	660	Deep red

* Wavelength in nanometers

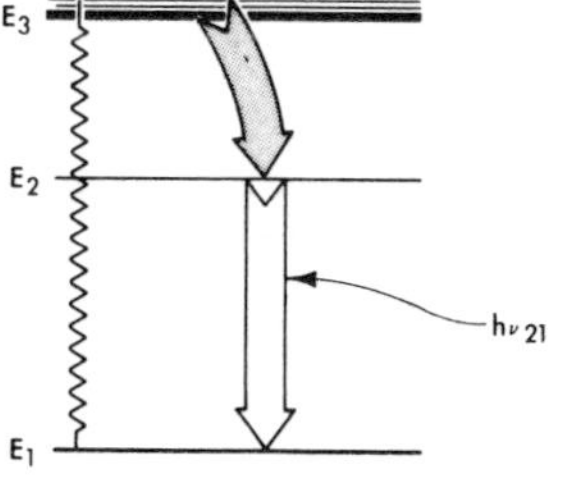

Fig. 2-13. Simplified diagrammatic representation of electronic transitions in a ruby laser.

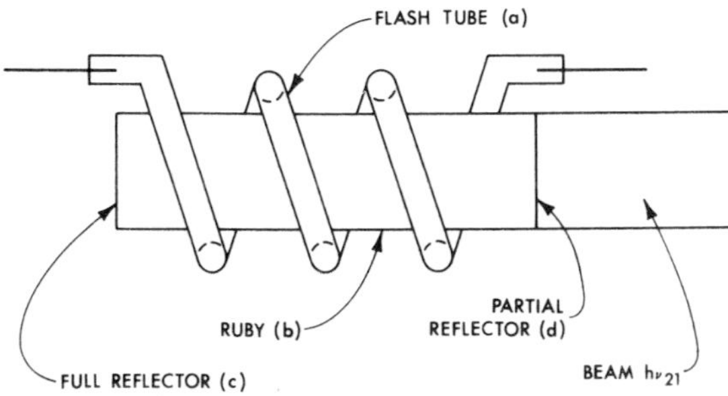

Fig. 2-14. Simplified diagram of a ruby laser.

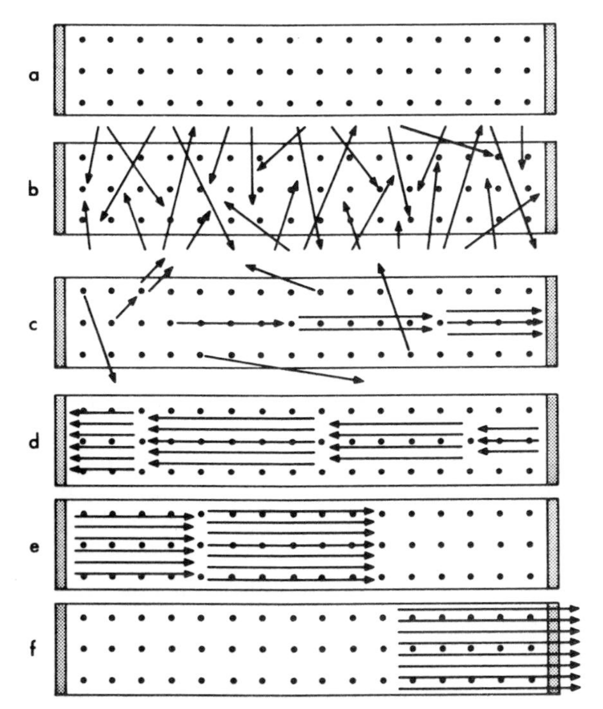

Fig. 2-15. Photon cascade in a solid laser.[13] Before the build-up begins, atoms in the laser crystal are in the ground state (a). Pumping light (arrows in b) raised most of the atoms to the excited state. The cascade (c) begins when an excited atom spontaneously emits a photon parallel to the axis of the crystal (photons emitted in other directions pass out of the crystal). The build-up continues (d) and (e) through thousands of reflections back and forth from the silvered surfaces at the ends of the crystal. When amplification is great enough, light passes out (f).*

* Adapted from *Scientific American*

can then move from E_2 to E_1 spontaneously and slowly emit fluorescent light, $h\nu_{21}$, or they can be stimulated to emit coherent light, $h\nu_{21}$. The full reflector (c) and the partial reflector (d) channel the coherent radiation, $h\nu_{21}$, until it has built up enough to emit coherent light $h\nu_{21}$ through (d). The fact that this light has been reflected many times by parallel mirrors ensures that it is well collimated. The electrons are then available for further pumping. See Fig. 2-15.

Gas Laser. In a solid laser there are three requirements—a material which reacts energetically to light, a population inversion generated by pumping in energy at the proper energy level and a growth of the internal energy caused by the reflection of photons within the solid. While the same requirements are met in a gas laser, two other characteristics are available—strong, narrow spectral lines and unequal emission at different energy levels. An example of such a gas laser is that containing a mixture of helium and neon. See Fig. 2-16. Helium is used as the energizing gas since it has a level of energy at which it can lose energy only by collision. This level corresponds to the level at which neon radiates energy in the form of red light. By energizing helium in a gas discharge inside a cavity whose ends are reflecting and containing both helium and neon, helium transfers energy by collision with neon. The excited neon emits photons which begin to amplify by cascading between the two reflecting surfaces until the internal energy is so large that the losses through the partially

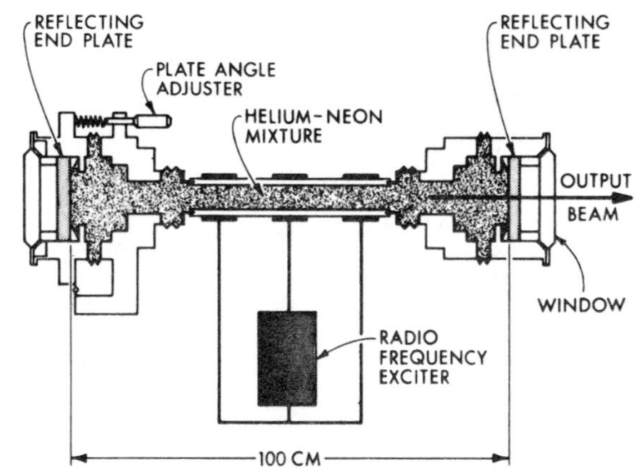

Fig. 2-16. Structure of CW helium-neon gas laser, showing essential parts.* Operation of a gas laser depends on the right mixture of helium and neon gases to provide an active medium. Radio frequency exciter puts energy into the medium. The output beam is built up by repeated passes back and forth between reflecting end plates.[13]

* Adapted from *Scientific American*

transmitting mirror become equal to the internal gains and the laser becomes saturated.

Semiconductor Laser. A third type of laser utilizes a semiconducting solid material where the electron current flowing across a junction between p- and n-type material produces extra electrons in the conduction band. These radiate upon their making a transition back to the valence band or lower-energy states. If the junction current is large enough there will be more electrons near the edge of the conduction band than there are at the edge of the valence band and a population inversion may occur. To utilize this effect, the semiconductor crystal is polished with two parallel faces perpendicular to the junction plane. The amplified waves may then propagate along the plane of the junction and are reflected back and forth at the surfaces. See Fig. 2-17.

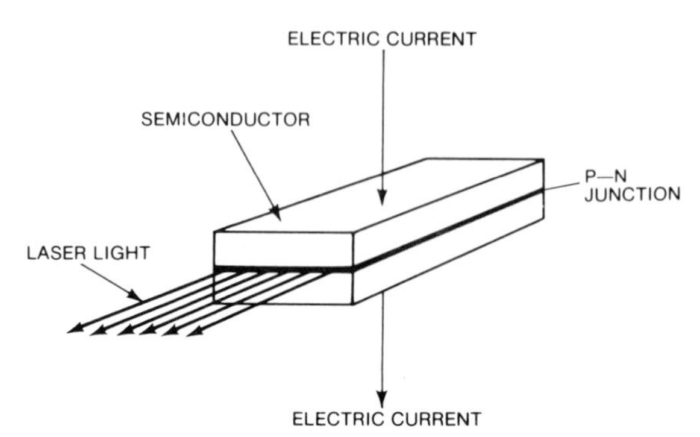

Fig. 2-17. Highly schematic diagram of a semiconductor (gallium arsenide) laser.

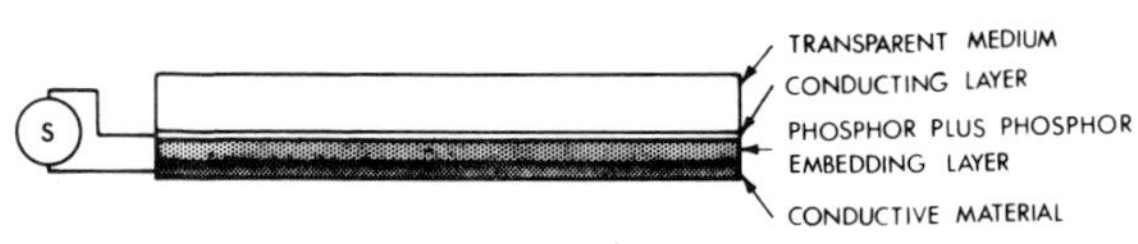

Fig. 2-18. Diagrammatic cross section of an electroluminescent lamp.

Electroluminescence

Certain special phosphors convert alternating current energy directly into light without using an intermediate step, such as the processes in a gas discharge, by utilizing the phenomenon of electroluminescence.

Electroluminescent Lamps (AC Capacitive). An electroluminescent lamp is composed of a two-dimensional area conductor (transparent or opaque) on which a dielectic-phosphor layer is deposited. A second two-dimensional area conductor of transparent material is deposited over the dielectric-phosphor mixture.

An electric field is established between the two conductors with the application of a voltage across the two-dimensional (area) conductors. Under the influence of such an applied alternating electric field, some electrons in the electroluminescent phosphor are excited. During the return of these electrons to their ground or normal state the excess energy is radiated as visible light.

Fig. 2-18 shows a cross-sectional view of an electroluminescent lamp. Fig. 2-19 gives the properties of some electroluminescent phosphors.

The color of light emitted by the electroluminescent lamp is dependent on frequency, while the luminance is dependent on frequency and voltage. These effects vary depending upon the specific phosphor.

The efficacy of electroluminescent devices is low compared to incandescent lamps. It is of the order of a few lumens per watt.

Fig. 2-19. Properties of Some Electroluminescent Phosphors

Material	Activators	Color of Light
Cubic zinc sulfide	Copper (low), lead	Blue
Cubic zinc sulfide	Copper (high), lead	Green
Cubic zinc sulfide	Copper (high), lead, manganese	Yellow
Hexagonal zinc sulfide	Copper (very high)	Green
Hexagonal zinc sulfide	Copper (very high), manganese	Yellow
Zinc sulfo selenide	Copper	Green to yellow
Zinc cadmium sulfo selenide	Copper	Yellow to pink

Light Emitting Diodes. Light emitting diodes (LED), also called solid state lamps (SSL), produce light by electroluminescence when low-voltage direct current is applied to a suitably doped crystal containing a p-n junction. The phenomenon has been observed as early as 1923 in naturally occurring junctions but was not considered practical due to the low luminous efficacy in converting electric energy to light. Recently it was discovered that under certain conditions the conversion was significant.

The efficacy is dependent upon the visible energy generated at the junction and losses due to reabsorption when light tries to escape through the crystal. Due to the high index of refraction of most semiconductors, light is re-reflected back from the surface into the crystal and highly attenuated before finally exiting. The term used to express this ultimate measurable visible energy is "external" efficacy. While external efficacies are moderate, internal efficacies are calculated to be very high.

For more information see Section 8.

Cathodoluminescence. Cathodoluminescence is light emitted when a substance is bombarded by an electron beam from a cathode, as in cathode-ray and television picture tubes.

Miscellaneous Luminescence Phenomena

Galvanoluminescence. Galvanoluminescence is light which appears at either the anode or cathode when solutions are electrolyzed.

Crystalloluminescence. Crystalloluminescence (lyoluminescence) is observed when solutions crystallize and is believed to be due to the rapid reformation of molecules from ions. The light intensity increases upon stirring which might be due to some *triboluminescence.*

Chemiluminescence. Chemiluminescence (oxyluminescence) is the production of light during a chemical reaction at room temperatures. True chemiluminescences are oxidation reactions involving valence changes.

Thermoluminescence. Thermoluminescence (heat) is exhibited by some materials when slightly heated. In all cases of thermoluminescence the effect is dependent upon some previous illumination or radiation of the crystal. Diamonds, marble apatite, quartz and fluorspar are thermoluminescent.

Triboluminescence. Triboluminescence (piezoluminescence) is light produced by shaking, rubbing, or crushing crystals. Triboluminescent light may result from unstable light centers previously exposed to some source or radiation, such as, light, X-rays, radium and cathode rays; centers not exposed to previous radiation but characteristic of the crystal itself; or electrical discharges from fracturing crystals.

Sonoluminescence. Sonoluminescence is the light which is observed when sound waves are passed through fluids. Luminescence occurs when fluids are completely shielded from an electrical field and is always connected with cavitation (the formation of gas or vapor cavities in a liquid). It is believed the minute gas bubbles of cavitated gas develop a considerable charge as their surface increases. When they collapse their capacity decreases and their voltage rises until a discharge takes place in the gas causing a faint luminescence.

Radioluminescence. Radioluminescence is light emitted from a material under bombardment from alpha rays, beta rays, gamma rays or X-rays.

LIGHT DETECTION

The most universally used detector of light is the human eye (see Section 3). Other common detectors are photovoltaic cells, photoconductor cells, photoelectric tubes, photodiodes, phototransistors, and photographic film.

Photovoltaic Cells

Photovoltaic cells commonly include selenium barrier-layer cells and silicon or gallium arsenide photodiodes operated in the photovoltaic or unbiased mode. They depend upon the generation of a voltage as a result of the absorption of a photon. The cell is comprised of a metal plate coated with a semiconductor material, such as selenium on iron as the p-type material and cadmium oxide as the semitransparent n-type material. Upon exposure to light, electrons liberated from the semiconductor are trapped at the interface unless there is an external circuit provided through which they may escape. They thus convert radiant energy to electric energy which can be used directly or amplified to drive a microammeter (see Fig. 2-20). The cells can be filtered to correct their spectral response so that the meter can be calibrated in units of illuminance. Factors such as response time, fatigue, temperature effects, linearity, stability, noise and magnitude of current influence the choice of cell and circuit for a given application.

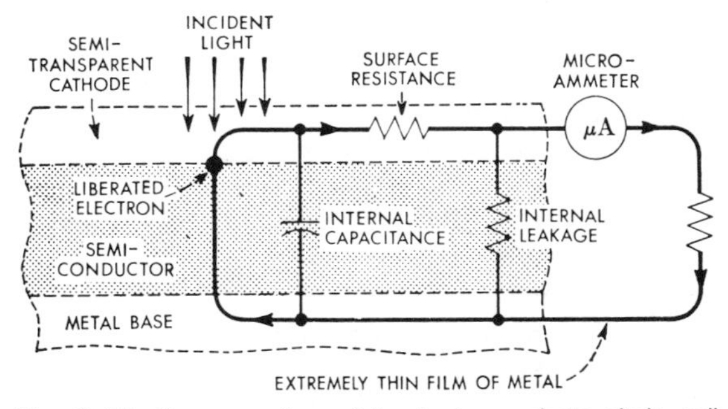

Fig. 2-20. Cross section of barrier-layer photovoltaic cell showing motion of photoelectrons through microammeter circuit.

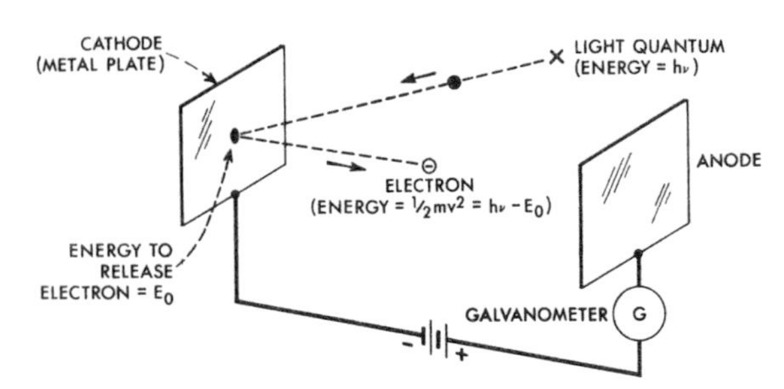

Fig. 2-21. By the photoelectric effect, electrons may be liberated from illuminated metal surfaces. In an electric field these will flow to an anode and create an electric current which may be detected by means of a galvanometer.

Photoconductor Cells

Photoconductor cells depend upon the resistance of the cell changing directly as a result of photon absorption. These detectors use materials, such as, cadmium sulfide, cadmium selenide and selenium. Cadmium sulfide and cadmium selenide are available in transparent resin or glass envelopes and are suitable for low illuminance levels ($<10^{-4}$ lux or $<10^{-5}$ footcandles).

Photoelectric Tubes

The emission of electrons from a surface when bombarded by sufficiently energetic photons is known as the *photoelectric effect.* If the surface is connected as a cathode in an electric field (see Fig. 2-21) the liberated electrons will flow to the anode creating a photoelectric current. An arrangement of this sort may be used as an illuminance meter and can be calibrated in lux or footcandles.

It has been found that the photoelectric current in vacuum varies directly with the illuminance level over a wide range (spectral distribution, polarization and cathode potential remaining the same). In gas-filled tubes the response is linear over only a limited range. If the radiant energy is polarized, the photoelectric current will vary as the orientation of the polarization is changed (except at normal incidence).

Photodiodes and Phototransistors

Photodiodes or "junction photocells" are based on solid-state p-n junctions that react to external stimuli such as light. Conversely, if properly constructed, they can emit light (see Light Emitting Diodes). In a photosensitive diode the reverse saturation current of the junction increases in proportion to the illuminance. Such a diode can, therefore, be used as a sensitive detector of light and is particularly suitable for indicating extremely short pulses of radiation because of its very fast response time.

Phototransistors operate in a manner similar to photodiodes but, because they provide an additional amplifier effect, they are many times more sensitive than simple photodiodes.

OPTICAL CONTROL[15]

Optical control may be provided in a number of ways, all of which are applications of one or more of the following phenomena: reflection, refraction, polarization, interference, diffraction, diffusion and absorption.

Reflection and Reflectors

Reflection is the process by which a part of the light falling on a medium leaves that medium from the incident side. Reflection may be specular, spread, diffuse, or compound, and selective or non-selective. Reflection from the front of a transparent plate is called first-surface reflection and that from the back second-surface reflection. Refraction and absorption by supporting media are avoided in first-surface reflection.

Specular Reflection. If a surface is polished, it reflects specularly; that is, the angle between the reflected ray and the normal to the surface will equal the angle between the incident ray and the normal as shown in Fig. 2-22. If two or more rays are reflected, they may produce a virtual, erect or inverted image of the source. A lateral reversal of the image occurs when an object is reflected in an odd number of plane mirrors.

Specular Reflectors. Examples of specular reflectors are:

1. Smooth polished, anodized and electroplated metals, and first-surface silvered glass or plastic mirrors. Internally reflectorized lamps utilize first-surface reflection where the incident light strikes the thin metal reflecting layer without passing through the glass, as shown in Fig. 2-23b.

Light reflected from the upper surface of a transparent medium, such as glass plate, as in Figs. 2-23a and c, also is an example of first-

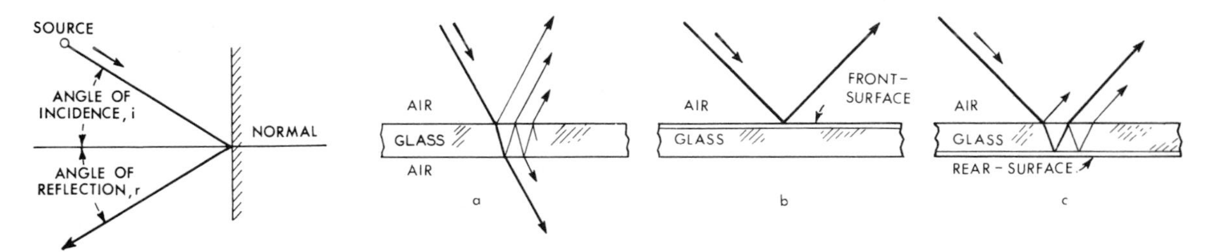

Fig. 2-22. The law of reflection states that the angle of incidence *i* = angle of reflection *r*.

Fig. 2-23. Reflections from (a) a transparent medium, such as clear plate glass and (b) from front-surface and (c) rear-surface mirrors.

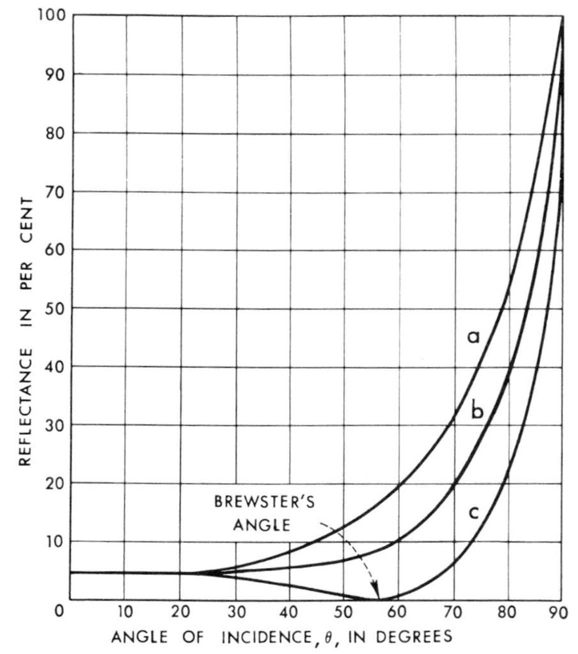

Fig. 2-24. Effect of angle of incidence and state of polarization on per cent of light reflected at an air-glass* surface: a. Light that is polarized in the plane of incidence. b. Nonpolarized light. c. Light that is polarized in plane perpendicular to plane of incidence.

* For spectacle crown glass, n = 1.523.

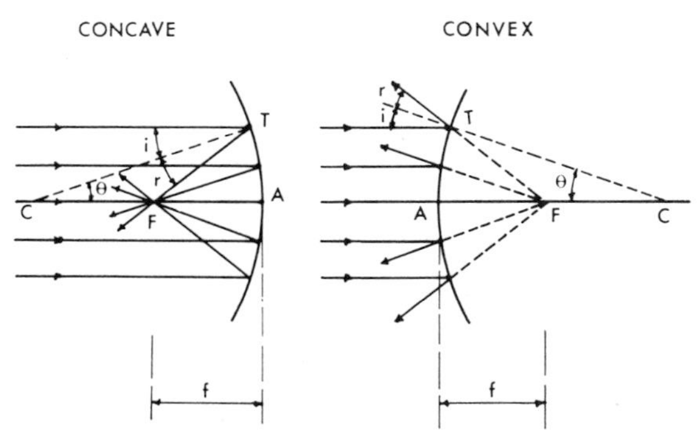

Fig. 2-25. Focal point and focal length of curved surfaces.

surface reflection. As shown in Fig. 2-24, less than 5 per cent of the incident light is reflected at the first surface unless it strikes the surface at wide angles from the normal. The sheen of silk and the shine from smooth or coated paper are images of light sources reflected in the first surface.

2. Rear-surface mirrors. Some light, the quantity depending on the incident angle, is reflected by the first surface. The rest goes through the transparent medium to the rear-surface mirror and is reflected back through, as shown in Fig. 2-23c, parallel to the ray reflected by the first surface.

Reflection from Curved Surfaces. Fig. 2-25 shows the reflection of a beam of light by a concave surface and by a convex surface. A ray of light striking the surface at *T* obeys the law of reflection (see Fig. 2-22), and by taking each ray separately, the paths of the reflected rays may be constructed.

In the case of parallel rays reflected from a concave surface, all the rays can be directed through a common point *F* by properly designing the curvature of the surface. This is called the *focal point.* The *focal length* is *f* (*FA*).

Spread Reflection. If a reflecting surface is figured in any way (corrugated, deeply etched or hammered) it spreads any rays it reflects; that is, an incident ray is spread out into a cone of reflected rays, as shown in Fig. 2-26b.

Spread Reflectors. Depolished metals and similar surfaces reflect individual rays at slightly different angles but all in the same general direction. These are used where smooth beam and moderate control are required.

Corrugated, brushed, dimpled, etched, or pebbled surfaces consist of small specular surfaces in irregular planes. Brushing the surface spreads the image at right angles to the brushing. Pebbled, peened or etched surfaces produce a random patch of highlights. These are used where beams free from striations and filament images are required; widely used for sparkling displays.

The angle through which reflections are spread can be controlled by proper peening, for which equations describing peen radius and depth are available.[21] Shot or sand blasting and etching may cause serious losses in efficiency.

Diffuse Reflection. If a material has a rough surface or is composed of minute crystals or pigment particles, the reflection is diffuse. Each ray falling on an infinitesimal particle obeys the law of reflection, but as the surfaces of the particles are in different planes, they reflect the light at many angles, as shown in Fig. 2-26c.

Diffuse Reflectors. Flat paints and other matte finishes and materials reflect at all angles and exhibit little directional control. These are used where wide distribution of light is desired.

Compound Reflection. Most common materials are compound reflectors and exhibit all three reflection components (specular, spread and diffuse) to varying degrees. In some, one or two components predominate, as shown in Fig. 2-27. Specular and narrowly spread reflections (usually surface reflections) cause the "sheen" on etched aluminum and semigloss paint.

Diffuse-Specular Reflectors. Porcelain enamel, glossy synthetic finishes, and other surfaces with a shiny transparent finish over a matte base exhibit no directional control except for the specularly reflected ray that is shown in Fig. 2-27a, which usually amounts to a total of 5 to 15 per cent of the incident light.

Total Reflection. Total reflection of a light ray at a surface of a transmitting medium (see Fig. 2-28) occurs when the angle of incidence (i) exceeds a certain value such that its sine equals or exceeds n_2/n_1 (ratio of indices of refraction). If the index of refraction of the first medium (n_1) is greater than that of the second medium (n_2), sin r will become unity when sin i is equal to n_2/n_1. At angles of incidence greater than this critical angle (i_c) the incident rays are reflected totally, as in Fig. 2-28. In most glass total reflection occurs whenever sin i is greater than 0.66, that is, for all angles of incidence greater than 41.8 degrees (glass to air). Both edge lighting and efficient light transmission through rods and tubes are examples of total (internal) reflection.[23,24]

When light, passing through air, strikes a piece of ordinary glass ($n_2/n_1 \approx 1.5$) normal to its surface, about 4.5 per cent is reflected from the upper surface and about 4 per cent from the

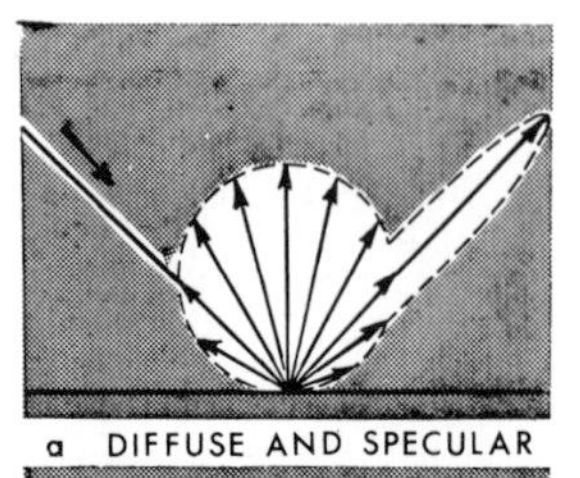

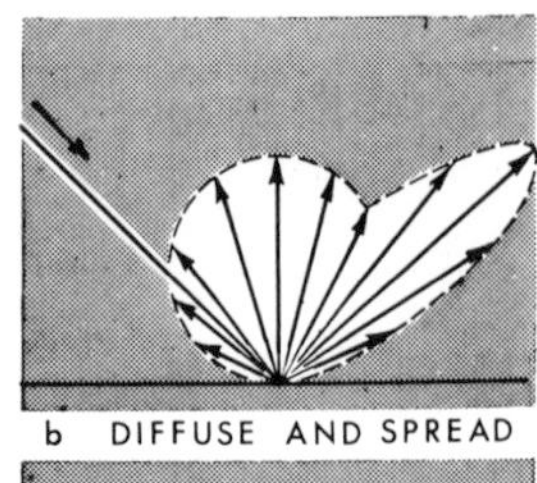

Fig. 2-27. Examples of compound reflection: (a) diffuse and specular; (b) diffuse and spread; (c) specular and spread.

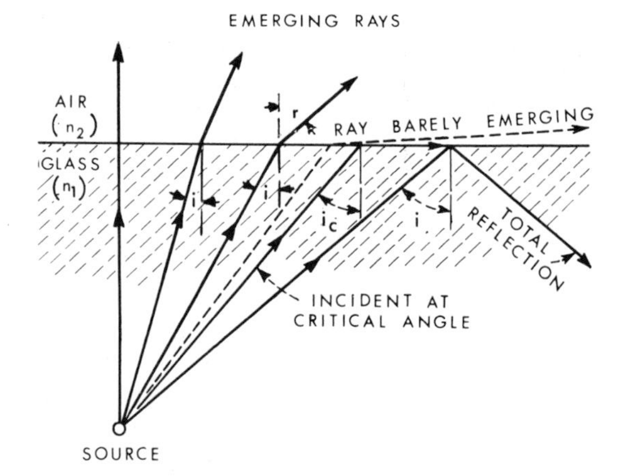

Fig. 2-28. Total reflection occurs when sin r = 1. The critical angle i_c, varies with the media.

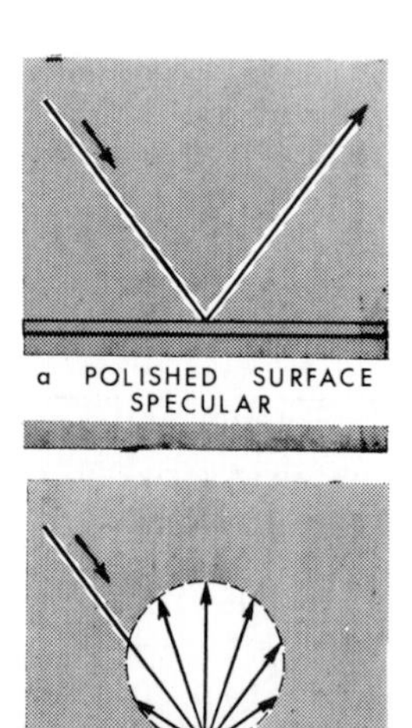

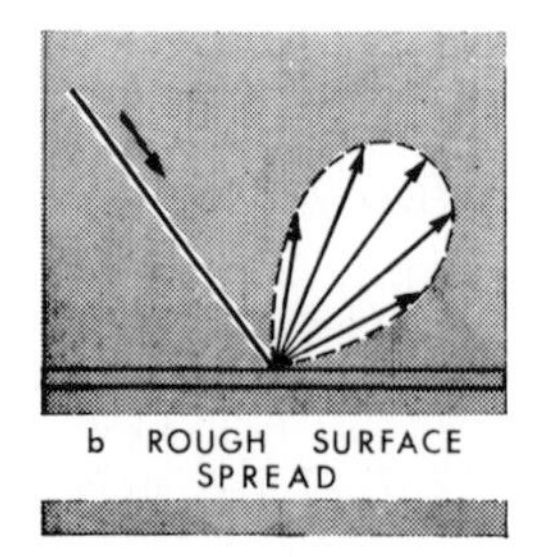

Fig. 2-26. The type of reflection varies with different surfaces: (a) polished surface (specular); (b) rough surface (spread); (c) matte surface (diffuse).

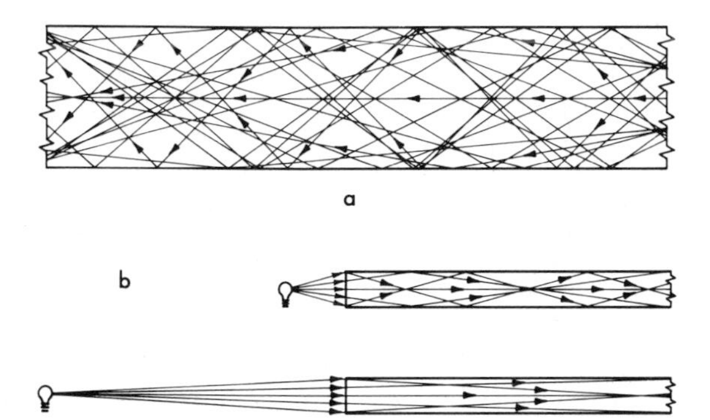

Fig. 2-29. Representation of light transmission through a single fiber of a fiber optics system showing internal reflections (a), and effect of light source location on collimation of light (b).

lower surface. Approximately 90 per cent of the light is transmitted and some absorbed. The proportion of reflected light increases as the angle of incidence is increased. See Fig. 2-24.

Fiber Optics. Fiber optics[18] is a term given to that phase of optical science concerned with thin, cylindrical glass or plastic fibers of optical quality. Light entering one end of the fiber is transmitted to the other end by a process of total internal reflection. See Fig. 2-29. Applications for a single fiber are more theoretical than practical. Therefore, a large number of fibers (from 100 to 1,000,000) are clustered together to form a bundle. The ends of the bundle are bonded together, ground and polished. In order to prevent light leaking from one fiber to another, each is insulated with a glass coating of lower refractive index than that of the fiber. The entire bundle is encased in a flexible tubing to protect the fibers. Fiber bundles are of two major types: coherent and non-coherent. The first is used for transmitting images and each individual fiber is carefully oriented with respect to its neighbors in the entire bundle. Non-coherent bundles have random fiber locations in the bundle, but are suitable for transmitting light between points.

Refraction and Refractors

A change in the velocity of light (speed of propagation, not frequency) occurs when a ray leaves one material and enters another of greater or less optical density. The speed will be reduced if the medium entered is more dense and increased if it is less dense.

Except when light enters at an angle normal to the surface of a medium of different density, the change in speed always is accompanied by a bending of the light from its original path at the point of entrance, as shown in Fig. 2-30. This is known as refraction. The degree of bending depends on the relative densities of the two substances, on the wavelength of the light, and on the angle of incidence, being greater for large differences in density than for small. The light is bent toward the normal to the surface when it enters a more dense medium and away from the normal when it enters a less dense material.

When light is transmitted from one medium to another, each ray follows the law of refraction. When rays strike or enter a new medium, they may be broken up and scattered in many directions because of irregularities of the surface, such as fine cracks, mold marks, scratches or changes in contour or because of foreign deposits of dirt, grease or moisture.

Snell's Law. The law of refraction (Snell's Law) is expressed:

$$n_1 \sin i = n_2 \sin r$$

where n_1 = the index of refraction of the first medium
i = the angle the incident light ray forms with the normal to the surface
n_2 = the index of refraction of the second medium
r = the angle the refracted light ray forms with the normal to the surface.

When the first medium is air, of which the index of refraction usually is assumed to be 1 (correct to three decimal places but actually the index for a vacuum) the formula becomes:

$$\sin i = n_2 \sin r$$

The two interfaces of the glass plate shown in Fig. 2-30 are parallel and therefore the entering

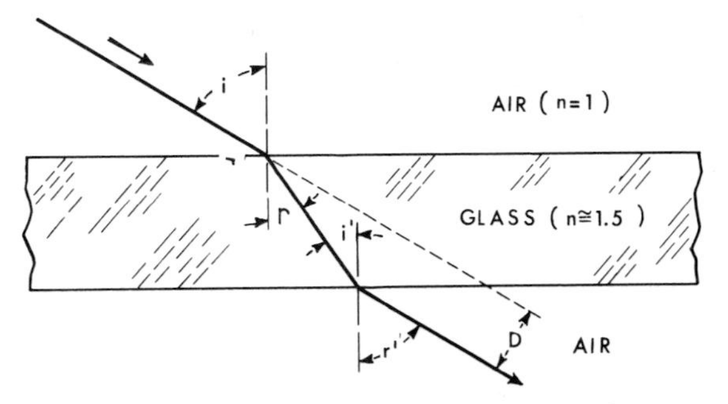

Fig. 2-30. Refraction of light rays at a plane surface causes blending of the incident rays and displacement of the emergent rays. A ray passing from a rare to a denser medium is bent toward the normal to the interface, while a ray passing from a dense to a rarer medium is bent away from the normal.

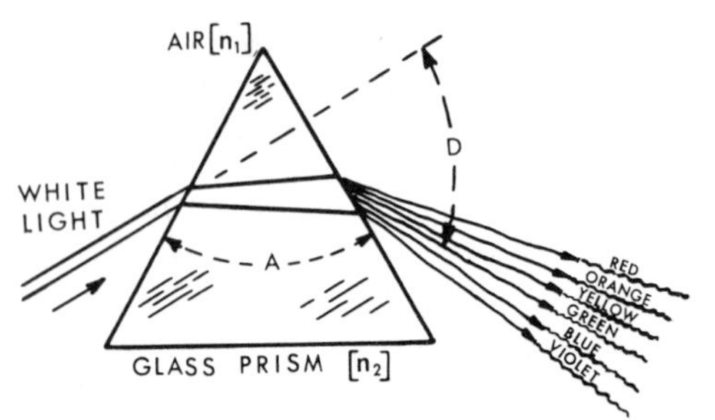

Fig. 2-31. White light is dispersed into its component colors by refraction when passed through a prism. The angle of deviation, *D*, (illustrated for green light) varies with wavelength.

and emerging rays also are parallel. The rays are displaced from each other (distance *D*) because of refraction.

Examples of Refraction. A common example of refraction is the apparent bending of a straw at the point where it enters the water in a drinking glass. Although the straw is straight, light rays coming from that part of the straw under water are refracted when they pass from the water into the air and appear to come from higher points. These irregularities cause irregular refraction of transmitted rays and distortion of the images of objects at which the rays originate.

Prismatic light directors, such as shown in Fig. 2-32, a and b, may be designed to provide a variety of light distributions using the principles of refraction. Lens systems controlling light by refraction are used in automobile headlights, and in beacon, floodlight and spotlight Fresnel lenses as shown in Fig. 2-32.

Prisms. Consideration of Snell's Law

$$n_2 = \frac{\sin i}{\sin r} = \frac{\text{velocity of light in air}}{\text{velocity in prism}}$$

suggests, since the velocity of light is a function of the index of refraction of the media involved and also of wavelength, that the exit path from a prism will be different for each wavelength of incident light and for each angle of incidence. See Fig. 2-31. This orderly separation of incident light into its spectrum of component wavelengths is called *dispersion.*

Refracting Prisms. The degree of bending of light at each prism surface is a function of the refractive indices of the media and the prism angle (*A* in Fig. 2-31). Light can be directed accurately within certain angles by having the proper angle between the prism faces.

In the design of refracting equipment, the same general considerations of proper flux distribution hold true as for the design of reflectors. Following Snell's Law of refraction, the prism angles can be computed to provide the proper direction of the light rays from the source. For most commercially available transparent material like glass or plastics, the index of refraction used lies between 1.4 and 1.6.

Often by proper placement of the prisms, it is possible to limit the prismatic structure to only one surface of the refractor, leaving the other surface entirely smooth for easy maintenance. The number and the sizes of prisms used are governed by several considerations. Among them are ease of manufacture and convenient maintenance of lighting equipment in service. A large

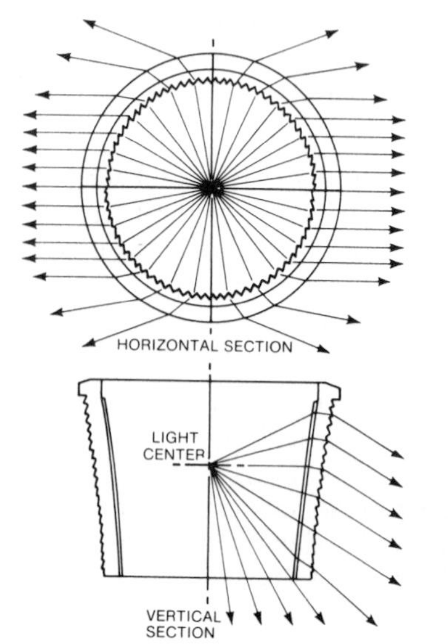

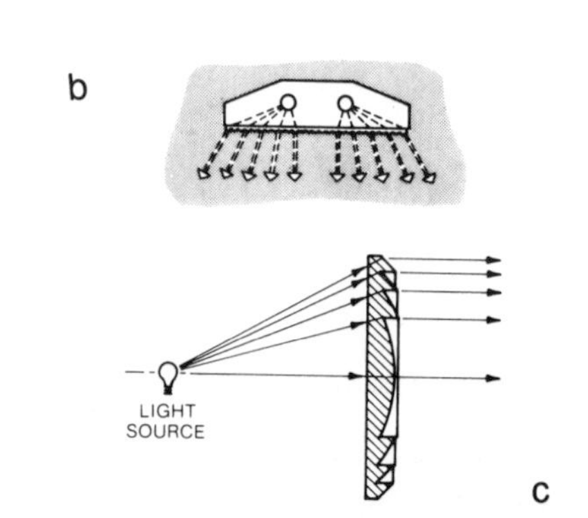

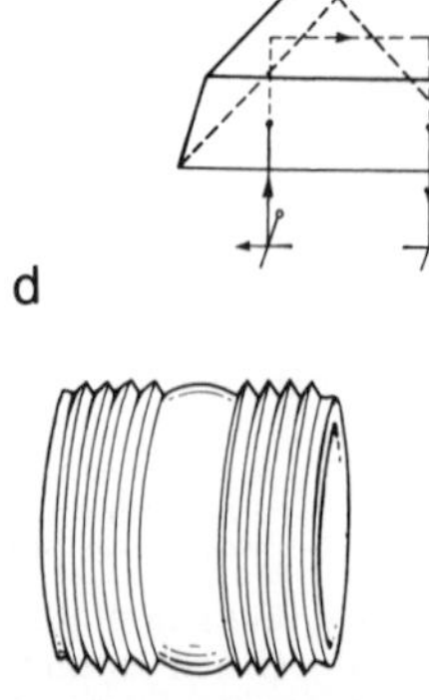

Fig. 2-32. Optical systems utilizing the refractive properties of prisms and lenses: a. Street lighting unit in which the outer piece controls the light in vertical directions (concentrating the rays into a narrow beam at about 75 degrees from the vertical) and the inner piece redirects the light in the horizontal plane. The result is a "two-way" type of candlepower distribution. b. Prismatic lens for fluorescent lamp luminaire intercepts as much light as possible, redirecting part from the glare zone to more useful directions. c. Cylindrical and flat Fresnel lenses. d. Reflecting prism.

number of small prisms may suffer from prisms rounding in actual manufacture; on the other hand, small prisms produce greater accuracy of light control. Refracting prisms are used in headlight lenses, refracting luminaires, etc.

Ribbed and Prismed Surfaces. These can be designed to spread rays in one plane or scatter them in all directions. These surfaces are used in luminaires, footlight lenses, luminous elements, glass blocks, windows and skylights.

Reflecting Prisms. These reflect light internally, as shown in Fig. 2-32d, and are used in luminaires and retrodirective markers. Performance quality depends on flatness of reflecting surfaces, accuracy of prism angles, elimination of black surface dirt in optical contact with the surface and elimination (in manufacture) of prismatic error.

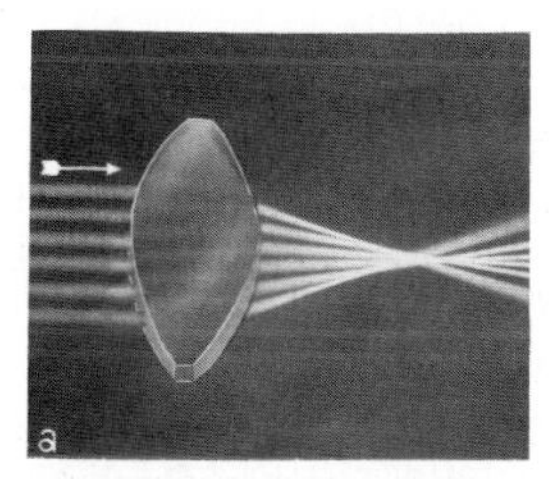

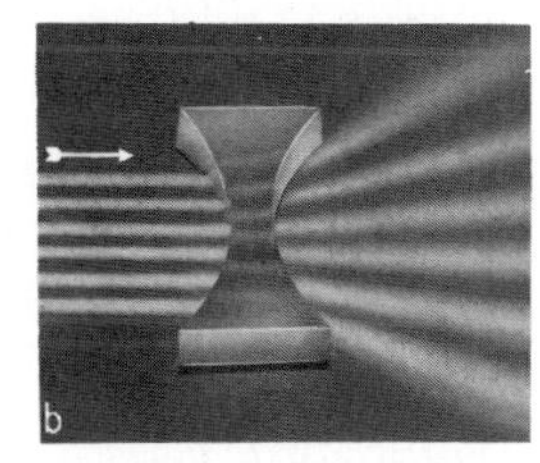

Fig. 2-33. Ray path traces through lenses: a. positive, b. negative.

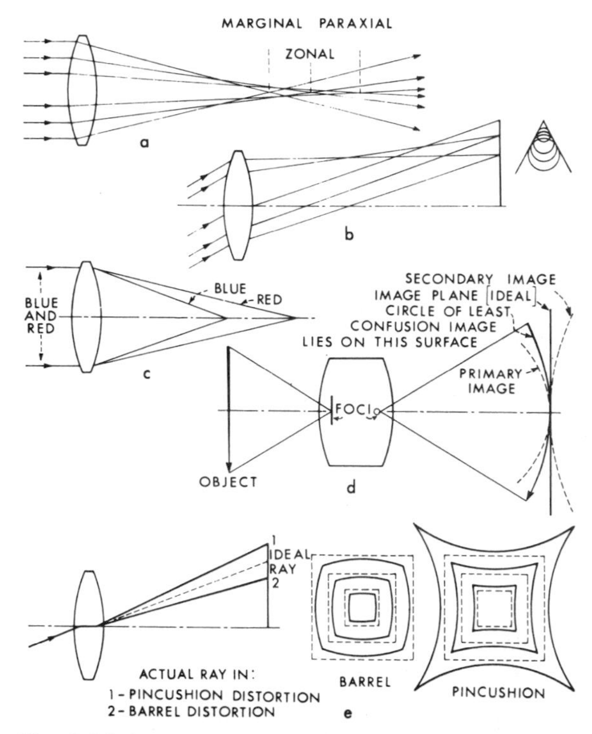

Fig. 2-34. Lens aberrations. a. Spherical aberration: conversion at different focal points of parallel rays at varying distances from the axis of a lens. b. Coma: difference in the lateral magnification of rays passing through different zones of a lens. c. Chromatism: a difference in focal length for rays of different wavelengths. d. Astigmatism and curvature: existence in two parallel planes of two mutually perpendicular line foci and a curved image plane. e. Distortion: a difference in the magnification of rays passing through a lens at different angles.

Lenses. Positive lenses form convergent beams and real inverted images as in Fig. 2-33a. Negative lenses form divergent beams and virtual, inverted images as in Fig. 2-33b.

Fresnel Lenses. Excessive weight and cost of glass in large lenses used in illumination equipment can be reduced by a method developed by Fresnel, as shown in Fig. 2-32c. The use of lens surfaces parallel to those replaced brings about a reduction in thickness and the optical action is approximately the same. Although outside prisms are slightly more efficient, they are likely to collect more dust. Therefore, prismatic faces often are formed on the inside.

Lens Aberrations. There are, in all, seven principal lens aberrations: spherical, coma, axial and lateral chromatism, astigmatism, curvature, and distortion. See Fig. 2-34. Usually they are of little importance in lenses used in common types of lighting equipment. The simpler the lens system, the more difficult is the correction of the aberrations.

Transmission and Transmitting Materials

Transmission is a characteristic of many materials: glass, plastics, textiles, crystals and so forth. The luminous transmittance τ of a material is the ratio of the total emitted light to the total incident light; it is affected by reflections at each surface of the material, as explained in Fig. 2-24, and by absorption within the material. Fig. 2-35 lists characteristics of several materials.

Bouguer's or Lambert's Law. Absorption in a clear transmitting medium is an exponential function of the thickness of the medium traversed:

$$I = I_0 \tau^x$$

where I = intensity of transmitted light
I_0 = intensity of light entering the medium after surface reflection
τ = transmittance of unit thickness
x = thickness of sample traversed.

Optical density (D) is the common logarithm of the reciprocal of transmittance:

$$D = \log_{10} (1/\tau).$$

Fig. 2–35. Reflecting and Transmitting Materials

Material	Reflectance * or Transmittance† (per cent)	Characteristics
		Reflecting
Specular		
Mirrored and optical coated glass	80 to 99	Provide directional control of light and brightness at specific viewing angles. Effective as efficient reflectors and for special decorative lighting effects.
Metallized and optical coated plastic	75 to 97	
Processed anodized and optical coated aluminum	75 to 95	
Polished aluminum	60 to 70	
Chromium	60 to 65	
Stainless steel	55 to 65	
Black structural glass	5	
Spread		
Processed aluminum (diffuse)	70 to 80	General diffuse reflection with a high specular surface reflection of from 5 to 10 per cent of the light.
Etched aluminum	70 to 85	
Satin chromium	50 to 55	
Brushed aluminum	55 to 58	
Aluminum paint	60 to 70	
Diffuse		
White plaster	90 to 92	Diffuse reflection results in uniform surface brightness at all viewing angles. Materials of this type are good reflecting backgrounds for coves and luminous forms.
White paint**	75 to 90	
Porcelain enamel**	65 to 90	
White terra-cotta**	65 to 80	
White structural glass	75 to 80	
Limestone	35 to 65	
		Transmitting
Glass‡		
Clear and optical coated	80 to 99	Low absorption; no diffusion; high concentrated transmission. Used as protective cover plates for concealed light sources.
Configurated, obscure, etched, ground, sandblasted, and frosted	70 to 85	Low absorption; high transmission; poor diffusion. Used only when backed by good diffusing glass or when light sources are placed at edges of panel to light the background.
Opalescent and alabaster	55 to 80	Lower transmission than above glasses; fair diffusion. Used for favorable appearance when indirectly lighted.
Flashed (cased) opal	30 to 65	Low absorption; excellent diffusion. used for panels of uniform brightness with good efficiency.
Solid opal glass	15 to 40	Higher absorption than flashed opal glass; excellent diffusion. Used in place of flashed opal where a white appearance is required.
Plastics		
Clear prismatic lens	70 to 92	Low absorption; no diffusion; high concentrated transmission. Used as shielding for fluorescent luminaires, outdoor signs and luminaires.
White	30 to 70	High absorption; excellent diffusion. Used to diffuse lamp images and provide even appearance in fluorescent luminaires.
Colors	0 to 90	Available in any color for special color rendering lighting requirements or esthetic reasons.
Marble (impregnated)	5 to 30	High absorption; excellent diffusion; used for panels of low brightness. Seldom used in producing general illumination because of the low efficiency.
Alabaster	20 to 50	High absorption; good diffusion. Used for favorable appearance when directly lighted.

* Specular and diffuse reflectance.
** These provide compound diffuse-specular reflection unless matte finished.
† Inasmuch as the amount of light transmitted depends upon the thickness of the material and angle of incidence of light, the figures given are based on thicknesses generally used in lighting applications and on near normal angles of incidence.
‡ See also Fig. 6–1.

Spread Transmission. Spread transmission materials offer a wide range of textures. They are used for brightness control as in frosted lamp bulbs, in luminous elements where accents of brilliance and sparkle are desired, and in moderately uniform brightness luminaire enclosing globes. Care should be used in placing lamps to avoid glare and spotty appearance.

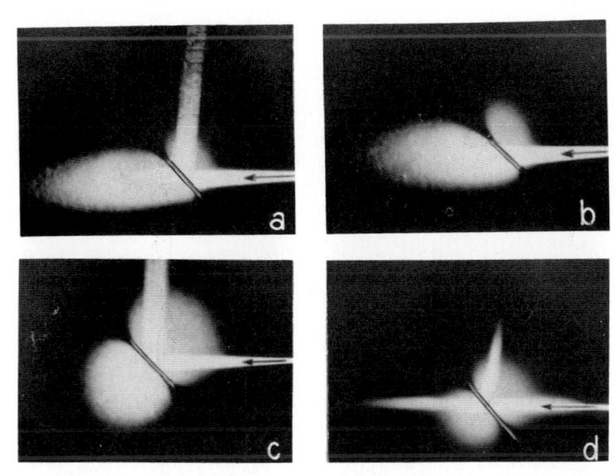

Fig. 2-36. a. Spread transmission of light incident on *smooth* surface of figured, etched, ground, or hammered glass samples. b. Spread transmission of light incident on *rough* surface of the same samples. c. Diffuse transmission of light incident on solid opal and flashed opal glass, white plastic or marble sheet. d. Mixed transmission through opalescent glass.

Fig. 2-36a shows a beam of light striking the smooth side of a piece of etched glass. In Fig. 2-36b the frosted side is toward the source, a condition that with many ground or otherwise roughened glasses results in appreciably higher transmittance. For outdoor use the rough surface usually must be enclosed to avoid excessive dirt collection.

Diffuse Transmission. Diffusing materials scatter light in all directions, as shown in Fig. 2-36c. White, opal and prismatic plastics and glass are widely used where uniform brightness is desired.

Mixed Transmission. Mixed transmission is a result of a spectrally selective diffusion characteristic exhibited by certain materials such as fine opal glass, which permits the regular transmission of certain colors (wavelengths) while diffusing other wavelengths. This characteristic in glass varies greatly, depending on such factors as its heat treatment, composition and thickness, and the wavelengths of the incident light.

Polarization

Unpolarized light consists of visible electromagnetic waves having transverse vibrations of equal magnitude in an infinite number of planes, all of which contain the line representing the direction of propagation. See Fig. 2-37. In explaining the properties of polarized light, it is common to resolve the amplitude of the vibrations of any light ray into components vibrating in two orthogonal planes each containing the light ray. These two principal directions are usually referred to as the horizontal and vertical vibrations. The horizontal component of light is the summation of the horizontal components of the infinite number of vibrations making up the light ray. When the horizontal and vertical components are equal, the light is unpolarized. When these two components are not equal, the light is partially or totally polarized as shown in Fig. 2-37.

The percentage polarization of light from a source or luminaire at a given angle is defined by the following relation:[19]

$$\text{Per cent vertical polarization} = \frac{I_v - I_h}{I_v + I_h} \times 100$$

where I_v and I_h are the intensities of the vertical and horizontal components of light, respectively, at the given angle.

Reference to vertically polarized light or horizontally polarized light can be misleading in that it suggests that all light waves vibrate either horizontally or vertically. A better notation would be to refer to light at a given instant as consisting of one component vibrating in a horizontal plane and another component vibrating in a vertical plane. A general notation would identify the light components in terms of two reference planes as shown in Fig. 2-38. One plane is the plane of the task at the point of the incident light ray; and the second plane is the plane of incidence or that plane perpendicular to the plane of the task and containing the incident light ray. Then the two components of light would be referred to as the parallel component or component in the plane of incidence and the perpendicular component. This notation would

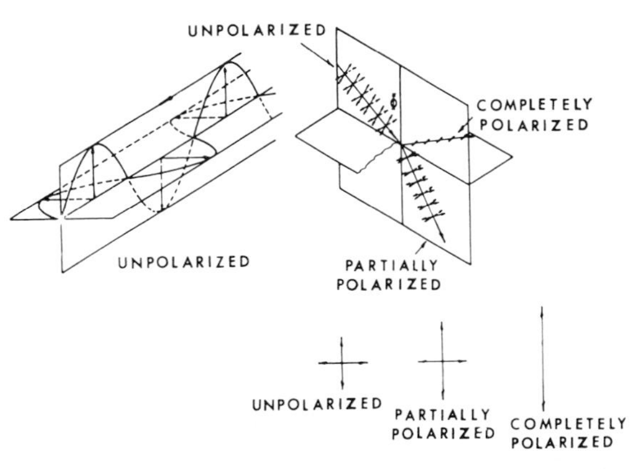

Fig. 2-37. Graphic representations of polarized and unpolarized light.

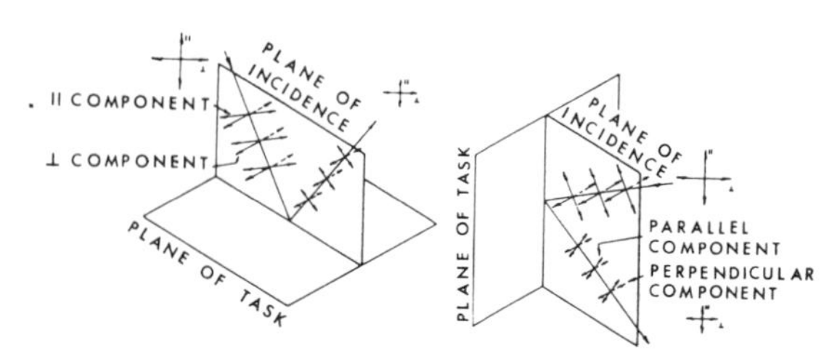

Fig. 2-38. Reference planes of task.

apply to any task position and would not have ambiguity with space references.

Polarized light can be produced in four ways: (1) scattering, (2) birefringence, (3) absorption and (4) reflection and refraction.

Scattering is applicable to daylighting; that is, light from a clear blue sky is partially polarized due to the scattering of light by dust particles in the air.

The *birefringence* or double refraction property of certain crystals can be utilized to achieve polarization. However, the size of these crystals limits this technique to scientific applications and is not suitable to general lighting.

Polarization by *absorption* can be achieved by using dichroic polarizers. These polarizers absorb all of the light in one particular plane and transmit a high percentage of the light in a perpendicular plane. High percentage of polarization can be obtained by this method but at a loss of total luminous transmittance. This type of polarizer is commonly used in sunglasses where it is oriented to transmit the vertical component of light while suppressing the horizontal component.

Light may be polarized by utilizing the *reflection* characteristics of dielectric materials. When light is reflected from a glass surface, it is partially polarized; *i.e.*, a larger percentage of the horizontal component is reflected than of the vertical component. See Fig. 2-24. At approximately 57 degrees, or Brewster's angle, the reflected light is composed of only the horizontal component. See Fig. 2-24. However, for one surface only, 15 per cent of the incident horizontal component is reflected. The light transmitted through a plate at this angle is made up of the remaining portion of the horizontal component and all the vertical component of the original beam. The resulting light is partially polarized. See Fig. 2-39. As additional glass plates are added to the system, more and more of the horizontal component is reflected and the transmitted light is more completely vertically polarized. A stack of glass plates, as shown in Fig. 2-40, then becomes a method of producing polarization, and the polarizing effect is greatest at Brewster's angle. The percentage polarization is less at all other angles and is zero for a light ray at normal incidence. Polarization by this method can be obtained by arranging glass or plastic flakes in a suitable material.

Interference

When two light waves come together at different phases of their vibration, they combine to make up a single wave whose amplitude equals the sum of the amplitudes of the two. Fig. 2-41 shows interference. Part of the incident light *ab* is first reflected as *bc*. Part is refracted as *bd*, which again reflects as *de*, and finally emerges as *ef*. If waves *bc* and *ef* have appreciable width of wave fronts, they will overlap and interfere. Optical interference coatings have been used for many years in cameras, projectors and other

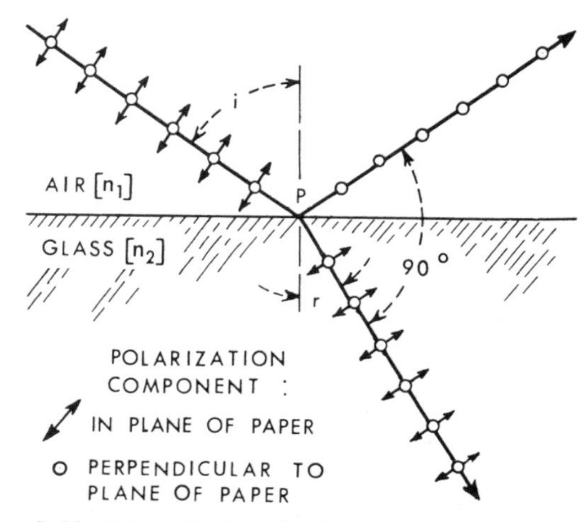

Fig. 2-39. Polarization by reflection at a glass-air surface is at a maximum when the sum of the angle of incidence *i* plus the angle of refraction *r* equals 90 degrees.

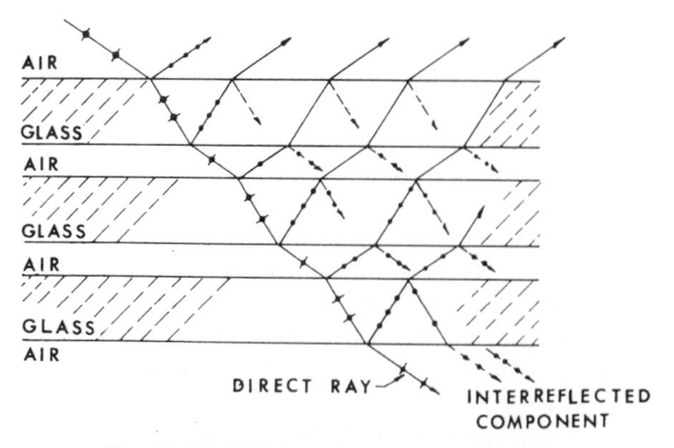

Fig. 2-40. Principle of multilayer polarizers.

optical instruments, and can reduce reflections from transmitting surfaces, separate heat from light, transmit or reflect light according to color, increase reflections from reflectors, or perform other light control functions. Naturally occurring examples of interference are soap bubbles and oil slicks. Also, many birds, insects and fish get their irridescent colors from interference films. The application of interference coatings can significantly increase the reflectance of reflectors or transmittance of luminaire glass or plastic enclosures.

Low Reflectance Films. Dielectric optical interference films are applied to surfaces to reduce reflectance, increase transmittance, and consequently improve contrast relationships. Films a quarter wavelength thick with an index of refraction between that of the medium surrounding the glass and that of the glass are used. The hardest and most permanent films are those of magnesium fluoride condensed on the transmitting surface after thermal evaporation in vacuum.

The usual, uncoated, 4 per cent reflection at air-glass surfaces may be reduced to less than 0.5 per cent at each filmed surface at normal incidence, as a result of the canceling interference between the waves reflected at the air-to-film and film-to-glass surfaces.

Diffraction

When a wave front is obstructed partially, as by the edge of a reflector or a louver, the shadow cast by the reflector or louver may be "sharp" or "soft," depending on the geometrical relationship and size of the source, reflector and illuminated surface. When a series of fuzzy and ill-defined shadows is produced, this phenomenon is known as diffraction.[17]

Diffusion

Diffusion is the breaking up of a beam of light and the spreading of its rays in many directions by irregular reflection and refraction from microscopic crystalline particles, droplets, or bubbles within a transmitting medium, or from microscopic irregularities of a reflecting surface. Perfect diffusion seldom is attained in practice but sometimes is assumed in calculations in order to simplify the mathematics. See Figs. 2-26c and 2-27.

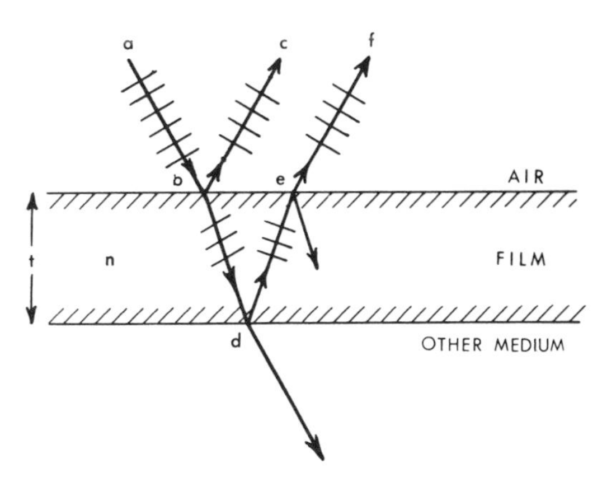

Fig. 2-41. Interference.

Absorption

Absorption occurs when a light beam passes through a transparent or translucent medium or meets a dense body such as an opaque reflector surface. If the intensity of all wavelengths of the light passing through a transparent body is reduced by nearly the same amount, the substance is said to show general absorption. The absorption of certain wavelengths of light in preference to others is called selective absorption. Practically all colored objects owe their color to selective absorption in some part of the visible spectrum with resulting reflection and/or transmission in other selected parts of the spectrum.

REFERENCES

1. Richtmyer, F. K., Kennard, E. H. and Lauritsen, T.: *Introduction to Modern Physics*, McGraw-Hill Book Company, New York, 1969. Born, M.: *Atomic Physics*, rev. by R. J. Blin-Stoyle and J. M. Radcliffe, 8th ed., Hafner Publishing Company, New York, 1969. Born, M., and Wolf, E.: *Principles of Optics; Electromagnetic Theory of Propagation, Interference and Diffraction of Light*, Pergamon Press, New York, 1970. Elenbaas, W.: *Light Sources*, Crane, Russak & Company, New York, 1973.
2. Maxwell, J. C.: *A Treatise on Electricity and Magnetism*, Dover Publications, New York, 1960.
3. Forsythe, W. E.: *Measurement of Radiant Energy*, McGraw-Hill Book Company, New York, 1937.
4. Forgan, B. W.: "Solar Constants and Radiometric Scales," *App. Opt.*, Vol. 16, p. 1628, 1977.
5. Elenbaas, W.: *Fluroescent Lamps and Lighting*, The Macmillan Company, New York, 1959.
6. Waymouth, J. F.: *Electric Discharge Lamps*, MIT Press, Cambridge, Massachusetts, 1971.
7. Fonda, G. R. and Seitz, F.: *Preparation and Characteristics of Solid Luminescent Materials*, John Wiley & Sons, New York, 1948. Leverenz, H. W.: *Introduction to Luminescence of Solids*, John Wiley & Sons, New York, 1950.
8. Harvey, E. N.: *A History of Luminescence*, American Philosophical Society, Philadelphia, 1957.

9. Wachtel, A.: "The Effect of Impurities on the Plaque Brightness of a 3000 K Calcium Halosphosphate Phosphor," *J. of Electro-Chem. Soc.*, Vol. 105, pp. 256–60, May, 1958.
10. Ivey, H. F.: *Electroluminescence and Related Effects*, Academic Press, New York, 1963.
11. Brotherton, M.: *Masers and Lasers; How They Work, What They Do*, McGraw-Hill Book Company, New York, 1964.
12. Harvey, A. F.: *Coherent Light*, John Wiley & Sons, New York, 1971.
13. Lengyel, B. F.: *Introduction to Laser Physics*, John Wiley & Sons, New York, 1966.
14. Goodeve, C. F.: "Relative Luminosity in the Extreme Red," *Proc. Roy. Soc.* (London), Vol. A155, p. 664, July, 1936.
15. Committee on Light Control and Equipment Design of the IES: "IES Guide to Design of Light Control," Parts I and II, *Illum. Eng.*, Vol. LIV, p. 722, November and p. 778, December, 1959.
16. Halliday, D. and Resnick, R.: *Physics*, John Wiley & Sons, Inc., New York, 1977.
17. Hardy, A. C. and Perrin, F. H.: *The Principles of Optics*, McGraw-Hill Book Co., Inc., New York, 1932.
18. *Light Wires*, Bausch & Lomb Publication No. D-2045.
19. Committee on Testing Procedures for Illumination Characteristics of the IES: "Resolution on Reporting Polarization," *Illum. Eng.*, Vol. LVIII, p. 386, May, 1963.
20. Jolley, L. B. W., Waldram, J. M., and Wilson, G. H.: *The Theory and Design of Illuminating Engineering Equipment*, Chapman & Hall, Ltd., London, 1930.
21. Elmer, W. B.: *The Optical Design of Reflectors*, John Wiley & Sons, Inc., New York, 2nd Edition, 1979.
22. Elmer, W. B.: "The Optics of Reflectors for Illumination," *IEEE Trans., Ind. Appl.*, Vol. IA-19, p. 776, September/October, 1983.
23. Whitehead, L. A., Nodwell, R. A. and Curzon, F. L.: "New Efficient Light Guide for Interior Illumination," *Appl. Opt.*, Vol. 21, p. 2755, August 1982.
24. Whitehead, L. A.: "Simplified Ray Tracing in Cylindrical Systems," *Appl. Opt.*, Vol. 21, p. 3536, October, 1982.

Light and Vision

SECTION 3

Vision is dependent on light. It is the role of those responsible for lighting design to provide an environment in which people, through the sense of vision, can function effectively, efficiently and comfortably. To have the ability to predict human behavior as a function of lighting, however restricted that behavior may be, it is necessary to know the physical, physiological and psychological components of the behavior and how lighting interacts with each. Although a great deal is known about how simple visual stimuli produce simple visual behaviors, little is known about human response to even moderately complex scenarios.

Optometrists and ophthalmologists are responsible for optimizing the human visual system; designers of the lighting are responsible for optimizing the visual environment, and for doing so without material waste. Both groups are dependent on data from basic and applied scientists from many disciplines who are supported by agencies, such as the Lighting Research Institute, government, universities and industry. Together, they continue toward the development of better lighting practice based on a solid foundation of knowledge.

This section attempts to highlight some of the basic interactions between light and vision. It is intended to provide some fundamental data which the designer of the lighting may find useful, and to call attention to the types of considerations that may be necessary if one is to design lighting for optimal visual performance and comfort.

THE STRUCTURE AND FUNCTION OF THE EYE

The eye is a complex sensory organ which maintains the spatial and temporal relationships of objects in visual space and converts the light energy it receives into electrical signals for processing by the brain. The eye can be divided into optical components (the cornea, crystalline lens, pupil and intraocular humors) and neurological components (the retina and optic nerve). See Fig. 3–1.

Optical Components

A thin film of *tears* is the first optical component of the eye. This film is important because it cleans the surface of the eye, starts the optical refraction (light bending) process necessary for focusing objects, and smooths out small imperfections in the surface of the subsequent refracting medium, the cornea. The *cornea* covers the transparent anterior one-fifth of the eyeball. With the tear layer, it forms the major refracting component of the eye and gives the eye about 70 per cent of its power. The *crystalline lens* provides most of the remaining 30 per cent of the refracting power. The *ciliary muscles* have the ability to change the curvature of the lens and thereby adjust the power of the eye, when needed, in response to changing object distances or certain types of refractive errors; this change in power is called *accommodation* (see page 3–7).

The *aqueous humor* and *vitreous humor* help maintain the shape of the globe and provide nutrients to the nonvascular structures within the eye.

The transmittance of the eye varies with wavelength and with age.[1] In young eyes, the cornea absorbs most of the incident radiation shorter than 300 nanometers while the crystalline lens effectively filters out wavelengths shorter than 380 nanometers (see Fig. 3–2). Accordingly, the *retina* receives radiation in the range between 380 to 950 nanometers with little attenuation. Beyond 950 nanometers, transmittance is variable with major absorptance in the infrared water bands. There is very little infrared radiation passed beyond 1400 nanometers.

NOTE: References are listed at the end of each section.

Because of the excellent transmittance of the eye to infrared, such radiation can be dangerous to the retina and sources with a substantial infrared radiance should not be viewed directly.

In the visible range of the spectrum, the optics of the eye transmit more light at long wavelengths (the red end) than at short wavelengths (the blue end). On the average, some 70 to 85 per cent of the visible spectrum reaches the retina in young eyes.[2] As one ages, there is a general reduction in the transmittance at all wavelengths combined with a marked reduction (greater than four times) in short wavelength transmittance due primarily to yellowing of the crystalline lens (see Fig. 3-3).[3]

In addition to the absorptance of light by the optical media, the efficiency of light for visual perception can be reduced by reflection from the tear-cornea surface and by scatter within the eye.[4] Because the eye suffers primarily from large particle scatter, only part of the scatter is wavelength dependent; therefore, the amount of scattered light within the eye decreases slightly with wavelength. In young eyes, some 25 per cent of the scattered light is produced by the cornea,[5, 6] another 25 per cent by the fundus[7] and the rest from the lens and the vitreous humor. The aqueous humor produces little, if any, scattered light. The amount of scattered light in the eye increases with age. Almost all of this increase is due to changes in the lens.[4]

Neurological Components

The posterior 4/5ths of the eyeball is enclosed by three layers of tissue:

1. The *sclera*—the outermost covering of the globe which is continuous with the cornea and which protects the eye's contents.
2. The *choroid*—a highly vascular tissue that contains the blood supply to much of the eye.
3. The *retina*—the innermost layer of the eyeball which receives the radiant energy and converts it into electrical signals that are sent to the brain.

Radiation which reaches the retina can be absorbed by photopigments located in the outer segments of the retinal receptors—the rods and cones. This absorbed energy is then converted into neural electrochemical signals within the receptor and are then communicated to subsequent neurons in the retina.

The receptors are grouped into functional classes on the basis of the photopigment they contain as well as by their morphological characteristics (see Fig. 3-4). There is one class of rods and three classes of cones. Each of the three cone classes are differentiated primarily on the basis of their photopigments. These photopigments can be conveniently characterized by their peak wavelength sensitivities (see Fig. 3-5). Thus, these three cone classes have been called long, middle and short wavelength sensitive cones. They are also referred to as the red, green and blue cones, respectively, even though these photopigments, by themselves, do not give rise to our perceptions of color.

Rods and cones are not evenly distributed throughout the retina (see Fig. 3-6). The greatest concentration of cones occurs at the fovea. The number of cones falls off dramatically just outside this area until the cone population reaches a relatively constant density across the rest of the retina. The three cone classes are apparently equally distributed throughout the retina except that the short wavelength cones are sparse or nonexistent in the fovea. Rods are not found in the fovea either. Their peak density is found some 20 degrees from the fovea.

A person's relative sensitivity to various wavelengths is determined, to a first approximation, by the absorption spectra of the photoreceptors combined with the transmittance of the preretinal optics. This relative spectral sensitivity changes depending upon the total amount of light reaching the retina. At dim levels the rod system determines sensitivity, and at bright levels the cone system dominates. Figure 3-7

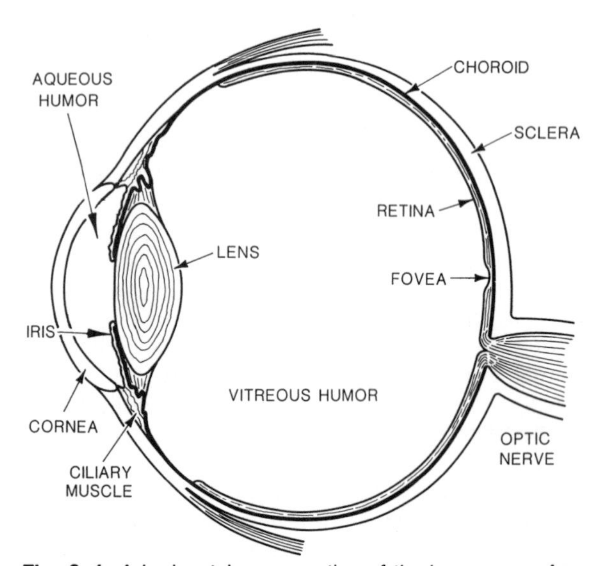

Fig. 3-1. A horizontal cross-section of the human eye. Approximate length from cornea to cone layer of retina is 24 millimeters. Thickness of choroid is about 0.05 millimeters and the sclera 1.0 millimeters.

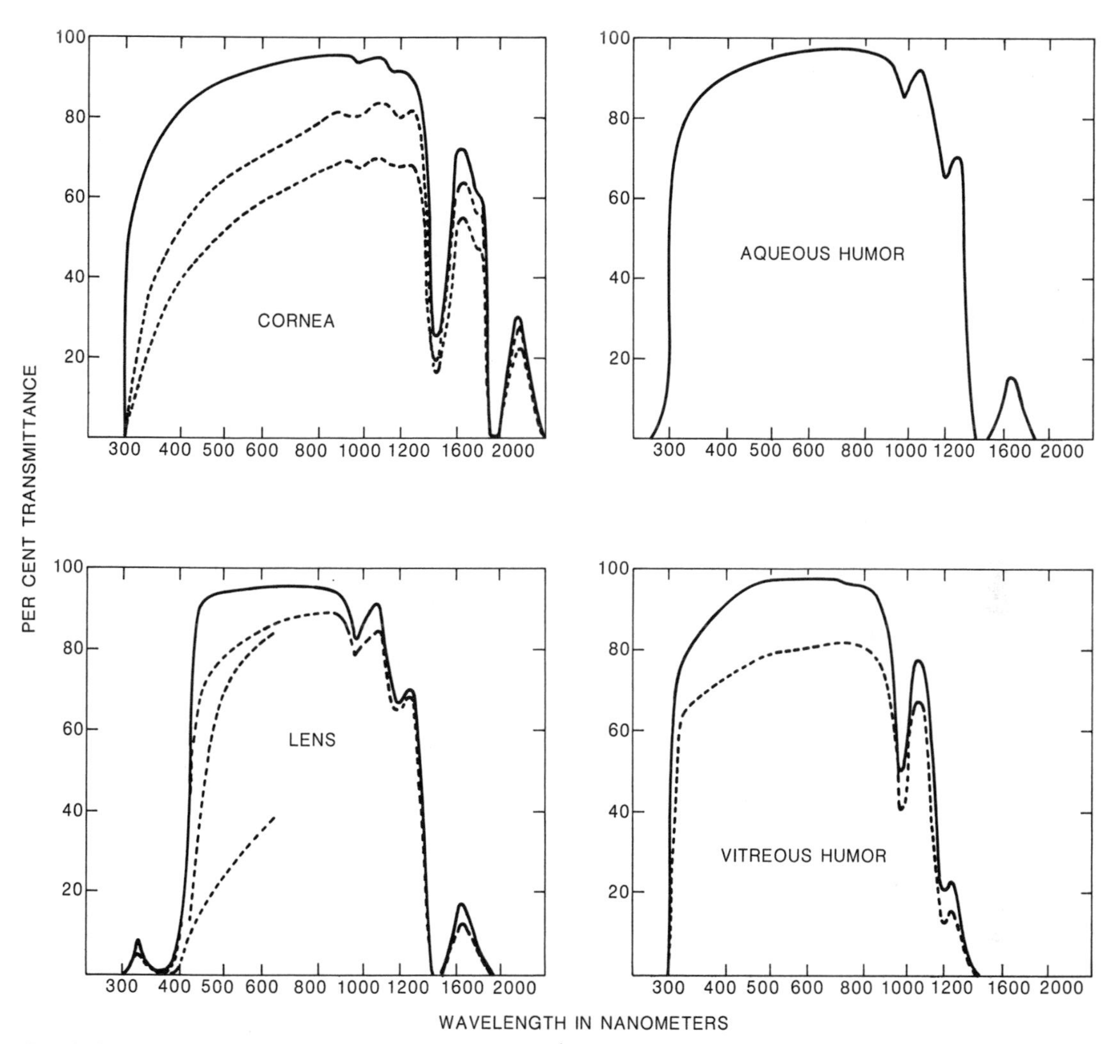

Fig. 3-2. Transmission properties of human ocular media.[1] The solid curves refer to the total light transmitted through the media. The dashed curves refer to the direct, unscattered components only. Where more than one dashed curve is shown the lower are for older eyes.

gives the relative luminous efficiency functions for scotopic (rod) and photopic (cone) vision. (Tabulated values for scotopic and photopic functions may be found in Section 1 and other references.[7]) It should be noted that there is a gradual transition between the scotopic and photopic systems as light level is changed. The luminous efficiency of the visual system in this, so called, *mesopic* region is a poorly understood combination of the scotopic and photopic systems.

An interesting perceptual phenomenon associated with this transition between scotopic and photopic vision is known as the *Purkinje shift.* The relative brightness of colored objects can change because of the differences in the luminous spectral sensitivities of the scotopic and

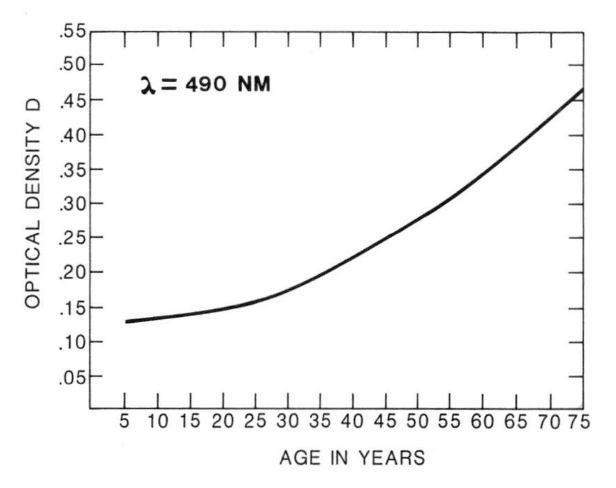

Fig. 3-3. The optical density of the human lens as a function of age.[3] $D = \log \frac{1}{t}$ where t = total internal transmittance.

photopic systems. Blue-green objects (those having high reflectances near 505 nanometers) that may appear dimmer than yellow-green objects (those having high reflectances near 555) under photopic conditions may actually appear brighter (but achromatic) under scotopic conditions.

Neither the rod nor cone system, under common levels of retinal illuminance, is sensitive to radiations longer than about 770 nanometers and, accordingly, these wavelengths are not seen even though there is retinal irradiance as far as 1400 nanometers. Wavelengths shorter than about 380 nanometers are absorbed by the preretinal media and are therefore not "seen" (see material above). The term "light" is applied only to those radiations which reach the retina and stimulate vision; therefore, "light" is restricted to wavelengths between 380 and 770 nanometers. Aphakic individuals, that is, those who have no crystalline lens within their eye, can often detect radiations as short as 310 nanometers and, in that specific case, perhaps the range of wavelengths included as light should extend below 380 nanometers.

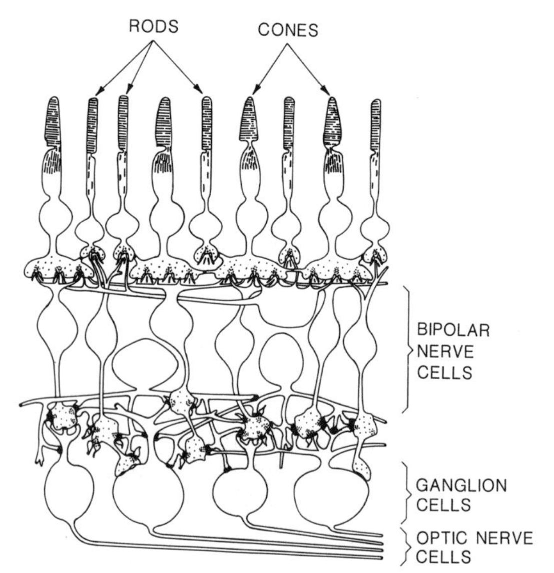

Fig. 3-4. Simplified diagram of the connections among neural elements in the retina. The regions where the cells are contiguous are synapses. The direction of the light is from the bottom up in this diagram.

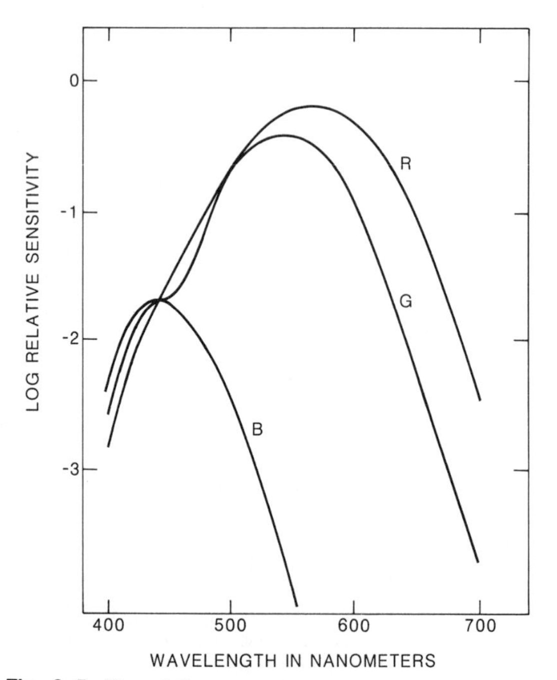

Fig. 3-5. The relative spectral sensitivity curves of the three cone types: long (R), middle (G) and short (B).[9,10]

Sensitivity to light is also determined by the neural organization of the retina. The output from the photoreceptors is recoded by a series of cellular processes in the inner layers of the retina and is transmitted to the brain by the axons of the ganglion cells (the optic nerve, see Fig. 3-4) as pulse-trains of variable frequency. To a first approximation, the ratio of receptors to each ganglion cell determines the absolute sensitivity of that cell. This ratio varies considerably across the retina.

The fovea centralis has a very small receptor-to-ganglion cell ratio; as such, the sensitivity to light at the fovea is low. As one moves out from the fovea, the ratio increases and the sensitivity to light is also increased. The rod system, which predominates in the retinal periphery, has a very high receptor-to-ganglion cell ratio and, therefore, is very sensitive to light. At night, the rod system is operative while the less sensitive cone systems remain below threshold. See Fig. 3-6.

Although a small receptor-to-ganglion cell ratio is a handicap for seeing at low light levels, it permits excellent spatial localization and is responsible for the ability of the foveal region of the retina to resolve fine details. Thus, the system seems to trade off absolute sensitivity for spatial resolution.

For tasks which require the recognition of detail the visual system performs with maximum efficiency when the target to be viewed falls on the *primary line of sight* and its image falls on the fovea centralis. *Visual acuity*, the ability of the eye to resolve fine details, decreases rapidly as the retinal image of the target becomes displaced from the fovea (see Fig. 3-8).

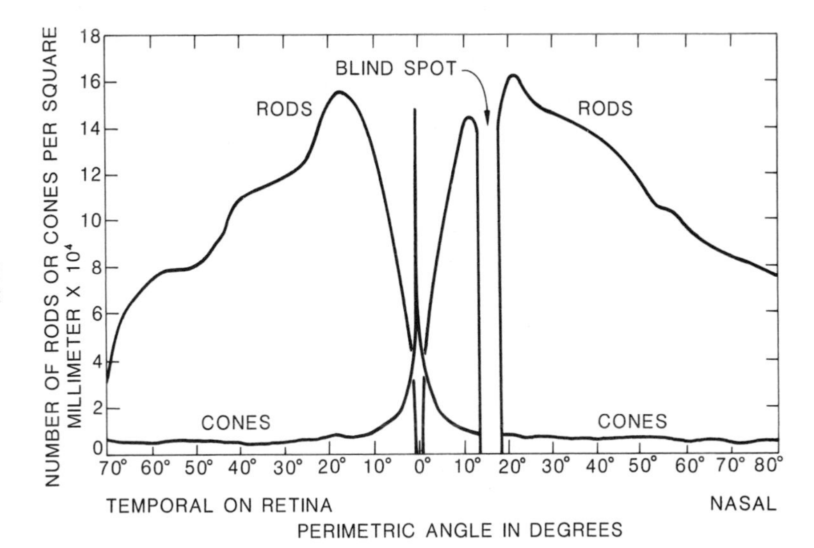

Fig. 3-6. The distribution of rods and cones in the retina. The O-degree point represents the fovea centralis.[15]

Adaptation

The human eye can process information over an enormous range of luminances (about twelve log units). The visual system changes its sensitivity to light, a process called *adaptation*, so that it may detect the faintest signal on a dark night and yet not be overloaded by the high brightnesses of a summer beach scene. Adaptation involves three major processes:

1. Change in Pupil Size. The iris constricts and dilates in response to increased and decreased levels of retinal illumination. Iris constriction is larger and faster (about 0.3 second) than dilation (about 1.5 seconds).[11] There are wide variations in pupil sizes between individuals and for a particular individual at different times. Thus, for a given luminous stimulus, some uncertainty is associated with an individual's pupil size unless it is measured. In general, however, the range in pupil size diameter for young people may be considered to be between two millimeters for high levels and eight millimeters for low levels of retinal illumination. This change in pupil size in response to retinal illumination can only account for a 1.2 log unit change in sensitivity to light. Older people tend to have smaller pupils under comparable conditions.

2. Neural Adaptation. This is a fast (less than one second) change in sensitivity produced by synaptic interactions in the visual system.[12] Neural processes account for virtually all the transitory changes in sensitivity of the eye where cone photopigment bleaching has not yet taken place (discussed below). In other words, at luminance values commonly encountered in electrically lighted environments—below about 600 candelas per square meter [60 candelas per square foot]. Because neural adaptation is so fast and is operative at moderate light levels, the sensitivity of the visual system is typically well-adjusted to the interior scene. Only under special circumstances in interiors (*e.g.*, glancing out a window or directly at a bright light source before looking back at a task) would the capabilities of rapid neural adaptation be exceeded. Under these conditions, and in situations associated with exteriors, neural adaptation will not be completely able to handle the changes necessary for efficient visual function.

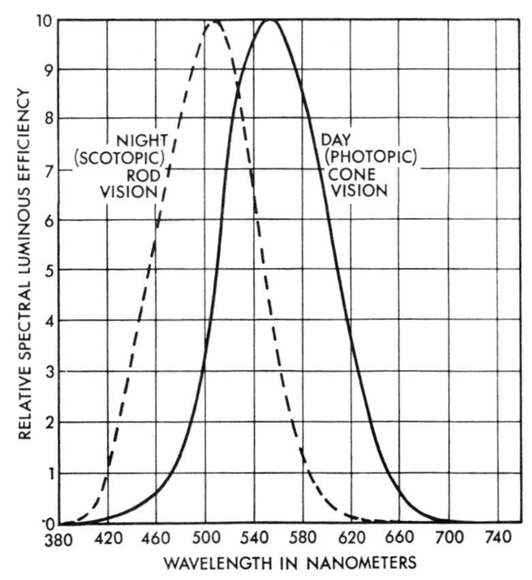

Fig. 3-7. The relative spectral sensitivity for photopic (cone) and scotopic (rod) vision.

3. Photochemical Adaptation. The retinal receptors (rods and cones) contain pigments which, upon absorbing light energy, change composition and release electrons which provide, after processing, an electrical signal to the brain. There are believed to be four photopigments in the human eye, one in the rods (rhodopsin) and one each in the cones (called by some erythrolabe, chlorolabe and cyanolabe). When light is absorbed, the pigment breaks down into an unstable aldehyde of vitamin A and a protein (opsin) and gives off energy that generates signals which are relayed to the brain and interpreted as light. In the dark, the pigment is regenerated and is again available to receive light.

The sensitivity of the eye to light is largely a function of the percentage of unbleached pigment. Under conditions of steady luminance, the concentration of photopigment is in equilibrium; when the luminance is changed, pigment is either bleached or regenerated to reestablish equilibrium. Because the time required to accomplish the photochemical reactions is finite, changes in the sensitivity lag behind the stimulus changes. The cone system adapts much more rapidly than does the rod system; even after exposure to high levels of luminance, the cones will regain nearly complete sensitivity in ten to twelve minutes while the rods will require 60 minutes (or longer) to fully dark-adapt (see Fig. 3-9).[13]

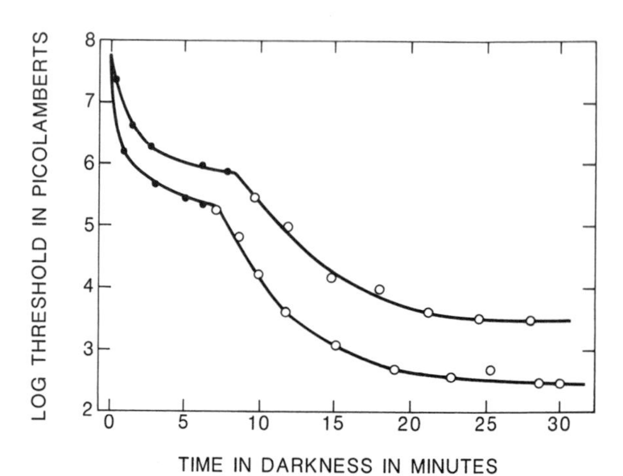

Fig. 3-9. The increase in sensitivity to light (decrease in threshold) as a function of time in the dark after exposure to a bright light. Sensitivity is measured at a point 7 degrees from the line of sight.[13] The two curves represent the extremes of the normal range of observers.

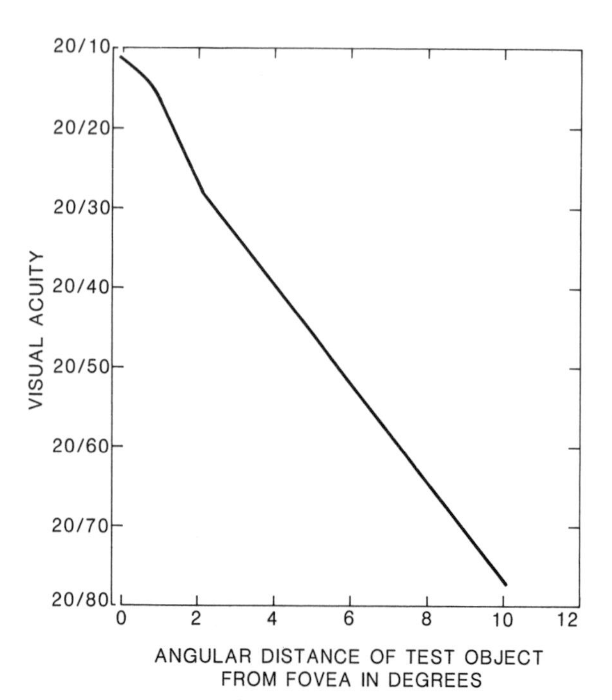

Fig. 3-8. Relation of visual acuity to angular distance of the test object from the fovea.[16]

4. Transient Adaptation. Transient adaptation is a phenomenon associated with reduced visibility after viewing a higher or lower luminance than that of the task.[14] If recovery from transient adaptation is fast (less than one second), neural processes are accounting for the change. If recovery is slow (longer than one second), some changes in the photopigments have taken place. Transient adaptation is usually insignificant in interiors, but can be a problem in brightly lighted exteriors where photopigment bleaching has taken place. The reduced visibility after entering a dark movie theatre from the outside on a sunny day is an illustration of this latter effect.

Eye Movements

It is the function of the oculomotor systems of the eye to position the lines of sight of the two eyes so that they are pointed at the object of regard. If the image of the target does not fall at the fovea, acuity for that target may be reduced. Additionally, if both lines of sight are not aimed at the object, the target may be seen as double (diplopia).

Eye movements can be classified into the following general categories:

1. *Micronystagmus*—very small movements (10 to 60 minutes of arc) of the eyes that occur when observers consider themselves to be fixating steadily in a given direction. Movements consist of high frequency tremors, drifts and flicks.

2. *Saccades*—high velocity movements, usually generated to move the line of sight from one target to another, or to make it join up with a moving target. Velocities may range up to 1000 degrees per second depending upon the distance moved. Saccadic eye movements have a latency of 150 to 200 milliseconds which limits the number of fixations that can be made in a given time period; four or five fixations per second would be a maximum. Vision is substantially reduced during saccadic movements. Saccades are also referred to as fixational jump movements.
3. *Pursuits*—smooth eye movements used to follow a moving target after a saccade has been used to make the line of sight join up with a moving target. The pursuit system cannot follow targets at high velocities. This may cause the eyes to lag behind a rapidly moving object. When speeds exceed 20 to 30 degrees per second, detection and acuity abilities decrease (see Fig. 3-10).[17] Binocular pursuit and jump movements, which involve objects in a frontal plane, are referred to as *versions* because when both eyes are involved, the two eyes make equal movements in the same direction; there is no change in the angle of convergence. See Fig. 3-11.
4. *Compensatory movements*—movements of the eyes to compensate for the movement of the head so that eyes continue to fixate the same target while the head is turning.
5. *Binocular movements*—although the movements described above can be observed in a single eye while the opposite eye is covered, the covered eye will generally mimic the uncovered

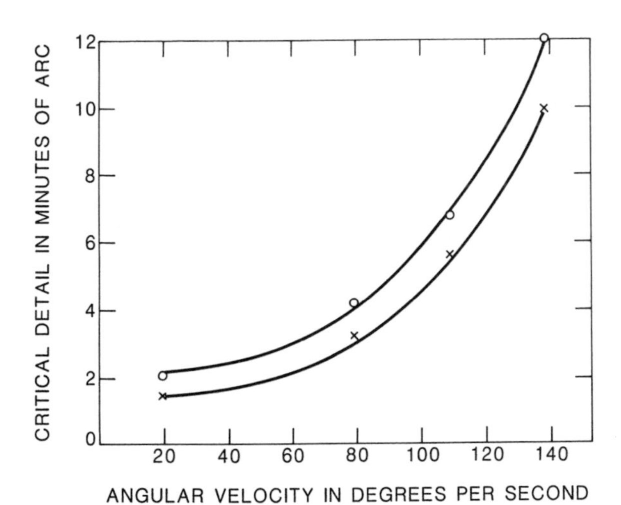

Fig. 3-10. The mean dynamic visual acuity threshold values of nine subjects obtained in the horizontal (O) and vertical (×) planes of pursuit[17].

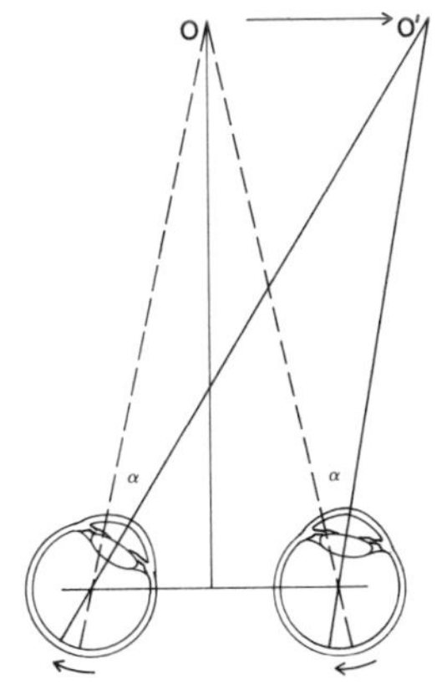

Fig. 3-11. In a version movement, as the target moves from *O* to *O'*, the angle (α) between the eyes remains constant.

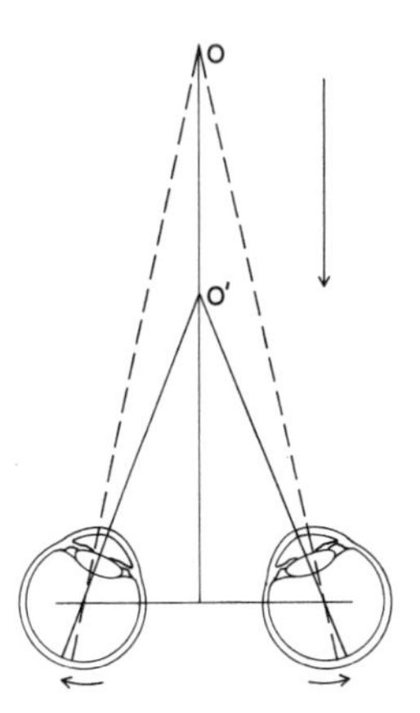

Fig. 3-12. In a vergence movement, as the target moves from *O* to *O'*, the angle between the eyes changes.

eye so that most of the movements of the fixing eye are correlated with conjugate or corresponding movements in the covered eye.

Vergence movements are movements of the two eyes which keep the primary lines of sight converged on a target or which may be used to switch fixation from a target at one distance to a new target at a different distance (see Fig. 3-12). Vergences can occur as a jump movement or can smoothly follow a target moving in a fore and aft direction. When the primary lines of sight drift apart so that they fail to converge at the intended point of fixation, vergence movements play a major role in making the eyes converge on the target. These movements include also cyclovergence movements of the eyes around their primary lines of sight.

Accommodation

As the distance between the viewer and the target is decreased, the refracting power of the eyes must be increased to maintain a clear image on the retina. Such a change is called accommodation and is accomplished by a change in the shape and position of the crystalline lens within the eye (see Fig. 3-13).

Accommodation is always a response to a target located on or near the fovea. It is used to bring a defocused target into focus or to change focus from one target to another at a different distance. It may be gradually changed to keep a target which is moving in a fore and aft direction in focus. Any condition, either physical or physiological, that handicaps the fovea will adversely affect accommodative ability.

Accommodative function decreases rapidly with age such that, by the mid-forties, most individuals can no longer see clearly at normal near-working distances and may need optical assistance. By age 60, there is very little accommodative ability remaining in most of the population (see Fig. 3–14). The need for optical aids can be minimized by appropriate lighting design.

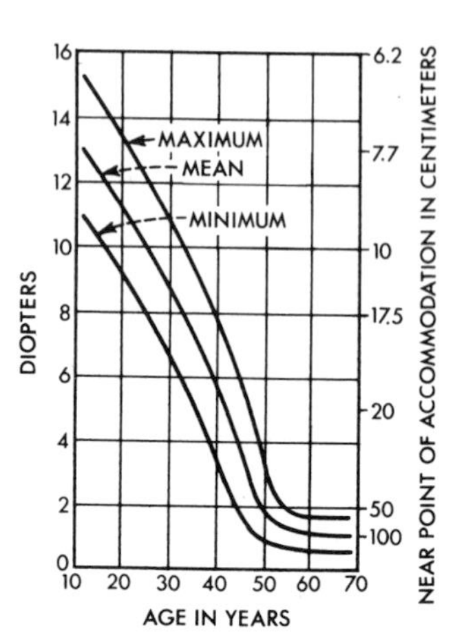

Fig. 3–14. The decrease of the amplitude of accommodation with age.

Color Vision

The ability to discriminate among wavelengths of light is due to a combination of photochemical and neurological processes. Signals from the three cone types are coded in the retina and the lateral geniculate body (in the brain) into chromatic and achromatic information. As a first-order model of color and brightness, chromatic information is a result of a subtraction of receptor signals, while the achromatic information is a result of the addition of receptor signals. However, many experiments using various testing procedures and stimuli demonstrate that this is an oversimplification of how the visual system processes light. Chromatic, achromatic, spatial and temporal information are nonlinearly combined to give final perceptions of light and color. For example, equal luminance colored lamps may have different apparent brightnesses because of the interaction between achromatic and chromatic channels.[18] Figure 3–15 is a proposed model of how the visual system combines the information from these various channels to produce human perceptions.

The normal visual system varies in its ability to discriminate among wavelengths. There are three peaks of maximum wavelength discrimination in the center of the visible spectrum; discrimination falls off rapidly at the spectral extremes.[19] Likewise, the ability to discriminate hue from white is wavelength dependent. Monochromatic colors from the ends of the visual spectrum are more easily discriminated from white because they are more saturated than those wavelengths at the center of the spectrum.[20] The ability to discriminate nonspectral colors is also related to their chromaticness.[21]

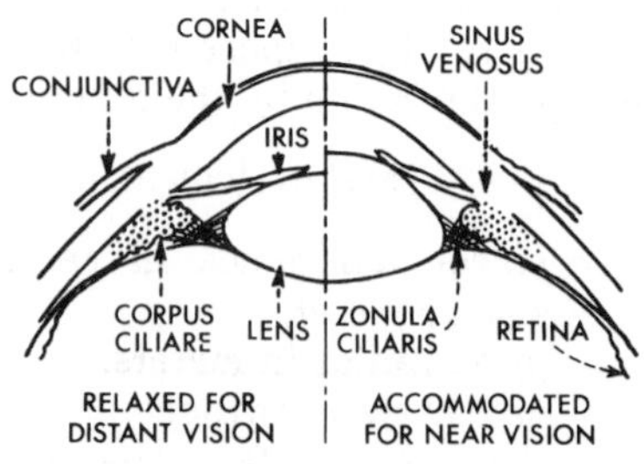

Fig. 3–13. Change in the form and position of the lens during accommodation.

Generally, color discrimination is best in the fovea and decreases toward the periphery where there are fewer cones. However, color discrimination for small fields—20 minutes of arc or less—presented to the fovea will not be good because there are no short wavelength (blue) cones in the center of the fovea. This effect is known as small field tritanopia.[23]

About eight per cent of the male population is color deficient to some degree and may be limited in their ability to perform certain tasks requiring color discrimination. Additionally, certain types of color deficiencies modify the achromatic sensitivity of the eye and reduce the effective brightness of middle and long wavelength lights; the long wavelength loss in sensitivity often can be severe.

The Field of View

The extent of the visual field seen by an individual when looking straight ahead must be divided into monocular and binocular portions. The monocular field is generally considered to be approximately 90 degrees temporally, 60 degrees nasally (depending on the prominance of the nose), 70 degrees inferiorly (restricted by the cheek), and 50 degrees superiorly (restricted by the brow). In addition, the field of each eye contains a blind spot (scotoma) corresponding to the area of the retina where the optic nerve and retinal blood vessels leave the globe and where there are no photoreceptors. The blind spot is located approximately 16 degrees temporal to the line of sight and is elliptically shaped, 8 degrees high by 5 degrees wide.

The monocular visual fields overlap to form a combined binocular field, the central 120 degrees of which is seen by both eyes (see Fig. 3–16). Under binocular conditions it is difficult to detect the blind spots because of the retinal overlap in the opposite eye. It is interesting to note that people are usually unaware of the blind spots, even under monocular conditions, because neurological processes "fill in" the blind spot. Even cortical scotomas resulting from stroke or insult to the cortex may go unnoticed because of this neurological phenomenon.

The visual field will vary considerably depending upon the facial anatomy, state of adaptation, stimulus size used to measure it, stimulus duration and color.

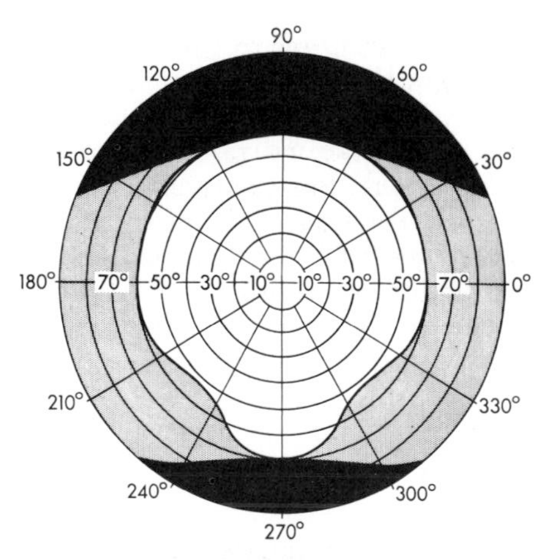

Fig. 3–16. The normal field of view of a pair of human eyes. The white central portion represents the region seen by both eyes. The gray portions, right and left, represent the regions seen by the respective eyes alone. The cut-off by the eyebrows, cheeks and nose is shown by the dark areas.

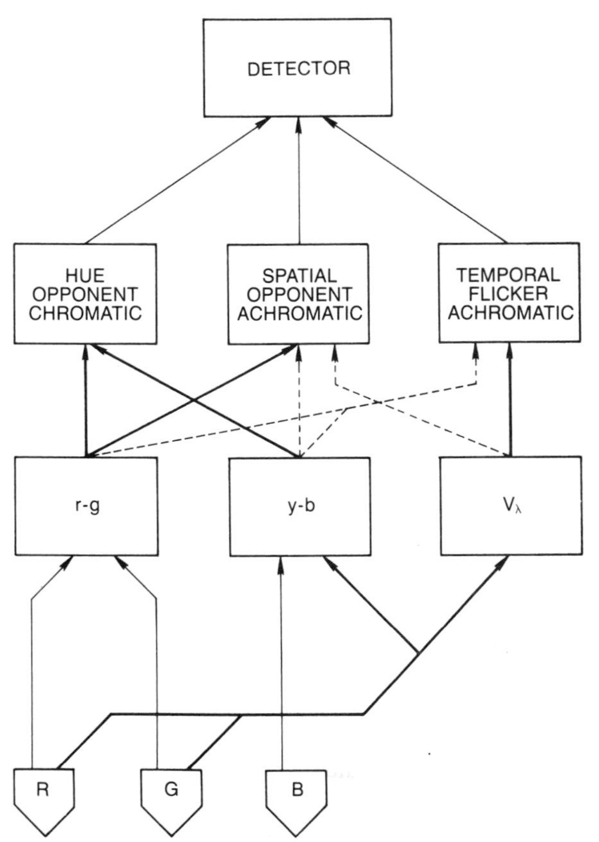

Fig. 3–15. A proposed model for connections in the visual system. Information from the first-stage photoreceptors (*R*, *G*, and *B*) goes to mechanisms that sum or subtract input to give achromatic and chromatic information, respectively. Subsequent "cortical analysis" mechanisms receive multiple inputs from the second stage. Such a model attempts to qualitatively describe some of the nonlinearities in the visual system that have been discovered using stimuli that vary in several dimensions (spatial, temporal, chromatic, and achromatic).[24]

RETINAL IMAGE

Retinal Illuminance

The retina is sensitive to the flux density (illuminance) of light falling onto it. In daylight conditions, and for all but very small or briefly presented objects, brightness sensation is monotonically related to the retinal illuminance produced by that object. However, flux density by itself does not determine the perception of brightness. Rather, it is the *relative* flux density, or contrast, between adjacent areas of the retina that determine these impressions. A patch of retina with an invariant retinal illuminance may appear bright or dark depending upon the illuminance falling on adjacent areas.

Luminances in object space can be related to illuminance at the retina by the function:

$$E_r = (L)\pi(g^2)(t)(\cos\theta)/k^2 \qquad (3\text{–}1)$$

where L = object luminance in candelas per square meter.
g = pupil radius in meters.
t = ocular transmittance.
θ = displacement from the primary line of sight.

k = 0.015.
E_r = retinal illuminance in lumens per square meter.

It can be seen that individuals with different pupil sizes or different transmission characteristics may have significantly different visual effects from identical objects. These individual variations must be considered when predictions of visual behavior are made solely from luminance descriptions of the scene.

Blur

An object can be thought of as an array of points, each of which radiates light toward the eye. If that object is imaged onto the retina by a theoretically perfect optical system, the rays from each point in object space will focus at a point in image space (see Fig. 3–17a). However, if the optical system of the eye does not focus the object's rays at the retina, the image of each object point will no longer be a point, but rather will form a patch of light (see Fig. 3–17b). The image will be blurred and, when the blur is great enough, the image will be perceived as "blurred."

Blurring of the image will result in a defocussed image of each point, an overlap of the images of adjacent points and, consequently, a reduction in perceived contrast when the objects are small.

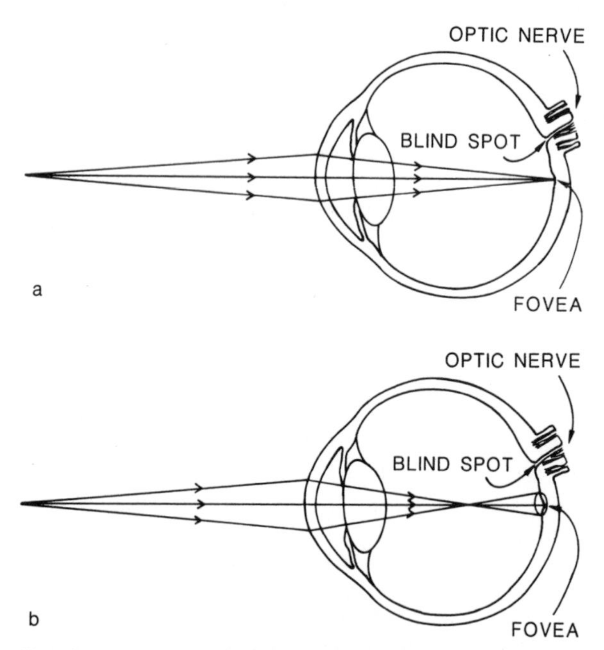

Fig. 3–17. (a) In a perfectly focused optical system, a point source or object is imaged as a point on the retina. (b) The retinal image of a point source or object in an eye with too much power for its length is no longer a point. Such an eye is called *myopic* or "near-sighted."

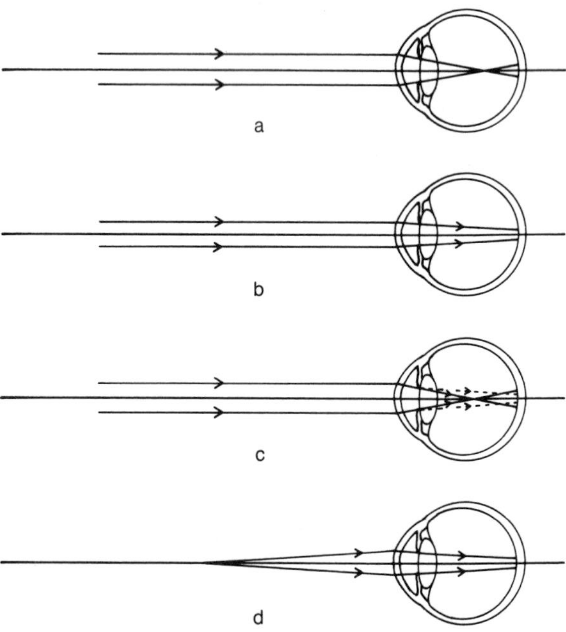

Fig. 3–18. The relationship between the image of a point object and the retina in the common refractive errors. (a) In myopia, the image forms in front of the retina. (b) In hyperopia, the image forms behind the retina. (c) In astigmatism, multiple foci are formed due to different powers in the various meridians of the eye. (d) In presbyopia, accommodation is sufficiently limited so that near objects focus behind the retina.

Blur can be caused by several factors in the human eye:

1. *Uncorrected refractive errors*—The eyes do not focus a sharp image on the retina (see Fig. 3–18).
   a. *Myopia*: object points at a distance focus in front of the retina (see Fig. 3–18a).
   b. *Hyperopia*: object points at a distance focus behind the retina (see Fig. 3–18b).
   c. *Astigmatism*: object points do not focus at a point in any plane (see Fig. 3–18c). This condition is caused by different powers in the various meridians of the eye.
   d. *Presbyopia*: near objects focus behind the retina (see Fig. 3–18d). Caused by a loss of accommodative ability, usually as one ages.
2. *Aberrations*—even when the eye is perfectly corrected for refractive errors, a residual blur remains due to the existence of spherical and chromatic aberrations.
   a. *Spherical aberration*: occurs because light rays which enter the periphery of the cornea are refracted more than those which enter through the central zones (see Fig. 3–19a). Thus, the light in retinal image is partially

redistributed over a larger retinal area than would be the case in an aberration-free system. The amount and type of spherical aberration varies with the state of accommodation (see Fig. 3-20).[25]

b. *Chromatic aberration*: occurs because shorter wavelengths (blues) are refracted more than are the longer wavelengths (reds) (see Fig. 3-21). As in spherical aberration, the results of the different foci causes blur (see Fig. 3-19b).

While these aberrations (and others) are of theoretical interest to the lighting designer, they are partially compensated by the structure of the eye and can usually be neglected in practical lighting design. They may, however, be important in certain specialized applications, such as color processing, work under reduced illuminances where pupil sizes may be large, and rendering from color display terminals.

3. *Diffraction*—regardless of whether rays are in focus or not, there is always a certain amount of blur due to the diffraction of light. This determines the ultimate resolving power when the eye is in best focus, but the blur is not large enough to be perceived.

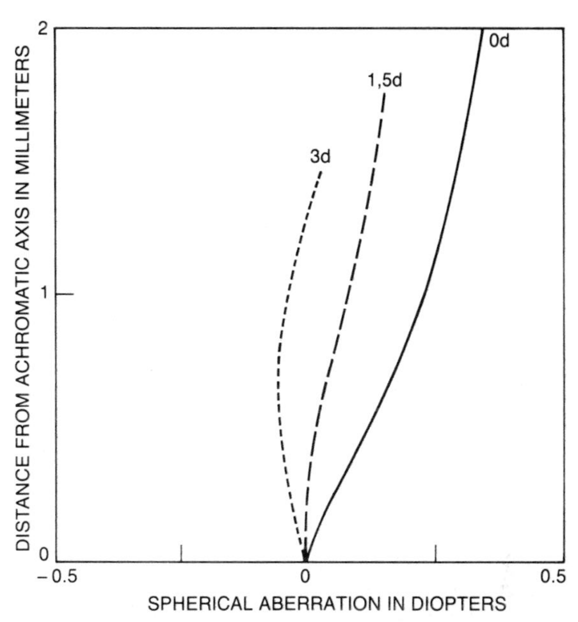

Fig. 3-20. Spherical aberration of the human eye based upon the average of ten observers. The amount of spherical aberration (in diopters) is on the abscissa (positive when under corrected) and the distance from the achromatic axis is on the ordinate. The solid line corresponds to the unaccommodated eye; the dashed line corresponds to the average eye with 1.5 diopters accommodation; and the dotted line to 3.0 diopters accommodation.[25]

Veiling Luminance (Disability Glare)

If the eye were a perfect optical system, the "blurredness" of a scene could be substantially accounted for by the degree to which the image on the retina was out of focus. In reality, however, the ocular media contain many inhomogenieties which scatter the incident light and, accordingly, reduces the theoretical contrast of even a perfectly focused retinal image. This reduction in contrast occurs because light intended for the primary image is scattered to adjacent retinal areas while light intended for the adjacent areas is scattered onto the primary image.

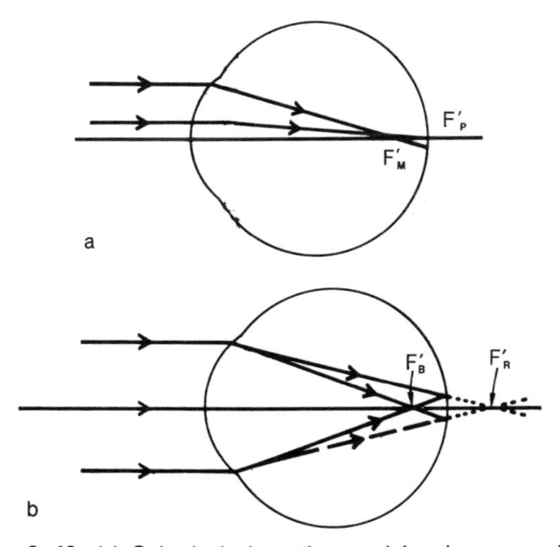

Fig. 3-19. (a) Spherical aberration: peripheral rays are focused in front of rays entering the eye near the center of the pupil. (b) Longitudinal chromatic aberration: because the eye's index of refraction is greater for short wavelengths (blues) than for long wavelengths (oranges and reds), the eye focuses short wavelengths in front of long wavelengths.

Because the reduction in contrast of an image from scattered light can be mimicked by adding a uniform "veil" of luminance to the object, the effect is considered to be equivalent to a *veiling luminance*; it is also called *disability glare*.

Although veiling luminance is more commonly thought of as coming from discrete sources of glare, such as that from on-coming automobile headlamps, every luminous point in space acts as a source of stray light for nearby points, lines, and borders and reduces their contrast.

Because the visibility of objects on or near the line of sight are usually of concern, it is useful to understand the effect of surrounding areas on the appearance of objects which are viewed foveally. Several investigators[27-30] have analyzed the role of glare source luminance and distance

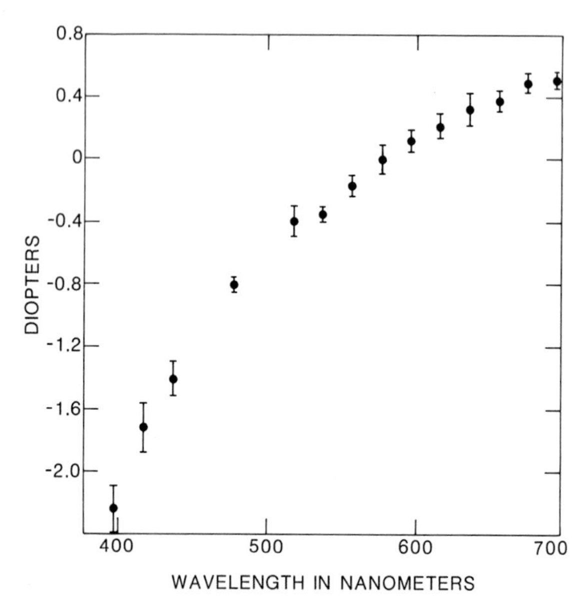

Fig. 3–21. The power necessary to correct an eye for the difference in refraction of wavelengths between 400 nanometers and 700 nanometers (zero is arbitrarily set at 589 nanometers).[26]

from the primary object of regard as producers of disability glare; while they have each derived slightly different functions, they generally take the form:[27]

$$L_v = 9.2\ E_o/\theta(\theta + 1.5) \qquad (3\text{–}2)$$

where: L_v = equivalent veiling luminance in candelas per square meter.
E_o = illuminance from the glare source at the eye in lux.
θ = angle between the primary object and the glare source in degrees.

Although the constants and the form of the function vary from observer to observer, for small angles the function must always adhere to the following form:

$$L_v = k_1 E_o/f(\theta) \qquad (3\text{–}3)$$

This is essential for integrating the effects at a given point produced by stray light from all other points.

DISCOMFORT GLARE

Discomfort glare is a sensation of annoyance or pain caused by high or nonuniform distributions of brightness in the field of view. Discomfort glare may be accompanied by disability glare (see above), but is a distinctly different phenomenon.

While the cause of disability glare is well known (intraocular light scatter), the mechanism of discomfort glare is less well understood. Laboratory studies[31,32] have related discomfort glare to pupillary activity, but such data are as yet insufficient to be applied in engineering practice. Consequently, most of the assessments of discomfort glare are based on consideration of the size, luminance, and number of glare sources, their locations in the field of view, and the background luminance. Most measurements of discomfort produced by single glare sources have been performed by determining the luminance (L) just necessary to cause discomfort—a threshold criterion termed the "Borderline between

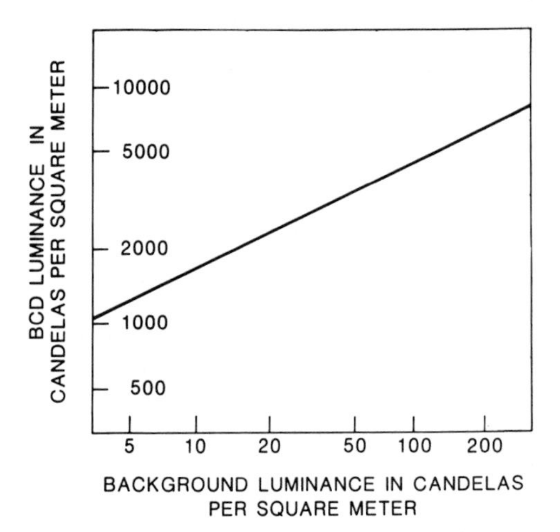

Fig. 3–22. The luminance of a glare source at discomfort threshold (BCD) as a function of the background luminance (light to which the eyes are adapted) – moderate background luminances. $L_b = 302\ F^{0.44}$.

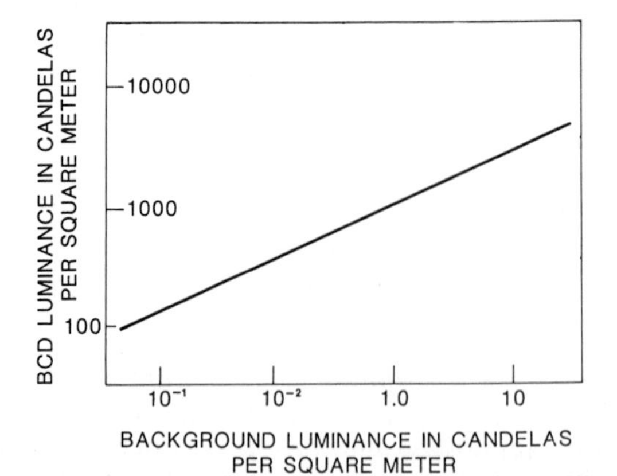

Fig. 3–23. The luminance of a glare source at discomfort threshold (BCD) as a function of the background luminance (light to which the eyes are adapted)—low luminance backgrounds. $L_b = 529\ F^{0.44}$.

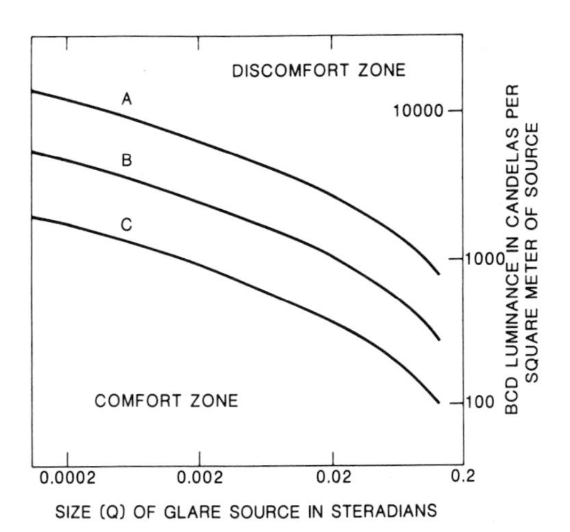

Fig. 3-24. The effect of glare source on the BCD at three levels of background luminance (L_b). A = 343 candelas per square meter, B is 1/10 A, C is 1/100 A. Larger sources result in a lower BCD. Sources are located on the primary line of sight—moderate background luminance.

Comfort and Discomfort" (BCD).[33] These investigations have included the effect of adaptation luminance (F) on BCD at the moderate levels of luminance likely to be encountered in interior spaces. Fig. 3-22 shows the results for a source subtending 0.0011 steradian (approximately two degrees) on the line of sight. Data from another investigation,[34] performed as a parallel study at low adaptation luminances, are presented in Fig. 3-23. Both sets of data fit the function:

$$L = cF^{0.44} \qquad (3\text{-}4)$$

where $c = 302$ in one study[33] and 529 in the other.[34]

The difference in the value of the constant is significant and may be due to both the different levels of background luminance at which the experiments were performed and to the differences among the subject populations. Care must be taken to utilize the appropriate function for the conditions (that is, the data of Fig. 3-22 for interior applications and those from Fig. 3-23 for night driving, for example).

Figs. 3-24 and 3-25 show the effect of glare source size on BCD luminance for various size sources. Once again, it is clear that there are differences between the higher and lower luminance conditions and that caution must be used in applying the results in a general fashion.

As the glare source is moved away from the primary line of sight the BCD luminance increases. Data for locations at various eccentricities and along several meridians is presented in Fig. 3-26 for a background luminance of 34.3 candelas per square meter.[33]

The data taken at moderate luminances, combined with considerations such as the effect of multiple sources, have been used to develop the IES-VCP (visual comfort probability) system described in the Application Volume and Section 9 of this Volume.

DESCRIPTION OF THE VISUAL DISPLAY

Visual displays (visual stimuli, visual tasks) can be specified in terms of quantities that are

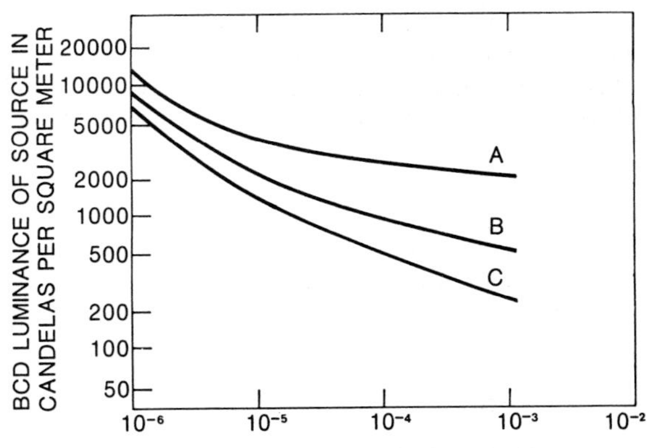

Fig. 3-25. The effect of glare source on the BCD at three levels of background luminance (L_b). A = 0.34 candelas per square meter, B is 1/10 A, C is 1/100 A. Larger sources result in a lower BCD. Sources are located on the primary line of sight—low luminance background.

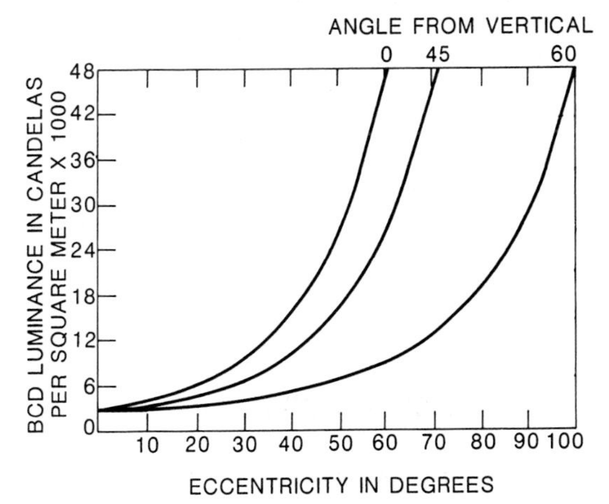

Fig. 3-26. BCD luminance for glare sources at various angular distances from the line of sight along several meridians. Background luminance was 34.3 candelas per square meter.

measurable in object space. Some of these quantities are:

1. Luminance of the object.
2. Luminance of the background (adapting luminance).
3. Contrast (calculated from 1 and 2 above).
4. Spectral distributions of the object and background.
5. Size (in units of linear measure or visual angle).
6. Duration.
7. Temporal frequency characteristics.
8. Location relative to the line of sight.
9. Movement in the field of view.
10. Nonuniformities of luminance in the object and the background.

The visibility of any target is a function of all of the above parameters; additionally, cognitive factors such as attention, expectation and habituation will greatly affect the detectability and recognition of objects.

The lighting designer can exert a most direct effect on the first four parameters listed above with opportunities to influence and affect the remainder in many cases. In discussions that follow both the object and the background have a uniform luminance.

An object may be differentiated from its background because it differs either in luminance or in color; that is, there may be either a luminance contrast or a chromatic contrast (the term "luminance" is used instead of "brightness" because most objects are measured with photoelectric photometers today). Except in the case of self-luminous objects, both types of contrast are a function of the reflectance properties of the scene and of the incident illumination.

Luminance Contrast

Luminance contrast is defined in several ways:

$$C = \frac{L_d - L_b}{L_b} \tag{3-5}$$

where L_d = luminance of detail
L_b = luminance of background

$$C = \frac{L_g - L_l}{L_g} \tag{3-6}$$

where L_g = greater luminance
L_l = lesser luminance

$$C = \frac{L_g - L_l}{L_l} \tag{3-7}$$

where L_g = greater luminance
L_l = lesser luminance

$$M = \frac{L_{max} - L_{min}}{L_{max} + L_{min}} \tag{3-8}$$

where L_{max} = maximum luminance
L_{min} = minimum luminance

Equation 3–5 will result in contrasts that range between 0 and 1 for objects that are darker than their backgrounds and 0 and ∞ for objects that are brighter than their backgrounds. This equation is recommended by some because the contrasts so determined represent relative visibilities.

Equation 3–6 results in contrasts between 0 and 1 for all objects whether brighter or darker than their backgrounds. It is especially applicable in a situation like a bipartite pattern in which neither of the areas on the two sides of the border can be identified as object or background. It also applies to square wave gratings where neither the dark nor the bright bars can be designated as objects as opposed to backgrounds.

Equation 3–7 is an alternative to equation 3–6 which is commonly used when the object is brighter than the background.

The quantity defined by equation 3–8 is often called contrast but is usually, and more properly, called *modulation*. It applies to periodic patterns, such as gratings, which have one maximum and one minimum in each cycle.

When the contrast of an object is specified, one must know which definition is being used.

When an object and its background are illuminated by a single source and the luminances are dependent on their reflectances, the contrast is not affected by changing the intensity of the source. A target that consists of a disk (50 per cent reflectance) on a background (90 per cent reflectance) will have a contrast (equation 3–5) of 0.44 regardless of the illuminance on the scene.

$$C = \left|\frac{L_d - L_b}{L_b}\right|$$

$$= \left|\frac{E(\rho_d) - E(\rho_b)}{E(\rho_b)}\right| = \left|\frac{\rho_d - \rho_b}{\rho_b}\right|$$

The contrast is determined by the reflectances.

It should be noted that, because reflectances are rarely uniform in all directions, the contrast may be changed by varying the position of the observer or that of the source. This situation is described by saying that changing the position of the observer or of the source affects the *contrast rendition*.

Chromatic Contrast

Visual targets that have the same luminance as the background may still be discerned by color information. These equal luminance chromatic constrasts are generally less distinct than achromatic contrasts,[35,36] but under some conditions they can be quite visible. This is especially true if there is a sharp, spatial discontinuity between the two colors. The distinctness of such a chromatic border can be estimated by matching it to the contrast of an achromatic border. Compared in this way, chromatic contrasts range from below achromatic contrast threshold to about 0.30. A model for chromatic contrast has been developed using the responses of the long and middle wavelength-sensitive cones.[37]

Because color contributes to the perceptions of scenes, predictions of visibility based only upon achromatic luminances can be wrong.[38] These errors can arise not only because of chromatic contrast (as discussed above), but also because perceived brightness is modified by color information. As described previously, brightness is not based only upon achromatic luminance. The input from the color channels can also modulate luminance information to make borders more or less distinct. Roughly, the larger the difference in saturation between the target and the background, the larger the discrepancy between the perceived contrast and the estimated contrast based only on achromatic luminance.

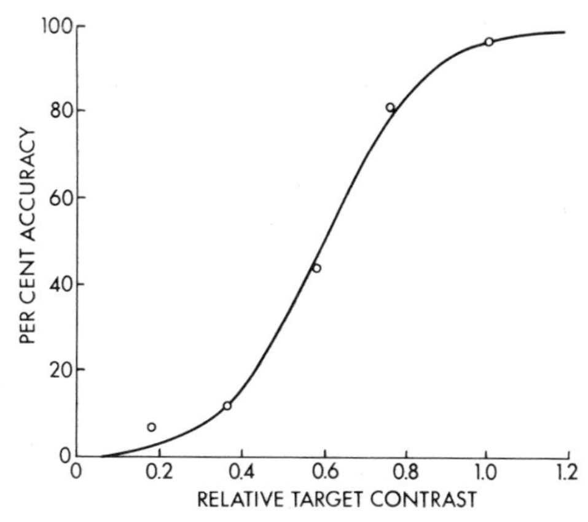

Fig. 3-27. A "frequency of seeing" function. As contrast is increased, the number of times it is seen relative to the number of times it is presented increases to a maximum of 100 per cent.

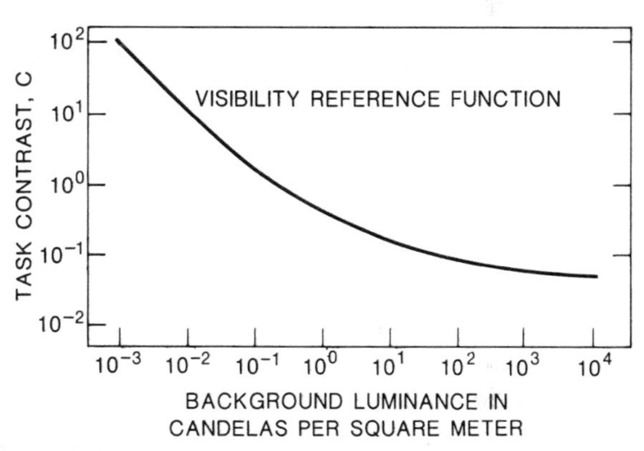

Fig. 3-28. Plot of the Visibility Reference Function representing task contrast required at different levels of task background luminance to achieve threshold visibility for a 4-minute luminous disk exposed for ⅕ second.

VISUAL ABILITY

Contrast Detection

Contrast detection is the basic task from which all other visual behaviors are derived. The visual system gives virtually no useful information when the retina is uniformly illuminated, but is highly specialized to inform about luminous discontinuities and gradients in the visual field.

The simplest visual function is one in which a small change in luminance must be detected in an otherwise uniform surround. This function has been studied in great detail[39] and has related the probability of detecting a small round test object on a uniform background to the contrast and the luminance of the background (see Fig. 3-27).

As the contrast (defined as in Equation 3-5) is raised, the probability of seeing a test object increases until, at a certain contrast, the object can be detected 100 per cent of the time. The contrast at which the object can be detected 50 per cent of the time is called the *threshold contrast* and will vary among individuals, with the duration of exposure of the object, with the size of the object, and with the luminance of the background.

The change in threshold contrast as a function of background luminance is shown in Fig. 3-28. It represents a specific group of observers and a particular test object on a uniform background. This curve has been used by the IES to illustrate a fundamental relationship between object detection and the luminance of the background. It is called the *Visibility Reference Function.*

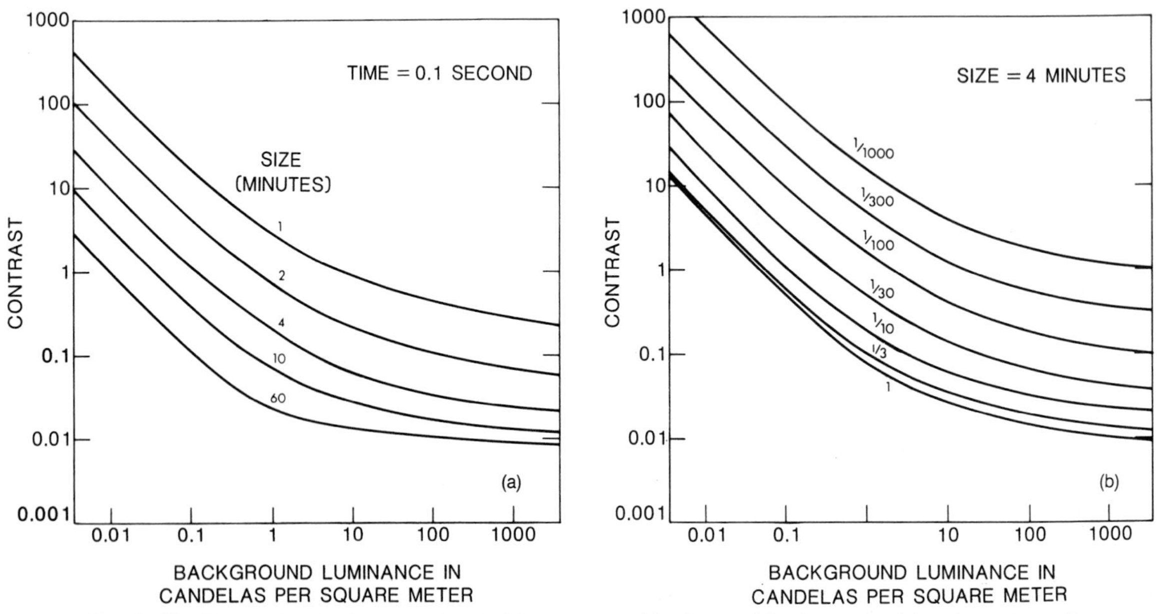

Fig. 3–29. The relationship between threshold contrast and background luminance for (a) various sizes of object and (b) various exposure times.

The shape of the contrast threshold curve is relatively invariant for simple detection tasks in controlled environments. Furthermore, there is some evidence that the curve reasonably predicts visual performance for more complex functions such as the recognition of letters or words.[40]

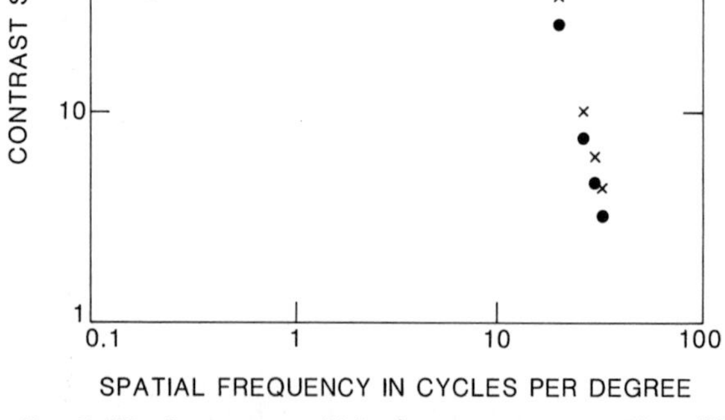

Fig. 3–30. Contrast sensitivity for square-wave gratings (X) and sine-wave gratings (●). The space-average luminance of the grating was 500 candelas per square meter.[41] Contrast sensitivity is the reciprocal of the contrast necessary to just detect the orientation of the gratings.

Size and exposure duration each affect the contrast necessary to detect a simple target (see Figs. 3–29 and 3–30).

Visual Acuity

While contrast detection may be considered the simplest visual function, visual acuity is somewhat more complex. The word "acuity" is often used to describe the visibility of fine details involved in various kinds of displays. Several different kinds of acuity are recognized.

1. *Resolution acuity*—the ability to detect that there are two stimuli, rather than one, in the visual field. It is measured in terms of the smallest angular separation, measured at the eye, between the two stimuli that can be present and still have the targets seen as separate.
2. *Recognition acuity*—the ability to correctly identify a visual target, such as differentiating between a "G" and a "C." Usually, but not always, it is measured in terms of the smallest size target, in angular measure, that can be discriminated. Visual acuity testing performed using letters, such as done clinically, is a form of recognition acuity.

Recognition acuity can involve very complex perceptual processing and often cannot be simply related to resolution acuity. Although detecting the difference between an "O" and a "C" may be identical to resolution acuity, differentiating between a circle and an octagon is a more complex task.

Fig. 3-31 illustrates how visual angle is determined; the "E" and "C" in Fig. 3-32 could be measured in terms of either their critical detail (as shown) or in terms of their overall angular size.

Visual acuity, like detection, varies with exposure duration and luminance (retinal illuminance). See Figs. 3-33 and 3-34.

Temporal Resolution

Just as the visual system responds to contrasts in space, it also responds to contrasts in time. Because of the movement of targets across the retina, either because of eye movements or movements of the object itself, virtually all spatial contrasts can also involve temporal contrasts.

At detection threshold, the visual system summarizes flashes of light over brief periods of time, *i.e.*, for brief flashes (less than 100 milliseconds, see Fig. 3-35), luminance (L) can be traded for duration (t), a relationship known as Block's law:

$$(L)(t) = \text{constant} \qquad (3\text{-}9)$$

For flashes longer than 100 to 200 milliseconds, threshold becomes independent of time. That is, threshold is solely a function of stimulus luminance.

Repetitive square-wave flashes at frequencies between two and twenty hertz may have suprathreshold brightnesses greater than the same

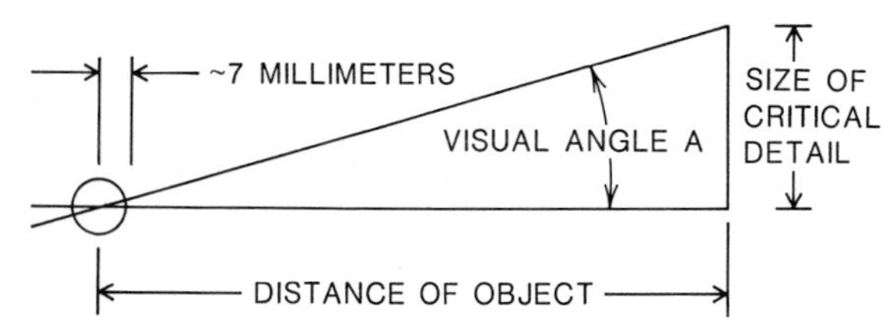

Fig. 3-31. Visual angle is the angle subtended by an object (or some important detail of an object) at the nodal points of the eye. It is usually measured in minutes of arc.

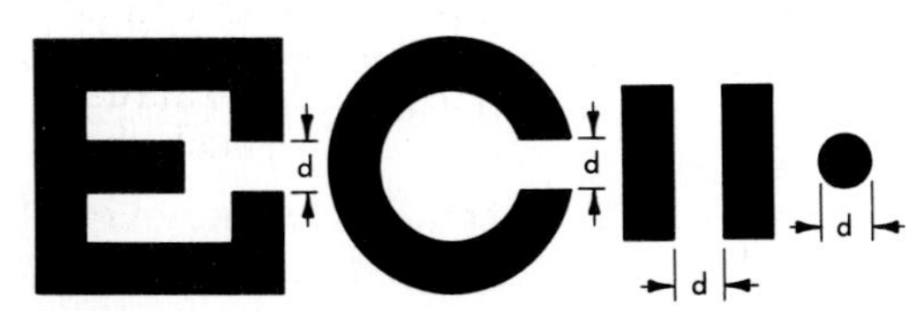

Fig. 3-32. Commonly used test objects for determining size discrimination and visual acuity. The critical detail is represented by "d."

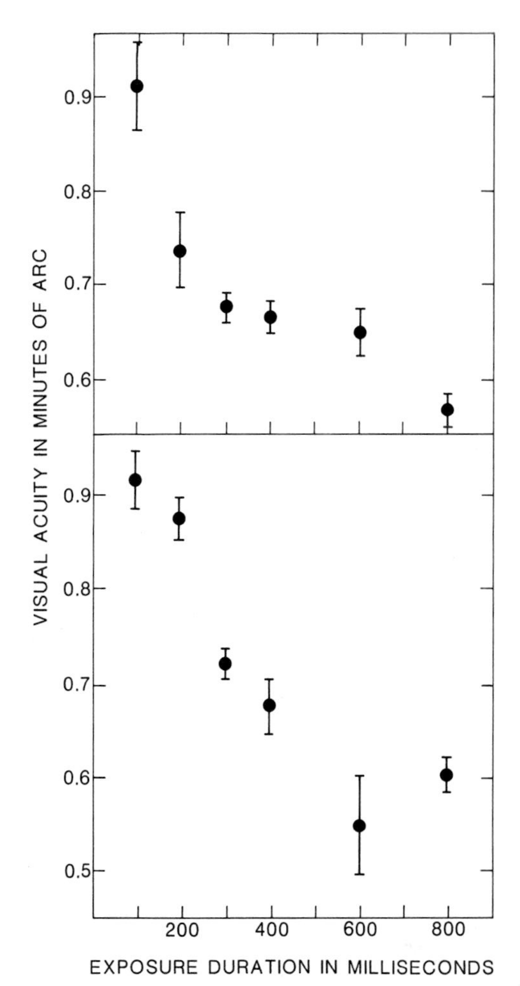

Fig. 3-33. The minimum angle of resolution versus exposure duration for two different subjects. Visual acuity is found to improve with exposure durations out to about 400 milliseconds and then reaches a plateau. The vertical bars indicate the 50 per cent fiducial limits of the means of the fitted frequency-of-seeing curves.[42]

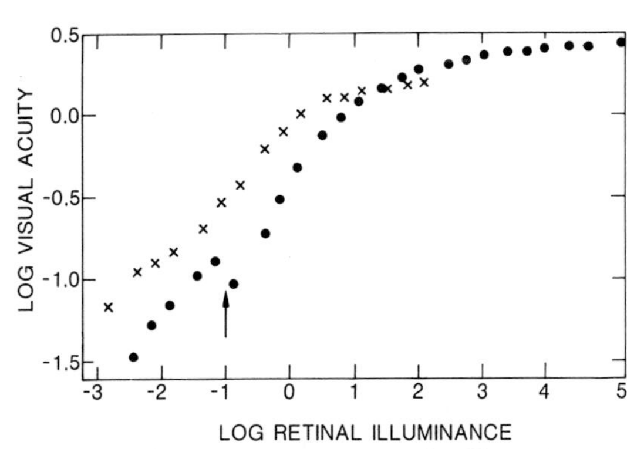

Fig. 3-34. Effect of illuminance on acuity: (X) grating target; (●) Landolt ring target. Arrow indicates point at which subject began using central fixation when viewing Landolt ring.[43]

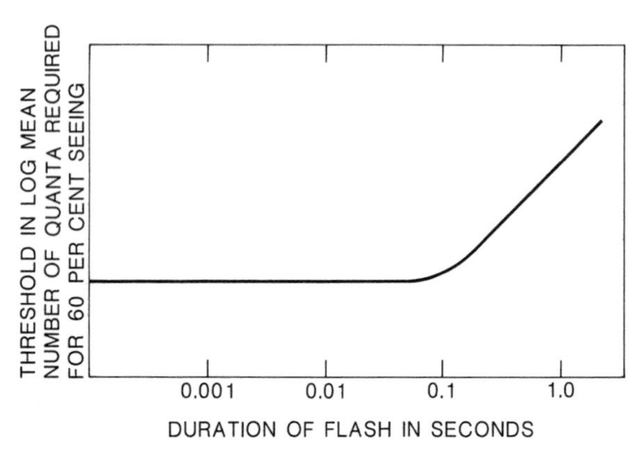

Fig. 3-35. Total light required for seeing a flash as a function of the duration of the flash.[44]

stimulus presented continuously.[45] This is known as *brightness enhancement.* Even single pulses of light may, under the proper stimulus conditions, appear brighter than a steady light.[47]

As a flashing stimulus is increased in frequency, it will eventually reach a point where it will be perceived as steady rather than as intermittent; this is the *Critical Flicker Fusion Frequency* (CFF). The frequency at which the fusion occurs will vary with stimulus size, shape, retinal location, spectral distribution, luminance, modulation depth and state of adaptation. Fig. 3-36 shows the relationship of CFF to adaptation luminance for centrally fixated test objects of several different sizes. CFF rarely exceeds 60 hertz even under optimal conditions.

For an intermittent stimulus, the modulation necessary to just detect the flicker varies with the temporal frequency of the stimulus[48] (see Fig. 3-37) such that frequencies in the mid-range (6 to 10 hertz) are easier to detect than are stimuli presented at either a slower or faster rate. Functions such as in Fig. 3-37 are known as DeLange curves and may be called temporal modulation transfer functions similar to spatial modulation transfer functions (see Fig. 3-30 for comparison). To be just detectable, the rate of change of a flash whose luminance increases with time must be greater for a short flash than for a long flash (see Fig. 3-38).[49]

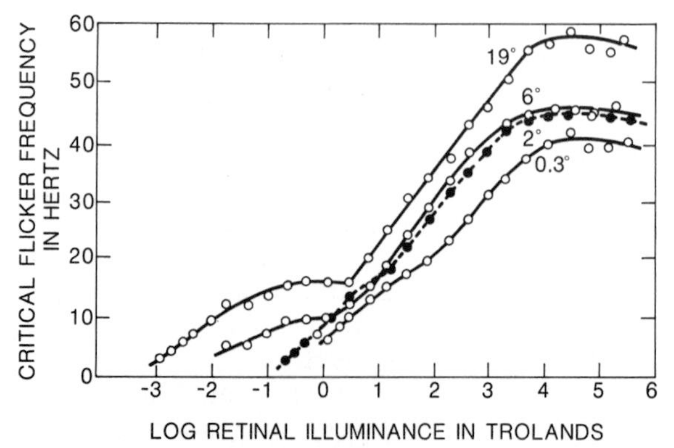

Fig. 3-36. Critical flicker fusion frequency as a function of source area and retinal illuminance.[46]

Knowledge of flicker effects is most helpful when considering such problems as fluorescent lamp flicker, signal detection and animated signs.

Visibility

In this section Visibility (or Visibility Level, *VL*) is defined according to CIE document 19/2.1.[50] The reader should not infer that this definition is necessarily the same as "visibility" used in the general sense, or that there are no other ways that "visibility" might be defined. To aid the reader in discriminating between usages, a capital "*V*" will be used with referring to Visibility as defined in the CIE document,[50] and a lower case "*v*" will be used when referring to visibility as used in the vernacular.

Definition of Visibility. In the case of a particular display, such as a small disk on a darker surround, visibility can be defined in terms of contrast. The contrast (C) is defined as

$$C = \frac{L_D - L}{L} \qquad (3\text{-}10)$$

where L_D is the luminance of the disk and L is the luminance of the background. The contrast can be reduced to the threshold of visibility by reducing the difference between L and L_D until the threshold is reached. At this level the contrast is called the threshold contrast $\bar{C}$. Visibility (V) can be defined as the ratio of C to $\bar{C}$.

$$V = C/\bar{C} \qquad (3\text{-}11)$$

Ideally, the Visibility of the particular display can be measured with a visibility meter. These instruments usually superimpose a patch of veiling luminance L_v on the viewed field and, at the same time, reduce the luminance of both the disk and the background by a factor, F, such that

$$L_V + FL = L \qquad (3\text{-}12)$$

At threshold

$$\bar{C} = \frac{L_D F - LF}{FL + L_V} = F\left(\frac{L_D - L}{L}\right) = FC \qquad (3\text{-}13)$$

Hence $$V = 1/F \qquad (3\text{–}14)$$

The visibility meter can also be used, again ideally, to measure the Visibility of any arbitrary display such as printed letters, pictures or three-dimensional arrays. The display may have sharp or blurred luminous or colored borders. In the case of small targets on large backgrounds, background luminance is naturally defined. No theoretically sound definition exists for background luminance in complex displays having many luminous or colored edges. Two alternatives exist: the background luminance may be arbitrarily defined as either the mean luminance of the display,[51] or as the luminance of the area immediately adjacent to the target of interest.[51]

It must be noted that measurement of Visibility with an ideal visibility meter would not have to include photometric assessment of contrast. However, no complex optical system like a visibility meter has perfect optical fidelity. Small targets will have less apparent contrast when viewed through the visibility meter at maximum transmission than they will when viewed naturally. Consequently, the visual threshold of the target, especially a small target, estimated by the visibility meter will be artificially high. In equation (3–14) the value of F will be too large, and, thus, the estimate of V will be too small.

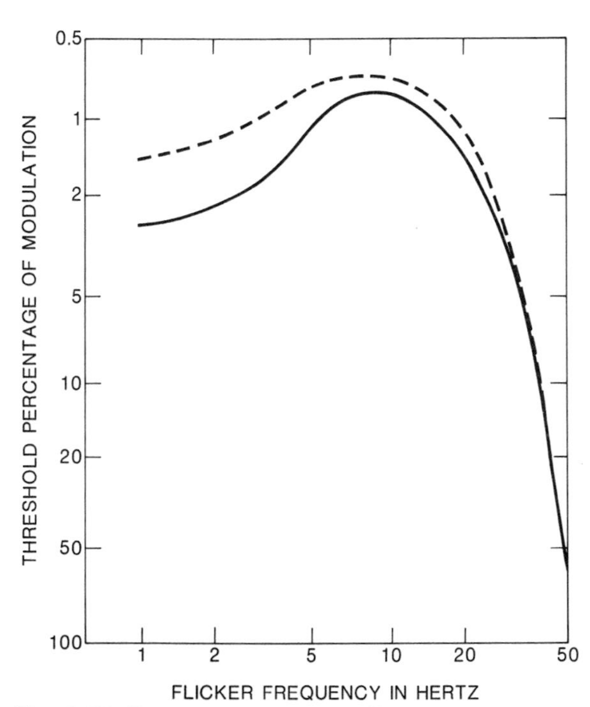

Fig. 3-37. Temporal modulation transfer functions for small flickering fields against a nonflickering background whose luminance is the same as the average of the flickering field. The dash curve is for a field with sharp edges and the solid one for a field with blurred edges.[48]

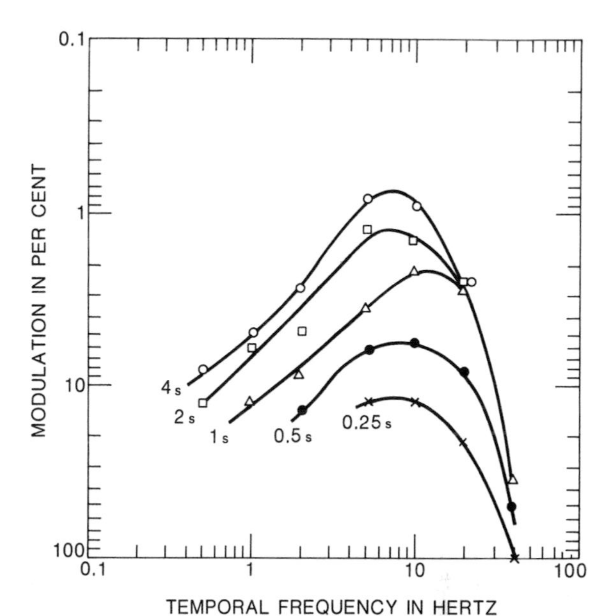

Fig. 3-38. Flicker thresholds plotted as a function of the frequency ("De Lange" curves) for various stimulus durations (in seconds).[49]

Measurement of Visibility. As previously noted, visibility is a complex function of physical stimulus variables, physiological processes, and psychological predispositions and attitudes, all of which interact to determine the appearance of each display. Over the years, there have been many definitions of visibility, but most have recognized that visibility indicates something about how easy or difficult a target is to see. Accordingly, attempts have been made to measure target difficulty in terms that would be meaningful to lighting designers. Visibility meters have helped to solve this problem.

While there are several different types of visibility meters in terms of design, mechanical operation and ease of use they all act to reduce the contrast of a scene until some pre-determined part of the scene (called the critical detail) is just visible; *i.e.*, at its threshold. The amount of reduction in contrast necessary to bring the detail to threshold is taken as a measure of the difficulty of the task; a scene with "high" visibility will require a large amount of contrast reduction, while a "low" visibility scene will require little contrast reduction.

In practice, the actual task detail difficulty is referenced to the contrast of a standard target which is of equal difficulty—difficulty is measured in terms of the amount to which the target is above its own threshold contrast. This reference contrast is called the *equivalent contrast* ($\tilde{C}$). A visibility meter must be used to determine $\tilde{C}$ for realistic stimuli.

The Relation of Visibility to the Luminance of the Background. The relation for a 4-minute disk on a darker background has been well established. As the luminance of the background increases the contrast threshold decreases and the visibility of a disk of a given contrast increases (see Fig. 3–28).

The relationship between Visibility (V) and the luminance of the background (L) is given by the following formula:

$$V = 16.846 \left[\left(\frac{1.639}{L}\right)^{0.4} + 1\right]^{-2.5} C \quad (3\text{–}15a)$$

where C is the contrast between the 4-minute disk and its background.

To use this formula it is necessary to make an assessment of C.

The ratio between the Visibility of any given display and the Visibility $\overline{V}$ of a 4-minute test object of a given contrast $\hat{C}$ is assumed to be a fixed value which holds at all luminance levels:

$$V = K\overline{V} \quad (3\text{–}15b)$$

Hence for any display:

$$V = 16.846\, K\hat{C} \left[\left(\frac{1.639}{L}\right)^{0.4} + 1\right]^{-2.5} \quad (3\text{–}15c)$$

Therefore, for any display it is possible to measure the Visibility with a theoretically perfect visibility meter at any given level of illuminance and solve equation (3–15c) for the constant ($16.846\, K\hat{C}$). Then equation (3–15c) can be used to predict the Visibility at any other level of luminance.

RELATIONSHIPS AMONG LIGHTING, VISIBILITY AND TASK PERFORMANCE

The basic question that has faced the illuminating engineer since the beginning of lighting design as an integral component of structures is "how much light is needed to see?" It has been known since the mid-1800's that, in general, it is easier to perform simple tasks as the luminance level of a visual environment becomes higher. However, it was not until the 1930's that the effect of lighting level on specific industrial and office tasks was investigated. In England, the problem was approached by measuring the performance on actual industrial tasks under varying lighting conditions;[52] from such field data, it was possible to assess the relative increase in task performance (in most cases restricted to their visual components) for a given increase in luminance. For years, the British lighting recommendations were based on that work.

In the United States, that approach was used[53] to assess the relative importance of lighting levels on the speed of reading. Data from that study, too, were limited to specific tasks and could be generalized to other situations only with great caution.

In 1937[54] and, more recently in 1959,[39] a different approach was chosen to address the problem of determining lighting levels. The assumption was that if the basic processes that governed visual performance were understood, one could predict the response behavior from a knowledge of stimulus parameters and could avoid the immense problem of having to study each task independently. This approach was first published by the CIE[55] as a method of determining illuminance levels, and for many years was the foundation of the IES method of prescribing illumination. Primarily through the work of Blackwell, the Visibility concepts were expanded to include factors other than those described above. As more information was accumulated, the system was revised to reflect data on age[56] (see Fig. 3–39), off-foveal viewing[57] (see Fig. 3–40) and other variables.

In recent years, the IERI supplemented its studies of visibility with studies of performance such as had been performed previously.[52,53] Blackwell, in particular, has tried to relate per-

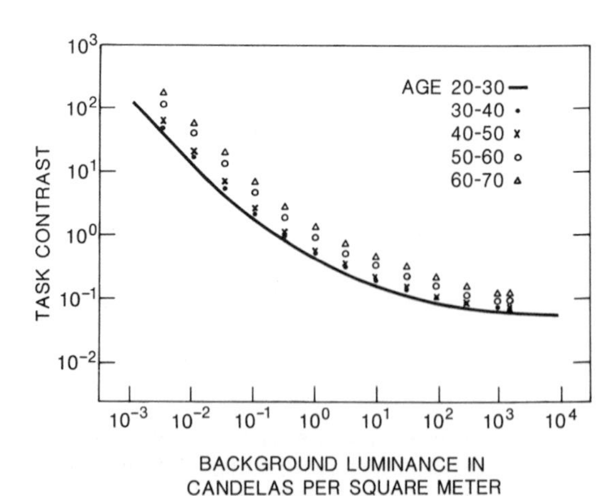

Fig. 3–39. Effect of age on required task contrast for task performance.[56]

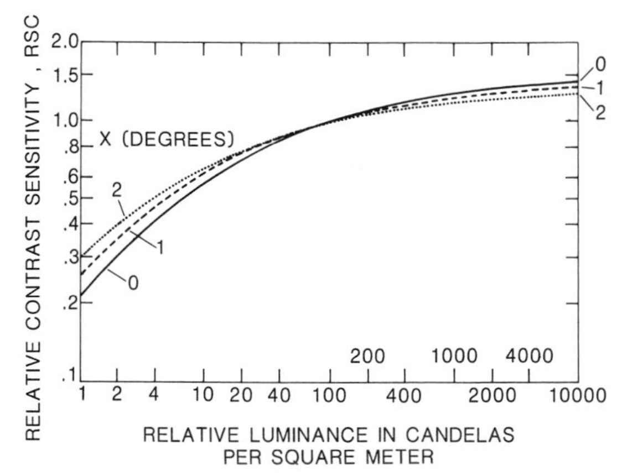

Fig. 3-40. Effect of eccentricity of task location (from the line of sight) upon contrast sensitivity. The curves are normalized at a background luminance of 100 candelas per square meter.

formance to visibility so that there could be tradeoffs between increasing contrast, size and sharpness, on the one hand, and illuminance on the other.[50] Emphasis has been given to the effect of contrast enhancement by controlling the luminous intensity distribution of individual luminaires and arrangement of a number of them in a given environment for workers located at specific locations and with specific orientations.

Blackwell has concentrated on developing a task which can simulate the effect of manipulating such variables as distance of critical detail from the line of sight, and the simultaneous processing of bits of information, etc., and showing the relation of these variables to performance.

Smith and others[58] have carried out studies of performance and levels of luminance using tasks which closely simulate tasks performed in schools and offices with the hope that these data can be used to better understand the effect of lighting on various types of tasks in various environments. These kinds of data can also be used to reduce the uncertainty about the relation between cost of lighting and improvements in performance at the higher levels of luminance.

Attention is being paid also to rating techniques in which subjects make judgments about the adequacy of the lighting for the performance of a task without making actual assessments of performance. The same techniques are being used to investigate other features of an environment such as spaciousness.[59]

Future work is more likely to bridge fundamental research in visual science with applied illuminating engineering. Such a scheme would give the lighting specialist a better appreciation of how the visual system responds under a variety of situations. Nevertheless, at this time, human visual performance can be confidently predicted on only the simplest tasks and under a restricted set of conditions. More information is needed on how visual performance varies above threshold, how visual performance changes with time and how nonvisual factors interact with the sense of sight.

Because of these uncertainties in the Visibility concept, the IES has currently chosen to provide an illuminance selection procedure that is based upon consensus.* This procedure is described in the *Application Volume.* Like the Visibility approach, it is primarily based upon visual criteria, *i.e.*, on illuminances that experienced designers have found to be adequate for the variations in requirements among age groups and task characteristics. It is an oversimplified solution to a difficult and complex problem; however, used with intelligence and judgment, along with the principles of good lighting design, it should aid the designer by giving a qualitative indication of the more important factors that affect seeing in the lighted environment.

VISUAL RANGE OF LUMINOUS SIGNALS

Basic Principles

It is the illuminance at an observer's eye, produced by a light, that largely determines if the light will be seen. The illuminance E produced at a distance x by a source of luminous intensity I in an atmosphere having a transmissivity (transmittance per unit distance) T is:

$$E = IT^x/x^2 \qquad (3\text{-}16)$$

If the illuminance E, at the eye, is greater than E_m, the minimum perceptible (or threshold) illuminance, the light will be visible. The distance at which E is equal to E_m is designated as V, the visual range of the light. Then:[60]

$$E_m = IT^V/V^2 \qquad (3\text{-}17)$$

* The procedure was developed as a guide in selecting illuminance levels for different tasks. Thus, it was not developed to provide mandatory IES illuminance requirements. A given illuminance level derived from the procedure does not mean that the lighting system is satisfactory with respect to visibility, energy conservation, luminance ratios, quality of light or any other criterion. Nor does it mean that it is a good compromise between various criteria. If a lighting designer arrives at some value based upon the particular situation and upon past experience (rather than by the illuminance selection procedure), that value may be more valid.

Equation 3-17 is generally known as *Allard's law*.[61, 62]

Equations 3-16 and 3-17 are strictly applicable only when the luminance of the background is small compared to the average luminance of the light.[61, 63] Otherwise equation 3-16 becomes:

$$E = [I - (L - L')A]T^x/x^2 \qquad (3\text{-}18)$$

where L is the luminance of the background of the light, L' is the average luminance of the unlighted projector, both in candelas per unit area, and A is the area of the entire projector projected on a plane normal to the line of sight.

Both L and L' are measured in the direction of the line of sight from a position near the light.

The quantity $(L - L')A$ is the intensity required of the light to make its average luminance equal to that of the background. The visual range of the light is determined by the net intensity, that is, the difference between the measured intensity of the light and this intensity. Typically, the term $(L - L')A$ has a significant effect on the visual range of a signal light only under daylight conditions when the light is dimmed or when the light has a low average luminance in the direction of view.

Effects of the Atmosphere. The atmosphere is never perfectly transparent. Hence, unless the viewing distance is short, atmospheric losses may have a significant effect upon the illuminance at the observer's eye. In fog the law of diminishing returns takes effect at relatively short distances. For example, if the transmissivity is 0.01 per mile (light fog), a light with an intensity of 100 candelas will produce an illuminance of one milecandle at a distance of 1 mile; an intensity of 40,000 candelas is required to produce that illuminance at 2 miles; and 9,000,000 candelas is required at 3 miles.

Equations 3-16, 3-17 and 3-18 can be best solved by using an iterant process with a computer or programmable hand-calculator or by graphical methods. In Fig. 3-41 curves relating distance and the ratio of intensity to illuminance are shown for several values of atmospheric transmissivity. These curves may be used with any consistent set of units; for example, I in candelas, D in miles, E in lumens per square mile, and T per mile.

The clarity of the atmosphere may be conveniently expresssed by a distance defined as the meteorological optical range, the distance at which the transmittance of the light path for light of color temperature of 2700 K is equal to 0.05. This distance corresponds closely with the maximum distance at which dark objects may be observed through such an atmosphere in the daytime.[64, 65]

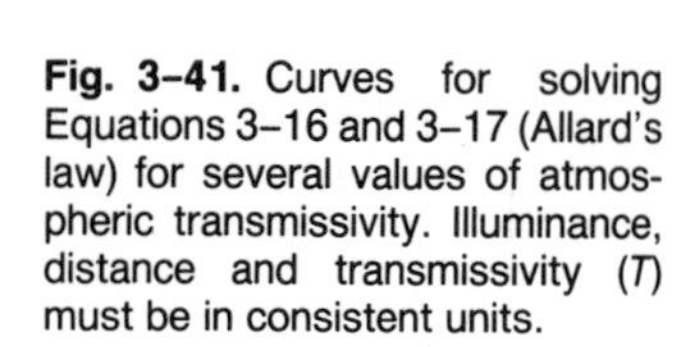

Fig. 3-41. Curves for solving Equations 3-16 and 3-17 (Allard's law) for several values of atmospheric transmissivity. Illuminance, distance and transmissivity (T) must be in consistent units.

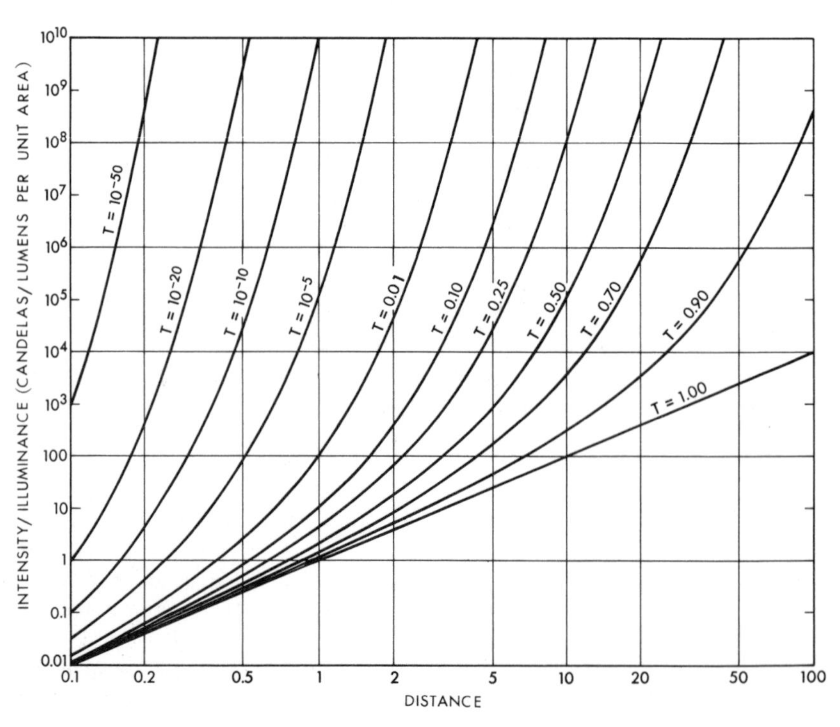

Fig. 3-42. Transmissivities and Extinction Coefficients for Various Weather Conditions

Code Number	Weather	Maximum Meterorological Optical Range (kilometers)	Minimum Extinction Coefficient (per meter)	Maximum Transmissivity		
				(per kilometer)	(per statute mile)	(per nautical mile)
9	Exceptionally clear	50+	0.00006−	0.94+	0.91+	0.89+
8	Very clear	50	0.00006	0.94	0.91	0.89
7	Clear	20	0.00015	0.86	0.79	0.76
6	Light haze	10	0.00030	0.74	0.62	0.57
5	Haze	4	0.00075	0.47	0.30	0.25
4	Thin fog	2	0.0015	0.22	0.090	0.062
3	Light fog	1	0.0030	0.050	0.0081	0.0039
2	Moderate fog	0.5	0.0060	0.0025	0.000065	0.000015
1	Thick fog	0.2	0.015	3.1×10^{-7}	3.4×10^{-11}	9.0×10^{-13}
0	Dense fog	0.05	0.060	9.5×10^{-27}	1.3×10^{-42}	6.5×10^{-49}
	Very dense fog	0.03	0.10	4.3×10^{-44}	1.6×10^{-70}	4.8×10^{-81}
	Exceptionally dense fog	0.015	0.20	1.8×10^{-87}	2.6×10^{-140}	2.3×10^{-161}

The distance V_o at which a large black object can be seen against the horizon sky is given by the relation:[61, 66]

$$\epsilon = e^{-\sigma V_o} \qquad (3\text{–}19a)$$

or

$$\epsilon = T^{V_o} \qquad (3\text{–}19b)$$

where σ is the extinction coefficient, T is the transmissivity, and ϵ is the minimum perceptible contrast, or threshold contrast, of the observer. A value of 0.05 is considered representative of the daylight contrast threshold of a meterological observer.[64]

Values of meteorological optical ranges, extinction coefficient, and transmissivity for various descriptors are given in Fig. 3-42.

Equation 3-19 is a particular case of *Koschmieder's law.*[61,66,67] If the object is not black, equation 3-19 becomes

$$\epsilon = [(L_o - L_H)/L_H]T^V \qquad (3\text{–}20a)$$

or

$$\epsilon = C_oT^V \qquad (3\text{–}20b)$$

where L_o is the luminance of the object, L_H is the luminance of the horizon sky, V is the visual range of the object, and C_o is the inherent contrast between the object and the sky. (Note that ϵ may be either positive or negative, having the same sign of $L_o - L_H$.) Equation 3-20 applies to artificially lighted as well as naturally lighted objects.

If the object, or area, is viewed against a background other than the horizon sky equation 3-20 becomes[61]

$$\epsilon = \frac{[L_o - L_H - (L_b - L_H)T^d]T^V}{[(L_b - L_H)T^{(d+V)} + L_H]} \qquad (3\text{–}21)$$

where L_b is the inherent luminance of the background and d is the distance between the object and its background.

Equations 3-19 and 3-20 may be used without significant error in computing the visual range of objects, or area sources, viewed against a terrestrial background if the distance, d, between the object and its background exceeds one half of V_o of equation 3-19.

Equations 3-16 through 3-21 are based upon the assumption that the transmittance of the atmosphere is independent of wavelength throughout the visible portion of the spectrum. In clean fogs and in rain this assumption is usually valid. However, in smoke or dust there may, on occasion, be significant differences with the transmittance of red light being greater than the transmittance of blue light, and these equations must be applied wavelength by wavelength.[68]

For example, equation 3-16 takes the form

$$E_m = \sum_{380}^{830} I_\lambda V(\lambda)[T(\lambda)]^x \Delta\lambda / x^2 \qquad (3\text{–}22)$$

where I_λ is the spectral radiant intensity of the source.

Threshold Illuminance. The threshold illuminance is not a constant. It is a function of the luminance of the background of the light, the position of the light in the field of view, the angular size and shape of the light, its color, and, if it is not steady-burning, its flash characteristics. In addition, the observer's knowledge of the position of the light and his time for search have a significant influence on the threshold.

Fig. 3-43 shows the relation between threshold illuminance and background luminance for steady-burning, white, point sources for about 98 per cent probability of detection. The threshold illuminance values shown are applicable only when the observer knows precisely where to look for the light. Even if the illuminance is twice the values shown the light will be hard to find. The illuminance values must be increased by a factor of 5 to 10 if the light is to be easy to find.[69]

These increases in illuminance are applicable only when the observer is looking for the light signal. Much greater increases are needed if the light signal is to attract the attention of an observer who is not searching for it. Factors of 100 to 1000 are not excessive.[70]

The break in the curve represents the changes from cone to rod vision. At low background luminances, the threshold illuminance for cone vision remains essentially unchanged as indicated by the broken line at the left. The horizontal portion of the curve represents most night seeing conditions since a light used as a signal is usually observed by looking directly at it; hence cone, not rod, vision is used. Moreover, it is doubtful whether those engaged in transport, with the possible exception of lookouts on ships, ever reach the state of dark adaptation required for rod vision.

Representative background luminances are in Fig. 3–44. It should be noted that the luminance of the night sky in the vicinity of cities and airports seldom falls below 0.003 candela per square meter because of the effects of man-made sources. Note also that, unless there are glare sources in the field of view, it is probably necessary to consider only the background in the immediate vicinity of the light.[71]

Source Size. The threshold illuminance values shown in Fig. 3–43 are applicable to sources

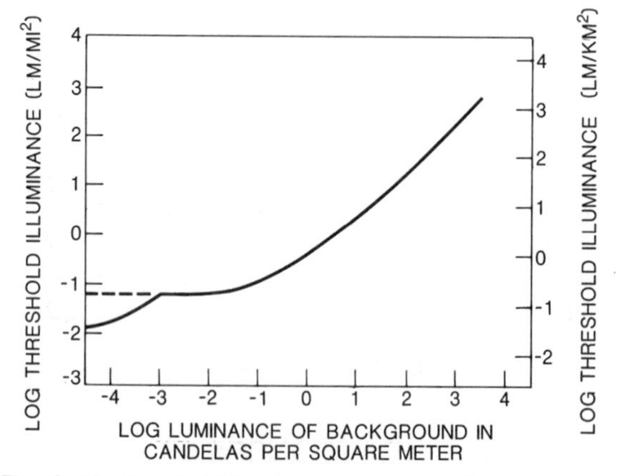

Fig. 3–43. Threshold illuminance at the eye from an achromatic (white) point source for about 98 per cent probability of detection as a function of the background luminance.

Fig. 3–44. Representative Luminance Values of Various Backgrounds Against Which Luminous Signals are Viewed

Background	Representative Luminance — Candelas per Square Meter
Horizon sky	
Overcast, no moon	0.00003
Clear, no moon	.0003
Overcast, moon	.003
Clear, moonlight	.03
Deep twilight	.3
Twilight	3
Very dark day	30
Overcast day	300
Clear day	3000
Clouds, sun-lighted	30000
Daylight fog	
Dull	300–1000
Typical	1000–3000
Bright	3000–16000
Ground	
On overcast day	30–100
On sunny day	300
Snow, full sunlight	16000

which are in effect point sources. Fig. 3–45 shows the maximum diameter of a source which may be considered a point source.[72] Most signal lights behave as point sources. Approximate thresholds for sources which are too large to be considered point sources may be obtained by multiplying the thresholds obtained from Fig. 3–43 by the size factors given in Fig. 3–46.[73]

Figs. 3–45 and 3–46 apply only to threshold and near threshold viewing. Recent work has shown that the intensity of a red traffic signal light required to produce optimum recognition under bright daylight conditions is independent of source size for sources subtending up to 16.5 minutes of arc.[74]

Colored Light Signals

The threshold for the identification of the color of a light signal is about the same as the cone-threshold for detection of that light. At the spectral extremes these threshold values are nearly identical, but at the center of the visible spectrum the cone-threshold for detection may be as much as a half a log unit lower. In the luminance range between absolute rod-threshold and the cone-threshold for identification of color (about three log units) no perception of hue can be made. This luminance range is known as the photochromatic interval.[75] Thus, color discrimi-

nation should not be required when signal light luminances are in this interval.

As noted before, color perception of very small lights may not be good due to small field tritanopia. For this reason, color identification is rarely required for signal lights that have to be seen at great distances. Lights of larger areal extent should be used when color discrimination is necessary. Traffic lights, for example, are usually of sufficient areal extent when color judgments are required (*i.e.*, near a road intersection). Similarly, atmospheric pollutants may selectively attenuate certain regions of the spectrum. In such circumstances the apparent hue of the signal light may change, and color discrimination may be impaired.

Flashing Light Signals

The luminance of a steady light at threshold will be less than the peak luminance of a brief light flash at threshold.[76] It is convenient to evaluate light flashes in terms of the intensity of a steady light of the same color, size and shape that will produce the same criterion response. This intensity is known as the effective luminous intensity of the flash. Although alternate equations are available,[77] the effective intensity may be computed from the relation:[78]

$$I_e = \left(\int_{t_1}^{t_2} I dt\right) \Big/ (a + t_2 - t_1) \quad (3\text{-}23)$$

where I_e is the effective intensity, I is the instantaneous intensity, and t_1 and t_2 are the times in seconds of the beginning and end of that part of

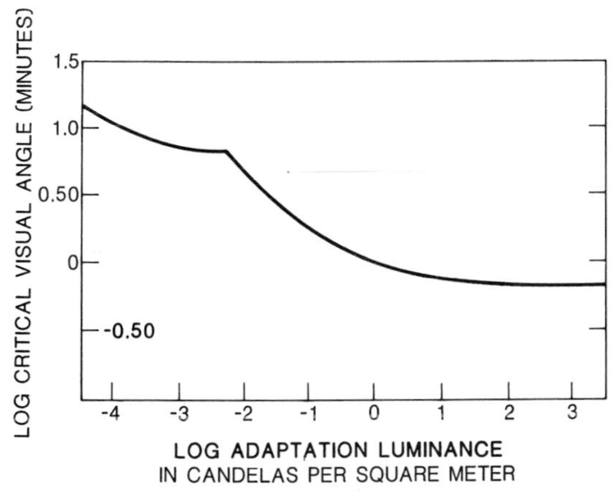

Fig. 3-45. Maximum angular diameter of a circular source which can be considered strictly as a point source.

Fig. 3-46. Size Factors for Sources Other Than Point Sources

Ratio of Source Diameter to Viewing Distance	Size Factor	
	Night	Day
0.0005	1.0	1.0
0.001	1.0	1.2
0.003	1.1	2.5
0.005	1.4	4.9
0.01	2.5	20.0

the flash when the value of I exceeds I_e. This choice of times maximizes the values of I_e.

Methods have been developed whereby the maximum value of I_e can be obtained in two or three steps. "Cut and try" methods are not necessary.[78]

The term a is a function of the effective illuminance produced by the flashing source at the eye. The value of a decreases as the illuminance increases above threshold. Various values of a have been suggested from time to time as the most applicable to a particular signal problem. In the United States a value of 0.2 is usually used for specification purposes and, to promote uniformity, as the nighttime threshold value.[78] This value was chosen as an interim value until a value appropriate to the search situation is determined by definitive experimental work. Thus, by definition

$$I_e = \left(\int_{t_1}^{t_2} I dt\right) \Big/ (0.2 + t_2 - t_1) \quad (3\text{-}24)$$

If the duration of the flash is less than about one millisecond, the effective intensity is given by:

$$I_e = 5 \int I dt \quad (3\text{-}25)$$

where the integration is performed over the entire flash cycle. The effective intensity can then be measured directly using simple electronic integration.

As noted previously, repetitive flashes may appear brighter than comparable steady lights of the same luminance due to brightness enhancement. Consequently, flickering lights are often used to increase the conspicuity of signal lights.

REFERENCES

1. Boettner, E. A. and Wolter, J. R.: "Transmission of the Ocular Media," *Invest. Ophthalmol.*, Vol. 1, p. 776, 1962.

2. Said, F. S. and Weale, R. A.: "The Variation with Age of the Spectral Transmissivity of the Living Human Crystalline Lens," *Gerontol.*, Vol. 3, p. 213, 1959.
3. Coren, S. and Girgus, J. S.: "Density of Human Lens Pigmentation: *In Vivo* Measures over an Extended Age Range," *Vis. Res.*, Vol. 12, p. 343, 1972.
4. Wolf, E. and Gardiner, J. S.: "Studies on the Scatter of Light in the Dioptric Media of the Eye as a Basis of Visual Glare," *Arch. Ophthal.*, Vol. 74, p. 338, September, 1965.
5. Vos, J. J. and Boogaard, J.: "Contribution of the Cornea to Entoptic Scatter," *J. Opt. Soc. Am.*, Vol. 53, p. 869, 1963.
6. Boynton, R. M. and Clarke, F. J. J.: "Sources of Entoptic Scatter in the Human Eye," *J. Opt. Soc. Am.*, Vol. 54, p. 111, 1964.
7. Wyszecki, G. and Stiles, W. S.: *Color Science: Concepts and Methods, Quantitative Data and Formulae*, 2nd ed., John Wiley & Sons, New York, 1982.
8. Vos, J. J.: "Contribution of the Fundus Oculi to Entoptic Scatter," *J. Opt. Soc. Am.*, Vol. 53, p. 1449, 1963.
9. Smith, V. C. and Pokorny, J.: "Spectral Sensitivity of the Foveal Cone Photopigments Between 400 and 500 nm," *Vis. Res.*, Vol. 15, p. 161, 1975.
10. Boynton, R. M.: *Human Color Vision*, Holt, Rinehart and Winston, New York, 1979.
11. Bouma, H.: *Receptive Systems Mediating Certain Light Reactions of the Pupil of the Human Eye*, Philips Research Report, Supplement No. 5, Eindhoven, Netherlands, 1965.
12. Dowling, J. A.: "The Site of Visual Adaptation," *Sci.*, Vol. 155, p. 273, January 20, 1967.
13. Hecht, S. and Mandelbaum, J.: "The Relation Between Vitamin A and Dark Adaptation," *J. Am. Med. Assoc.*, Vol. 112, p. 1910, 1939.
14. Boynton, R. M. and Miller, N. D.: "Visual Performance Under Conditions of Transient Adaptation," *Illum. Eng.*, Vol. LVII, p. 541, August, 1963. Boynton, R. M., Rinalducci, E. J. and Sternheim, C.: "Visibility Losses Produced by Transient Adaptational Changes in the Range from 0.4 to 4000 Footlamberts," *Illum. Eng.*, Vol. 64, p. 217, April, 1969. Boynton, R. M., Corwin, T. R. and Sternheim, C.: "Visibility Losses Produced by Flash Adaptation," *Illum. Eng.*, Vol. 65, p. 259, April, 1970.
15. Pierenne, M. H.: *Vision and the Eye*, 2nd ed., Chapman and Hall, Ltd., London, 1967.
16. Ludvigh, E. and Miller, J. W.: "Study of Visual Acuity during the Ocular Pursuit of Moving Test Objects, I. Introduction," *J. Opt. Soc. Am.*," Vol. 48, p. 799, 1958.
17. Miller, J. W.: "Study of Visual Acuity during the Ocular Pursuit of Moving Test Objects, II. Effects of Direction of Movement, Relative Movement, and Illumination," *J. Opt. Soc. Am.*, Vol. 48, p. 803, 1958.
18. Alman, D. H.: "Errors in the Standard Photometric System when Measuring the Brightness of General Illumination Light Sources," *J. Illum. Eng. Soc.*, Vol. 71, No. 1, p. 55, October, 1977.
19. Bedford, R. E. and Wyszecki, G.: "Wavelength Discrimination for Point Sources," *J. Opt. Soc. Am.*, Vol. 48, p. 129, 1958.
20. Wright, W. D.: *Researches on Normal and Defective Color Vision*, Henry Kimpton, London, 1946.
21. Robertson, A. R.: "Color Differences," *Die Farbe*, Vol. 29, p. 273, April 6, 1981.
22. Boynton, R. M.: *Human Color Vision*, Holt, Rinehart and Winston, New York, 1979.
23. Brindley, G. S.: *Physiology of the Retina and Visual Pathway*, Edward Arnold, Ltd., London, 1970.
24. Ingling, C. R.: "Luminance and Opponent Color Contributions to Visual Detection and to Temporal and Spatial Integration: Comment," *J. Opt. Soc. Am.*, Vol. 68, No. 8, p. 1143, 1978.
25. Ivanoff, A.: "About Spherical Aberration of the Eye," *J. Opt. Soc. Am.*, Vol. 46, No. 10, p. 901, October, 1956.
26. Lewis, A. L., Katz, M. and Ochrlein, C.: "A Modified Achromatizing Lens," *Am. J. Optom. Physio. Opt.*, Vol. 59, p. 909, 1982.
27. Fry, G. A.: "A Re-evaluation of the Scattering Theory of Glare," *Illum. Eng.*, Vol. XLIX, p. 98, February, 1954.
28. Holladay, L. L.: "Fundamentals of Glare and Visibility," *J. Opt. Soc. Am.*, Vol. 12, p. 271, 1926.
29. Holladay, L. L.: Action of Light Source in the Field of View on Lowering Visibility," *J. Opt. Soc. Am.*, Vol. 14, p. 1, 1927.
30. Stiles, W. S.: "The Effect of Glare on the Brightness Difference Threshold," *Proc. Roy. Soc.*, Vol. B104, p. 322, London, 1929.
31. Fugate, J. M. and Fry, G. A.: "Relation of Changes in Pupil Size to Visual Discomfort," *Illum. Eng.*, Vol. LI, p. 537, July, 1956.
32. Fry, G. A. and King, V. M.: "The Pupillary Response and Discomfort Glare," *J. Illum. Eng. Soc.*, Vol. 4, p. 307, July, 1975.
33. Luckiesh, M. and Guth, S. K.: "Brightness in the Visual Field at Borderline Between Comfort and Discomfort (BCD)," *Illum. Eng.*, Vol. XLIV, p. 650, November, 1949.
34. Putnam, R. C. and Faucett, R. E.: "The Threshold of Discomfort Glare at Low Adaptation Levels," *Illum. Eng.*, Vol. XLVI, p. 505, October, 1951.
35. Kaiser, P. K., Herzberg, P. A. and Boynton, R. M.: "Chromatic Border Distinctness and its Relation to Saturation," *Vis. Res.*, Vol. 11, p. 953, 1971.
36. Boynton, R. M. and Kaiser, P. K.: "Vision: The Additivity Law Made to Work for Heterochromatic Photometry with Bipartite Fields," *Sci.*, Vol. 161, p. 366, 1968.
37. Tansley, B. W. and Boynton, R. M.: "Chromatic Border Distinctness: The Role of Red- and Green-Sensitive Cones," *Vis. Res.*, Vol. 18, p. 683, 1978.
38. Kaiser, P. K.: "Luminance and Brightness," *Appl. Opt.*, Vol. 10, p. 2768, 1971.
39. Blackwell, H. R.: "Contrast Thresholds of the Human Eye," *J. Opt. Soc. Am.*, Vol. 36, p. 624, 1946. Blackwell, H. R.: "Development and Use of a Quantitative Method for Specification of Interior Illumination Levels on the Basis of Performance Data," *Illum. Eng.*, Vol. LIV, p. 317, June, 1959.
40. Guth, S. K. and McNelis, J. F.: "Visual Performance: A Comparison in Terms of Detection of Presence and Discrimination of Detail," *Illum. Eng.*, Vol. 63, p. 32, January, 1968. McNelis, J. F. and Guth, S. K.: "Visual Performance—Further Data on Complex Test Objects," *Illum. Eng.*, Vol. 64, p. 99, February, 1969.
41. Campbell, F. W. and Robson, J. G.: "Application of Fourier Analysis to the Visibility of Gratings," *J. Physiol.*, Vol. 197, p. 551, London, 1968.
42. Baron, W. S. and Westheimer, G.: "Visual Acuity as a Function of Exposure Duration," *J. Opt. Soc. Am.*, Vol. 63, p. 212, 1973.
43. Shlaer, S.: "The Relation Between Visual Acuity and Illumination," *J. Gen. Physiol.*, Vol. 21, p. 165, 1937.
44. Sperling, H. G. and Jolliffe, C. L.: "Intensity Time Relationship at Threshold for Spectral Stimuli in Human Vision," *J. Opt. Soc. Am.*, Vol. 55, p. 191, 1965.
45. Bartley, S. H.: "Intermittent Photic Stimulation at Marginal Intensity Levels," *J. Psychol.*, Vol. 32, p. 217, 1951. Bartley, S. H.: "Subjective Brightness in Relation to Flash Rate and the Light-Dark Ratio," *J. Exper. Psychol.*, Vol. 23, p. 313, 1938.
46. Hecht, S. and Smith, E. L.: "Intermittent Stimulation by Light: V. The Relation Between Intensity and Critical Frequency for Different Parts of the Spectrum," *J. Gen. Physiol.*, Vol. 19, p. 979, 1936.
47. Katz, M. A.: "Brief Flash Brightness," *Vis. Res.*, Vol. 4, p. 361, 1964.
48. Kelly, D. H.: "Flickering Patterns and Lateral Inhibition," *J. Opt. Soc. Am.*, Vol. 59, p. 1361, 1969.
49. Van Der Wildt, G. J. and Rijsdijk, J. P.: "Flicker Sensitivity Measured with Intermittent Stimuli: I. Influence of the Stimulus Duration on the Flicker Threshold," *J. Opt. Soc. Am.*, Vol. 69, p. 660, 1979.
50. CIE Committee TC-31: "An Analytic Model for Describing the Influence of Lighting Parameters upon Visual Performance," *CIE Publication No.* 19/2.1, Commission Internationale de l'Eclairage, Paris, 1981.
51. Blackwell, H. R.: "Development of Procedures and Instruments for Visual Task Evaluation," *Illum. Eng.*, Vol. 65, p. 267, April, 1970.
52. Weston, H. C.: "The Relation Between Illumination and Visual Efficiency—The Effect of Brightness Contrast," *Ind. Health Res. Board Rep.*, Vol. 87, p. 35, Great Britain Medical Research Council, 1945. Weston, H. C.: "The Relation Between Illumination and Industrial Efficiency, 1. The Effect of Size of Work," *Jt. Rep. Indus. Health Res. Board Illum. Res. Comm.* (D.S.I.R.), London, HMSO, 1935. Weston, H. C.: *Light, Sight, and Efficiency*, Lewis and Co., London, 1949.
53. Tinker, M. A.: "Brightness Contrast, Illumination Intensity, and Visual Efficiency," *Am. J. Optom.*, Vol. 36, p. 221, 1959.
54. Luckiesh, M. and Moss, F. M.: *The Science of Seeing*, Van Nostrand, New York, 1937.
55. CIE Committee TC-3.1: "A Unified Framework of Methods for Evaluating Visual Performance Aspects of Lighting," *CIE Publication No. 19*, Commission Internationale de l'Eclairage, Paris, 1972.
56. Blackwell, O. M. and Blackwell, H. R.: "Visual Performance

Data for 156 Normal Observers of Various Ages," *J. Illum. Eng. Soc.*, Vol. 1, p. 3, October, 1971. Blackwell, O. M. and Blackwell, H. R.: "Individual Responses to Lighting Parameters for a Population of 235 Observers of Varying Ages," *J. Illum. Eng. Soc.*, Vol. 9, p. 205, July, 1980.

57. Blackwell, H. R. and Scott, D. E.: "Analysis of Visual Performance Data Obtained in a Landolt Ring Task Without Response Limitation," *J. Illum. Eng. Soc.*, Vol. 2, p. 445, July, 1973. Blackwell, H. R. and Blackwell, O. M.: "Population Data for 140 Normal 20–30 Year Olds for Use in Assessing Some Effects of Lighting upon Visual Performance," *J. Illum. Eng. Soc.*, Vol. 9, p. 158, April, 1980.

58. Smith, S. W. and Rea, M. S.: "Proofreading Under Different Levels of Illumination," *J. Illum. Eng. Soc.*, Vol. 8, p. 47, October, 1978. Smith, S. W.: "Performance of Complex Tasks Under Different Levels of Illumination: Part I—Needle Probe Task," *J. Illum. Eng. Soc.*, Vol. 5, p. 235, July, 1976. Smith, S. W.: "Is There an Optimum Light Level for Office Tasks?" *J. Illum. Eng. Soc.*, Vol. 7, p. 255, July, 1978.

59. Flynn, J. E., Spencer, T. J., Martyniuk, O. and Hendrick, C.: "Interim Study of Procedures for Investigating the Effect of Light on Impressions and Behavior," *J. Illum. Eng. Soc.*, Vol. 3, p. 87, October, 1973. Flynn, J. E.: "A Study of Subjective Responses to Low Energy and Non-Uniform Lighting Systems," *Light. Des. Appl.*, Vol. 7, p. 6, February, 1977. Flynn, J. E. and Spencer, T. J.: "The Effects of Light Source Color on User Impression and Satisfaction," *J. Illum. Eng. Soc.*, Vol. 6, p. 167, April, 1977. Flynn, J. E., Hendrick, C., Spencer, T. and Martyniuk, O.: "A Guide to Methodology Procedures for Measuring Subjective Impressions in Lighting," *J. Illum. Eng. Soc.*, Vol. 8, p. 95, January, 1979.

60. Reynaud, L.: "Memoir upon the Illumination and Beaconage of the Coasts of France," Paris, 1864 (Translation—Hains, Peter C., Government Printing Office, Washington, DC, 1876).

61. Middleton, W. E. K.: *Vision Through the Atmosphere*, University of Toronto Press, Toronto, 1952.

62. Douglas, C. A. and Booker, R. L.: *Visual Range: Concepts, Instrumental Detemination, and Aviation Applications*, National Bureau of Standards Monograph 159, Washington, DC, 1977.

63. Kevern, G. M.: "Effect of Source Size upon Approach Light Performance," *Illum. Eng.*, Vol. XLV, p. 96, February, 1950.

64. Douglas, C. A. and Young, L. L.: "Development of a Transmissimeter for Determining Visual Range," Civil Aeronautics Administration Technical Development Report No. 47, Washington, DC, 1945.

65. "Measurement of Visibility," *Guide to Meteorological Instrument and Observing Practices*, Chapter 10, World Meteorological Organization, 1971.

66. Koschmieder, H.: "Theorie der Horizontalen Sichtweite," *Beitr. Phys. Freien Atm.* 12, 33, and 171, 1924. Duntley, S. Q.: "The Reduction of Apparent Contrast by the Atmosphere," *J. Opt. Soc. Am.*, Vol. 38, p. 179, February, 1948.

67. Duntley, S. Q., Gordon, J. I., Taylor, J. H., White, C. T., Boileau, A. R., Tyler, J. E., Austin, R. W. and Harris, J. L.: "Visibility," *Appl. Opt.*, Vol. 3, p. 549, May, 1964.

68. Zuev, V. E.: "Atmospheric Transparency in the Visible and the Infrared, (*Prozrachnost' atmosfery dlya vidimykh i infrakrasykh lnchei*)," *Izdatel' stv "Sovetskoe Radio,"* Moskva, 1966,—Translated into English by Z. Lerman, Israel Program for Scientific Translations, Jerusalem, 1970. Available from National Technical Information Service, Springfield, Virginia 22151 as TT69-55102.

69. Tousey, R. and Koomen, M. J.: "The Visibility of Stars and Planets During Twilight," *J. Opt. Soc. Am.*, Vol. 43, p. 177, March, 1953.

70. Breckenridge, F. C. and Douglas, C. A.: "Development of Approach- and Contact-Light Systems," *Illum. Eng.*, Vol. XL, p. 785, November, 1945.

71. Knoll, H. A., Tousey, R., and Hulbert, E. O.: "Visual Thresholds of Steady Point Sources of Light in Fields of Brightness from Dark to Daylight," *J. Opt. Soc. Am.*, Vol. 36, p. 480, August, 1946.

72. Blackwell, H. R.: "Contrast Thresholds of the Human Eye," *J. Opt. Soc. Am.*, Vol. 36, p. 624, November, 1946.

73. de Boer, J. B.: "Visibility of Approach and Runway Lights," *Philips Res. Rep.*, Vol. 6, p. 224, 1951.

74. Fisher, A. J. and Cole, B. L.: "The Photometric Requirements of Vehicle Traffic Signal Lanterns," *Proc. Aust. Road Res. Board*, Vol. 7, No. 5, 1974.

75. Stiles, W. S., Bennett, M. G. and Green, H. N.: "Visibility of Light Signals with Special Reference to Aviation Lights," A. R. C. Reports and Memoranda No. 1793, H. M. Stationery Office, London, 1937. Hill, N. E. G.: "The Recognition of Colored Light Signals which are Near the Limit of Visibility," *Proc. Phys. Soc.*, Vol. 59, p. 560, 1947. Hill, N. E. G.: "The Measurement of the Chromatic and Achromatic Thresholds of Colored Point Sources against a White Background," *Proc. Phys. Soc.*, Vol. 59, p. 574, 1947.

76. Blondel, A. and Rey, J.: "The Perception of Lights of Short Duration at Their Range Limits," *Trans. Illum. Eng. Soc.*, Vol. VII, p. 625, November, 1912. The paper is a complete account in English of the material contained in two papers published in the *J. de Phy. et Radium*, Vol. 1, Series 5, pp. 530, 643, 1911. Projector, T. H.: "Effective Intensity of Flashing Lights," *Illum. Eng.*, Vol. LII, p. 630, December, 1957.

77. Schmidt-Clausen, J. J.: "A Comparison of Different Methods for the Determination of the Effective Luminous Intensity of Signal Lights in the Form of Multiple Pulses," *CIE J.*, Vol. 1, No. 1, p. 18, 1982.

78. Committee on Aviation of the IES: "IES Guide for Calculating the Effective Intensity of Flashing Signal Lights," *Illum. Eng.*, Vol. LIX, p. 747, November, 1964. Douglas, C. A.: "Computation of the Effective Intensity of Flashing Lights," *Illum. Eng.*, Vol. LII, p. 641, December, 1957.

Measurement of Light and Other Radiant Energy

SECTION 4

Progress in a branch of science or engineering is dependent, to a very large extent, on the ability to measure the quantities associated with that subject. Thus, each advance in measuring technique means a broadening of knowledge.

The measurement of light is called photometry, and the basic instrument employed is known as a photometer. The earlier photometers depended on visual appraisal as the means of measurement. These are now used only rarely, having been superseded by modern nonvisual "physical" photometers giving higher accuracy and easier use. Physical photometers utilize radiant energy incident upon a receiver producing measurable electrical quantities.

Spectral Luminous Efficiency Curve. In general, light measurements with physical instruments are useful only if they indicate reliably how the eye would react to a certain stimulus. In other words, such an instrument should be sensitive to the spectral power distribution of light in the same way as the eye. Because of very substantial differences between individual pairs of eyes in the latter respect, the CIE standard observer response curve has been established (also known as the eye sensitivity curve). See Figs. 1–6 and 4–4. Therefore, the sensitivity characteristics of a physical receiver should be the equivalent of this standard observer. The required match is usually accomplished by adding filters between the light-sensitive element and the light source. See page 4–4.

Photopic and Scotopic Vision. All ordinary photometric measurements are assumed to be made with the observer's eye in the photopic or light-adapted state. When the luminance of the surface to be measured is much below 0.034 candela per square meter (0.003 candela per square foot), the eye can no longer be regarded as light-adapted. Most measurements of fluorescent and phosphorescent materials are made in the scotopic (dark-adapted) range, or the region between the scotopic and photopic. Because of the change in the spectral response of the eye at these low levels, it is necessary to take special measures in evaluating the results of such measurements.[1]

NOTE: References are listed at the end of each section.

Measurable Quantities

As indicated in Fig. 4–1, many characteristics of light, light sources, lighting materials and lighting installations, may be measured. In addition to the characteristics covered in Fig. 4-1, most of the standard photometric terms that apply to the material in this section are also defined in Section 1. Further information on ultraviolet, visible light and infrared are covered in the Application Volume.

The measurements of most general interest are:

1. Illuminance.
2. Luminance.
3. Luminous intensity.
4. Luminous flux.
5. Contrast.
6. Color (appearance and rendering).
7. Spectral distribution.
8. Electrical characteristics.
9. Radiant energy.

Fig. 4-1. Some Measurable Characteristics of Light, Light Sources and Lighting Materials

Characteristic	Dimensional Unit	Equipment	Technique
		Light	
Wavelength	Nanometer†	Spectrometer	Laboratory
Color	None	Spectrophotometer and colorimeter	Laboratory
Flux density (illuminance)	Lumen per unit area (lux and footcandle)‡	Photometer	Laboratory or field
Orientation of polarization	Degree (angle)	Analyzing Nicol prism	Laboratory
Degree of polarization*	Per cent (dimensionless ratio)	Polarization photometer	Laboratory
		Light Sources	
Energy radiated	Joule per square meter	Calibrated radiometer	Laboratory
Color temperature	Kelvin (K)	Colorimeter or filtered photometer	Laboratory or field
Luminous intensity	Candela§	Photometer	Laboratory or field
Luminance	Candela per unit area§	Photometer or luminance meter	Laboratory or field
Spectral power distribution	Watts per nanometer	Spectroradiometer	Laboratory
Power consumption	Watt	Wattmeter, or voltmeter and ammeter for dc, and unity power factor ac circuits	Laboratory or field
Light output (total flux)	Lumen	Integrating sphere photometer	Laboratory
Zonal distribution	Lumen or candelas§	Distribution or goniophotometer	Laboratory
		Lighting Materials	
Reflectance	Per cent (dimensionless ratio)§	Reflectometer	Laboratory or field
Transmittance	Per cent (dimensionless ratio)§	Photometer	Laboratory or field
Spectral reflectance and transmittance	Per cent (at specific wavelengths)§	Spectrophotometer	Laboratory
Optical density	Dimensionless number§	Densitometer	Laboratory

* Committee on Testing Procedures for Illumination Characteristics of the IES: "Resolution on Reporting Polarization," *Illum. Eng.*, Vol. LVIII, p. 386, May 1963.

† See Fig. 1-11 for conversion factors for units of length.

‡ See Fig. 1-7 for illuminance conversion factors.

§ See Fig. 1-1 for standard units, symbols and defining equations for fundamental photometric and radiometric quantities.

BASIS OF PHOTOMETRY[2]

Principles

Photometric measurements frequently involve a consideration of the inverse-square law (which is strictly applicable only for point sources) and the cosine law.

Inverse-Square Law. The inverse-square law (see Fig. 4-2a) states that the illumination E at a point on a surface varies directly with the candlepower I of the source, and inversely as the square of the distance d between the source and the point. If the surface at the point is normal to the direction of the incident light, the law may be expressed as follows:

$$E = \frac{I}{d^2}$$

This equation holds true within one-half per cent when d is at least five times the maximum dimension of the source (or luminaire) as viewed from the point on the surface.

Cosine Law. The Lambert cosine law (see Fig. 4-2b) states that the illuminance of any surface varies as the cosine of the angle of incidence. The angle of incidence θ is the angle between the normal to the surface and the direction of the incident light. The inverse-square law and the cosine law can be combined as follows:

$$E = \frac{I}{d^2} \cos \theta$$

Cosine-Cubed Law. A useful extension of the cosine law is the "cosine-cubed" equation (see Fig. 4-2c). By substituting $h/\cos \theta$ for d, the above equation may be written:

$$E = \frac{I \cos^3 \theta}{h^2}$$

Standards[3]

Standards of candlepower, luminous flux and color are established by national physical labo-

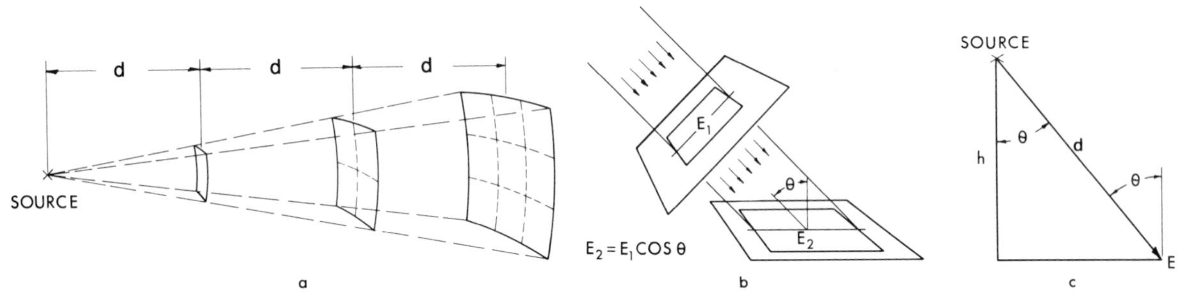

Fig. 4-2. a. The inverse-square law illustrating how the same quantity of light flux is distributed over a greater area, as the distance from source to surface is increased. b. The Lambert cosine law showing that light flux striking a surface at angles other than normal is distributed over a greater area. c. The cosine-cubed law explaining the transformation of the formula.

ratories. Various types of standards may be used in photometric laboratories.

Primary Standards. A primary standard, reproducible from specifications, is usually found only in a national physical laboratory. See Section 1.

Secondary Standards. Secondary standards are usually derived directly from primary standards. A secondary standard must be prepared using precise electrical and photometric equipment, and is usually maintained at a national physical laboratory. See Section 1.

They are used in photometric laboratories throughout the industry.

Preservation of the rating is of prime importance; accordingly, a standard will be used as seldom as possible. Such a standard is used, handled and stored with the greatest care. Such lamps will be found wherever photometric accuracy warrants the highest attainable precision.

Reference Standards. Reference standard is an alternate term for a secondary standard.

Working Standards. Working standards may be prepared for everyday use, because of the cost of the reference standards. A laboratory prepares its own working standards for use in calibrating a photometer. Even a working standard, however, is not used during the conduct of a test, except where a direct comparison is necessary.

General Methods

Photometric measurements, in general, make use of the basic laws of photometry, mentioned on page 4-2, applied to readings from visual photometric comparison or photoelectric instruments. Various types of procedures are designated as follows:

Direct Photometry. Direct photometry consists of the simultaneous comparison of a standard lamp and an unknown light source.

Substitution Photometry. Substitution photometry consists of the sequential evaluation of the desired photometric characteristics of a standard lamp and an unknown light source in terms of an arbitrary reference.

Relative Photometry. To avoid the use of standard lamps, the relative method is widely applied. Relative photometry consists of the evaluation of the desired photometric characteristic based upon an assumed lumen output of the test lamp. Alternately, it may be the measurement of one uncalibrated light source in terms of another.

Photometry of Non-Steady-State Sources. It is sometimes necessary to measure the output of sources which are non-steady or cyclic, in which case extreme care should be taken.[4]

Means of Attenuation. In making photometric measurements, it often becomes necessary to reduce the luminous intensity of a source in a known ratio, to bring it within the range of the measuring instrument. A rotating sector disk with one or more angular apertures is one means of doing this. If such a disk is placed between a source and a surface, and rotated at such speed that the eye perceives no flicker, the effective luminance of the surface is reduced in the ratio of the time of exposure to the total time (Talbot's law). The reduction is by the factor $\theta/360$ degrees, where θ is the total angular aperture in degrees. The sector disk has advantages over many filters. It is not affected by a change of characteristics over a period of time. It reduces

Fig. 4-3. Portable photoelectric photometers: (a) pocket-size color- and cosine-corrected illuminance meter; (b) small luminance/illuminance meter (shown in illuminance mode); (c) paddle-type illuminance/irradiance meter; (d) Freund luminance spot meter; (e) Pritchard luminance photometer; (f) combined illuminance, luminance and irradiance photometer; (g) small, hand-held luminance meter.

luminous flux without changing its spectral composition. Sector disks should not be used with light sources having cyclical variation in output.[5]

Various types of neutral filters of known transmittance are also used for attenuation. Wire mesh or perforated metal filters are perfectly neutral, but have a limited range. Partially silvered mirrors have high reflectance, and the reflected light must be controlled to avoid errors in the photometer. When this type of filter is perpendicular to the light source photometer axis, serious errors may be caused by multiple reflections between the filter and receiver surface. These may be avoided by mounting the filter at a small angle (not over 3 degrees) from the perpendicular, and at a sufficient distance from the receiver surface to throw reflections away from the photometric axis. In this canted position, care must be taken not to reflect light from adjacent surfaces on to the receiver. Also, it is difficult to secure completely uniform transmission over all parts of the surface.

So-called "neutral" glass filters are seldom neutral and transmission characteristics should be checked before use. In general, they have a characteristic high transmittance in the red region, and low in the blue, and spectral correction filters may be required. However, this type of filter varies in transmittance with ambient temperature, as do many other optical filters.

"Neutral" gelatin filters are quite satisfactory, although not entirely neutral. Some have a small seasoning effect, losing neutrality over a period of time. They must be protected by being cemented between two pieces of glass and watched carefully for loss of contact between the glass and gelatin. Filters should not be stacked together unless cemented, due to errors which may be created by interference between surfaces.

With modern metering techniques, electronic alterations can be accomplished to keep the output of a receiver-amplifier combination in range of linearity and readability.

PHOTOMETERS

A photometer is a device for measuring radiant energy in the visible spectrum. Various types of physical instruments consisting of a radiant energy sensitive element and appropriate measuring equipment are used to measure ultraviolet and infrared radiant energy. When used with a filter to correct their response to the CIE standard observer, they measure visible light and are usually called physical photometers.

In general, photometers may be divided into two classifications: (1) laboratory photometers which are usually fixed in position and yield results of highest accuracy, and (2) portable photometers of lower accuracy for making photometric measurements in the field or outside the laboratory. These in turn may be grouped according to function, such as photometers to measure luminous intensity (candlepower), luminous flux, illuminance, luminance, light distribution, light reflectance and transmittance, color, spectral distribution and visibility.

Illuminance Photometers

In recent years, visual photometric methods have largely been supplanted commercially by physical methods; however, visual methods, because of their simplicity, are still used in educational laboratories for demonstrating photometric principles, and the less routine types of photometric measurements.[6–10]

Photoelectric photometers[11] may be divided into two classes: those employing solid state devices such as photovoltaic and photoconductive cells; and those employing photoemissive tubes, which require considerable additional equipment for operation.

Photovoltaic Cell Meters.[12] A photovoltaic cell is one that directly converts radiant energy into electrical energy. It not only provides a small current which is approximately proportional to the incident illumination, but also produces a small electromotive force capable of forcing this current through a low resistance circuit. Photovoltaic cells provide much larger currents than do photoemissive cells, and can directly operate a sensitive instrument such as a microammeter or galvanometer. However, photovoltaic cells depart from linearity of response at higher levels of incident illumination as the resistance of circuit to which they are connected increases; therefore, for precise results, the external circuitry and metering should apply nearly zero impedance across the photocell.

Some of the portable illuminance meters in use today consist of a photovoltaic cell, or cells, connected to a meter calibrated directly in lux or footcandles. See Fig. 4–3. However, with modern solid state electronic devices, operational amplifiers have been used successfully to amplify the output of photovoltaic cells, and the condition which produces most favorable linearity between cell output and incident light is automatically achieved, namely, by reducing the potential difference across the cell to zero. The amplifier power requirements are small, and easily supplied by small batteries. In addition, digital readouts may be conveniently used to eliminate the ambiguities inherent in deflection type instruments.[14]

Spectral Response. The spectral response of photovoltaic cells is quite different from that of the human eye, and color-correcting filters are usually employed.[15] As an example, Fig. 4–4 illustrates the degree to which a typical commercially corrected selenium photovoltaic cell, as commonly used in illuminance meters, approximates the standard spectral luminous efficiency curve. Cells of a given type vary considerably in this respect, and for precision laboratory photometry each cell should be individually color-corrected.

The importance of color correction can be illustrated by comparing the "human eye" match under illumination generated by a monochromatic source. For example, if a predominately blue light source is used, the majority of the visible energy is concentrated near 465 nanometers. It can be seen in Fig. 4–4 that the relative eye and filtered receptor responses are approximately 10 and 15 percent, respectively. This represents a 50 percent differential and indicates that this photoreceptor could read as much as 50 per cent high under that blue light source. Care should be taken to correct for this difference.

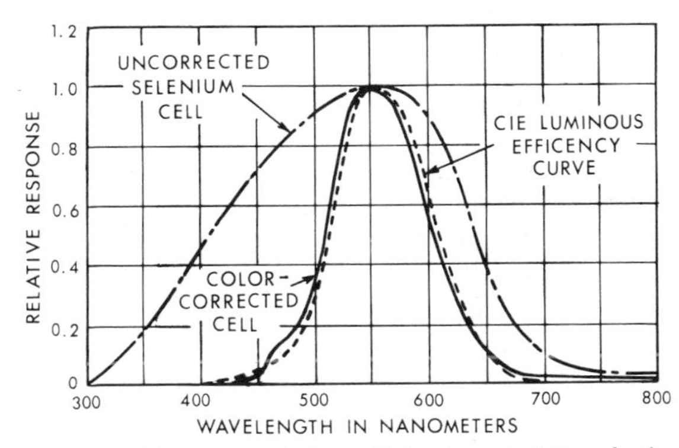

Fig. 4-4. Average spectral sensitivity characteristics of selenium photovoltaic cells, compared with CIE spectral luminous efficiency curve.

Transient Effects. The output of photovoltaic cells, when exposed to a constant illumination, will require a short finite rise time to reach a stable output and thereafter may decrease slightly over a period of time due to fatigue.[16] Rise times for silicon cells typically are considerably shorter than for selenium cells.

Effect of Angle of Incidence (Cosine Effect). Part of the light reaching a photovoltaic cell at high angles of incidence is reflected by the cell surface and the cover glass, and some may be obstructed by the rim of the case. The resultant error increases with angle of incidence, and in measuring illuminance where an appreciable portion of the flux comes at wide angles, an uncorrected meter may read as much as 25 per cent below the true value. The cells used in most illuminance meters are now provided with diffusing covers or some other means of correcting the light-sensitive surface to approximate the true cosine response. The component of illuminance contributed by single sources at wide angles of incidence may be determined by orienting the cell perpendicular to the direction of the light, and multiplying the reading thus obtained by the cosine of the angle of incidence. Other methods have been proposed. The possibility of cosine error must be taken into consideration in some laboratory applications of photovoltaic cells. One satisfactory solution to the problem consists in placing an opal diffusing acrylic plastic disk with a matte surface over the cell. At high angles of incidence, however, it reflects the light specularly so that the readings are too low. This can be compensated by allowing light to reach the cell through the edges of the plastic. The readings at very high angles will then be too high but can be corrected by using a screening ring.[17]

Effect of Temperature.[18] Wide temperature variations, either high or low, affect the performance of photovoltaic cells, particularly where the external resistance of the circuit is high. Prolonged exposure to temperatures above 50 °C (120 °F) will permanently damage selenium cells. Silicon cells are considerably less affected by temperature. Measurements at high temperatures and at high illuminance levels should, therefore, be made rapidly to avoid overheating the cell. Hermetically sealed cells provide greater protection from the effects of temperature and humidity. When using photovoltaic cells at other than their calibrated temperature, conversion factors may be employed, or means provided to maintain cell temperature in the vicinity of 25 °C (77 °F).

Effect of Cyclical Variation of Light. When electric discharge sources are operated on alternating current power supplies, precautions should be taken with regard to the effect of frequency on photocell response.[19] In some cases, these light sources may be modulated at several thousand hertz. Consideration should then be given to whether the response of the cell is exactly equivalent to the Talbot's Law response of the eye for cyclic varying light. Due to the internal capacitance of the cell, it cannot always be assumed that its dynamic response exactly corresponds to the mean value of the illuminance. It has been found that a low or zero-resistance circuit is the most satisfactory for determining the average intensity of modulated or steady state light sources with which photovoltaic cell instruments are generally calibrated. Although a microammeter or galvanometer appears to register a steady photocell current, it may not be receiving such a current, and may be actually receiving a pulsating current which it integrates since its natural period of oscillation is long compared to the pulses. Meters are available which average over a period of time, which eliminates the effect of cyclic variation.

Photometer Zeroing. It is important to check photometer zeroing prior to taking measurements. If an analog meter is used, this requires setting the needle to zero. For any type of equipment using an amplifier, it may be necessary to zero both the amplifier and the dark current. Where possible, it should be verified that the meter remains correctly zeroed when the photometer scale is changed. Alternately, any deviation from zero under dark current conditions may be measured and subtracted from the light readings.

Electrical Interference. With modern day electronic instrumentation, extreme care should be taken to eliminate interference induced in the leads between the cell and the instrumentation. This may be achieved by filter networks, shielding, grounding or combinations of the above.

Photomultiplier Tube Meters. Photoelectric tubes produce current when radiant energy is received on a photoemissive surface and then amplified by a phenomenon known as secondary emission. They require a high voltage to operate (2000 to 5000 volts) and will require an amplifier to provide a measurable signal. The resultant current may be measured by a deflection meter, oscillograph or digital output device. "Dark current", or current flowing through the device while it is in absolute darkness, must be compensated for in the circuitry or subtracted from the lighted tube output. Meters employing this device are usually extremely sensitive.

Because of the construction of such meters they are sensitive to shock and strong magnetic fields. Shocks will damage the photomultiplier tube and magnetic fields will alter the calibration of the instrument. Also the device is temperature sensitive and it is recommended to operate the tube at or below room ambient temperature. As with other photoelectric devices, the photomultiplier spectral response curve does not match the human eye; therefore, color correcting filters are required. Because of the large number of photomultiplier types available, the manufacturer should supply the proper optical filter for the particular photomultiplier of interest.

When the photomultiplier tube is used in conjunction with an optical lens system a luminance meter can be constructed of high sensitivity and broad range. A number of such meters are available (see Fig. 4–3).

Luminance Photometers

The basic principles discussed earlier relating to photometers for the measurement of illuminance apply equally well to those for the measurement of luminance. Luminance meters consist essentially of a photoreceptor in front of which is positioned a means of focusing an image of the object of interest onto the photoreceptor. By suitable optics, therefore, the luminance of a certain size spot when cast onto the receptor will generate an electrical signal which is dependent upon the object luminance. This signal can be measured, and assuming the necessary calibration has been performed, a reading is produced which is a direct measure of luminance. Usually an eyepiece is provided such that the user is able to see the general field of view through the instrument.

By interchanging the lens system in front of the photoreceptor, different fields of view, and therefore different sizes of measurement area may be achieved. This may vary from areas subtending a few minutes of arc up to several degrees.

Photoreceptors may be selenium, but typically in modern instruments they are silicon cells or photomultipliers. The meter reading may be analog or digital and either built into the meter itself or remote. Amplifier controls for zeroing and scale selection usually will be provided, as may be optical filters for color work or scale selection by means of neutral density filters.

Freund Brightness Spot Meter.[21] This is a photoelectric photometer for measuring the luminance of small areas (Fig. 4–3). Models available cover 1/4, 1/2 or 1 degree field of view. A beam splitter allows a portion of the light from the objective lens to reach a reticule viewed by the eyepiece. The remainder of the light is reflected on to the field operative in front of the photomultiplier tube, and the output of the tube after amplification is read on a microammeter with a scale calibrated in candelas per square meter or footlamberts. One of the filters provided with the instrument approximately corrects the response of the photomultiplier to the standard spectral luminous efficiency curve. Full scale deflection is produced by 10^{-1} to 10^{7} candelas per square meter.

Pritchard Photometer. This is a high sensitivity precision photomultiplier photometer with interchangeable field apertures covering fields from 2 arc minutes to 3 degrees in diameter (Fig. 4–3). Full scale sensitivity ranges are from 10^{-4} to 10^{8} candelas per square meter.The readings of the light being measured are free from the effects of polarization since there are no internal reflections of the beam. The spectral response of each photometer is individually measured and the filters to best match it to the standard spectral luminous efficiency curve are determined and inserted. Filters are also in-

Fig. 4–5. Visual task photometer (VTP).

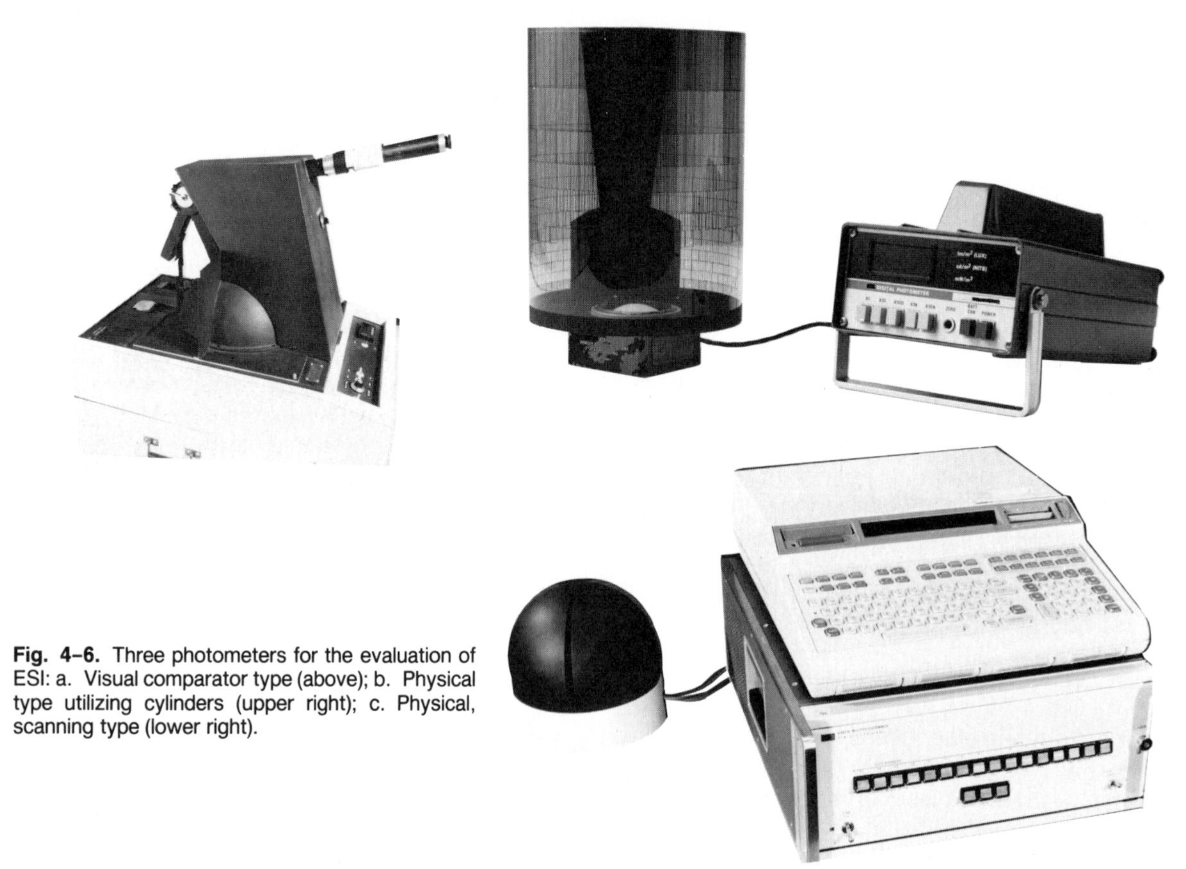

Fig. 4-6. Three photometers for the evaluation of ESI: a. Visual comparator type (above); b. Physical type utilizing cylinders (upper right); c. Physical, scanning type (lower right).

cluded to permit evaluation of polarization and color factors.[22]

ESI Photometers

Equivalent sphere illumination (ESI) can be measured both visually and physically. Visual measurements are made by visual comparisons between a task viewed in the measured (actual) environment and the task viewed in a luminous sphere, using a visibility meter (see Section 3). Physical ESI measurements are based on the algorithms described beginning on page 9-36. The only parameters that must be measured in situ are L_b and L_t. All physical ESI devices measure, in one way or another, L_b and L_t. The measuring devices discussed here are presented in chronological order of development.

Visual Task Photometer (VTP). This photometer is the basic, reference instrument against which all others are compared.[23] ESI is not measured directly; L_b and L_t are measured and ESI subsequently calculated. The task to be evaluated is mounted on a target shifter and a telephotometer is aimed at it from the desired viewing angle. The task and telephotometer (usually mounted on a cart) are then positioned so that the task is in the location where the measurement is to be made, and the telephotometer is facing the proper direction of view. The standard body shadow (attached to the telephotometer) shades the task in a manner similar to an actual observer. L_b is measured. Then the shifter is activated to bring the target into view of the telephotometer and L_t is measured. See Fig. 4-5.

Visual ESI Meter. The visual ESI meter[24] consists of an optical system, variable luminous veil, target carrier, luminous sphere, illuminance meter (inside the sphere) and body shadow (see Fig. 4-6a). A task is placed on the target carrier and viewed through the optical system. The contrast of the task is then reduced to threshold by adjusting a variable luminous veil. Field luminance is automatically kept constant so as not to alter the adaptation luminance of the observer's eye. The task is then carried inside the sphere and the optical system adjusted until the target is again at threshold; task visibility is the same inside the sphere as it was outside. The illuminance in the sphere is measured to directly determine ESI.

Physical ESI Meters. Two devices are available that do not rely on the actual presence of a task for their precision. Instead, they use numerical data which represents the task's reflectance characteristics—bidirectional reflectance distribution functions (BRDF's).

One meter utilizes cylinders[25] that represent an optical analogy of the VTP. See Fig. 4-6b. There are two cylinders used per measurement, one which represents a task's L_b, called the background cylinder, and one which represents $L_b - L_t$, called the difference cylinder. These two parameters can be used to calculate ESI in place of L_b and L_t alone. Each cylinder has its own body shadow. A cosine corrected illuminance probe is placed where the measurement is desired. The background cylinder is placed atop the probe and oriented in the appropriate viewing direction. Background illuminance is then recorded. The background cylinder is replaced by the difference cylinder, oriented in the same direction and the difference illuminance recorded. ESI is then calculated from the background and difference illuminance readings.

The other meter, a scanning luminance meter,[26] consists of a narrow field luminance probe attached to a motorized scanning apparatus and a minicomputer to control scanning, store BRDF data and perform calculations (see Fig. 4-6c). To use this instrument to measure ESI, the probe is positioned at the desired location and the minicomputer instructed to begin scanning. Luminances are multiplied by their appropriate BRDF's to determine L_b and L_t. The minicomputer then calculates ESI directly. The scanning luminance meter has the capability of rotating the task in any viewing direction, and of determining the ESI on different tasks with only one set of scanning measurements.

Task Photometry. The bidirectional reflectance distribution functions (BRDF's) used with the physical meters are obtained by illuminating a task from a particular direction and by viewing the task from some other unique direction. A visual task photometer is used to perform these measurements. The VTP used is the same as that used for ESI measurements except that a collimated light source is included which can be positioned anywhere on a hemisphere over the task (see Fig. 4-7). Thus the task is illuminated from each azimuth and declination angle (usually in five-degree increments) and the reflectance measured at each angle. The collection of bidirectional reflectance data for the task and its background form the bidirectional reflectance distribution function (BRDF)[27].

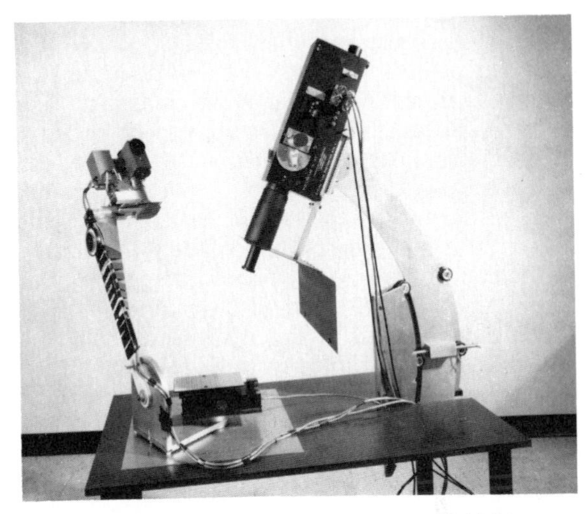

Fig. 4-7. Visual task photometer and collimated light source arrangement for measurement of BRDF's.

Reflectometers

Reflectometers are photometers used to measure reflectance of materials or surfaces in specialized manners. They measure diffuse, specular and/or total reflectance.[28] Those designed to determine specular reflectance are referred to as *glossmeters.* A popular type of reflectometer uses

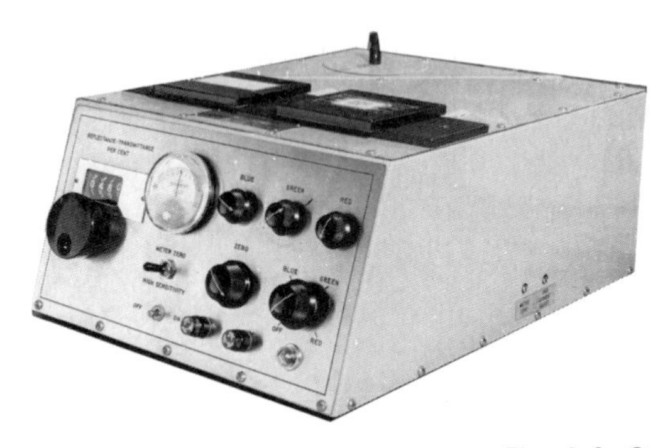

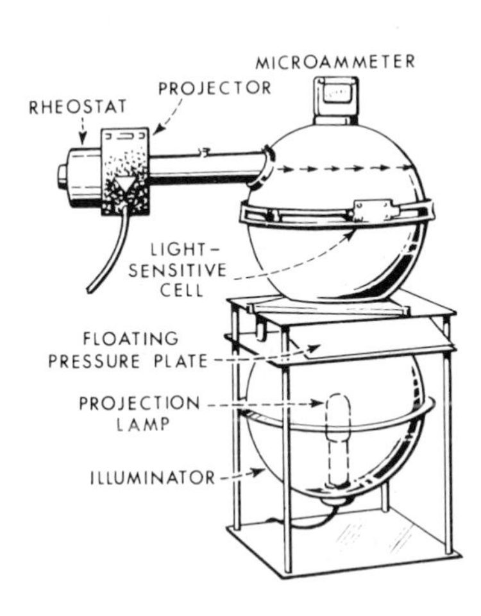

Fig. 4-8. Gardner reflectometer (left) and the Baumgartner light cell reflectometer showing arrangement for transmittance measurement (right).

a collimated beam and photovoltaic cell. See Fig. 4–8. The beam source and cell are mounted in a fixed relationship in the same housing. The housing has an aperture through which the beam travels. This head or sensor is set on a standard reflectance reference with the aperture against the standard. The collimated beam strikes the standard at a 45-degree angle. The photovoltaic cell is constructed so that it measures the light reflected at 0-degrees from the standard. The instrument is then adjusted to read the value stated on the standard. The sensor then is placed on the surface to be measured and the reading recorded. Two cautions are recommended. The standard used should be in the range of the value expected of the surface to be measured. Also, if the area to be considered is large, several measurements should be taken and averaged to obtain a representative value.

Another type[28] of instrument, the Baumgartner reflectometer, see Fig. 4–8, measures both total reflectance and diffuse transmittance. It consists of two spheres, two light sources and two photovoltaic cells. The upper sphere is used alone for the measurement of reflectance. The sample to be measured is placed over an opening at the bottom of the sphere, a collimated beam of light is directed on it at about 30 degrees from the normal, and the total reflected light, integrated by the sphere, is measured by two cells mounted in the sphere wall. The tube carrying the light source and the collimating lenses is then rotated so that the light is incident on the sphere wall, and a second reading taken. The test sample is in place during both measurements, so that the effect on both readings of the small area of the sphere surface it occupies is the same.[29] The ratio of the first reading to the second is the reflectance of the sample for the conditions of the test. Test specimens of translucent materials should be backed by a nonreflecting diffuse material.

Transmittance for diffuse incident light is measured by using the light source in the lower sphere, and taking readings with and without the sample in the opening between the two spheres. The introduction of the sample will change the characteristics of the upper sphere. Correction must be made to compensate for the error thus introduced.[30, 31]

Various instruments are available for measuring such properties as specular reflectance and the gloss characteristics of materials.[32] For example, an instrument similar to that described above for the measurement of diffuse reflectance may be used, except that the cell is fixed at 45 degrees on the opposite side of the sample to the light source, thus measuring the specularly reflected beam. The angle subtended by the photocell to the sample will affect the reading obtained, and appropriate recommendations are available.

Radiometers

Radiometers (sometimes called radiometric photometers) are used to measure radiant power over a wide range of wavelengths which may include the ultraviolet, visible or infrared spectral regions. They may employ detectors (receivers) which are nonselective in wavelength response or which give adequate response in the desired wavelength band. Nonselective detectors, *i.e.*, those which response varies little with wavelength, are thermocouples, thermopiles, bolometers and pyroelectric detectors. One class of wavelength-selective detectors is photoelectric in nature and includes photoconductors, photoemissive tubes, photovoltaic cells and solid-state sensors, such as photodiodes, phototransistors and other junction devices.

The over-all response of such detectors can be modified by using appropriate filters to approximate some desired function. For example, they can be color corrected by means of a filter to duplicate the standard luminous efficiency curve in the visible range or to level a detector's response to radiant power over some band of wavelengths. The corrections must compensate for any selectivity in the spectral response of the optical system. Special care must be exercised to eliminate a detector's response to radiation lying outside the desired range of interest.

When a monochromator is used to disperse the incoming radiation one can determine the radiant power in a very small band of wavelengths. Such an instrument is called a *spectroradiometer* and is used to determine the spectral power distribution (SPD), *i.e.*, the radiant power per unit wavelength as a function of wavelength, of the radiation in question. The SPD is fundamental: from it one can determine in an absolute manner radiometric, photometric and colorimetric properties of the radiation. The advent of digital processing has greatly facilitated both the measurement and the use of the SPD.[33]

The range of spectral response generally depends on the nature of the detector. Photomultiplier tubes extend from 125 to 1100 nanometers.[34] Various types of silicon photodiodes cover the range from 200 to 1200 nanometers.[34] In the infrared are intrinsic germanium (0.9 to 1.5 micrometers), lead sulfide (1.0 to 4.0 micrometers), indium arsenide (1.0 to 3.6 micrometers), indium antinomide (2.0 to 5.4 micrometers), various types of doped germanium (zinc, 2.0 to 40 micrometers), and mercury cadmium telluride (1.0 to 13 micrometers).[35] The response of nonselective detectors spans from near ultraviolet to 30 micrometers and beyond.[35]

The electrical output of detectors (voltage, current or charge) is very small, thus special precautions are often required to achieve acceptable signal levels, signal-to-noise ratios, and response times (for rapidly varying signals). Photon counting and charge integration techniques are used for extremely low radiation levels.[36]

In all radiometric work it is of the utmost importance to avoid stray radiation, and great care must be taken to be sure it is excluded. This is difficult, because such stray radiation is not visible to the eye, and a surface appearing black to the eye may actually be an excellent reflector of radiant energy outside the visible spectrum. Often the unwanted radiation can be absorbed by an appropriate filter. Sometimes such a high flux of power must be removed that the absorption filter heats to the point of breaking or such that its transmittance for other desired wavelengths is altered. In these cases one must use nonabsorbing interference filters.

Since radiated flux of some wavelengths is dispersed or absorbed by a layer of air between the radiator and detector, consideration must be given to the placement of the source and detector, and to the medium surrounding them.

Spectrophotometers

Photometry is the measurement of power in the visible spectrum, weighted according to the visual response curve of the eye; however, when the power is measured as a function of wavelength, the measurement is referred to as spectrophotometry. Its applications extend from precise quantitative chemical analysis to the exact determination of the physical properties of matter. In illuminating engineering, spectrophotometry is important in the determination of spectral transmittance and spectral reflectance. It also is applied to the measurement of the spectral emittance of lamps, in which case it is referred to as *spectroradiometry*. This form of measurement commonly will cover not only the visible portion of the spectrum, but also the ultraviolet[37] and near-infrared wavelengths.

Instruments used for performing the measurements mentioned above are called spectrophotometers and spectroradiometers, respectively. These instruments consist basically of a device called a *monochromator* which separates or disperses the various wavelengths of the spectrum by means of prisms or gratings, and a suitable receptor which measures the power contained within a certain wavelength range of the dispersed light. If the spectrum is examined visually rather than by a photoreceptor, the instrument is referred to as a *spectroscope*.

In the visible spectrum, the only fundamental means of examining a color for analysis, standardization and specification is by means of spectrophotometry. In addition, it is the only means of color standardization that is independent of material color standards (always of questionable permanence), and independent of the abnormalities of color vision existing among even so-called normal observers.

Colorimetric data may be computed from spectrophotometry by means of the methods of measuring and specifying color in the CIE standard observer and coordinate system for colorimetry. See Section 5.

Commercial development of spectrophotometers has served to extend the wavelength range from about 200 to 2500 nanometers, to make them automatic recording and to add tristimulus integration. Recent developments utilizing self-scanned silicon photodiode arrays provide nearly instantaneous determination of spectral power distribution.

Basic Equipment Types

Optical Bench Photometers. Optical bench photometers are used for the calibration of instruments for illumination measurement. They provide a means for mounting sources and photocells in proper alignment and a means for easily determining the distances between them. If the source is of known candlepower the inverse-square law is used to compute illuminance, provided that the source to detector distance is at least 5 times the maximum source dimension.

Distribution Photometers. Luminous intensity (candlepower) measurements are made on a distribution photometer which may be one of the following types:

1. Goniometer and single cell.
2. Fixed multiple cell.
3. Moving cell.
4. Moving mirror.

All types of photometers have advantages and disadvantages. The significance attached to each advantage or disadvantage is very dependent on other factors, such as available space and facilities, polarization requirements, economic considerations, etc.

Goniometer and Single Cell. The light source is mounted on a goniometer which allows the source to be rotated about both horizontal and vertical axis. The candlepower is measured by a single fixed cell.

There are several different versions of goniometers. Each is related to the type of luminaire being photometered and the facilities in which it is located. With the use of computers the coordinate system of one goniometer system can be easily changed to another coordinate system,[38] thus the universality of data reporting becomes more practical. Fig. 4-9 show two types of goniometer systems, known as Type A and Type B.

Fixed Multiple Cell Photometer. Numerous individual photocells are positioned at various angles around the light source under test. Readings are taken on each photocell to determine the candlepower distribution. See Fig. 4-10.

Moving Cell Photometer. This device consists of a photocell which rides on a rotating boom or arc shaped track, where the light source is centered in the arc traced by the cell. Readings are collected with the cell positioned at the desired angular settings. Sometimes a mirror is placed on a boom to extend the test distance. See Fig. 4-11.

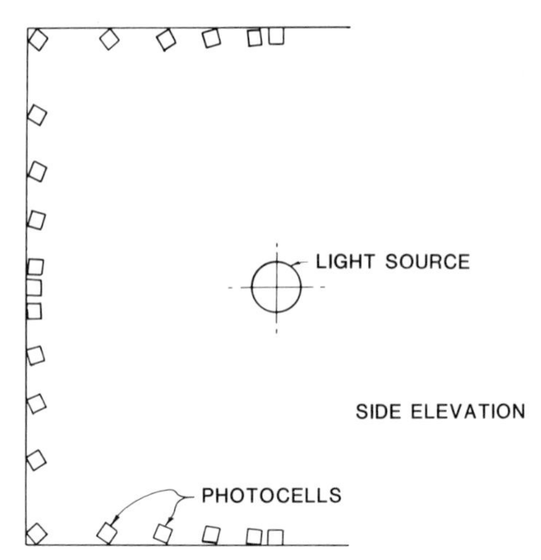

Fig. 4-10. Schematic side elevation of a fixed multiple cell photometer.

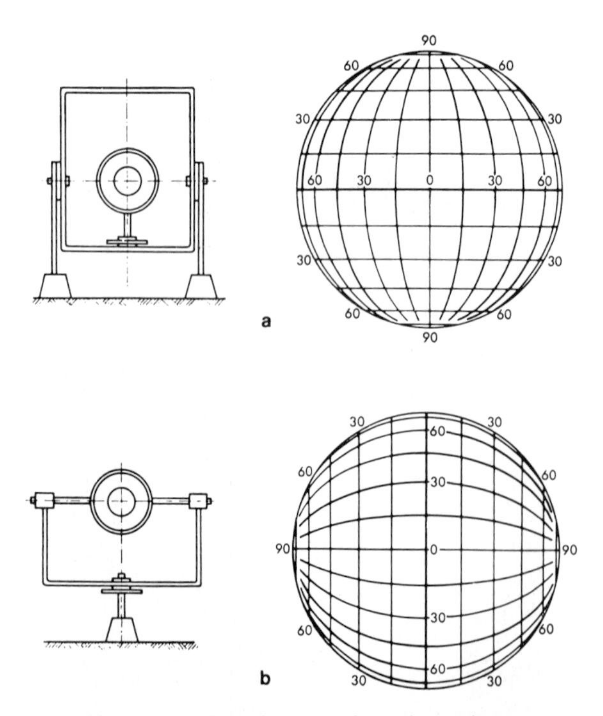

Fig. 4-9. a. Type A goniometer where the projector turns about a fixed horizontal axis and also about an axis which, in the position of rest, is vertical, and upon rotation follows the movement of the horizontal axis. b. Type B goniometer where the light source turns about a fixed vertical axis and also about a horizontal axis following the movement of the vertical axis. The grid lines shown represent the loci traced by the photocell as the goniometer axes are rotated.

Moving Mirror Photometer. In this type of photometer the mirror rotates around the light source, reflecting the candlepower to a single cell. Readings are taken at each desired angle as the mirror moves to that location. See Fig. 4-12.

Integrating Sphere Photometer. The total luminous flux from a source (lamp or luminaire) is measured by some form of integrator, the most common one being the Ulbricht[39] sphere. See Fig. 4-13. Other geometric forms are sometimes used.[40] The theory of the integrating sphere as-

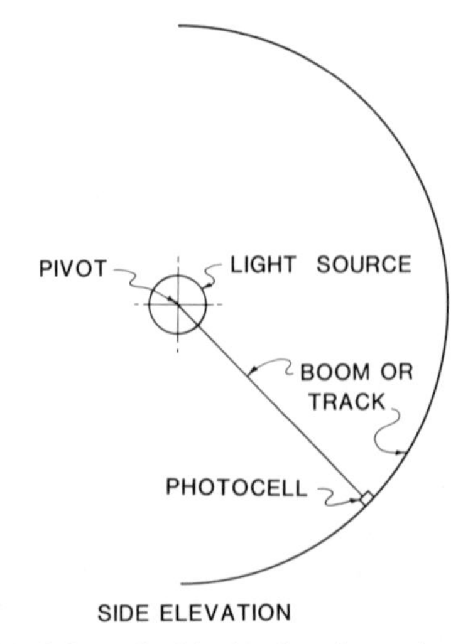

Fig. 4-11. Schematic side elevation of a moving cell photometer.

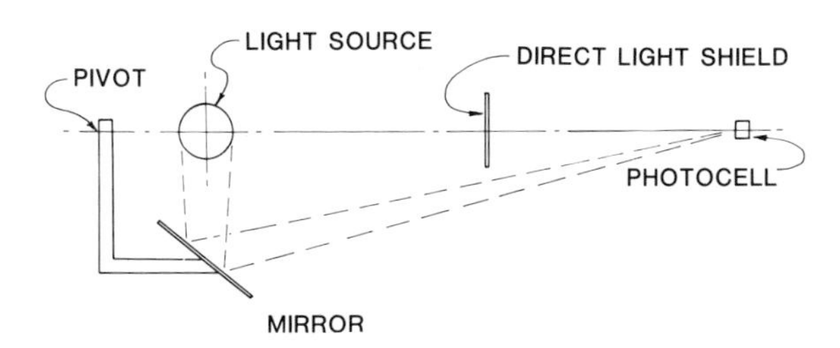

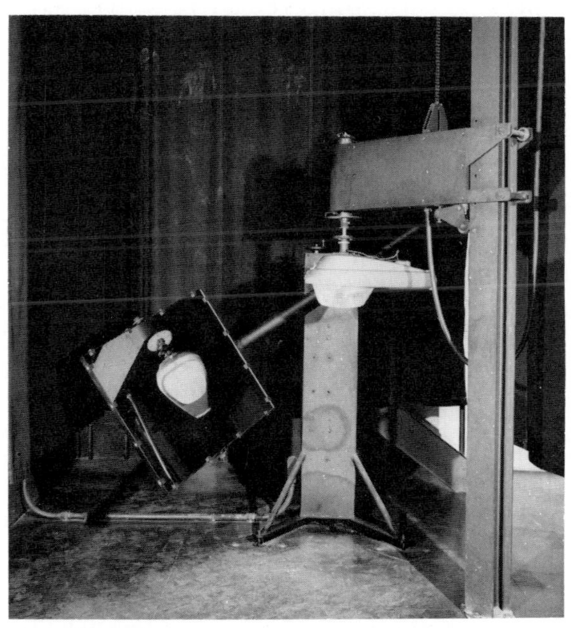

Fig. 4-12. Schematic diagram and photograph of a moving mirror photometer.

sumes an empty sphere whose inner surface is perfectly diffusing and of uniform non-selective reflectance. Every point on the inner surface then reflects to every other point, and the illuminance at any point is, therefore, made up of two components; the flux coming directly from the source, and that reflected from other parts of the sphere wall. With these assumptions, it follows that the illuminance, and hence the luminance, of any part of the wall due to *reflected light only* is proportional to the total flux from the source, regardless of its distribution. The luminance of a small area of the wall, or the luminance of the outer surface of a diffusely transmitting window in the wall, carefully screened from direct light from the source, but receiving light from other portions of the sphere, is therefore a relative measurement of the flux output of the source. The presence of a finite source, its supports, electrical connections, the necessary shield, and the aperture or window, are all obvious departures from the basic assumptions of the integrating sphere theory. The various elements entering into the considerations of a sphere, as an integrator, make it generally undesirable to use a sphere for absolute measurement of flux unless various correction factors are applied[41], but do not detract in the least from its use when a substitution method is employed.[42]

LABORATORY MEASUREMENTS[43-46]

Precision and accuracy of laboratory measurements results only from a combination of several elements: perhaps the most important is properly trained personnel. Others are control of electrical power, good electrical and light measuring instruments, electrical stability of test lamps, and meticulous care to avoid sources of error. Errors frequently arise from insecure luminaire mounting; stray light; infrequent checks of meters, cells and standards; inadequate warm-up of discharge lamps and ballasts; and failure to recognize inherent nonsymmetry of sources and luminaires.[46]

The availability of a room providing an adequate test distance is often a problem since long ranges are sometimes needed for sealed beam, floodlight and projector lamp measurements. The use of mirrors to permit folding the effective

Fig. 4-13. Integrating (Ulbricht) sphere used at the Canadian Standards Association. It is used for directly obtaining total light output of lamps and luminaires.

test distance may help considerably. Temperature and air circulation control are serious requirements for discharge lamp photometry. Cleanliness, provision for suitable electric power, and storage space for lamps, luminaires and delicate instruments are requirements that scarcely need mentioning.

Special techniques and current procedures are described in the pertinent IES and other testing and measurement guides found in the references at the end of this section.

Electrical Measurements of Incandescent Filament Lamps[47]

It is often desirable to determine the electrical characteristics of light sources in connection with photometric measurements. Where additional information is required the reader is referred to any of the many texts or handbooks on electrical engineering.[48]

It is mandatory that test lamps be electrically stable before attempting any accurate electrical measurement. To accomplish this it is necessary to season them in accordance with established procedures.[49]

Any analog types of electrical instruments used should be selected to have current and voltage ratings corresponding to the circuit conditions to be encountered and to give indications at half-scale or above. The use of these instruments at lower scale indications make the accuracy of measurement questionable.[50]

Corrections for instrument currents or voltage drops must be taken into account since the light output of incandescent filament lamps is a high order exponential function of current and voltage. Instruments should be selected that have ¼-per cent full scale accuracy or better to give measurements of high accuracy and precision.

Instrumentation—Direct Current Circuits.

Current. When the light source current is measured by means of an ammeter, all of the current to that source should go through the ammeter. When any voltmeter current is included in the ammeter indication, a correction must be made to obtain true lamp current.

Voltage. The voltage applied to the light source is measured by means of a voltmeter. To avoid correction for voltage drop in the ammeter, the voltmeter is generally connected directly across the load. Often, separate voltage leads are connected to the base of the lamp through special lamp holders to avoid voltage drop resulting from socket-to-lamp connections.

Precise Method. For the measurement of direct currents and voltages, the deflection potentiometer,[51] equipped with suitable multipliers and shunts, provides a rapid and accurate method of measurement. Adjustment of a single dial gives an approximate balance of the unknown value against a standard cell. The residual unbalance is read from the deflection of a sensitive millivoltmeter, which replaces the galvanometer of the null potentiometer.

The initial cost of a deflection potentiometer is considerably greater than a voltmeter and ammeter; but speed of operation, and increased accuracy, often justify the expenditure.

High precision can also be obtained with modern digital voltmeters (DVM) having accuracies of 0.1 per cent of reading and better. These instruments draw negligible power from the circuit being measured compared to moving element instruments.

Power. Power may be computed by multiplying together the observed values of current and voltage.

Instrumentation—Alternating Current Circuits.

Current. Ammeters for use in ac circuits may be self-contained or used in connection with a current transformer, in order to adapt the current rating of the instrument to the actual current in the lamp circuit.

Voltage. Voltmeters for use in ac circuits may be self-contained or used with either a multiplier or a potential transformer. With the voltmeter connected directly across the load, the current taken by the voltmeter is included in the ammeter indication, and the power consumed by the voltmeter is included in any ammeter indication. As in the case of direct-current measurements, the voltmeter current is computed by dividing the voltmeter indication in volts by the voltmeter impedance in ohms. The power taken by the voltmeter is computed in the same manner as that taken by the wattmeter potential circuit, and a correction made to the wattmeter indication.

Care must be exercised in selecting ac instruments compatible with ac wave shape. Non-sinusoidal ac wave shape may require true rms measuring instruments for highest accuracy. All ac instruments must be suitable for the frequency of the supply power to the lamp or be calibrated for the test frequency. Some instruments sense average currents but are calibrated in rms values on the assumption of sine wave usage.

Power. In ac circuits the power consumed by incandescent lamps can be determined by the volt-ampere product or by a wattmeter. Care should be taken to correct the wattage read for the power consumed by the instruments. For example, if the voltmeter is connected directly across the lamp the current drawn by the voltmeter must be subtracted from the ammeter reading. The current is easily computed by dividing the actual voltage read by the scale resistance. The power is best determined by using a wattmeter. If one is used, the current coil of the wattmeter is connected into the line in a manner similar to an ammeter and the voltage or potential circuit of the wattmeter connected directly across the load, as for a voltmeter. Under this condition, the wattmeter indication will include the power taken by the wattmeter voltage circuit and a correction should be made. The correction is determined by computation and is equal to the square of the load (lamp) voltage divided by the resistance of the potential circuit of the wattmeter. Certain wattmeters are designed to compensate automatically for the power taken by the wattmeter potential coil. In such cases the wattmeter is referred to as a "compensated wattmeter"[52] and when so arranged, no correction need be made.

Precise Method. High precision is attainable on ac circuits using modern digital voltmeters (DVM) having accuracies of 0.1 per cent of reading or better. These devices should read true rms values and draw negligible power from the circuit being measured.

Electrical Measurements of Electric Discharge Lamps and Circuits

All electric discharge lamps have negative volt-ampere characteristics and must therefore be operated in conjunction with internal or external current limiting devices, such as resistors or reactors. These are described in Section 8. Generally all discharge lamps are measured on ac circuits using rms indicating instruments, although the meters may not necessarily respond to rms input.

Electric discharge lamp circuit measurements may involve lamps or ballasts, but in many cases the two are inseparable and the combined operation must be considered. Ballasts used commercially, because of normal manufacturing tolerances, supply lamps with some variation in voltage and current characteristics which affect the electrical input and the light output of lamps. To promote uniformity of testing, the International Electrotechnical Commission (IEC), working through the American National Standards Institute (ANSI), the Canadian Standards Association (CSA), and similar national standardizing bodies throughout the world, has or is establishing standardized testing procedures for determining the electrical characteristics for most of the common arc discharge lamps.

Where international standards have not been established, national standards are used.

Testing Procedures—Discharge Lamps. Lamp parameters are influenced by many factors and detailed accepted testing procedures are described in the appropriate IES guides.[53-55] Some of the more important conditions affecting lamp and ballast testing are discussed below.

A. ***Ambient Temperature.*** If a lamp is operated within an enclosure, such as an Ulbricht sphere, the air temperature within the enclosure becomes the ambient. An ambient temperature of 25 °C plus or minus 1 °C should be maintained for fluorescent lamps. High intensity discharge lamps are less sensitive to temperature and for them ±5 °C is satisfactory.

B. ***Drafts.*** Extreme caution should be exercised to reduce air movement over the surface of the lamps, especially for fluorescent and to a lesser extent for high intensity discharge lamps.

C. ***Lamp Position.*** Unless tests are designed to meet special conditions, fluorescent lamps should be tested in a horizontal position. High intensity discharge lamps may be tested in any position unless specifically designed for a certain position, in which case they should be tested in that position.

D. ***Lamp Connections.*** Fluorescent lamps of the hot cathode bipin type should be operated with the same two pins connected to the operating circuit for all conditions during tests.[56]

High intensity discharge lamp circuit connections to the center contact should always be the high voltage side of the line supply.

E. ***Lamp Stabilization.*** Before any measurements are taken, the lamp should be operated until the performance characteristics are stable. If the nature of the test requires seasoned lamps, a minimum of 100 hours operation is recommended.[49] The position of the lamp during stabilization should be the same as the position in which the lamp is to be measured.

If lamps are warmed up on one ballast and then transferred to a different ballast for measurement, an additional period of burning in the measurement circuit is usually necessary to bring the lamp to equilibrium.

No lamp which shows swirling or other abnormal behavior should be considered to be stabilized for measurement purposes.

F. ***Power Supply.*** The ac wave shape should be such that the rms summation of the harmonic components does not exceed three per cent of the fundamental.

The input impedance, looking into the power source from the lamp and ballast, should not exceed ten per cent of the ballast impedance.

Unless automatic voltage regulation within 0.1 per cent is available, constant checking and readjustment are necessary. If a static type voltage stabilizer is used, it is particularly important to check the wave shape for the harmonic components.

G. ***Ballasts.*** Rating tests of lamps should be conducted with the lamp operating in series with the assigned Reference Ballast or variable linear reactor. If no Reference Ballast specification has been issued for the lamp under test, a reactor complying with the general lamp requirements should be used.

Special tests may be made with other than specified Reference Ballasts. The results of these tests are not directly comparable with data taken on a Reference Ballast.

H. ***Rating Measurements of Fluorescent Lamps with Continuously Heated Electrodes.*** Lamps using the rapid start principle are designed to be operated with the electrode continuously heated by separate windings within the ballast. Because of the complexity of circuit conditions, measurements of these lamps, operated in conjunction with a series ballast, may not exactly represent the values when the lamp is operated by a rapid start ballast.

I. ***Test Circuits.*** A typical measurement circuit is shown in Fig. 4-14. Since the line voltage may be insufficient for lamp starting or operation, a step-up as well as a variable transformer has been shown.

In the case of fluorescent lamps, a preheat lamp has been shown; the starting circuit is omitted for instant-start lamps. The procedure for testing lamps designed for operation with continuously heated electrodes requires special attention.[53]

Switches are shown for removing instruments from the circuit. Those used to short circuit current coils should have very low resistance and must be maintained to meet this requirement. The Reference Ballast or reactor and the accompanying variable resistor can be set for impedance and power factor, with or without the am-

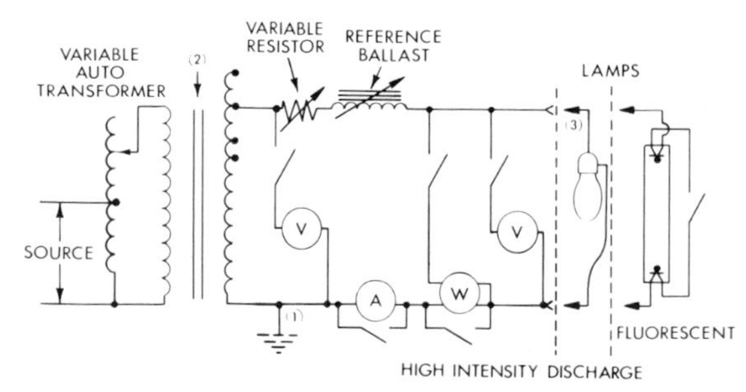

Fig. 4-14. Basic measurement circuit for electric discharge lamps and ballasts. Note:

(1) This side of circuit should be the one connected to ground if a grounded supply system is used.
(2) It may be necessary to insert the step-up transformer shown at this point in order to obtain the open circuit voltage required by the Reference Ballast for certain sizes of lamps. A double-wound transformer (as illustrated) should be used.
(3) Center contact on base.

meter and wattmeter current coils included. Either procedure is satisfactory, but instrument corrections must be determined for the method used.

J. ***Instrumentation.***[50] The voltage across an electric discharge lamp has a distorted wave shape, hence the instruments used must be of a type whose deflection responds to true rms values.

The combined impedance of instruments connected in parallel with the lamp should not draw more than three per cent of the rated lamp current. The voltage drop across instruments connected in series with the lamp should be less than two per cent of the rated lamp voltage. Where necessary, corrections should be made to lamp current and voltage readings.

High precision is attainable using modern digital voltmeters (DVM) having accuracies of 0.1 per cent of reading or better. These devices should read true rms values. Very little power is drawn from the measurement circuit when using these instruments.

Because of the transient voltages (in the order of 1500 to 4500 volts) experienced when turning these lamps on and off suitable protection should be provided for the semiconductor components.

Testing Procedures—Ballasts. Ballast parameters are influenced by many factors and detailed accepted testing procedures are described in the appropriate ANSI standards.[57] Some of the more important conditions affecting ballast testing are discussed below.

A. ***Voltage Range.*** For most tests, ballasts should be operated at their rated primary voltage. If they are rated for a range, the voltage should be the center of the range; the center for 110–125 should be 118 volts.

B. ***Reference Lamps.*** Some tests on ballast specify that the ballast shall be operating a Reference Lamp. These are seasoned lamps which, when operated under stated conditions with the specified Reference Ballast, operate at values of lamp volts, amperes and watts each within plus or minus 2.5 per cent of the values established by the appropriate existing or proposed specifications (See J.)

C. ***Open Circuit Voltage.*** This measurement is necessary only for ballasts containing a transformer.

Series-sequence fluorescent ballasts are so designed that two lamps operate in series. The open circuit voltage of each of the lamp positions should be measured with an operable lamp in the other position.

Fluorescent ballasts for lamps with continuously heated cathodes have four terminals which connect to the lamp. Open circuit voltage should be measured between the two giving the highest voltage with all cathode heating windings loaded with the appropriate dummy resistance load.

Open circuit voltage for high pressure sodium lamps should be measured with the starter inoperative.

D. ***Electrode Heating Voltage.*** On ballasts for use with lamps having continuously heated electrodes, the electrode heating voltages are measured with the electrode windings loaded with the specified dummy load.

E. ***Short Circuit Current (High Intensity Discharge Ballasts).*** An ammeter is inserted in the circuit in place of the lamp, and the short circuit current of the ballast measured.

F. ***Starting Current.***

Fluorescent Instant Start Ballasts. A resistor and ammeter, in series, with a total resistance equivalent to the value specified in the appropriate standard[57] should be used instead of the lamp.

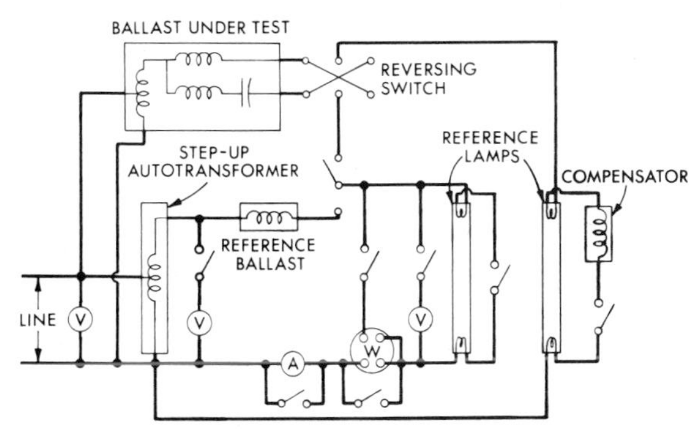

Fig. 4-15. Circuit for comparing fluorescent lamp performance with two different ballasts.

High Intensity Discharge Ballasts. The secondary circuit should be short circuited.

G. ***Electrode Preheating Current (Fluorescent Preheat Ballasts).*** This measurement is made with an ammeter connected in series with the lamp electrodes while the lamp is maintained in the preheat condition.

H. ***Fluorescent Ballast Output.***

Preheat and Instant Start. Specifications are in terms of the power delivered to a Reference Lamp operated by the ballast under test, as compared to the power delivered to the same lamp by the appropriate Reference Ballast.

The general circuit for comparing fluorescent lamp performance on commercial and Reference Ballasts is shown in Fig. 4-15.

Continuously Heated Electrodes. Specifications are in terms of the light output of a Reference Lamp operated by the ballast under test, as compared with the light output of the same Reference Lamp when operated by the appropriate Reference Ballast.

I. ***Fluorescent Ballast Regulation.*** This measurement involves the relative lamp power input and light output at 90 per cent and 110 per cent of rated ballast input voltage.

J. ***Fluorescent Lamp Current.***

Reference Lamps. The current of a Reference Lamp should be measured on both the ballast under test and the Reference Ballast.

Lamps with Continuously Heated Electrodes.[58] Unless the internal connections of the ballasts are accessible, measurement of lamp current requires special instrumentation to provide the vector summation of currents in the two leads to an electrode.

Photometric Measurements of Incandescent Filament Lamps[2,47]

The photometric characteristics of bare incandescent filament lamps are usually determined in conjunction with electrical measurements. Lamp testing procedures, circuits and instrumentation are similar, and the same general precautions should be observed during photometry as in making electrical measurements (see page 4-14). These lamps (except series types) are usually measured at rated voltage.

Reference standard lamps[3,59] may be purchased from the National Bureau of Standards or established laboratories. Lamp standards are usually rated for lumens at a current or voltage a little below their rating to improve the length of burning time without important change in the assigned values. Care should be taken to maintain correct color temperature and/or lamp temperature. Nickel plated bases are normally used on these standards to reduce corrosion and high resistance problems over their long life.

Working standards are usually made by comparison with reference standards. They should have the loops of filament supports closed firmly around the filament to avoid the possibility of random short circuiting of a portion of the filament by the support. They should be adequately seasoned and selected by successive comparisons with reference standards for stability. All standards should be handled carefully to avoid exposure to electrical and mechanical shocks. Exposure to current or voltage above the standardized value may alter lamp ratings. Power should be applied slowly, taking at least 10 seconds to go from zero to rated power.

Lamps may be measured on an optical bench photometer if candlepower in an oriented direction, or mean horizontal candlepower, is desired. Lamps standardized for unidirectional measurements are usually marked to indicate the orientation. A common practice is to inscribe a circle and vertical line on opposite sides of the bulb. The standardized direction is from the circle toward the line, when they are centered on each other, looking toward the receiver.

Most routine photometric measurements of incandescent filament lamps require total light output and are made in a sphere (see Fig. 4–13). Details of operation are described on pages 4–12 and 4–13. Strict substitution procedures should be followed, or corrections made for any unavoidable departures.

A standard should be selected of about the same physical size, lumen output, and color temperature as the lamp under test. The unknown lamp should be measured in the sphere in the same position as the standard.

If photometric readings are to be taken during lamp life, some special precautions should be observed. Readings to determine lamp depreciation are usually taken at 70 per cent rated lamp life. At this time, some blackening of the bulb may have occurred, hence, a blackened unknown lamp replaces a clear standard in the sphere, violating strict substitution. The rather important errors which this may introduce can be avoided by using a third lamp shielded from the light measuring device and the test lamps. This third lamp, commonly called the "absorption," "comparison" or "sub-standard" lamp, is lighted and consecutive measurements made with the test and standard lamps in the test position but not in operation. The difference in the two readings indicates the amount of light absorbed by the blackened test lamp. The same general procedure can be followed in most cases where the characteristics of the integrator are altered during the test by the introduction of light absorbing elements.

Photometric Measurements of Electric Discharge Lamps[2,60]

The photometric characteristics of electric discharge lamps are usually determined in conjunction with electrical measurements. Lamp testing procedures, circuits and instrumentation are similar and the same general precautions should be observed during photometry as in making electrical measurements (see page 4–15). The substitution method is normally employed for photometric measurements. Details concerning photometric measurements will be found in the appropriate IES guides.[53–55]

Photometric Equipment.

A. ***Ballasts.*** When a lamp is being photometered for rating purposes it should be operating on the appropriate Reference Ballast.[57] If no standard exists, the ballast used should comply with the general lamp requirements.[57] Photometric measurements of fluorescent lamps burning on commerical ballasts should be made with the ballast operating at rated input voltage, and measurements of high intensity discharge lamps made with the lamp operating at rated wattage. The ballast should be operated long enough to reach thermal equilibrium. The use of commercial ballasts should conform to the procedures given in the appropriate standards.[57]

B. ***Detectors.*** Detectors used should be selected for linearity and absence of fatigue. Spectral response should be corrected physically, by means of filters or special instrumentation,[61] to the standard spectral luminous efficiency curve (see Fig. 4–4). Spectral calibration data may also be applied to the measured data.

C. ***Standard Lamps.*** These should have characteristics similar to the lamp under test with respect to light output, physical size, shape and spectral distribution.

D. ***Integrating Sphere Photometer.***[39,40] The sphere diameter should be 1.5 meters (5 feet) for high intensity discharge lamps and at least 1.2 times the length of the lamp for straight lamps; the area of the light source should not exceed 2 per cent of the interior surface of the sphere, unless strict substitution methods are

employed. Precautions should be taken to prevent supports and baffles from absorbing light or affecting the lamp temperature. There should be no selective spectral absorption by the sphere paint, sphere window or internal supports. The lamp should be positioned in the center of the sphere, but not remain so long as to raise ambient temperature. See Fig. 4-13. The light measuring device should receive no direct light from the lamp.

There have been advances in the formulation of substances for coating the interiors of spheres —both highly diffuse and substantially uniform reflectance are of importance. Present literature shows much less near ultraviolet and violet absorption, characteristics that become meaningful in the classical Ulbricht equation and implications related to biologically significant radiant energy.[62]

E. ***Candlepower Distribution Photometer.***[63,64] The lamp is mounted in open air with the distance between receiver and lamp at least five times lamp length, or 3 meters (ten feet), whichever is greater. Except as stipulated below, the lamp should be operated in the same burning position as associated with the luminaire and be held stationary during measurement since any movement may disturb stabilization causing erroneous test results.[53-55] Total light output can be computed if the lumen-candlepower ratio is known, or if strict substitution is practiced. The candlepower values of a lamp are established by multiplying the test readings by the photometric calibration constant. Total light output of a lamp is established by summing the products of the candlepower values and the appropriate zonal lumen constants. Care should be taken to exclude stray light, to control ambient temperature and drafts, and to reduce the effects of light absorbing or reflecting materials. For fluorescent measurements the lamp is mounted in a horizontal position and measurements taken normal to the axis of the lamp. To provide the greatest accuracy these measurements must be taken at several points about the axis of the lamp by rotating the lamp around its axis between each set of measurements. For high intensity discharge lamp measurements, especially those of metal halide lamps, the lamp should be placed in its designed operating position. If this is vertical, measurements may be made while the lamp is rotating slowly on its longitudinal axis. Holding the lamp stationary is most desirable for stabilization and accuracy. If the lamp is to be operated in any other position, it must be held stationary during measurement since the light distribution from a high intensity discharge lamp is a function of arc position, which is influenced by gravity.

Color Appearance of Light Sources. For measurement of color appearance of light source, see page 5-17.

Photometric Measurements of Reflector-Type Lamps[65]

For purposes of identification, a reflector-type lamp is defined as a lamp having a reflective coating applied to the reflector part of a bulb, the reflector being specifically contoured for control of the luminous distribution. Included are pressed or blown lamps such as PAR and ER lamps plus other lamps with optically contoured reflectors. Excluded are: lamps of standard bulb shape to which an integral reflector is added, such as silvered-bowl and silvered-neck lamps; lamps designated for special applications, such as automotive headlamps and picture projection lamps, for which special test procedures are already established; lamps having translucent coatings, such as partially phosphorcoated mercury lamps; and reflector fluorescent lamps.

Intensity Distribution. Several different methods of making intensity distribution measurements may be used depending on the type of lamp and the purpose of the test. The *photometric center of lamp* normally should be taken as center of bulb face disregarding any protuberences or recesses in face center, and the *test distance* should be great enough so that the inverse-square law applies. Generally, by convention, the *receptor* should subtend an angle of one square degree.

The intensity distribution of a normally *circular beam* is commonly represented by an average curve in the plane of the *beam axis* (defined as the axis around which the average distribution is substantially symmetrical—beam axis and photometric axis are adjusted to coincide). The curve is obtained either by taking measurements with the lamp rotating about the beam axis, or by averaging a number of curves (at least three) taken in planes at equally spaced azimuthal intervals about the axis.

The intensity distribution of a lamp whose beam is nominally oval or rectangular in cross section is not adequately represented by one average curve. For some lamps two curves through the beam axis, one in the plane of each axis of symmetry, may supply sufficient information. The necessary number of traverses, their distribution within the beam, and the intervals between individual readings vary considerably with the type of lamp; sufficient measurements should be made to adequately describe the average distribution pattern.

When reflector-type lamps are considered for a specific application, test results will be most readily comparable when in the same form as that for equipment used for the same application. For example, when a direct performance comparison of a reflector lamp with floodlighting luminaires is desired, the lamp should be tested according to approved floodlight testing procedures.[66] The same is true for indoor luminaire applications. Although frowned upon by convention, beam patterns that reveal typical nonuniformity are sometimes of paramount interest.

Total Flux Measurements.[67] Total flux may be obtained by direct measurement in an integrating sphere or by calculation from intensity distribution data. Because of the high intensity spot produced by most reflector-type lamps, special precautions should be taken when using an integrating sphere. One possible position for the test lamp in the sphere is with its base close to the sphere wall and the beam aimed through the sphere center, thus distributing the flux over as large an area of the sphere as possible. An appropriate baffle should be placed between the light source and the receptor.

When reflector-type standards are available the calibration of the sphere follows the usual substitution procedure and for maximum accuracy the standard lamps should be of the same type as the test lamps which are used.

Beam and Field Flux. Beam and field flux may be calculated from an average intensity distibution curve or from an isocandela diagram. Of particular interest is the flux contained within the limits of 50 per cent and 10 per cent respectively of the maximum intensity. *Beam angle* is designated as the total angular spread of the cone intercepting the 50 per cent of maximum intensity. *Field angle* is designated as the total angular spread of the cone intercepting the 10 per cent of maximum intensity.

Life Performance Testing of Lamps

Life tests are performed on a very small portion of the product under consideration. Under such conditions test program planning, sampling techniques, and data evaluation become especially important. Helpful guidance can be found in a variety of documents.[68]

It is recognized that it is not practical to test lamps under all of the many variables that occur in service hence, specific reproducible procedures must be included in the test experiment plan.

Incandescent Lamp Life Testing.[69,70] Life tests of incandescent filament lamps may be divided into two classes, *rated-voltage* and *over-voltage tests.*

A. ***Rated Voltage.*** Lamps are operated in the specified burning position at labeled voltage or current held within plus or minus 0.25 per cent of rated value. Sockets should be designed to assure good contact with lamp bases, and the racks should not be subjected to excessive shocks or vibration. If lamps are removed for interim photometric readings, great care should be taken to avoid accidental filament breakage. Sockets should be slightly lubricated because the vibration of a "squeak", as the lamp is removed or replaced, may be sufficient to break a filament which has been rendered brittle by burning.

B. ***Over Voltage Tests.*** Lamp life is shortened by the application of voltage in excess of rated. The mathematical relationship between voltage and lamp life is shown in Section 8. By means of extreme over voltage life testing, sometimes called "high forced testing," lamp life may be shortened so that an evaluation of 1000-hour lamps may be obtained in an eight hour day. The exponents are empirical and require many comparison tests at rated voltages to determine them. This type of accelerated testing introduces additional uncertainties which may cause appreciable errors; hence, the results are only approximate.

Electric Discharge Lamp Life Testing.[71–73] Tests should be made on ac, except when a lamp is designed for dc, and the power supply should have a voltage wave shape in which the harmonic content does not exceed 3 per cent of the fundamental. The line voltage may fluctuate as much as plus or minus 5 per cent from the rated ballast input without any appreciable effect on discharge lamp life. For this and other reasons there is no known method of accelerated life testing. As a result, and because of their long rated life, it normally requires from 18 to 60 months to evaluate life.

There are a number of factors which make the evaluation of life characteristics difficult.

A. ***Auxiliaries.*** Since an electric discharge lamp must be operated with auxiliaries which often affect lamp life they must be selected to conform with the requirements of the appropriate guides, test methods and specifications.

B. ***Test Cycles.*** An on-off cycle is normally employed to "simulate" field conditions. The commonly accepted cycles are 3 hours on and 20 minutes off for fluorescent lamps and 11 hours on and one hour off for high intensity discharge lamps, although others are in use. It it known that increasing the rapidity of this cycling (*i.e.*, burning lamps for a shorter period of time between outages) will materially shorten lamp life, but correlation with the standard cycle is not sufficiently accurate to predict life on the standard cycle.

C. ***Environment.*** The effects of vibration, shock, room temperature, drafts, etc., are extremely variable and unless they are closely controlled a wide deviation in test results may be obtained.

Photometry of Luminaires

The purpose of photometering a luminaire is to accurately determine and report the light distribution and characteristics of the luminaire that will most adequately describe its performance. Characteristics such as candlepower distribution, zonal lumens, efficiency, luminances, beam widths and typing are necessary in designing, specifying and selecting lighting equipment. Photometric data is essential in deriving and developing additional application information.

The information that follows is only to serve as rudimentary guide in outlining the photometry of luminaires. Specific photometric guides and practices are referenced and should be consulted to obtain the detailed testing procedure for each type of luminaire. The "IES General Guide to Photometry,"[2] and the "IES Practical Guide to Photometry,"[44] provide information covering general photometric practices, equipment and related information. Each specific type of luminaire, *e.g.*, indoor lighting, task lighting, floodlighting, streetlighting, etc., require different testing procedures. However, there are several general requirements that should be met in all tests. The luminaire to be tested should be: (1) typical of the unit it is to represent; (2) clean and free of defects (unless it is the purpose of the test to determine the effects of such conditions); (3) equipped with lamps of size and type recommended for use in service; and (4) equipped with the light source in the recommended operating position for service. If the location of the source in a beam-producing luminaire is adjustable, it should be positioned as recommended to obtain such beam as desired in service.

To provide an accurate description of the characteristics of the materials used in the manufacture of a luminaire, measurements should be made of the reflectances of reflecting surfaces where applicable.

Luminaires should be tested in a controlled environment under controlled conditions. The photometric laboratory temperature should be held steady. Typically, for fluorescent photometry, where lamps are sensitive to temperature variations, the room temperature should be held to 25 °C ± 1 °C. Power supplies should be regulated and free of distortion to minimize any effects of line voltage variations. Test rooms should be painted black and/or provided with sufficient baffling to minimize or eliminate extraneous and reflected light during testing.

For accurate measurements, the distance between the luminaire and the light sensor should be great enough that the inverse-square law applies. The minimum test distance is governed by the dimensions of the luminaire. This distance should not be less than 3 meters (ten feet), and at least five times the maximum dimension of the luminous area of the luminaire. For maximum precision the test distance should be measured from the center of the apparent source to the surface of the light sensor. However, from a practical standpoint, the following rules should be followed: (1) for recessed, coffered and totally direct luminaires, the test distance should be measured to the plane of light opening (plane of the ceiling); (2) for luminous sided luminaires the test distance should be measured to the geometric center of the lamps; (3) for suspended luminaires (a) if the light center of the lamp(s) is within the bounds of the reflector and there is no refractor, the test distance should be measured to the plane of the light opening, (b) if the light center of the lamp(s) does not fall within the bounds of the reflector and there is no refractor, the test distance should be measured to the light center of the lamp(s), and (c) if a refractor is attached then the test distance should be measured to the geometric center of the refractor.

General Lighting Luminaires (Candlepower Distribution). For specific information on testing general lighting luminaires, the following IES Guides should be consulted: Photometric Testing of Fluorescent Luminaires,[74] Photometric Testing of Filament Type Luminaires for General Lighting Service,[75] Approved Method for Photometric Testing of Indoor Luminaires

Using High Intensity Discharge Lamps[76] and Reporting General Lighting Equipment Engineering Data.[77]

The basic measurement made in a photometric test of a luminaire is the luminous intensity in specified planes and angles. The resulting candlepower distribution is used to determine zonal lumens, efficiency and average luminances. It is therefore essential that sufficient data be taken to adequately describe the candlepower distribution and the luminaires total light output.

Luminaires having a symmetrical distribution can be photometered in five to twelve equally spaced planes and these planes averaged. Most fluorescent luminaires are photometered in five planes per quadrant in diagonally opposite quadrants and the results of the two quadrants averaged to give the five-plane data. To adequately describe a highly asymmetric luminaire it may be necessary to photometer in planes at ten degrees (or less) intervals.

The distribution in each vertical plane is determined by taking readings at 10-degree (or less) intervals. To facilitate calculation of lumens by zones, candlepower measurements are usually taken at mid-zone angles; *i.e.*, 5°, 15°, 25°, etc. for 10-degree intervals. If the luminaire is of the beam forming type, the candlepower measurements should be made at closer intervals in the beam forming area. If visual comfort probability or ESI calculations are to be made, it is recommended that candlepower measurements be made at least at every five degrees in the vertical plane, and preferably every 2½ degrees.

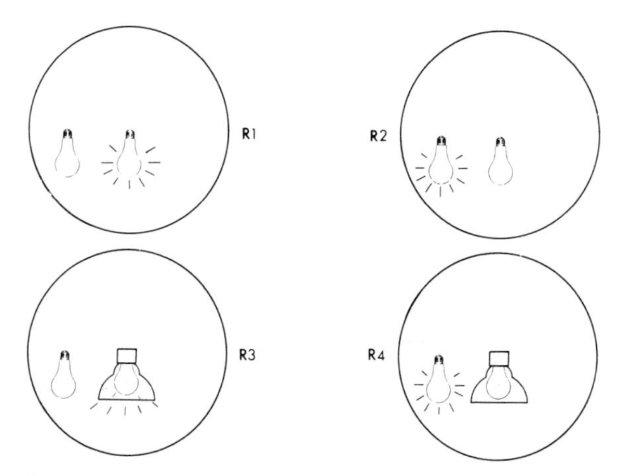

Fig. 4-16. Luminaire and lamp positions within an integrating sphere for the relative method of determining luminaire efficiency. The diameter of the sphere should be at least twice the maximum dimension of the luminaire to be photometered.

Most luminaires are photometered using the relative method. In this method lamps are photometered the same as a luminaire. From these readings a total quasi-lumen value can be calculated using the zonal lumen method. The lumen rating of the lamp is divided by this value to give a constant or factor for that lamp on that photometer. The readings taken on the luminaire can be converted to candlepower values that would be obtained at that point if the lamp were furnishing rated lumens, by multiplying by that constant.

The candlepower distribution data is generally presented in tabular form on the test report

Fig. 4-17. Constants for Use in the Zonal Method of Computing Luminous Flux from Candlepower Data. (1, 2, 5 and 10 Degree Zones)

1 Degree Zones		2 Degree Zones		5 Degree Zones		10 Degree Zones	
Zone Limits (degrees)	Zonal Constant	Zone Limits (degrees)	Zonal Constant	Zone Limits (degrees)	Zonal Constant	Zone Limits (degrees)	Zonal Constant
0–1	0.0009	0–2	0.0038	0–5	0.0239	0–10	0.095
1–2	.0029	2–4	.0115	5–10	.0715	10–20	.283
2–3	.0048	4–6	.0191	10–15	.1186	20–30	.463
3–4	.0067	6–8	.0267	15–20	.1649	30–40	.628
4–5	.0086	8–10	.0343	20–25	.2097	40–50	.774
5–6	.0105	10–12	.0418	25–30	.2531	50–60	.897
6–7	.0124	12–14	.0493	30–35	.2946	60–70	.993
7–8	.0143	14–16	.0568	35–40	.3337	70–80	1.058
8–9	.0162	16–18	.0641	40–45	.3703	80–90	1.091
9–10	.0181	18–20	.0714	45–50	.4041		
				50–55	.4349		
				55–60	.4623		
				60–65	.4862		
				65–70	.5064		
				70–75	.5228		
				75–80	.5351		
				80–85	.5434		
				85–90	.5476		

Fig. 4–18. Constants for Converting Beam Candlepower of Projector Type Luminaires (Searchlights, Floodlights, Spotlights) into Lumens
(0.1 × 0.1 to 10 × 10 degree steps)

Spacing 0.1° Vertical

Horizontal Angle and Setting	K	Horizontal Angle and Setting	K	Horizontal Angle and Setting	K
0.1° Horizontal					
0.05	$0.^{5}3046$	2.75	$0.^{5}3043$	5.45	$0.^{5}3032$
0.15	3046	2.85	3042	5.55	3032
0.25	3046	2.95	3042	5.65	3031
0.35	3046	3.05	3042	5.75	3031
0.45	3046	3.15	3042	5.85	3030
0.55	3046	3.25	3041	5.95	3030
0.65	3046	3.35	3041	6.05	3029
0.75	3046	3.45	3041	6.15	3029
0.85	3046	3.55	3040	6.25	3028
0.95	3046	3.65	3040	6.35	3028
1.05	3046	3.75	3040	6.45	3027
1.15	3046	3.85	3039	6.55	3026
1.25	3045	3.95	3039	6.65	3026
1.35	3045	4.05	3039	6.75	3025
1.45	3045	4.15	3038	6.85	3024
1.55	3045	4.25	3038	6.95	3024
1.65	3045	4.35	3037	7.05	3023
1.75	3045	4.45	3037	7.15	3023
1.85	3045	4.55	3037	7.25	3022
1.95	3044	4.65	3036	7.35	3021
2.05	3044	4.75	3036	7.45	3021
2.15	3044	4.85	3035	7.55	3020
2.25	3044	4.95	3035	7.65	3019
2.35	3044	5.05	3034	7.75	3018
2.45	3043	5.15	3034	7.85	3018
2.55	3043	5.25	3033	7.95	3017
2.65	3043	5.35	3033		
0.2° Horizontal					
0.1	$0.^{5}6092$	2.9	$0.^{5}6085$	5.5	$0.^{5}6065$
0.3	6092	3.1	6084	5.7	6063
0.5	6092	3.3	6083	5.9	6061
0.7	6092	3.5	6081	6.1	6058
0.9	6091	3.7	6080	6.3	6056

Spacing 0.1° Vertical *Continued*

Horizontal Angle and Setting	K	Horizontal Angle and Setting	K	Horizontal Angle and Setting	K
0.2° Horizontal *Continued*					
1.1	6091	3.9	6079	6.5	6054
1.3	6091	4.1	6077	6.7	6051
1.5	6090	4.3	6076	6.9	6049
1.7	6089	4.5	6074	7.1	6046
1.9	6089	4.7	6073	7.3	6044
2.1	6088	4.9	6071	7.5	6041
2.3	6087	5.1	6069	7.7	6038
2.5	6087	5.3	6067	7.9	6035
2.7	6086				
0.4° Horizontal					
0.2	$0.^{4}1219$	3.0	$0.^{4}1217$	5.8	$0.^{4}1212$
0.6	1218	3.4	1216	6.2	1211
1.0	1218	3.8	1216	6.6	1210
1.4	1218	4.2	1215	7.0	1210
1.8	1218	4.6	1215	7.4	1208
2.2	1217	5.0	1214	7.8	1207
2.6	1217	5.4	1213		
0.6° Horizontal					
0.3	$0.^{4}1827$	3.3	$0.^{4}1825$	5.7	$0.^{4}1819$
0.9	1827	3.9	1824	6.3	1816
1.5	1827	4.5	1822	6.9	1814
2.1	1827	5.1	1821	7.5	1810
2.7	1826				
0.8° Horizontal					
0.4	$0.^{4}2437$	3.6	$0.^{4}2432$	6.0	$0.^{4}2424$
1.2	2436	4.4	2429	6.8	2420
2.0	2435	5.2	2427	7.6	2416
2.8	2434				
1.0° Horizontal					
0.5	$0.^{4}3046$	3.5	$0.^{4}3041$	6.5	$0.^{4}3027$
1.5	3045	4.5	3037	7.5	3020
2.5	3043	5.5	3032		

Spacing 0.2° Vertical

Horizontal Angle and Setting	K	Horizontal Angle and Setting	K	Horizontal Angle and Setting	K
0.2° Horizontal					
0.1	$0.^{4}1219$	2.9	$0.^{4}1217$	5.5	$0.^{4}1213$
0.3	1219	3.1	1217	5.7	1212
0.5	1218	3.3	1217	5.9	1212
0.7	1218	3.5	1216	6.1	1212
0.9	1218	3.7	1216	6.3	1211
1.1	1218	3.9	1216	6.5	1211
1.3	1218	4.1	1215	6.7	1210
1.5	1218	4.3	1215	6.9	1210
1.7	1218	4.5	1215	7.1	1209
1.9	1218	4.7	1214	7.3	1209
2.1	1218	4.9	1214	7.5	1208
2.3	1218	5.1	1214	7.7	1208
2.5	1217	5.3	1213	7.9	1207
2.7	1217				
0.4° Horizontal					
0.2	$0.^{4}2437$	3.0	$0.^{4}2434$	5.8	$0.^{4}2425$
0.6	2437	3.4	2433	6.2	2423
1.0	2437	3.8	2432	6.6	2421
1.4	2436	4.2	2430	7.0	2419
1.8	2436	4.6	2429	7.4	2417
2.2	2435	5.0	2428	7.8	2414
2.6	2434	5.4	2426		
0.6° Horizontal					
0.3	$0.^{4}3655$	3.3	$0.^{4}3649$	5.7	$0.^{4}3637$
0.9	3655	3.9	3647	6.3	3633
1.5	3654	4.5	3644	6.9	3629
2.1	3653	5.1	3641	7.5	3624
2.7	3651				
0.8° Horizontal					
0.4	$0.^{4}4874$	3.6	$0.^{4}4864$	6.0	$0.^{4}4848$
1.2	4872	4.4	4858	6.8	4840
2.0	4870	5.2	4854	7.6	4832
2.8	4868				

Fig. 4–18. *Continued*

Horizontal Angle and Setting	K	Horizontal Angle and Setting	K	Horizontal Angle and Setting	K
Spacing 0.2° Vertical *Continued*					
1.0° HORIZONTAL					
0.5	$0.^{4}6092$	3.5	$0.^{4}6082$	6.5	$0.^{4}6054$
1.5	6090	4.5	6074	7.5	6040
2.5	6086	5.5	6064		
Spacing 0.4° Vertical					
0.2° HORIZONTAL					
0.1	$0.^{4}2437$	2.9	$0.^{4}2434$	5.5	$0.^{4}2426$
0.3	2437	3.1	2434	5.7	2425
0.5	2437	3.3	2433	5.9	2424
0.7	2437	3.5	2432	6.1	2423
0.9	2437	3.7	2432	6.3	2422
1.1	2437	3.9	2431	6.5	2421
1.3	2436	4.1	2431	6.7	2420
1.5	2436	4.3	2430	6.9	2419
1.7	2436	4.5	2430	7.1	2418
1.9	2436	4.7	2429	7.3	2417
2.1	2435	4.9	2428	7.5	2416
2.3	2435	5.1	2427	7.7	2415
2.5	2434	5.3	2426	7.9	2414
2.7	2434				
0.4° HORIZONTAL					
0.2	$0.^{4}4874$	3.0	$0.^{4}4867$	5.8	$0.^{4}4849$
0.6	4874	3.4	4865	6.2	4846
1.0	4874	3.8	4863	6.6	4842
1.4	4872	4.2	4861	7.0	4838
1.8	4872	4.6	4858	7.4	4834
2.2	4870	5.0	4855	7.8	4829
2.6	4869	5.4	4852		
0.6° HORIZONTAL					
0.3	$0.^{4}7310$	3.3	$0.^{4}7299$	5.7	$0.^{4}7274$
0.9	7310	3.9	7294	6.3	7267
1.5	7308	4.5	7288	6.9	7258
2.1	7306	5.1	7282	7.5	7248
2.7	7302				

Horizontal Angle and Setting	K	Horizontal Angle and Setting	K	Horizontal Angle and Setting	K
Spacing 0.4° Vertical *Continued*					
0.8° HORIZONTAL					
0.4	$0.^{4}9747$	3.6	$0.^{4}9728$	6.0	$0.^{4}9694$
1.2	9746	4.4	9719	6.8	9679
2.0	9742	5.2	9707	7.6	9662
2.8	9736				
1.0° HORIZONTAL					
0.5	$0.^{3}1218$	3.5	$0.^{3}1216$	6.5	$0.^{3}1211$
1.5	1218	4.5	1215	7.5	1208
2.5	1217	5.5	1212		
Spacing 0.6° Vertical					
0.2° HORIZONTAL					
0.1	$0.^{4}3655$	2.9	$0.^{4}3650$	5.5	$0.^{4}3638$
0.3	3655	3.1	3650	5.7	3637
0.5	3655	3.3	3649	5.9	3636
0.7	3655	3.5	3649	6.1	3635
0.9	3655	3.7	3648	6.3	3634
1.1	3655	3.9	3647	6.5	3632
1.3	3654	4.1	3646	6.7	3631
1.5	3654	4.3	3645	6.9	3629
1.7	3654	4.5	3644	7.1	3628
1.9	3653	4.7	3643	7.3	3626
2.1	3653	4.9	3642	7.5	3625
2.3	3653	5.1	3641	7.7	3622
2.5	3652	5.3	3640	7.9	3621
2.7	3652				
0.4° HORIZONTAL					
0.2	$0.^{4}7310$	3.0	$0.^{4}7301$	5.8	$0.^{4}7273$
0.6	7310	3.4	7299	6.2	7268
1.0	7310	3.8	7295	6.6	7262
1.4	7308	4.2	7292	7.0	7256
1.8	7307	4.6	7288	7.4	7250
2.2	7306	5.0	7283	7.8	7243
2.6	7303	5.4	7278		

Horizontal Angle and Setting	K	Horizontal Angle and Setting	K	Horizontal Angle and Setting	K
Spacing 0.6° Vertical *Continued*					
0.6° HORIZONTAL					
0.3	$0.^{3}1097$	3.3	$0.^{3}1095$	5.7	$0.^{3}1091$
0.9	1097	3.9	1094	6.3	1090
1.5	1096	4.5	1093	6.9	1089
2.1	1096	5.1	1092	7.5	1087
2.7	1095				
0.8° HORIZONTAL					
0.4	$0.^{3}1462$	3.6	$0.^{3}1459$	6.0	$0.^{3}1454$
1.2	1462	4.4	1458	6.8	1452
2.0	1461	5.2	1456	7.6	1449
2.8	1460				
1.0° HORIZONTAL					
0.5	$0.^{3}1828$	3.5	$0.^{3}1824$	6.5	$0.^{3}1816$
1.5	1827	4.5	1822	7.5	1812
2.5	1826	5.5	1819		
Spacing 0.8° Vertical					
0.2° HORIZONTAL					
0.1	$0.^{4}4874$	2.9	$0.^{4}4867$	5.5	$0.^{4}4851$
0.3	4874	3.1	4867	4.7	4850
0.5	4874	3.3	4866	5.9	4848
0.7	4874	3.5	4865	6.1	4846
0.9	4874	3.7	4864	6.3	4845
1.1	4874	3.9	4862	6.5	4842
1.3	4872	4.1	4862	6.7	4841
1.5	4872	4.3	4860	6.9	4838
1.7	4872	4.5	4859	7.1	4837
1.9	4871	4.7	4858	7.3	4834
2.1	4870	4.9	4856	7.5	4833
2.3	4870	5.1	4854	7.7	4830
2.5	4869	5.3	4853	7.9	4828
2.7	4869				
0.4° HORIZONTAL					
0.2	$0.^{4}9747$	3.0	$0.^{4}9734$	5.8	$0.^{4}9698$
0.6	9747	3.4	9730	6.2	9691

Horizontal Angle and Setting	K	Horizontal Angle and Setting	K	Horizontal Angle and Setting	K
Spacing 0.8° Vertical *Continued*					
0.4° HORIZONTAL *Continued*					
1.0	9747	3.8	9726	6.6	9683
1.4	9744	4.2	9722	7.0	9675
1.8	9743	4.6	9717	7.4	9667
2.2	9741	5.0	9710	7.8	9658
2.6	9738	5.4	9704		
0.6° HORIZONTAL					
0.3	$0.^{3}1462$	3.3	$0.^{3}1460$	5.7	$0.^{3}1455$
0.9	1462	3.9	1459	6.3	1453
1.5	1462	4.5	1458	6.9	1452
2.1	1461	5.1	1456	7.5	1450
2.7	1460				
0.8° HORIZONTAL					
0.4	$0.^{3}1949$	3.6	$0.^{3}1946$	6.0	$0.^{3}1939$
1.2	1949	4.4	1944	6.8	1936
2.0	1948	5.2	1941	7.6	1932
2.8	1947				
1.0° HORIZONTAL					
0.5	$0.^{3}2437$	3.5	$0.^{3}2432$	6.5	$0.^{3}2421$
1.5	2436	4.5	2429	7.5	2416
2.5	2435	5.5	2426		
Spacing 1.0° Vertical					
0.1° HORIZONTAL					
0.05	$0.^{4}3046$	2.75	$0.^{4}3043$	5.45	$0.^{4}3032$
0.15	3046	2.85	3042	5.55	3032
0.25	3046	2.95	3042	5.65	3031
0.35	3046	3.05	3042	5.75	3031
0.45	3046	3.15	3042	5.85	3030
0.55	3046	3.25	3041	5.95	3030

Horizontal Angle and Setting	K	Horizontal Angle and Setting	K	Horizontal Angle and Setting	K
Spacing 1.0° Vertical *Continued*					
0.1° HORIZONTAL *Continued*					
0.65	3046	3.35	3041	6.05	3029
0.75	3046	3.45	3041	6.15	3029
0.85	3046	3.55	3040	6.25	3028
0.95	3046	3.65	3040	6.35	3028
1.05	3046	3.75	3040	6.45	3027
1.15	3046	3.85	3039	6.55	3026
1.25	3045	3.95	3039	6.65	3026
1.35	3045	4.05	3039	6.75	3025
1.45	3045	4.15	3038	6.85	3024
1.55	3045	4.25	3038	6.95	3024
1.65	3045	4.35	3037	7.05	3023
1.75	3045	4.45	3037	7.15	3023
1.85	3045	4.55	3037	7.25	3022
1.95	3044	4.65	3036	7.35	3021
2.05	3044	4.75	3036	7.45	3021
2.15	3044	4.85	3035	7.55	3020
2.25	3044	4.95	3035	7.65	3019
2.35	3044	5.05	3034	7.75	3018
2.45	3043	5.15	3034	7.85	3018
2.55	3043	5.25	3033	7.95	3017
2.65	3043	5.35	3033		
0.2° HORIZONTAL					
0.1	$0.^{4}6092$	2.9	$0.^{4}6085$	5.5	$0.^{4}6065$
0.3	6092	3.1	6084	5.7	6063
0.5	6092	3.3	6083	5.9	6061
0.7	6092	3.5	6081	6.1	6058
0.9	6092	3.7	6080	6.3	6056
1.1	6092	3.9	6079	6.5	6054
1.3	6090	4.1	6077	6.7	6051
1.5	6090	4.3	6076	6.9	6049
1.7	6089	4.5	6074	7.1	6046
1.9	6089	4.7	6073	7.3	6044

Horizontal Angle and Setting	K	Horizontal Angle and Setting	K	Horizontal Angle and Setting	K
Spacing 1.0° Vertical *Continued*					
0.2° HORIZONTAL *Continued*					
2.1	6088	4.9	6071	7.5	6041
2.3	6087	5.1	6069	7.7	6038
2.5	6087	5.3	6067	7.9	6035
2.7	6086				
0.4° HORIZONTAL					
0.2	$0.^{3}1219$	3.0	$0.^{3}1217$	5.8	$0.^{3}1212$
0.6	1218	3.4	1216	6.2	1211
1.0	1218	3.8	1216	6.6	1210
1.4	1218	4.2	1215	7.0	1210
1.8	1218	4.6	1215	7.4	1208
2.2	1217	5.0	1214	7.8	1207
2.6	1217	5.4	1213		
0.6° HORIZONTAL					
0.3	$0.^{3}1828$	3.3	$0.^{3}1825$	5.7	$0.^{3}1819$
0.9	1828	3.9	1824	6.3	1816
1.5	1827	4.5	1822	6.9	1814
2.1	1827	5.1	1821	7.5	1810
2.7	1826				
0.8° HORIZONTAL					
0.4	$0.^{3}2437$	3.6	$0.^{3}2432$	6.0	$0.^{3}2424$
1.2	2436	4.4	2430	6.8	2420
2.0	2436	5.2	2427	7.6	2416
2.8	2434				
1.0° HORIZONTAL					
0.5	$0.^{3}3046$	3.5	$0.^{3}3041$	6.5	$0.^{3}3027$
1.5	3045	4.5	3037	7.5	3020
2.5	3043	5.5	3032		
Spacing 2° Vertical					
2° HORIZONTAL					
1	$0.^{2}122$	31	$0.^{2}104$	61	$0.^{3}59$

Fig. 4-18. *Continued*

Horizontal Angle and Setting	K	Horizontal Angle and Setting	K	Horizontal Angle and Setting	K
Spacing 0.2° Vertical *Continued*					
0.2° HORIZONTAL *Continued*					
3	122	33	102	63	55
5	121	35	100	65	51
7	121	37	097	67	48
9	120	39	095	69	44
11	120	41	092	71	40
13	119	43	089	73	36
15	118	45	086	75	32
17	116	47	083	77	27
19	115	49	080	79	23
21	114	51	077	81	19
23	112	53	073	83	15
25	110	55	070	85	11
27	108	57	066	87	06
29	107	59	063	89	02
5° HORIZONTAL					
2.5	$0.^{2}3046$	32.5	$0.^{2}2570$	62.5	$0.^{2}1406$

Horizontal Angle and Setting	K	Horizontal Angle and Setting	K	Horizontal Angle and Setting	K
Spacing 0.2° Vertical *Continued*					
5° HORIZONTAL *Continued*					
7.5	3020	37.5	2416	67.5	1166
12.5	2970	42.5	2246	72.5	0918
17.5	2906	47.5	2060	77.5	658
22.5	2814	52.5	1856	82.5	396
27.5	2702	57.5	1638	87.5	134
10° HORIZONTAL					
5	$0.^{2}6066$	35	$0.^{2}4986$	65	$0.^{2}2572$
15	5876	45	4306	75	1576
25	5576	55	3494	85	0530
Spacing 5° Vertical					
5° HORIZONTAL					
2.5	$0.^{2}760$	32.5	$0.^{2}642$	62.5	$0.^{2}352$

Horizontal Angle and Setting	K	Horizontal Angle and Setting	K	Horizontal Angle and Setting	K
Spacing 5° Vertical *Continued*					
5° HORIZONTAL *Continued*					
7.5	755	37.5	604	67.5	291
12.5	744	42.5	562	72.5	229
17.5	726	47.5	514	77.5	165
22.5	704	52.5	463	82.5	099
27.5	676	57.5	409	87.5	033
10° HORIZONTAL					
5	0.015165	35	0.012465	65	$0.^{2}6430$
15	14690	45	10765	75	3940
25	13790	55	08735	85	1325
5	0.0304	35	0.0249	65	0.0129
15	294	45	214	75	076
25	276	55	174	85	026

* Note: Small numbers following the decimal point indicate number of zeros following the decimal point but before the numbers shown.

sheet. Data for a lens or indoor luminaire is given in five planes. Three distributions are usually presented in graphical or polar distribution curve form, and these three vertical planes being: parallel to the lamps, 45 degrees to parallel, and perpendicular to the lamps.

General Lighting Luminaires (Luminance). Either before or after the photometric test, while the lamps are installed and stabilized, the maximum luminance of the luminaire should be measured at the angles specified in the appropriate guide and at the shielding angles. The measurements may be in candelas per square meter, candelas per square inch or footlamberts. The readings should be taken both crosswise and lengthwise in the case of fluorescent luminaires or luminaires with an asymmetric distribution. The projected area of the measurement field should be circular and of 6.45 square centimeters (one square inch area). Care must be taken that luminance measurements are related to the lumen output of the lamps, and thus luminance measuring instrument must be calibrated against the test lamps, employing the fundamentals of the techniques described above.

If average luminance values are desired, they can be calculated by using the candlepower measurements obtained from the test data. By definition, luminance is the luminous intensity (candlepower) of any surface in a given direction per unit of projected area of the surface viewed from that direction. See Reference 78.

General Lighting Luminaires (Total Lumen Output). The total light output of the luminaire, needed to establish its efficiency in terms of the lumen output of lamps with which it is equipped, can be determined in a spherical integrating photometer or by computations from the candlepower distribution data.

If it is to be measured in a sphere, the efficiency can be determined by the relative method. See Fig. 4-16. First lamps are mounted in the center of the sphere and a reading taken. A reading is then taken on a lamp mounted at some other point in the sphere. The luminaire is then mounted in the center of the sphere and a reading is taken on the luminaire. Then another reading is taken on the other lamp mounted in the sphere. The efficiency is then calculated as follows:

$$\text{Per Cent Efficiency} = \frac{R_3 \times R_2}{R_1 \times R_4} \times 100$$

where
R_1 = Reading of lamp(s) in center of sphere
R_2 = Auxiliary lamp reading
R_3 = Luminaire reading
R_4 = Auxiliary lamp reading with luminaire in the sphere

It is understood that while a reading is being taken the other lamp must be extinguished. The sphere method is not considered to be normally as accurate as the method using candlepower distribution data, described below.

Candlepower distribution data is used to compute the luminous flux in any angular zone from nadir to 180 degrees. The product of the mid-zone candlepower and the zonal constant gives the zonal lumens. The summation of the zonal lumens multiplied by 100 and divided by the nominal lamp lumens gives the per cent efficiency.

Calculation of Luminaire Lumens from Candlepower Data. Constants useful in calculating luminous flux from candlepower data are given in Figs. 4-17 and 4-18. For computing luminous flux, multiply the average candlepower at the center of each zone by the zonal constant (see Fig. 4-17) equal to $2\pi(\cos\theta_1 - \cos\theta_2)$.

The constants in Fig. 4-18 are computed for candlepower measurements of projector-type luminaires made on a goniometer of the type shown in Fig. 4-9b. If the measurements have been made with the type as shown in Fig. 4-9a the constants may be used by interchanging the vertical and horizontal angular arguments, *i.e.*, by substituting the word "vertical" wherever "horizontal" appears. The zonal constants for Fig. 4-18 were computed as equal to $\phi\pi(\sin\theta_2 - \sin\theta_1)/180$, where ϕ is the vertical interval and θ_1 and θ_2 are the limits of the horizontal interval. If a goniometer of the type in Fig. 4-9a is used the constant is equal to $\theta\pi(\sin\phi_2 - \sin\phi_1)/180$, where θ is the horizontal interval and ϕ_1 and ϕ_2 are the limits of the vertical interval. See Reference 62 for additional zonal constants.

If a number of constants are to be calculated for the same interval, the following shortcut method is a time-saver but is completely accurate.

For the first formula above, let θ_m be the mid-zone angle and let P equal ½ of the zone interval. The formula becomes:

$$4\pi \sin P\ (\sin\theta_m)$$

Since the zone width is often the same for a series of constants, the first term will simply be multiplied successively by the sines of the mid-zone angles.

For the second formula above, let θ_m be the

median angle on the horizontal interval. The formula then becomes:

$$2\pi \frac{\theta}{180} \sin P\ (\cos\theta_m)$$

Floodlight-Type Luminaires. The following applies to floodlighting equipment having a total beam spread (divergence) of more than 10 degrees. For specific information on testing this type of equipment, consult the "IES Approved Method for Photometric Testing of Floodlights Using Incandescent Filament or Discharge Lamps."[66] For lighting equipment having a beam spread less than 10 degrees, see the "IES Guide for Photometric Testing of Searchlights."[79]

The classification of floodlights is based upon the beam width of the floodlight on the horizontal and vertical axes of the floodlight. The classification is designated by NEMA type numbers.[80] For symmetrical beams the floodlight type is defined as the average of the horizontal and vertical beam spreads. For asymmetrical beam floodlights the type is designated by the horizontal and vertical beam spreads in that order, *e.g.*, a floodlight with a horizontal beam spread of 75 degrees (Type 5) and a vertical beam of 35 degrees (Type 3) would be designated as a Type 5 × 3 floodlight.

Stray light may be defined as that light emitted by the floodlight which is outside the floodlight beam as defined by the beam classification. In some instances stray light may be useful in illumination or detrimental to vision depending upon its magnitude and direction. When it is desired to determine the amount and direction of stray light, it is necessary to make measurements as far horizontally and vertically as the readings have significant values in relation to the measuring system.

If the light center of the test lamp (if more than one lamp, the geometric center of lamp light centers) is not enclosed by the reflector, the floodlight should be mounted on the goniometer so that the light center of the lamp is at the goniometer center. If the lamp light center is within the reflector, the floodlight should be positioned so that the center of the reflector opening coincides with the goniometer center.

Either the direct or the relative method of photometry may be used for floodlights, but the relative method lends itself particularly well since cumulative errors may be reduced, and maintenance of standards of luminous intensity and flux are not necessary. In the latter method, relative candlepower readings of the test lamp alone made on a distribution photometer, and of the lamp-floodlight combination made on the floodlight photometer are taken with the lamp operating under identical electrical conditions in both tests.

Beams produced by projector-type luminaires utilizing specular reflectors and filament lamps are likely to be non-uniform in intensity. To average out such variations in the beams, an integrating device should be used which will integrate the illumination over one square degree [524 millimeters by 524 millimeters at 30-meter test distance (20.94 inches by 20.94 inches at 100-foot test distance)] or over a circular degree at a distance of 30 meters (100 feet).

The method of taking candlepower readings is to traverse the beam with such angular spacings as to give approximately 100 reading stations uniformly spaced throughout the beam. For beam limit definition see Reference 81. By interpolating between these readings an isocandela diagram may be plotted on rectangular co-ordinates (see Fig. 9-53, page 9-45). The lumens in the beam may be computed using the constants in Fig. 4-18.

The information usually reported for floodlights includes the following: NEMA type, horizontal and vertical beam distribution curves, maximum beam luminous intensity, average maximum beam luminous intensity, beam spread in both horizontal and vertical directions, beam flux, beam efficiency, total floodlight flux and total efficiency. In addition, the report should indicate whether the data was obtained on a type A or a type B goniometer.[81]

Roadway Luminaires. A guide has been prepared to provide test procedures and methods of reporting data to promote the uniform evaluation of the optimal performance of roadway luminaires using incandescent filament and high intensity discharge lamps.[82]

Luminaires selected for test should be representative of the manufacturer's typical product. A test distance of 7.6 meters (25 feet) should be sufficient for most beam forming luminaires. Photometric test distance is, in general, defined as the distance from the goniometer center to the surface of the test plate or photosensitive element, taking into account the distance to and from any mirror or mirrors that may be used.

The number of planes explored during photometric measurements should be determined by the symmetry or irregularity of distribution and by the end results desired from the test. The number of vertical angles at which readings are taken will depend on how the readings are to be used. If an isocandela diagram is to be plotted, readings may have to be taken at close intervals, especially if values are changing rapidly. Increased usage of computers to provide comprehensive evaluation of luminaires and lighting application designs requires that readings be taken at vertical angle intervals through the beam section not exceeding 2½ degrees.

For luminaires having a distribution that is symmetric about a vertical axis (IES Type V), readings may be taken in one vertical plane while the entire luminaire is being rotated at a speed of 60 rpm or less. The lamp may be rotated within a stationary optical assembly provided readings are taken in not less than ten vertical planes and averaged. The performance of metal halide lamps may suffer from rotation and consideration should be given to photometry of these lamps without rotation. If neither luminaire nor lamp is rotated, average of the readings taken in ten equally spaced vertical planes should suffice.

For luminaires having a distribution that is symmetric about a single vertical plane (IES Types II, III, IV and II 4-way) readings may be taken in vertical planes that are ten degrees apart. Due to the method used in data processing, it may be advantageous to divide laterally into 10 degree zones and measure at the mid-zone angle. Averages may be made of the readings taken at corresponding angles on the opposite sides of the plane of symmetry. Any computations that are to be performed may then be done on one side of the plane of symmetry, using averaged data.

For luminaires having a distribution that is symmetric about two vertical planes (IES Type I), readings may be taken as directly above, but computations may be performed in one quadrant of the sphere.

For luminaires having a distribution that is symmetric about four vertical planes (IES Type I 4-way), readings may be taken as above, but computations may be performed in one octant of the sphere.

For luminaires having an asymmetric type distribution, readings may be taken in vertical planes that are ten degrees apart. Since there is no symmetry, any computations performed should be done without averaging.

Sufficient data should be obtained to allow classification of the light distribution in accordance with recommended practice (see the Application Volume) as well as to provide an isolux (isofootcandle) diagram, utilization efficiency, and the total and four quadrant efficiencies.

Luminaire-Lamp-Ballast Operating Factor. It is commonly assumed that a ballast operated at its rated input voltage delivers rated wattage to a lamp and that a lamp operated at its rated wattage delivers its rated lumen output. In many instances this may not be the case. Therefore, a procedure has been developed[83] to provide a factor to be applied to high intensity discharge luminaire photometric data to adjust them to the specific combination of luminaire, lamp type and ballast used in a system. By repetitive tests, this procedure may be used to determine variations of system performance exclusive of lamp variations. Such a factor can be applied specifically to lumens, candelas, and lux (footcandles) as they appear on photometric data sheets.

Two possibilities are recognized: the first, (LLB_1) in which the lamp is used in the operating position for which it is rated; the second (LLB_2), in which the lamp is operated in a position other than the one for which it is rated. The factors determined relate to equipment (luminaire, lamp, ballast combination) operating factor under initial conditions unless otherwise specified.

The procedures essentially involve a relative light output measurement using the selected lamp in the test luminaire with its test ballast operated at rated supply voltage, after operating conditions have stabilized, and a relative light output measurement with the lamp switched to a *reference ballast* without being extinguished, holding the rated input voltage specified for the *reference ballast*. The *equipment operating factor* is the ratio of the light output measurements made in the first measurement to that made in the second.

Projector Photometry[79]

The illuminance from searchlights, beacons or other highly collimating luminaires, if measured at distances greater than a certain minimum,

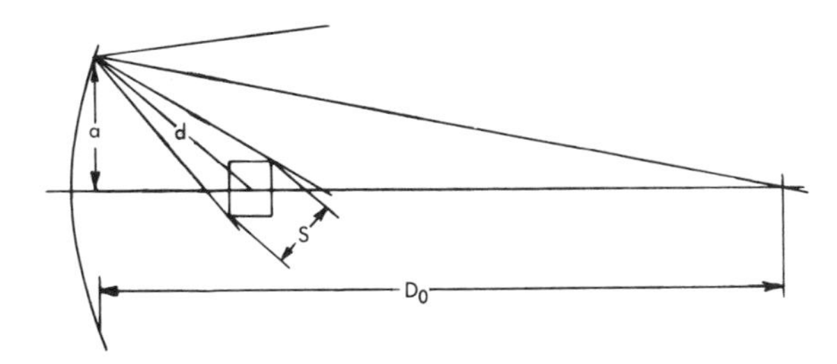

Fig. 4-19. Diagram showing distances and dimensions used to determine minimum inverse-square distance.

Fig. 4-20. Typical goniometer for indoor photometric range.[86] Two rotary tables of the type used in large machine tools are incorporated into the goniometer to provide two of the three rotations available. For horizontal distributions either the rotary table on which the outer frame is mounted or the inner table on which the test equipment is mounted may be used. This makes for flexibility and is especially useful for obtaining polar angle distributions. The pinion gear in the vertical drive may be disengaged readily and the inner frame with the test unit mounted can then be balanced with the adjustable counterweights. After balancing, a small constant torque is applied to the inner frame through the use of the pulley and weight at the side, thus eliminating backlash.

obeys the inverse-square law. This minimum distance is a function of the focal length of the reflector, the diameter of the reflector aperture and the diameter of the smallest element of the light source (arc stream or filament). This minimum distance point is called the "beam crossover" point and is the point where the optic is seen to be completely flashed. Only at distances greater than this does the inverse-square law apply. The minimum inverse-square distance may be calculated by using the following general formula:

$$L_0 = \frac{ad}{Ks}$$

where (see Fig. 4-19)

L_0 = Minimum distance for optic under consideration

a = Distance from the optical axis to the outermost flashed point when viewed from a distant point on the optical axis.

d = Distance from centroid of light source to same outermost flashed point as used to determine a.

s = Diameter of smallest element of light source. For example, the arc stream width of an arc source or one coil of a multi-coil filament lamp.

K = Constant equal to 500 when a, d, and s are in millimeters and L_0 is in meters; and 6 when a, d and s are in inches and L_0 is in feet.

The above calculation determines the minimum inverse-square distance based on ideal light sources and axial measurements and, therefore, should be considered approximate. In practice, the range used should be much in excess of this calculated distance to insure conformance. Methods for using shorter ranges and "zero length photometry" have been devised, however, the full length range is preferable for highest accuracy. These methods and a fuller discussion of minimum inverse-square distance calculations may be found in References 79, 84 and 85.

Goniometers. The usual method for measuring angles in candlepower distribution measurements is to mount the test equipment on a goniometer, keeping the photometric equipment fixed.[86] Basically goniometers provide means for mounting the equipment, for rotating it around the two axes (horizonal and vertical) and for measuring the angles of rotation. (See Figs. 4-9 and 4-20).

When the searchlight is of unusual size and weight it may be necessary to utilize its own mounting and goniometric facilities for the photometric work or to hold the searchlight fixed and traverse the beam by moving the photometric receptor.

Since searchlights may have a total beam spread of less than one degree and may furthermore be of unusual weight, the mechanical requirements of the goniometer are severe. Suitable rigidity, freedom from backlash, and accuracy of angular measurements are prime requirements. There should be provision for accurate angular settings of the order of 0.1 degree. In special cases higher orders of accuracy may be required.

Photometric Ranges. Indoor ranges are in general preferable, but are frequently impracticable because of the lengths required. For photometry on the relatively shorter indoor ranges, proper photometric procedures must be followed. See References 44, 79 and 87. Outdoor ranges suitable for daytime measurements require much more attention to methods of reducing stray light, minimizing atmospheric disturbances and correcting for atmospheric transmission. Range sites should be selected where the terrain is flat and homogeneous. The range should be as high off the ground as practicable. Ranges should not be located where atmospheric disturbances occur regularly or where dust or moisture are prevalent, and corrections must be made for other than perfect atmospheric transmission. Stray light should be minimized by a suitable system of diaphragms. Any remaining stray light should be measured and subtracted from all readings.

Correction for Atmospheric Transmission. The absorption of light by moisture, smoke and dust particles, even in an apparently clear atmosphere, may introduce considerable errors in measurements.[88] Therefore, it is desirable to measure the atmospheric transmittance before and after the test has been made. A standard reference projector is frequently employed, and is calibrated either by repeated observations in the clearest weather, when the atmospheric transmittance may be accurately estimated or independently measured, or by laboratory measurement methods.

Photometers for Searchlight Measurements.[84] The equipment required for photometric measurements is basically similar in most respects to that required for other types of photometric measurements.

Automatic recording photometers are frequently employed in order to obtain rapid measurements or to obtain statistically valid quantities of data. Multi-element oscillographs are particularly valuable since they facilitate simultaneous recording of searchlight parameters, such as voltage and current, along with the basic photometric data.

Luminance Measurements[89]

Lamps and Luminaires. Luminance measurements of lamps and luminaires should be made by either the absolute or relative method. With the absolute method, reference standards must be available for equipment calibration. In practice the relative method is generally used.

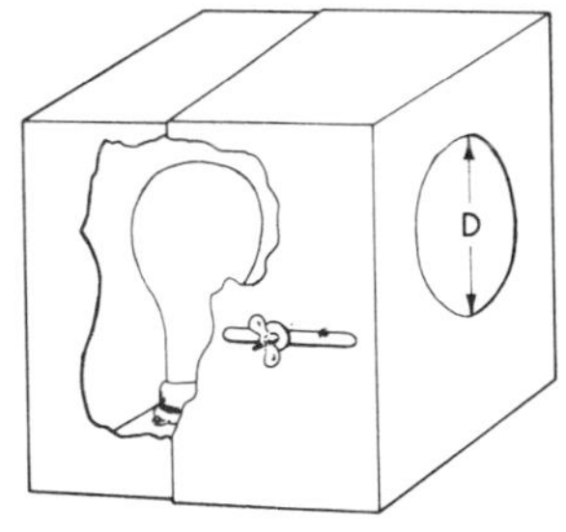

Fig. 4-22. An adjustable source of luminance, useful when calibrating luminance measuring devices having apertures of different shapes. D is about 76 millimeters (3 inches).

The published luminances of non-asymmetric fluorescent lamps are computed from the rated lumen output of the lamp according to the following formula:[90]

$$L_{avg} \text{ (candelas per square meter or candelas per square inch)} = \frac{K \times \text{Total Lamp Lumens}}{\text{Lamp Diameter} \times \text{Lamp Luminous Length}}$$

where L_{avg} is the average luminance of the full width of the lamp at its center. The diameter and length are expressed in meters or inches and K is:

for 1.22-meter (48-inch) T-12 lamps $K = \dfrac{1.09}{9.25}$

1.83-meter (72-inch) T-12 lamps $K = \dfrac{1.07}{9.28}$

2.44-meter (96-inch) T-12 lamps $K = \dfrac{1.05}{9.28}$

To compute approximate luminance L_θ of a fluorescent lamp at any angle with the lamp axis, the following formula is used:

$$L_\theta = \frac{\text{Total Lamp Lumens}}{K_\theta \times \text{Lamp Diameter} \times \text{Lamp Luminous Length} \times \sin\theta}$$

where K_θ is as shown in Fig. 4-21.

In the laboratory, an accurate means of establishing lamp luminance is secured by construction of a baffled collimating tube 0.6 to 1.0 meters

Fig. 4-21. Values at Various Angles of the Lamp: Candlepower Ratio K_θ for Preheat-Starting Types of Fluorescent Lamps*

Angle (degrees)	0	10	20	30	40	50	60	70	80	90
K_θ	—	172.0	46.0	24.7	17.1	13.3	11.3	10.1	9.5	9.25

* Average for 15-, 20-, 30-, 40-, and 100-watt lamps.

(2 to 3 feet) long, with a receptor at one end, and at the other end a rectangular aperture having one dimension equal to the lamp diameter, and the second dimension such that the product of the two dimensions equals 645 square millimeters (one square inch).[91] This is placed against the lamp near its midpoint so that the entire diameter of the lamp fills the aperture. The deflection of the indicating equipment is directly proportional to L_{avg}. The instrument is thus calibrated when a lamp of known lumen output is viewed, and direct comparisons can be made with lamps of similar diameters. If the lumen output is assumed to be the rated value, the instrument is calibrated for the relative method so that all subsequent luminance measurements are related to rated lumen output. Techniques also are available for the relative method of calibration for incandescent and high intensity discharge sources.

For measuring luminance of a luminaire, an instrument having a circular aperture is preferable, so that rotation of the instrument or angular position of luminaire will not affect readings. It is convenient, therefore, to use another collimating tube with an aperture 29 millimeters (1⅛-inches) in diameter, or luminance meters with circular fields of view, encompassing 645 square millimeters (1 square inch) at the measuring distance, and calibrated against a "working standard" which can be similar to the device shown in Fig. 4-22. In use, luminance of the device's diffusing face is adjusted such that the deflection of the indicating instrument used with the collimating tube equipped with a rectangular aperture is the same as that indicated when the same aperture is placed against the reference fluorescent lamp. The collimator with circular aperture is then substituted and the deflection noted—then this will be the deflection for the lamp luminance. All luminaire luminance readings made with this circular aperture are then directly related to the lamp luminance.

An alternate method is the use of a luminance meter to make direct measurements of the luminaire luminance. The characteristics of the luminance meter should be such that the field of measurement corresponds to a projected area of 645 square millimeters (one square inch) normal to its axis at the distance of measurement. This may be achieved by the use of an appropriate lens system and a measurement distance which may be determined from the manufacturer's specifications. This technique has the advantage of the observer being able to view through the luminance meter and see the exact area being measured. It does not suffer from the disadvantage of the collimating tube which arises due to the possibility of the tube casting a shadow on the area being measured.

Meter Calibration. A relative reading luminance meter may be used, with methods of meter calibration using either a calibrated fluorescent lamp or a diffusing test plate.

Where luminance meters are calibrated by means of a diffusing test plate with a known illuminance, the characteristics of the test plate for specific conditions of incidence, spectral power distribution of the source, and angle of view must be known. For a perfectly diffusing reflecting test plate the luminance is uniform at all angles, and its value in footlamberts is the product of the reflectance and the illuminance in footcandles. However, no surface is perfectly diffusing, and serious error can result from the assumption that this simple relationship holds. It is essential that the characteristics (luminance vs. angle of view) of the test plate be known, and taken into consideration in the calibration.

Spectroradiometric Measurements[93]

Spectroradiometric measurements of light sources can be reported in relative terms or in absolute values, which for spectral irradiance values are generally watts per unit area and wavelength band.

Basically there are two ways to determine the spectral properties of sources. One method is to use a broad band detector in conjunction with specific filters which define the spectral response region desired. The second method, and the one described below, is to determine the complete spectral power distribution of the source using a monochromator (see page 4-11) and a detector. In making measurements, it should be realized that if a spectroradiometric measurement is to give meaningful results, the measuring setup, the equipment used, and the technique employed must satisfy strict requirements.[94]

Measurement Methods. Measurement involves the comparison of two light sources, a reference source of known spectral power distribution and a test source whose spectral power distribution is to be determined. The two sources can be compared wavelength by wavelength or they can be measured by scanning throughout the spectrum.

When the sources are compared wavelength by wavelength, the two sources are operated simultaneously. This approach necessitates discontinuous wavelength scanning, pausing at each measurement point. It also requires that the reference and test source not drift over the lengthy measurement period.

When the sources are measured throughout the spectral region of interest, the measurement sequence usually comprises a measurement of the standard followed by the measurement of the test source. Either continuous or discontinuous wavelength scanning may be adopted. When this method is used, the spectral properties of the measuring system, the monochromator, detector, and so forth, must remain constant throughout the measurement period. Instrumental drift should be checked by remeasuring the standard source.

The data may be recorded by an operator, but for greater efficiency and to reduce the possibility of errors, it is better if the results are automatically recorded.[95,96] Some form of immediate presentation of the data in readable form is desirable to aid in the recognition of malfunction in the measurement system.

The presentation of the data is usually in the form of a series of numbers which describe the spectral power distribution of a light source over a particular spectral range. For the type of instruments described here, employing a monochromator-detector system, a large number of readings or a continuous spectral curve describe the spectral power distribution of the lamp. The observed data is a function of the spectral power distribution of the source, the spectral transmission of the optical system, the spectral bandwidth of the monochromator, and the spectral responsivity of the detector. Since the last three functions are properties of the spectroradiometer and should be constant, their effect is cancelled by comparing the data from the standard source with that from the test source. The accuracy of the final spectral power distribution is dependent on the accuracy of the calibration curve and the spectral data of the test lamp.

The spectral power distribution from an incandescent lamp can be adequately represented by a series of measurements at 10-nanometer intervals across the spectrum. In this situation the bandwidth of the monochromator is not too critical. Problems can arise when this technique is used to evaluate discharge lamp spectra which contain a number of emission lines. One way to alleviate this problem is to take more readings at closer intervals and to use a narrower bandwidth if the line structure of the source is important. A common approach is to reduce the measurement interval of each wavelength step to approximately 2 nanometers. If the sampling interval is greater than the bandwidth, gaps will be left in the spectrum. If the sampling interval is smaller than the bandwidth, overlapping will exist. The method recommended for general use involves scanning the spectrum continuously from one end to the other at a uniform speed and integrating the signal from the detector. This approach effectively converts the spectral power distribution into a curve under which the height of the curve is proportional to the power emitted by the source. In principle, this method does not require that the continuous scanning interval and the bandwidth be exactly equal, but the best resolution is obtained when the two are approximately equal. Calculation of chromaticity, which is one of the most critical applications of spectroradiometric data, indicates that for almost any type of source, a 2-nanometer bandwidth will give values of x and y accurate to 0.001, which is quite adequate for most purposes. It should be noted, however, that measurement precision may be limited by source variability.

Calorimetry of Luminaires

A thermal testing method has been developed for compiling data on air cooled heat transfer luminaires.[97] In that method, the entire laboratory room in which the calorimeter is located becomes a part of the calorimetric system. (A calorimeter is a device used to measure the thermal energy distribution of luminaires; primarily those luminaires which handle either air, water or both.) The room must be controlled closely with respect to temperature, air motion and relative humidity. Varying conditions can materially affect the results of the calorimetric measurements. The room conditions should be as follows: the temperature should be controlled at 25 °C ± 0.3 °C; the velocity of the air in the space containing the calorimeter should be held constant and not exceed 0.15 meters per second (30 feet per minute); the relative humidity should be held constant at any convenient value between 20 and 50 per cent; and the room should be located in an area that will not be affected by external conditions.

The selection of a calorimeter type is determined by the specific purpose of the device and the degree to which its conditions can be controlled. Three types of calorimeters are: the zero heat-loss calorimeter, a calorimeter so constructed to compensate for the heat transfer through its walls; the calibrated heat-loss calorimeter,* a box in which the heat loss can be determined by dissipating a measured quantity of energy in the plenum; and the continuous fluid-flow calorimeter, a modification of the zero heat-loss calorimeter consisting of a heavily insulated heat exchanger installed over the luminaire.

Precision instrumentation is needed to measure temperature (thermometers, thermocouples, thermistors and resistance elements), pressure (manometers, micromanometers, draft gages and swinging vane type gages), mass flow rate of air and water, electrical quantities, and light output.

Each luminaire that is to be tested for energy distribution should first be photometered in accordance with accepted procedures (see page 4-21). During calorimetry, a light meter or photovoltaic cell and microammeter or similar instrumentation should be installed at luminaire nadir, not less than 300 millimeters (one foot) from the bottom of the enclosure. The distance at which the cell should be mounted below the luminaire should be governed by the distance required for the cell to see the full luminous area of the luminaire. It is necessary that precautions be taken to prevent the cell from seeing luminous flux other than that transmitted by the luminaire under test. This position must not be changed during test.

The data to be recorded and reported should include: description and size of luminaire; mode of operation; test conditions—space and plenum temperatures; relative light output from a 25 °C base with the luminaire operated in free air outside the calorimeter; energy to space; energy removed by exhaust air stream; and exhaust air temperature as functions of exhaust flow rate. Energy and relative light output values should be plotted as ordinates vs. air volumes as abscissa.

FIELD MEASUREMENTS—INTERIORS

In evaluating an actual lighting installation in the field it is necessary to measure or survey the quality and quantity of lighting in the particular environment. To help do this, the IES has developed a uniform survey method of measuring and reporting the necessary data.[98] The results of these uniform surveys can be used alone or with other surveys for comparison purposes, can be used to determine compliance with specifications, and can be used to reveal the need for maintenance, modification or replacement.

Field measurements apply only to the conditions that exist during the survey. Recognizing this it is very important to record a complete detailed description of the surveyed area and all factors that might affect results, such as: interior surface reflectances, lamp type and age, voltage, and instruments used in the survey.

In measuring illuminance, cell type instruments should be used which are cosine and color corrected. They should be used at a temperature above 15 °C (60 °F) and below 50 °C (120 °F), if possible. Before taking readings, the cells should be exposed to the approximate illuminance to be measured until they become stabilized. This usually requires 5 to 15 minutes. Casting a shadow on the light sensitive cell should be avoided while reading the instrument. A high intensity discharge or fluorescent system must be lighted for at least one hour before measurements are taken to be sure that normal operating output has been attained. In relatively new lamp installations, at least 100 hours of operation of a gaseous source should elapse before measurements are taken. With incandescent lamps, seasoning is accomplished in a shorter time (20 hours or less for common sizes).

Illuminance Measurements—Average

Determination of Average Illuminance on a Horizontal Plane from General Lighting Only. The use of this method in the types of areas described should result in values of average illuminance within 10 per cent of the values that would be obtained by dividing the area into 0.6-meter (2-foot) squares, taking a reading in each square and averaging.

The measuring instrument should be positioned so that when readings are taken, the surface of the light sensitive cell is in a horizontal plane and 760 millimeters (30 inches) above the floor. This can be facilitated by means of a small portable stand of wood or other material that will support the cell at the correct height and in the proper plane. Daylight may be excluded during illuminance measurements. Readings can be taken at night or with shades, blinds or other opaque covering on the fenestration.

Regular Area With Symmetrically Spaced Luminaires in Two or More Rows. (See Fig. 4-23a.) (1) Take readings at stations r-

* The approved IES calorimeter is of the calibrated heat-loss type.

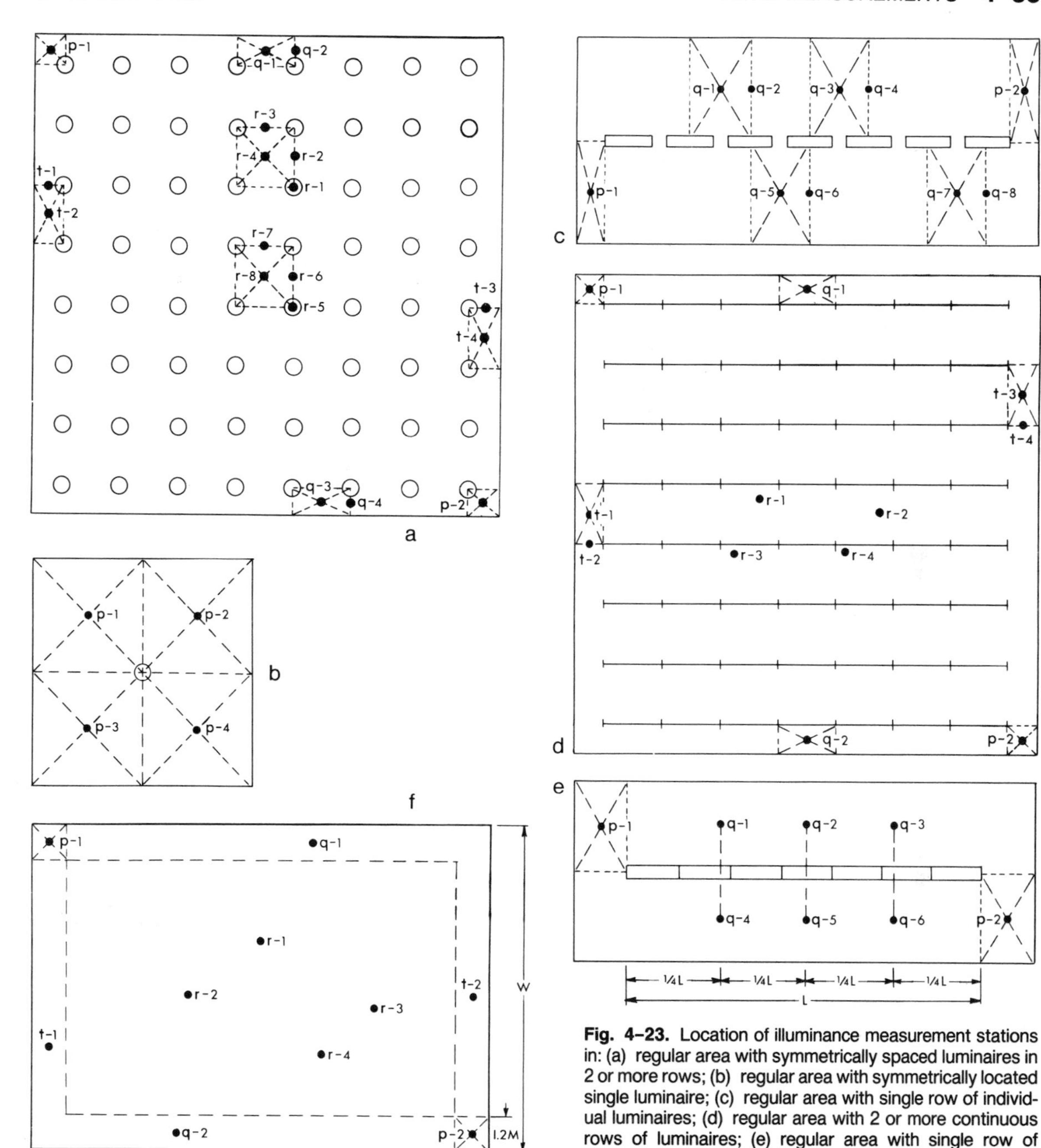

Fig. 4-23. Location of illuminance measurement stations in: (a) regular area with symmetrically spaced luminaires in 2 or more rows; (b) regular area with symmetrically located single luminaire; (c) regular area with single row of individual luminaires; (d) regular area with 2 or more continuous rows of luminaires; (e) regular area with single row of continuous luminaires; (f) regular area with luminous or louverall ceiling.

1, r-2, r-3, and r-4 for a typical inner bay. Repeat at stations r-5, r-6, r-7 and r-8 for a typical centrally located bay. Average the 8 readings. This is R in the equation directly below. (2) Take readings at stations q-1, q-2, q-3 and q-4 in two typical half bays on each side of room. Average the 4 readings. This is Q in the equation below. (3) Take readings at stations t-1, t-2, t-3 and t-4 in two typical half bays at each end of room. Average the 4 readings. This is T in the equation. (4) Take readings at stations p-1 and p-2 in two typical corner quarter bays. Average the 2 readings. This is P in the equation. (5) Determine the average illuminance in the area by solving the equation:

$$\text{Average Illuminance} = \frac{R(N-1)(M-1) + Q(N-1) + T(M-1) + P}{NM}$$

where

N = number of luminaires per row;
M = number of rows.

Regular Area With Symmetrically Located Single Luminaire. (See Fig. 4-23b.) (1) Take readings at stations p-1, p-2, p-3 and p-4 in all 4 quarter bays. Average the 4 readings. This is P, the average illuminance in the area.

Regular Area With Single Row of Individual Luminaires. (See Fig. 4-23c.) (1) Take readings at stations q-1 thru q-8 in 4 typical half bays located two on each side of the area. Average the 8 readings. This is Q in the equation directly below. (2) Take readings at stations p-1 and p-2 for two typical corner quarter bays. Average the 2 readings. This is P in the equation. (3) Determine the average illuminance in the area by solving the equation:

$$\text{Average Illuminance} = \frac{Q(N-1)+P}{N}$$

where N = number of luminaires.

Regular Area With Two or More Continuous Rows of Luminaires. (See Fig. 4-23d.) (1) Take readings at stations r-1 thru r-4 located near the center of the area. Average the 4 readings. This is R in the equation directly below. (2) Take readings at stations q-1 and q-2 located at each midside of the room and midway between the outside row of luminaires and the wall. Average the 2 readings. This is Q in the equation below. (3) Take readings at stations t-1 thru t-4 at each end of the room. Average the 4 readings. This is T in the equation. (4) Take readings at stations p-1 and p-2 in two typical corners. Average the 2 readings. This is P in the equation. (5) Determine the average illuminance in the area by solving the equation:

$$\text{Average Illuminance} = \frac{RN(M-1) + QN + T(M-1) + P}{M(N+1)}$$

where

N = number of luminaires per row;
M = number of rows.

Regular Area With Single Row of Continuous Luminaires. (See Fig. 4-23e.) (1) Take readings at stations q-1 thru q-6. Average the 6 readings. This is Q in the equation directly below. (2) Take readings at stations p-1 and p-2 in typical corners. Average the 2 readings. This is P in the equation below. (3) Determine the average illuminance in the area by solving the equation:

$$\text{Average Illuminance} = \frac{QN+P}{N+1}$$

where N = number of luminaires.

Regular Area With Luminous or Louverall Ceiling. (See Fig. 4-23f.) (1) Take readings at stations r-1 thru r-4 located at random in the central portion of the area. Average the 4 readings. This is R in the equation directly below. (2) Take readings at stations q-1 and q-2 located 0.6 meter (2 feet) from the long walls at random lengthwise of the room. Average the 2 readings. This is Q in the equation below. (3) Take readings at stations t-1 and t-2 located 0.6 meter (2 feet) from the short walls at random crosswise of the room. Average the 2 readings. This is T in the equation below. (4) Take readings at stations p-1 and p-2 located at diagonally opposite corners 0.6 meter (2 feet) from each wall. Average the 2 readings. This is P in the equation. (5) Determine the average illuminance in the area by solving the equation:

$$\text{Average Illuminance} = \frac{R(L-8)(W-8) + 8Q(L-8) + 8T(W-8) + 64P}{WL}$$

where

W = width of room;
L = length of room.

Illuminance Measurements—Point

With task, general and supplementary lighting in use, the illuminance at the point of work should be measured with the worker in his normal working position. The measuring instrument should be located so that when readings are

Fig. 4-24. Form for Tabulation of Point Illuminance Measurements

Work Point	Description of Work Point	Height Above Floor	Plane (horizontal, vertical, or inclined)	Illuminance	
				Total (general + supplementary)	General Only
1—(max.)					
2—(min.)					
3—					
4—					
5—					

Fig. 4-25. Form for Tabulation of Luminance Measurements

Work Point Location*	Luminance					
	A	B	C	D	E	F
Luminaire at 45° above eye level						
Luminaire at 30° above eye level						
Luminaire at 15° above eye level						
Ceiling, above luminaire						
Ceiling, between luminaires						
Upper wall or ceiling adjacent to a luminaire						
Upper wall between two luminaires						
Wall at eye level						
Dado						
Floor						
Shades and blinds						
Windows						
Task						
Immediate surroundings of task						
Peripheral surroundings of task						
Highest luminance in field of view						

* Describe locations A thru F.

taken, the surface of the light sensitive cell is in the plane of the work or of that portion of the work on which the critical visual task is performed—horizontal, vertical or inclined. Readings as shown in Fig. 4-24 should be recorded.

Luminance Measurements

Luminance surveys may be made under actual working conditions and from a specified work point location with the combinations of daylight and electric lighting facilities available. Consideration should be given to sun position and weather conditions, both of which may have marked effect on luminance distribution. All lighting in the area, task, general and supplementary, should be in normal use. Work areas used only in the daytime should be surveyed in the daytime; work areas used both daytime and nighttime should preferably have two luminance surveys made under the two sets of conditions, as the luminance distribution and the possible comfort or discomfort will differ markedly at these times. Nighttime surveys should be made with shades drawn. Daytime surveys should be made with shades adjusted for best control of daylight.

On a floor plan sketch of the area, an indication should be made of which exterior wall or walls, if any, were exposed to direct sunlight during the time of the survey by writing the word "Sun" in the appropriate location. Readings should be taken, successively, from the worker's position at each work point location A, B, C, etc. and luminance readings from each location recorded as shown in Fig. 4-25.

FIELD MEASUREMENTS—OUTDOOR

In roadway and many floodlight installations light is projected in a direction forming a large angle of incidence with the surface to be lighted, and each unit must be adjusted carefully to produce the best utilization and quality of illumination. For an accurate evaluation of this type of installation, special care must be taken in the measurement of the resultant illumination. A summary of appropriate IES guides follows, but the full guides should be consulted before making an actual survey.[99-101]

Preparation for the Survey. (1) Inspect and record the condition of the luminaires (globes, reflectors, refractors, lamp positioning, etc.). In the case of roadway lighting, make sure luminaires are level and their lateral placement is correct. Unless the purpose of the test is to check depreciation or actual in-service performance, all units should be cleaned and new lamps installed. New lamps should be seasoned properly.[82] While inoperative lamps are readily noticed in roadway installations, they can easily be overlooked in large floodlighting systems. If these lamps are not replaced for the field survey, proper consideration must be given when evaluating the test. (2) Record the mounting height of the luminaires. (3) Record the location of the poles, the number of units per pole, the wattage of the lamps and other pertinent data. Check these data against the recommended layout; a small change in the location or adjustment of the luminaires may make a considerable difference in the resultant illuminance. (4) Determine and record the hours of burning of the installed lamps. (5) Record the atmospheric conditions. Because of the effect of adverse atmospheric conditions, the survey should be made only when the atmosphere is clear. Extraneous light produced by a store, service station, or other lights in the vicinity, requires careful attention in street lighting tests. (6) Because of the influence of the electrical circuit operating conditions on lamp light output, it is usually necessary to know precisely the electrical circuit operating conditions at the luminaires in the system at the time the photometric measurements are being made. At night, during the hours when the luminaires will normally be used, record the voltage at the lamp socket with all of the lamps operating. The voltage at the main switch may be measured provided allowance is made for the voltage drop to the individual units. If discharge lamps are installed, record the input voltage to the ballast at the ballast terminals. Discharge lamps should be operated at least one half hour to reach normal operating conditions before measurements are made.

Survey Procedures. Measurements should be made with a recently calibrated, color- and cosine-corrected photometer capable of being leveled for horizontal measurements or positioned accurately for other measurement planes as required. The photometer should be selected for its portability and repeatability of measurements at any point of the scale which is used. If required by the spectral characteristic of the light source in the system being measured, appropriate corrections should be made to each reading.

1. For roadways, because of the difficulty and expense of making field measurements of pavement luminance (this requires the closure of at least one traffic lane for substantial periods of time), it has been found that those who utilize a computerized design procedure using point calculations can easily generate the horizontal illuminance level at each of the pavement luminance measurement points recommended. As a check on the performance of the lighting system, it is only necessary to measure the illuminance at these points.[99]

2. For roadway signs, the minimum and maximum illuminance levels are determined by scanning the sign face. Additional illuminance measurements are taken at specific locations according to the sign size. Luminance measurements are also made for both externally and internally illuminated signs.[101]

3. For sports installations,[100] the sports area, or that portion of the area under immediate consideration, should be divided into test areas of approximately 5 per cent of the total area and readings should be taken in the center of each area. Where illuminance for color television is involved,[102] multiple readings should be taken at each station: one reading with the meter cell tilted 15 degrees from vertical in the direction of each camera location, and 0.9 meters (36 inches) above ground level, unless otherwise specified for the particular sports activity, and a final reading with the cell in the horizontal position.

4. Readings should be made at each test station with repeat measurements at the first station frequently enough to assure stability of the system and repeatability of results. Readings should be reproducible within 5 per cent. Enough readings should be taken so that additional readings in similar locations will not change the average results significantly. Care should be exercised while taking readings to avoid casting shadows on the receptor of the measuring instrument, and also by standing far enough away from the receptor, especially when wearing light colored clothes, to prevent light from the source from being reflected onto it.

REFERENCES

1. Wright, W. D.: *Photometry and the Eye*, Hatton Press, Ltd., London, pp. 31, 80, 123, 124, 1949.
2. Walsh, J. W. T.: *Photometry*, 3rd ed., Constable and Co., Ltd., London, pp. 120–173, 1958. Committee on Testing Procedures of the IES: "IES General Guide to Photometry," *Illum. Eng.*, Vol. L, p. 147, March, 1955.
3. Walsh, J. W. T.: *Photometry*, 3rd ed., Constable and Co., Ltd., London, pp. 174, 1958.
4. Walsh, J. W. T.: *Photometry*, 3rd ed., Constable and Co., Ltd., London, pp. 235–237, 1958. Projector, T. E.: "Effective Intensity of Flashing Lights," *Illum. Eng.*, Vol. LII, p. 630, December, 1957. Douglas, C. A.: "Computation of the Effective Intensity of Flashing Lights," *Illum. Eng.*l, Vol. LII, p. 641, December, 1957. Lash, J. D. and Prideaux, G. F.: "Visibility of Signal Lights," *Illum. Eng.*, Vol. XXXVIII, p. 481, November, 1943. Preston, J. S.: "Note on the Photoelectric Measurement of the Average Intensity of Fluctuating Light Sources," *J. Sci. Insts.*, Vol. 18, p. 57, April, 1941. Schuil, A. E.: "The Effect of Flash Frequency on the Apparent Intensity of Flashing Lights Having Constant Flash Duration," *Trans. Illum. Eng. Soc.* (London), Vol. XXXV, p. 117, September, 1940. Neeland, G. K., Laufer, M. K., and Schaub, W. R.: "Measurement of the Equivalent Luminous Intensity of Rotating Beacons," *J. Opt. Soc. Amer.*, Vol. 28, p. 280, August 1938. Blondel, A. and Rey, J.: "The Perception of Lights of Short Duration at Their Range Limits," *Trans. Illum. Eng. Soc.*, Vol. VII, p. 625, November, 1912.
5. Walsh, J. W. T.: *Photometry*, 3rd ed., Constable and Co., Ltd., London, p. 223, 1958. Lewin, I., Baker, G. A. and Baker, M. T.: "Developments in High Speed Photometry and Spectroradiometry," *J. Illum. Eng. Soc.*, Vol. 8, p. 214, July, 1979.
6. Kaufman, J. E. and Christensen, J. F.: *IES Lighting Handbook*, 5th ed., Illuminating Engineering Society, New York, 1972.
7. Kingsbury, E. F.: "A Flicker Photometer Attachment for the Lummer-Brodhun Contrast Photometer," *J. Franklin Inst.*, August, 1915. Guild, J.: "A New Flicker Photometer for Heterochromatic Photometry," *J. Sci. Instrum.*, March, 1924. Feree, C. E. and Rand, G.: "Flicker Photometry," *Trans. Illum. Eng. Soc.*, Vol. XVIII, p. 151, February, 1923.
8. Little, W. F. and Estey, R. S.: "The Use of Color Filters in Visual Photometry," *Trans. Illum. Eng. Soc.*, Vol. XXXII, p. 628, June, 1937. Johnson, L. B.: "Photometry of Gaseous-Conductor Lamps," *Trans. Illum. Eng. Soc.*, Vol. XXXII, p. 646, June, 1937.
9. Sharp, C. H. and Little, W. F.: "Compensated Test Plate for Illumination Photometers," *Trans. Illum. Eng. Soc.*, Vol. X, p. 727, November, 1915. Little, W. F.: "Practical Hints on the Use of Portable Photometers," *Trans. Illum. Eng. Soc.*, Vol. X, p. 766, November, 1915. Morris, A., McGuire, F. L., and Van Cott, H. P.: "Accuracy of Macbeth Illuminometer as a Function of Operator Variability, Calibration, and Sensitivity," *J. Opt. Soc. Amer.*, Vol. 45, p. 525, July, 1955.
10. Taylor, A. H.: "A Portable Reflectance Meter," *Illum. Eng.*, Vol. LV, p. 614, November, 1960.
11. Weibel, W. A.: "Portable Electric Photometers—A Survey," *Light. Des. Appl.*, Vol. 5, p. 5, August, 1975.
12. "Report of Committee on Portable Photoelectric Photometers,"

Trans. Illum. Eng. Soc., Vol. XXXII, p. 379, April, 1937. Lange, B., *Photoelements and Their Application*, Reinhold Publishing Corp., New York, May, 1938.
13. Projector, T. H., Laufer, M. K., and Douglas, C. A.: "An Improved 'Zero-Resistance' Circuit for Photo-Cell Photometry," *Rev. Sci. Instrum.*, April, 1944. Barbrow, L. E.: "A Photometric Procedure Using Barrier-Layer Photocells," *J. Res. N. B. S.*, Vol. 25, p. 703, December, 1940.
14. Horton, G. A.: "Electronic Instrumentation in Light Measurements," *Illum. Eng.*, Vol. 64, p. 701, December, 1969.
15. Fogle, M. E.: "New Color Corrected Photronic Cells for Accurate Light Measurements," *Trans. Illum. Eng. Soc.*, Vol. XXXI, p. 773, September, 1936. Parker, A. E.: "Measurement of Illumination From Gaseous Discharge Lamps," *Illum. Eng.*, Vol. XXXV, p. 883, November, 1940. Preston, J. S.: "The Relative Spectral Response of the Selenium Rectifier Photocell in Relation to Photometry and the Design of Spectral Correction Filters," *J. Sci. Instrum.*, Vol. 27, p. 135, May, 1950.
16. Elvegard, E., Lindroth, S., and Larson, E.: "The Drift Effect in Selenium Photovoltaic Cells," *J. Opt. Soc. Amer.*, Vol. 28, p. 33, February, 1938. Houston, R. A.: "The Drift of the Selenium Barrier-Layer Photo-Cell," *Phil. Mag.*, June, 1941. Preston, J. S.: "Fatigue in Selenium Rectifier Photocells," *Nature*, Vol. 153, p. 680, June, 1944.
17. "Report of the Committee on Portable Photoelectric Photometers," *Trans. Illum. Eng. Soc.*, Vol. XXXII, p. 379, April, 1937. Goodbar, I.: "New Procedure to Measure Accurately Illumination at Large Angles of Incidence, with a Barrier-Layer Cell," *Illum. Eng.*, Vol. XL, p. 830, November, 1945. Morton, C. A.: "Cosine Response of Photocells and the Photometry of Linear Light Sources," *Light and Lighting* (London), Vol. XXVIII, p. 157, November, 1945. Buck, G. B., II: "Correction of Light-Sensitive Cells for Angle of Incidence and Spectral Quality of Light," *Illum. Eng.*, Vol. XLIV, p. 293, May, 1949. Dows, C. L.: "Illumination Measurements with Light Sensitive Cells," *Illum. Eng.* Vol. XXXVII, p. 103, February, 1942. Pleijel, G., and Longmore, J.: "A Method of Correcting the Cosine Error of Selenium Rectifier Photocells," *J. Sci. Instrum.*, Vol. 29, p. 137, May, 1952.
18. Atkinson, J. R., Campbell, N. R., Palmer, E. H., and Winch, G. T.: "The Accuracy of Rectifier-Photoelectric Cells," *Proc. Phys. Soc.* (London), November, 1936. Also see ref. 12.
19. MacGregor-Morris, J. T. and Billington, R. M.: "The Selenium Rectifier Photocell: Its Characteristics and Response to Intermittent Illumination," *J. Inst. Elec. Eng.* (London), Vol. 78, p. 435, October, 1936. Zworkin, V. K. and Ramberg, E. G.: *Photoelectricity and Its Application*, John Wiley & Sons, New York, p. 211, 1949. Gleason, P. R.: "Failure of Talbot's Law for Barrier-Layer Photocells," *Phys. Rev.*, Vol. 45, p. 745, 2nd ser., 1934. Walsh, J. W. T.: *Photometry*, 3rd ed., Constable and Co., Ltd., London, p. 98, 107, 1958. Lange, B., *Photoelements and Their Application*, Reinhold Publishing Co., New York, p. 151, 1938.
20. *RCA Photomultiplier Manual*, Booklet PT-61, RCA Electronic Components, Harrison, NJ, 1970.
21. Freund, K.: "Design Characteristics of a Photoelectric Brightness Meter," *Illum. Eng.*, Vol. XLVIII, p. 524, October, 1953.
22. Horton, G. A.: "Evaluation of Capabilities and Limitations of Various Luminance Measuring Instruments," *Illum. Eng.*, Vol. LX, p. 217, April, 1965. Eastman, A. A.: "Contrast Determination with the Pritchard Telephotometer," *Illum. Eng.*, Vol. LX, p. 179, April, 1965. Spencer, D. E.: "Out of Focus Photometry," *J. Opt. Soc. Amer.*, Vol. 55, p. 396, April, 1965. Spencer, D. E. and Levin, R. E.: "On the significance of Photometric Measurements," *Illum. Eng.*, Vol. LXI, p. 196, April, 1966.
23. Blackwell, H. R., Helms, R. N. and DiLaura, D. L.: "Application Procedures for Evaluation of Veiling Reflections in Terms of ESI: III Validation of a Prediction Method for Luminaire Installations," *J. Illum. Eng. Soc.*, Vol. 2, p. 284, April, 1973.
24. Ngai, P. Y., Zeller, R. D. and Griffith, J. W.: "The ESI Meter—Theory and Practical Embodiment," *J. Illum. Eng. Soc.*, Vol. 5, p. 58, October, 1975.
25. DiLaura, D. L. and Stannard, S. M.: "An Instrument for the Measurement of Equivalent Sphere Illumination," *J. Illum. Eng. Soc.*, Vol. 7, p. 183, April, 1978.
26. Green, J. D.: "A Practical Direct Reading ESI Meter for Field Use," *J. Illum. Eng. Soc.*, Vol. 9, p. 247, July, 1980.
27. Blackwell, H. R. and DiLaura, D. L.: "Application Procedures for Evaluation of Veiling Reflections in Terms of ESI: II Gonio Data for the Standard Pencil Task," *J. Illum. Eng. Soc.*, Vol. 2, p. 254, April, 1973.
28. Committee on Testing Procedures of the IES: "IES Approved Method of Reflectometry," *J. Illum. Eng. Soc.*, Vol. 3, p. 167, January, 1974. Baumgartner, G. R.: "A Light-Sensitive Cell Reflectometer," *Gen. Elec. Rev.*, Vol. 40, p. 525, November, 1937.
29. Taylor, A. H.: "Errors in Reflectometry," *J. Opt. Soc. Amer.*, Vol. 43, p. 51, February, 1953.
30. McNichols, H. J.: "Absolute Methods in Reflectometry," Research Paper No. 3, *J. Res. Bur. Stand.*, p. 29, 1928. Hunter, R. S.: "A Multipurpose Photoelectric Reflectometer," Research Paper RP 1345, *J. Res. Nat. Bur. Stand.*, p. 581, November, 1940. McNicholas, H. J.: "Equipment for Measuring the Reflective and Transmissive Properties of Diffusing Media," Research Paper RP 704. *J. Res. Nat. Bur. Stand.*, p. 211, August, 1934. Hunter, R. S.: "A New Goniophotometer and Its Applications," *Bulletin No. 106*, Henry A. Gardner Laboratory, Inc., Bethesda 14, Maryland, December, 1951. Sharp, C. H. and Little, W. F.: "Measurement of Reflection Factors," *Trans. Illum. Eng. Soc.*, Vol. XV, p. 802, December, 1920.
31. Taylor, A. H.: "A Simple Portable Instrument for the Absolute Measurement of Reflection and Transmission Factors," Sci. Paper No. 405, *Bull. Bur. Stand.*, p. 421, July 28, 1920.
32. Dows, C. L. and Baumgartner, G. R.: "Two Photo-Voltaic Cell Photometers for Measurement of Light Distribution," *Trans. Illum. Eng. Soc.*, Vol. XXX, p. 476, June, 1935. "Permanent Gloss Standards," *Illum. Eng.* Vol. XLV, p. 101, February, 1950. Hunter, R. S.: *A New Goniophotometer and Its Applications*, Bulletin No. 106, Gardner Laboratory, Incorporated, Bethesda, Maryland, December, 1951. Spencer, D. E. and Gray, S. M.: "On the Foundations of Goniophotometry," *Illum. Eng.*, Vol. LV, p. 228–229, April, 1960. Nimeroff, I.: "Analysis of Goniophotometric Curves," *J. Res. of NBS*, RP2335, Vol. 48, p. 441–447, June, 1952. *Standard Recommended Practice for Goniophotometry of Objects and Materials*, ANSI/ASTM E167 - 77, American Society for Testing and Materials, Philadelphia, PA.
33. Field, H. P.: "Digital Processing in Optical Radiometry," *Electro-Opt. Sys. Des.*, Vol. 7, p. 78, November, 1975.
34. Cunningham, R. C.: "Silicon Photodiode or Photomultiplier Tube?," *Electro-Opt. Sys. Des.*, Vol. 6, p. 21, August, 1974.
35. Bode, D. E.: "Optical Detectors," *Handbook of Lasers*, p. 171, Chemical Rubber Company, 1971. Bode, D. E.: "Infrared Detectors," *Applied Optics and Optical Engineering*, Vol. VI, p. 323, Academic Press, New York, 1980.
36. Weekes, P.: "Photon Counting—Notes on a Basic System," *Electro-Opt. Sys. Des.*, Vol. 9, p. 30, June, 1977. Morton, G. A.: "Photon Counting," *Appl. Opt.*, Vol. 7, p. 1, 1968.
37. Committee on Testing Procedures of the IES: "IES Guide for Measurement of Ultraviolet Radiation from Light Sources," *J. Illum. Eng. Soc.*, Vol. 8, p. 53, October, 1978.
38. McCullock, J. H. and McCullock, H.: "Floodlight Photometry Without Special Photometer and Without Tipping Luminaire," *Illum. Eng.*, Vol. LXII, p. 243, April, 1967.
39. Rosa, E. B. and Taylor, A. H.: "Theory Construction and Use of the Photometric Integrating Sphere," Sci. Paper No. 447, *Bull. Bur. Stand.*, September 26, 1921. Walsh, J. W. T.: *Photometry*, 3rd ed., Constable and Co., Ltd., London, p. 257, 1958. Buckley, H.: "The Effect of Non-Uniform Reflectance of the Interior Surface of Spherical Photometric Integrators," *Trans. Illum. Eng. Soc.* (London), Vol. XLI, p. 167, July, 1946. Hardy, A. C. and Pineo, O. W.: "The Errors Due to the Finite Size of Holes and Sample in Integrating Spheres," *J. Opt. Soc. Amer.*, Vol. 21, p. 502, August, 1931. Gabriel, M. H., Koenig, C. F., and Steeb, E. S.: "Photometry, Parts I and II," *Gen. Elec. Rev.*, Vol. 54, pp. 30, 23, September and October, 1951.
40. Weaver, K. S. and Shackleford, B. E.: "The Regular Icosahedron as a Substitute for the Ulbricht Sphere," *Trans. Illum. Eng. Soc.*, Vol. XVIII, p. 290, March, 1923.
41. Procedures for the Measurement of Luminaire Flux of Discharge Lamps and for Their Calibration as Working Standards, CIE Pub. No. 25, Paris, 1973.
42. Walsh, J. W. T.: *Photometry*, Constable and Co., Ltd., London, 3rd ed., p. 265, 1958.

43. Committee of Testing Procedures of IES: "IES General Guide to Photometry," *Illum. Eng.*, Vol. L, p. 201, April, 1955. Stephenson, H. F.: "The Equipment and Functions of an Illumination Laboratory," *Trans. Illum. Eng. Soc.* (London), Vol. XVII, p. 1, January, 1952.
44. Committee on Testing Procedures of the IES: "IES Practical Guide to Photometry," *J. Illum. Eng. Soc.*, Vol. 1, October, 1971.
45. Walsh, J. W. T.: *Photometry*, Constable and Co., Ltd., London, 3rd ed., p. 486, 1958.
46. Levin, R. E.: "The Photometric Connection," *Light. Des. Appl.*, Vol. 12, Part 1, p. 28, September, 1982; part 2, p. 60, October, 1982; Part 3, p. 42, November, 1982; and Part 4, p. 16, December 1982.
47. Committee on Testing Procedures of the IES: "IES Approved Method for Electrical Photometric Measurements of General Service Incandescent Filament Lamps," *J. Illum. Eng. Soc.*, Vol. 11, p. 23, October, 1981.
48. Fink, D. G.: *Standard Handbook for Electrical Engineers*, 11 ed., McGraw Hill Book Co., Inc., New York, 1978.
49. Committee on Testing Procedures of the IES: "IES Guide to Lamp Seasoning," *J. Illum. Eng. Soc.*, Vol. 7, p. 144, January, 1978.
50. Committee on Testing Procedures of the IES: "IES Guide for the Selection, Care, and Use of Electrical Instruments in the Photometric Laboratory," *J. Illum. Eng. Soc.*, Vol. 9, p. 77, January, 1980.
51. Brooks, H. B.: "Deflection Potentiometers for Current and Voltage Measurements," *Nat. Bur. of Stand. Tech. News Bull.*, p. 395, 1912.
52. Karapetoff, V. and Dennison, B. C.: *Experimental Electric Engineering*, Vol. 1, 4th ed., John Wiley and Sons, New York, p. 132, 1933. Gabriel, M. H., Koenig, C. F., and Steeb, E. S.: "Photometry, Parts I and II," *Gen. Elec. Rev.*, Vol. 54, pp. 30, 23, September and October, 1951. John Fluke Mfg. Co., Inc., Seattle 99, Wash. "Electronic Volt-Ampere-Wattmeter with Negligible Circuit Burden."
53. Committee on Testing Procedures of the IES: "IES Approved Method for the Electrical and Photometric Measurements of Fluorescent Lamps," *J. Illum. Eng. Soc.*, Vol. 11, p. 16, October, 1981.
54. Committee on Testing Procedures of the IES: "IES Approved Method for Photometric Measurements of High-Intensity Discharge Lamps," *J. Illum. Eng. Soc.*, Vol. 4, p. 229, April, 1975.
55. Testing Procedures Committee of the IES: "IES Approved Method for the Electrical and Photometric Measurments of Low Pressure Sodium Lamps," *J. Illum. Eng. Soc*, Vol. 12 p. 210, July, 1983.
56. Little, W. F. and Salter, E. H.: "The Measurement of Fluorescent Lamps and Luminaires," *Illum. Eng.*, Vol. XLII, p. 217, February, 1947.
57. "Specifications for Fluorescent Lamp Ballasts," ANSI C82.1-1977. "Methods of Measurement of Fluorescent Lamp Ballasts," ANSI C82.2-1983. "Specifications for Fluorescent Lamp Reference Ballasts," ANSI C82.3-1983. "Specifications for High-Intensity-Discharge Lamp Ballasts (Multiple-Supply Type)," ANSI C82.4-1978. "Specifications for High-Intensity-Discharge Lamp Reference Ballasts," ANSI C82.5-1983. "Methods of Measurement of Mercury Lamp Ballasts and Transformers," ANSI C82.6-1980, American National Standards Institute, New York.
58. Miller, J. H.: "A Vector Sum Ammeter," *Weston Eng. Notes*, p. 7, December, 1954.
59. Teele, R. P.: "Gas-Filled Lamps as Photometric Standards," *Trans. Illum. Eng. Soc.*, Vol. XXV, p. 78, January, 1930. Knowles-Middleton, W. E. and Mayo, E. G.: "Variation in the Horizontal Distribution of Light from Candlepower Standards," *J. Opt. Soc. Amer.*, Vol. 41, p. 513, August, 1951. Winch, G. T.: "Recent Developments in Photometry and Colorimetry," *Trans. Illum. Eng. Soc.* (London), Vol. XXI, p. 91, May, 1956. Also see ref. 3.
60. Winch, G. T.: "Photometry and Colorimetry of Fluorescent and Other Discharge Lamps," *Trans. Illum. Eng. Soc.* (London), Vol. XI, p. 107, June, 1946. Voogd, J.: "Physical Photometry," *Philips Tech. Rev.*, Vol. 4, p. 260, September, 1939. Winch, G. T.: "The Measurement of Light and Color," *Proc. Inst. Elec. Eng.* (London), p. 452, June, 1949.
61. Dickey, J. W., Waltman, D. J. and Hardesty, G. K. C.: "A Programmed-Response, Spectral-Scanning Photometric System," Naval Ship R & D Center, Washington, D.C., Report 6-135, December, 1971, and U.S. Patent, Hardesty, G. K. C., et al., (same title), #3,634,694, January, 1972.
62. "Biologically Related Units of the International System of Units (SI)," *Optical Radiation News*, National Bureau of Standards, Washington, D.C., January, 1977.
63. Franck, K. and Smith, R. L.: "A Photometric Laboratory for Today's Light Sources," *Illum. Eng.*, Vol. XLIX, p. 287, June, 1954. Baumgartner, G. R.: "New Semi-Automatic Distribution Photometer and Simplified Calculation of Light Flux," *Illum. Eng.*, Vol. XLV, p. 253, April, 1950.
64. Baumgartner, G. R.: "Practical Photometry of Fluorescent Lamps and Reflectors," *Illum. Eng.*, Vol. XXXVI, p. 1340, December, 1941.
65. Committee on Testing Procedures of the IES: "IES Approved Method for Photometric Measuring and Reporting Tests on Reflector-Type Lamps," *J. Illum. Eng. Soc.*, Vol. 11, p. 130, April, 1982.
66. Committee on Testing Procedures of the IES: "IES Approved Method for Photometric Testing of Floodlights Using Incandescent Filament or Discharge Lamps," *Illum. Eng.*, Vol. 66, p. 107, February, 1971.
67. Walsh, J. W. T.: *Photometry*, 3rd ed., Constable and Co., Ltd., London, p. 257, 1958.
68. "Experimental Statistics," *Handbook 91*, Section 3, National Bureau of Standards. "Recommended Practice for Probability Sampling of Materials," ASTM E105-58, American Society for Testing Materials. "Evaluation of Test Data," *Handbook 91*, Section 2, National Bureau of Standards.
69. Lewinson, L. J.: "The Interpretation of Forced Life Tests of Incandescent Electric Lamps," *Trans. Illum. Eng. Soc.*, Vol. XI, p. 815, November, 1916. Lewinson, L. J. and Millar, P. S.: "The Evaluation of Lamp Life," *Trans. Illum. Eng. Soc.*, Vol. VI, p. 774, November, 1911. Purcell, W. R.: "Saving Time in Testing Life of Incandescent Lamps," *Proc. AIEE*, Vol. 68, p. 617, July, 1949.
70. Committee on Testing Procedures of the IES: "IES Approved Method for Life Testing of General Lighting Incandescent Filament Lamps," *J. Illum. Eng. Soc.*, Vol. 8, p. 152, April, 1979.
71. Committee on Testing Procedures of the IES: "IES Approved Method for Life Performance Testing of Fluorescent Lamps," *J. Illum. Eng. Soc.*, Vol. 11, p. 213, July, 1982.
72. Committee on Testing Procedures of the IES: "IES Approved Method for Life Testing of High-Intensity Discharge Lamps," *J. Illum. Eng. Soc.*, Vol. 10, p. 211, July, 1981.
73. Committee on Testing Procedures of the IES: "IES Approved Method of Life Testing of Low Pressure Sodium Lamps," *J. Illum. Eng. Soc.*, Vol. 12, p. 215, July 1983.
74. Committee on Testing Procedures of the IES: "IES Approved Method for Photometric Testing of Indoor Fluorescent Luminaires," *J. Illum. Eng. Soc.*, Vol. 2, p. 70, October, 1972.
75. Committee on Testing Procedures of the IES: "Photometric Testing of Filament Type Luminaires for General Lighting Service," *Illum. Eng.*, Vol. LXII, p. 587, October, 1967.
76. Committee on Testing Procedures of the IES: "IES Approved Method for Photometric Testing of Indoor Luminaires Using High-Intensity Discharge Lamps," *J. Illum. Eng. Soc.*, Vol. 3, p. 160, January, 1974.
77. Committee on Testing Procedures of the IES: "IES Guide for Reporting General Lighting Equipment Engineering Data," *J. Illum. Eng. Soc.*, Vol. 1, p. 175, January, 1972, and "Addendum to IES Guide for Reporting General Lighting Equipment Engineering Data," *J. Illum. Eng. Soc.*, Vol. 5, p. 243, July, 1976.
78. Committee on Testing Procedures of the IES: "Determination of Average Luminance of Luminaires," *J. Illum. Eng. Soc.*, Vol. 1, January, 1972.
79. Committee on Testing Procedures of the IES: "IES Guide for Photometric Testing of Searchlights," *J. Illum. Eng. Soc.*, Vol. 13, July, 1984.
80. *Outdoor Floodlighting Equipment*, NEMA Standard Publication No. FA1-1973 (Reaffirmed 1979), National Electrical Manufacturers Association, Washington, DC.
81. Committee on Testing Procedures of the IES: "IES Approved Method for Photometric Testing of Floodlights Using Incan-

descent Filament or Discharge Lamps," *Illum. Eng.*, Vol. 66, p. 107, February, 1971. Committee on Equipment Performance Ratings, Joint IES-SMPTE: "Recommended Practice for Reporting Photometric Performance of Incandescent Filament Lighting Units Used in Theatre and Television Production," *Illum. Eng.*, Vol. LIII, p. 516, September, 1958. CIE Technical Committee TC 2.4: *Photometry of Floodlights*, Commission Internationale de L'Eclairage, 1979.

82. Committee on Testing Procedures of the IES: "IES Approved Method for Photometric Testing of Roadway Luminaires," *J. Illum. Eng. Soc.*, Vol. 7, p. 132, January, 1978.

83. Committee on Testing Procedures of the IES: "IES Approved Method for Determining Luminaire-Lamp-Ballast Combination Operating Factors for High Intensity Discharge Luminaires," *Illum. Eng.*, Vol. 65, p. 718, December, 1970.

84. Chernoff, L.: "Photometry of Projectors at the National Bureau of Standards," *Nat. Bur. of Stand. Tech. Note No. 198*, December, 1963. Johnson, J.: "Zero Length Searchlight Photometry System," *Illum. Eng.*, Vol. LVII, p. 187, March, 1962.

85. Frederiksen, E.: "Unidirectional-Sensitive Photometer," *Light Light.*, Vol. 60, p. 46, February, 1967.

86. Breckenridge, F. C. and Projector, T. H.: "Construction of a Goniometer for Use in Determining the Candlepower Characteristics of Beacons," CAA Technical Development Report No. 39, February, 1944. Projector, T. H.: "Versatile Goniometer for Projection Photometry," *Illum. Eng.*, Vol. XLVIII, p. 192, April, 1953.

87. Committee on Testing Procedures of the IES: "IES General Guide to Photometry," *Illum. Eng.*, Vol. L, p. 201, April, 1955. Stephenson, H. F.: "The Equipment and Functions of an Illumination Laboratory," *Trans. Illum. Eng. Soc.* (London), Vol. XVII, p. 1, January, 1952.

88. Douglas, C. A. and Young, L. L.: "Development of a Transmissiometer for Determining Visual Range." *CAA Tech. Dev. Rep. No. 47*, U. S. Dept. of Commerce. Knowles-Middleton, W. E.: *Vision Through the Atmosphere*, Univ. of Toronto Press, 1952. Committee on Instruments and Measurements of the IES, Annual Report of "Part II—Description of Method for Measuring Atmospheric Transmission," *Illum. Eng.*, Vol. XXXVIII, p. 515, November, 1943.

89. Committee on Testing Procedures of the IES: "IES Guide for Measurement of Photometric Brightness (Luminance)," *Illum. Eng.*, Vol. LVI, p. 457, July, 1961.

90. Linsday, E. A.: "Brightness of Cylindrical Fluorescent Sources," *Illum. Eng.*, Vol. XXXIX, p. 23, January, 1944.

91. Horton, G. A.: "Modern Photometry of Fluorescent Luminaires," *Illum. Eng.*, Vol. XLV, p. 458, July, 1950.

92. Levin. R. E., "On Fluorescent Photometry," *J. Illum. Eng. Soc.*, Vol. 12, p. 218, July 1983.

93. Committee on Testing Procedures of the IES: "IES Guide to Spectroradiometric Measurements," *J. Illum. Eng. Soc.*, Vol. 12, p. 136, April, 1983.

94. Jones, O. C.: *Spectroradiometric Measurement of Light Sources*, CIE Report, International Commission on Illumination, C/O National Bureau of Standards, Washington, D.C. 20234, April, 1981.

95. Elby, J. E.: "Computer Based Spectroradiometer System," *Appl. Opt.*, Vol. 9, April, 1970.

96. Lewin, I., Baker, G. A. and Baker, M. T.: "Developments in High Speed Photometry and Spectroradiometry," *J. Illum. Eng. Soc.*, Vol. 8, p. 214, July, 1979.

97. Committee on Testing Procedures of the IES: "IES Approved Guide for the Photometric and Thermal Testing of Air Cooled Heat Transfer Luminaires," *J. Illum. Eng. Soc.*, Vol. 8, p. 57, October, 1978.

98. Lighting Survey Committee of the IES: "How to Make a Lighting Survey," *Illum. Eng.*, Vol. LVII, p. 87, February, 1963.

99. Committee on Testing Procedures of the IES: "IES Guide for Photometric Measurement of Roadway Lighting Installations," *J. Illum. Eng. Soc.*, Vol. 4, p. 154, January, 1975. Committee on Roadway Lighting of the IES: "American National Standard Practice for Roadway Lighting," *J. Illum. Eng. Soc.*, Vol. 12, p. 146, April, 1983.

100. Committee on Testing Procedures of the IES: "IES Guide for Photometric Measurements of Area and Sports Lighting Installations," *J. Illum. Eng. Soc.*, Vol. 4, p. 60, October, 1964.

101. Committee on Testing Procedures of the IES: "IES Guide for Photometric Measurements of Roadway Sign Installations," Vol. 5, p. 244, July, 1976.

102. Committee on Sports and Recreational Areas of the IES: "Interim Report—Design Criteria for Lighting Sports Events for Color Television Broadcasting," *Illum. Eng.*, Vol. 64, March, 1969.

103. Committee on Testing Procedures of the IES: "IES Guide for Calibration of Photoelectric Control Devices," *J. Illum. Eng. Soc.*, Vol. 4, p. 69, October, 1974.

Color

SECTION 5

Architects, engineers, interior and industrial designers, colorists and color stylists, and lighting designers all have a need to understand color. To satisfy this need, this section has been prepared to increase mutual understanding among those responsible for creating the environment and making it visible and visually functional.

Electromagnetic radiant energy provides a physical stimulus that enters the eye and causes the sensation of color. The spectral characteristics of the stimulus are integrated by the visual system and cannot be differentiated without the use of an instrument. Because the color and the color rendering properties of light sources are becoming increasingly important in the design of an illuminated environment, today's designer of lighting needs a good working knowledge of the vocabulary and practices of modern color science.

The esthetic use of color to produce a pleasing interior requires coordination between the interior designer and the person designing the lighting. Each needs to know how to use color to help provide the desired brightness levels and distributions. Today's lighting designer is faced not only with a choice of color in light sources, but with wide variations in color rendering properties of light sources that may be identical in color.

To provide lighting designers with a basis for their studies in color, the IES committees have developed several reports[1-3,5] that are used as the general plan for this section. In addition, the section concludes with examples of several fields of special applications. A few other sections contained brief discussions of color, with specialized applications (see Index).

NOTE: References are listed at the end of each section.

BASIC CONCEPTS OF COLOR

Color Terms

In Section 1 color terms are carefully defined to provide a way of distinguishing between several commonly confused meanings of the term *color*. Whether one makes strict use of the definitions or not, an understanding of the purpose and need for the differentiations that are made is basic to an understanding of the subject.

The *perceived object color*, the color perceived as belonging to an object, is something perceived instantaneously. It is so common an experience that many persons find it hard to understand why color is not simple to explain in a few easy lessons. But a color perception results from the interaction of many highly complex factors such as: the characteristics of the object, the incident light and the surround; the viewing direction; observer characteristics; and the observer's adaptation. Characteristics of object, light, surround and observer may vary both spectrally and directionally, each in a different manner. The observer may vary in regard to time of seeing, what was seen last, or how attention was focussed in relation to time of seeing. Unless the circumstances of a former situation with which the layman, interior designer, or lighting designer may be familiar, are similar enough in all important respects, a new situation cannot be responded to by reference to past experience alone. Laymen may cope with a new situation by making certain assumptions, or by limiting themselves to the use of conditions with which they are familiar. But lighting designers may not do this if they are to deal with all types of architectural situations, with all types of light sources, and with requirements that will fit new or specialized situations.

The color of an object, or *object color*, as distinct from the *color perceived as belonging to that object*, is defined as the color of light reflected or transmitted by an object when it is illuminated by a standard light source. For this purpose, a CIE* standard observer, using standardized conditions of observation, must also be assumed.

Color, as distinct from *color of an object*, or *color perceived as belonging to an object*, is defined as the characteristic of light by which an observer may distinguish between patches of light of the same size, shape and structure. It reduces itself to a basic description of light in terms of amounts of radiant power at the different wavelengths of the visually effective spectrum, which for most practical purposes is considered to extend from 380 to 770 nanometers. (To identify colors due in part to fluorescent dyes activated by energy in the ultraviolet, it is necessary in specifying the spectral distribution of a light source to extend the wavelength range beyond that which is visually effective, down to 300 nanometers in the ultraviolet, particularly for sources that are intended to reproduce daylight.) Identical colors are produced not only by identical spectral power distributions but also by many different spectral power distributions depending upon their relative visual effectiveness. Such different spectral distributions are called *metamers* (see page 5-16).

The word *color* often is used to cover the three meanings as discussed above. When assumed standard conditions are the same, then there is less need for distinguishing between the three meanings, the *color*, *color of object*, and *perceived color of object*. If lighting designers are to handle new problems in color, including new light sources that may vary widely in spectral distributions, they must know the basic differences between the meanings of color under the above conditions, and keep these distinctions in mind even when using the one term *color* to cover all three situations.

The term *color temperature* is widely used, and often misused, in illumination work. It relates to the color of a completely radiating (blackbody) source at a particular temperature and of light sources that color match such a body. The *color temperature of a light source* is the absolute temperature of a blackbody radiator having a color equal to that of the light source. The *correlated color temperature (of a light source)* is the absolute temperature of a blackbody whose color most nearly resembles that of the light source.

* Commission Internationale de l'Eclairage (International Commission on Illumination).

Defective Color Vision

About 8 per cent of males and 0.5 per cent of females have color vision that differs from that of the majority of the population. These people are usually called "color-blind" although very few (about 0.003 per cent of the total population) can see no color at all.[6] Most color-blind people can distinguish yellows from blues, but confuse reds and greens. They should be excluded from any color measurements or color evaluation procedures that use visual judgments.

Color Rendering

Color rendering is a general expression for the effect of a light source on the color appearance of objects in conscious or subconscious comparison with their color appearance under a reference light source.

The color rendering properties of a light source cannot be assessed by visual inspection of the source, or by a knowledge of its color.[7] For this purpose, full knowledge of its spectral power distribution is required. Viewed in succession under lamps that look quite alike but are different in spectral distribution, objects may look entirely different in color. An extreme case would be a group of objects seen under a pair of color-matched low pressure sodium and yellow fluorescent lamps. Objects that in daylight look red, yellow, green, blue and purple, would appear quite different under these two lamps. In the case of the sodium lamp most objects would lose their daylight appearance, appearing more or less as grays, from light to very dark, near-black. Under the yellow fluorescent lamp, more hues could be recognized, but the color of objects would still vary considerably from their daylight color.

Methods of measuring and specifying color rendering properties of light sources depend on the color appearance of objects under a reference, or standard, light source compared with the appearance of the same objects under the test source (see page 5-24).

Basis for Measurements

If *color* is the characterisic of light by which a human observer distinguishes patches of light, and if *light* is visually evaluated radiant energy, then color may be computed by combining physical measurements of radiant power, wavelength by wavelength, with data on how an observer matches colors. The color-matching characteristics of the internationally adopted CIE standard observers, defined by the tristimulus values of an equal power spectrum, are provided in Figs. 5-1 and 5-6. These *spectral tristimulus values* are often called *color-matching functions.* With data for a standard observer and the spectroradiometric measurement of a light source, the color of that light source can be calculated. Thus spectroradiometry becomes a tool for color measurement. Measurements of radiant power are physical, while evaluation of radiant power by a human observer, based solely on perceptions, is psychological. Visual evaluations, tied down to numbers through measurements made for standardized conditions of test, provide psychophysical methods of measurement.

Visual evaluation of appearance of objects and light sources may be in terms derived wholly from one's perceptions. One convenient and useful set of terms describing these perceptions is *hue, brightness* and *colorfulness.*[8] Hue is the attribute according to which an area appears to be similar to one, or to proportions of two, of the

Fig. 5-1. Spectral Tristimulus Values for Equal Spectral Power Source

a. CIE 1931 Standard Observer

Wavelength (nanometer)	$\bar{x}(\lambda)$	$\bar{y}(\lambda)$	$\bar{z}(\lambda)$	Wavelength (nanometer)	$\bar{x}(\lambda)$	$\bar{y}(\lambda)$	$\bar{z}(\lambda)$
380	0.0014	0.0000	0.0065	580	0.9163	0.8700	0.0017
385	0.0022	0.0001	0.0105	585	0.9786	0.8163	0.0014
390	0.0042	0.0001	0.0201	590	1.0263	0.7570	0.0011
395	0.0076	0.0002	0.0362	595	1.0567	0.6949	0.0010
400	0.0143	0.0004	0.0679	600	1.0622	0.6310	0.0008
405	0.0232	0.0006	0.1102	605	1.0456	0.5668	0.0006
410	0.0435	0.0012	0.2074	610	1.0026	0.5030	0.0003
415	0.0776	0.0022	0.3713	615	0.9384	0.4412	0.0002
420	0.1344	0.0040	0.6456	620	0.8544	0.3810	0.0002
425	0.2148	0.0073	1.0391	625	0.7514	0.3210	0.0001
430	0.2839	0.0116	1.3856	630	0.6424	0.2650	0.0000
435	0.3285	0.0168	1.6230	635	0.5419	0.2170	0.0000
440	0.3483	0.0230	1.7471	640	0.4479	0.1750	0.0000
445	0.3481	0.0298	1.7826	645	0.3608	0.1382	0.0000
450	0.3362	0.0380	1.7721	650	0.2835	0.1070	0.0000
455	0.3187	0.0480	1.7441	655	0.2187	0.0816	0.0000
460	0.2908	0.0600	1.6692	660	0.1649	0.0610	0.0000
465	0.2511	0.0739	1.5281	665	0.1212	0.0446	0.0000
470	0.1954	0.0910	1.2876	670	0.0874	0.0320	0.0000
475	0.1421	0.1126	1.0419	675	0.0636	0.0232	0.0000
480	0.0956	0.1390	0.8130	680	0.0468	0.0170	0.0000
485	0.0580	0.1693	0.6162	685	0.0329	0.0119	0.0000
490	0.0320	0.2080	0.4652	690	0.0227	0.0082	0.0000
495	0.0147	0.2586	0.3533	695	0.0158	0.0057	0.0000
500	0.0049	0.3230	0.2720	700	0.0114	0.0041	0.0000
505	0.0024	0.4073	0.2123	705	0.0081	0.0029	0.0000
510	0.0093	0.5030	0.1582	710	0.0058	0.0021	0.0000
515	0.0291	0.6082	0.1117	715	0.0041	0.0015	0.0000
520	0.0633	0.7100	0.0782	720	0.0029	0.0010	0.0000
525	0.1096	0.7932	0.0573	725	0.0020	0.0007	0.0000
530	0.1655	0.8620	0.0422	730	0.0014	0.0005	0.0000
535	0.2257	0.9149	0.0298	735	0.0010	0.0004	0.0000
540	0.2904	0.9540	0.0203	740	0.0007	0.0002	0.0000
545	0.3597	0.9803	0.0134	745	0.0005	0.0002	0.0000
550	0.4334	0.9950	0.0087	750	0.0003	0.0001	0.0000
555	0.5121	1.0000	0.0057	755	0.0002	0.0001	0.0000
560	0.5945	0.9950	0.0039	760	0.0002	0.0001	0.0000
565	0.6784	0.9786	0.0027	765	0.0001	0.0000	0.0000
570	0.7621	0.9520	0.0021	770	0.0001	0.0000	0.0000
575	0.8425	0.9154	0.0018	775	0.0001	0.0000	0.0000
580	0.9163	0.8700	0.0017	780	0.0000	0.0000	0.0000
	Totals				21.3714	21.3711	21.3715

b. CIE 1964 Supplementary Observer

Wavelength (nanometer)	$\bar{x}_{10}(\lambda)$	$\bar{y}_{10}(\lambda)$	$\bar{z}_{10}(\lambda)$	Wavelength (nanometer)	$\bar{x}_{10}(\lambda)$	$\bar{y}_{10}(\lambda)$	$\bar{z}_{10}(\lambda)$
380	0.0002	0.0000	0.0007	580	1.0142	0.8689	0.0000
385	0.0007	0.0001	0.0029	585	1.0743	0.8256	0.0000
390	0.0024	0.0003	0.0105	590	1.1185	0.7774	0.0000
395	0.0072	0.0008	0.0323	595	1.1343	0.7204	0.0000
400	0.0191	0.0020	0.0860	600	1.1240	0.6583	0.0000
405	0.0434	0.0045	0.1971	605	1.0891	0.5939	0.0000
410	0.0847	0.0088	0.3894	610	1.0305	0.5280	0.0000
415	0.1406	0.0145	0.6568	615	0.9507	0.4618	0.0000
420	0.2045	0.0214	0.9725	620	0.8563	0.3981	0.0000
425	0.2647	0.0295	1.2825	625	0.7549	0.3396	0.0000
430	0.3147	0.0387	1.5535	630	0.6475	0.2835	0.0000
435	0.3577	0.0496	1.7985	635	0.5351	0.2283	0.0000
440	0.3837	0.0621	1.9673	640	0.4316	0.1798	0.0000
445	0.3867	0.0747	2.0273	645	0.3437	0.1402	0.0000
450	0.3707	0.0895	1.9948	650	0.2683	0.1076	0.0000
455	0.3430	0.1063	1.9007	655	0.2043	0.0812	0.0000
460	0.3023	0.1282	1.7454	660	0.1526	0.0603	0.0000
465	0.2541	0.1528	1.5549	665	0.1122	0.0441	0.0000
470	0.1956	0.1852	1.3176	670	0.0813	0.0318	0.0000
475	0.1323	0.2199	1.0302	675	0.0579	0.0226	0.0000
480	0.0805	0.2536	0.7721	680	0.0409	0.0159	0.0000
485	0.0411	0.2977	0.5701	685	0.0286	0.0111	0.0000
490	0.0162	0.3391	0.4153	690	0.0199	0.0077	0.0000
495	0.0051	0.3954	0.3024	695	0.0138	0.0054	0.0000
500	0.0038	0.4608	0.2185	700	0.0096	0.0037	0.0000
505	0.0154	0.5314	0.1592	705	0.0066	0.0026	0.0000
510	0.0375	0.6067	0.1120	710	0.0046	0.0018	0.0000
515	0.0714	0.6857	0.0822	715	0.0031	0.0012	0.0000
520	0.1177	0.7618	0.0607	720	0.0022	0.0008	0.0000
525	0.1730	0.8233	0.0431	725	0.0015	0.0006	0.0000
530	0.2365	0.8752	0.0305	730	0.0010	0.0004	0.0000
535	0.3042	0.9238	0.0206	735	0.0007	0.0003	0.0000
540	0.3768	0.9620	0.0137	740	0.0005	0.0002	0.0000
545	0.4516	0.9822	0.0079	745	0.0004	0.0001	0.0000
550	0.5298	0.9918	0.0040	750	0.0003	0.0001	0.0000
555	0.6161	0.9991	0.0011	755	0.0002	0.0001	0.0000
560	0.7052	0.9973	0.0000	760	0.0001	0.0000	0.0000
565	0.7938	0.9824	0.0000	765	0.0001	0.0000	0.0000
570	0.8787	0.9556	0.0000	770	0.0001	0.0000	0.0000
575	0.9512	0.9152	0.0000	775	0.0000	0.0000	0.0000
580	1.0142	0.8689	0.0000	780	0.0000	0.0000	0.0000
	Totals				23.3294	23.3324	23.3343

Fig. 5-2. Spectral Power Distributions of CIE Standard Illuminants

Wavelength (nanometer)	Standard Illuminants					
	A	B	C	D_{55}	D_{65}	D_{75}
300	9			0.2	0.3	0.4
310	14			21	33	51
320	19	0.2	0.1	112	202	298
330	27	5	4	206	371	549
340	36	23	26	239	399	573
350	47	54	66	278	449	627
360	61	93	123	306	466	630
370	78	148	203	343	521	703
380	98	218	313	326	500	667
390	121	304	450	381	546	700
400	147	402	601	610	828	1019
410	177	507	765	686	915	1119
420	210	615	932	716	934	1128
430	247	711	1067	679	867	1031
440	287	786	1154	856	1049	1212
450	331	831	1178	980	1170	1330
460	378	859	1169	1005	1178	1324
470	429	895	1176	999	1149	1273
480	482	926	1177	1027	1159	1268
490	539	939	1146	981	1088	1178
500	599	916	1065	1007	1094	1066
510	661	882	972	1007	1078	1137
520	725	871	920	1000	1048	1087
530	791	897	931	1042	1077	1104
540	859	943	970	1021	1044	1063
550	929	982	999	1030	1040	1049
560	1000	1000	1000	1000	1000	1000
570	1072	998	972	972	963	956
580	1144	982	929	977	958	942
590	1217	965	885	914	887	870
600	1290	953	852	944	900	872
610	1363	958	840	951	896	861
620	1436	970	837	942	877	836
630	1508	982	836	904	833	787
640	1580	994	834	923	837	784
650	1650	1011	838	889	800	748
660	1720	1021	835	903	802	743
670	1788	1020	820	939	823	754
680	1854	1011	798	900	783	716
690	1919	988	762	797	697	639
700	1983	964	725	828	716	651
710	2044	936	688	848	743	681
720	2104	904	649	702	616	564
730	2161	870	612	793	699	642
740	2217	845	584	850	751	692
750	2270	829	562	719	636	586
760	2321	824	552	528	464	426
770	2370	831	553	759	668	614
780	2417			718	634	583
790	2461			729	643	591
800	2503			673	595	547
810	2543			587	520	479
820	2581			650	574	529
830	2616			683	603	555
	Chromaticity coordinates (CIE 1931 System)					
x	0.4476	0.3484	0.3101	0.3324	0.3127	0.2990
y	0.4074	0.3516	0.3162	0.3475	0.3290	0.3150

perceived colors red, orange, yellow, green, blue and purple. Brightness is the attribute according to which an area appears to be emitting, transmitting or reflecting more or less light. Colorfulness is the attribute according to which an area appears to exhibit more or less chromatic color; *i.e.*, departure from gray. The term *saturation*, which is colorfulness judged in proportion to brightness, is often more useful than colorfulness itself. For example, the image of a light source viewed in a sheet of glass is less bright and less colorful than the actual light source, but the saturation is the same. Thus the color appearance of a light source may be described in terms of hue, brightness and saturation. The perceived color of an object, as opposed to a light source, is often best described by *hue*, *lightness* and *perceived chroma*, where lightness and perceived chroma are brightness and colorfulness, respectively, judged in proportion to the brightness of a similarly illuminated area that appears to be white or highly transmitting. The utility of lightness and perceived chroma arises because the color of an object is usually judged relative, rather than absolute, to similarly illuminated surrounding areas.

Many widely used psychophysical methods for describing and specifying color show poor correlation with perceptual factors, and often these are converted to more meaningful visual terms, usually to a more uniform color spacing, of which the Munsell system[9] and the CIE 1976 Uniform Color Spaces[10,11] are prime examples.

CIE System of Color Specification

Basic CIE System.[12] This is a system originally recommended in 1931 by the CIE to define all metameric pairs (see page 5-16) by giving the amounts X, Y, Z, of three imaginary primary colors required by a standard observer to match the color being specified. These amounts may be calculated as a summation from the spectral compositions of the radiant power of the source (and/or of the color specimen), and the spectral tristimulus values for an equal power source (see Fig. 5-1). For example:

$$X = k \sum_{\lambda=380}^{\lambda=780} S(\lambda)\rho(\lambda)\bar{x}(\lambda)\Delta\lambda$$

where

$S(\lambda)$ is the spectral irradiance distribution of the source (see Fig. 5-2),

$\rho(\lambda)$ is the spectral reflectance of the specimen,

k is a normalizing factor,

$\bar{x}(\lambda)$ is the spectral tristimulus value from Fig. 5-1 with similar expressions for Y and Z wherein $\bar{y}(\lambda)$ and $\bar{z}(\lambda)$ respectively are substituted for $\bar{x}(\lambda)$.

The normalizing factor, k, may be assigned any arbitrary value provided it is kept constant throughout any particular application. Where only the relative values of X, Y, and Z are required, the value of k is usually chosen so that Y has the value 100.0. For object colors the normalizing factor is usually given the value:

$$k = \frac{100}{\Sigma S(\lambda)\bar{y}(\lambda)\Delta\lambda}$$

According to this normalization, the value of Y is the luminous reflectance expressed as a percentage. In the special case where the absolute values of $S(\lambda)\Delta\lambda$ are given (*e.g.*, in watts), it is convenient to take $k = K_m = 683$ lumens per watt, whereby the value of Y gives the luminous flux in the stimulus in lumens. Here, the accepted symbolism for $S(\lambda)$ is Φ_e, λ (see Section 1).

Depending on the geometrical arrangement, the spectral reflectance, $\rho(\lambda)$, may be replaced by the spectral reflectance factor (in which case Y is the percentage luminous reflectance factor), or the spectral transmittance (in which case Y is the percentage luminous transmittance). For convenience, the rest of this Section will use only reflectance, with the understanding that one of the other quantities should be substituted whenever it is more appropriate.

To simplify the computation, normalized values of $\bar{x}(\lambda)S(\lambda)$ $\bar{y}(\lambda)S(\lambda)$ and $\bar{z}(\lambda)S(\lambda)$ for each of the CIE standard illuminants are given in steps of 10 nm in Fig. 5-3. These have been calculated from the values of $\bar{x}(\lambda)$, $\bar{y}(\lambda)$ and $\bar{z}(\lambda)$ in Fig. 5-1 and $S(\lambda)$ in Fig. 5-2. For each illuminant the values have been normalized so that $\Sigma\bar{y}(\lambda)S(\lambda) = 100.0$. This simplifies the computation further.

The fractions, $X/(X + Y + Z)$, $Y/(X + Y + Z)$, $Z/(X + Y + Z)$ are known as chromaticity coordinates, x, y, z respectively. Note that $x + y + z = 1$ whereby specification of any two fixes the third. By convention, chromaticity usually is stated in terms of x and y and plotted in a rectangular coordinate system as shown in Fig. 5-7. In this *chromaticity diagram*, the points representing lights of single wavelengths plot along a horseshoe-shaped curve called the *spectrum locus*. The line joining the extremities of the spectrum locus is known as the *purple boundary* and is the locus of the most saturated purples obtainable.

A sample calculation for determining the CIE coordinates x, y, and Y is shown in Fig. 5-4 for a deep red surface when illuminated by CIE illuminant C. In Fig. 5-4, column I is a listing of wavelengths in 10-nanometer steps, column II is a tabulation of spectral reflectance values for the deep red surface at each wavelength in column I, and column III lists the CIE tristimulus computational data from Fig. 5-3. By multiplying column II by column III and summing each tabulation in column IV, X, Y and Z values are determined. Then by using the three fractions in the previous paragraph, the chromaticity coordinates are determined. The percentage luminous reflectance is determined by multiplying the Y value by the normalizing factor $k = 100/\Sigma S(\lambda)\bar{y}(\lambda)\Delta\lambda = 1/1000$.

CIE standards include data for the 1931 Standard Observer (for angular subtense between 1° and 4°) as shown in Fig. 5-1a, the 1964 Supplementary Observer (for angular subtense greater than 4°) as shown in Fig. 5-1b, four standard sources (A representing tungsten at a color temperature of 2856 K; B representing direct sunlight with a correlated color temperature of approximately 4900 K; C representing average daylight with a correlated color temperature of approximately 6800 K; and D_{65} representing a phase of natural daylight at approximately 6500 K), and two supplementary standard sources (D_{55} representing a phase of natural daylight at approximately 5500 K and D_{75} representing a phase of natural daylight at approximately 7500 K). The spectral power distributions of these sources (known as CIE Standard Illuminants) are given in Fig. 5-2. Four alternative conditions for illuminating/viewing a test sample are specified: a) 45°/normal, b) normal/45°, c) diffuse/normal, and d) normal/diffuse (diffuse illuminating or viewing is usually achieved by the use of an integrating sphere). For an extended discussion of the calculation and application of CIE data, including extensive tables of quantitative data and methods of colorimetry, consult *Color Science.*[13]

Dominant Wavelength and Excitation Purity. Dominant wavelength and excitation purity are quantities more suggestive of color appearance of objects than a CIE x,y-specification, and may be determined on an x,y-diagram in relation to the spectrum locus and an assumed achromatic point (for object colors this is usually the point for the light source). See Fig. 5-7. The *dominant wavelength* of all colors whose x,y-coordinates fall on a straight line connecting the achromatic point with a point on the spectrum locus is the wavelength indicated at the intersection of that line with the spectrum locus. For some colors, the straight line from the achromatic point though the test chromaticity will

Fig. 5–3. CIE Tristimulus Computational Data for Several Illuminants Computed for the CIE 1931 Standard Observer

Wavelength (nanometers)	A			B			C			D_{65}			D_{55}			D_{75}		
	$\bar{x}(\lambda)S(\lambda)$	$\bar{y}(\lambda)S(\lambda)$	$\bar{z}(\lambda)S(\lambda)$	$\bar{x}(\lambda)S(\lambda)$	$\bar{y}(\lambda)S(\lambda)$	$\bar{z}(\lambda)S(\lambda)$	$\bar{x}(\lambda)S(\lambda)$	$\bar{y}(\lambda)S(\lambda)$	$\bar{z}(\lambda)S(\lambda)$	$\bar{x}(\lambda)S(\lambda)$	$\bar{y}(\lambda)S(\lambda)$	$\bar{z}(\lambda)S(\lambda)$	$\bar{x}(\lambda)S(\lambda)$	$\bar{y}(\lambda)S(\lambda)$	$\bar{z}(\lambda)S(\lambda)$	$\bar{x}(\lambda)S(\lambda)$	$\bar{y}(\lambda)S(\lambda)$	$\bar{z}(\lambda)S(\lambda)$
380	0.001	0.000	0.006	0.003	0.000	0.014	0.004	0.000	0.020	0.007	0.000	0.031	0.004	0.000	0.020	0.009	0.000	0.041
390	0.005	0.000	0.023	0.013	0.000	0.060	0.019	0.000	0.089	0.022	0.001	0.104	0.015	0.000	0.073	0.028	0.001	0.132
400	0.019	0.001	0.093	0.056	0.002	0.268	0.085	0.002	0.404	0.112	0.003	0.532	0.083	0.002	0.394	0.137	0.004	0.650
410	0.071	0.002	0.340	0.217	0.006	1.033	0.329	0.009	1.570	0.377	0.010	1.796	0.284	0.008	1.354	0.457	0.013	2.179
420	0.262	0.008	1.256	0.812	0.024	3.890	1.238	0.037	5.949	1.189	0.035	5.711	0.916	0.027	4.400	1.424	0.042	6.839
430	0.649	0.027	3.167	1.983	0.081	9.678	2.997	0.122	14.628	2.330	0.095	11.370	1.836	0.075	8.959	2.749	0.112	13.414
440	0.926	0.061	4.647	2.689	0.178	13.489	3.975	0.262	19.938	3.458	0.228	17.343	2.838	0.187	14.235	3.964	0.262	19.883
450	1.031	0.117	5.435	2.744	0.310	14.462	3.915	0.443	20.638	3.724	0.421	19.627	3.136	0.354	16.528	4.199	0.475	22.133
460	1.019	0.210	5.851	2.454	0.506	14.085	3.362	0.694	19.299	3.243	0.669	18.614	2.780	0.574	15.959	3.614	0.746	20.746
470	0.776	0.362	5.116	1.718	0.800	11.319	2.272	1.058	14.972	2.124	0.989	13.998	1.858	0.865	12.243	2.336	1.088	15.395
480	0.428	0.622	3.636	0.870	1.265	7.396	1.112	1.618	9.461	1.048	1.524	8.915	0.933	1.357	7.936	1.139	1.655	9.682
490	0.160	1.039	2.324	0.295	1.918	4.290	0.363	2.358	5.274	0.330	2.142	4.791	0.299	1.941	4.341	0.354	2.301	5.146
500	0.027	1.792	1.509	0.044	2.908	2.449	0.052	3.401	2.864	0.051	3.343	2.815	0.047	3.094	2.606	0.054	3.536	2.978
510	0.057	3.080	0.969	0.081	4.360	1.371	0.089	4.833	1.520	0.095	5.132	1.614	0.089	4.819	1.516	0.099	5.371	1.689
520	0.425	4.771	0.525	0.541	6.072	0.669	0.576	6.462	0.712	0.628	7.041	0.775	0.602	6.754	0.744	0.646	7.245	0.798
530	1.214	6.322	0.309	1.458	7.594	0.372	1.523	7.934	0.338	1.687	8.785	0.430	1.641	8.546	0.418	1.717	8.941	0.438
540	2.313	7.600	0.162	2.689	8.834	0.188	2.785	9.149	0.195	2.869	9.425	0.201	2.821	9.267	0.197	2.899	9.523	0.203
550	3.732	8.568	0.075	4.183	9.603	0.084	4.282	9.832	0.086	4.267	9.796	0.086	4.245	9.747	0.085	4.270	9.803	0.086
560	5.510	9.222	0.036	5.840	9.774	0.038	5.880	9.841	0.039	5.625	9.415	0.037	5.656	9.466	0.037	5.584	9.345	0.037
570	7.571	9.457	0.021	7.472	9.334	0.021	7.322	9.147	0.020	6.947	8.678	0.019	7.048	8.804	0.019	6.844	8.550	0.019
580	9.719	9.228	0.018	8.843	8.396	0.016	8.417	7.992	0.016	8.304	7.885	0.015	8.520	8.089	0.016	8.108	7.699	0.015
590	11.579	8.540	0.012	9.728	7.176	0.010	8.984	6.627	0.010	8.612	6.352	0.009	8.927	6.584	0.010	8.387	6.186	0.009
600	12.704	7.547	0.010	9.948	5.909	0.007	8.949	5.316	0.007	9.046	5.347	0.007	9.541	5.668	0.007	8.704	5.171	0.007
610	12.669	6.356	0.004	9.436	4.734	0.003	8.325	4.176	0.002	8.499	4.264	0.003	9.074	4.553	0.003	8.114	4.071	0.002
620	11.373	5.071	0.003	8.140	3.630	0.002	7.070	3.153	0.002	7.089	3.101	0.002	7.658	3.415	0.002	6.710	2.992	0.002
630	8.980	3.704	0.000	6.200	2.558	0.000	5.309	2.190	0.000	5.062	2.088	0.000	5.528	2.280	0.000	4.754	1.961	0.000
640	6.558	2.562	0.000	4.374	1.709	0.000	3.693	1.443	0.000	3.547	1.386	0.000	3.934	1.537	0.000	3.301	1.290	0.000
650	4.336	1.637	0.000	2.815	1.062	0.000	2.349	0.886	0.000	2.147	0.810	0.000	2.397	0.905	0.000	1.993	0.752	0.000
660	2.628	0.972	0.000	1.655	0.612	0.000	1.361	0.504	0.000	1.252	0.463	0.000	1.417	0.524	0.000	1.152	0.426	0.000
670	1.448	0.530	0.000	0.876	0.321	0.000	0.708	0.259	0.000	0.680	0.249	0.000	0.781	0.286	0.000	0.620	0.227	0.000
680	0.804	0.292	0.000	0.465	0.169	0.000	0.369	0.134	0.000	0.347	0.126	0.000	0.401	0.145	0.000	0.315	0.114	0.000
690	0.404	0.146	0.000	0.220	0.080	0.000	0.171	0.062	0.000	0.150	0.054	0.000	0.172	0.062	0.000	0.136	0.049	0.000
700	0.209	0.075	0.000	0.108	0.039	0.000	0.082	0.029	0.000	0.077	0.028	0.000	0.090	0.032	0.000	0.070	0.025	0.000
710	0.110	0.040	0.000	0.053	0.019	0.000	0.039	0.014	0.000	0.041	0.015	0.000	0.047	0.017	0.000	0.037	0.013	0.000
720	0.057	0.019	0.000	0.026	0.009	0.000	0.019	0.006	0.000	0.017	0.006	0.000	0.019	0.007	0.000	0.015	0.005	0.000
730	0.028	0.010	0.000	0.012	0.004	0.000	0.008	0.003	0.000	0.009	0.003	0.000	0.011	0.004	0.000	0.008	0.003	0.000
740	0.011	0.006	0.000	0.006	0.002	0.000	0.004	0.002	0.000	0.005	0.002	0.000	0.006	0.002	0.000	0.005	0.002	0.000
750	0.006	0.002	0.000	0.002	0.001	0.000	0.002	0.001	0.000	0.002	0.001	0.000	0.002	0.001	0.000	0.002	0.001	0.000
760	0.004	0.002	0.000	0.002	0.001	0.000	0.001	0.001	0.000	0.001	0.000	0.000	0.001	0.001	0.000	0.001	0.000	0.000
770	0.002	0.000	0.000	0.001	0.000	0.000	0.001	0.000	0.000	0.001	0.000	0.000	0.001	0.000	0.000	0.001	0.001	0.000
Total	109.828	100.000	35.547	99.072	100.000	85.223	98.041	100.000	118.103	95.018	100.000	108.845	95.655	100.000	92.102	94.954	100.000	122.520

Fig. 5-4. Determination from Spectrophotometric Curve of CIE Coordinates for a Surface Illuminated by Illuminant C

I Wavelength (nanometers)	II Reflectance, $\rho(\lambda)$	III CIE Data for Illuminant C (from Fig. 5-3) (multiplied by 1000)			IV (II × III)		
		$\bar{x}(\lambda)S(\lambda)$	$\bar{y}(\lambda)S(\lambda)$	$\bar{z}(\lambda)S(\lambda)$	$\rho(\lambda)\bar{x}(\lambda)S(\lambda)$	$\rho(\lambda)\bar{y}(\lambda)S(\lambda)$	$\rho(\lambda)\bar{z}(\lambda)S(\lambda)$
380	0.051	4		20			1
390	.051	19		89	1		5
400	.051	85	2	404	4		21
410	.051	329	9	1,570	17		80
420	.050	1,238	37	5,949	62	2	297
430	.050	2,997	122	14,628	150	6	731
440	.050	3,975	262	19,938	199	13	997
450	.047	3,915	443	20,638	184	21	970
460	.045	3,362	694	19,299	151	31	868
470	.044	2,272	1,058	14,972	100	47	659
480	.043	1,112	1,618	9,461	48	70	407
490	.041	363	2,358	5,274	15	97	216
500	.041	52	3,401	2,864	2	139	117
510	.041	89	4,833	1,520	4	198	62
520	.041	576	6,462	712	24	265	29
530	.041	1,523	7,934	388	62	325	16
540	.041	2,785	9,149	195	114	375	8
550	.042	4,282	9,832	86	180	413	4
560	.043	5,880	9,841	39	253	423	2
570	.050	7,322	9,147	20	366	457	1
580	.075	8,417	7,992	16	631	599	1
590	.145	8,984	6,627	10	1,303	961	1
600	.290	8,949	5,316	7	2,595	1,542	2
610	.465	8,325	4,176	2	3,871	1,942	1
620	.575	7,070	3,153	2	4,065	1,813	1
630	.623	5,309	2,190		3,308	1,364	
640	.648	3,693	1,443		2,393	935	
650	.667	2,349	886		1,567	591	
660	.683	1,361	504		930	344	
670	.699	708	259		495	181	
680	.713	369	134		263	96	
690	.725	171	62		124	45	
700	.739	82	29		61	21	
710	.749	39	14		29	10	
720	.762	19	6		14	5	
730	.775	8	3		6	2	
740	.785	4	2		3	2	
750	.791	2	1		2	1	
760	.795	1	1		1	1	
Sums		98,040	100,000	118,103	$X = 23,597$	$Y = 13,337$	$Z = 5,497$

$x = \frac{X}{X+Y+Z} = 0.556$, $y = \frac{Y}{X+Y+Z} = 0.314$, $z = \frac{Z}{X+Y+Z} = 0.130$, and percentage luminous reflectance = 13.3

strike the purple boundary rather than the spectrum locus. For these colors the line must be extended backwards from the achromatic point. The point where the extended line strikes the spectrum locus determines the *complementary wavelength* of such a color. *Excitation purity* is defined as the distance from the achromatic point expressed relative to the total distance in the same direction from the achromatic point to the spectrum locus or the purple boundary, as the case may be. An x,y-specification of any object color relates it only to the light source for which the object color is calculated; consequently, dominant wavelength and excitation purity of any object depend on the spectral composition of its illumination.

CIE Uniform Color Space System. Distances in the CIE (x,y)-diagram or X,Y,Z space do not correlate well with the perceived magnitudes of color differences. For this reason various transformations have been suggested that provide more uniform spacing.

In 1960 the CIE provisionally recommended

that whenever a diagram is desired to yield chromaticity spacing more uniform than the CIE (x,y)-diagram, a uniform chromaticity-scale diagram based on that described in 1937 by MacAdam[14] be used. The ordinate and abscissa of this (u,v)-diagram are defined as:

$$u = \frac{4X}{(X + 15Y + 3Z)} \text{ or, } u = \frac{4x}{(-2x + 12y + 3)}$$

$$v = \frac{6Y}{(X + 15Y + 3Z)} \text{ or, } v = \frac{6y}{(-2x + 12y + 3)}$$

To convert this to a three-dimensional metric system that is useful in studying color differences, the CIE added a recommendation developed for the purpose in 1963 by Wyszecki[15] that converts Y to a lightness index, W^*, by the relationship

$$W^* = 25Y^{1/3} - 17 \ (1 \leqq Y \leqq 100)$$

and converts the chromaticity coordinates (u,v) to chromaticness indices U^* and V^* by the relationships:

$$U^* = 13\ W^*\ (u - u_0)$$
$$V^* = 13\ W^*\ (v - v_0)$$

The lightness index, W^*, approximates the Munsell value function in the range of Y from 1 to 100 per cent. The chromaticity coordinates (u_0, v_0) refer to the nominally achromatic color (usually that of the source) placed at the origin of the (U^*, V^*)-system.

In 1976 the CIE[10,11] recommended two new uniform color spaces known as CIELUV and CIELAB. Although these give a more uniform representation of color differences and therefore supersede the $(U^*V^*W^*)$ space for most purposes, the latter is still used for the calculation of CIE color rendering indices (see page 5–24).

The three coordinates of CIELUV are L^*, u^* and v^*, defined by:

$$L^* = 116(Y/Y_n)^{1/3} - 16 \quad \text{for} \quad Y/Y_n > 0.008856$$
$$L^* = 903.29(Y/Y_n) \quad \text{for} \quad Y/Y_n \leqslant 0.008856$$
$$u^* = 13L^*\ (u' - u'_n)$$
$$v^* = 13L^*\ (v' - v'_n)$$

where

$$u' = 4X/(X + 15Y + 3Z)$$
$$v' = 9Y/(X + 15Y + 3Z)$$

and u'_n, v'_n and Y_n are the values of u', v' and Y of the nominally achromatic color (usually that of the source with $Y_n = 100$). The major change from the $(U^*V^*W^*)$ system is that v' is equal to 1.5 v. L^* is a minor modification of W^*, and u' is the same as u.

The three coordinates of CIELAB are L^*, a^* and b^*, defined by:

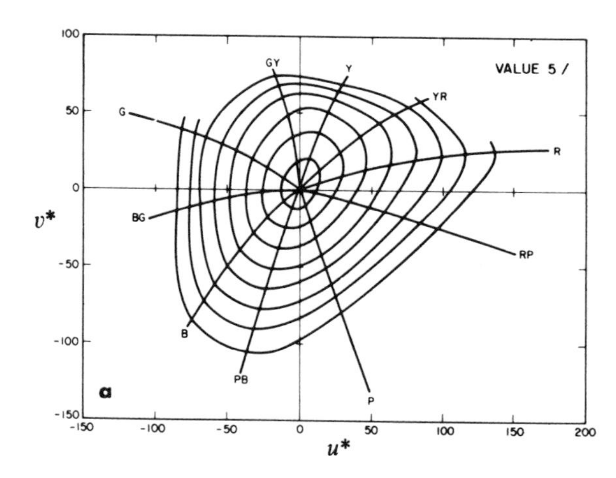

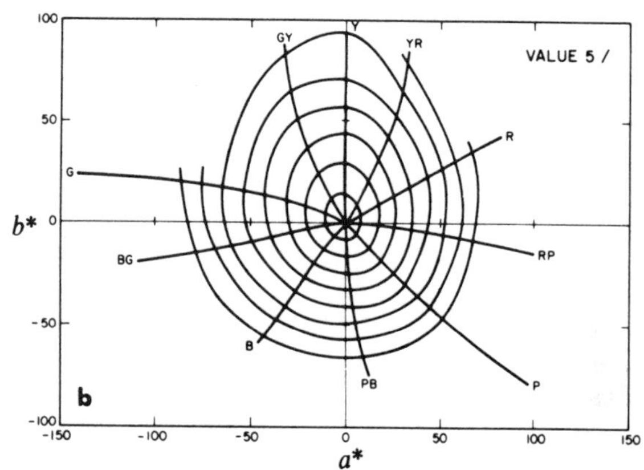

Fig. 5-5. Loci of constant Munsell hue and chroma: (a) plotted in the CIE 1976 u^* v^* diagram, and (b) plotted in the CIE 1976 a^* b^* diagram.[16]

$$L^* = 116f(Y/Y_n) - 16$$
$$a^* = 500\ [f(X/X_n) - f(Y/Y_n)]$$
$$b^* = 200\ [f(Y/Y_n) - f(Z/Z_n)]$$

where $f(q) = (q)^{1/3}$ for $q > 0.008856$

$f(q) = 7.787q + 16/116$ for $q \leq 0.008856$

with $q = X/X_n$, Y/Y_n or Z/Z_n

X_n, Y_n and Z_n are the values of X, Y and Z of the nominally achromatic color (usually that of the source with $Y_n = 100$). The lightness index, L^*, is the same for both CIELUV and CIELAB.

Loci of constant Munsell hue and chroma for value 5/ are plotted in u^*v^* and a^*b^* diagrams in Fig. 5–5.

These two uniform color spaces each have, associated with them, a color-difference formula by which a measure of the total color-difference between two object colors may be calculated. In the CIELUV system, the color difference is measured by

$$\Delta E^*_{uv} = [(\Delta L^*)^2 + (\Delta u^*)^2 + (\Delta v^*)^2]^{1/2}$$

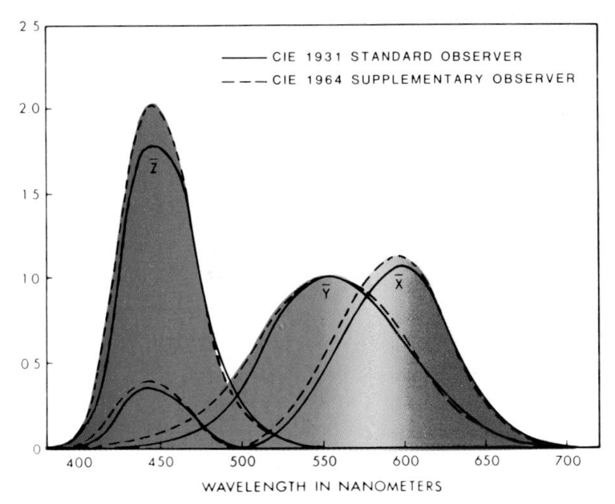

Fig. 5-6. Spectral tristimulus values for equal spectral power source. (Colors shown are approximate representations.)

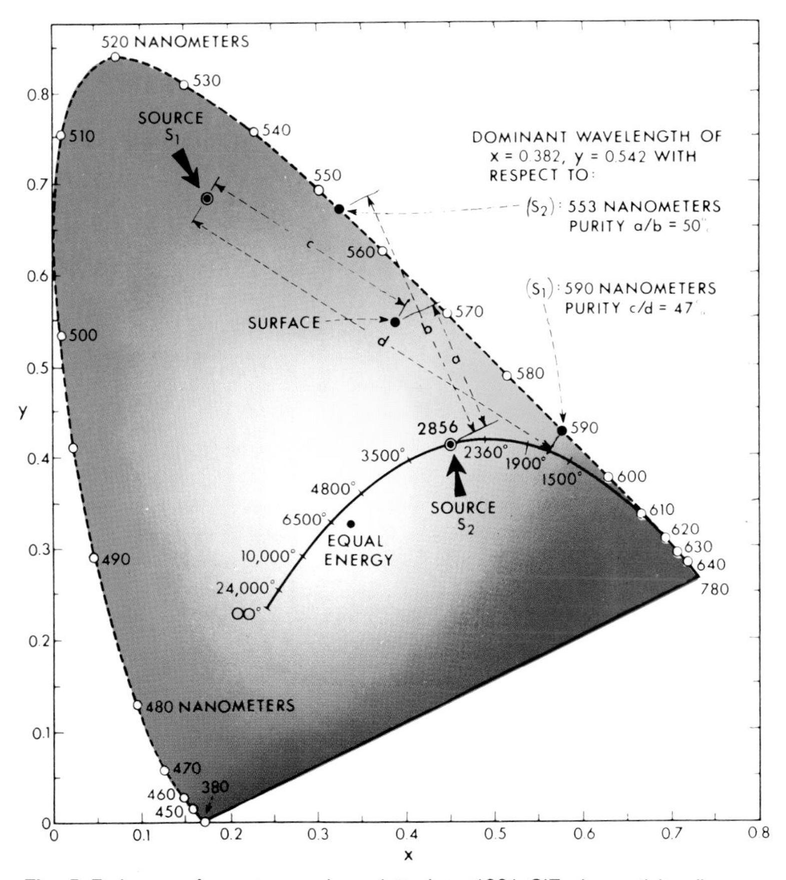

Fig. 5-7. Locus of spectrum colors plotted on 1931 CIE chromaticity diagram showing method of obtaining dominant wavelength and purity for different samples under different light sources. (The colors shown are only very approximate representations of the appearance of the various chromaticities to an observer adapted to D_{65}. In particular, due to the limited gamut of printing inks, nearly all the colors are much less saturated than the corresponding chromaticities would normally appear.)

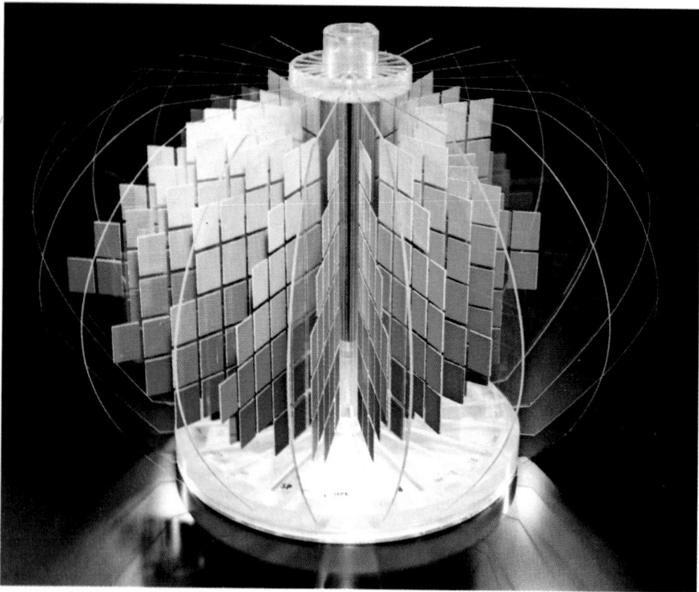

Fig. 5-8. Form of Munsell color solid is shown in the photograph at the left. Diagram below is a cut-away view showing notation scales of hue, value, and chroma (*e.g.*, 5Y 5/4), and the relation of constant hue charts to the three-dimensional representation.

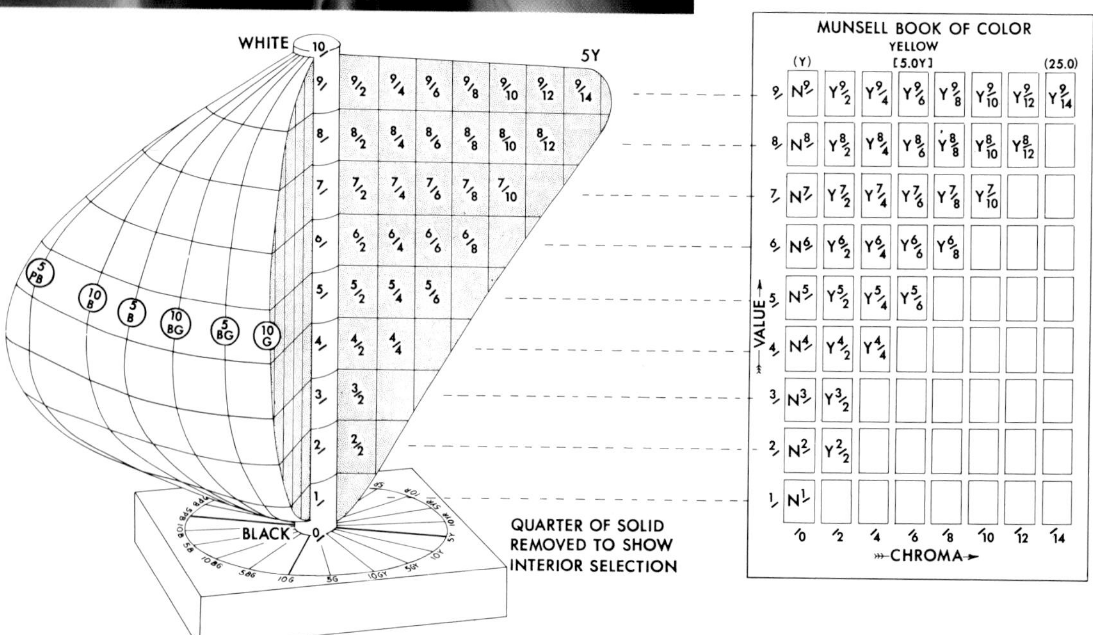

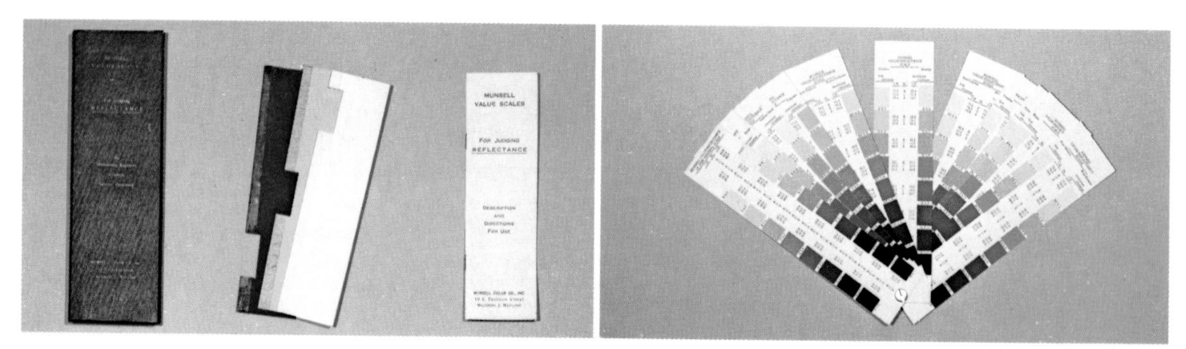

Fig. 5-9. Munsell value scales for judging reflectance.

In the CIELAB system, similarly, the color-difference is measured by

$$\Delta E^*_{ab} = [(\Delta L^*)^2 + (\Delta a^*)^2 + (\Delta b^*)^2]^{1/2}$$

These two formulas are useful for setting color tolerances in industrial situations. They are recommended by the CIE to unify practice which in the past has involved the use of 10 or 20 different color-difference formulas.

Psychometric Lightness, Chroma and Hue-Angle. Correlates of the subjective attributes lightness, perceived chroma and hue-angle can be derived from either CIELUV or CIELAB. They are known as psychometric lightness, psychometric chroma and psychometric hue-angle and are defined as follows:

psychometric lightness $= L^*$

psychometric chroma $= C^*_{uv} = (u^{*2} + v^{*2})^{1/2}$ or $C^*_{ab} = (a^{*2} + b^{*2})^{1/2}$

psychometric hue-angle $= h_{uv} = \tan^{-1}(v^*/u^*)$ or $h_{ab} = \tan^{-1}(b^*/a^*)$

Although these quantities are approximate correlates of the respective subjective attributes, the actual perceived color depends significantly on the viewing conditions such as the nature of the surround. The exact degree of agreement of these psychometric measures with the corresponding subjective attributes, even for standard daylight viewing conditions, has not been determined.

The geometrical relationships between these quantities are illustrated in Fig. 5–10.

Other Systems of Color Specification

Munsell System. This is a system of specifying color on scales of hue, value and chroma. These are exemplified by a collection of color chips forming an atlas of charts that show linear series for which two of the three variables are constant. See Fig. 5–8. The hue scale contains five principal and five intermediate hues. The value scale contains ten steps from black to white, 0 to 10. The chroma scale contains up to 16 or more steps from neutral (white, gray or black) to highly saturated. Each of the three scales is intended to represent equal visual intervals for a normal observer fully adapted to daylight viewing conditions (CIE source C) with gray to white surroundings. Under these conditions the hue, value and chroma of a color correlate closely with hue, lightness and perceived chroma of color perception; under other conditions the correlation is lost. It is only for daylight conditions that Munsell samples are expected to appear equally spaced. When problems of color adaptation are fully solved, it may be possible to calculate the change in appearance and spacing that takes place when samples are viewed under a light source of different spectral power distribution. Munsell notations are useful whether or not reference is made to Munsell samples. For daylight conditions the system represents uniform color spacing as adequately as anything that was available until recently. (A set of colors now sold by the Optical Society of America, 1816 Jefferson Pl., N.W., Washington, D.C. 20036 is believed to be more uniformly spaced, but is not intended for color specification[17].) The Munsell notation is written in hue, value/chroma order, *e.g.*, 5R 4/10. This is read "five red, 4 over 10". Colors of zero-chroma, which are known as neutral colors, are written N1/, N2/, etc., as shown in Fig. 5–8. One widely used approximation of equivalence between hue, value/chroma units is 1 value step = 2 chroma steps = 3 hue steps (when the hue is at chroma 5). For use as standards or in technical color control, collections of carefully standardized color chips in matte or glossy surface may be obtained from Munsell Color Company, 2441 North Calvert Street, Baltimore, Maryland 21218, in several different forms. Since 1943 the smoothed renotations for the system, recommended by the Optical Society of America's Colorimetry Committee, have been recognized as the primary standard for these papers. Instructions for obtaining Munsell notations by calculation, or by conversion through CIE are contained in several publications.[6,9,20] The relationship between Munsell value and CIE luminous reflectance is summarized in Fig. 5–11.

ISCC-NBS Method of Designating Colors. The Inter-Society Color Council—National Bureau of Standards method of designating colors appeared in its original form in 1939 as NBS Research Paper RP 1239. The second edition appeared in book form in 1955 as NBS Circular 553, usually called the *Color Names Dictionary* (CND). The first Supplement to the CND, called the *Centroid Color Charts* (1965),[21] provides useful low-cost color charts that illustrate with one-inch square samples the centroid color for as many (251) of the 267 color names in the system as could be matched at that time. Each of the names defines a block in color space. This method is distinguished from all others in that the boundaries of each name are given, rather than points. These boundaries are defined in Munsell notations. The method does not provide for pinpointing colors, but provides an un-

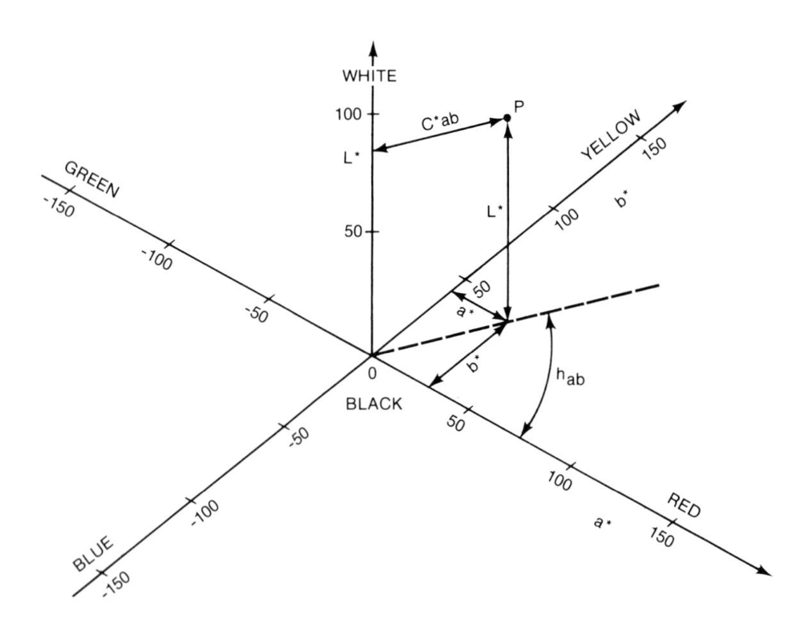

Fig. 5-10. CIE 1976 ($L^*a^*b^*$) uniform color space. The color represented by the point P has psychometric lightness = L^*, psychometric chroma = C^*_{ab}, and psychometric hue angle = h_{ab}. The geometrical arrangement of the CIE 1976 ($L^*u^*v^*$) uniform color space is similar.

derstandable color description. When close distinctions must be made among samples that might bear the same ISCC-NBS designation, other specifications such as CIE or Munsell should be used. The method is simple in principle: terms *light*, *medium* and *dark* designate decreasing degrees of lightness, and the adverb *very* extends the scale to *very light* and *very dark*; adjectives *grayish*, *moderate*, *strong* and *vivid* designate increasing degrees of saturation. These, and a series of hue names, used in both noun and adjective forms, are combined to form names for describing color in terms of its three perceptual attributes: hue, lightness and perceived chroma. A few adjectives are added to cover combinations of lightness and perceived chroma: *brilliant* for *light and strong*; *pale* for *light and grayish*; and *deep* for *dark and strong*. Hue names and modifiers are listed in Fig. 5-12.

The second Supplement to the CND, entitled *The Universal Color Language* (UCL), was published also in 1965.[22] The UCL serves as the means of updating the CND. It brings together all the well known color-order systems and methods of designating colors, and interrelates them in six correlated levels of fineness of color designation, each higher level indicating a finer division of the color solid. It follows closely and extends the original requirements of the ISCC-NBS Method of Designating Colors in the CND. The CND and the UCL have been published together as NBS Special Publication SP 440, with the UCL illustrated in color.[23]

Ostwald System. This is a system describing colors in terms of Color Content, White Content, and Black Content. It is usually exemplified by color charts in triangular form with Full Color, White and Black at the apexes providing a gray scale of White and Black mixtures, and parallel scales of constant White Content and constant Black Content as these grays are mixed with varying proportions of the Full Color. Twenty-four or more triangles form a collection of charts with color samples on each illustrating constant dominant wavelength (called hue), those colors lying parallel to the gray scale illustrating constant purity (called Shadow Series). The *Color*

Fig. 5-11. Relationship Between Munsell Value and Luminous Reflectance

Munsell Value	Luminous Reflectance* (per cent)
10.0	100.0
9.5	87.8
9.0	76.7
8.5	66.7
8.0	57.6
7.5	49.4
7.0	42.0
6.5	35.3
6.0	29.3
5.5	24.0
5.0	19.3
4.5	15.2
4.0	11.7
3.5	8.8
3.0	6.4
2.5	4.5
2.0	3.0
1.5	2.0
1.0	1.2
0	0

* Relative to a perfect diffuser.

Harmony Manual produced by the Container Corporation of America, but currently out of print, is an update and extension of the Ostwald System in which the visual spacing between adjacent samples has been smoothed and new samples added to fill empty spaces. The colored samples in the third edition are hexagonal with one side glossy and the other matte. The CIE *x,y*-coordinates for each chip in the first three editions of the *Color Harmony Manual* have been determined, and the notations for these chips may be transposed through CIE coordinates to Munsell notations or any other notations for which CIE data are available.

Correlation Among Methods. Frequently it is desirable to convert from one system of specification to another, or to convert or identify the color of samples on a chart or color card to terms of another. If the coordinates or samples of one system are given in CIE or Munsell terms, they may be converted or compared to any other system for which a similar conversion is available. Color charts of the German standard 6164 DIN system are provided with both CIE and Munsell equivalents. The Japanese standard system of color specification, JIS Z 8721-1958, is in terms of hue, value and chroma of the Munsell renotation system, according to the CIE *Y,x,y*-coordinates recommended by the Optical Society of America's 1943 subcommittee report.[9] The name blocks of the ISCC-NBS method are in terms of the Munsell renotation system with samples measured in CIE terms. Having a common conversion language helps promote international cooperation and understanding of the subject. Complete sets of CIE-Munsell conversion charts are contained in the OSA subcommittee report,[9] and are available from the Munsell Color Company. Many of the available conversions are referenced in a 1957 paper by Nickerson.[24] For more detailed descriptions of color systems or conversions, consult *Color in Business, Science and Industry.*[6]

Fig. 5-12. ISCC—NBS Standard Hue Names and Modifiers.

Hue Name	Abbreviation	Hue Name	Abbreviation
red	R	purple	P
reddish orange	rO	reddish purple	rP
orange	O	purplish red	pR
orange yellow	OY	purplish pink	pPk
yellow	Y	pink	Pk
greenish yellow	gY	yellowish pink	yPk
yellow green	YG	brownish pink	brPk
yellowish green	yG	brownish orange	brO
green	G	reddish brown	rBr
bluish green	bG	brown	Br
greenish blue	gB	yellowish brown	yBr
blue	B	olive brown	OlBr
purplish blue	pB	olive	Ol
violet	V	olive green	OlG

Hue Modifier	Abbreviation	Hue Modifier	Abbreviation
very pale	v.p.	moderate	m.
pale	p.	dark	d.
light grayish	l.gy.	very dark	v.d.
grayish	gy.	brilliant	brill.
dark grayish	d.gy.	strong	s.
blackish	bk.	deep	deep
very light	v.l.	very deep	v.deep
light	l.	vivid	v.

Color Temperature

Blackbody characteristics at different temperatures are defined by Planck's radiation law. See Section 1. The perceived colors of blackbodies at different temperatures depend on the state of adaptation of the observer. Fig. 5-18 gives an approximate illustration of the perceived colors at various color temperatures for various states of adaptation.

The locus of blackbody chromaticities on the *x,y*-diagram is known as the Planckian locus. Any chromaticity represented by a point on this locus may be specified by color temperature. Strictly speaking, color temperature should not be used to specify a chromaticity that does not lie on the Planckian locus. However, what is called the *correlated* color temperature (the temperature of the blackbody whose chromaticity most nearly matches that of the light source) is sometimes of interest. The correlated color temperature can be determined from diagrams[25] similar to the one shown in Fig. 5-13, either by graphical interpolation or by computer program.[26]

Equal *color differences* on the Planckian locus are more nearly expressed by equal steps of *reciprocal color temperature* than by equal steps of color temperature itself. The usual unit is the reciprocal megakelvin (MK^{-1}) so that reciprocal color temperature = 10^6 divided by color temperature in kelvins. The term "mired" (pronounced my'red), an abbreviation for micro-reciprocal-degree, was formerly used for the unit. A difference of one reciprocal megakelvin indicates approximately the same color difference anywhere on the color temperature scale above 1800 K; yet it varies from about 4 kelvins at 2000 K to 100 kelvins at 10,000 K.

Color temperature is a specification of chro-

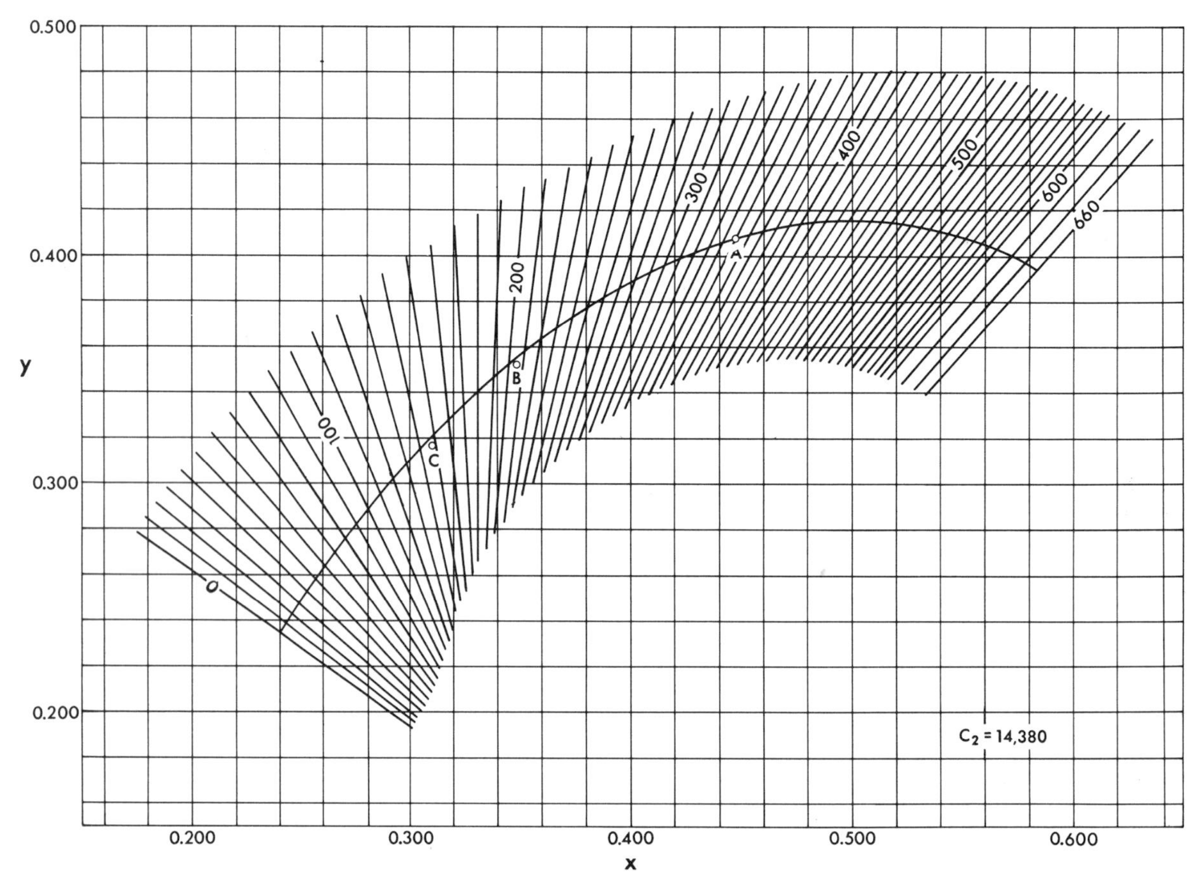

Fig. 5-13. CIE 1931 (*x*, y) chromaticity diagram showing isotemperature lines.[25] Lines of constant correlated color temperature are given at every 10 reciprocal megakelvins (formerly microreciprocal degrees).

maticity only. It does not represent the spectral power distribution of a light source. Chromaticities of many "daylight" lamps plot very close to the Planckian locus, and their colors may be specified in terms of correlated color temperature. However, this specification gives no information about their spectral power distributions which can, and often do, depart widely from that of daylight. The addition of light from two sources each having blackbody distribution but different color temperatures does not produce a blackbody mixture. Figs. 5-14 and 5-15 show spectral curves for Planckian distributions for different color temperatures, and distributions based on daylight[27] for several correlated color temperatures.

Most tungsten filament lamps approach the relative spectral power distribution of a blackbody quite closely. The color temperature of such lamps varies with the current passing through them. By varying the voltage across such a lamp a series of color temperatures can be obtained covering a very wide range up to about 3600 K.

Color Constancy and Adaptation[3, 28, 29]

A nonluminous colored object contributes to observed color by modifying the spectral power distribution of the light. The color of the light reflected or transmitted by the object when it is illuminated by a standard light source is known as the *object color* and can be calculated by assuming certain conventions (as in the CIE system). The color seen when the object is viewed normally in daylight is a mental phenomenon, and is referred to as the *perceived color* of the object. While there are exceptions, the perceived colors of objects, when illuminated by various sources, do not change as much as might be expected from the calculated difference in chromaticities. This phenomenon is known as "color constancy." Objects whose perceived colors change greatly when there is a wide change in illumination, as for example from daylight to incandescent filament light, are said to have unstable colors. It is important to remember that whereas the perceived color of an object may not change much with a change of light source color, the object color (as specified for example by CIE chromaticity coordinates) will change. For example, a piece of white paper will appear white under both incandescent light and daylight, but the object color will be quite different in the two cases because the paper, being spectrally neutral, will have almost the same chromaticity as the source in each case.

The impression that the perceived colors of most objects do not change greatly with the spectral power distribution of the light source is due primarily to a low degree of spectral selectivity in daylight and incandescent sources. Color constancy is affected by such factors as awareness of the illuminant, persistence of memory colors, consistence of attitude toward the object, and adaptation of the visual mechanism. Adaptation contributes to this relative color constancy because it tends to reduce to huelessness the chromatic response to the illumination. In other words, adaptation tends to counteract the shift of the color correlated with the shift of the source color, although there are cases where even slight residual shifts may be annoying or even intolerable. Such cases may be encountered with foodstuffs, displayed merchandise, or in the grading of various commercial products. It is these residual shifts that determine the color rendering properties of a source.

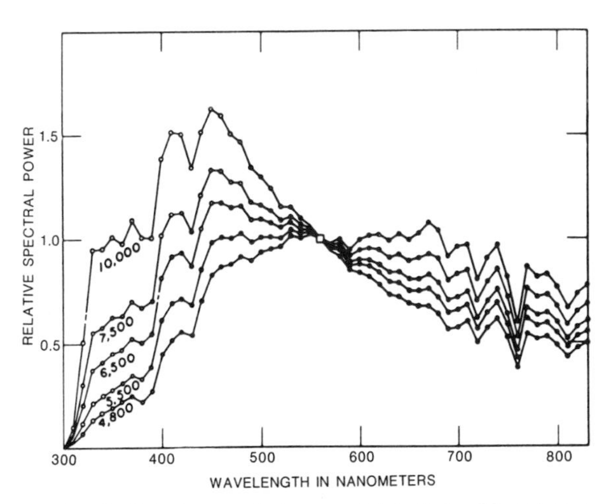

Fig. 5-15. Relative spectral power distributions of five phases of daylight for correlated color temperatures 4800, 5500, 6500, 7500 and 10,000 K.[27]

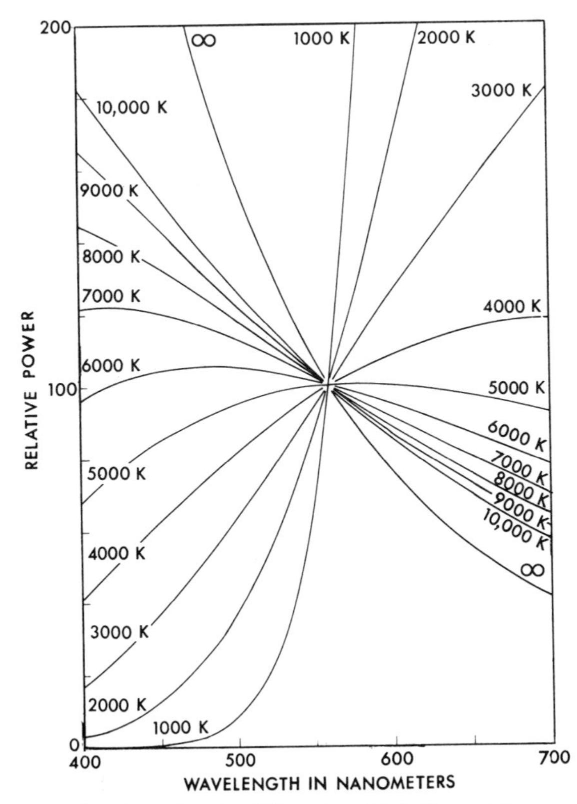

Fig. 5-14. Family of Planckian distribution curves.

The facts of color constancy and adaptation are not yet well enough known to make possible the computation of color rendering properties of a lamp with sufficient accuracy except when the reference or standard lamp is limited to the same color as the test lamp. When it becomes possible to compute the effects of constancy and adaptation so that the results agree accurately with the facts of experience, it will then be possible in calculating the color rendering properties of a lamp to take correctly into consideration differences in lamp color, as well as differences in spectral power distributions of lamps that have the same color. Nevertheless, as will be seen later, the CIE color rendering index does make an allowance for chromatic adaptation, even though the allowance is known not to be perfectly accurate.

Color Contrast

The appearance of chromatic areas is affected markedly by adjacent areas, particularly if one surrounds the other. Thus a central chromatic patch appears brighter or less gray if surrounded by a sufficiently large and relatively dark area, but dimmer or more gray if surrounded by a relatively light area. Juxtaposed chromatic areas also produce shifts in hue and saturation. Hues shift in opposite directions, tending to make them complementary, and saturations tend to shift away from each other, magnifying the saturation difference. In the general case there tends to be a simultaneous complex shift with respect to all three attributes.

Metameric and Conditional Color Matches

If two lights are visually indistinguishable because they have the same spectral compositions, they are said to form a spectral (or nonmetameric) match. A spectral match is an unconditional match; nobody can tell them apart. Two lights may, however, be visually indistinguishable in spite of having quite different spectral power distributions. Such a color match is said to be metameric and the lights to be metamers. In the CIE system, the computed match is identified by application of color matching functions which show that the tristimulus values for one light are identical to those for the other. If the lights are viewed by an observer characterized by different color-matching functions, the match may be upset. All metameric matches are therefore conditional matches. Even if one observer cannot, another one may be able to tell them apart.

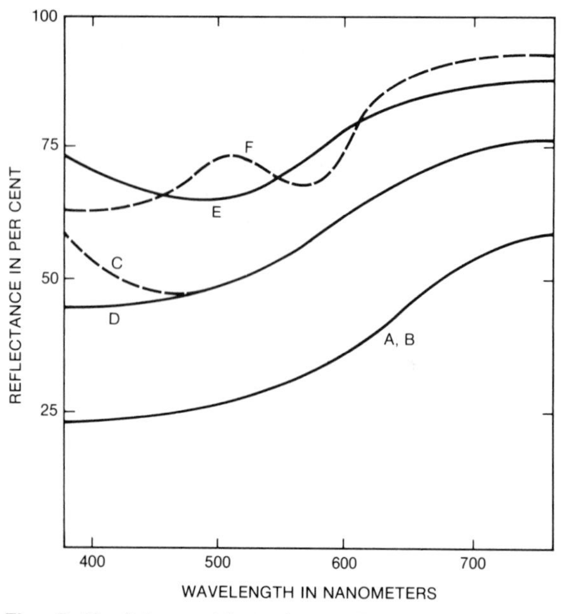

Fig. 5-16. Color matches. A and B are nonmetameric matches, *i.e.*, will match for any observer under any light source. C and D will match for any observer under a source with no power at wavelengths less than 500 nanometers, but will not match for some observers under a source that does have some power in the wavelength region 380 to 480 nanometers. C and D produce a source-conditional match, but they do not form a metameric pair, any more than A and B do, because the matching beams do not have different spectral compositions. E and F may form a metameric pair for some source-observer combination.

Metameric lights look alike in spite of the different spectral compositions (*e.g.*, incandescent filament compared to warm white fluorescent). The metameric character of such a match sometimes will be revealed by looking at a spectrally selective object and noting that the object is of different color when illuminated by the two lights. This is illustrated in Fig. 5-19.

Objects with identical spectral reflectance distributions (see samples A and B, Fig. 5-16) are said to produce an unconditional match. They match to everyone no matter what source is used. If, however, the color-matched reflected light comes from identically illuminated objects whose reflectances have different wavelength dependencies (see samples E and F, Fig. 5-16), the match is metameric. Substitution of another light source, or another observer, may upset the match; so objects that can produce a metameric match, though identically illuminated, may be said to produce a match that is both observer-conditional and source-conditional. Such a match is illustrated in Fig. 5-20.

It has sometimes been wrongly argued that, because the presence of metamerism always corresponds to a conditional match, a metameric match is the same as a conditional match. Not all conditional matches, however, are metameric. Fig. 5-16 shows the reflectance curves of two samples, C and D, that have different colors if the light source contains a significant amount of power of wavelength between 380 and 480 nm, but produce a spectral match if the power of the source is confined to wavelengths greater than 500 nm. Samples C and D thus form a source-conditional match; but no metamerism is involved, because there is no source for which light reflected from C and D have the same color but are spectrally different.

There is a necessary, though not sufficient, condition that must be satisfied by the spectral reflectances of objects that, identically illuminated, may produce metameric reflected lights. First, the two reflectance curves must be different in some part of the visible spectrum, or the reflected lights will be a spectral rather than a metameric match. Second, to be a color match the two objects must reflect equal amounts of the incident light, and this means that the curves must cross at least once within the visible spectrum. Third, the two objects must not differ in the yellow-blue sense; but if the curves cross at only one wavelength within the visible spectrum, a yellow-blue difference is implied. Therefore, the curves must cross at least at two wavelengths. Finally, the two objects must not differ in the purple-green sense; but if the curves cross at only two wavelengths, a purple-green difference is implied. Therefore, the reflectance curves of objects capable of producing a metameric match for some combination of trichromatic observer and source must cross at three wavelengths or more in the visible spectrum. Samples E and F, Fig. 5-16, have this property, and so for some source-observer combination might produce a metameric pair of reflected lights, *i.e.*, might match. For a discussion of the exact conditions under which a certain number of intersections is required, see Stiles and Wyszecki.[30]

COLORIMETRY OF LIGHT SOURCES

Color of a Light Source

The perceived color of a light source is its appearance apart from its brightness, geometry and time variations. The physical stimulus that causes the sensation of color is the spectral power distribution of the radiant flux from the source. The actual sensation occurs within the observer and depends not only on the stimulus itself but also on a host of other factors, such as, the nature of other stimuli in the field of view, the nature of other stimuli viewed in the recent past, the size and shape of the stimulus, and many more. Two sources may have the same color appearance and yet have entirely different relative spectral power distributions (*i.e.*, they may produce a metameric match). But if they have the same relative spectral power distributions they must have the same color appearance, all other factors being constant.

In measuring the color of a light source one makes use of the CIE 1931 or 1964 standard observer functions. (See Fig. 5-1 and Fig. 5-6.) Usually the "color" is expressed in terms of tristimulus values (X, Y, Z) or chromaticity coordinates (x,y). It can also be expressed as color temperature if the chromaticity falls on the blackbody locus.

Measurement of Light Source Color

The fundamental determination of chromaticity of light sources is by means of spectroradiometry.[31-38] Here the spectral power distribution of the source is measured and the chromaticity computed using the CIE 1931 or 1964 standard observer tristimulus functions. The spectroradiometric curve is fundamental to determinations of chromaticity, color temperature, color rendering and other psychophysical attributes of light sources. Current chromaticity standards in the United States for four types of fluorescent lamps are based on spectroradiometrically determined assignments of the National Bureau of Standards. See reference 39 and Fig. 5-17. Incandescent lamp color standards, on the other hand, are based on correlated color temperature comparisons. No standards exist for mercury, sodium, xenon or metal halide lamps. Colorimetry of light sources can be made directly (*i.e.*, without computation) by using calibrated photoelectric colorimeters, visual tristimulus colorimeters[6], or by visual color comparisons.

Color from Spectroradiometry

All of the information needed to determine the chromaticity, color temperature, and color rendering index of a light source is found in the spectral power distribution curve of the source. Where visual phenomena such as color are involved the curve usually is plotted between the wavelengths 380 and 770 nanometers and should be plotted at sufficiently narrow wavelength intervals and be based on sufficiently narrow bandwidths to show the desired resolution.

Chromaticity is calculated by the methods described earlier (see page 5-5) and has been greatly simplified through the use of modern computers, which in some systems have been tied directly to the spectroradiometer output. Several such systems are now available commercially.

Photoelectric Colorimeters

Photoelectric colorimeters are used extensively for measuring lamp chromaticity. Such instruments produce electrical responses to light of various wavelengths that are intended to be proportional to the computed values of the CIE system's X, Y, Z. The IES "Practical Guide to Colorimetry of Light Sources"[1] describes the use of this type of instrument and the precautions required. For example, due to the mismatch between the spectral sensitivity of the available detectors and the CIE 1931 tristimulus functions, it is necessary to have a lamp standard with a spectral power distribution very close to that of each test lamp to be measured. It is not sufficient just to have a standard of the same chromaticity—a close spectral match is also required. Through the use of operational amplifiers, logic circuitry and, more recently, microprocessors, these instruments often read chromaticity directly.

Visual Colorimeters

The eye is a remarkably sensitive color difference detector and can be used effectively to determine when a color match is made between a test light source and a comparison source consisting of a combination of three known "primary" sources. An instrument for facilitating such a comparison and for determining the proportions of the primaries needed to effect the match is called a visual tristimulus colorimeter.[6] The proportions of the primaries needed can be converted into CIE tristimulus values or chromaticity coordinates.

Direct Visual Comparisons

The purpose of evaluating the chromaticity or color appearance of light sources is usually to keep lamps of a given type (*e.g.*, cool white fluorescent lamps) within specified limits so that they appear approximately the same. Therefore, if no more quantitative procedure is available, it is reasonable to obtain judgments of color appearance by qualified observers and abide by their average findings. It is important to establish that the observers are normal in their color vision and preferably familiar with color comparison. The ability of observers to discriminate small color differences is by no means equal, even if color-defective persons are excluded. Color matching experience is very helpful, but that experience should be accumulated upon a basis of known color differences upon which there is agreement. There should be assurance that the color judgment is normal. A good test for detecting color-defective observers is the Hardy-Rand-Rittler test published by the American Optical Company. The Matching Color Aptitute Test, prepared by the Inter-Society Color Council (ISCC) and distributed by the Federation of Societies for Coatings Technology,* seems to offer considerable promise in selecting individuals with innate ability to distinguish small color differences. The test involves matching color chips and rating the results.

The technique of matching light sources by eye is important. Lamps to be compared should be adjacent to each other, and neither the standard nor the sample should be the end lamp; *i.e.*, there should be lamps at each side of those being judged. These outside lamps should be similar in color to those being judged, and all lamps involved should have about the same luminance.

* 1315 Walnut Street, Philadelphia, PA 19107.

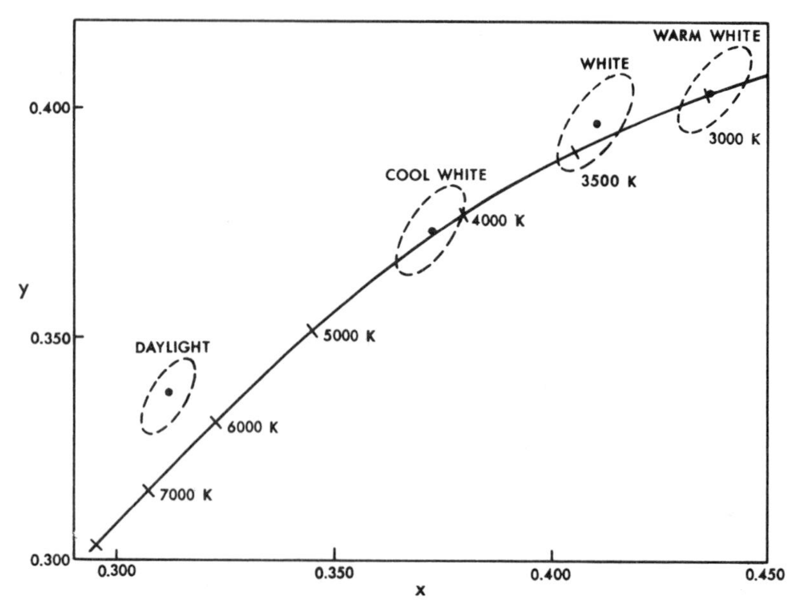

Fig. 5-17. ANSI colorimetric standards for color of fluorescent lamps. For each class of lamp the chromaticity must lie within the elliptical area shown.

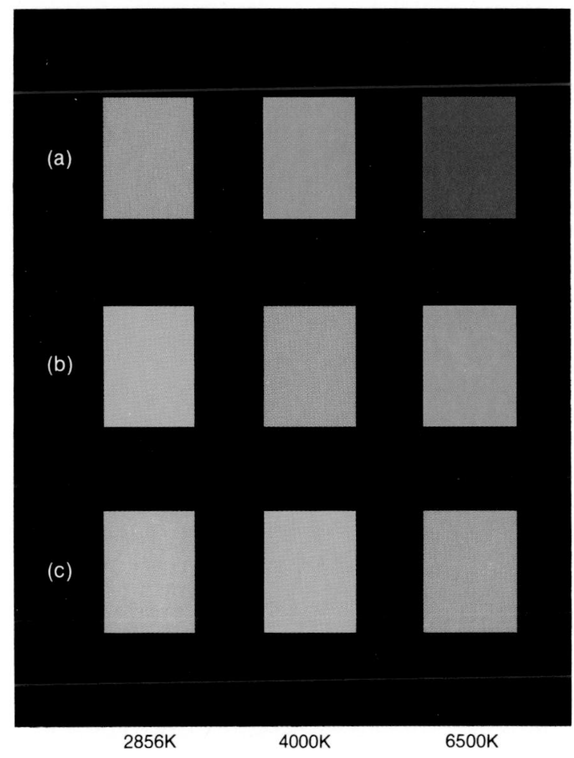

Fig. 5-18. Illustration of appearance of sources of different color temperatures after different chromatic adaptations: (a) after adaptation to 2856 K; (b) after adaptation to 4000 K; and (c) after adaptation to 6500 K. (Colors shown are approximate representations.)

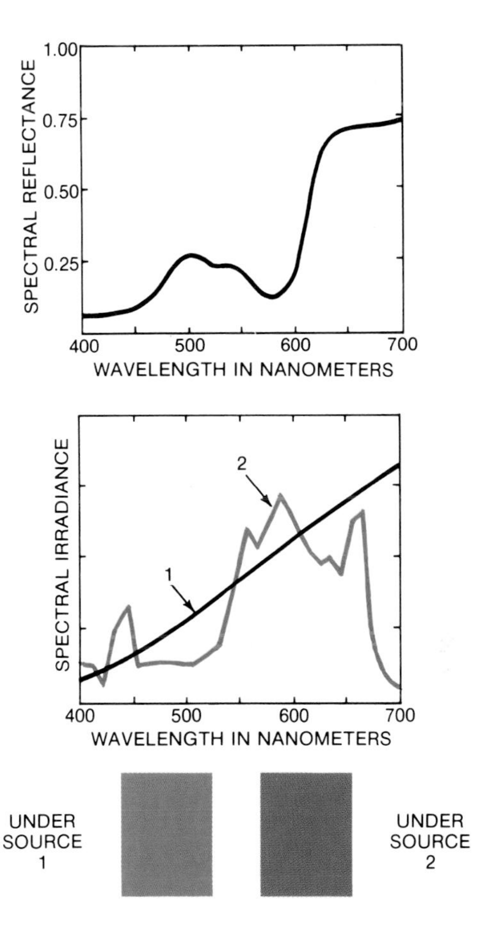

Fig. 5-19. Illustration of effect of metamerism of light sources. The two light sources have different spectral power distributions but are the same color. An object, such as the one whose spectral reflectance curve is shown, may have different color appearances when illuminated by one or the other source. (The actual colors shown are for illustration only. They are not intended to give an accurate representation of the color appearance.)

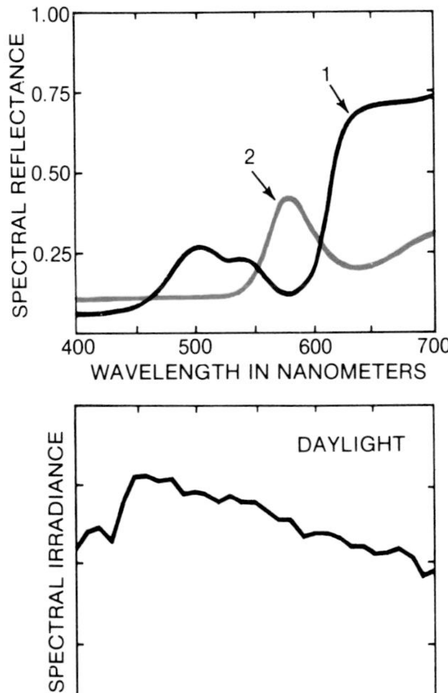

Fig. 5-20. Illustration of metamerism of reflecting objects. The objects whose spectral reflectance curves are shown have the same color when illuminated by daylight but are quite different in color when they are illuminated by an incandescent lamp. (The actual colors shown are for illustration only. They are not intended to give an accurate representation of the color appearance.)

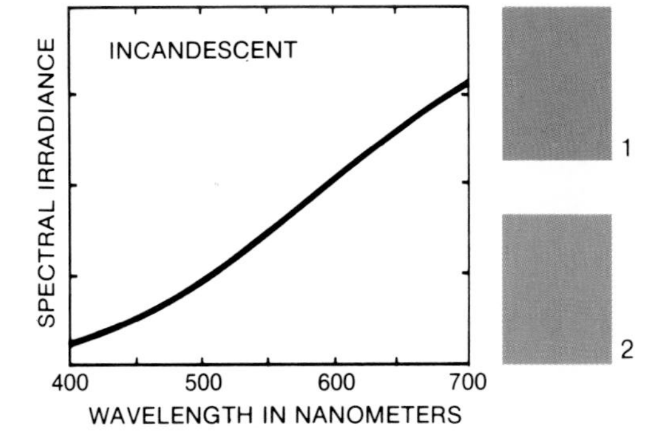

Fig. 5-21. Specification of ANSI Safety Colors Viewed Under CIE Standard Illuminant C*

Color Name	Munsell Notation	CIE Specification			ISCC-NBS Name
		x	y	Y	
Safety Red	7.5R 4.0/14	0.5959	0.3269	12.00	Vivid Red
Safety Orange	5.0YR 6.0/15	0.5510	0.4214	30.05	Vivid Orange
Highway Brown	5.0YR 2.75/5	0.4766	0.3816	5.52	Moderate Brown
Safety Yellow	5.0Y 8.0/12	0.4562	0.4788	59.10	Vivid Yellow
Safety Green	7.5G 4.0/9	0.2110	0.4120	12.00	Strong Green
Safety Blue	2.5PB 3.5/10	0.1690	0.1744	9.00	Strong Blue
Safety Purple	10.0P 4.5/10	0.3307	0.2245	15.57	Strong Reddish Purple
Safety White	N9.0/	0.3101	0.3162	78.70	White
Safety Gray	N5.0/	0.3101	0.3162	19.80	Medium Gray
Safety Black	N1.5/	0.3101	0.3162	2.02	Black

* Since this figure is a printed reproduction, the colors shown are approximations and should never be used in place of the actual samples.

Fig. 5-22. Specifications of Test Colors Used in Calculating the CIE Color Rendering Index*†
(Calculated for CIE Standard Illuminant C.)

Test Color Number	Munsell Notation	CIE Specification			ISCC-NBS Name
		x	*y*	*Y*	
1.	7.5 R 6/4	0.375	0.331	29.9	Light grayish red
2.	5 Y 6/4	0.385	0.395	28.9	Dark grayish yellow
3.	5 GY 6/8	0.373	0.464	30.4	Strong yellow green
4.	2.5G 6/6	0.287	0.400	29.2	Moderate yellowish green
5.	10 BG 6/4	0.258	0.306	30.7	Light bluish green
6.	5 PB 6/8	0.241	0.243	29.7	Light blue
7.	2.5 P 6/8	0.284	0.241	29.5	Light violet
8.	10 P 6/8	0.325	0.262	31.5	Light reddish purple
9.	4.5 R 4/13	0.567	0.306	11.4	Strong red
10.	5 Y 8/10	0.438	0.462	59.1	Strong yellow
11.	4.5 G 5/8	0.254	0.410	20.0	Strong green
12.	3 PB 3/11	0.155	0.150	6.4	Strong blue
13.	5 YR 8.4	0.372	0.352	57.3	Light yellowish pink (Caucasian complexion)
14.	5 GY 4/4	0.353	0.432	11.7	Moderate olive green (leaf green)

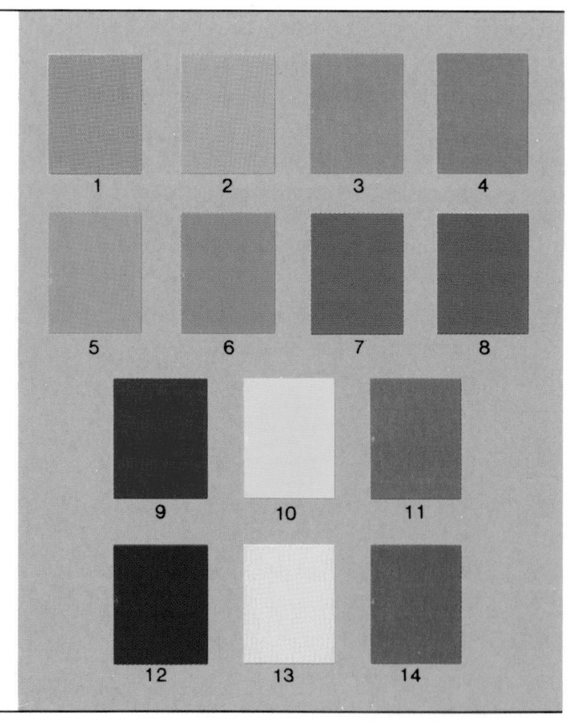

* The definitive specifications are in terms of spectral radiance factor.[4]
† Since this figure is a printed reproduction, the colors shown are approximations and should never be used in place of the actual samples.

Because the background may affect the observer's judgment due to the color of the reflected light, the background should be nonselective (gray) and of matte finish. The center portion of the lamp is normally used for color judgment. The observer should inspect the standard and the unknown lamps from at least two directions, and at grazing angles.

The observer should have the ability to detect small color differences and to determine the direction of any difference in hue that may be important. For example, if the test lamp is somewhat bluer than the standard, the addition of yellow light to the test lamp may be indicated, and a skilled observer can sometimes judge the percentage of yellow light that should be added.

Observations, combined with colorimetric data, can be useful if the limitations of both are well understood. Two light sources of different spectral power distributions may have exactly the same appearance to one normal observer, yet they may appear somewhat different to another normal observer. Colorimetric data may indicate an exact match, and yet a normal observer may disagree, and vice versa. It is often expedient to rely on an averaged consensus concerning a particular match (or lack of it) and to bear in mind that it is impossible in all cases to suit everyone's sense of a good match.

USE OF COLOR

Reflectance

Every object reflects some fraction of the light incident on it. Depending on the geometrical arrangement, the appropriate measure may be total luminous reflectance, regular (specular) luminous reflectance, diffuse luminous reflectance, luminous reflectance factor or luminance factor. In the CIE System, the Y tristimulus value of a reflecting object is the appropriate one of these quantities expressed as a percentage. For convenience, only per cent luminous reflectance will be used in the following discussion.

The luminous reflectance scale is not visually uniform between 0 and 100 per cent, black to white. An object that reflects 50 per cent does not look halfway between black and white, but looks much lighter than halfway. On the other hand, the purpose of the Munsell value scale is to illustrate equal visual steps for a given set of standard conditions. Thus, under daylight conditions, and for a light gray surround, a Munsell value of 5 should look about halfway in lightness between the black and white end points of the scale. Yet the luminance factor of a value of 5 sample is only about 20 per cent. In a condensed table, Fig. 5-11, reflectance and Munsell value units are related. A Munsell color value of 7 is called a light color, yet it reflects less than half—actually only 42 per cent—of the light it receives. This is an important point for lighting designers to consider, for unless all the colors in the color scheme of a room layout are very light, well over 50 per cent of the light is absorbed. If 5 value colors are used, as much as 80 per cent of the incident light may be absorbed. With practice in the use of a Munsell value scale, or in the special set of Munsell scales developed for lighting and interior designers illustrated in Fig. 5-9 (page 5-10) one may learn to estimate value rather accurately, and convert to luminous reflectance by means of Fig. 5-11. Value-reflectance conversion tables for every Munsell value step of 0.1 are available in several publications[6, 9, 20] or may be calculated by:[6]

$$1.0257Y = 1.2219V - 0.23111V^2 + 0.23951V^3 - 0.021009V^4 + 0.0008404V^5$$

where
Y = per cent luminous reflectance relative to a perfect diffuser
V = Munsell value

The factor 1.0257 converts Y to a formerly used scale on which smoked magnesium oxide had the value of 100.

The luminous reflectance of spectrally nonselective white, gray and black objects remains constant for all light sources, but the luminous reflectances of colored objects will differ in accordance with the spectral power distribution of the light source. For example, with illumination from incandescent sources which are relatively rich in the red and yellow portions of the spectrum and poor in the blue end, yellow objects appear lighter and blue objects darker than they do under daylight illumination; under blue sky the reverse would be true. On the special set of Munsell scales for judging reflectance, shown in Fig. 5-9, reflectances of each sample are given for three light sources; CIE "A" at 2856 K, CIE D_{65} at 6500 K, and cool white fluorescent at 4300 K.

Light walls and ceilings, whether neutral or chromatic, are much more efficient than dark walls in conserving energy and distributing light uniformly. Step by step changes studied by Brainerd and Massey in 1942[40] have been reported in terms of footcandles and coefficients of utilization and are shown in Fig. 5-23. Mathematical analyses by Moon[41] of effect of wall colors on illuminance and luminance ratios in cubical rooms show that an increase of wall reflectance by a factor of 9 could result in an increase in illuminance by a factor of about 3. Moon[41] has also published much information concerning spectral and colorimetric characteristics of material used in room interiors.

When neutral and chromatic surfaces of equal reflectance are equally and directly illuminated, they will have equal luminances. But by *interreflections* in a room, the light reaching a working surface will have undergone several reflections from ceiling and walls, and greater illumination will result from use of chromatic surfaces than from neutral surfaces. The calculated amount of improvement is shown in Fig. 5-24 for yellow, blue-green, and pink walls, each in comparison to gray walls of the same paired reflectances. The greater the number of interreflections, the greater the advantage. Spencer,[42] O'Brien,[43] and Jones and Jones[44] have published basic studies in this field. Spencer has analytically solved the color shift due to interreflection in an infinite room and in a finite rectangular room, and O'Brien has developed and used computer methods and results to provide charts and tables to aid designers in making detailed predictions of illuminance and luminance distributions in rooms, a prerequisite for solving the problem for color interreflections. In France, Barthès has published experimental measurements for a model room.[45] In Japan, Krossawa[46] has computed data on a closed surface painted with a uniform color and derived a general empirical formula on a color shift due to interreflections for different colors. Yamanaka and Nayatani[47] have compared results for computed and actual rooms, and consider the agreements to be quite satisfactory under the conditions. Gradually the data based on such studies will reach a form in which the practicing lighting designer can use them. Meanwhile designers should understand the general principles so that color change by room interreflections can be taken into consideration in planning a lighting layout.

Color Schemes—Choosing Suitable Colors

No set of simple rules can allow for tastes of different people, or for different conditions and changing fashions. However, the following suggestions provide a place to start:

1. *Ceilings* are assumed to be white, or slightly tinted. (Note: Some hospital ceilings are treated as a fifth wall for the supine patient.)
2. *Walls, floors and other structural elements*, which can be changed only after a long projected life or period of use must be considered first.
3. *Smaller areas*, machinery or furniture, need only blend or contrast with walls and floors.
4. Color schemes, represented by material, surfacing or paint (or coating) samples, should be assembled and evaluated under lighting conditions closely duplicating those under which the scheme will be used. This will avoid considerably the problem of significant color shifts and of metameric matches (see page 5-16).
5. As major surfaces can contribute considerably to the useful distribution of light by reflection and interreflection, luminous reflectance (Munsell value) should be high, if appropriate for the purpose of the room or space.
6. The *prime purpose* of the color scheme needs consideration (it may be seeing efficiency in a schoolroom, dignity in a church, a sense of well being for a factory, atmosphere of excitement for a circus, quiet for an office).

STEP	CEILING			WALLS		FLOOR		FURNITURE	
	HEIGHT	COLOR	ρ*	COLOR	ρ*	COLOR	ρ*	COLOR	ρ*
START	12 FT	WHITE	30						
1	10 FT		65	WHITE AND GRAY	40				
2	10 FT		85			DARK RED	12	DARK OAK	28
3	10 FT	CREAM	85						
4	10 FT		85	GREEN	72	WHITE	85		
5	10 FT		85					BLOND	50
6	10 FT		85			WHITE AND RUSSET	70		

*REFLECTANCE

LUX → 0 200 400 600

AVERAGE ILLUMINANCE
COEFFICIENT OF UTILIZATION

CU → 0 .10 .20 .30 .40 .50 .60

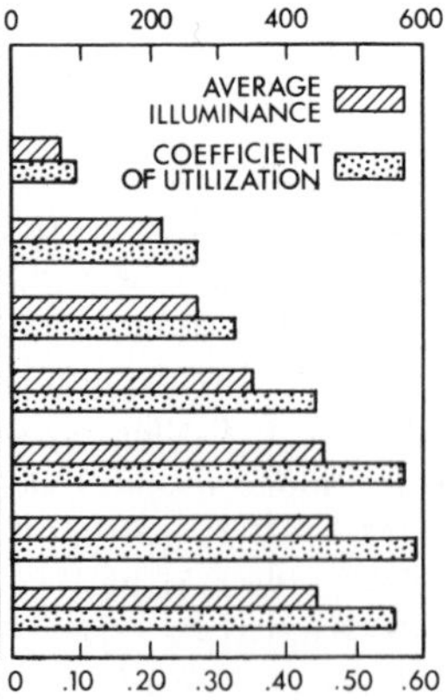

Fig. 5-23. Variation of illuminance and utilization coefficient with color scheme. Results of this historic study, made in 1942, are still valid. The luminaire had a general diffuse distribution.[40]

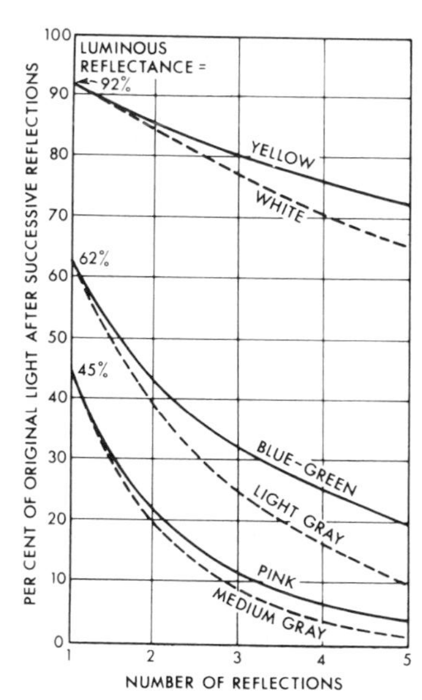

Fig. 5-24. Comparison of interreflection efficiencies of colored and neutral surfaces having the same luminous reflectance.

7. *Limitations* may exist in which redecoration schemes must be built around existing colors of carpeting or other flooring, draperies, or furniture.

The 1962 report of the Color Committee,[2] has three useful color charts. The first provides scales of hue, value and chroma to help in understanding color terminology used in interior design. The second provides a series of 66 color chips arranged to show strong versus weak chromas, and warm versus cool colors, with reflectances and Munsell notations for each sample, of colors used for interior surfaces. The third illustrates a 10-sample hue circle of typical wall colors at 60 per cent reflectance, and three sample color scheme selections.

Color schemes usually are variations of basic plans classified as monochromatic (single-hued), complementary, adjacent or analogous, split complementary, or triads. The dominant character usually is determined by the largest area, and in three-hued schemes this usually is the least saturated. A large pattern, strong in value contrast, makes a room seem smaller; a small pattern, in gentle contrast and high reflectance can make a room appear larger. Absence of pattern can provide the illusion of maximum space. The effects of strong contrasts of color or pattern are similar to each other: they are stimulating, make people restless, and make time seem longer. They are effective for corridors, places of entertainment, entrance halls, public washrooms, quick lunch counters, and other locations where it is desired that people spend a short time, but are undesirable, as for example, in hospitals. Gentle contrast is restful and makes time seem shorter. The play of molded form and texture can add interest; contrasts of natural wood, brick, stone and woven materials add interest to smooth painted walls.

Although personal tastes in color vary with climate, nationality, age, sex and personality, there is almost universal agreement to call yellows, yellow-reds, reds, and red-purples warm colors, and to call greens, blue-greens, blues, and purple-blues cool colors. All grayed colors approach neutrality of character, whether from the warm or cool side.

Apparent size and position of objects are affected by color. High chroma warm colors usually are the most advancing, with cool colors the most receding. Lowering the chroma reduces the effect on position. Light colors make objects appear larger; conversely, dark colors make objects appear smaller.

Selection Guide

By considering factors such as warmth, spaciousness, and excitement level, it is possible to determine a suitable dominant color and degree of contrast. These considerations may be dealt with in four steps[2] to help decide on values, hues, chromas and contrasts.

Step 1: This step helps decide *value*, that is, how light or dark a color scheme should be. If a high illuminance level is necessary, colors with a high reflectance should be used. Dark color combinations tend to produce luminance ratios that are unsatisfactory for efficient seeing. For areas in which illuminances of 750 lux [75 footcandles] or higher are recommended, the dominant values should be kept high, with reflectances of 40 to 60 per cent—the higher reflectance where the task is critical. Where lower illuminance levels are recommended, around 300 lux [30 footcandles], lower values may be introduced, at reflectances of 35 to 60 per cent. For still lower levels, the dominant values may be even lower, with reflectances for large areas down to 15 to 35 per cent. Where the best feasible visibility at low illuminances must be the goal, as for example in a parking garage, light colored (high value) walls and ceilings are recommended.

Step 2: This step helps decide the *hue*. Use warm, exciting, advancing colors where rooms have northern exposure, cool temperatures, and low noise element; where the room is too large and has smooth textures; where there is light physical exertion, time exposure is short, and stimulating atmosphere is required; and where lamps are cool fluorescent. Use cool, restful, receding colors for rooms with southern exposure, warm temperatures, and high noise element; for small rooms with rough texture; where physical exertion is heavy, time exposure is long, and a restful atmosphere is desired; and where lamps are incandescent or warm fluorescent.

Step 3: This step helps decide the *chroma.* Strong chromas are used primarily for advertising, display, accents and food merchandising; grayed colors of low chroma are primarily used for fashion areas, general interiors and merchandising. Use strong chromas if time exposure is short, responsibility low, lively atmosphere desired, noise level is low, and sense of taste or smell unimportant. Use grayed, low chroma colors if time exposure is long, responsibility high, atmosphere of dignity desired, noise level is high and sense of taste or smell important.

Step 4: This step helps decide the amount of *contrast.* Contrast is obtained by using light with dark, low with high chromas, and hues that are complementary. Consider use of little or no contrast if time exposure is long, room size is small, dignified atmosphere is required, and wall surfaces are textured. Consider use of strong contrast if time exposure is short, room size is large, lively or exciting atmosphere is desired, and wall surfaces are flat.

These recommendations represent a consensus from working knowledge of designers and architects, based on field experience. As long as results from scientific investigations do not contradict them, they remain generally accepted. The lighting designer must realize that in some cases the interior designer and architect may decide on a different approach and overrule common practice for reasons of their own or of their client.

Safety Colors

Safety colors are used to indicate the presence of a hazard or safety facility such as an explosive hazard or a first aid station. These are carefully developed colors that are specified in American National Standard Z53.1-1979.[48] The background around these safety colors should be kept as free of competing colors as possible and the number of other colors in the area should be kept to a minimum. These colors should be illuminated by a source to levels which will permit positive identification of the color and the hazard or situation which the color identifies, but which will not distort the color and, therefore, the message the color identification conveys.

The specification of the colors is given in Fig. 5–21. Lighting designers must be aware that these color specifications are based on illuminant C. The colors will be recognizable under daylight and/or conventional incandescent and fluorescent light sources which have a relatively broad spectrum; but high intensity discharge light sources, which have come into use in industry and may be widely used in the future, render some colors differently than the sources mentioned above. They may cause some confusion, especially at illuminances of 5 lux [0.5 footcandle] and lower, which are not uncommon in industrial spaces. Possible solutions are given in reference 49.

Color tolerance charts showing the safety colors and their tolerance limits are available from Hale Color Consultants, 1220 Bolton Street, Baltimore, MD 21217.

Chromaticity and Illuminance

The chromaticity of the light source should be matched to the illuminance level.[50] From experience it has been found that at low illuminances a "warm light" (<3300 K) is usually preferred, but the color temperature of the light source should increase as the illuminance increases.[51,52]

COLOR RENDERING

As discussed previously under Basic Concepts, lamps cannot be assessed for color rendering properties by visual inspection of the lamps themselves. To provide a color rendering index that will in all cases represent the facts, it is necessary to have accurate and precise spectroradiometric measurements of light sources. It is also necessary to understand the mechanisms of color vision, particularly chromatic adaptation. Knowledge in this area is still incomplete. If such facts as now can be ascertained are considered to have a sufficient degree of accuracy and precision, and the application is held within prescribed limits, it is possible to provide a useful, though not complete, answer. These recommendations are based on the following assumptions.

The color shift that occurs when an object is observed under different light sources may be classified in three ways—as a *colorimetric shift*, an *adaptive shift*, or a *resultant color shift* in which the first two are combined. To understand the subject it is extremely important that the three concepts be separately understood.

1. *Colorimetric shift* is the difference between the color (luminance and chromaticity, *e.g.*, Y,x,y) of an object illuminated by a non-standard source and the color of the same object illuminated by the standard source, usually measured on a scale appropriate for color differences.

2. *Adaptive color shift* is the difference in the perceived color of an object caused solely by chromatic adaptation.

3. *Resultant color shift* is the difference between the perceived color of an object illuminated by a non-standard source and that of the same object illuminated by the standard source for specified viewing conditions. The conditions usually are that the observer shall have normal color vision and be adapted to the environment illuminated by each source in turn. Resultant color shift is the resultant of colorimetric and adaptive color shifts.

Colorimetric shift can be determined using standard CIE conventions, but determination of adaptive shift requires some assumptions about the effects of chromatic adaptation.

CIE Test Color Method

The CIE recommends a test color method for measuring and specifying the color rendering properties of light sources.[4] It rates lamps in terms of a color rendering index (CRI), that is limited to consideration of the degree of colorimetric shift of a test object under a test lamp in comparison to its color under a standard lamp of the same correlated color temperature. The indices are based on a general comparison of the lengths of chromaticity-difference vectors on the CIE 1960 uniform chromaticity u,v-diagram. The rating consists of a General Index, R_a, which is the mean of the Special Indices, R_i, for a set of eight test-color samples that have been found adequate to cover the hue circuit. (See Fig. 5–22). This may be supplemented by Special Indices based on special-purpose test samples. Unless otherwise specified the reference light source for sources with a correlated color temperature below 5000 K is a Planckian radiator of the same correlated color temperature (see Fig. 5–14). For 5000 K and above the reference source is one of a series of spectral energy distributions of daylight based on reconstituted daylight data[27] developed from daylight measurements made in Enfield, England; Rochester, N. Y.; and Ottawa, Canada (see Fig. 5–15). Tables of colorimetric data are included in the CIE recommendations for Planckian radiators up to 5000 K, and on these reconstituted daylight curves from 5000 K to infinity, for eight general and six special test-color samples.

The current version of the CIE method is basically the same as an earlier version but with a better correction for the adaptive color shift. A paper by Nickerson and Jerome[53] on the earlier version provides a working text and formulas, discusses the meaning of the index, and shows applications to a number of lamps. The 1962 IES report[3] discusses in more detail the problems involved in its more than 10-year study of the subject. It indicates some of the problems, particularly those of chromatic adaptation, that remain to be solved before an all-purpose, completely satisfactory method can be established for rating a lamp, regardless of its color, against a single standard (probably daylight). Because of these problems, the index is not an absolute figure. For example, a 6500 K daylight lamp and a 3000 K warm white lamp having equal General Color Rendering Indices close to 100 will differ about as much as their respective reference illuminants, in this case the CIE phase of daylight D_{65} and the 3000 K Planckian radiator. These reference illuminants are—compared with each other—different in their color rendering. This will also apply to the two lamps tested, even though they have the same General Color Rendering Index.[4]

CIE ratings are in terms of a single index, R_a, but to provide more information on the color rendering properties of a lamp, it is recommended that this be accompanied by a listing of the eight special indices on which the rating is based. Since the eight test samples cover the hue circuit, this makes it possible to obtain a record of the relative colorimetric shift of the different hues under the test lamp. Plotting the chromaticity difference vectors provides even more information, for this indicates the direction, as well as the degree, of colorimetric shift that is involved.

The color rendering properties of a lamp with an index of 100 are identical to those of a standard reference lamp at least for these eight test-color samples. A lamp with an index of 50 has color rendering properties that shift the colors of objects on the average about as much as they are shifted by a standard fluorescent warm white lamp at 3000 K in comparison to the colors of the same objects under an incandescent reference lamp at 3000 K. Indices below 50 indicate lamps under which the average colorimetric shift is relatively greater than this. Fig. 5-25 shows the entire basis for the CIE index for the case where the test and reference lamp have the same chromaticity. At the top, on the CIE 1960 uniform chromaticity diagram, there is a plot of the eight test samples illuminated by the reference lamp and by a test lamp with a high index, R_a = 95, and at the bottom there is the same information for a lamp of very low index, R_a = 18. The difference in color rendering properties of the two lamps is shown by the differences in the colorimetric shift for the eight test colors, red through a reddish-purple, under each lamp in turn. The shifts under lamp *A* are very small; under lamp *B* they are very large. Yet the lamps themselves have the same chromaticity. For the case where the test and reference sources have different chromaticities, the adaptive color shift has to be taken into account as well. Details of how to do this are recommended by the CIE.[4]

If two lamps differ in R_i by about 5 units, the colors of test sample *i* rendered by the two lamps will be just perceptibly different under the best conditions, provided that the directions of the color shifts are nearly the same. No such simple rule can be given for R_a. It is obtained as the mean of eight R_i values, and even when two lamps have exactly the same R_a, differences of about 5 units or more in one or more of the R_i's may still be possible, so that their color rendering properties will be different for the object colors in question. Where the R_a values are close to 100, the R_i values are unlikely to show variations large enough to result in noticeable color differences. But, as the value of R_a decreases from 100, the corresponding special indices R_i show increasing spread. Ratings are illustrated in Fig. 5-26 for a number of typical lamps. The best color rendering lamps not only must have a high index, but also have the least variation in special indices for the different hues that are used as test samples. The closer one comes to an optimum color-matching lamp, the tighter must be these tolerances. As shown in Fig. 5-26, several lamps, including xenon and tungsten, have indices approaching the theoretical maximum of 100. Special fluorescent lamps have been reported to have similarly high indices.

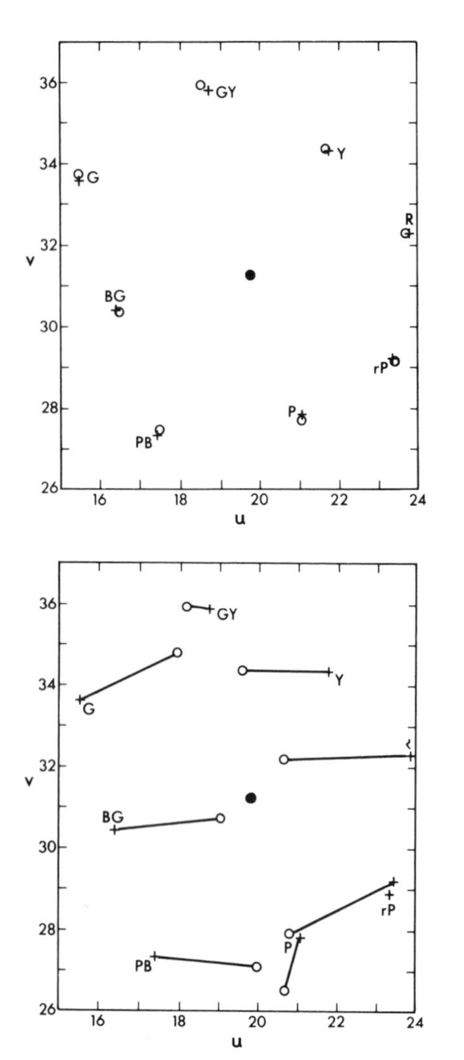

Fig. 5-25. The entire basis for CIE Color Rendering Index, R_a, is illustrated in this one figure which shows a portion of the CIE 1960 uniform chromaticity diagram. For test lamp A (upper) the index is 95, with each of the test colors (O) almost the same as under the reference (+); for test lamp B (lower) the index is 18, with each of the test colors (O) a long way from its color under the reference (+). Eight standard test colors, R, Y, GY, etc., are illustrated. When differences between test and reference lamp results are small, the index is high; when differences are large, the index becomes low. The filled circle (●) indicates the chromaticity of the test and reference lamps.

Visual Clarity

The concept of "visual clarity" has been used in several studies[54,57] to indicate a preferential appearance of scenes containing colored objects when illuminated by certain sources. "Visual clarity" seems to be a combination of various factors, such as: perceived color and contrast, color rendering, color discrimination, color preference and border sharpness.

Fig. 5-26. Color and Color Rendering Characteristics of Common Light Sources.*

Test Lamp Designation	CIE Chromaticity Coordinates		Correlated Color Temperature (Kelvins)	CIE General Color Rendering Index	CIE Special Color Rendering Indices, R_i													
	x	y		R_a	R_1	R_2	R_3	R_4	R_5	R_6	R_7	R_8	R_9	R_{10}	R_{11}	R_{12}	R_{13}	R_{14}
Fluorescent Lamps																		
Warm White	.436	.406	3020	52	43	70	90	40	42	55	66	13	−111	31	21	27	48	94
Warm White Deluxe	.440	.403	2940	73	72	80	81	71	69	67	83	64	14	49	60	43	73	88
White	.410	.398	3450	57	48	72	90	47	49	61	68	20	−104	36	32	38	52	94
Cool White	.373	.385	4250	62	52	74	90	54	56	64	74	31	−94	39	42	48	57	93
Cool White Deluxe	.376	.368	4050	89	91	91	85	89	90	86	90	88	70	74	88	78	91	90
Daylight	.316	.345	6250	74	67	82	92	70	72	78	82	51	−56	59	64	72	71	95
Three-Component A	.376	.374	4100	83	98	94	48	89	89	78	88	82	32	46	73	53	95	65
Three-Component B	.370	.381	4310	82	84	93	66	65	28	94	83	85	44	69	62	68	90	76
Three-Component C	.448	.408	2857	80	78	96	75	61	81	87	89	74	18	89	51	69	79	80
Simulated D_{50}	.342	.359	5150	95	93	96	98	95	94	95	98	92	76	91	94	93	94	99
Simulated D_{55}	.333	.352	5480	98	99	98	96	99	99	98	98	96	91	95	98	97	98	98
Simulated D_{65}	.313	.325	6520	91	93	91	85	91	93	88	90	92	89	76	91	86	92	91
Simulated D_{70}	.307	.314	6980	93	97	93	87	92	97	91	91	94	95	82	95	93	94	93
Simulated D_{75}	.299	.315	7500	93	93	94	91	93	93	91	94	91	73	83	92	90	93	95
Mercury, clear	.326	.390	5710	15	−15	32	59	2	3	7	45	−15	−327	−55	−22	−25	−3	75
Mercury, improved color	.373	.415	4430	32	10	43	60	20	18	14	60	31	−108	−32	−7	−23	17	77
Metal Halide, clear	.396	.390	3720	60	52	84	81	54	60	83	59	5	−142	68	55	78	62	88
Xenon, high pressure arc	.324	.324	5920	94	94	91	90	96	95	92	95	96	81	81	97	93	92	95
High pressure sodium	.519	.418	2100	21	11	65	52	−9	10	55	32	−52	−212	45	−34	32	18	69
Low pressure sodium	.569	.421	1740	−44	−68	44	−2	−101	−67	29	−23	−165	−492	20	−128	−21	−39	31
DXW Tungsten Halogen	.424	.399	3190	100	100	100	100	100	100	99	100	100	100	99	100	100	100	100

* Lamps representative of the industry are listed. Variations from manufacturer to manufacturer are likely. A high positive value of R_i indicates a small color difference for sample *i*. A low value of R_i indicates a large color difference.

LIGHT SOURCES FOR SPECIAL PURPOSE APPLICATIONS

Data for Daylight Standards. Standard specifications for the color and spectral quality of light sources for special applications have been adopted in several fields. Some have been adopted formally, as in the graphic arts field, by the IES; others, as for photographic prints and transparencies, by the American National Standards Institute; and still others by the American Society for Testing and Materials. Other less formal standard specifications are recommended, as those for textiles by the American Association of Textile Chemists and Colorists, and for diamond grading by the Gemological Institute of America. In many of these, the target standard is the spectral quality of daylight at around 7500 K, sometimes used with an additional lamp at lower color temperature, sometimes with the addition of ultraviolet. Information on the spectral quality and color of daylight itself is well summarized in papers by Nayatani and Wyszecki,[59] by Macbeth and Reese,[60] and by Judd, MacAdam, and Wyszecki.[27] See Fig. 5-27.

These and other daylight measurements had led to adoption by the CIE of target curves for daylight based on a family of reconstituted spectral distribution curves of typical daylight[27] such as are shown in Fig. 5-15. In the CIE method for specifying color rendering the reconstituted curves for daylight are used as the reference sources for correlated color temperatures above 5000 K.

Artificial Daylighting. Specifications for the best artificial daylighting for accurate work include: a large source of relatively low luminance; duplication of color of a moderately overcast north sky; and more light for inspecting dark colored samples than light colored samples.

Levels of at least 800 lux [80 footcandles] for light samples (40 per cent reflectance and over), and up to 3000 lux [300 footcandles] for dark (6 per cent or less) samples are indicated.

The color specification of an artificial daylight source for commercial color appraisal should be aimed at the best obtainable duplicate of preferred natural daylight conditions, and today this requires lamps that have a color rendering index above 90.

Inspection for suitability of color of materials to be used in daylight (as by a customer in a retail store) usually requires a less rigid approximate duplication of the spectral power distribution of natural daylight, because larger object-color variations are tolerable, but even in this case the target for a color-rendering index should be close to 90 for lamp colors in the daylight range. The normal eye adapts to rather large changes in the chromaticity of a light source, and if the color rendering properties of the two lamps are sufficiently good, the apparent colors of objects often will remain approximately constant for different color temperatures.

Preference of Textile Color Matchers

Data obtained by the Inter-Society Color Council indicate that illumination and color temperature combinations of natural daylight are preferred by textile color matchers as shown in Fig. 5–28. At 1000 lux [100 footcandles] the minimum color temperature preferred is close to

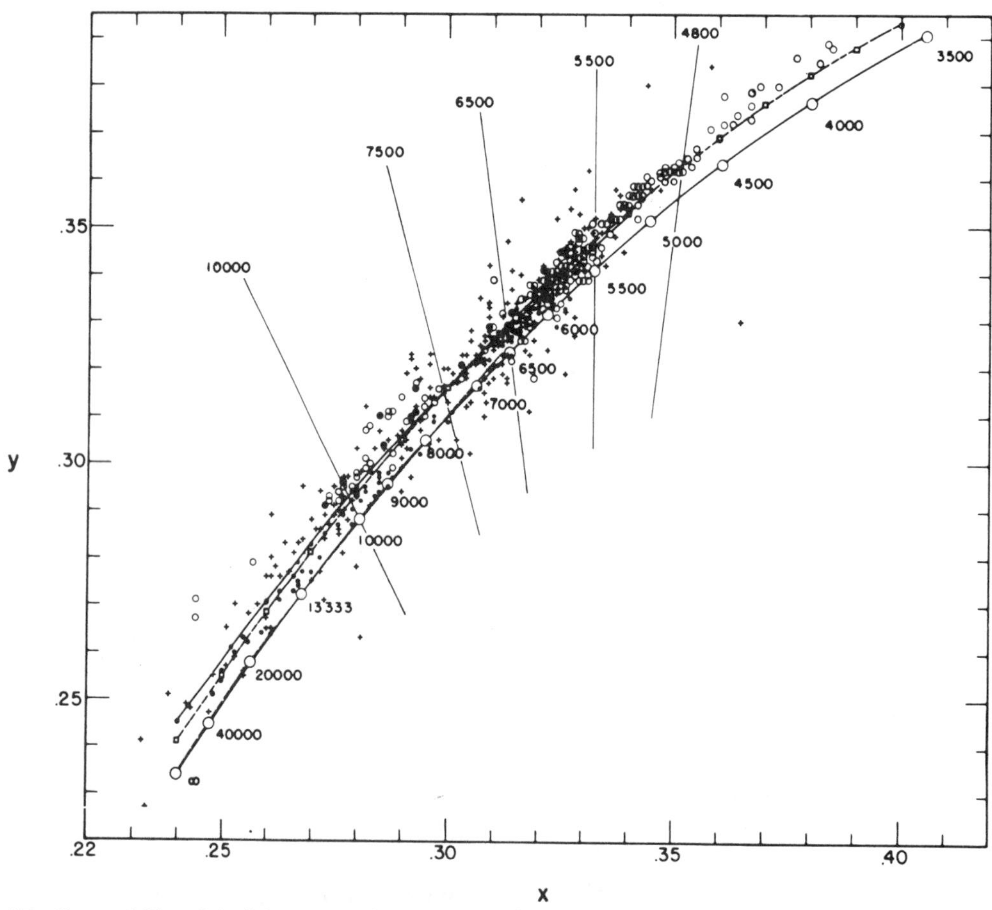

Fig. 5–27. Chromaticities of daylight compared to the locus of chromaticities implied by the Planck radiation law (open circles connected by solid lines).[13] The temperatures of the blackbody radiator corresponding to the points at the center of these open circles are indicated in kelvin. The straight lines intersecting this locus at 4800, 5500, 6500, 7500, and 10,000 K correspond to lines of constant correlated temperature. The central locus of the chromaticity range for daylight found by Nayatani and Wyszecki is shown by solid circles connected with a solid line; that found by Chamberlin, Lawrence and Belbin, is shown by a dotted line. Chromaticity points computed from the measured spectral distribution curves are indicated by open circles (Condit and Grum), by crosses (Henderson and Hodgkiss), and by solid circles (Budde). The locus of chromaticities taken to be typical of daylight conforms to the relation $y = 2.870x - 3.000x^2 - 0.275$, and is shown by squares connected by a dotted line.[27]

7500 K, ranging upward to a maximum above 25,000 K. The preferred color temperature may drop to 5700 K when the illumination is 3000 lux [300 footcandles][61] or more.

Color Photography

Color photography is based on the same general principle as color vision (see Section 3). The normal human retina has three kinds of light sensors, known as cones, that are sensitive to light in three different regions of the spectrum. Color film has three layers that are sensitive to light in three regions, but the spectral sensitivities of films are not identical to those of the visual cones; therefore, the colors of photographs may differ somewhat from the colors of the objects photographed. Lights or objects that match visually may be depicted in different colors on the photograph, or different original colors may be rendered as matching. Color films are designed to give satisfactory color rendition for a specific kind of illumination, usually daylight or incandescent filament light having a correlated color temperature of either 3200 K or 3400 K. Flash lamps are designed to produce light having a spectral quality like daylight. The American National Standards Institute (ANSI) and the International Organization for Standardization (ISO) have adopted standards specifying illumination for testing the sensitivities of films, the color contribution of camera lenses, methods of testing flash equipment, and conditions for viewing transparencies and prints. See Section 11 in the *1981 Application Volume* and references 62 through 67.

Color Television

Considerations similar to those for color photography apply to lighting for color television. The suitability of a light source for illuminating an indoor or outdoor scene to be televised depends on the spectral power distribution of the light source and on the color analysis characteristics of the television cameras being used.[68] See also Section 11 in the *1981 Application Volume.*

Lighting of Classing Rooms for Color Grading

A large scale application of artificial daylighting to replace natural daylighting in color grading is in the field of raw cotton classification. Studies made in the U. S. Department of Agriculture and reported to the IES as early as 1939 provided the basis for specifying no less than 600 to 800 lux [60 to 80 footcandles] on the classing table, and color and spectral quality as close as possible to that of daylight of a moderately overcast sky, about 7500 K. Following World War II, a combination of daylight and blue fluorescent lamps, plus incandescent, was developed to match the color of 7500 K with a spectral power as close as possible to that of daylight. Since then there have been improvements, first by combining the phosphors of the daylight and blue lamps into a single fluorescent lamp and using with it low voltage incandescent lamps to increase their life for use with the long-life fluorescent lamp, and in 1956 by the production of fluorescent lamps with much improved spectral power distributions. The profusion of mercury lines is still a problem in fluorescent lamps, but except for these lines it is possible to produce a fluorescent lamp with a spectral power distribution very close to that of daylight. The advantage of greater efficacy and lower cost over the filtered-tungsten lamps first used in classing rooms has made it the accepted lighting standard in practically all modern cotton classing rooms in the United States as well as in many other countries.

In cotton classing rooms it is found that the geometry of the luminaires, as well as the spectral power distribution, needs control in order to minimize glare, for as the illuminance is increased this factor becomes increasingly important. Control of surrounding conditions is very important, and instructions, used by the United States Department of Agriculture, are included in ASTM D-1684-81.[69]

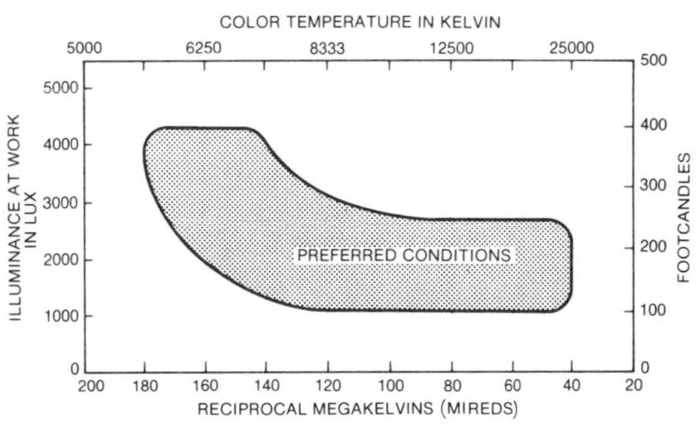

Fig. 5-28. Tests conducted under the direction of the Inter-Society Color Council show the characteristics of preferred daylight illumination conditions for color matching, grading and classing.

Lighting for Graphic Arts Color Work

Recommendations for lighting of color appraisal of reflection-type materials in the graphic arts industry[70] are given in Section 9 in the *1981 Application Volume.*

There are four basic tasks in graphic arts color work, and in these, the precision of the lighting requirements is not necessarily the same. The spectral power distribution of CIE sources D5000 and D7500 are recommended, however, as the design standards for the various tasks. These two standards are specified to serve as a basis for making the specifying color corrections and to establish a common understanding between customer and supplier. The recommendations for the four tasks are as follows:

Task 1. When determining the original color selection, or artist's choice of colors, the copy should be evaluated under commonly used light sources as well as under a standard source; many undesirable choices would be avoided if original color selections are checked under lighting conditions in which they will ultimately be used.

Task 2. When matching color of basic materials, or selecting pigments, two light sources of widely different color should be used to detect metamerism. One should be CIE source D7500 (an approximation to overcast north sky daylight); the other may be an incandescent filament lamp.

Task 3. When visually appraising the color quality of the printed result against the original master, *i.e.*, proofs versus copy, CIE source D5000 is specified.

Task 4. When visually appraising color uniformity of production sheets, the use of CIE source D7500 is specified in the United States.[70]

In each case, the standard sources are to have CIE general color rendering indices, R_a, above 90. Geometrical, background and surround conditions are specified to avoid glare and undesirable reflections.

Control of Color

Color Selection. The problem may simply be one of selection, as for example that faced by a housewife about to choose from an assortment of meat at the meat dealer's, or of fruit or vegetables at the grocer's, or from an assortment of dress or upholstery fabrics, paints, or wallpapers at a department store; the decision will be based on the appearance of the object on display and upon the customer's estimate of its probable appearance under the conditions most likely to be encountered in use. The conditions of display and use differ more often than they coincide, something that should be corrected by the lighting designer as rapidly as lamps with better color rendering properties become available.

Color Matching. Visual color matches satisfactory for many purposes may usually be assured by the simple expedient of checking the match under each of two widely different light sources of continuous power distribution. A practical application of this general method is used in dye houses. When samples are to be matched they are viewed under two light sources selected at or near wide extremes of daylight color temperatures.

A perfect match under all conditions will be obtained only in materials having matching spectrophotometric curves. Even then, if their surface textures are not the same (smooth paint and rough textiles, for example), their appearance may vary slightly depending on the angle at which they are illuminated and viewed.

Color Grading. The market value of many things—raw cotton, tobacco, fruit, vegetables, furs, textiles, and so forth—varies with their colors over a very wide range. In some instances such products are accepted or rejected on the basis of color specifications or standards. They may be separated according to nearly imperceptible color differences into a number of "standard" grades, each of which may have a different market value.

Color grading of a group of materials known to have similar spectral reflectance characteristics may not always require close duplication of the spectral power distribution of daylight. This use, however, differs from the other uses listed because large departures from the spectral power distribution of daylight are allowable as long as they yield in undiminished amount the object-color differences characteristic of daylight inspection.

Whenever the light source magnifies the characteristic differences so that they may more easily be detected, departures from the spectral power distribution of daylight may be desirable. If yellow samples are to be examined, a light source rich in energy in the blue portion of the spectrum, where the spectral reflectances of yellow samples are apt to differ most widely, will enable the observer to discriminate differences more easily than when using a source deficient in the blue portion of the spectrum. When blue samples are to be examined, the reverse is true, *i.e.*, a light source rich in energy in the yellow portion of the spectrum, will facilitate discrimination. It should be remembered, however, that while *differences* may be revealed by such a method, the average daylight appearance of the samples may be quite distorted. To experienced color graders, duplication of the spectral distribution of the light with which they are most familiar permits them to take full advantage of their experience and makes conversion to new conditions much less difficult.

Color Control in a Lighting Installation. The artist, architect and lighting designer, after agreeing on a design having suitable decorative qualities and which at the same time will provide the proper quantity and quality of illumination, have the problem of transferring their plans to the location in question. This must be done by specifying to the contractor and builder, as well as to the furniture, wall covering, drapery, and paint manufacturers, what the restrictions are from the standpoint of colored surfaces.

Color Preference

Work has been done under grants from IERI, that is expected to add to our knowledge of color preferences in lighting. A few Helson reports[72–75] already are published in this and related fields, and reports[19, 71] on more extensive studies with fluorescent sources are available.

In his reports, Dr. Helson states that the pleasantness of object colors depends on the interaction of light source with background color and the hue, lightness, and saturation of the object color. The best background colors for enhancing the pleasantness of object colors were found to have either very high Munsell values (8/ or 9/) or very low values (2/ or 1/), and, with only one exception, very low or zero chroma. Color of background was found to be more important than spectral power distribution of light source. Neutrals rank high as background colors, but almost at the bottom for object colors. High chromas are preferred to low for object colors.

The chief single factor responsible for pleasant color harmonies was found to be lightness contrast between object and background colors. The greater the lightness contrast, the greater are the chances of good color combinations. Lightness contrast is important for pattern vision and this may be why this factor is the most important single factor in good color combinations. The influences of hue and saturation difference cannot be stated simply. A certain amount of variety, change, differentiation, or contrast is pleasant; sameness, monotony and repetition tend to be unpleasant. Configurations of colors should contain some variations in hue, lightness and saturation, and over a period of time different configurations of colors should be employed to prevent satiation by over-familiar patterns of stimulation. Color preferences may differ due to such factors as function, size, configuration, climate and socio-cultural factors.

REFERENCES

1. Committee on Testing Procedures of the IES: "Practical Guide to Colorimetry of Light Sources," *J. Illum. Eng. Soc.*, Vol. 5, p. 177, April, 1976.
2. Color Committee of the IES: "Color and the Use of Color by the Illuminating Engineer," *Illum. Eng.*, Vol. LVII, p. 764, December, 1962.
3. Subcommittee on Color Rendering of the Light Sources Committee of the IES: "Interim Method of Measuring and Specifying Color Rendering of Light Sources," *Illum. Eng.*, Vol. LVII, p. 471, July, 1962.
4. CIE Committee TC-3.2: *Method of Measuring and Specifying Color Rendering Properties of Light Sources,* 2nd Edition, CIE Publication No. 13.2, Paris, 1974.
5. Color Committee of the IES: "Color and Illumination" (tentative title), an expanded and revised monograph based in part on the 1961 report, "Color and the Use of Color by the Illuminating Engineer" (see reference 2), to be published.
6. Judd, D. B. and Wyszecki, G.: *Color in Business, Science and Industry*, 3rd Edition, John Wiley & Sons, Inc., New York, 1975.
7. Nickerson, D.: "Light Sources and Color Rendering," *J. Opt. Soc. Amer.*, Vol. 50, p. 57, January, 1960.
8. Hunt, R. W. G.: "Colour Terminology," *Color Res. Appl.*, Vol. 3, p. 79, 1978.
9. Newhall, S. M., Nickerson, D., and Judd, D. B.: "Final Report of the OSA Subcommittee on Specifying of the Munsell Colors," *J. Opt. Soc. Amer.*, Vol. 33, p. 385, July, 1943.
10. CIE Committee TC-1.3: *Recommendations on Uniform Color Spaces-Color-Difference Equations-Psychometric Color Terms,* CIE Publication No. 15, Supplement No. 2, Paris, 1978.
11. Robertson, A. R.: "The CIE 1976 Color-Difference Formulae," *Color Res. Appl.*, Vol. 2, p. 7, 1977.
12. CIE Committee TC-1.3: *Colorimetry*, CIE Publication No. 15, Paris, 1971.
13. Wyszecki, G. and Stiles, W. S.: *Color Science*, John Wiley and Sons, Inc., New York, 1982.
14. MacAdam, D. L.: "Projective Transformations of ICI Color Specifications," *J. Opt. Soc. Amer.*, Vol. 27, p. 294, 1937.
15. Wyszecki, G.: "Proposal for a New Color-Difference Formula," *J. Opt. Soc. Amer.*, Vol. 53, p. 1318, 1963.
16. "CIE Recommendations on Uniform Color Spaces, Color Spaces, Color-Difference; Equations, and Metric Color Terms," *Color Research and Application*, Vol. 2, No. 1, p. 10, Spring, 1977.
17. Uniform Color Scales Committee: *Uniform Color Scales*, Optical Society of America, Washington, D.C., 1977.
18. Committee on Colorimetry of Optical Society of America: *The Science of Color*, Optical Society of America, Washington, D. C., 1963.
19. Judd, D. B.: "Choosing Pleasant Color Combinations," *Light. Des. Appl.*, Vol. 1, p. 31, August, 1971.
20. American Society for Testing Materials: *Method for Specifying Color by the Munsell System D 1535 - 80*, American Society for Testing and Materials, Philadelphia, Pa., 1980.

21. Kelly, K. L. and Judd, D. B.: *The ISCC-NBS Centroid Color Charts*, SRM #2106, The Office of Standard Reference Materials, National Bureau of Standards, Washington, D. C., 1965.
22. Kelly, K. L.: "The Universal Color Language," *Color Engineering*, Vol. 3, p. 16, March–April, 1965.
23. Kelly, K. L. and Judd, D. B.: *Color: Universal Language and Dictionary of Names*, National Bureau of Standards Special Publication 440, Superintendent of Documents, US Government Printing Office, Washington, D. C., 1976 and 1978.
24. Nickerson, D.: "Horticultural Color Chart Names with Munsell Key," *J. Opt. Soc. Amer.*, Vol. 47, p. 619, July, 1957.
25. Kelly, K. L.: "Lines of Constant Correlated Color Temperature Based on MacAdam's (u,v) Uniform Chromaticity Transformation of the CIE Diagrams," *J. Opt. Soc. Amer.*, Vol. 53, p. 999, August, 1963.
26. Robertson, A. R.: "Computation of Correlated Color Temperature," *J. Opt. Soc. Amer.*, Vol. 58, p. 1528, 1968.
27. Judd, D. B., MacAdam, D. L., and Wyszecki, G.: "Spectral Distribution of Typical Daylight as a Function of Correlated Color Temperature," *J. Opt. Soc. Amer.*, Vol. 54, p. 1031, August, 1964: Summary in *Illum. Eng.*, Vol. LX, p. 271, April, 1965.
28. Evans, R. M.: *An Introduction to Color*, John Wiley & Sons, Inc. New York, 1948.
29. Burnham, R. W., Hanes, R. M., and Bartleson, C. J.: *Color: A Guide to Basic Facts and Concepts*, John Wiley & Sons, Inc. New York, 1963.
30. Stiles, W. S. and Wyszecki, G.: "Intersections of the Spectral Reflectance Curves of Metameric Object Colors," *J. Opt. Soc. Amer.*, Vol. 58, p. 32, January, 1968.
31. Committee on Testing Procedures of the IES: "IES Guide to Spectroradiometric Measurements," *J. Illum. Eng. Soc.*, Vol. 12, p. 136, April, 1983.
32. Jones, O. C. and Moore, J. R.: *The Spectroradiometry of Light Sources*, National Physical Laboratory (UK) Report DES 70, 1981.
33. Thorington, L., Parascandola, J., and Schiazzano, G.: "Chromaticity and Color Rendition of Light Sources from Fundamental Spectroradiometry," *Illum. Eng.*, Vol. LX, p. 227, April, 1965.
34. Sanders, C. L. and Gaw, W.: "A Versatile Spectroradiometer and Its Application," *Appl. Opt.*, Vol. 6, p. 1639, October, 1967.
35. Hammond, H. K., Holford, W. L. and Kuder, M. L.: "Ratio Recording Spectroradiometer," *J. Res. NBS*, Vol. 64C, p. 151, 1960.
36. Brown, R. L.: "Direct Recording Spectroradiometer for Light Sources," *Illum. Eng.*, Vol. LXI, p. 230, April, 1966.
37. Eby, J. E.: "A Computer-Based Spectroradiometer System," *Appl. Opt.*, Vol. 9, p. 888, April, 1970.
38. Heaps, W. L.: "Automatic Recording Spectroradiometer System," *Appl. Opt.*, Vol. 10, p. 2045, September, 1971.
39. *Specifications for the Chromaticity of Fluorescent Lamps*, ANSI C78.376-1969 (R1980), American National Standards Institute, New York, 1980.
40. Brainerd, A. A. and Massey, R. A.: "Salvaging Waste Light for Victory," *Illum. Eng.*, Vol. XXXVII, p. 738, December, 1942.
41. Moon, P.: "Wall Materials and Lighting," *J. Opt. Soc. Amer.*, Vol. 31, p. 723, December, 1941.
42. Spencer, D. E. and Sanborn, S. E.: "Interflections and Color," *J. Franklin Inst.*, Vol. 252, p. 413, November, 1961.
43. O'Brien, P. F.: "Lighting Calculations for Thirty-Five Thousand Rooms," *Illum. Eng.*, Vol. LV, p. 215, April, 1960.
44. Jones, B. F. and Jones, J. R.: "A Versatile Method of Calculating Illumination and Brightness," *Illum. Eng.*, Vol. LIV, p. 113, February, 1959.
45. Barthès, M. E.: "Etudes Expérimental de Mm. Tarnay et Barthès pour la Mise au Point d'une Méthode de Calcul de Point de Couleur de la Lumière Reçue par le Plan Utile dans un Local a Parois Colorées," *Bull. Soc. Franc. Elec.*, Ser. 7, Vol. VII, p. 546, September, 1957.
46. Krossawa, R.: "Color Shift of Room Interior Surfaces Due to Interreflection," *Die Farbe*, Vol. 12, p. 117, 1963.
47. Yamanaka, T. and Nayatani, Y.: "A Note on Predetermination of Color Shift Due to Interreflection in a Colored Room," *Acta Chromatica*, Vol. 1, p. 111, October, 1964.
48. *Safety Color Code for Marking Physical Hazards*, ANSI Z53.1-1979, American National Standards Institute, New York, 1979.
49. Color Committee of the IES: "Potential Misidentification of Industrial Safety Colors with Certain Lighting," *Light. Des. Appl.*, Vol. 10, p. 20, November 1980. Glass, R. A., *et al*: "Some Criteria for Colors and Signs in Workplaces," NBSIR 83-2604, National Bureau of Standards, Washington, D.C., 1983.
50. CIE Committee TC-4.1: *Guide on Interior Lighting*, CIE Publication No. 29, Paris, 1975.
51. Kruithof, A. A.: "Tubular Luminescence Lamps for General Illumination," *Philips Tech. Rev.*, Vol. 6, p. 65, 1941.
52. *IES Lighting Handbook, 1981 Application Volume*, Illuminating Engineering Society of North America, New York, 1981, Fig. 2-11.
53. Nickerson, D. and Jerome, C. W.: "Color Rendering of Light Sources: CIE Method of Specification and Its Application," *Illum. Eng.*, Vol. LX, p. 262, April, 1965.
54. Aston, S. M. and Bellchambers, H. E.: "Illumination, Colour Rendering and Visual Clarity," *Light. Res. Technol.*, Vol. 1, No. 4, p. 259, 1969.
55. Bellchambers, H. E. and Godby, A. C.: "Illumination, Colour Rendering and Visual Clarity," *Light. Res. Technol.*, Vol. 4, No. 2, p. 104, 1972.
56. Lemons, T. M. and Robinson, A. V.: "Does Visual Clarity Have a Meaning for IES Illumination Recommendations for Task Lighting?" *Light. Des. Appl.*, Vol. 6, p. 24, November, 1976.
57. Thornton, W. A. and Chen, E.: "What Is Visual Clarity?" *J. Illum. Eng. Soc.*, Vol. 7, p. 85, January, 1978.
58. Boyce, P. R.: "Investigation of the Subjective Balance between Illuminance and Lamp Colour Properties," *Light. Res. Technol.*, Vol. 9, p. 11, 1977.
59. Nayatani, Y. and Wyszecki, G.: "Color of Daylight from North Sky," *J. Opt. Soc. Amer.*, Vol. 53, p. 626, May, 1963.
60. Macbeth, N. and Reese, W. B.: "Color Matching," *Illum. Eng.*, Vol. LIX, p. 461, June, 1964.
61. Nickerson, D.: "The Illuminant in Textile Color Matching," *Illum. Eng.*, Vol. XLIII, p. 416, April, 1948.
62. "Color," a volume in the *Life Library of Photography*, Time-Life Books, New York, 1976.
63. "Lighting," *Encyclopedia of Practical Photography*, Vol. 9, American Photographic Book Publishing Co., Garden City, NY, 1978.
64. SPSE Handbook of Photographic Science and Engineering, Society of Photographic Scientists and Engineers, John Wiley and Sons, New York, 1973, Sections 1, 2, 4, 7 and 18.
65. Spencer, D. A.: *Color Photography in Practice*, Focal Press, New York, 1966.
66. Hunt, R. W. G.: *The Reproduction of Colour*, Fountain Press, London, 1967.
67. McCamy, C. S.: "A Nomograph for Selecting Light Balancing Filters for Camera Exposure of Color Films," *Photogr. Sci. and Eng.*, Vol. 3, pp. 302–304, 1959.
68. Taylor, E. W.: "The Television Consistency Index Formulation and Preliminary Tests," *IEE Proc.*, Vol. 129, Pt. A, No. 7, p. 454, September, 1982.
69. *Recommended Practice for Lighting Cotton Classing Rooms for Color Grading*, ASTM D184-81, American Society for Testing and Materials, Philadelphia, 1981.
70. *Viewing Conditions for the Appraisal of Color Quality and Color Uniformity in the Graphic Arts*, ANSI PH2.32-1972 (R1981) American National Standards Institute, New York, 1981.
71. Helson, H. and Lansford, T.: "The Role of Spectral Energy of Source and Background Color in the Pleasantness of Object Colors," *Appl. Opt.*, Vol. 9, p. 1513, July, 1970.
72. Helson, H.: "Color and Vision," *Illum. Eng.*, Vol. XLIX, p. 92, February, 1954.
73. Helson, H.: "Color and Seeing," *Illum. Eng.*, Vol. L, p. 271, June, 1955.
74. Helson, H., Judd, D. B. and Wilson, M.: "Color Rendition with Fluorescent Sources of Illumination," *Illum. Eng.*, Vol. LI, p. 329, April, 1956.
75. Helson, H.: "Role of Sources and Backgrounds on Pleasantness of Object Colors," a paper presented at the IES National Technical Conference, New York, September, 1965.

SECTION 6

Luminaires

A luminaire is a complete lighting unit consisting of one or more lamps (light sources) together with the parts designed to distribute the light, to position and protect the lamps and to connect the lamps to the power supply. This section primarily deals with the "parts" rather than lamps. For the latter see Section 8, Light Sources. Also, for luminaire data for calculation purposes, such as coefficients of utilization values, see Section 9, Lighting Calculations.

CONSIDERATIONS IN DESIGNING AND SELECTING LUMINAIRES[1]

Some of the factors to be considered in luminaire design and selection include: (1) codes and standards relative to the construction and installation of luminaires, (2) physical and environmental characteristics of luminaires, (3) electrical and mechanical considerations that affect luminaires, (4) thermal considerations relative to the light source, the environment and the luminaire, (5) safety considerations as they apply to all areas of luminaire design and (6) economic factors which may affect the design and selection of luminaires.

Codes and Standards

Local codes, national codes, international codes, federal standards, professional standards and manfacturers' standards relate to specific requirements which must be met in the construction and installation of a luminaire. Standards usually relate to minimum requirements (safety, construction or performance) which may be exceeded to provide a better product.

Note: References are listed at the end of each section.

Some codes and standards deal with fire and safety (electrical, mechanical and thermal); others relate to performance (photometric) and construction (materials and finishes). They will vary to some extent depending on geographic location and end-use of the equipment. Conformance to the appropriate set of specifications is often determined by certified laboratory tests. Certification is often denoted by an identifying label. Local inspection agencies may or may not rely on conformance to national, federal or industrial codes and standards.

Local Codes. Local codes are normally, but not always, patterned after national codes. Information regarding local codes may be obtained from electrical inspection departments. Several other code jurisdictions may apply to regional and state codes. Building Officials and Code Administrators International (BOCA), Southern Building Code Congress International (SBCC) and International Conference of Building Officials (ICBO), all promulgate codes applicable to lighting installations. BOCA issues the Basic Building Code, ICBO issues the Uniform Building Code, and SBCC issues the Standard Building Code. All contain sections which control the use of materials in luminaires and luminous ceilings. Most state, county and city code jurisdictions operate under one of them.

National Codes. The NEC (National Electrical Code), the CEC (Canadian Electrical Code), and similar ones in most major countries throughout the world, state specific electrical requirements which must be met by all electrical equipment, including luminaires. They have been developed by safety protection and inspection agencies in conjunction with fire protection agencies. Although based on good manufacturing practices, these codes may vary considerably in many respects. The National Electrical Code may or may not be accepted in total by local agencies throughout the United States.

National and International Standards. The UL (Underwriters Laboratories, Inc.), the CSA (Canadian Standards Association), and similar ones in other countries publish minimum safety standards for electrical and associated products which are in conformance with the respective electrical codes of their country. They have testing laboratories to which equipment must be submitted for listing. Most manufacturers design luminaires to meet these standards.

Industry Standards. Industry standards are published by professional societies, associations and institutes, in most cases, utilizing national technical committees. Represented on these committees are inspection and safety protection agencies, manufacturers, professionals and consumers. Conformance to such standards may be desirable or specified, but are otherwise not binding on the designer. Such organizations include:

ASTM (American Society for Testing and Materials)
CBM (Certified Ballast Manufacturers)
IEEE (Institute of Electrical and Electronic Engineers)
IES (Illuminating Engineering Society of North America)
NEMA (National Electrical Manufacturers Association)
ANSI (American National Standards Institute)

Federal Standards (U.S. Government). Standards written by the following and similar federal agencies are almost exclusively for their own requirements. They are rarely specified for non-federal commercial applications.

BU Ships (Bureau of Ships)
CAB (Civil Aeronautics Board)
FAA (Federal Aviation Administration)
GSA (General Services Administration)
USCG (U.S. Coast Guard)
CE (Corps of Engineers)

Manufacturers' Standards. Since codes and standards deal primarily with saftey and performance but not with quality, the specifier should be aware of the quality standards practiced by the manufacturer.

Luminaire Characteristics

The luminaire and light source to be employed for a given application depend upon many factors. In addition to the illumination aspects (luminance, glare, uniformity, illuminance, etc.), consideration should be given to appearance, color of light, heating effect, noise level, efficiency, life and economics.

Reflector Design Considerations. Most lamps do not act as point sources in the size of reflectors into which they are placed. Physically large lamps such as phosphored mercury require larger reflectors to control light to a given degree than do small filament sources. Generally, the larger the source, the larger the reflector required for equivalent control.

Secondary effects of reflector or housing design can often be detrimental to the performance of a luminaire. As an example, if reflected energy is concentrated on the lamp, lamp parts may fail or reduced lamp life may be experienced (as in the case of high pressure sodium sources); if the beam is concentrated on a lens front, the glass may fail from thermal stress.

When fluorescent lamps are used in confining units, two effects take place. The build-up of heat in the lamp compartment raises bulb wall temperature, which may reduce light output. See Fig. 8-34. As lamps are moved closer, a mutual heating occurs and, beyond this, light which might be redirected out of the unit is trapped between lamps, or between lamps and reflector. Both conditions lower luminaire efficiency.

Lamping and Lamp Position. Lamp operating position is important. Many lamps are designed only for base up, base down, or some other specified operating position. Ignoring such limitations will normally result in unsatisfactory lamp performance or lamp life, or both.

A basic consideration is that of ready lamp insertion and removal. Recognition should be made of possible lamp changing devices that might be used and space allowed for clearance.

Effects of Radiant Energy. Consideration must be given to the effects of lamp energy in the nonvisible regions on luminaire materials and performance. Some plastics, for example, can be altered by ultraviolet energy. Some reflector materials are excellent for visible light but absorb infrared, thus creating thermal problems.

If the intended lamp can present a hazard to people or objects under some conditions of operation (for example, some high intensity discharge lamps operated with broken envelopes), protective devices should be designed into the luminaire or lamp.

Lamp Wattages. Lamp dimensions are rarely the sole criteria in determining luminaire size. Careful consideration must be given to adequate heat dissipation to insure normal lamp life and luminaire performance.

Luminaire Efficiency. This is normally a function of physical configuration and the selection of materials used. It should be recognized that many materials will change to some extent with use.

Appearance. The luminaire designer must coordinate technical, safety and economic considerations with the final luminaire appearance. Where the lighting is primarily functional, performance has maximum importance. Design efforts will probably be concentrated on reflectors, refractors and shielding elements.

Decorative luminaires are often selected because of their appearance. In this case, not unlike a piece of jewelry, they may serve to complement or accent a decorative scheme. It may be desirable to sacrifice optimum performance in order to attain pleasing proportions and shapes. Sometimes both can be coordinated into a single luminaire.

Color, texture and form all play an important role in the attempt to achieve either of the above goals.

Glare. Light sources and luminaires are potential glare sources. The degree of brightness control to be designed into a luminaire depends upon the intended use of the luminaire and luminous environment within which it will be used. Frequently, design compromises are required between visual comfort, utilization and esthetics.

Systems exist to evaluate the potential glare from luminaires. See the Application Volume. These establish criteria commonly used to guide the luminaire designer in determining acceptable limits of maximum and average luminances. A thorough understanding of the principles involved is essential.

The problem of glare and its control should be recognized for both interior and exterior luminaires.

Thermal Distribution. The integration of lighting units and the air handling and/or architectural aspects of a building greatly influence the basic construction of a luminaire. Some materials used in luminaires can be good reflectors of light and good absorbers of far infrared. For reflectance data see the index.

Ventilation and Circulation. Air movement through the lamp compartment of a luminaire may result in the lowering of light output due to accumulation of dust and dirt, or it may maintain the lumen output by the cleaning action of air moving past the lamps.

In fluorescent heat transfer luminaires a reverse air flow is sometimes used to trap dirt and dust before it enters the lamp compartment. Consideration must also be given to the effect air currents have on bulb wall temperature since this affects light output. See Section 8. Ambient temperature affects the striking voltage of all discharge lamps.[2]

Acoustical. There are two acoustical problems associated with lighting equipment (usually more critical in spaces that are intended to maintain exceptionally low ambient sound levels). The first is the introduction of sound, generated at the unit. The second problem involves sound transmission from room to room via the luminaire and plenum.

Undesirable sound generation is sometimes a problem with fluorescent or other discharge lamps ballasted with electromagnetic or solid state devices. Luminaires can transmit this sound to the rest of the space and, in some cases, add luminaire vibration to it. Large, flat surfaces and loose parts tend to amplify the condition. Steps taken to minimize transmission of sound from the ballast to the luminaire may affect heat transfer characteristics.

Where luminaires are used as air supply or air return devices, the air controlling surfaces should be designed with full consideration for air noise. In this case, there are well accepted criteria for permissible sound levels.[3,4]

If luminaires are designed for recessing, thought should be given to the possibility of sound transfer from one room to another through the luminaire and then through the plenum. This can be minimized by making the luminaire as "leak-proof" as possible. Here, too, there is an established testing and rating procedure.[5,6]

Where low noise levels are necessary, consideration should be given to the possibility of remotely mounting the ballasting equipment or using light sources having inherently quieter operation of its control devices.

Vibration. Incandescent lamps are normally made in the smaller sizes up to 150 watts for vibration service and up to 500 watts for rough service. Where more severe vibration is present or larger wattages are required, vibration resisting sockets or shock absorbing mountings may be used. For high bay mounting on a building's steel structure, a spring steel loop of proper size and tension for the specific luminaire's weight usually solves the most difficult vibration problems.

High intensity discharge lamps are less fragile than most incandescent lamps and will, consequently, stand more vibration. If vibration is severe then shock absorbers should be used.

For fluorescent luminaires that may be installed in vibration areas, spring loaded lampholders should be used—rather than those of the twist type. There are conditions of very high frequencies such as turbine deck areas where special lampholders may be required.

Radiation Interference. Electromagnetic radiation from gaseous discharge lamps, especially of the fluorescent type, and auxiliary components such as starters and phase control devices may be sufficient to cause interference with nearby radios, television receivers, sound amplifying equipment, electro-cardiograph devices, sensitive electronic equipment, and certain military radar and tracking equipment. This interference is transmitted in two ways: (1) by direct radiation through the face of the luminaire and (2) by conduction through ballast and supply line.

To eliminate direct radiation the luminaire should be entirely metal enclosed, except for the light transmitting opening. Normal tolerances on the fit of parts are acceptable but there should be no appreciably large open holes in any of the metal parts and the electrical service should be brought in through grounded conduit or shielded cable. Conducted interference can be isolated by proper line filters located at the luminaire. See page 8-31 on Radio Interference.

Life and Maintenance. The life of any luminaire is dependent primarily upon the luminaire's ability to withstand the environmental conditions in which it is installed. To ensure reasonable life, appropriate treatment of all materials should be considered. Where conditions are such that electrolytic action may persist, the use of dissimilar metals and high copper content alloys of aluminum should be restricted.

Maintenance may be a problem unless proper consideration is given to minimizing luminaire light loss factors at the time the luminaire is designed. Ventilation, gasketing and filtering, in addition to the selection of materials to diminish the effect of dust and dirt, should be utilized. Ease of relamping and of access to auxiliary components are important considerations as well.

Environmental Conditions

Ambient Temperature. The most obvious influence of ambient temperatures is its effect on the starting and operating of fluorescent lamps. While the effect of ambient temperature on other sources is not visually apparent, the effect of high and low temperatures on other luminaire components should be considered. Excess heat generated by internal heat sources such as lamps, and in some cases ballasts, must be transferred to the outer surfaces in the environment.

Surface mounted luminaires and recessed equipment, frequently operating in non-ventilated plenums, are often exposed to high ambients. Some industrial areas have shown 65 °C (150 °F), or even higher, to be realistic. Refrigerated storage areas, and in some climates unheated storage areas, attain temperatures of −29 °C (−20 °F) and frequently lower. A given luminaire should be expected to operate efficiently through these extremes.

Incandescent Filament Lamps. An incandescent filament operates at a very high temperature, near 2620 °C (4750 °F) for a 200-watt lamp. Normal changes in ambient temperature will not appreciably affect its light output or life.

Fluorescent Lamps. The performance of fluorescent lamps is affected by ambient temperature and draft. See page 8-28. In totally enclosed luminaires, at normal room ambients, fluorescent lamps usually operate at temperatures above their optimum. Designs which "bottle-up" lamp and ballast energy may reduce light output. Ventilation or other means of heat dissipation are helpful and should be considered.

Where it is not possible to reduce temperature by luminaire design, the use of components with higher temperature ratings may be indicated.

High Intensity Discharge Lamps. Ambient temperature is not normally a significant factor in the light output of conventional high intensity discharge lamps. Arc-tube temperatures are such that modest variations due to varying ambients will not make much difference. Starting requirements are sometimes affected at low temperatures and ballasts meeting such requirements must be used. Special auxiliary equipment also may be required at elevated temperatures. The relationships between temperatures and service life of wiring and component parts must be studied. There are UL and National Electrical Code requirements which apply in some cases.

Wet Locations. The National Electrical Code requires that luminaires for wet locations be so labeled by the manufacturer under authority from the Underwriters Laboratories. Booklet UL57 covers the minimum design requirements of such luminaires in detail. Article 680 of the National Electrical Code[7] specifically covers swimming pools. See also National Electrical Code, Article 410-4.

Corrosive Locations. In some areas, lighting equipment is subject to unusual* corrosive atmospheres. Many outdoor locations subject to a windborne corrosive agent, abrasives, or salt water, have corrosive action that can rapidly depreciate exposed or unprotected surfaces.

The characteristics of all materials should be carefully considered where possible corrosive or severe local atmospheric conditions (as near salt water) can be anticipated.

Hazardous Locations. All equipment for hazardous locations must be labeled by the Underwriters Laboratories for the environment in which it is to be used or installed. To obtain approval for such labeling, the construction must comply with *Standards for Safety,* "Electric Lighting Fixtures,"[2] and "Electric Lighting Fixtures for Use in Hazardous Locations."[8] General requirements for hazardous locations are covered in articles 500 through 517 of the National Electrical Code.[7]

Some industries have additional requirements. For example, flour mills and elevators are covered by the *Electrical Code of the Mill Mutual Fire Prevention Bureau,* Two N. Riverside Plaza, Chicago, IL 60606.

There are also more restrictive local codes, such as in New York City, Chicago and Los Angeles. For Marine use, the Underwriters Laboratories Booklet UL595 should be consulted.[9]

Classification. The design of lighting equipment for hazardous locations is a highly specialized topic. It is imperative that the designer be completely familiar with the latest versions of the applicable documents and related writings.

The National Electrical Code classifies the various hazardous locations by the type of hazard (Class I, Class II, etc.) and by the degree of probability of hazard (Division 1 and Division 2).[7] There are also Canadian equivalents,[10] requiring labeling by the Canadian Standards Association (CSA).

"Enclosed and Gasketed" luminaires are not recognized by UL as being suitable for hazardous locations.

Acoustics.† Room acoustics and sound level sometimes determine the type of luminaire to be utilized. In some areas which require very low ambient noise levels, discharge luminaires require the use of very quiet ballasts and assembly methods which prevent sound build-up.[3,4] These applications may warrant the incorporation of acoustic properties and material into the luminaire proper, the design of recessed equipment, or the remote location of the ballast equipment.

Vibration.† There are spaces within a building where the operation of equipment creates a condition of high vibration. The solution to this problem is discussed under Luminaire Characteristics.

Radiation.† Certain areas of buildings, especially laboratories with specialized equipment, may emit radiation which will have a detrimental effect on the luminaire. These are special cases, and the possible effects of fading or color change of paints, and degradation of plastics should be studied. In these cases, special consideration should be given to selection of luminaire materials.

Air Movement. Outdoor luminaires should be designed to withstand wind-loading to which they might be subjected. In indoor locations, the air pressure on a fluorescent troffer lens or a luminous ceiling component can dislodge a diffuser when a door to a small office is suddenly opened.

Dirt. The effect that dirt will have on a luminaire configuration will depend on the intended end-use of the luminaire. (For further details on Lighting Maintenance, see the Application Volume).

Miscellaneous Environmental Effects. A complete listing of the possible effects of environment on luminaire design is impossible. The designer, however, should be alert to the possibilities of many others not specifically covered, such as, the effect of ultraviolet energy on outdoor components, thermal shock, extremes in humidity and foreign substances in the atmosphere.

Electrical Considerations

Sockets and lampholders, wire, ballasts and other controls are parts in electrical circuits of luminaires. All gaseous discharge lamps require some form of ballast or control equipment to provide adequate starting voltage, and to limit the current after the lamp has started.

Fluorescent Lamp Circuits. Each type of lamp requires a specific ballast circuit. See Section 8.

* In some cases, these might also be classified as "Hazardous Locations" and require National Electrical Code Class I equipment.
† See also Luminaire Characteristics, page 6-2.

Ballast Quality. Rated lamp life and light output for fluorescent lamps are directly dependent on the ballast's ability to meet specified limits set by the American National Standards Institute (ANSI) Standards.* Ballasts not meeting ANSI Standards, can reduce lamp life as much as 50 per cent and light output by more than 30 per cent.

The life of ballasts made with 105 °C (Class A) insulation is approximately 45,000 hours (continuous operation not exceeding 105 °C). Ballast coil temperatures will not exceed the 105 °C rating in properly designed ballasts when luminaires maintain a maximum ballast case temperature of 90 °C. Other factors affecting ballast life are input voltage, luminaire heat dissipation characteristics, luminaire mounting and environment. Field data indicates a 12 year median ballast life for a duty cycle of 16 hours per day, 6 days per week, or 5000 hours per year. This life rating considers the fact that ballast warm-up time allows the ballast to operate at peak temperatures 12 out of the 16 hours per day or 3750 hours per year.

Ballast safety standards are set by Underwriters Laboratories. Ballast case temperature is limited to 90 °C with the ballast mounted in a luminaire and the luminaire in a 25 °C ambient.

Excessive ballast operating temperatures cause ballast insulation deterioration resulting in short life and possible actuation of the ballast protective device. A convenient rule of thumb is that every 10 °C increment above 90 °C results in a 50 per cent reduction in life. Luminaire design should prevent undesirable heat rise within the luminaire.

Ballast hum is a natural result of an electromagnetic device. Various ballast circuits produce varying noise levels and the ballast manufacturer should be consulted for specific sound ratings. The lighting equipment designer must consider the most common application of his product when selecting specific ballast types.

Remote locations of ballasts may involve complications of wiring, voltage, thermal considerations and code restrictions. (See the National Electrical Code[7] for details.)

Ballast Protection—Class P Ballasts. The National Electrical Code requires that protected ballasts be used in fluorescent luminaires (except reactor types). Protected ballasts are called "Class P" ballasts and are listed by UL as such. All new fluorescent luminaires must use UL listed Class P ballasts in order to carry the UL label. The intent is to limit the maximum temperature a ballast case can reach in a luminaire under both normal and abnormal conditions.

UL requires that the protector open within two hours after the ballast case temperature has reached 110 °C. The ballast-luminaire combination must be such, thermally, that the luminaire can be installed as intended without the ballast protector being actuated.

The maximum ballast case hot spot temperature should not exceed 90 °C when the luminaire is in service.

High Frequency Operation. Most fluorescent lamps can be operated on higher frequencies and with increased luminous efficacy (see Section 8.) As system frequency is increased, ballasting devices can be smaller, lighter and more efficient.

Low Temperature Operation. Fluorescent lamps require higher open circuit voltages for low temperature starting. Low temperature ballasts should be used where ambient temperatures are below 10 °C.

Gaseous Discharge Lamp Circuits. Mercury, metal halide or high pressure sodium ballasts (see Section 8) are usually classified as:

1. *Reactor*—consists of a series reactor to limit lamp current. Line voltage must be high enough to start lamp, and line regulation must be within ±5 per cent.
2. *Autotransformer*—used when line voltage is higher or lower than required starting voltage for the lamp; it is combined with a reactor to limit lamp current and a capacitor for power factor correction. Primary taps can be provided to extend line voltage range. Line voltage regulation limited to ±5 per cent of tap voltage range.
3. *Constant Wattage or Regulator Type*—consists of special high power factor lead circuit to provide improved lamp current regulation over an extended line voltage range. Typical designs for mercury lamps allow a line voltage variation of ±13 per cent with a lamp wattage variation of only ±2 per cent. With this type of ballast the operating line current is always higher than the starting or open circuit line currents.
4. *Lead Circuit Autotransformers*—also called Constant Wattage Autotransformers or Lead Peaked Autotransformers. These are similar to the Constant Wattage ballasts except line and lamp wattage regulation is lower and the ballast is an autotransformer rather than an isolated transformer. Typical performance allows a line

* Such ballasts bear the CBM/ETL label which means that they have met the Certified Ballast Manufacturers' requirements which includes the ANSI C82 standard and have been tested and certified by ETL Testing Laboratories, Inc.

voltage variation of ±10 per cent with a lamp wattage variation of from 5 to 10 per cent.

Sockets and Lampholders. For *incandescent lamp sockets,* maximum wattage ratings should not be exceeded and careful consideration should be given to the following:

1. Non-metallic housings and paper insulating liners should only be used for low wattage lamps in well ventilated luminaires.
2. Porcelain sockets should be used in recessed or totally enclosed luminaires in spotlights.
3. Screwshells of aluminum or copper should be used only for low wattage applications.
4. Nickel plated brass or stainless steel screwshells should be used for high heat conditions.
5. High heat sockets should be used for tungsten-halogen and other lamps of high wattage in small envelopes.
6. See Underwriters Laboratories and Canadian Standards Association standards for sockets to be used for outdoor luminaires.

For *fluorescent lampholders,* a good connection, mechanical and electrical, between lamp pins and contacts is imperative, particularly for rapid start and trigger start circuits. Silver plated contacts reduce electrical contact resistance. Positive lamp seating is imperative to prevent destructive arcing at contact points.

Wire.

Total current. Wire smaller than No. 18 AWG may not be used for line voltages of 120 V or higher.

Voltage. Thermoplastic insulation is limited to 1000V, Type AF wire limited to 250V.

Temperature on insulation. 60 °C to 200 °C ratings are generally available.

Application will determine use of stranded or solid wire. Braided insulation usually is required where flexing is expected. Third (grounding) conductor and three-prong plug may be required for cord sets. Chain suspended luminaires may require grounding wires.

Codes, both national and local, should be carefully consulted before specifying a given wire size and insulation. There are many special requirements, for example, supply leads for a recessed luminaire must be No. 14 AWG minimum regardless of how little current they will be required to carry.

Starters. Protective starters for fluorescent lamp operation provide additional safety by disconnecting lamps which have failed. Other starter types include thermal, glow discharge, manual reset and automatic reset. See Section 8. Choice is usually dictated by specifier.

Mechanical Considerations

Tolerances. It is necessary to establish suitable dimensional tolerances for both the component parts and the final assembly. This includes allowances for the different rates of thermal expansion of the various parts. An industry standard[11] defines tolerances of recessed luminaires. Surface and pendant mounted luminaires may have less critical tolerances. Particular attention should be given to such components as spinnings, plastics, glass and lamp tolerances.* Plenum space available should be considered in case of recessed luminaires.

Strength. Strength considerations will depend on the luminaire type. All luminaires should have housings of sufficient rigidity to withstand normal handling and installation. Luminaires intended for outdoor use should incorporate mounting and design features suitable to withstand high winds, rain and snow accumulation.

Recessed luminaires in poured concrete should have an enclosure of suitable strength, tightness and rigidity for the application.

Surface mounted luminaires should be strong enough so that they will not excessively distort when mounted on uneven ceilings.

Suspended luminaires should have adequate strength to minimize vertical sag between supports as well as lateral distortion and twist. Provision must be made for attachment of supports at suitable locations. Vibration should also be considered.

Local conditions such as in earthquake areas, sometimes require conformity with special local codes. Local codes should be investigated for these rulings.

Component Support. Lampholders and sockets should be held in place to prevent movement and to insure maintenance of proper spacing lengthwise for good lamp contact in fluorescent luminaires. To hold large, high wattage incandescent and high intensity discharge lamps, gripping sockets are often helpful. Ballasts should be securely fastened to the luminaire housing for good thermal contact.

* See the applicable American National Standards Institute (ANSI) standards.

Enclosure Fastening. Luminaire enclosures should be designed to incorporate a secure method for retention in the luminaire. Attention must be given to wind, rain, vibration and shock. (See also Environmental Considerations.) *Tamper-proof fastenings* may be required in correctional institutions and mental hospitals. *Wire guards* may be required around exposed lamps in accessible locations.

Assembly. Factory assembly should be as complete as possible for minimum field assembly. Refer also to Canadian Standard Association (CSA) and Underwriters Laboratories (UL) requirements. Instruction sheets should describe how components are to be assembled and installed on the job site.

Ultimate Mounting. Mounting components should satisfy the requirements of UL, CSA, local codes and end-use environment. Local codes should be considered with regard to power supply, continuous row wiring and grounding of the luminaire. Recessed luminaires intended for use in forced air return plenums may be subject to special requirements. Instructions and warnings should be marked in a readily discernible and permanent manner. The text and physical dimensions may be governed by code requirements.

Maintenance. Ease of maintenance is an important consideration. Instructions for maintenance and correct lamping should be provided to prevent damage to and insure continuous maximum utilization of the luminaire. Electrical components should be replaceable without removing the luminaire from its mounting. Where ceiling plenums are not otherwise accessible, provision must be made for access through the luminaire to splice boxes connecting the recessed luminaire to the branch circuit.

Size. The size of a luminaire is controlled by components such as lamps, ballasts and reflectors. Mounting limitations and building modules can also determine size.

Thermal Considerations

Heat is a natural by-product of most light sources and the amount of heat generated by a particular source depends on its luminous efficacy, total wattage consumed, and type of source. Ballasts and transformers may also contribute heat. Environmental temperature is an added consideration.

Code Requirements. The National Electrical Code,[7] Canadian Electrical Code,[10] and their respective testing laboratories set specific temperature test limits for electrical components and critical areas immediately surrounding a luminaire. Much of their testing is performed at an ambient temperature of 25 °C (77 °F). Code requirements usually do not relate to performance characteristics and thermal considerations are regarded from a safety standpoint.

Thermal Environment. A theoretical thermal environment of 25 °C (77 °F) is usually not adequate. Temperatures at the luminaire are frequently higher and ambient temperatures in ceiling plenums of 60 °C (140 °F) are not uncommon. In industry ambient temperatures may reach as high as 65 °C (150 °F). In cold storage areas, temperatures may be −29 °C (−20 °F) or lower. High or low temperature differentials will directly affect the thermal dissipating capabilities of a luminaire. A luminaire may be expected to operate without breakdown or reduced component life through both extremes.

Very high or low ambient temperatures may cause electrical components to fail. For example, contact of hot glass with cold air or water may result in breakage and excess heat may cause thermoplastics to distort. Satisfactory performance at the luminaire interior temperature is a major consideration in selecting components and finishes.

Supply Voltage Effect. Higher than nominal line voltages usually cause higher temperatures within electrical components. Lower voltages do not necessarily produce lower temperatures.

Luminaire Components. Ballasts and other auxiliary equipment, sockets and wires have definite safe operating temperature limits defined by UL, CSA, ANSI and the manufacturer. Special electrical components may have to be specified for extreme temperatures.

Metal components and their finishes may be affected by temperature within a luminaire. Thought must also be given to metals as conductors of heat either to or from component parts.

Glass and plastic components should be very carefully chosen to prevent cracking, shattering, deformation or other deterioration. This may be either or both long and short term.

Expansion or contraction of components due to thermal changes should be considered. Different coefficients of expansion between materials in intimate contact can lead to serious problems.

Safety Considerations

The need for attention to safety considerations, particularly in the design stage, cannot be overemphasized. Usually, if the design meets applicable code requirements, it will be in conformance with the following criteria.

Electrical. The safe transfer of electrical energy from the source to the light control equipment is of utmost importance. In this context important wiring considerations are:

1. Current carrying capacity of the conductors.
2. Insulation rating of the conductors.
3. Grounding. A ground conductor is often required for portable lighting equipment. Permanently installed lighting equipment should be electrically grounded to eliminate potential shock hazards.
4. Temperature rating of the conductors.
5. Connections to the junction boxes. Where supplied, connections must be in conformance with local codes.
6. Wire termination color coding in conformance with code requirements for safe field installation.
7. Mechanical strength and flexing requirements.

Safety interlocks are advisable in equipment where high open circuit voltages are present for protection during servicing and are often advisable in damp areas and basements.

In those applications where required, fuses and thermal protectors must be included in the design of the equipment.

Splices, clearances and sockets should conform to applicable UL, CSA and ANSI requirements.

Low voltage units should be considered for outdoor applications.

Thermal. Luminaires should be designed to safely meet the requirements of the thermal environment and use. If the unit is to be firmly mounted, and will not be handled during operation, it can operate at a higher temperature than a unit which must be held or moved by hand. (See also Thermal Considerations.)

Mechanical. Mechanical design of light control equipment for safety involves consideration of the components of the equipment, both individually and collectively, and may include the effect of vibration, weight, wind-load, shock, impact, snow and ice. Materials, castings, sheets and extrusions should meet the unit requirements for strength as follows:

1. Lenses, diffusers or louvers should be chosen to withstand the expected mechanical loading (wind or vibration) and secured to the lighting equipment in such a way that they are not a hazard by falling.
2. Reflector assemblies should be securely fastened to the mounting members of the equipment.
3. The method of attaching the ballast should permit field replacement while still retaining the same positive holding means.
4. Trunnions or other mounting assemblies should support the weight of the equipment, plus any normal loading such as wind, vibration and shock, without permanent sag, settling or damage.
5. Pipe stems should have adequate wall thickness for full threads. Thin wall tubing should not be used for threaded connections. Breakable couplings are required in some applications such as airport stem mounted runway lights or standards for street lighting luminaires.

Environmental. Hazardous area requirements for safety are covered by Underwriters Laboratories Booklet UL844[8] and also under Environmental Considerations in this Section.

There are, in addition, some general safety considerations to be given to the environment when designing a luminaire and its equipment. The unit must withstand the environmental conditions without its mounting breaking loose. In outdoor applications, exposure to rain, snow, water, ice and condensation may necessitate the use of waterproof seals and weep holes.

The nameplate or other permanent marking should include data on the lamp to be used, the voltage to be supplied, safe operating conditions, catalog number, and other pertinent information. A complete instruction sheet should be included.

Economic Considerations

The choice of components and performance level used in designing a luminaire is governed by its intended end use. The design may have to be altered to provide the most economical product to meet a given set of design requirements.

Ready-Made Components. Standard components are generally less expensive than those made only on special order or in low volume. Luminaires which use all commercially available parts tend to be less distinctive than those made of specially built parts.

Fabrication. Frequently decisions will be dependent on the equipment available to manufacture the product. Many lighting companies are specialists in sheet metal fabrication and work in that area, others specialize in castings and use those extensively. If a part can be completely fabricated in one piece, a casting may prove to be cheaper; however, if additional pieces are to be attached, sheet metal materials may be spot welded whereas castings must be jointed by screws, rivets, adhesives or similar methods.

Inventory and Packaging. In the economics of a company and the services offered on customer orders, inventory becomes a major consideration. Many types of lighting equipment offer multiple lenses, or choice of finish, or variation in trim design on a common housing.

In some instances large inventories of finished, but not assembled, parts are desired by a manufacturer. This provides the flexibility of assembling these items in the proper combination to fill any requirement. The difficulty with this plan is the time delay necessary to assemble saleable items from stocked parts. To overcome this weakness, the method of packaging can be varied. It is best to check the Underwriters Laboratories or CSA requirements when considering packaging of subassemblies instead of complete units.

The most desirable, but not necessarily the most economical stocking for ease in filling an order, would be to have a complete unit in one carton. There is less chance of error in shipment or breakage from handling in this type of stocking, but considerable inventory is required.

Tooling. Generally, the amount of money available for the purchase of tooling for a given product is dependent on the anticipated volume of sales for that item. Proper selection of tooling may also reduce unit labor costs and/or improve product acceptance.

Standards vs. Specials. The term "standard fixture" usually indicates the cataloged and warehoused luminaires manufactured by a company which are available from stock. All expenses of producing the unit such as engineering, tooling, photometric testing and UL listing are included in the manufacturing costs for a relatively large volume over an extended period of time. The fixed costs are spread over the maximum number of luminaires to obtain the most economic cost per unit. A special unit is not usually cataloged or warehoused.

If the same type of unit is made on a "special basis," the expenses of producing the unit may be chargeable to the items on the specific order, therefore, a special may be more costly to manufacture.

Finishing. Decisions on finishes are frequently based on the type and amount of equipment available in the manufacturing facility; other factors are: volume, quality of finish, appearance, durability, and amount of light control desired.

Material is available in a variety of preapplied finishes; such as, anodized aluminum, porcelain enamel, baked enamel, and plated sheet steel. Textured surfaces of adhesive vinyl or embossed patterns may be desirable. In the use of these materials, special attention should be given to the methods of fabricating and joining which are necessary due to the nature of the finish. Equipment required to apply porcelain coatings is much more expensive than required for baked enamel.

Protective finishes require investigation into the end-use of the product and its location. A protective finish must be compatible with the base material and strongly adhere to it. It should act as a barrier to attack or degradation from elements surrounding it.

Fasteners. If it is not essential that parts be disassembled, use of permanent fastening by welding, cementing or riveting may be used. If components are to be removed for installation or maintenance, the use of screws or nuts and bolts should be considered. The use of snap fasteners and clips formed of sheet metal or wire has become common. These devices can be designed to perform multiple functions.

MATERIALS USED IN LUMINAIRES[12]

The materials most commonly used in luminaires are glass, plastics, metals and applied finishes and coatings. Each of these materials is briefly described here along with comments on manufacturing processes and techniques. The use of these materials in lighting as well as some general information to aid in the selection and evaluation of the most suitable light control materials is also discussed in this section. For more specific applications the material manufacturer's data should always be consulted.

Glass

Glass is an inorganic product of fusion which has been cooled to a rigid condition without crystallizing. Chemically, glasses are mixtures of oxides. The oxides most commonly used in glass composition are silicon, boron, aluminum, lead, sodium, magnesium, calcium and potassium.

The chemical and physical properties of glasses, such as color, refractive index, thermal expansion, hardness, corrosion resistance, dielectric strength and elasticity, are obtained by varying composition, heat treatment and surface finish. Colors produced for high temperature glass are limited.

Glasses which are important to illumination can be classified into several groups with characteristic properties.

Soda-Lime. Soda-lime glasses (or lime glasses), used for window glass, lamp envelopes, lens covers and cover glasses (tempered), etc., are easily hot-worked, and are usually specified for service where high heat resistance and chemical stability are not required.

Lead-Alkali. Lead-alkali glasses (or lead glasses) are used for electric light bulb stems, neon sign tubing and certain optical components. They are useful because of their good hot-workability, high electrical resistivity and high refractive indices. They will not withstand high temperatures or sudden temperature changes.

Borosilicate. Borosilicate glasses are used for refractors, reflectors, lenses, sealed-beam lamp parts, etc. because of their high chemical stability, high heat shock resistance and excellent electrical resistivity. They may be used at higher temperatures (about 230 °C) than soda-lime or lead glasses. They are not as convenient to fabricate as either soda-lime or lead-alkali glasses.

Aluminosilicate. Aluminosilicate glasses are used where high thermal shock resistance is required. They have good chemical stability, high electrical resistivity, and a high softening temperature enabling use at moderately high temperatures (about 400 °C).

Ninety-Six Per Cent Silica. Ninety-six per cent silica glasses are used where high operating temperatures are required. They may be regularly used at 800 °C. They are also useful because of their extremely high chemical stability, good transmittance to ultraviolet and infrared radiation, and resistance to severe thermal shock. They are considerably more expensive than soda-lime and borosilicates, and are more difficult to fabricate. Due to their low coefficient of thermal expansion, these glasses cannot be tempered to increase mechanical strength.

Vitreous Silica. Vitreous silica is a glass composed essentially of SiO_2. It is used for lamp envelopes where high temperature operation and excellent chemical stability are required. It has high resistance to severe thermal shock, high transmission to ultraviolet, visible, and infrared radiation, and excellent electrical properties. However, due to its low coefficient of thermal expansion, it cannot be tempered to increase mechanical strength. Depending on the method of manufacture, this glass may be known as fused silica, synthetic fused silica, or fused quartz.

Manufacturing Techniques. Glass may be formed by several techniques: pressing, blowing, rolling, drawing, centrifugal casting, and sagging. These operations may be performed by hand methods or by automatic machines, depending on volumes.

After being formed, many glass products must be annealed to relieve excessive stress. Also, additional finishing operations may be required, such as cutting, grinding, polishing and drilling. Further treatment, such as tempering or chemical strengthening, may be required to obtain the desired physical properties.

The finishes for glass surfaces are varied in nature, depending on the forming technique. The surfaces may be subsequently altered by chemical etching, sandblasting or shot blasting, polishing, staining and coating. These operations are used to obtain reflection, control radiation or make a surface electrically conductive.

The function of glass in lighting may be divided into the following general categories:

1. Control of light and other radiant energy.
2. Protection of the light source.
3. Safety.
4. Decoration.

These functions may be combined for a particular application. Typical properties of glasses are shown in Fig. 6-1.

Plastics

Plastics generally are high molecular weight organic compounds that can be, or have been, changed by application of heat and pressure, or by pressure alone, and once formed retain their shape under normal conditions.

They can be broadly classified as thermoplas-

Fig. 6-1. Properties of Lighting Glasses

Type of Glass	Color[a]	Coefficient of Thermal Expansion per °C[b] (× 10^{-7})	Upper Working Temperatures °C(°F) (Mechanical Considerations Only)				Thermal Shock[e,f] Resistance Plates 15 cm x 15 cm x 0.64 cm (6 in x 6 in x ¼ in) thick	Impact Abrasion Resistance[g]	Density (grams per cc)	Young's Modulus		Poisson's Ratio[h]	Refractive Index (589.3 nm)
			Annealed		Tempered								
			Normal Service[c,e]	Extreme Limit[d,e]	Normal Service[c,e]	Extreme Limit[d,e]				(10^6 kg/cm²)	(10^6 lb/in²)		
Soda-Lime	Clear	85–97	110 (230)	430–460 (806–860)	200–240 (392–464)	250 (482)	50	1.0–1.2	2.47–2.49	0.7–0.71	10–10.2	.24	1.512–1.514
Lead-alkali	Clear	85–91	110 (230)	370–400 (698–752)	220 (428)	240 (464)	45	.56	2.85–3.05	0.61–0.62	8.7–8.9	.22	1.534–1.56
Borosilicate	Clear	32–46	230 (446)	460–490 (860–914)	250–260 (482–500)	250–290 (482–554)	100–150	3.1–3.2	2.13–2.43	0.65–0.66	9.3–9.5	.20	1.474–1.488
Aluminosilicate	Clear	34–52	200 (392)	650 (1202)	400 (752)	450 (842)	100–150	2.0	2.43–2.64	0.87–0.89	12.5–12.7	.25	1.524–1.547
96% Silica	Clear	8	800–900 (1472–1652)	1090–1200 (1994–2192)			1000	3.5–3.53	2.18	0.67	9.6	.18	1.458
Vitreous Silica	Clear	5.5	1000 (1832)	1200 (2192)			1000	3.6	2.2	0.73	10.4	.16	1.458

[a] All glasses can be colored by the addition of metallic oxides that become suspended or dissolved in the parent glass usually without substantially changing its chemical composition or physical properties.
[b] From 0° to 300°C, in/in/°C, or cm/cm/°C.
[c] Normal Service: no breakage from excessive thermal shock is assumed.
[d] Extreme Limits: depends on the atmosphere in which the material operates. Glass will be very vulnerable to thermal shock; tests should be made before adopting final designs.
[e] These data approximate only.
[f] Based on plunging sample into cold water after oven heating. Resistance of 100°C means no breakage if heated to 110°C and plunged into water at 10°C. Tempered samples have over twice the resistance of annealed glass.
[g] Data show relative resistance to sandblasting.
[h] Value applies to only one glass of group.

tic or thermosetting. Thermosplastic resins may be repeatedly softened and hardened by heating and cooling. No chemical change takes place during such actions. Thermosetting resins cannot be softened and reshaped once they have been heated and set since their chemical structure has changed. Some of the commercially important thermoplastics are: acrylonitrile butadiene styrenes (ABS), acrylics, cellulosics, acetals, fluorocarbons, nylons, polyethylenes, polycarbonates, polypropylenes, polystyrenes and vinyls. In the thermosetting group resins of importance are epoxies, melamines, phenolics, polyesters, ureas and silicones.

Most resins, whether thermoplastic or thermosetting, can be processed into structural or low ratio expanded foams. Among the important properties of many foams are: stress free parts, improved insulation characteristics and lighter weight, and sometimes greater strength and toughness than the unfoamed form of the material.

Fillers and reinforcing agents are frequently added to plastics to obtain improved heat resistance, strength, toughness, electrical properties, chemical resistance, and to alter formability characteristics. Some of the fillers and reinforcements in general use are: aluminum powder, asbestos, calcium carbonate, clay, cotton fibers, fibrous glass, graphite, nylon, powdered metals and wood flour.

The basic compounds from which today's plastics are produced are obtained from such sources as air, water, natural gas, petroleum, coal and salt. The involved chemicals are reacted in large closed vessels under controlled heat and pressure with the aid of catalysts. The resultant solid product is then subjected to such further operations as: reduction of particle size; addition of fillers, softeners and modifiers; and conversion of form to granules, pellets, film, etc.

If the plastic surfaces need protection from ultraviolet energy, a film compounded from a material resistant to this energy is sometimes laminated to the surface facing the energy source.

The forming and converting of plastics into end products is a highly specialized field. Some of the more important processes are: injection molding (fluid plastic forced under pressure into a controlled temperature mold); compression molding (resin placed in mold and the cavity filled by application of heat and pressure); blow molding (a thin cylinder of plastic is placed in a mold and inflated by air pressure to conformity with the mold cavity); extrusion (fluid plastic is screw driven through a die); thermoforming (a sheet, heated to a state of limpness, is draped over a mold and forced into close conformity with the mold by pressure or vacuum); spray coating (a solution or emulsion is sprayed on a prepared surface); machining (solid plastic is shaped by basic wood or metal working operations); rotational molding (resin is charged into a mold rotated in an oven, centrifugal forces distribute the resin and the heat melts and fuses the charge to the shape of the cavity); cold stamping (stamping and/or forming of cold plastic sheet); and ultrasonic welding (bonding of plastics by conversion of sonic vibrations to heat).

Lighting uses of important resins and properties of plastics used in lighting are shown in Figs. 6-2 and 6-3, respectively.

Steel

In the fabrication of lighting equipment, steel serves primarily in a structural capacity. Sheet steel, while having the greater strength and lower cost needed for a large volume material, must be processed additionally with platings or applied coatings before it can serve as a light controlling medium. Many grades and types of sheet steel are available and should be selected for the proper use. In certain instances, other forms of steel are used in lighting equipment and many

Fig. 6-2. Lighting Uses of Important Resins

Resin	Uses
Thermoplastics	
Acrylics	louvers, formed light diffusers, prismatic lenses, diffusers, film
Cellulosics	sign faces, vacuum formed diffusers, globes, light shades, light supports
Flexible vinyls	gaskets, wire coating
Nylons	electro-mechanical parts, wire insulation, coil forms
Polycarbonates	insulators, globes, diffusers, antivandalism street lighting globes
Polyethylenes	(high density)—wire coatings, housings
Polyethylenes	(low density)—formed light diffusers, blow molded globes
Polystyrenes	same as acrylic
Rigid vinyls	formed lighting diffusers, corrugated sheet for luminous ceilings
Thermosetting	
Melamines	switches, insulators
Phenolics	wire connectors, switches, sockets, shades
Polyesters	(glass reinforcing)—shades, reflector housings, diffusers
Ureas	louvers, lamp holders, shades
Filled reinforced plastics	insulators, reflectors, housings, globes, light shades, switch bases

of these sizes of bars, rods, etc., are available readily from steel warehouses.

Sheet steel used in lighting equipment is of three basic types: *hot rolled steel*, *cold rolled steel* and *porcelain enameling sheets.*

Hot rolled steel, because of the rolling process at elevated temperatures, carries an oxide coating or mill scale. It is not normally used where a smooth appearance is desired. The scale can be removed in an acid bath (pickling) but the surface still is somewhat rough.

Cold rolled steel is used primarily because of its smooth surface appearance. It is also available in thinner gauges than hot rolled steel. It is usually obtained by pickling hot rolled steel coil and further reducing the thickness without elevating the temperatures.

Porcelain enameling sheets are similar to hot rolled steel or cold rolled steel, but have very low carbon content. The low carbon content is required, because porcelain is normally fused at temperatures which would cause cold rolled steel

Fig. 6-3. Properties of Plastics Used in Lighting

Materials	Common Forms										Reasons for Use and General Data**										
	Castings	Compression Moldings	Extrusions	Fiber	Film	Foam	Injection Moldings	Sheet	Reinforced Plastic Moldings	Industrial Laminates	Chemical Resistance	Colorability	Flammability Rating*	Flexibility	High Dielectrics	Low Moisture Absorption	Clarity	Strength and Rigidity	Toughness	Effect of Ultraviolet†	Resistance to Heat °C Continuous
Thermoplastics																					
ABS‡			X				X	X			X	X	HB-VO		X			X	X	NP	60–110
Acetals‡			X				X				X	X	HB		X			X	X	C	80
Acrylics	X		X	X			X	X			X	X	HB				X	X		N	60–90
Acrylic-styrene copolymers			X				X	X			X	X	HB		X	X	X	X		NP-SL	80–95
Cellulose acetates			X	X	X		X	X				X	HB-VO		X		X	X	X	SL	65–105
Cellulose acetate butyrates			X		X		X	X				X	HB		X		X	X	X	NL	60–105
Cellulose propionates			X		X		X	X				X	HB		X		X	X	X	SL	70–105
TFE-fluorocarbons			X		X			X			X		VO		X	X				N	260
Nylons‡			X	X	X		X	X			X		HB-VO		X			X	X	CO	80–150
Polycarbonates‡			X				X				X	X	HB-VO			X		X	X	SL	120
Polyethylenes			X	X	X	X	X	X			X	X	HB-VO	X	X	X		X	X	NP	80–120
Polypropylenes‡			X	X	X		X	X			X	X	HB-VO		X	X		X	X	NP	120–160
Polystyrenes‡	X		X	X	X		X	X			X	X	HB-VO		X		X	X	X§	NP	75–95
Styrene acrylonitrile copolymers‡			X	X			X	X			X	X	HB-VO		X		X	X		NP	80–95
Vinyls	X		X	X	X	X	X	X			X	X	HB-VO	X	X	X		X	X	SL	65–95
Polysulfones			X			X	X				X	X	V1-VO		X			X	X		150–175
Thermosetting Plastics																					
Epoxies	X					X			X	X	X	X	HB-VO		X	X		X		SL	120–290
Melamines		X							X	X	X	X	VO		X	X		X	X	SL	100
Phenolics	X	X				X			X	X	X		HB-VO		X			X	X	D	120
Polyesters (other than molding compounds)	X			X	X				X	X	X	X	HB-VO		X		X		X	SL	150–210
Polyesters		X							X	X			HB-VO		X	X				SL	150–175
Silicones	X	X				X			X	X			HB-VO		X	X				SL-N	315
Ureas		X									X	X	V1-VO							G	75

** For general guidance only—Evaluate suitability of specific compounds upon detailed data from manufacturer.

*Per UL 94. Because of the complexity of flammability ratings, exact information on specific compounds should be obtained from the supplier of the plastic resins.

† NP-needs protection, C-chalks, N-none, SL-slight, NL-nil, CO-colors, D-darkens, G-grays.

‡ Available in glass filled forms which, in general, yield parts with improved toughness, higher rigidity, significantly higher heat resistance and slower burning rates.

§ Rubber modified form.

to mill scale beneath the porcelain coating, creating an unsatisfactory finish.

Each of these steels is available from the mill in several grades. The requirements of a material for spinning are different from that of a material which will be deep drawn or simply formed. Therefore, grades are designed for each of these requirements.

Steel sheet is also available with any of a number of different finish treatments already applied. These types of finish include: galvanized sheet, prepainted sheet, aluminum clad sheet and plastic coated sheet. Pre-plated sheets are also available including chrome, brass and copper.

Steel sheet used in lighting is often referred to only in terms of gauge thickness. The metal gauge most commonly used is the Manufacturer's Standard Gauge for Steel Sheets.

Aluminum

Aluminum is a non-ferrous, corrosion-resistant, lightweight, non-magnetic metal having good thermal and electrical conductivity. It is high in the electro-chemical series and resists attacks by either air or water because of the formation of an invisible protective covering of aluminum oxide.

Uses in Lighting. Aluminum is used in lighting for: structural parts such as tubes or poles, fitters, holders, housings, channels, hardware, mechanical parts and trim; and light controlling surfaces such as reflectors, louvers, baffles and decorative surfaces.

Aluminum is also used as a reflecting surface when vaporized on glass and plastic, and as a paint when in fine powder form and suspended in a suitable liquid vehicle.

An aluminum reflector can have a high-permanence, high reflectance, diffuse or semi-diffuse surface of graduated brightness; the value of the brightness depending on the intensity and angle at which the light reaches the surface and the angle at which the surface is viewed.

Types of Aluminum. Aluminum is used in its near-pure state or may be alloyed by the addition of other elements to improve its mechanical, physical and chemical properties. Silicon, iron, copper, manganese, magnesium, chromium, nickel and zinc are the most common elements used. Aluminum alloys may be cast, extruded and rolled as shapes or sheets. In sheet form, aluminum is available as reflector, homogeneous, clad and lighting.

Finishes. The final finish on aluminum parts will depend on service requirements but structural members often require only cleaning. Aluminum may be etched, polished, brushed, plated, anodized, vacuum coated with a dielectric, color anodized, brightened, plastic coated with or without vaporization, coated with clear or dye lacquers, finished with baked or porcelain enamel, or some combination of these finishes. Reflector finishes may range from diffuse, such as baked enamel or etched surfaces, to highly specular such as polished, anodized or coated surfaces. Aluminum paint, made of aluminum as a fine powder suspended in a suitable liquid vehicle, has found wide use as an attractive and practical finish for many surfaces.

Processing. Anodizing, an electrochemical process, is used to form a protective surface of aluminum oxide of thickness greater than 100 times that formed naturally in air. The aluminum oxide surface is smooth, continuous, inseparable, and has a particle hardness with a Mohs' value of 9. Anodizing combined with chemical or electrochemical brightening provides surface finishes of uniformly high reflectance and permanence.

A high purity aluminum must be used if a clear, colorless transparent, high reflectance oxide surface is required. Impurities and alloying materials will result in lower reflectance, cloudiness, dullness or streaking of the oxide surface.

Prefinished anodized sheet is available for the forming of simple reflector elements.

In the anodizing process, colored surfaces may be obtained by depositing dyes or pigments within the open pores of the aluminum oxide just before the final sealing of the surface. A wide range of colors and tints are available.

Physical Characteristics. Properties for several alloys of aluminum that may be of interest to the designer are shown in Fig. 6-4. The types and values shown are intended as typical illustrations. New alloys are being developed frequently, and any contemporary listing becomes rapidly out-of-date.

Other Metals

Stainless Steel. Stainless steel includes those iron-base alloys which contain sufficient chromium to render them corrosion resistant. The classification "stainless" is usually reserved for those steels having 12 to 30 per cent chromium;

those with greater than 30 per cent are classed as heat-resisting alloys and not as stainless.

The family of stainless steels may be divided into three main groups:

Straight Chrome Group. The steels in this group are all magnetic. They may exhibit characteristics of rusting on exposure to corrosive atmospheres; however, the rusting is only a superficial film and acts as a barrier to further corrosive action. These are identified as AISI Type 400 Series.

Chrome-Nickel Group. These steels are non-magnetic. They do not exhibit the characteristics of rusting, but some alloys may not be satisfactory in certain corrosive atmospheres. These steels are designated as AISI Type 300 Series.

Chrome-Nickel-Manganese Group. These steels are non-magnetic. In this group, manganese substitutes for part of the nickel. These steels do not exhibit the characteristics of rusting, but some alloys may not be satisfactory in certain corrosive atmospheres. These steels are designated as AISI Type 200 Series.

Stainless steels are widely used in luminaires intended for installation outdoors or in other corrosive atmospheres. Some applications of stainless steel are for housings, springs, latches, mounting straps, hinges, fittings, fasteners and lampholder screw shells.

Copper. Copper is used extensively for the conductors, bus bars and associated switchgear necessary for the distribution and control of electrical energy used for lighting. Copper is ductile, malleable, flexible and fairly strong, and may be formed by a variety of standard machines and proceses.

Non-Ferrous Alloys. Bronze, an alloy of copper and tin, and brass, an alloy of copper and zinc, are often used in specialty luminaires where appearance of the attractive color is a prime consideration. A more utilitarian use of bronze and brass is in luminaires for marine use where strength and resistance to salt water corrosion are highly important.

Chromium copper or beryllium copper are often used for conducting springs, contacts, and similar highly stressed members that have to be formed in manufacture. Parts are shaped soft and then strengthened by heat treatment.

FINISHES USED IN LUMINAIRES[12]

A finish is the final treatment given to the surface of a material in the course of manufacture to render it ready for use. Three major purposes for finishes on lighting equipment are: the control of light, the protection of material and the enhancement of appearance. In addition, there are several special applications such as flame retardant and color stabilizing treatments.

Types of Finishes

Finishes are classified both by the method of application and by the kind of material applied. The three basic types are coatings, laminates and chemical conversion finishes.

Coatings can be separated into four general classes as organic, ceramic, metallic and others.

Organic Coatings.

Lacquers may be clear, transparent or opaque and will cure rapidly at room temperatures. They may be used for decoration or protection.

Fig. 6-4. Typical Physical Properties of Aluminum

Type	Alloy	Federal Specification Number	Average Coefficient of Thermal Expansion*	Specific Gravity**	Thermal Conductivity at 25°C (CGS units)	Reflectance (per cent)
Specular, processed sheet	#12 Reflector sheet		23.6	2.71	0.53	80–95
Diffuse, processed sheet	#31 Reflector sheet		23.6	2.71	0.53	75–80
Mill finish sheet	#1100-H14	QQ-A-561c	23.6	2.71	0.53	70
Extruded	#6061-T4	QQ-A-270a	23.4	2.7	0.37	
Extruded	#6063-T4	QQ-A-274	23.4	2.7	0.46	
Extruded	#6463-T4		23.4	2.7	0.52	
Cast, sand, or permanent	#43-F	QQ-A-371c	22.1	2.69	0.34	
Cast, sand, or permanent	#214-F	QQ-A-371c	22.3	2.89	0.29	
Cast, sand (heat treat)	#220-T4	QQ-A-371c	24.7	2.57	0.29	
Cast, die	#360	QQ-A-591a-2	20.9	2.64	0.27	
Cast, die	#380	QQ-A-591a-2	20.9	2.72	0.23	

* μcm/cm·°C, 20 to 100°C.
** Also weight in g/cm³.

Enamels are pigmented coatings and are applied for protection, decoration or reflectance. They cure by oxidation (air or forced drying) or polymerization (baking or catalytic action) and result in tougher finishes than lacquers.

Baked clear coatings, sometimes called baking lacquers, are used for decoration and protection. They cure by polymerization (baking or catalytic action).

Organisols and plastisols are usually applied by dipping and spraying. These plastic dispersion coatings offer good exterior corrosion as well as scratch and abrasion resistance and are also able to conceal many surface defects.

Ceramic Coatings. Ceramic coatings, including porcelain enamels, are fired on glass and metals at temperatures in excess of 540 °C. Primary features include high resistance to corrosion, good reflectance and easy maintenance on metals; and reduction of brightness and increase of diffusion on glass.

Metallic Coatings.

Electrochemical deposition, commonly called electroplating, causes a second metal to be deposited over the first by means of an electrolytic action. Zinc, cadmium or nickel is used to provide protection. Brass and silver plated finishes are primarily decorative.

Vacuum metallizing, consists of vaporizing a metal, usually aluminum, in a vacuum chamber and depositing it on surfaces of plastic, glass or metal. Finishes of high specular reflectance are obtained and can be used for either light control or decorative purposes.

Dip and spray coatings, such as galvanizing and sherardizing, deposit a second metal to protect the base metal against corrosion.

Other coatings. Semiconductors, such as silicon and germanium, and inorganic dielectrics, such as magnesium fluoride and titanium dioxide, are vacuum deposited on such materials as glass, aluminum and plastics. The coatings of interest for lighting uses are multilayered and less than one micrometer thick.

Laminates. This type of finish is created by bonding a thin layer to a base material, such as a plastic film to sheet metal. The laminate can be a decorative material or it can be a light-controlling material.

Chemical Conversion Finishes. Anodizing converts an aluminum surface by an anodic process to aluminum oxide which has outstanding protective qualities against corrosion and abrasion. The resultant finish may be clear or can be dyed in a variety of colors.

Typical characteristics of finishes are indicated in Fig. 6-5. Because of the great number of possible variations in composition and application of all types of finishes, numerical values are not shown and relative gradings only are used. For more details on finishes, technical assistance should be obtained from suppliers and available literature.

REFLECTOR DESIGN[21]

The design of reflector contour is an extensive subject because the possible shapes for a particular application are almost limitless. However, the end use usually limits the choice.

For design purposes, reflector contours can be divided into two classes: *basic contours* and *general contours.* Basic contours may be defined as those which are mathematically predictable as to action and can be designed mathematically. General contours are those required to satisfy many candlepower (intensity) distribution curves, but which do not conform to any of the basic contours.

Basic Reflector Contours

Basic contours which are used very frequently are the conic section and the spherical reflector.

A conic section is by definition the locus of a point whose distance from a fixed point is in a constant ratio to its distance from a fixed straight line. The fixed point is called the *focus* of the conic section; the fixed line is its *directrix.* The constant ratio is the *eccentricity* of the conic section. If the eccentricity, e, of the conic section is equal to one, the section is called a parabola; if e is less than one, an ellipse. A third conic section, the hyperbola, occurs when e is greater than one.

Parabolic Reflectors. An inherent property of the parabola is its ability to redirect a ray of light originating at the focal point in a direction parallel to the axis. The proof of the property is shown in Fig. 6-6 where $A'A''$ is tangent to the curve at A, BA is perpendicular to DD', and $BA = FA$. Assuming a point source at the focus of a perfect parabolic mirror, all light from the source striking the mirror would be redirected as a beam

Fig. 6-5. Properties of Finishes

Type of Finish	Method of Application[a]	Principal Uses[b]	Colors Possible	Character of Reflected Light	Per Cent Reflectance[d]	Resistance[c]				Stability[c]	Flammability
						Heat	Corrosion	Abrasion	Impact		
Organic coatings											
Lacquers	D, B, S	A, P	Colorless or any color	Mixed to diffuse	10–90	F	F	P	F	F	Slow burn
Emulsions	D, B, S	A, P	All colors	Mixed to diffuse	10–90	G	G	G	G	G	Slow burn
Enamels	D, B, S	A, P, R	All colors	Mixed to diffuse	10–90	G	G	G	G	G	Slow burn
Baked clear coatings	D, B, S	A, P	Colorless, clear color	Diffuse to specular	0	G	G	G	G	G	Slow burn
Organisols	D, S	A, P	All colors	Mixed to diffuse	10–90	F	E	G	G	F	None
Ceramic coatings											
Vitreous enamels	D, S	A, P, R	All colors	Diffuse to specular	10–90	E	E	E	P	E	None
Ceramic enamels	D, S, B	A, R	All colors	Mixed to specular	10–90	E	E	E	P	E	None
Metallic coatings											
Chrome plate	Electrochemical	A, P	Fixed; depending on color of plated metal	Specular to diffuse	60–88	E	E	E	E	E	None
Nickel plate	Electrochemical	A, P		Specular to diffuse	55	E	G	E	E	E	None
Cadmium plate	Electrochemical	P		Specular to diffuse	85	G	G	F	P	E	None
Brass plate	Electrochemical	A		Specular to diffuse	55–80	P	P	F	P	F	None
Silver plate	Electrochemical	A, R		Specular	85–95	P	P	F	P	E	None
Laminates	Laminate	A, P, R	All colors of metallic effects	Mixed	10–90	Depends on nature of laminate					Slow burn
Conversion coatings											
Anodized aluminum	Electrochemical	A, P, R	Natural aluminum (or a wide variety of colors)	Diffuse to specular	60–90	E	E	E	E	E	None
Vacuum metalizing	Vacuum chamber	A, R	Natural aluminum (or a wide variety of colors)	Specular	10–70	Depends on nature of protective coating					None
Vacuum deposition	Vacuum chamber	A, R, T	Colorless to interference effects	Diffuse to specular	0–99	E	E	E	E	E	None

[a] D—dip, S—spray, B—brush.
[b] A—appearance, P—protection, R—reflectance, T—transmittance.
[c] P—poor, F—fair, G—good, E—excellent.
[d] Depends upon color.

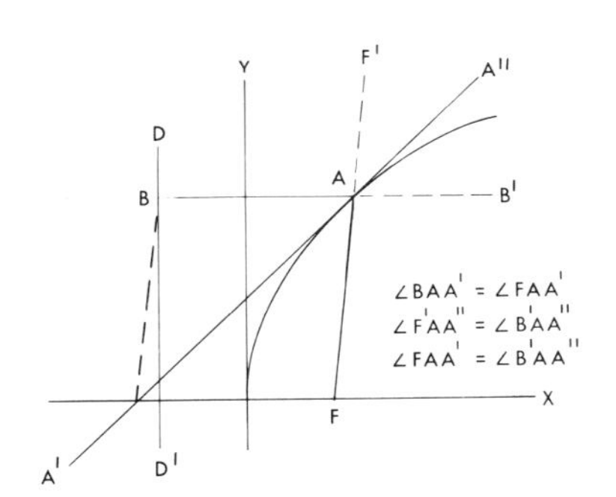

Fig. 6-6. Parabolic conic section. *DD'* is the directrix. *F* is the focus.

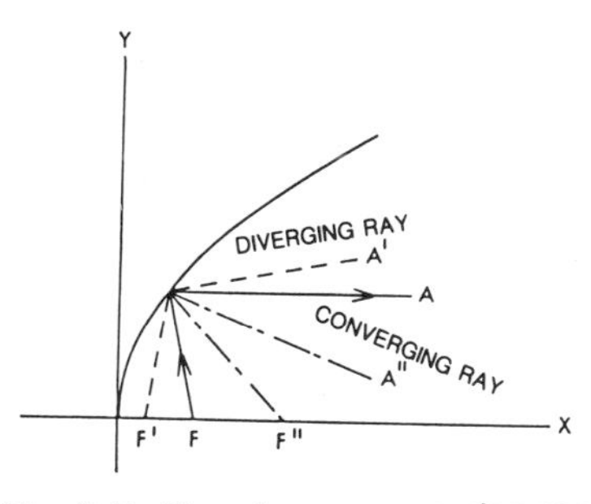

Fig. 6-7. Diverging or converging rays from point source on axis of parabolic reflector, but behind or ahead of focal point *F*.

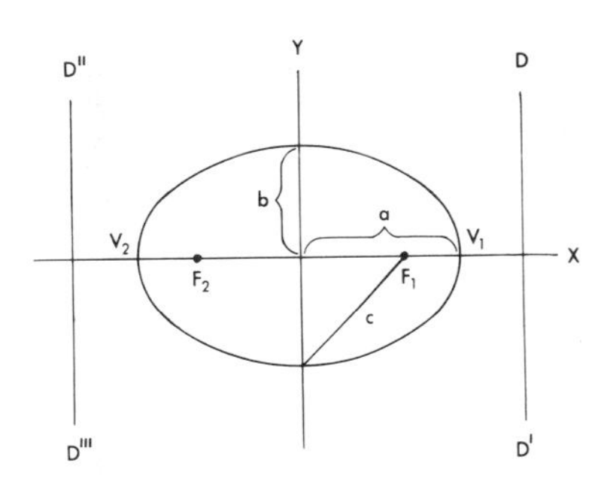

Fig. 6-8. The two foci, directrices, and axes of an ellipse.

of light parallel to the axis. The ideal conditions of the perfect mirror and the point source cannot be reached in practice nor, in most cases, would this be desirable. The further conditions deviate from the ideal, the greater will be the deviation of the light from a parallel beam. Formulas have been derived expressing the light divergence from shallow mirrors when sources of various shapes are used. Fig. 6-7 illustrates the action of a point source lying on the axis of the parabola but ahead of or behind the focal point.

Ellipsoidal Reflectors. Ellipsoidal reflectors are an efficient means of producing beams of controlled divergence and for collecting light to be controlled by a lens or lens system. The ellipse, Fig. 6-8, can be described by the following equation:

$$\frac{x^2}{a^2} + \frac{y^2}{b^2} = 1$$

Useful working equations are:

$$y = \pm \frac{b}{a} \sqrt{a^2 - x^2}$$

$$x = \pm \frac{a}{b} \sqrt{b^2 - y^2}$$

$$b^2 = a^2 - \left(\frac{\overline{F_1F_2}}{2}\right)^2$$

Fig. 6-9 illustrates the action of a perfect complete ellipsoidal mirror with a light source at F_1 and a ray of light striking the mirror at P_1, passing through F_2 to P_2, and being reflected again through F_1 to P_3 and on to P_4. If the mirror is "chopped off" at F_2 (the conjugate focus) or at a point closer to F_1, all of the light from the theoretical source at F_1 will leave the mirror either directly or after one reflection. By moving the opening back along the major axis, the beam limits of the reflected light are narrowed. When the plane of the opening reaches the center of the ellipse, the maximum angle of the reflected light coincides with the angle of the direct light from the source.

Hyperbolic Reflectors. The diverging beam typical of the ellipsoidal reflector can also be produced by a reflector having a hyperbolic contour. The main difference, as shown in Fig. 6-10, is that the hyperbolic reflector will form a virtual image F_2 behind the focus, whereas the ellipsoidal reflector produces a real image in front of the focus. The equation of a hyperbola is:

$$\frac{x^2}{a^2} - \frac{y^2}{b^2} = 1.$$

Spherical Reflectors. A spherical reflector can be considered as a special form of the ellipsoidal reflector where the two foci are coincident. Any light leaving the source located at the focus would return and pass through the same point. This has obvious disadvantages when dealing with practical sources since the concentration of energy can often have a damaging effect on the source or the bulb wall surrounding it. The principle is, however, often used in projecting devices to increase the amount of light collected by a lens as shown in Fig. 6-11.

General Reflector Contours

For many applications, reflectors can be designed which are mathematically predictable as to contour and action. A more general problem is that of determining the contour of a reflector which does not meet these conditions. Usually there are two known factors; an approximate candlepower distribution curve for the finished reflector, and the material from which it is to be manufactured.

Diffuse Reflecting Surfaces. In the rare case of surfaces with perfectly diffuse reflection, the contour of the reflector will have very little effect on the distribution of light. The candlepower distribution curve of such a reflector (after that portion due to the bare lamp has been deducted) would be very nearly a circle, with the maximum ordinate normal to the plane of the opening. The distribution curve would be even closer to a circle except that all portions of the reflector would not be uniformly illuminated. A unit strip of area near the lower edge of the reflector, being farther from the source, would receive less light than a similar strip higher in the reflector. Since these lower portions are the parts directing light out at the upper angles, the candlepower falls off faster than the cosine of the angle. In this case, there is little that can be done by the designer in the way of light control except by aiming the entire reflector.

Reflectors with Specular and Semi-Specular Surfaces. General contours for specular reflectors are usually obtained through graphical rather than strictly mathematical methods. The problem consists of determining what reflector shape is necessary to redirect luminous flux from the lamp into the proper directions to achieve a predetermined candlepower distribution curve.

The basic steps in one method of determining reflector contour are as follows:[22-24]

1. Select the width, in degrees, of zones to be considered. From the required candlepower distribution curve (polar curve), calculate the lu minous flux required in each zone (candlepower × zonal constant for each zone). See Fig. 4-17, Section 4. Tabulate.

2. Calculate the luminous flux emitted by the bare lamp in each zone (using the polar curve of the bare lamp and zonal constants), and tabulate.

3. Find the reflected lumens needed in each zone by subtracting the bare lamp lumens (data from Step 2) from the required luminous flux (data from Step 1) in those zones where no reflection will take place, *i.e.*, from nadir up to cut-off. Tabulate.

4. Decide upon the general action of the reflector. There are, in general, four basic actions of the reflectors as shown in Fig. 6-12. For a given cut-off, the forms shown in Fig. 6-12 (*c* and *d*) usually require a very large reflector. The form Fig. 6-12b has the disadvantage that much of the light is passed through the lamp bulb. For most cases, the form Fig. 6-12a results in the smallest reflector and redirects the least light back through the lamp.

5. Plot a curve of reflected flux obtained in Step 3. Starting at 0 degrees, show cumulative sums of lumens required to be reflected into each zone from nadir up to cut-off, see Fig. 6-13. Similarly, plot a curve of available lamp flux (from Step 2 data). Starting at cut-off, show how many lumens are incident on the reflector progressively from cut-off to 180 degrees (Fig. 6-13b). Since all flux considered here must be reflected from the reflector surface, the available lamp lumens must first be multiplied by the reflectance of the surface and by other loss factors. It is convenient to work with rectangular coordinates, plotting the lumens along the horizontal axis and degrees along the vertical axis. Plot the reflected flux curve (Fig. 6-13a) to the same scale and directly below the available lamp flux (bare lamp flux corrected for losses) curve. If the reflector action illustrated in Fig. 6-12 (b or d) was selected, the horizontal (lumen) scales of the two curves will be in the same direction. If Fig. 6-12 (a and c) was selected, the scales will be reversed as shown in Fig. 6-13. Take intercepts at intervals (one for each zone) along the reflected flux curve (Fig. 6-13a) and project upward to the available lamp flux curve as shown

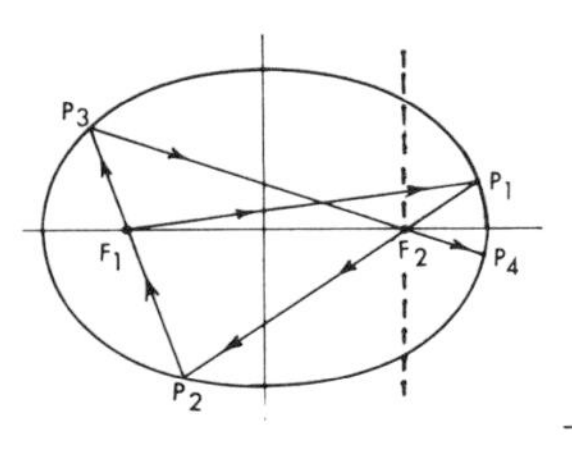

Fig. 6-9. The action of a perfect ellipsoidal mirror.

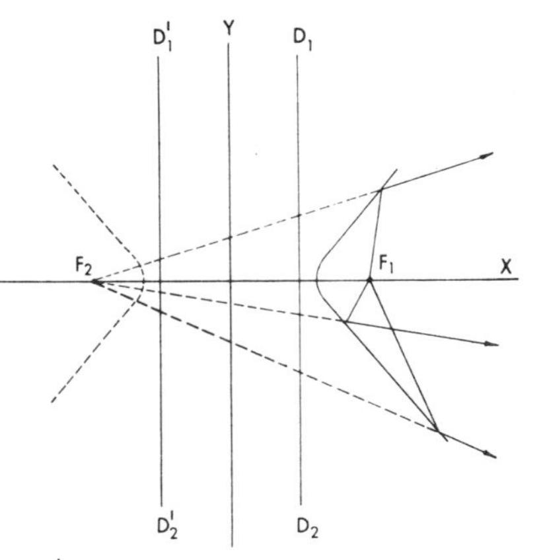

Fig. 6-10. Hyperbolic reflector action.

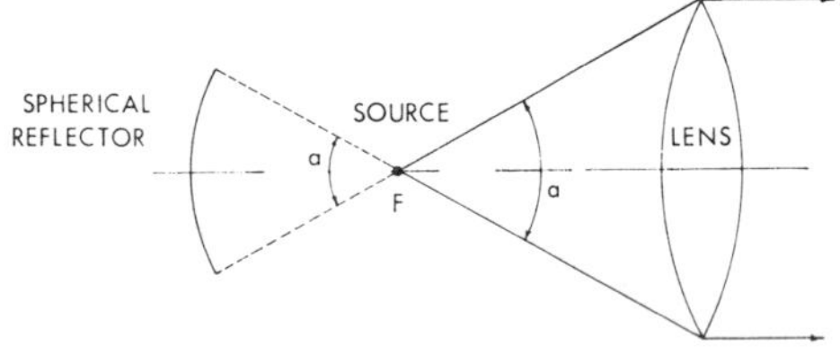

Fig. 6-11. Projecting device with a spherical reflector.

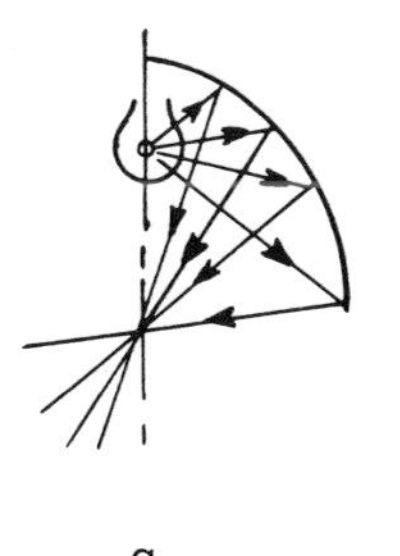

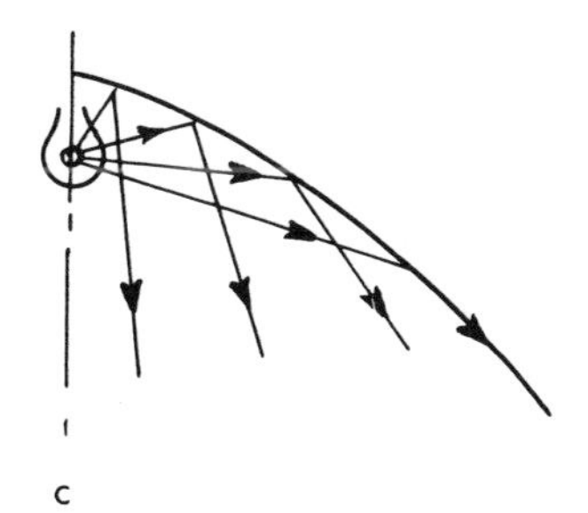

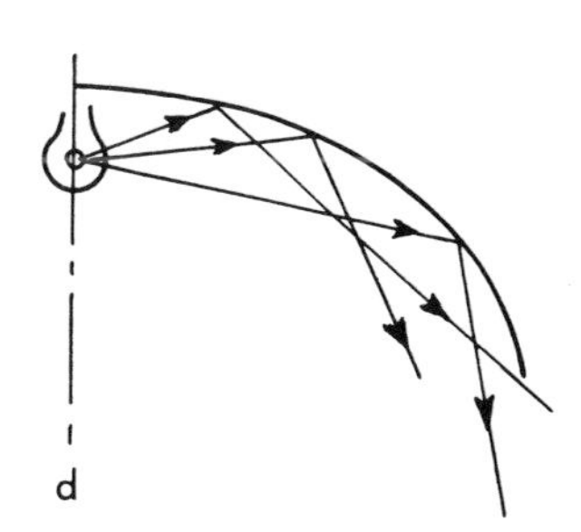

Fig. 6-12. Four basic reflector actions from most compact design to least.

in Fig. 6-13. At the point where each intercept cuts the available lamp flux curve (Fig. 6-13b), project horizontally to degrees on the vertical axis. The spacings between intercepts on the vertical axis indicate how large an angular, zonal segment of the reflector is required to direct enough light at a particular angle to satisfy the requirements of the reflected flux curve.

6. Lay out the reflector contour by starting at the point nearest the lamp and progressively drawing small, straight segments (one segment for each zone). Each segment must be of the required length, and placed in the correct position for the flux striking it to be reflected in the amount and direction determined in Step 5. Obtain the final contour by drawing a smooth curve tangent to the segments.

The procedure outlined above assumes a point source and true specular reflection. The effect of a normal filament would be to smooth out the final distribution, rounding off sharp points. The effect will be more noticeable as the reflector gets smaller in comparison with the source.

Departures from true specularity will have a similar effect which will increase as the proportion of diffuse to total reflection increases. With semispecular surfaces, the effect of specularity of the surface increases as the angle of incident light increases. Hence, for accurate control of the light, it is well to keep the angle of incidence as large as possible for semi-specular surfaces or to use highly specular high-reflectance surfaces when multiple reflection exists.

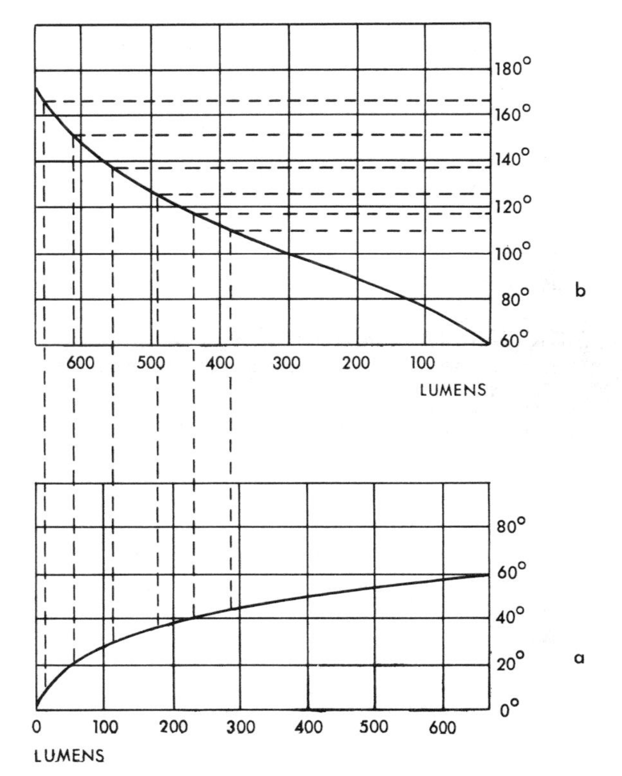

Fig. 6-13. Reflection plan of available lamp lumens (shown for 60-degree cut-off). a. Reflected flux curve. b. Incident or available flux curve (corrected for losses).

REFRACTOR DESIGN[21]

Optic Surface to Redirect Light Rays

A common problem, illustrated in Fig. 6-14, is the determination of an optic surface to cause the eventual refraction of a light ray from a source to a required direction. Initially, five elements of this problem are known:

1. Exit surface S, which is the side of the lens away from the source. This surface could be flat, spherical, cylindrical or irregularly shaped; the basic fundamentals being the same.
2. Optic surface T to which an optical configuration must be designed in order to create the required distribution of light rays. Surface T is normally parallel to exit surface S.
3. The location of the light source.
4. The incident ray from the source to the lens.

5. The required emergent ray. The angle θ of the optic surface T' which will change the incident ray to the direction of the emergent ray is the unknown quantity in this problem.

The first step is to draw all the known components of the problem: exit surface S, optic surface T, incident ray, emergent ray required and the source. Note that the surface T is a dashed line indicating that at the end of the complete design this surface will no longer be smooth, but will be configurated with optic elements.

To point A, the intersection of the emergent ray and exit surface S, draw the normal to exit surface S; with point A as center strike arc n_1 with radius proportional to n_1, the refractive index of the outside medium (probably air where $n_1 = 1$). See Refraction, page 2-17. Again with a point A as center, strike arc n_2 with radius proportional to the refractive index of the optical medium. Through point W, the intersection of arc n_1 and the emergent ray required, draw line C parallel to the normal to exit surface S. Through the intersection of arc n_2 and line C, which is indicated as point X, draw the construction line to point A on exit surface S. Extended, this becomes a refracted ray through the optic surface T. Further extended, it intersects arc n'_2 at point Y. The radius of arc n'_2 is proportional to the refractive index of the optical medium. Note that the center of the arc, point V, lies on the optic surface which has not yet been established. This fact indicates that a few trials are necessary for this solution. The initial steps are repeated until point V coincides with the intersection of the incident ray and optic surface T'. With point V as center, strike arc n'_1, whose radius is proportional to the refractive index of the outside medium. Draw line D between point Y and point Z, the intersection of arc n'_1 and the incident ray. The normal to the optic surface T' is then drawn parallel to line D and intersecting point V. The optic surface T' is then a line perpendicular to this normal.

All rays at an internal angle to the normal greater than critical angle ($\sin \theta = n_1/n_2$) are internally reflected from the inside of the surface.

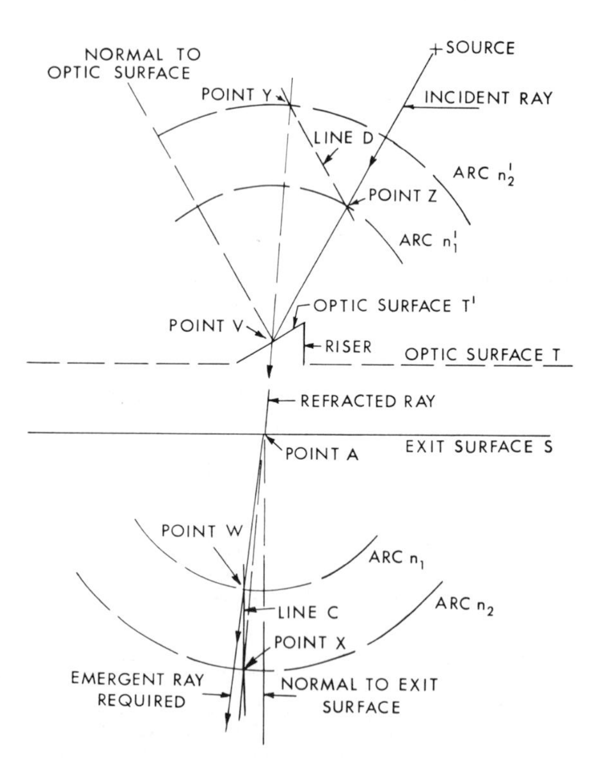

Fig. 6-14. Design of optic to redirect ray.

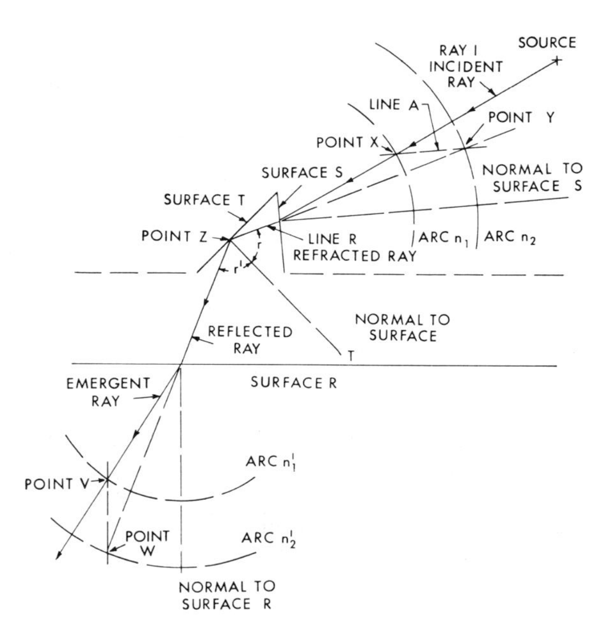

Fig. 6-15. Path of a light ray through a refraction-reflection system.

Refraction-Reflection System

Total internal reflection is used extensively in lens design when rays must be bent at an angle larger than is possible to accomplish with refraction alone. Total internal reflection problems can be solved graphically by methods similar to those used in solving refraction problems.

Fig. 6-15 illustrates the solution of a typical problem involving both refraction and total internal reflection. A ray of light from the source passes through the system in the following manner:

1. Ray 1 is refracted at surface S.
2. Refracted ray becomes incident to surface T at an angle greater than the critical angle of the medium.
3. Total internal reflection occurs at surface T so that $r' = r$.
4. Reflected ray becomes incident on surface R.
5. Ray is refracted at surface R and emerges from system.

The graphical method of determining the refracted rays has been described in the previous section. The graphical method of determining the reflected ray is accomplished by constructing the angle of reflected ray to normal equal to the angle of the incident ray to normal.

LUMINOUS AND LOUVERED CEILING DESIGN

Luminous ceiling systems employ translucent media with lamps installed in the cavity above that media. Luminous ceiling systems employing highly diffusing media produce about the same results as well designed indirect lighting from conventional suspended luminaires (or coves), but with generally greater uniformity of ceiling luminance than is possible with the latter method. When luminous ceilings employ prismatic materials they give generally higher utilization than those utilizing diffusing materials; they also have better direct glare control.

Either glass or plastic of proper optical characteristics may be used as the translucent medium. This material is usually supported in a structural framework and is formed in panels or rolls of a size which is convenient for handling. The diffusing material should have the highest possible transmittance consistent with adequate concealment of the lamps. Experience indicates that a material should be used having a transmittance of about 50 per cent and as low an absorptance as possible. Prismatic materials should have as high a transmittance as possible with the degree of light control desired. In general, such materials will have an effective transmittance of 70 to 75 per cent.

Ideally, the cavity above the diffuser should be of just sufficient depth to obtain the proper relation between spacing of lamps and vertical distance from the diffuser to produce acceptable uniform brightness. The luminous efficiency of the system decreases as the depth of the cavity is increased in respect to its length and width. The cavity should be relatively unobstructed and all surfaces painted in a white paint of at least 80 per cent reflectance. All dust leaks in cavity walls and ceilings should be sealed. Where deep heavily obstructed cavities occur, it is wise to consider the possibility of furring down the cavity ceiling to a reasonable level. The cost of such construction may be less than the additional cost of equipment, power and maintenance required to force an otherwise inefficient system to deliver a given illuminance level.

Successful luminous ceilings have been designed using practically all types of lamps, however fluorescent are generally used. A typical arrangement of lamps within the cavity is shown in Fig. 6-16. For uniform brightness appearance over a diffusing ceiling, the spacing of the lamps should not exceed 1.5 to 2 times their height above the diffuser. Where the plenum is shallow, the plenum ceiling painted white, and the lamps mounted against the plenum ceiling, S/L can be greater than is the case when the plenum above the lamps is deep, or dark, or obstructed. Also, when it is necessary to use units with reflectors such as those in Fig. 6-17, the ratio S/L should not exceed 1.5. With ceilings of prismatic materials, S/L should rarely exceed 1.0 if streaks of light and dark are to be avoided.

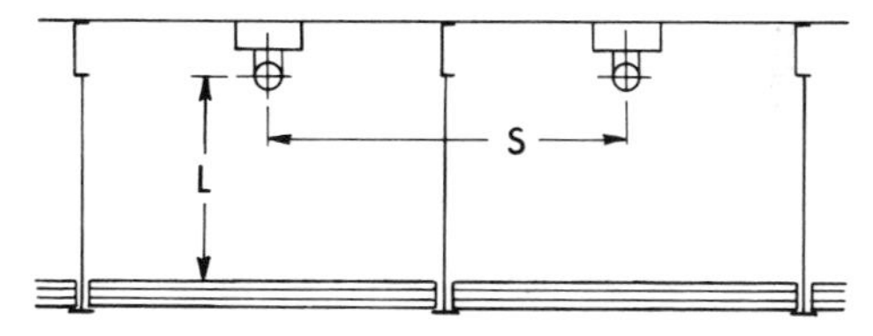

Fig. 6-16. Section of a portion of a luminous ceiling. The distance S should not exceed 1.5 to $2L$.

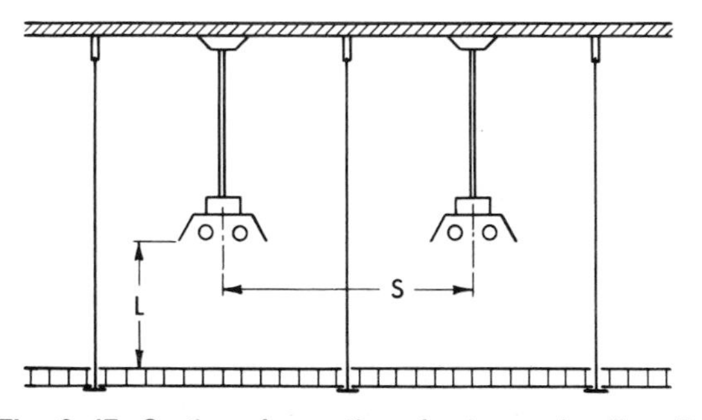

Fig. 6-17. Section of a portion of a louvered ceiling. For translucent louver varies the ratio S to L should not be over 1.5 to 1 and for opaque louvers, not over 1.5 to 1, depending upon the louver finish.

For greater uniformity of brightness over the entire ceiling, it is recommended that the lamp rows be spaced closer together at the sides of the cavity. The ends of the rows should be carried all the way to the ends of the cavity. In some instances it is desirable to install extra lamps between the rows at the end of the cavity, or across the ends of the rows, to increase the luminance at this point.

Luminous ceiling systems are often used for the dual purpose of providing light and concealing the visual clutter of beams and building services. Aside from the reduction in system efficiency caused by these devices, there is the possibility that they will cast unsightly shadows on the translucent medium. It is well to place the prime sources of light below, or at right angles to, such obstructions wherever possible.

Obtaining proper acoustical treatment of a room with a luminous ceiling can be difficult. In many cases, it is better to obtain the required results by treating other room surfaces. Generally this means wall surfaces in small rooms, and floors in large ones.

Louvered ceiling systems consist of an open network of translucent or opaque vanes through which light pases from the lighting recess above to the area below. The louver cells can be made in a number of shapes and sizes, depending upon mechanical design relationships and desired architectural effects, with various degrees of shielding. Fluorescent light sources are generally used with this type of ceiling. Here, the lighted appearance of the louvers is dependent on the conditions within the cavity, the spacing of the lamps, the height of the lamps above the louvers, the size of the louver cells, and the characteristics of the louver blades themselves. The transmittance (if any) of the blades, and both the type of reflectance (specular, spread or diffuse) and the amount (per cent) are important. Louvers of the parabolic wedge type have yet different requirements.

Since characteristics and requirements differ so greatly, it is difficult to give definite recommendations. In general, it should be remembered that wide spacings of luminaires above the louvers will result in streaks of varied brightness, and shadow patterns on the blades themselves which may or may not be desirable. Close spacings may result in a "flat" appearance of the louver.

As a guide, in unobstructed, shallow plenums with white-painted surfaces and ceiling-mounted strip luminaires, the ratio of spacing to height above the louver should not exceed 1.5 to 1 for translucent louvers, and 1 to 1 for opaque louvers. See Fig. 6–17. For other conditions, closer spacings are the rule.

Pipes, ducts and beams should be considered in their relationship to the louvered ceiling system in order to avoid the forming of unsightly shadows.

The lighting equipment should be mounted high enough above the louvers to permit ease of access and relamping. Means for removing the louvers in sections small enough for ease of handling facilitates the relatively infrequent cleaning necessary.

The acoustical effect of louvers depends on the cell size and the louver material. In most cases, however, it is best to ignore the louvers, and apply acoustical material to the ceiling of the plenum, or to other room surfaces.

The efficiency and effective reflectance of luminous and louvered ceilings vary with the cavity proportions and reflectances, the type of lighting equipment used, the reflectance and transmittance of the ceiling material, the shielding angle of louvers, and the amount and kind of obstruction within the cavity. Fig. 9–62 gives coefficients of utilization for typical ceilings of these types. Performance data for other ceilings can be determined by photometric tests or by zonal-cavity calculations for coupled enclosures. The ceiling

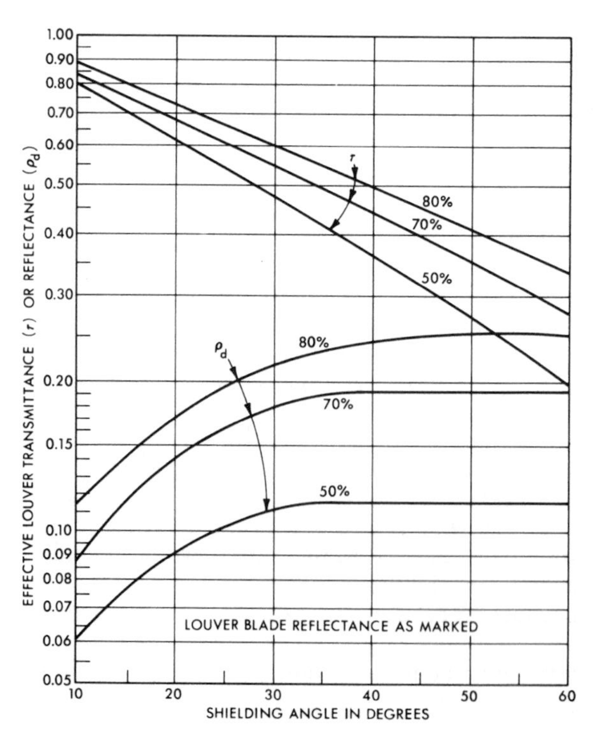

Fig. 6-18. Graph for obtaining effective transmittance and reflectance for square-cell opaque louvers.

cavity reflectance for a room with a luminous or louvered ceiling can be found by

$$\rho_{eff} = \rho_d + \tau^2 \left(\frac{\rho_{CC}}{1 - \rho_d \rho_{CC}} \right)$$

where

τ = ceiling material transmittance
ρ_d = ceiling material reflectance
ρ_{CC} = effective reflectance of ceiling cavity
ρ_{eff} = effective reflectance of luminous ceiling seen from below.

Effective transmittance and reflectance for opaque louvers can be estimated using the curves of Fig. 6–18.

Louvers made from translucent materials vary tremendously in performance, and cannot be accurately evaluated by Fig. 6–18. An approximation can be made, however, by measuring the effective transmittance of the louver in diffuse light, and using it along with the shielding angle to find the equivalent louver blade reflectance from Fig. 6–18. Using this value, the effective reflectance can be estimated from Fig. 6–18.

COVES

Coves may be for illumination or the creation of a decorative effect. In most instances, an efficient lighting system and a more pleasing brightness pattern are obtained if the brightness of the ceiling and the brightness of the wall above the cove are relatively uniform in appearance. Usually these objectives are more easily attained if the cove is not located too close to the ceiling. The brightness of the wall above the cove and that of the ceiling near the cove are produced largely by direct illumination from the light source and from reflections between the cove, wall and ceiling. The brightness of the ceiling at the far end of the room is largely dependent upon the general room surface reflectances. To attain more uniform wall and ceiling brightness, it is recommended that the lamps be mounted away from the wall and that the wall adjacent to the source be shielded in part from the direct radiation from the lamp. The channel serves this purpose very well. In general, the further the cove is located from the ceiling, the further the lamp and channel should be located from the wall. See Fig. 6–19. If the cove is 300 millimeters (12 inches) from the ceiling (H), tests have shown that the lamp center should be 60 millimeters (2½ inches) or more from the wall (S); if the cove is 380 to 500 millimeters (15 to 20 inches) from the ceiling, the lamp center should be 90 millimeters (3½ inches) or more from the wall; and for coves that are located 530 to 760 millimeters (21 to 30 inches) from the ceiling, it is recommended that the lamp center be 110 millimeters (4½ inches) or more from the wall.

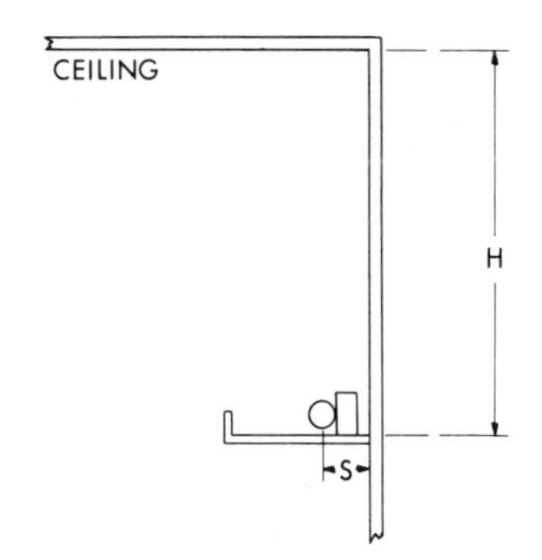

Fig. 6–19. Section through a fluorescent cove.

Reflectors can be used in fluorescent coves to obtain very pleasing results when the reflector is aimed about 20 degrees above the horizontal.

REFERENCES

1. Committee on Light Control and Equipment Design of the IES: "IES Guide to Design of Light Control," Part IV, *Illum. Eng.*, Vol. 65, p. 479, August, 1970.
2. "Electrical Lighting Fixtures," *Stand. for Safety*, UL57, Underwriters Laboratories, Inc., Chicago, 12th Edition, March, 1982.
3. *Precision Methods for the Determination of Sound Power Levels of Descrete-Frequency and Narrow-Band Noise Sources in Reverberation Rooms*, ANSI S1.32-1980, American National Standards Institute, New York, 1980. *ARI Standard for Sound Rating of Non-Ducted Indoor Air-Conditioning Equipment*, ARI 350–82, Air-Conditioning and Refrigeration Institute, Arlington, Va., 22209, 1982.
4. *ASHRAE Handbook, Systems Volume*, American Society of Heating, Refrigerating and Air-Conditioning Engineers, Atlanta, Georgia, 1984.
5. "Measurement of Room-to-Room Sound Transmission through Plenum Air Systems," No. ADC AD-63, Air Diffusion Council, Chicago, 1963.
6. "Tentative Method of Test by the Two-Room Method," Acoustical Materials Association, New York, March, 1969.
7. *National Electrical Code*, ANSI/NFPA 70–1984, National Fire Protection Association, Boston, Massachusetts, 1984.
8. "Electric Lighting Fixtures for Use in Hazardous Locations," *Stand. for Safety*, UL844, Underwriters Laboratories, Inc., Chicago, 9th Edition, November, 1983.
9. "Marine-Type Electric Lighting Fixtures," *Stand. for Safety*, UL595, Underwriters Laboratories, Inc., Chicago, 5th Edition, January, 1982.
10. *Canadian Electrical Code*, Canadian Standards Association, Ottawa, Canada, 1982.
11. "Fluorescent Luminaires," NEMA Standard LE1-1974, (R-1980) National Electrical Manufacturers Association, Washington, D.C., 1980.
12. Committee on Light Control and Equipment Design of the IES:

"IES Guide to Design of Light Control," Part III, *Illum. Eng.*, Vol. LXII, p. 483, August, 1967.
13. Shand, E. B.: *Glass Engineering Handbook*, McGraw-Hill Book Co., Inc., New York, 1958.
14. Phillip, C. J.: *Glass the Miracle Maker*, Pitman Publishing Co., New York, 1948.
15. Morey, G. W.: *Properties of Glass*, Reinhold Publishing Co., New York, 1938.
16. Condon, E. U. and Odishaw, H.: *Handbook of Physics*, McGraw-Hill Book Co., New York, Part 8, Chapter 7, "Glass," by Stookey, S. D. and Maurer, R. D., 2nd Edition, 1967.
17. *Metals Handbook*, Eighth Edition, Volume 1, "Properties and Selection of Metals," American Society for Metals, Metals Park, Oh., 8th Edition, 1961.
18. Baer, E.: *Engineering Design for Plastics*, Krieger Publishing Co., Huntington, NY, 1975.
19. Koppes, W. F.: *Metal Finishes Manual*, National Association of Architectural Metal Manufacturers, Chicago, 1964.
20. Wernick, S. and Pinner, R.: *Surface Treatment and Finishing of Aluminum and Its Alloys*, Second Edition, Robert Draper, Ltd., Teddington, England, 1959.
21. Committee on Light Control and Equipment Design of the IES: "IES Guide to Design of Light Control," Parts I and II, *Illum. Eng.*, Vol. LIV, p. 722, November and p. 778, December, 1959.
22. Jolley, L. B. W., Waldram, J. M. and Wilson, G. H.: *The Theory and Design of Illuminating Engineering Equipment*, Chapman & Hall, Ltd., London, 1930.
23. Elmer, W. B.: *The Optical Design of Reflectors*, John Wiley & Sons, Inc., New York, 2nd Edition, 1979.
24. Elmer, W. B.: "The Optics of Reflectors for Illumination," *IEEE Trans. Ind. Appl.*, Vol. IA-19, p. 776, September/October, 1983.

Daylighting

SECTION
7

Daylight, skillfully employed, provides the architect with one of his most effective modes of esthetic expression and means of energy conservation. Considerations of site, orientation, proportion and fenestration are all influenced or determined by the degree of importance attached to the utilitarian and esthetic aspects of daylighting in the structure.[1]

DAYLIGHT

To use daylight to advantage, the following design factors should be taken into account:
1. Variations in the amount and direction of the incident daylight.
2. Luminance and luminance distribution of clear, partly cloudy and overcast skies.
3. Variations in sunlight intensity and direction.
4. Effect of local terrain, landscaping and nearby buildings on the available light.

Daylight Source and Distribution

The daily and seasonal motions of the sun with respect to a particular building location produce a regular and predictable pattern of gradual variation in the amount and direction of available light. Superimposed, also, is a variable pattern caused by less regular changes in the weather, particularly to the degree of cloudiness.

The variability of daylight throughout the U.S. is given in Figs. 7–1 and 7–2. These figures show the number of clear and cloudy days per year. The number of partly cloudy days would be the total number of days in the year less the sum of the clear and cloudy days found from Figs. 7–1 and 7–2, respectively. For example, if San Diego, California was the city of interest, the approximated number of partly cloudy days would be found to be

$$365 - (200 + 60) = 105 \text{ days}$$

The Sun as a Light Source

The sun is an abundant source of radiant energy; however, about half of this energy reaches the earth's surface as sunlight or visible radiation. The other half of the radiant solar energy contains invisible shorter wavelength components (ultraviolet) and invisible longer wavelength components (infrared). When absorbed, virtually all the radiant energy from the sun is converted to heat, whether this energy happens to be visible or invisible. Therefore, solar energy, sunlight and solar heat are merely different names for radiant solar energy. The amount of usable visible energy in the solar spectrum varies depending on the depth of the atmosphere the light traverses. It depends on the elevation of the sun above the horizon and on variable atmospheric conditions such as moisture and dust.

The rotation of the earth about its own axis, as well as its revolution about the sun, produces a continual apparent motion of the sun with respect to any reference point on the earth's surface. The position of the sun with respect to such a reference point at any instant, is usually expressed in terms of two angles—the *solar altitude*, which is the vertical angle of the sun above the horizon, and the *solar azimuth*, which is usually taken as the horizontal angle of the sun from the due South Line. See Fig. 7–3.

The illuminance produced on an exterior surface by the sun is influenced by the altitude angle

NOTE: References are listed at the end of each section.

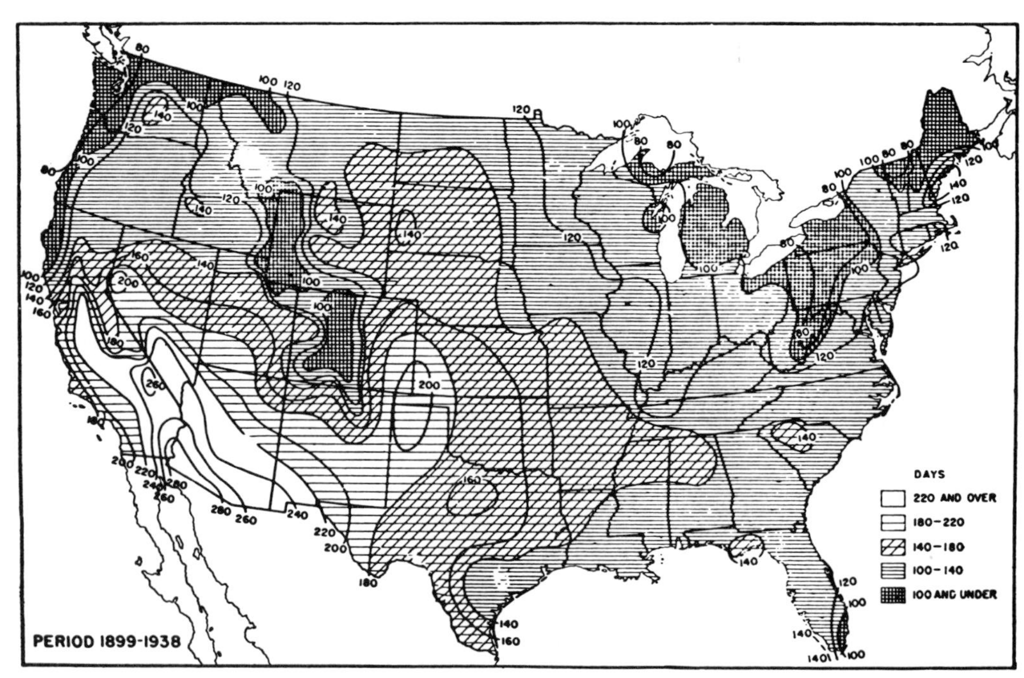

Fig. 7-1. Average number of clear days in the United States.

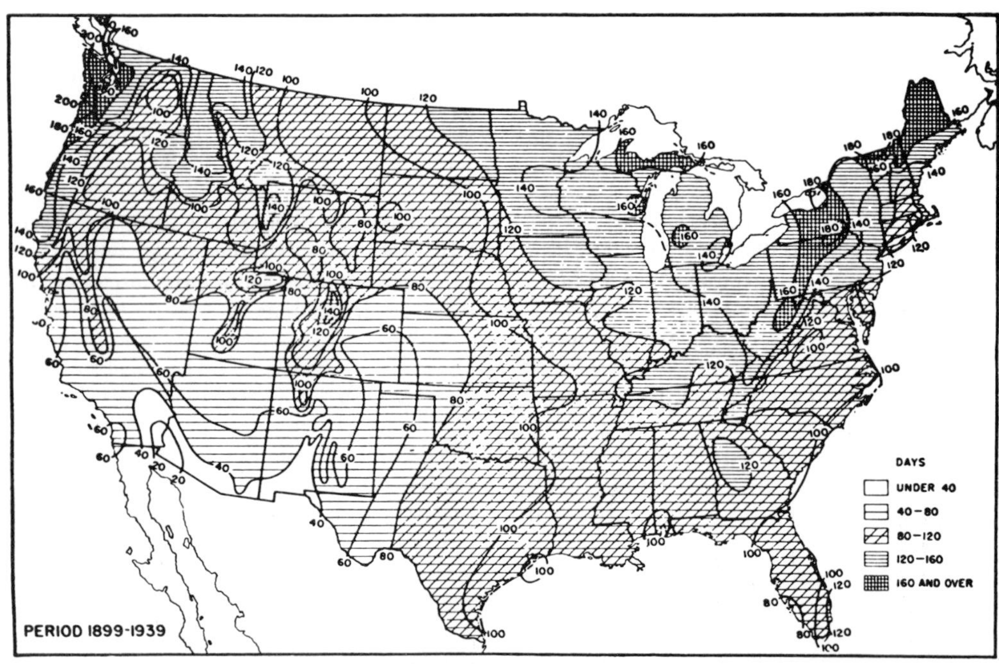

Fig. 7-2. Average number of cloudy days in the United States.

of the sun, the amount of haze and dust in the atmosphere and the angle between the incident sunlight and the surface on which the sunlight falls. Data on solar azimuth and altitude are shown in Fig. 7-4.

The Sky as a Light Source

In daylighting design, three conditions should be considered:

1. Overcast sky.

2. Clear sky.
3. Partly cloudy sky.

The amount of light received from an overcast sky reaching the windows of a building depends on the cloud pattern of the sky; the cloud pattern defining the luminous distribution. The luminance and luminous distribution vary with the geographical location, time, density and uniformity of the overcast.[2] "Uniformly" overcast skies are 2½ to 3 times as bright overhead as at the horizon. However, for design purposes, a single value representing equivalent uniform sky luminance may be used.

On clear days, the sky luminance varies with the position of the sun.[3] Except in the immediate vicinity of the sun, a clear sky is normally brighter near the horizon than overhead. For design purposes, the concept of equivalent sky luminance may also be used for clear skies. In daylight calculations for clear days, sky light only is included on non-sun exposures, while light from both sun and sky is included in calculations for sun exposures. Daylight calculations for partly cloudy skies are analogous to clear sky calculations. Only light from the sky is included on non-sun exposures, while light from both sun and sky is included in calculations for sun exposures. The concept of equivalent sky luminances may also be used for partly cloudy skies.

The Ground as a Light Source

Light reflected from the ground, or from other exterior surfaces, is important in daylighting design. As with other light sources, it may require brightness control. On sunny elevations, the light reflected from the ground may represent ten to fifteen per cent of the total daylight reaching a window area. It frequently exceeds this proportion with light-colored, sandy soils, light-colored vegetation, or snow cover. On non-sun exposures, the light reflected from the ground accounts for more than half the total light reaching the window.

The direction from which the ground light is received at the fenestration is such that it can be utilized most effectively in the interior of the room, particularly at points well removed from the window area. Furthermore, ground light may be under the partial control of the architect or engineer to a considerable extent. By use of light-colored ground surfacing materials near the building, the daylight incident on the window areas and penetrating the inner portions of the rooms inside the building can be increased significantly.

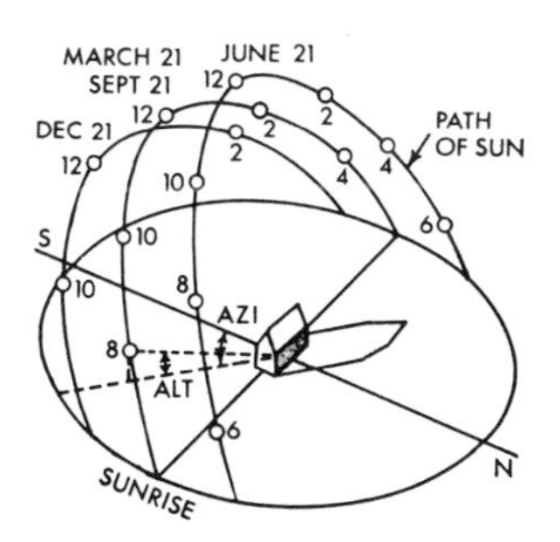

Fig. 7-3. Solar altitude and azimuth, 42°N. latitude.

DAYLIGHT AVAILABILITY[4]

Daylight availability is the amount of daylight available from the sun and the sky at a specific location, time, date and sky condition. Measurements of daylight illumination by various researchers in numerous worldwide locations over the past 60 years have resulted in very similar mean trend curves. The available daylight is determined from a set of these statistical mean trend curves.

Frequent use of the curves, or tables used to express daylight availability, can result in a feeling that the numbers represent instantaneous values, rather than mean design values. This is a mistake for it is not unusual for instantaneous values to be more than twice or less than half of mean design values.

Availability as a Function of Solar Position

Figs. 7-5 through 7-12 provide daylight availability as a function of solar altitude and solar azimuth with respect to the solar meridian. Figs. 7-5 and 7-6 give the direct sunlight availability for clear and partly cloudy skies, respectively. There is no graph of the direct sunlight from a cloudy (overcast) sky since it is assumed that the sun is obscured. Figs. 7-7 through 7-9 give the illuminances on vertical surfaces from clear, partly cloudy and overcast skies, and Figs. 7-10 through 7-12 give the illuminances on horizontal surfaces. In the latter figures, the half-sky horizontal illuminance is the component of horizontal illuminance from the same part of the sky as produced the vertical sky illuminance in Figs. 7-7 through 7-9.

Fig. 7-4. Solar Altitude and Azimuth for Different Latitudes

	Date	Solar Time* AM: 6 PM: 6	AM: 7 PM: 5	AM: 8 PM: 4	AM: 9 PM: 3	AM: 10 PM: 2	AM: 11 PM: 1	Noon

	Date	6	7	8	9	10	11	Noon
ALTITUDE	June 21	12	24	37	50	63	75	83
	Mar.-Sept. 21	—	13	26	38	49	57	60
	Dec. 21	—	—	12	21	29	35	37
AZIMUTH	June 21	111	104	99	92	84	67	0
	Mar.-Sept. 21		83	74	64	49	28	0
	Dec. 21	—		54	44	32	17	0

	Date	6	7	8	9	10	11	Noon
ALTITUDE	June 21	13	25	37	50	62	74	79
	Mar.-Sept. 21	—	12	25	36	46	53	56
	Dec. 21	—	—	9	18	26	31	33
AZIMUTH	June 21	110	103	95	90	78	58	0
	Mar.-Sept. 21		82	72	61	46	26	0
	Dec. 21	—	—	54	43	30	16	0

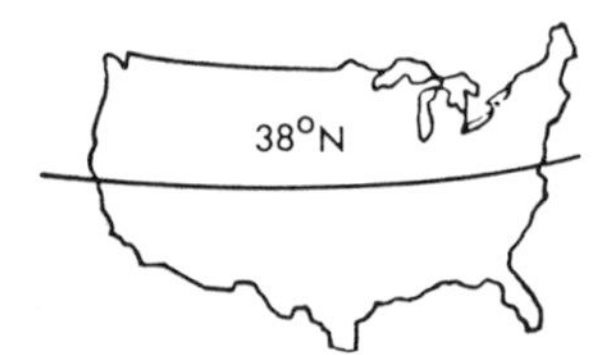

	Date	6	7	8	9	10	11	Noon
ALTITUDE	June 21	14	26	37	49	61	71	75
	Mar.-Sept. 21	—	12	23	34	43	50	52
	Dec. 21	—	—	7	16	23	27	28
AZIMUTH	June 21	109	101	90	83	70	46	0
	Mar.-Sept. 21		81	71	58	43	24	0
	Dec. 21	—	—	54	43	30	16	0

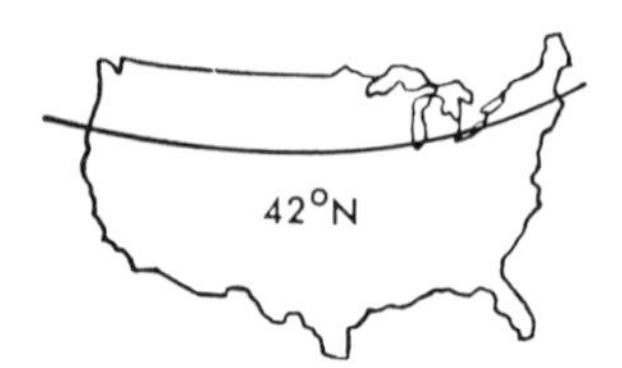

	Date	6	7	8	9	10	11	Noon
ALTITUDE	June 21	16	26	38	49	60	68	71
	Mar.-Sept. 21	—	11	22	32	40	46	48
	Dec. 21	—	—	4	13	19	23	25
AZIMUTH	June 21	108	99	89	78	63	39	0
	Mar.-Sept. 21		80	69	56	41	22	0
	Dec. 21	—	—	53	42	29	15	0

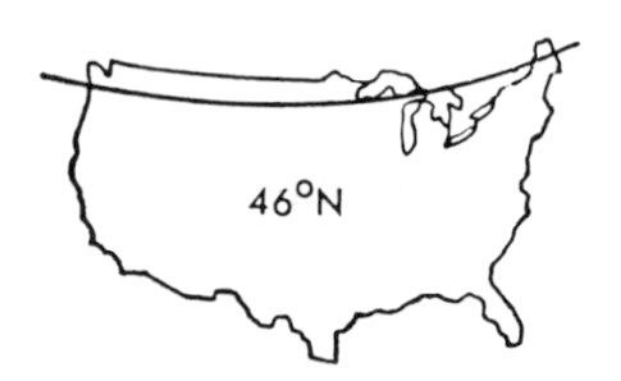

	Date	6	7	8	9	10	11	Noon
ALTITUDE	June 21	17	27	37	48	57	65	67
	Mar.-Sept. 21	—	10	20	30	37	42	44
	Dec. 21	—	—	2	10	15	20	21
AZIMUTH	June 21	107	97	88	74	58	34	0
	Mar.-Sept. 21		79	67	54	39	21	0
	Dec. 21	—	—	52	41	28	14	0

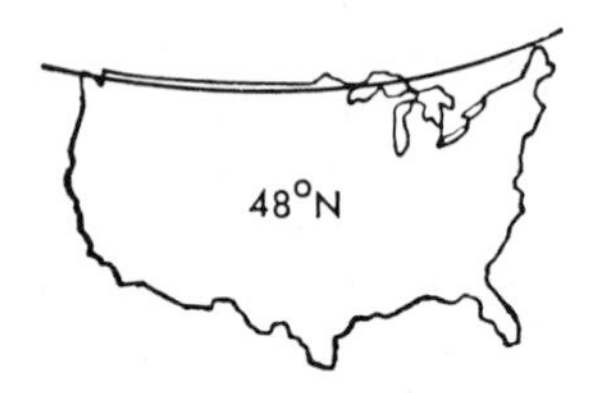

	Date	6	7	8	9	10	11	Noon
ALTITUDE	June 21	17	27	37	47	56	63	65
	Mar.-Sept. 21	—	10	20	29	36	40	42
	Dec. 21	—	—	1	8	14	17	19
AZIMUTH	June 21	106	95	85	72	55	31	0
	Mar.-Sept. 21		79	67	53	38	20	0
	Dec. 21	—	—	52	41	28	14	0

* Time measured by the daily motion of the sun. Noon is taken as the instant in which the center of the sun passes the observer's meridian.

Example. The available daylight on an east-facing window in San Diego, California, on December 21 at 10:00 a.m. solar time, for a clear day, is to be determined.

San Diego is at 33 degrees north latitude. From Fig. 7-4, the solar altitude is determined to be 26 degrees and the solar azimuth 30 degrees (using the nearest latitude of 34 degrees north). Since the window is facing east, the solar azimuthal angle with respect to the window is 90 − 30 = 60 degrees. The morning sun on a clear day would be shining on an east-facing window; thus, Fig. 7-5 is used for the direct solar component which is 37 kilolux. The daylight from the sky is determined from Fig. 7-7 and is 10 kilolux. Thus the total daylight on the window is 47 kilolux.

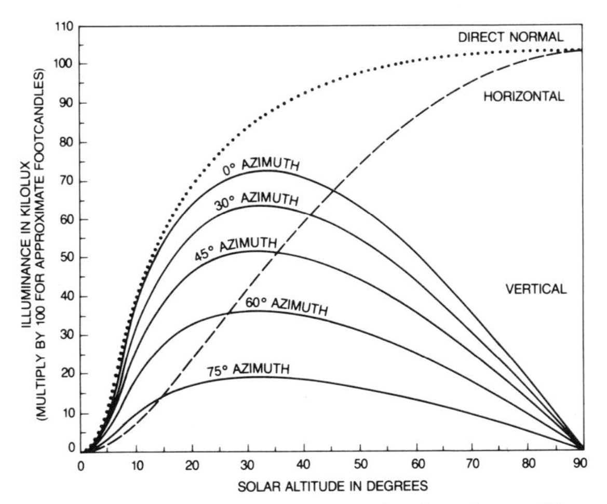

Fig. 7-5. Illuminance from the sun under clear sky conditions as a function of solar altitude and azimuth.[4]

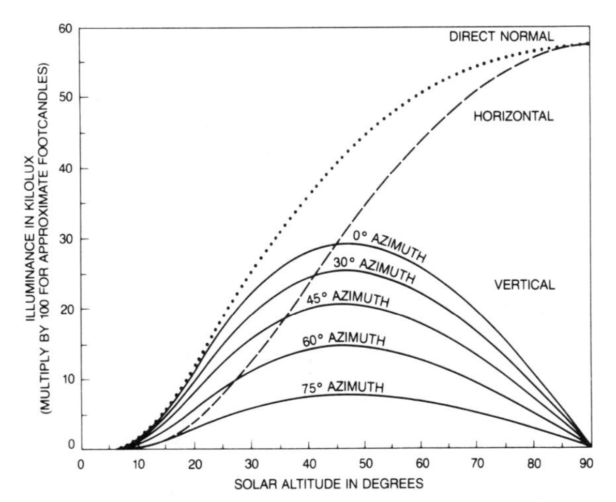

Fig. 7-6. Illuminance from the sun under partly cloudy sky conditions as a function of solar altitude and azimuth.[4]

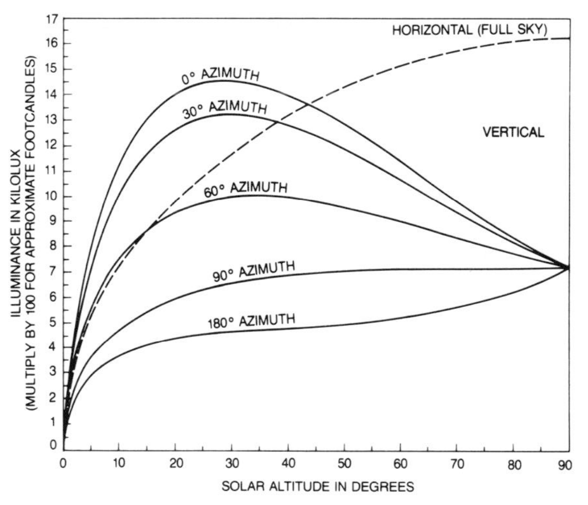

Fig. 7-7. Illuminance on vertical surfaces from clear sky conditions as a function of solar altitude and azimuth.[4]

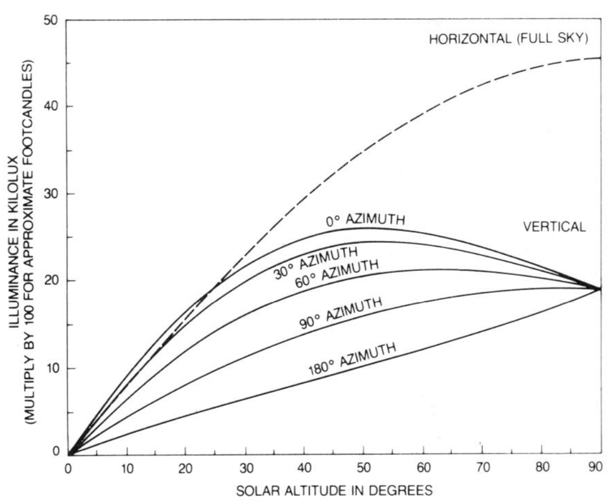

Fig. 7-8. Illuminance on vertical surfaces from partly cloudy sky conditions as a function of solar altitude and azimuth.[4]

Equivalent Sky Exitance

One technique used to estimate daylight illuminance is based on the concept of equivalent sky exitance. Equivalent exitance for north, south, east, and west skies are given in Figs. 7-13 through 7-16. Note that for overcast skies, the equivalent exitance is independent of cardinal orientation. For a point in a room that only sees a small portion of the sky through a window or a skylight, equivalent sky exitance is a useful concept for estimating the daylight illuminance at that point. For instance, it can be used in conjunction with Figs. 9-27 and 9-28 to calculate the illuminance from a rectangular window or skylight.

Availability as a Function of Site Location and Orientation

Although Figs. 7-5 through 7-12 provide a general method of determining daylight availability, it is possible to eliminate the intermediate steps of determining solar altitude and solar azimuth if site location, orientation, time, date and sky condition are predetermined. Figs. 7-17 through 7-20 provide daylight availability directly for many common situations. Exterior horizontal illuminances on clear, partly cloudy and overcast days can be determined from Fig. 7-17. Exterior vertical illuminances on a clear day for the four cardinal directions can be found in Fig. 7-18. Likewise, partly cloudy exterior vertical illuminances are given in Fig. 7-19, and overcast day exterior vertical illuminances in Fig. 7-20.

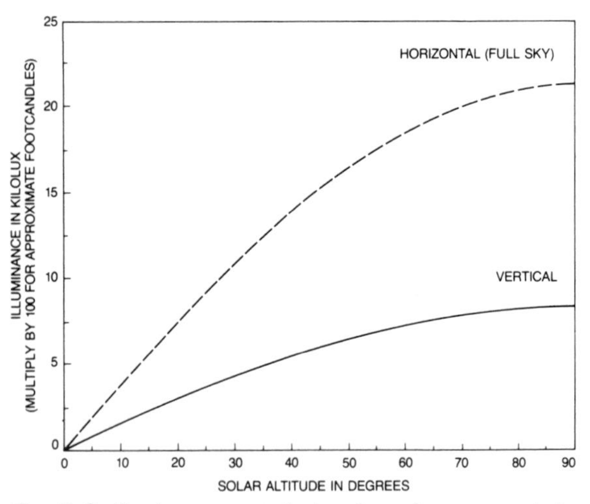

Fig. 7-9. Illuminance on vertical surfaces from overcast sky conditions as a function of solar altitude.[4]

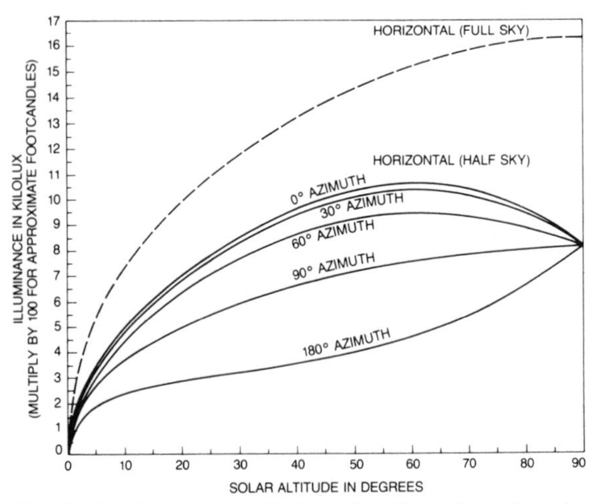

Fig. 7-10. Illuminance on horizontal surfaces from clear sky conditions as a function of solar altitude and azimuth.[4]

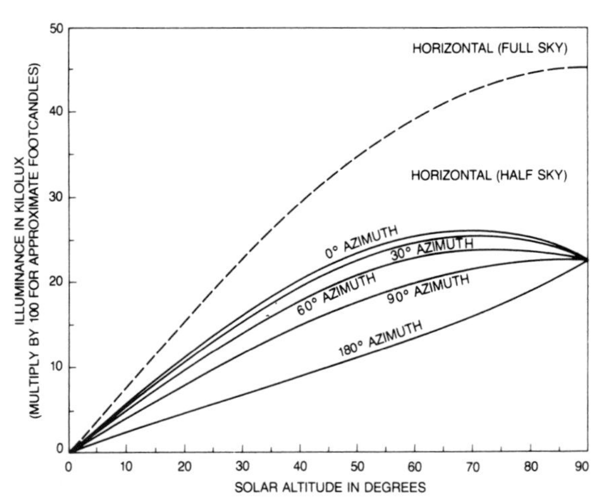

Fig. 7-11. Illuminance on horizontal surfaces from partly cloudy sky conditions as a function of solar altitude and azimuth.[4]

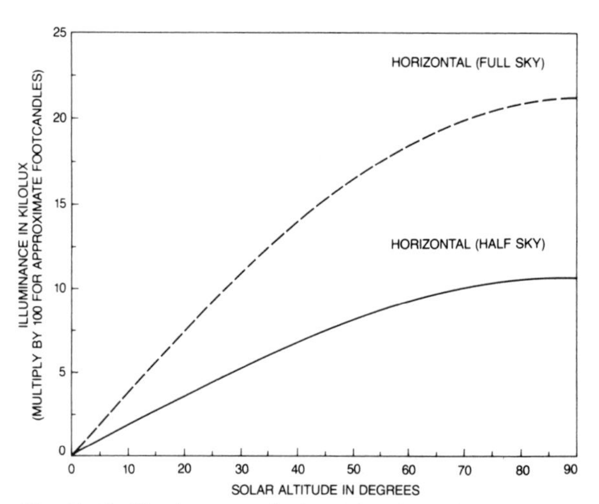

Fig. 7-12. Illuminance on horizontal surfaces from overcast sky conditions as a function of solar altitude.[4]

Example. The available daylight on a clear day in San Diego, California, on an east-facing window at 10:00 a.m. solar time on December 21 is to be determined.

Fig. 7-18 provides the exterior vertical illuminance facing east on a clear day. San Diego is at 33 degrees north latitude; the closest latitude in the figure is 34 degrees which is sufficiently close. At 10:00 a.m. on December 21, Fig. 7-18 shows a direct daylight component (sunlight) of 37 kilolux, a diffuse daylight component (sky component) of 10 kilolux for a total illuminance of 47 kilolux.

Fig. 7-13. Equivalent North Sky Exitance in Kilolumens per Square Meter (Multiply by 100 for Lumens per Square Foot*)

Latitude (degrees north)	December 21					March and September 21					June 21				
	8 AM	10 AM	Noon	2 PM	4 PM	8 AM	10 AM	Noon	2 PM	4 PM	8 AM	10 AM	Noon	2 PM	4 PM
Clear Day															
30	7	8	8	8	7	11	11	11	11	11	16	14	13	14	16
34	6	8	8	8	6	11	11	10	11	11	16	14	13	14	16
38	5	7	7	7	5	11	10	10	10	11	15	13	12	13	15
42	5	7	7	7	5	10	10	10	10	10	15	13	12	13	15
46	3	6	6	6	3	10	9	9	9	10	14	12	12	12	14
50	0	5	5	5	0	9	9	9	9	9	14	12	11	12	14
Partly Cloudy Day															
30	7	13	15	13	7	17	23	25	23	17	28	33	35	33	28
34	5	12	14	12	5	16	22	23	22	16	28	32	33	32	28
38	4	10	12	10	4	15	20	22	20	15	27	30	32	30	27
42	3	9	10	9	3	14	19	20	19	14	26	29	30	29	26
46	1	7	9	7	1	13	17	18	17	13	25	28	23	28	25
50	0	5	7	5	0	12	16	17	16	12	24	26	27	26	24
Overcast Day															
30	3	8	10	8	3	7	13	15	13	7	10	15	17	15	10
34	3	7	9	7	3	7	12	14	12	7	10	15	17	15	10
38	2	6	8	6	2	7	12	13	12	7	10	15	16	15	10
42	1	6	7	6	1	6	11	13	11	6	10	15	16	15	10
46	1	5	6	5	1	6	10	12	10	6	10	14	16	14	10
50	0	3	5	3	0	5	9	11	9	5	10	14	15	14	10

* More exact multiplier is 92.9.

Fig. 7-14. Equivalent South Sky Exitance in Kilolumens per Square Meter (Multiply by 100 for Lumens per Square Foot*)

Latitude (degrees north)	December 21					March and September 21					June 21				
	8 AM	10 AM	Noon	2 PM	4 PM	8 AM	10 AM	Noon	2 PM	4 PM	8 AM	10 AM	Noon	2 PM	4 PM
Clear Day															
30	16	25	28	25	16	16	21	22	21	16	14	15	15	15	14
34	15	25	28	25	15	17	22	24	22	17	14	16	16	16	14
38	14	24	28	24	14	17	23	25	23	17	14	17	17	17	14
42	12	23	27	23	12	17	23	26	23	17	15	18	19	18	15
46	9	22	26	22	9	17	24	27	24	17	15	19	20	19	15
50	0	21	25	21	0	17	24	27	24	17	16	20	21	20	16
Partly Cloudy Day															
30	15	38	47	38	15	24	43	52	43	24	24	36	42	36	24
34	13	35	45	35	13	23	43	53	43	23	25	38	44	38	25
38	10	32	41	32	10	23	43	53	43	23	26	39	46	39	26
42	7	28	37	28	7	22	42	52	42	22	27	40	48	40	27
46	3	24	33	24	3	21	41	51	41	21	27	42	50	42	27
50	0	20	28	20	0	20	40	49	40	20	28	43	51	43	28
Overcast Day															
30	3	8	10	8	3	7	13	15	13	7	10	15	17	15	10
34	3	7	9	7	3	7	12	14	12	7	10	15	17	15	10
38	2	6	8	6	2	7	12	13	12	7	10	15	16	15	10
42	1	6	7	6	1	6	11	13	11	6	10	15	16	15	10
46	1	5	6	5	1	6	10	12	10	6	10	14	16	14	10
50	0	3	5	3	0	5	9	11	9	5	10	14	15	14	10

* More exact multiplier is 92.9.

Fig. 7-15. Equivalent East Sky Exitance in Kilolumens per Square Meter (Multiply by 100 for Lumens per Square Foot*)

Latitude (degrees north)	December 21					March and September 21					June 21				
	8 AM	10 AM	Noon	2 PM	4 PM	8 AM	10 AM	Noon	2 PM	4 PM	8 AM	10 AM	Noon	2 PM	4 PM
Clear Day															
30	19	20	15	10	5	27	22	15	11	7	27	22	14	11	8
34	18	19	14	10	5	26	22	15	11	7	28	22	14	11	8
38	16	19	14	9	4	26	22	15	11	7	28	22	14	11	8
42	14	18	14	9	4	25	22	15	10	7	28	22	14	11	8
46	11	17	13	8	3	25	22	15	10	6	28	22	14	11	8
50	0	15	12	7	0	24	21	15	10	6	27	22	15	11	8
Partly Cloudy Day															
30	18	31	26	16	6	38	45	34	22	11	47	51	37	26	15
34	15	28	24	15	5	36	44	33	22	11	48	51	37	26	16
38	12	25	22	13	3	35	42	32	21	10	48	50	37	26	16
42	8	22	19	11	2	33	40	31	20	10	48	49	36	25	16
46	4	18	17	9	1	31	37	29	19	9	48	48	35	25	16
50	0	15	14	8	0	29	35	28	18	8	47	47	35	25	15
Overcast Day															
30	3	8	10	8	3	7	13	15	13	7	10	15	17	15	10
34	3	7	9	7	3	7	12	14	12	7	10	15	17	15	10
38	2	6	8	6	2	7	12	13	12	7	10	15	16	15	10
42	1	6	7	6	1	6	11	13	11	6	10	15	16	15	10
46	1	5	6	5	1	6	10	12	10	6	10	14	16	14	10
50	0	3	5	3	0	5	9	11	9	5	10	14	15	14	10

* More exact multiplier is 92.9.

Fig. 7-16. Equivalent West Sky Exitance in Kilolumens per Square Meter (Multiply by 100 for Lumens per Square Foot*)

Latitude (degrees north)	December 21					March and September 21					June 21				
	8 AM	10 AM	Noon	2 PM	4 PM	8 AM	10 AM	Noon	2 PM	4 PM	8 AM	10 AM	Noon	2 PM	4 PM
Clear Day															
30	5	10	15	20	19	7	11	15	22	27	8	11	14	22	27
34	5	10	14	19	18	7	11	15	22	26	8	11	14	22	28
38	4	9	14	19	16	7	11	15	22	26	8	11	14	22	28
42	4	9	14	18	14	7	10	15	22	25	8	11	14	22	28
46	3	8	13	17	11	6	10	15	22	25	8	11	14	22	28
50	0	7	12	15	0	6	10	15	21	24	8	11	15	22	27
Partly Cloudy Day															
30	6	16	26	31	18	11	22	34	45	38	15	26	37	51	47
34	5	15	24	28	15	11	22	33	44	36	16	26	37	51	48
38	3	13	22	25	12	10	21	32	42	35	16	26	37	50	48
42	2	11	19	22	8	10	20	31	40	33	16	25	36	49	48
46	1	9	17	18	4	9	19	29	37	31	16	25	35	48	48
50	0	8	14	15	0	8	18	28	35	29	15	25	35	47	47
Overcast Day															
30	3	8	10	8	3	7	13	15	13	7	10	15	17	15	10
34	3	7	9	7	3	7	12	14	12	7	10	15	17	15	10
38	2	6	8	6	2	7	12	13	12	7	10	15	16	15	10
42	1	6	7	6	1	6	11	13	11	6	10	15	16	15	10
46	1	5	6	5	1	6	10	12	10	6	10	14	16	14	10
50	0	3	5	3	0	5	9	11	9	5	10	14	15	14	10

* More exact multiplier is 92.9.

Fig. 7-17. Exterior Horizontal Illuminance in Kilolux (Multiply by 100 for Footcandles*)

Latitude (degrees north)	Component	December 21			March and September 21			June 21		
		8 AM 4 PM	10 AM 2 PM	Noon	8 AM 4 PM	10 AM 2 PM	Noon	8 AM 4 PM	10 AM 2 PM	Noon
Clear Day†										
30	Direct	9	42	55	34	72	87	52	86	99
	Diffuse	8	12	13	11	14	15	13	15	16
	Total	17	54	68	45	86	102	65	101	115
34	Direct	5	35	48	32	68	82	52	86	98
	Diffuse	7	11	12	11	14	15	13	15	16
	Total	12	46	60	43	82	97	65	101	114
38	Direct	3	29	41	29	64	77	53	84	96
	Diffuse	6	10	12	11	14	15	13	15	16
	Total	9	39	53	40	78	92	66	99	112
42	Direct	1	22	33	27	59	71	53	83	94
	Diffuse	5	10	11	10	13	14	13	15	16
	Total	6	32	44	37	72	85	66	98	110
46	Direct	0	16	25	24	54	65	53	80	91
	Diffuse	4	9	10	10	13	14	13	15	16
	Total	4	25	35	34	67	79	66	95	107
50	Direct	0	10	18	21	49	59	52	78	87
	Diffuse	0	8	9	10	12	13	13	15	15
	Total	0	18	27	31	61	72	65	93	102
Partly Cloudy Day‡										
30	Direct	0	13	20	9	33	44	19	44	55
	Diffuse	9	22	27	20	34	39	27	40	45
	Total	9	35	47	29	67	83	46	84	100
34	Direct	0	9	16	8	30	40	20	44	54
	Diffuse	7	20	25	19	33	38	27	40	45
	Total	7	29	41	27	63	78	47	84	99
38	Direct	0	6	12	7	27	36	20	43	52
	Diffuse	6	17	22	18	31	36	28	39	44
	Total	6	23	34	25	58	72	48	82	96
42	Direct	0	4	8	6	24	32	20	42	50
	Diffuse	4	15	19	17	29	34	28	39	43
	Total	4	19	27	23	53	66	48	81	93
46	Direct	0	2	5	4	20	28	20	40	48
	Diffuse	2	12	16	16	27	32	28	38	42
	Total	2	14	21	20	47	60	48	78	90
50	Direct	0	1	2	3	17	24	19	38	45
	Diffuse	0	10	13	15	25	29	27	37	41
	Total	0	11	15	18	42	53	46	75	86
Overcast Day§										
30	Direct	0	0	0	0	0	0	0	0	0
	Diffuse	4	11	13	9	16	18	13	19	21
	Total	4	11	13	9	16	18	13	19	21
34	Direct	0	0	0	0	0	0	0	0	0
	Diffuse	4	9	12	9	15	18	13	19	21
	Total	4	9	12	9	15	18	13	19	21
38	Direct	0	0	0	0	0	0	0	0	0
	Diffuse	3	8	10	9	15	17	13	19	21
	Total	3	8	10	9	15	17	13	19	21
42	Direct	0	0	0	0	0	0	0	0	0
	Diffuse	2	7	9	8	14	16	13	18	20
	Total	2	7	9	8	14	16	13	18	20
46	Direct	0	0	0	0	0	0	0	0	0
	Diffuse	1	6	8	8	13	15	13	18	20
	Total	1	6	8	8	13	15	13	18	20
50	Direct	0	0	0	0	0	0	0	0	0
	Diffuse	0	5	6	7	12	14	13	17	19
	Total	0	5	6	7	12	14	13	17	19

* More exact multiplier is 92.9. ‡ Atmospheric Extinction Coefficient = 0.80. § Typical nonprecipitative minimum.

† Atmospheric Extinction Coefficient = 0.21. Atmospheric extinction coefficient is the reciprocal of the optical air mass times the natural logarithm of the ratio of the extraterrestrial direct normal solar illuminance to the sea level direct normal solar illuminance. Optical air mass is the ratio of the sea level path length through the atmosphere toward the sun to the sea level path toward the zenith.

Fig. 7-18. Exterior Vertical Illuminance on a Clear Day* in Kilolux (Multiply by 100 for Footcandles†)

Latitude (degrees north)	Component	December 21					March and September 21					June 21				
		8 AM	10 AM	Noon	2 PM	4 PM	8 AM	10 AM	Noon	2 PM	4 PM	8 AM	10 AM	Noon	2 PM	4 PM
							Facing North									
30	Direct	0	0	0	0	0	0	0	0	0	0	10	0	0	0	10
	Diffuse	3	4	4	4	3	6	6	5	6	6	8	7	7	7	8
	Total	3	4	4	4	3	6	6	5	6	6	18	7	7	7	18
34	Direct	0	0	0	0	0	0	0	0	0	0	6	0	0	0	6
	Diffuse	3	4	4	4	3	5	5	5	5	5	8	7	6	7	8
	Total	3	4	4	4	3	5	5	5	5	5	14	7	6	7	14
38	Direct	0	0	0	0	0	0	0	0	0	0	3	0	0	0	3
	Diffuse	3	4	4	4	3	5	5	5	5	5	8	7	6	7	8
	Total	3	4	4	4	3	5	5	5	5	5	11	7	6	7	11
42	Direct	0	0	0	0	0	0	0	0	0	0	0	0	0	0	0
	Diffuse	2	3	3	3	2	5	5	5	5	5	7	6	6	6	7
	Total	2	3	3	3	2	5	5	5	5	5	7	6	6	6	7
46	Direct	0	0	0	0	0	0	0	0	0	0	0	0	0	0	0
	Diffuse	2	3	3	3	2	5	5	5	5	5	7	6	6	6	7
	Total	2	3	3	3	2	5	5	5	5	5	7	6	6	6	7
50	Direct	0	0	0	0	0	0	0	0	0	0	0	0	0	0	0
	Diffuse	0	3	3	3	0	5	5	4	5	5	7	6	6	6	7
	Total	0	3	3	3	0	5	5	4	5	5	7	6	6	6	7
							Facing South									
30	Direct	26	64	74	64	26	20	42	50	42	20	0	5	11	5	0
	Diffuse	8	12	14	12	8	8	11	11	11	8	7	8	8	8	7
	Total	34	76	88	76	34	28	53	61	53	28	7	13	19	13	7
34	Direct	20	63	75	63	20	21	46	55	46	21	0	11	18	11	0
	Diffuse	8	12	14	12	8	8	11	12	11	8	7	8	8	8	7
	Total	28	75	89	75	28	29	57	67	57	29	7	19	26	19	7
38	Direct	13	61	75	61	13	23	50	60	50	23	0	17	25	17	0
	Diffuse	7	12	14	12	7	8	11	12	11	8	7	8	9	8	7
	Total	20	73	89	73	20	31	61	72	61	31	7	25	34	25	7
42	Direct	5	57	72	57	5	24	53	64	53	24	1	23	31	23	1
	Diffuse	6	12	14	12	6	8	12	13	12	8	7	9	9	9	7
	Total	11	69	86	69	11	32	65	77	65	32	8	32	40	32	8
46	Direct	0	51	68	51	0	25	56	68	56	25	5	28	38	28	5
	Diffuse	5	11	13	11	5	8	12	13	12	8	8	9	10	9	8
	Total	5	62	81	62	5	33	68	81	68	33	13	37	48	37	13
50	Direct	0	41	60	41	0	25	58	70	58	25	8	34	44	34	8
	Diffuse	0	10	12	10	0	8	12	14	12	8	8	10	11	10	8
	Total	0	51	72	51	0	33	70	84	70	33	16	44	55	44	16
							Facing East									
30	Direct	36	39	0	0	0	68	48	0	0	0	69	45	0	0	0
	Diffuse	10	10	7	5	3	13	11	7	5	4	14	11	7	6	4
	Total	46	49	7	5	3	81	59	7	5	4	83	56	7	6	4
34	Direct	28	37	0	0	0	67	48	0	0	0	69	45	0	0	0
	Diffuse	9	10	7	5	2	13	11	7	5	3	14	11	7	6	4
	Total	37	47	7	5	2	80	59	7	5	3	83	56	7	6	4
38	Direct	17	35	0	0	0	65	47	0	0	0	69	44	0	0	0
	Diffuse	8	9	7	5	2	13	11	7	5	3	14	11	7	6	4
	Total	25	44	7	5	2	78	58	7	5	3	83	55	7	6	4
42	Direct	6	32	0	0	0	63	46	0	0	0	69	44	0	0	0
	Diffuse	7	9	7	4	2	13	11	8	5	3	14	11	7	6	4
	Total	13	41	7	4	2	76	57	8	5	3	83	55	7	6	4
46	Direct	0	27	0	0	0	60	45	0	0	0	69	44	0	0	0
	Diffuse	5	8	6	4	1	12	11	8	5	3	14	11	7	6	4
	Total	5	35	6	4	1	72	56	8	5	3	83	55	7	6	4
50	Direct	0	22	0	0	0	57	44	0	0	0	69	44	0	0	0
	Diffuse	0	8	6	4	0	12	11	7	5	3	14	11	7	6	4
	Total	0	30	6	4	0	69	55	7	5	3	83	55	7	6	4

See footnotes on page 7–11

Continued on next page.

Fig. 7-18. *Continued*

Latitude (degrees north)	Component	December 21					March and September 21					June 21				
		8 AM	10 AM	Noon	2 PM	4 PM	8 AM	10 AM	Noon	2 PM	4 PM	8 AM	10 AM	Noon	2 PM	4 PM
								Facing West								
30	Direct	0	0	0	39	36	0	0	0	48	68	0	0	0	45	69
	Diffuse	3	5	7	10	10	4	5	7	11	13	4	6	7	11	14
	Total	3	5	7	49	46	4	5	7	59	81	4	6	7	56	83
34	Direct	0	0	0	37	28	0	0	0	48	67	0	0	0	45	69
	Diffuse	2	5	7	10	9	3	5	7	11	13	4	6	7	11	14
	Total	2	5	7	47	37	3	5	7	59	80	4	6	7	56	83
38	Direct	0	0	0	35	17	0	0	0	47	65	0	0	0	44	69
	Diffuse	2	5	7	9	8	3	5	7	11	13	4	6	7	11	14
	Total	2	5	7	44	25	3	5	7	58	78	4	6	7	55	83
42	Direct	0	0	0	32	6	0	0	0	46	63	0	0	0	44	69
	Diffuse	2	4	7	9	7	3	5	8	11	13	4	6	7	11	14
	Total	2	4	7	41	13	3	5	8	57	76	4	6	7	55	83
46	Direct	0	0	0	27	0	0	0	0	45	60	0	0	0	44	69
	Diffuse	1	4	6	8	5	3	5	8	11	12	4	6	7	11	14
	Total	1	4	6	35	5	3	5	8	56	72	4	6	7	55	83
50	Direct	0	0	0	22	0	0	0	0	44	57	0	0	0	44	69
	Diffuse	0	4	6	8	0	3	5	7	11	12	4	6	7	11	14
	Total	0	4	6	30	0	3	5	7	55	69	4	6	7	55	83

* Atmospheric extinction coefficient = 0.21.
† More exact multiplier is 92.9.

Fig. 7-19. Exterior Vertical Illuminance on a Partly Cloudy Day* in Kilolux (Multiply by 100 for Footcandles†)

Latitude (degrees north)	Component	December 21					March and September 21					June 21				
		8 AM	10 AM	Noon	2 PM	4 PM	8 AM	10 AM	Noon	2 PM	4 PM	8 AM	10 AM	Noon	2 PM	4 PM
								Facing North								
30	Direct	0	0	0	0	0	0	0	0	0	0	4	0	0	0	4
	Diffuse	3	7	8	7	3	8	12	13	12	8	14	17	17	17	14
	Total	3	7	8	7	3	8	12	13	12	8	18	17	17	17	18
34	Direct	0	0	0	0	0	0	0	0	0	0	2	0	0	0	2
	Diffuse	3	6	7	6	3	8	11	12	11	8	14	16	17	16	14
	Total	3	6	7	6	3	8	11	12	11	8	16	16	17	16	16
38	Direct	0	0	0	0	0	0	0	0	0	0	1	0	0	0	1
	Diffuse	2	5	6	5	2	7	10	11	10	7	13	15	16	15	13
	Total	2	5	6	5	2	7	10	11	10	7	14	15	16	15	14
42	Direct	0	0	0	0	0	0	0	0	0	0	0	0	0	0	0
	Diffuse	1	4	5	4	1	7	9	10	9	7	13	15	15	15	13
	Total	1	4	5	4	1	7	9	10	9	7	13	15	15	15	13
46	Direct	0	0	0	0	0	0	0	0	0	0	0	0	0	0	0
	Diffuse	1	4	4	4	1	6	9	9	9	6	13	14	14	14	13
	Total	1	4	4	4	1	6	9	9	9	6	13	14	14	14	13
50	Direct	0	0	0	0	0	0	0	0	0	0	0	0	0	0	0
	Diffuse	0	3	3	3	0	6	8	8	8	6	12	13	13	13	12
	Total	0	3	3	3	0	6	8	8	8	6	12	13	13	13	12

See footnotes on page 7-12

Continued on next page.

Fig. 7-19. *Continued*

Latitude (degrees north)	Component	December 21					March and September 21					June 21				
		8 AM	10 AM	Noon	2 PM	4 PM	8 AM	10 AM	Noon	2 PM	4 PM	8 AM	10 AM	Noon	2 PM	4 PM
Facing South																
30	Direct	1	19	28	19	1	5	19	25	19	5	0	3	6	3	0
	Diffuse	8	19	24	19	8	12	21	26	21	12	12	18	21	18	12
	Total	9	38	52	38	9	17	40	51	40	17	12	21	27	21	12
34	Direct	0	16	25	16	0	5	20	27	20	5	0	6	10	6	0
	Diffuse	6	18	22	18	6	12	21	26	21	12	12	19	22	19	12
	Total	6	34	47	34	6	17	41	53	41	17	12	25	32	25	12
38	Direct	0	13	22	13	0	5	21	28	21	5	0	9	14	9	0
	Diffuse	5	16	21	16	5	11	21	26	21	11	13	20	23	20	13
	Total	5	29	43	29	5	16	42	54	42	16	13	29	37	29	13
42	Direct	0	9	17	9	0	5	21	29	21	5	0	11	17	11	0
	Diffuse	3	14	19	14	3	11	21	26	21	11	13	20	24	20	13
	Total	3	23	36	23	3	16	42	55	42	16	13	31	41	31	13
46	Direct	0	5	13	5	0	5	21	29	21	5	2	14	20	14	2
	Diffuse	2	12	17	12	2	11	21	25	21	11	14	21	25	21	14
	Total	2	17	30	17	2	16	42	54	42	16	16	35	45	35	16
50	Direct	0	2	8	2	0	4	20	28	20	4	3	16	23	16	3
	Diffuse	0	10	14	10	0	10	20	25	20	10	14	21	26	21	14
	Total	0	12	22	12	0	14	40	53	40	14	17	37	49	37	17
Facing East																
30	Direct	2	12	0	0	0	17	22	0	0	0	26	23	0	0	0
	Diffuse	9	15	13	8	3	19	23	17	11	6	24	26	19	13	8
	Total	11	27	13	8	3	36	45	17	11	6	50	49	19	13	8
34	Direct	1	10	0	0	0	16	21	0	0	0	26	23	0	0	0
	Diffuse	7	14	12	7	2	18	22	17	11	5	24	25	18	13	8
	Total	8	24	12	7	2	34	43	17	11	5	50	48	18	13	8
38	Direct	0	7	0	0	0	14	20	0	0	0	26	23	0	0	0
	Diffuse	6	12	11	7	2	17	21	16	10	5	24	25	18	13	8
	Total	6	19	11	7	2	31	41	16	10	5	50	48	18	13	8
42	Direct	0	5	0	0	0	13	18	0	0	0	26	22	0	0	0
	Diffuse	4	11	10	6	1	16	20	15	10	5	24	25	18	13	8
	Total	4	16	10	6	1	29	38	15	10	5	50	47	18	13	8
46	Direct	0	3	0	0	0	11	17	0	0	0	26	22	0	0	0
	Diffuse	2	9	8	5	0	16	19	15	9	4	24	24	18	13	8
	Total	2	12	8	5	0	27	36	15	9	4	50	46	18	13	8
50	Direct	0	1	0	0	0	9	15	0	0	0	26	21	0	0	0
	Diffuse	0	7	7	4	0	15	18	14	9	4	24	24	17	12	8
	Total	0	8	7	4	0	24	33	14	9	4	50	45	17	12	8
Facing West																
30	Direct	0	0	0	12	2	0	0	0	22	17	0	0	0	23	26
	Diffuse	3	8	13	15	9	6	11	17	23	19	8	13	19	26	24
	Total	3	8	13	27	11	6	11	17	45	36	8	13	19	49	50
34	Direct	0	0	0	10	1	0	0	0	21	16	0	0	0	23	26
	Diffuse	2	7	12	14	7	5	11	17	22	18	8	13	18	25	24
	Total	2	7	12	24	8	5	11	17	43	34	8	13	18	48	50
38	Direct	0	0	0	7	0	0	0	0	20	14	0	0	0	23	26
	Diffuse	2	7	11	12	6	5	10	16	21	17	8	13	18	25	24
	Total	2	7	11	19	6	5	10	16	41	31	8	13	18	48	50
42	Direct	0	0	0	5	0	0	0	0	18	13	0	0	0	22	26
	Diffuse	1	6	10	11	4	5	10	15	20	16	8	13	18	25	24
	Total	1	6	10	16	4	5	10	15	38	29	8	13	18	47	50
46	Direct	0	0	0	3	0	0	0	0	17	11	0	0	0	22	26
	Diffuse	0	5	8	9	2	4	9	15	19	16	8	13	18	24	24
	Total	0	5	8	12	2	4	9	15	36	27	8	13	18	46	50
50	Direct	0	0	0	1	0	0	0	0	15	9	0	0	0	21	26
	Diffuse	0	4	7	7	0	4	9	14	18	15	8	12	17	24	24
	Total	0	4	7	8	0	4	9	14	33	24	8	12	17	45	50

* Atmospheric extinction coefficient = 0.80.
† More exact multiplier is 92.9.

Fig. 7-20. Exterior Vertical Illuminance Facing Any Direction on an Overcast Day* in Kilolux (Multiply by 100 for Footcandles†)

Latitude (degrees north)	Component	December 21					March and September 21					June 21				
		8 AM	10 AM	Noon	2 PM	4 PM	8 AM	10 AM	Noon	2 PM	4 PM	8 AM	10 AM	Noon	2 PM	4 PM
30	Direct	0	0	0	0	0	0	0	0	0	0	0	0	0	0	0
	Diffuse	2	4	5	4	2	4	6	7	6	4	5	8	8	8	5
	Total	2	4	5	4	2	4	6	7	6	4	5	8	8	8	5
34	Direct	0	0	0	0	0	0	0	0	0	0	0	0	0	0	0
	Diffuse	1	4	5	4	1	4	6	7	6	4	5	7	8	7	5
	Total	1	4	5	4	1	4	6	7	6	4	5	7	8	7	5
38	Direct	0	0	0	0	0	0	0	0	0	0	0	0	0	0	0
	Diffuse	1	3	4	3	1	3	6	7	6	3	5	7	8	7	5
	Total	1	3	4	3	1	3	6	7	6	3	5	7	8	7	5
42	Direct	0	0	0	0	0	0	0	0	0	0	0	0	0	0	0
	Diffuse	1	3	4	3	1	3	5	6	5	3	5	7	8	7	5
	Total	1	3	4	3	1	3	5	6	5	3	5	7	8	7	5
46	Direct	0	0	0	0	0	0	0	0	0	0	0	0	0	0	0
	Diffuse	0	2	3	2	0	3	5	6	5	3	5	7	8	7	5
	Total	0	2	3	2	0	3	5	6	5	3	5	7	8	7	5
50	Direct	0	0	0	0	0	0	0	0	0	0	0	0	0	0	0
	Diffuse	0	2	2	2	0	3	5	5	5	3	5	7	8	7	5
	Total	0	2	2	2	0	3	5	5	5	3	5	7	8	7	5

* Typical nonprecipative minimum.
† More exact multiplier is 92.9.

ARCHITECTURAL DESIGN

Building Sections

Unilateral. This design (see Fig. 7-21) lends itself to continuous fenestration and curtain wall construction. Window heads are usually placed close to the ceiling line. For good daylight distribution, the distance of the inner wall to the outdoor wall should be limited to 2 to 2½ times the room height measured from the floor to the head of the window. Extreme luminance at the window, due to sun or sky, should be reduced using shades or tinted glass.

Bilateral. Bilateral daylighting design (see Fig. 7-22) permits doubling the room width. The second set of windows often occupies only the upper part of the wall. A reflecting roof under the secondary windows, acts like ground light and contributes materially to the light entering the room. At least one set of windows faces a sun exposure, necessitating brightness control. Sloping ceilings sometimes employed with this design generally have little effect on quantity or distribution of illumination. High reflectance materials used in the ceiling, however, do contribute to the utilization of light entering the space.

Roof Monitor. This fenestration system is most frequently used in industrial design where a central high bay area is set between two low bay areas. See Fig. 7-23. Brightness controls may be necessary. The roof surfaces below the monitors should be reflecting surfaces for improved light distribution.

Clerestory. The additional fenestration on the roof facing in the same direction as the main window (see Fig. 7-24) aids in overcoming the room width limitations of the unilateral section. Brightness control must be used on some exposures; brightness control is not as prominent a problem as with bilateral designs. The roof adjacent to the clerestory window should be a reflecting-type roof.

Sawtooth. This fenestration is used principally in low roof, large area, industrial buildings. See Fig. 7-25. The windows usually face north in northern latitudes; brightness controls are not then required. Slanting the windows increases the admission of daylight, but may increase dirt collection on the glazing material, in addition to increased thermal stresses in the tinted glazing material.

Skylight. Modern skylights assume many forms and are widely used in contemporary architecture. See Fig. 7-26. These are domes, panels with integral sun and brightness control,

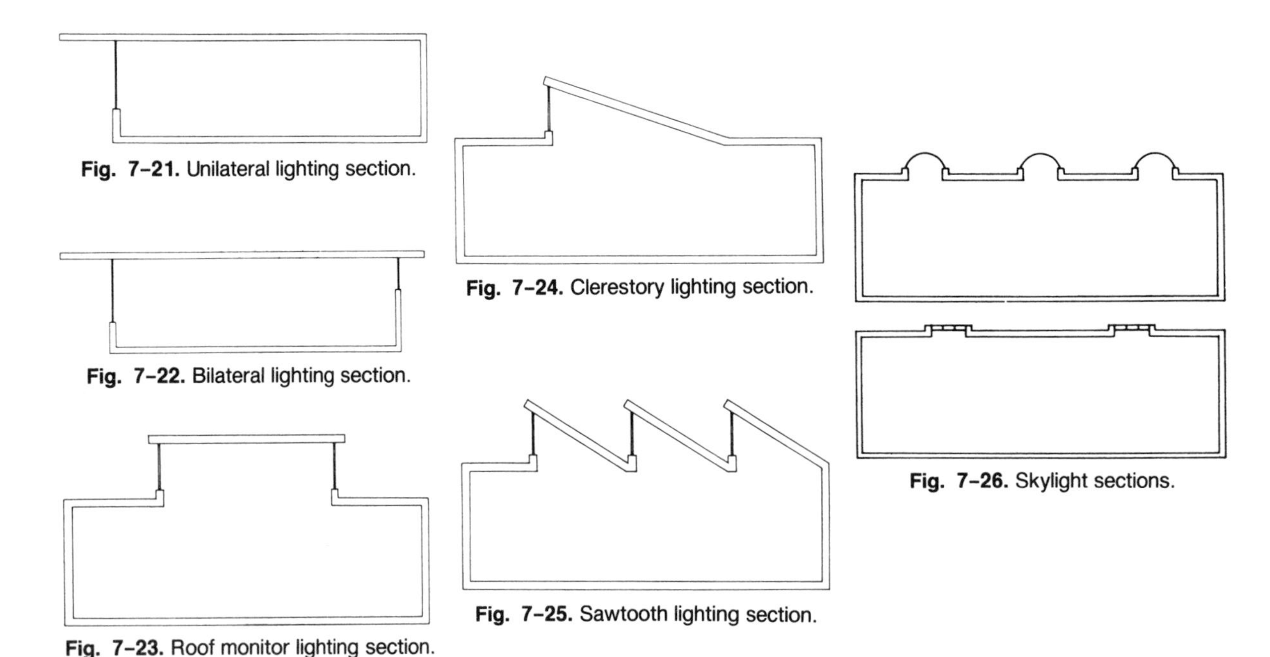

Fig. 7-21. Unilateral lighting section.

Fig. 7-22. Bilateral lighting section.

Fig. 7-23. Roof monitor lighting section.

Fig. 7-24. Clerestory lighting section.

Fig. 7-25. Sawtooth lighting section.

Fig. 7-26. Skylight sections.

panels of glass-fiber reinforced plastic, and louvered-louvers for heat and brightness control. In addition to effective light distribution, skylight design should be carefully detailed to provide for effective seals against moisture penetration and possible dripping from condensation. They also may be used to provide heat control and ventilation.

Window Orientation

In northern latitudes in the United States and Canada, windows oriented to the south are usually preferred. This orientation not only offers higher levels from daylight, particularly in the winter months, but permits the use of solar heat as an aid to room heating. As a general rule, east-west orientations present the most difficult problems of daylight control and complicate heating and cooling design.

Materials and Control Elements

The various materials and control elements used in daylighting are selected for their ability to transmit, absorb, reflect, diffuse or refract light. See Figs. 7-27 and 7-28.

Transparent (High Transmittance Materials). These include sheet, polished plate, float and molded glass, also rigid plastic materials and formed panels which transmit light without appreciably changing its direction or color and allow vision in either direction. Some of these materials are available as sealed double-glazed units which reduce conductive heat flow.

Transparent (Low Transmittance Materials). Low transmittance glasses and plastics offer a measure of brightness control which in-

Fig. 7-27. Transmittance Data of Glass and Plastic Materials

Material	Approximate Transmittance (per cent)
Polished Plate/Float Glass	80–90
Sheet Glass	85–91
Heat Absorbing Plate Glass	70–80
Heat Absorbing Sheet Glass	70–85
Tinted Polished Plate	40–50
Figure Glass	70–90
Corrugated Glass	80–85
Glass Block	60–80
Clear Plastic Sheet	80–92
Tinted Plastic Sheet	90–42
Colorless Patterned Plastic	80–90
White Translucent Plastic	10–80
Glass Fiber Reinforced Plastic	5–80
Double Glazed—2 Lights Clear Glass	77
Tinted Plus Clear	37–45
Reflective Glass*	5–60

* Includes single glass, double glazed units and laminated assemblies. Consult manufacturer's material for specific values.

creases as their transmittance is decreased. Reduction in radiant solar heat accompanies the reduction in visible light. During daylight hours with such materials, the view into the room is reduced while the view to the outdoors from the room is not noticeably affected. At night, the view into the room is apparent while the view from the room to the outdoors is reduced.

Reflective (High Reflectance, Low Transmittance Materials). Reflective glasses and plastics also offer a measure of brightness control which increases as their reflectance is increased. These materials act as one-way mirrors, depending on the ratios of the levels, indoors to outdoors. Also, they may reflect more heat while transmitting more light than non-reflective, low transmittance material.

Diffusing Materials. These include translucent and surface coated or patterned glass, plastics, and diffusing glass block. Transmittance is directionally non-selective. The amount of diffusion varies over a wide range depending on the material and surface. As a rule, transmittance and brightness decrease as diffusion increases. Some types may become excessively bright under sun exposure, requiring brightness control. The brightness of highly diffusing materials is nearly constant from all viewing angles.

Directional Transmitting Materials. These include prismatic surfaced glass and plastics to obtain the desired directional control of light and brightness. They are used in either horizontal or vertical panels.

Specularly Selective Transmitting Materials. These include the various heat absorbing and reflecting materials which are designed to pass most of the visible light but absorb or reflect a portion of the infrared radiation. The absorbed heat is then reradiated in approximately equal proportions inside and outside the building. Stained glass comes under this classification as it is selective in the visible portion of the spectrum. However, the primary use of stained glass is esthetic rather than as illumination control on visual tasks.

Overhangs. These shade the window from direct sunlight and reduce the luminance of the upper part of the window at a sacrifice in the amount of light reaching the room side. Overhangs of practical width do not provide complete shielding at all times. They can be designed to shade the window in summer and let solar heat in during the winter. In multi-story buildings, projecting balconies serve as overhangs.

Fig. 7-28. Reflectances of Building Materials and Outside Surfaces

Material	Reflectance (per cent)	Material	Reflectance (per cent)
Bluestone, sandstone	18	Asphalt (free from dirt)	7
Brick		Earth (moist cultivated)	7
light buff	48	Granolite pavement	17
dark buff	40	Grass (dark green)	6
dark red glazed	30	Gravel	13
Cement	27	Macadam	18
Concrete	40	Slate (dark clay)	8
Marble (white)	45	Snow	
Paint (white)		new	74
new	75	old	64
old	55	Vegetation (mean)	25
Glass			
clear	7		
reflective	20–30		
tinted	7		

Vertical Elements. Vertical opaque elements are effective on east and west walls as sun controls. Matte textures and suitable reflectances should be used. Combinations of vertical and horizontal elements as sun controls are most common in the tropics and southern states.

Shades and Draperies. These include opaque or diffusing shades and draperies for excluding or moderating daylight and solar energy. For darkening a room the material must be opaque and should tightly cover the entire window.

Louvers. These are widely used as shielding elements in daylighting design. The slats may be fixed or adjustable, horizontal or vertical. They may prevent entrance of direct sunlight and reduce radiant heat, while reflecting a high proportion of sun, sky and ground light into the interior. With fixed louvers, spacing and height of the slats should be determined to shield the light source at normal viewing angles. Overhangs for sun control are often made with louver elements so that more of the sky light can reach the windows. Louvers are also employed in top lighting arrangements, sometimes with two sets of slats set at right angles to form an egg crate. Matte textures and suitable reflectances should be used where possible.

Landscaping. Trees are effective shading devices when properly located with respect to the building and its fenestration. Deciduous trees provide shade and protect against sun glare during the warm months but allow the sun to reach the building during the winter. Deciduous vines

Fig. 7-29. Effect of Surface Reflectance—Room 9.1 × 9.8 × 3.7 Meters (30 × 32 × 12 Feet)—Unilateral Fenestration—1.8 × 9.8 Meters (6 × 32 Feet)—Directional Glass Block

Reflectance (per cent)			Relative Illuminance at Various Distances from Fenestration		
Walls	Floor	Ceiling	0.9 Meter (3 Feet)	4.6 Meters (15 Feet)	5.2 Meters (17 Feet)
6	6	6	6.57	2.66	1.00
28	28	28	8.55	3.62	1.60
62	28	62	11.98	5.75	3.16

on louvered overhangs or arbors provide a similar seasonal shade.

Exterior Reflecting Elements. Reflective pavements and similar treatment of the immediate terrain increase the amount of ground light entering the building. Reflecting materials or finishes on roofs below windows have the same effect.

Interior Reflecting Elements. Ceilings of higher reflectance contribute to the utilization of ground and sky light entering the space. Suitable reflectances of walls, floors and furnishings contribute to interreflections of light and to the luminance ratios required for visual tasks. It is important that the texture of all these surfaces be matte. See recommended reflectance values for various occupancies in appropriate Sections of the Application Volume, *e.g.*, Offices, Educational Facilities, Industries. The importance of interior reflectances to daylighting design is indicated in Fig. 7-29, showing increases in quantity and uniformity of daylight illumination produced by higher reflectances of the principal room surfaces.

BUILDING TYPES

Fenestration, the location of windows and openings in buildings: (1) provides for the admission, control and distribution of daylight;* (2) provides a distant focus for the eyes which relaxes the eye muscles; (3) eliminates the dissatisfaction many people experience in completely closed-in areas; (4) may provide for emergency exits or entrances; and (5) may be used for ventilation and energy conservation.

An adequate electric lighting system should always be provided because of the wide variation in daylight, from thousands of lux (footcandles) down to zero.

Effective daylighting complements a well designed electric lighting system. Such a combination can provide the recommended quantities and quality of lighting and also add desirable highlighting and variations.

Daylight illumination distribution is improved when light is introduced through more than one room surface. Unilateral fenestration limits the width of the space in which daylighting can be effective to about 2½ times the height from floor to window head. For wall fenestration, it is recommended that the light transmitting areas be continuous and preferably extend the full length of the room and to the ceiling. Piers, where required, should be of minimum width. When

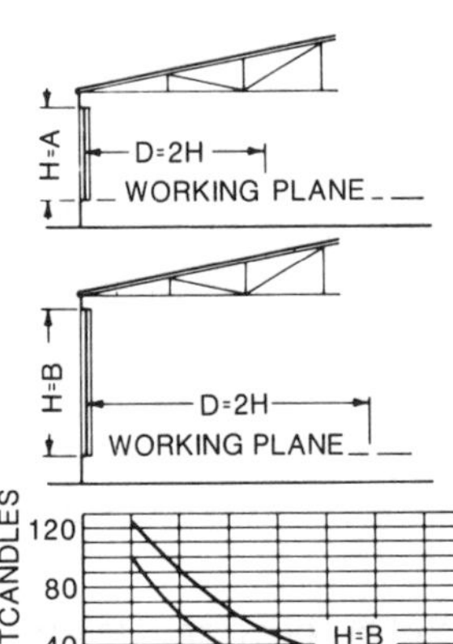

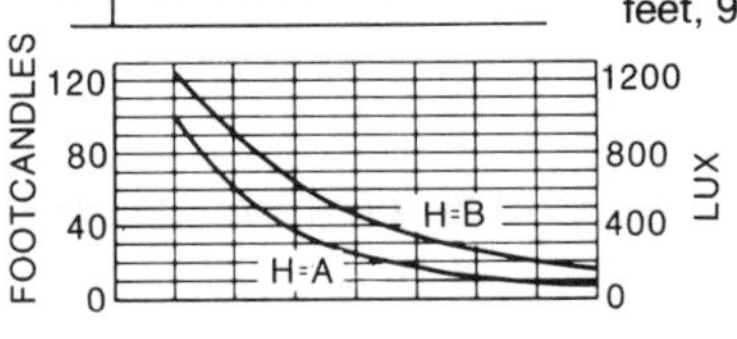

Fig. 7-30. Illuminance distribution from sidewall windows assuming no reflection, clean glass, 3430 candelas per square meter (343 candelas per square foot) uniform sky luminance, and window length equal to or greater than 5 times the window height. *A* = 2.87 meters (9 feet, 5 inches). *B* = 4.5 meters (14 feet, 9 inches).

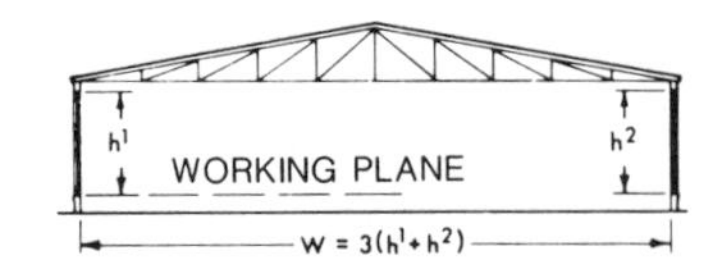

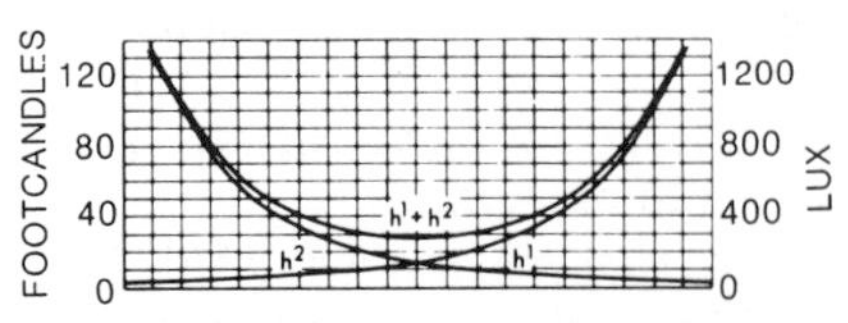

Fig. 7-31. Illuminance distribution from sidewall windows assuming no reflection, clean glass, 3430 candelas per square meter (343 candelas per square foot) uniform sky luminance, and window length equal to or greater than 5 times the window height. $h_1 = h_2 = 5.3$ meters (17 feet, 5 inches).

* Because light is only one portion of the solar spectrum, daylighting design inevitably affects the thermal environment of the interior space. For information on this important subject, refer to publications of the American Society of Heating, Refrigerating, and Air-Conditioning Engineers.

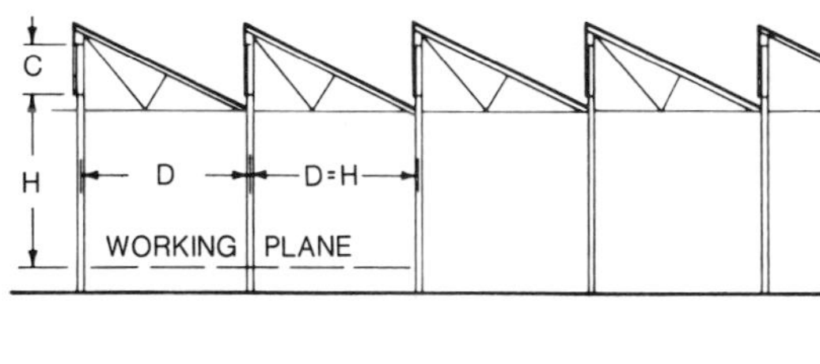

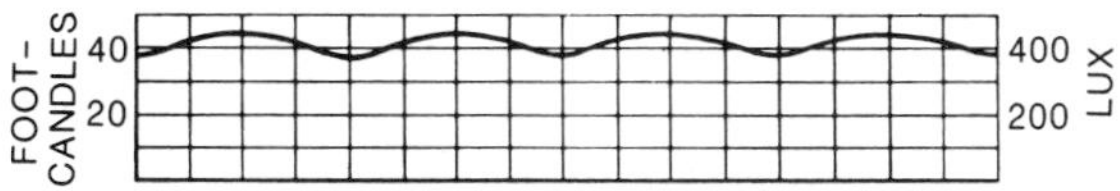

Fig. 7-32. Illuminance distribution from sawtooth windows assuming no reflection, clean glass, and 3430 candelas per square meter (343 candelas per square foot) uniform sky luminance; spacing equal to mounting height. $D = H = 6.1$ meters (20 feet). $C = 1.65$ meters (5 feet-5 inches).

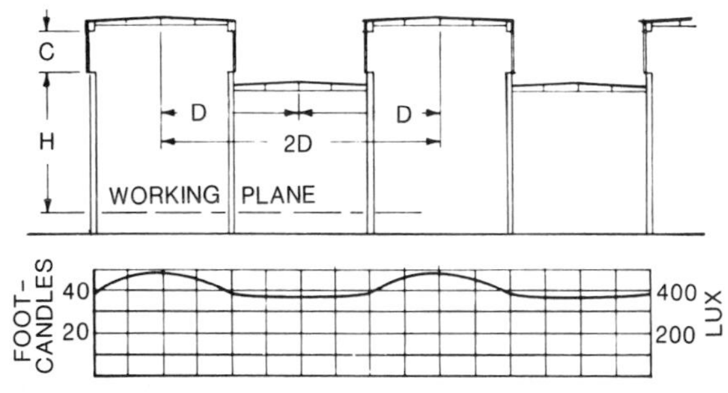

Fig. 7-33. Illuminance distribution from monitor windows assuming no reflection, clean glass, and 3430 candelas per square meter (343 candelas per square foot) uniform sky luminance; spacing equal to twice mounting height. $D = H =$ 6.1 meters (20 feet). $C = 1.65$ meters (5 feet-5 inches).

deep window reveals are necessary, they should be splayed.

Offices and Education Facilities. Detailed discussion of quantity and quality of illumination for the wide variety of visual tasks encountered in these occupancies will be found in the Application Volume. While the treatment for the various occupancies is from the viewpoint of electric lighting, the basic requirements of quantity, luminance ratios, and the reflectances of the principal architectural and work surfaces are the same whether the lighting be daylight, electric, or a combination of the two. Consideration of direct, reflected glare and veiling reflections are of equal importance whether the source be a window or a luminaire.

In offices and educational facilities, critical seeing is done over protracted periods of time from fixed positions involving frequent viewing of the fenestration. Direct and reflected glare from these sources should be minimized by brightness controls for all fenestration areas. In classrooms, to meet current educational practices, it is necessary to provide comfortable fenestration luminance for any orientation of the occupants throughout 360 degrees. Unfortunately, an accepted criterion for luminance limitation for vertical windows in the direct glare zone does not exist. However, the limitations of the luminance of luminaires established for classrooms and offices might well be used as a guide in the design of toplighting elements for buildings with these occupancies.

When light-directing panels are used, they should begin approximately 1.8 meters (six feet) above the floor. In conjunction with the light-directing panel, a vision strip should be incorporated below it. Provision should be made for the brightness control of this vision area, using draperies or blinds or other shading devices. Another design provides an eye-level window area of low transmission glass for outward vision and above an area of clear, high transmission glass to admit ground light. The sky brightness through the upper area may be shielded by louvers or overhangs.

Industrial. Detailed discussion of quantity and quality of illumination for the wide variety of visual tasks encountered in industry will be found in the Application Volume. A variety of fenestration treatments suited to industrial buildings and brightness control devices adapted to the problems encountered are described above. Figs. 7-30 through 7-33 show examples of illuminance distribution from typical industrial building sections.

MAINTENANCE

If the planned results of the daylighting system are to be permanently achieved, maintenance is required. This will involve periodic scheduled cleaning to meet the demands and criteria of the area enclosed. See Fig. 7-34 for typical light loss factors.

Fig. 7-34. Typical Light Loss Factors for Daylighting Design

Locations	Light Loss Factor Glazing Position		
	Vertical	Sloped	Horizontal
Clean Areas	0.9	0.8	0.7
Industrial Areas	0.8	0.7	0.6
Very Dirty Areas	0.7	0.6	0.5

DAYLIGHTING MEASUREMENTS

Daylighting surveys of illuminance and luminance may be made using established procedures with scale models prior to construction, or surveys may be made in actual buildings after construction.[5]

Survey Procedure. It is recommended that surveys be made by experienced persons during relatively representative daylight conditions. For surveys in a room, all electric lighting should be turned off, adjustments of all daylighting control devices should be noted, and exterior illuminance and luminance readings should be taken at the same time. Illumination values excluding direct sunlight or reflected light from the building wall may be obtained by placing black velvet or similar material in a position to shield the cell or test plate from the unwanted light flux. Vertical illuminance from the sun only may be obtained by subtracting the illuminance from the sky and ground from the total illuminance.[6]

Instruments. Luminance measurements may be taken with a visual comparison or color corrected photoelectric meter. Illuminance readings may be taken with a color and cosine corrected meter or with a luminance meter and test plate of known reflectance. Reflectances may be obtained with the luminance meter and test plate or by comparison with standard color chips.

DAYLIGHTING CALCULATIONS

Computation of Illuminance

Illuminance computation procedures involve: (1) determination of the daylight illuminance (sun, sky and ground light) incident on the fenestration, (2) the light flux actually entering the lighted space and (3) the illuminance produced on the surfaces of interest.

Exterior Illuminances. ***Horizontal Fenestration.*** For various days of the year and times of the day, the direct solar and diffuse sky illuminances on the horizontal plane may be found from Fig. 7-17 for a clear day, a partly cloudy day and an overcast day. Alternatively, if the solar altitude is known (or determined from Fig. 7-4), the horizontal illuminances can be read from Figs. 7-5 through 7-12.

Vertical Fenestration. For vertical fenestration that is facing in one of the four cardinal directions, and for various days of the year and times of the day, the direct solar and diffuse sky illuminances on the vertical plane may be found from Fig. 7-18 for a clear day, Fig. 7-19 for a partly cloudy day and Fig. 7-20 for an overcast day.

The direct solar and diffuse sky illuminances on vertical fenestration of any orientation may be obtained as follows: (1) from Fig. 7-4 determine the solar altitude and solar azimuth for the date and time, (2) determine the azimuth of the plane normal to the window, (3) add or subtract the two azimuths as appropriate to obtain the azimuthal angle of the plane normal to the window with respect to the solar meridian, and (4) read the vertical illuminances from Figs. 7-5 through 7-12.

The illuminance on vertical fenestration produced by light from the ground is dependent on ground exitance, which is the product of the exterior horizontal illuminance times the reflectance of the ground. A window receiving illumination from a uniformly bright ground of infinite extent has an illuminance equal, in lux or footcandles, to one-half the ground exitance in lumens per square meter or lumens per square foot, regardless of the height of the window above the ground. For ground of limited area, the ground illuminance on the window may be calculated by the methods outlined on page 9-18 for area sources.

Light Flux Entering a Room. Only a portion of the light flux incident on the fenestration area actually enters the lighted space. The gross fenestration area is reduced by the mechanical supports of the glazing material, such as mullions, window members or mortar joints. This reduction in transmission area can be readily determined.

There is also a transmission loss in the glazing material itself. Data on transmittance of glazing materials are included in Fig. 7-27. There is a further loss attributable to dirt collection on the glazing material (see Fig. 7-34), which should be included in computations as a light loss factor. Finally, there is absorption or reflection by daylight controls, such as shades, louvers, prismatic devices, overhangs, or by the light wells of toplighting (skylight) arrangements.

Daylight is generally abundant, and efficiency of transmission into the room can and must often be compromised with other factors, such as control of sunlight, heat gain or glare, in the overall daylighting design.

Interior Illuminances. The interior illuminances from windows or skylights are generally calculated by one of three methods: (1) rectangular source computation, (2) lumen method for toplighting (for skylights), or (3) lumen method for sidelighting (for windows).

Rectangular Source Computation. The interior illuminance from windows or skylights may be calculated by methods pertaining to rectangular area sources. No allowance is made in the calculations for inter-reflections. The two methods used assume either (1) a uniform luminance distribution or (2) a nonuniform overcast sky distribution.

Uniform Sky. If the sky luminance distribution is not known (or at least is known not to be overcast), a uniform sky is usually the best distribution to assume. This facilitates the use of the equivalent uniform sky exitance values found in Figs. 7-13 through 7-16. These figures do not include direct sunlight, which is assumed to be blocked by daylight controls. Alternatively, another uniform sky technique is more precise when the view of the sky through the fenestration from the calculation point is relatively small. The luminance of the sky seen through the center of the fenestration may be calculated according to the following equations:

(1) for clear skies[3]

$$L_\theta = L_z \frac{(1 - e^{-0.32/\cos\epsilon})(0.91 + 10e^{-3\delta} + 0.45\cos^2\delta)}{0.274(0.91 + 10e^{-3z\theta} + 0.45\cos^2 z\theta)}$$

where

L_θ = Luminance of sky position being considered
L_z = Luminance of the zenith
ϵ = Angular zenith distance of L_θ
δ = Angular distance of the sun from L_θ
$z\theta$ = Angular zenith distance of the sun

(2) for partly cloudy skies[4]

$$L_\theta = L_z \frac{(1 - e^{-0.80\cos\epsilon})(0.526 + 5e^{-3\delta/2})}{0.551\,(0.526 + 5e^{-3z\theta/2})}$$

where

L_θ = Luminance of sky position being considered
L_z = Luminance of the zenith
ϵ = Angular zenith distance of L_θ
δ = Angular distance of the sun from L_θ
$z\theta$ = Angular zenith distance of the sun

To determine the luminance of the zenith in the above equations, it is first necessary to determine the horizontal sky illuminance from Fig. 7-7 for the clear sky or Fig. 7-8 for the partly cloudy sky. Next the zenith luminance factor is determined from Fig. 7-35. The zenith luminance is the product of the zenith luminance factor times the horizontal sky illuminance divided by π.

The luminance of the sky position being considered can be converted to an equivalent exitance by multiplying the luminance by π. The area source methods discussed on page 9-18 can now be used to calculate the interior illuminance of the point. Also, the calculation should include the light reduction factors described under "Light Flux Entering a Room."

Example. A point in a room along the centerline of an east-facing window (in plain view) "sees" the sky through the center of the window at an angle of 40 degrees from zenith. What is the equivalent clear sky exitance at 10:00 a.m. solar time on December 21 if the site is at 34 degrees north latitude?

Method 1. From Fig. 7-15, the equivalent clear sky exitance for an east-facing window is 19 kilolumens per square meter.

Method 2. From Fig. 7-4, the solar altitude is determined to be 26 degrees and the solar azimuth 30 degrees. Thus the azimuthal angle (α) between the sun and the east-facing window is 60 degrees. The horizontal sky illuminance is read from Fig. 7-7 (or Fig. 7-17) and is 11 klx. By using linear interpolation on Fig. 7-35, the zenith luminance factor is calculated to be 0.473. Thus the zenith luminance L_z is $0.473 \times 11/\pi = 1.7$ kcd/m^2. To calculate the luminance of the sky from the clear sky luminance distribution:

$L_z = 1.7$ kcd/m^2

Fig. 7-35. Zenith Luminance Factors

Solar Altitude (degrees)	Solar Angular Zenith Distance (degrees)	Zenith Luminance Factor	
		Clear Sky	Partly Cloudy Sky
90	0	3.248	2.002
80	10	2.086	1.595
70	20	1.398	1.296
60	30	0.985	1.076
50	40	0.734	0.916
40	50	0.581	0.802
30	60	0.491	0.732
20	70	0.446	0.672
10	80	0.435	0.644
0	90	0.452	0.632

$\epsilon = 40$ degrees
$z\theta = 90° -$ solar altitude $= 90 - 26 = 64° = 1.1$ radian
$\delta =$ is determined from the relationship:

$$\cos\delta = \cos z\theta \cos\epsilon + \sin z\theta \sin\epsilon \cos\alpha$$

hence,

$$\delta = \text{arc cos} (\cos 64° \cos 40° + \sin 64° \sin 40° \cos 60°) = 51° = 0.9 \text{ radians}$$

Hence,

$$L_\theta = 1.7 \frac{(1 - e^{-0.32/\cos 40°})(0.91 + 10\, e^{-3(0.9)} + 0.45 \cos^2 51°)}{0.274\,(0.91 + 10e^{-3(1.1)} + 0.45 \cos^2 64°)}$$

Therefore,

$L_\theta = 2.7$ kilocandelas per square meter

Converting to equivalent exitance:

$M_\theta = \pi L_\theta = 8.5$ kilolumens per square meter

Comparison. The entire east sky provided an equivalent exitance of 19 kilolumens per square meter (Method 1), whereas the particular patch of sky "seen" in the center of the window had an equivalent exitance of only 9 kilolux per square meter (Method 2). If the sky seen through the window provides a view of nearly the entire east sky, the equivalent exitance would be close to 19 kilolumens per square meter. If only a small view of the sky could be seen, the equivalent exitance would be close to 9. The illuminance of the interior point can be calculated by the method described for area sources on page 9–18 once the equivalent exitance has been determined.

Nonuniform Overcast Sky. The luminance distribution of overcast skies can be approximated by the simple equation:

$$L_\theta = \frac{3}{7} \frac{E_h}{\pi} (1 + 2 \sin \theta)$$

where
$L_\theta =$ sky luminance at angle θ above the horizon
$E_h =$ illuminance on horizontal plane from unobstructed sky.

This equation can then be applied to the general method for the calculation of the illuminance from a rectangular source and new equations derived. Graphs of the results of this procedure are shown in Figs. 7–36 and 7–37 for the luminance distribution approximating an overcast sky. The results obtained from these curves when multiplied by the glass transmittance (Fig. 7–27), the equivalent sky exitance (Figs. 7–13 through 7–16), and the ratio of glass area to the area of the window give values which closely approximate test results for the conditions of dark walls and ceilings on an overcast day.

Lumen Method for Toplighting

For daylighting systems employing horizontal elements (skylights) at or slightly above roof level, a procedure very similar to the Zonal-Cavity Method for interior electric lighting design may be used to calculate the average horizontal illuminance on the work-plane. The basic formula is:

$$E_t = E_h \times \frac{A_t}{A_w} \times K_u \times K_m$$

where
$E_t =$ Average illuminance on the work-plane from skylights
$E_h =$ Horizontal illuminance on the exterior of the skylighting elements
$A_t =$ Gross area of the skylighting elements
$A_w =$ Area of the work-plane
$K_u =$ Utilization coefficient
$K_m =$ Light loss factor

The calculation procedure may be divided into four steps:

1. Determination of the horizontal illuminance on the exterior of the skylighting elements.
2. Determination of the net transmittances, direct and diffuse, of the skylighting elements.
3. Determination of the utilization coefficient and light loss factor.
4. Determination of:
   a. The average horizontal illuminance on the work-plane (if the number and size of the skylighting elements are specified), or
   b. The number and size of the skylighting elements (if the average horizontal illuminance is specified).

Step 1—Incident Horizontal Illuminance. The exterior horizontal illuminance (total of direct sun plus diffuse sky) incident on the skylights can be determined from Fig. 7–17 for clear, partly cloudy and overcast skies. Alternatively, through the use of Fig. 7–4, which relates latitude, date and time to solar altitude, the exterior horizontal illuminance can be read from Figs. 7–5 and 7–6 for the sun component, and Figs. 7–7 through 7–9 for the sky component.

Step 2—Net Transmittances. There are two transmittances to consider when dealing

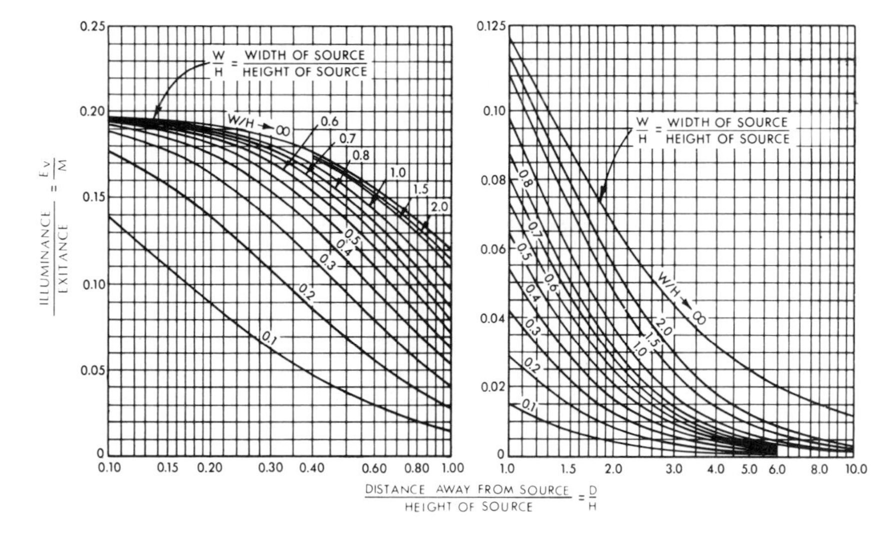

Fig. 7-36. Illuminance on a plane parallel to a rectangular nonuniform source (equivalent to an overcast sky).[7]

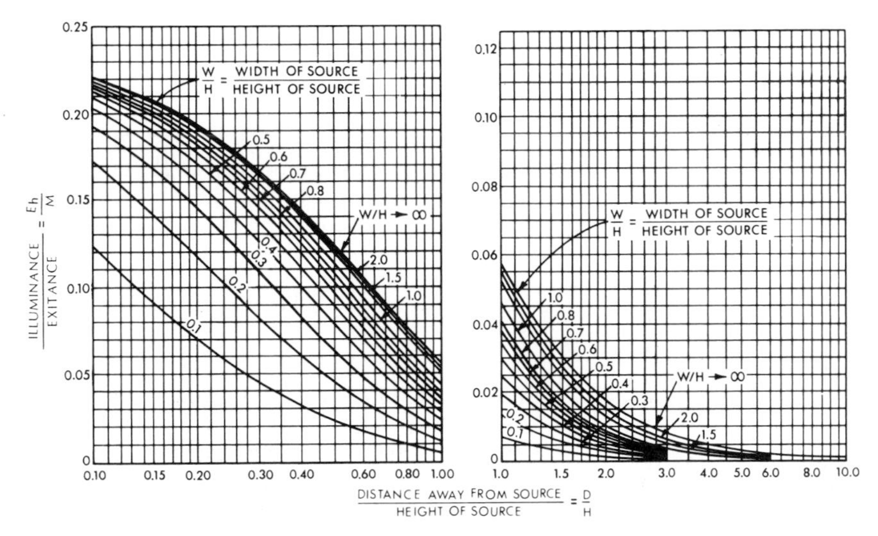

Fig. 7-37. Illuminance on a plane perpendicular to a rectangular nonuniform source (equivalent to an overcast sky).[7]

with skylights.[8] One is the direct transmittance (T_D), which is used whenever direct sunlight impinges on the skylight and which varies with the angle of incidence of the incoming solar radiation, decreasing rapidly for flat sheets at high values of this angle. The second is the diffuse transmittance (T_d), which is used with the sky illuminance and which does not vary with angle of incidence. Generally manufacturers provide transmittance data for flat sheets of their glass or plastic in the form of a single value of T_d and a curve showing the variation of T_D with angle.

Most skylights are domed, and this affects transmittance in three ways. First the process of doming significantly decreases sheet thickness at the center of the dome. Second, the angle of incidence of the direct sunlight varies over the dome's surface. Third, the dome, because it extends above the roof line, has greater light-gathering surface area than a flat sheet. The first of these factors can be included by modifying the flat sheet transmittance to:

$$T_{DM} = 1.25T_{FS}(1.18 - .416T_{FS})$$

where T_{DM} is dome transmittance and T_{FS} is flat sheet transmittance.

This equation does not change the transmittance of a transparent sheet (T_{FS} = 92 per cent) but increases the transmittance of a translucent sheet (T_{FS} = 44 per cent) by about 25 per cent, in conformity with what actually happens in practice. The second and third factors may be considered together by noting that the effect of doming[9] is to cause T_D to become constant within

10 per cent for all angles of incidence less than 70 degrees (sun angles greater than 20 degrees). Thus for most dome applications, a single number for T_D equal to its value at 0 degree angle of incidence can be used.

Because of energy considerations, most contemporary skylights are double domed, most often a transparent dome over a translucent dome. The over-all transmittance of such a unit may be obtained from[10]:

$$T = \frac{T_1 T_2}{1 - R_1 R_2}$$

where T_1 and T_2 are the transmittances of the individual domes, R_1 is the reflectance from the bottom side of the upper dome and R_2 is the reflectance from the top side of the lower dome. This formula takes into account the interflections between the two domes.

It remains to include the effect of any light well present between the dome and the ceiling plane of the room. The interflections within such a well will reduce the over-all transmittance of the skylighting system. First a well index (WI) is calculated from:

$$\text{WI} = \frac{h(w + l)}{2wl}$$

where h, w and l are well height, width and length, respectively.

Second, well efficiency (N_w) is obtained from Fig. 7-38, assuming well wall reflectance is known. Third, because a skylight has a non-light-transmitting frame to hold the dome in place, it is necessary to include the ratio of net to gross skylight area (R_a). Last, if any diffusers, lenses, louvers or other controls are present, their transmittances (T_c) must be included. Now the net transmittance (T_n) of the skylight-well system may be found from

$$T_n = T \times N_w \times R_a \times T_c$$

where T is either the diffuse or direct transmittance of the dome unit.

Step 3—Utilization Coefficient and Light Loss Factor. Room coefficients of utilization (RCU), indicating the fraction of the luminous flux entering the room that reaches the work-plane, are given in Fig. 7-39. These are based on a spacing to mounting height ratio of 1.5, a Lambertian distribution from the skylight, and a 20 per cent floor reflectance. For other values of floor reflectance, multiplying factors may be obtained from Fig. 9-40. The room cavity ratios (RCR) in Fig. 7-39 are obtained from:

$$\text{RCR} = \frac{5h_c \text{ (Room Length} \times \text{Room Width)}}{\text{(Room Length)} \times \text{(Room Width)}}$$

where h_c = room height − work-plane height

The utilization coefficient (K_u) is the fraction of the luminous flux incident on the skylights that reaches the work-plane. This is obtained from

$$K_u = \text{RCU} \times T_n$$

The light loss factor K_m is the product of room surface dirt depreciation (RSDD) and skylight dirt depreciation (SDD). It may be obtained from Fig. 7-34, or, in the absence of information about the dirtiness to be expected, a value of 0.7 may be assumed.

Step 4—Illuminance or Number and Size of Skylights. Let the number of skylights be N and the gross area of each skylight be A. Then the gross skylight area is $A_t = N \times A$.

On an overcast day, the average horizontal illuminance for a particular date, time and latitude is

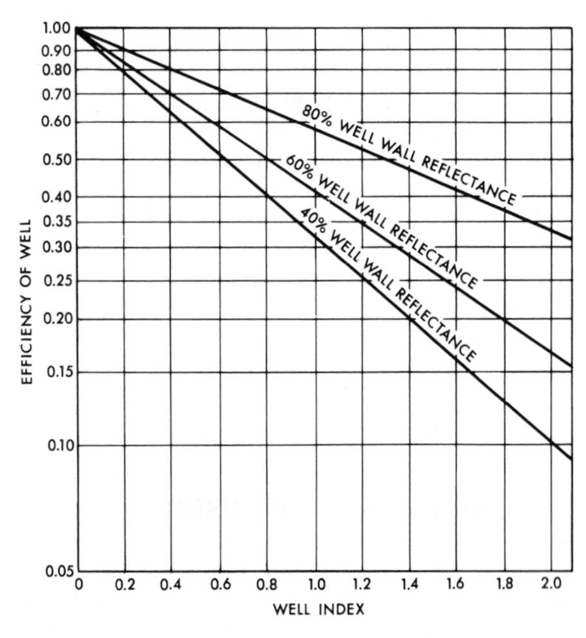

Fig. 7-38. Efficiency factors for various depths of light wells, based on well interflectance values where:

$$\text{Well Index} = \frac{\text{Well Height} \times \text{(Well Width + Well Length)}}{2 \times \text{Well Length} \times \text{Well Width}}$$

$$E_{to} = E_{ho} \times \frac{N \times A}{A_w} \times K_{ud} \times K_m$$

where the "o" denotes overcast and the "d" denotes that the diffuse transmittance is used in computing the utilization coefficient.

For a day with clear sky and direct sun, the illuminance equation becomes

$$E_{tcs} = E_{hc} \times \frac{N \times A}{A_w} \times K_{ud} \times K_m + E_{hs} \times \frac{N \times A}{A_w} \times K_{uD} \times K_m$$

where "cs" denotes clear sky plus direct sun, "c" denotes clear sky, "s" denotes direct sun, and "d" and "D" denote that the diffuse transmittance was used in computing the utilization coefficient for the clear sky contribution to illuminance and the direct transmittance was used in computing the utilization coefficient for the contribution of the direct sunlight.

For a partly cloudy sky and direct sun, use the same equation as for the clear sky and direct sun, except substitute "p" for "c". The partly cloudy daylight availability data for both the sky and sun components are to be used.

Example. Consider an office area 12 meters by 9 meters (40 feet by 30 feet) with a 3-meter (10-foot) ceiling height. Assume reflectances of 80 per cent ceiling, 50 per cent walls and 20 per cent floor.

The room is to be lighted by six 0.9-meter by 0.9-meter (3-foot by 3-foot) double-domed, transparent over translucent, skylights mounted on 230-millimeter (9-inch) high light wells whose wall reflectance is 60 per cent. Each skylight has a 40-millimeter (1.5-inch) wide retaining frame to hold the dome in place.

The transmittances of the flat sheets of plastic from which the dome is made are:

	T_d	T_D
transparent	.79	.92
translucent	.46	.45

The light loss factors are RSDD = .95 and SDD = .75.

Find the average horizontal illuminance on the work-plane at 10 a.m. on March or September 21 at 42 degrees north latitude for clear sky conditions.

Solution. *Step 1.* The horizontal illuminances on the exterior of the skylights may be found from Fig. 7-17.

Fig. 7-39. Room Coefficients of Utilization for Skylighting[11]
(Based on 20 per cent floor reflectance)

Ceiling Reflectance (per cent)	RCR	Wall Reflectance		
		50 Per Cent	30 Per Cent	10 Per Cent
80	0	1.19	1.19	1.19
	1	1.05	1.00	0.97
	2	0.93	0.86	0.81
	3	0.83	0.76	0.70
	4	0.75	0.67	0.60
	5	0.67	0.59	0.53
	6	0.62	0.53	0.47
	7	0.57	0.49	0.43
	8	0.54	0.47	0.41
	9	0.53	0.46	0.41
	10	0.52	0.45	0.40
50	0	1.11	1.11	1.11
	1	0.98	0.95	0.92
	2	0.87	0.83	0.78
	3	0.79	0.73	0.68
	4	0.71	0.64	0.59
	5	0.64	0.57	0.52
	6	0.59	0.52	0.47
	7	0.55	0.48	0.43
	8	0.52	0.46	0.41
	9	0.51	0.45	0.40
	10	0.50	0.44	0.40
20	0	1.04	1.04	1.04
	1	0.92	0.90	0.88
	2	0.83	0.79	0.76
	3	0.75	0.70	0.66
	4	0.68	0.62	0.58
	5	0.61	0.56	0.51
	6	0.57	0.51	0.46
	7	0.53	0.47	0.43
	8	0.51	0.45	0.41
	9	0.50	0.44	0.40
	10	1.49	0.44	0.40

$$E_{hc} = 13 \text{ klx} = 1200 \text{ fc}$$
$$E_{hs} = 59 \text{ klx} = 5500 \text{ fc}$$

Step 2. Next the net transmittance of the skylight is found. When the flat sheets are domed, their transmittances become:

	T_d	T_D
transparent	.84	.92
translucent	.57	.56

Combining them into a double dome yields $T_d = 0.51$, $T_D = 0.53$. Each well is 0.9 meters by 0.9 meters by 0.23 meters (3 feet by 3 feet by 0.75 foot) giving a well index of 0.25 and, from Fig. 7-38, a well efficiency of 0.80. With a 38 millimeter (1.5 inch) frame the ratio of net to gross skylight area is 0.84. Thus the net transmittances are:

$$T_d = .51 \times .80 \times .84 = .34$$

$$T_D = .53 \times .80 \times .84 = .36$$

Step 3. The room cavity ratio is calculated as 2.2. Entering Fig. 7–39 yields an RCU of 0.91. Then:

$$K_{ud} = 0.91 \times .34 = .31$$

$$K_{uD} = 0.91 \times .36 = .33$$

The light loss factor is:

$$K_m = .95 \times .75 = .71$$

Step 4. The average horizontal illuminance on the work-plane is:

$$E_{tcs} = 13000 \times \frac{6 \times .81}{108} \times .31 \times .71$$

$$+ 59000 \times \frac{6 \times .81}{108} \times .33 \times .71$$

$$= 750 \text{ lux}$$

$$= 70 \text{ footcandles}$$

Lumen Method for Sidelighting[1,12]

The calculation process used here for determining interior daylighting is based on a process of interpolation of test data from actual testing of an experimental building and, therefore, is limited to the general type of room, fenestration, openings and controls tested (*e.g.*, the top of the window is at the ceiling level and the sill is at desk height or 0.9 meter (36 inches). However, with the use of a little common sense, these parameters can be expanded to fit a wide variety of building designs.

Interior illuminance is determined for each of several conditions and then added together for final results. For instance, illuminances from windows and from skylights can be determined separately and then added together for final results. Illuminances from the sky and ground are determined separately and added for final results.

Standard Room Task Illuminance Points. The predicting of general daylighting has been simplified by referencing three points in a room—the "maximum," "middle," and "minimum" work-plane illuminance points, located on a line drawn perpendicular to the fenestration across the center of the room. The maximum point is 1.5 meters (5 feet) in from the fenestration, the middle point is in the center of the room, and the minimum point is 1.5 meters (5 feet) in from the wall opposite the fenestration. The height of the work-plane for all three points is 0.76 meters (30 inches) above the floor.

Calculation Steps. After the illuminance incident upon the vertical face of the fenestration has been determined, the net area of the window opening is calculated. The answer provides a factor for determining how much of the incident light will be allowed to pass through the opening. This is multiplied by a light loss factor accounting for dirt accumulation, and a transmission factor accounting for the characteristics of the glazing itself. The transmission of light may also be affected by draperies, shades or louvered screens and must be accounted for by another modifying transmission factor. Finally, the incoming daylight is reflected or absorbed from the room surfaces depending upon size and shape of the room, and reflectance of room surfaces.

$$E_p = E_i \times A_w \times \tau \times 10^* \; K_u$$

where

E_p = Work-plane illuminance at point P, in lux.
E_i = Illuminance from sky or ground incident on vertical windows, in lux.
A_w = Gross area of fenestration, in square meters.
τ = Net transmittance.
K_u = Utilization coefficient, which includes the effect of fenestration design, daylight controls, interior reflectances and room proportions ($C \times K$).

Step 1—Incident Vertical Illuminance. The exterior vertical illuminance incident on the window from the sky can be determined from Figs. 7–18 through 7–20 for a window facing one of the four cardinal directions. A more general method is to use Fig. 7–4 to find the solar altitude and solar azimuth, and by knowing the orientation of the window, the azimuthal angle between the sun and the normal to the window can be calculated. Figs. 7–7 through 7–9 can then be used to find the vertical illuminance on the window. Note that Figs. 7–5 and 7–6 are not used since the lumen method for sidelighting assumes that direct sunlight does not enter the room. Figs. 7–10 through 7–12 can be used to determine the horizontal illuminance at the window from the same half of the sky dome that produced the vertical illuminance in Figs. 7–7 through 7–9.

The exterior vertical illuminance on the window from the ground is calculated by finding the horizontal illuminance on the ground (Fig. 7–17)

* Omit when calculating illuminances in footcandles and using gross area in square feet.

from the sun and sky, and then multiplying the exterior horizontal illuminance by one-half of the ground reflectance (Fig. 7-28). Only one-half of the ground reflectance is used since the configuration factor between the window and the ground is 0.5.

Step 2—Gross Window Area. The entire area of the wall along the *length* of the room from sill height (0.9 meter or 36 inches above the floor) to ceiling height is assumed to be the window. It is this area which is used in the calculations.

Step 3—Net Transmittance. The light flux through the window is reduced by a number of factors which must be accounted for in the calculations. The net transmittance is the product of these factors: (1) the transmittance of the glazing (Fig. 7-27), (2) a light loss factor accounting for dirt accumulation (Fig. 7-34), (3) the net/gross window area factor accounting for the net glazing area of the gross window area, and (4) any other factors such as shades, drapes, etc. which reduce the transmittance of the window.

Step 4—Coefficient of Utilization (CU). For each source of light (overcast sky, clear sky, uniform sky or ground), there are at least two coefficients of utilization to be used: a "C" factor, for room length and width, and a "K" factor, for ceiling height and width (see Figs. 7-40 and 7-41).

(Note: If a diffuse window shade or drape is used, use the "uniform sky" coefficient table instead of overcast or clear sky; add the illuminance contributions on the windows from sky, sun and ground, and divide by 2.)

If venetian blinds are used, different "C" and "K" tables are used (see Fig. 7-41). In addition, a separate louver angle factor, for sky (Vs) and ground (Vg) light, is applied to the process. Rooms with overhang sun controls, when direct sunlight reaches the ground under the overhang, receive still another factor (Xf) (see Fig. 7-42).

There must be two separate calculations for each point in the room—for sky-source illuminance, and for ground-source illuminance, *i.e.* six separate calculations to determine the illuminance for the three points in the room. (If toplighting is used in the room, an additional separate calculation must be made and the results added to the above results.)

The coefficient of utilization factors were determined from simplified sky luminance distribution which were designated "overcast, clear or uniform." However, these descriptions are not always accurate. Here is a technique that can be used to determine the best category: (1) determine the vertical and horizontal half-sky illuminances from step 1, (2) compute the ratio of vertical to horizontal illuminance, and choose the appropriate coefficient of utilization tables based on Fig. 7-43.

Example: Which coefficient of utilization table should be used for a transparent window on a clear day when the sun is at zenith?

Fig. 7-7 gives a vertical illuminance of 7 kilolux and Fig. 7-10 gives a horizontal illuminance of 8 kilolux. The ratio of vertical to horizontal illuminance is 0.9. A comparison with Fig. 7-43 shows that use of either the overcast or uniform sky coefficient tables would give better result than use of the clear sky tables for this case.

Reflected Ground Light. Although the condition of the ground varies from site to site and, thus, is a part of the designed environment, in these calculations light reflected from the ground is considered as another light source. The light from the ground and from the sky must be separately determined. This is particularly critical where the sun shines directly upon a bright surface such as concrete or snow.

The daylight reflected onto the vertical plane from an infinite ground plane (12 meters (40 feet) beyond a one-story building, reflected ground light is of no significance) can be calculated by multiplying the illuminance on the ground from the sun and sky by the reflectance of the ground surface, divided by two.

If the ground plane involves more than one distinct area, such as grass and a concrete sidewalk or a gravel driveway, the reflected ground contribution to the illuminance incident upon the vertical plane of the fenestration can be estimated or approximated.

Rooms With Overhangs. To calculate the work-plane illuminance in a room with an overhang, use the "equivalent room" concept. The dimensions of the equivalent room are determined as follows:

Width. Add the width of the overhangs to the width of the room, as though the window wall were located at the outer edge of the overhang.

Length. The length (parallel to the window wall) of the equivalent room is determined by enlarging the actual room proportionally to the newly established equivalent width.

Window area. In rooms with overhangs, the window sill is disregarded in considering ground illumination and the window height is assumed to extend from the floor to the ceiling. (Actually, the typical sill height, being at or below the-

Fig. 7-40. Coefficients of Utilization: "C" and "K" Factors[1]
(Ceiling reflectance = 80 per cent; Floor reflectance = 30 per cent)

A. Illuminance from an overcast sky, without window controls

			C									K							
Room Length			6.1 M (20 FT)		9.1 M (30 FT)		12.2 M (40 FT)		Ceiling Height			2.4 M (8 FT)		3 M (10 FT)		3.7 M (12 FT)		4.3 M (14 FT)	
Wall Reflectance (per cent)			70	30	70	30	70	30	Wall Reflectance (per cent)			70	30	70	30	70	30	70	30
	Room Width M	FT								Room Width M	FT								
	6.1	20	.0276	.0251	.0191	.0173	.0143	.0137		6.1	20	.125	.129	.121	.123	.111	.111	.0991	.0973
MAX	9.1	30	.0272	.0248	.0188	.0172	.0137	.0131	MAX	9.1	30	.122	.131	.122	.121	.111	.111	.0945	.0973
	12.2	40	.0269	.0246	.0182	.0171	.0133	.0130		12.2	40	.145	.133	.131	.126	.111	.111	.0973	.0982
	6.1	20	.0159	.0117	.0101	.0087	.0081	.0071		6.1	20	.0908	.0982	.107	.115	.111	.111	.105	.122
MID	9.1	30	.0058	.0050	.0054	.0040	.0034	.0033	MID	9.1	30	.156	.102	.0939	.113	.111	.111	.121	.134
	12.2	40	.0039	.0027	.0030	.0023	.0022	.0019		12.2	40	.106	.0948	.123	.107	.111	.111	.135	.127
	6.1	20	.0087	.0053	.0063	.0043	.0050	.0037		6.1	20	.0908	.102	.0951	.114	.111	.111	.118	.134
MIN	9.1	30	.0032	.0019	.0029	.0017	.0020	.0014	MIN	9.1	30	.0924	.119	.101	.114	.111	.111	.125	.126
	12.2	40	.0019	.0009	.0016	.0009	.0012	.0008		12.2	40	.111	.0926	.125	.109	.111	.111	.133	.130

B. Illuminance from a clear sky without window controls

			C									K							
	Room Width M	FT								Room Width M	FT								
	6.1	20	.0206	.0173	.0143	.0123	.0110	.0098		6.1	20	.145	.155	.129	.132	.111	.111	.101	.0982
MAX	9.1	30	.0203	.0173	.0137	.0120	.0098	.0092	MAX	9.1	30	.141	.149	.125	.130	.111	.111	.0954	.101
	12.2	40	.0200	.0168	.0131	.0119	.0096	.0091		12.2	40	.157	.157	.135	.134	.111	.111	.0964	.0991
	6.1	20	.0153	.0104	.0100	.0079	.0083	.0067		6.1	20	.110	.128	.116	.126	.111	.111	.103	.108
MID	9.1	30	.0082	.0054	.0062	.0043	.0046	.0037	MID	9.1	30	.106	.125	.110	.129	.111	.111	.112	.120
	12.2	40	.0052	.0032	.0040	.0028	.0029	.0023		12.2	40	.117	.118	.122	.118	.111	.111	.123	.122
	6.1	20	.0106	.0060	.0079	.0049	.0067	.0043		6.1	20	.105	.129	.112	.130	.111	.111	.111	.116
MIN	9.1	30	.0054	.0028	.0047	.0023	.0032	.0021	MIN	9.1	30	.0994	.144	.107	.126	.111	.111	.107	.124
	12.2	40	.0031	.0014	.0027	.0013	.0021	.0012		12.2	40	.119	.116	.130	.118	.111	.111	.120	.118

C. Illuminance from a uniform ground without window controls

			C									K							
	Room Width M	FT								Room Width M	FT								
	6.1	20	.0147	.0112	.0102	.0088	.0081	.0071		6.1	20	.124	.206	.140	.135	.111	.111	.0909	.0859
MAX	9.1	30	.0141	.0112	.0098	.0088	.0077	.0070	MAX	9.1	30	.182	.188	.140	.143	.111	.111	.0918	.0878
	12.2	40	.0137	.0112	.0093	.0086	.0072	.0069		12.2	40	.124	.182	.140	.142	.111	.111	.0936	.0879
	6.1	20	.0128	.0090	.0094	.0071	.0073	.0060		6.1	20	.123	.145	.122	.129	.111	.111	.100	.0945
MID	9.1	30	.0083	.0057	.0062	.0048	.0050	.0041	MID	9.1	30	.0966	.104	.107	.112	.111	.111	.110	.105
	12.2	40	.0055	.0037	.0044	.0033	.0042	.0026		12.2	40	.0790	.0786	.0999	.106	.111	.111	.118	.118
	6.1	20	.0106	.0071	.0082	.0054	.0067	.0044		6.1	20	.0994	.108	.110	.114	.111	.111	.107	.104
MIN	9.1	30	.0051	.0026	.0041	.0023	.0033	.0021	MIN	9.1	30	.0816	.0822	.0984	.105	.111	.111	.121	.116
	12.2	40	.0029	.0018	.0026	.0012	.0022	.0011		12.2	40	.0700	.0656	.0946	.0986	.111	.111	.125	.132

D. Illuminance from the "uniform sky" without diffuse window shades

			C									K							
	Room Width M	FT								Room Width M	FT								
	6.1	20	.0247	.0217	.0174	.0152	.0128	.0120		6.1	20	.145	.154	.123	.128	.111	.111	.0991	.0964
MAX	9.1	30	.0241	.0214	.0166	.0151	.0120	.0116	MAX	9.1	30	.141	.151	.126	.128	.111	.111	.0945	.0964
	12.2	40	.0237	.0212	.0161	.0150	.0118	.0113		12.2	40	.159	.157	.137	.127	.111	.111	.0973	.0964
	6.1	20	.0169	.0122	.0110	.0092	.0089	.0077		6.1	20	.101	.116	.115	.125	.111	.111	.101	.110
MID	9.1	30	.0078	.0060	.0067	.0048	.0044	.0041	MID	9.1	30	.0952	.113	.105	.122	.111	.111	.110	.122
	12.2	40	.0053	.0033	.0039	.0028	.0029	.0024		12.2	40	.111	.105	.124	.107	.111	.111	.130	.124
	6.1	20	.0108	.0066	.0080	.0052	.0063	.0047		6.1	20	.0974	.111	.107	.121	.111	.111	.112	.119
MIN	9.1	30	.0047	.0026	.0042	.0023	.0029	.0020	MIN	9.1	30	.0956	.125	.103	.117	.111	.111	.115	.125
	12.2	40	.0027	.0013	.0022	.0012	.0018	.0011		12.2	40	.111	.105	.125	.111	.111	.111	.133	.124

Fig. 7–40. *Continued*

E. Illuminance from the ground, with window controls

C

Room Length			6.1 M (20 FT)		9.1 M (30 FT)		12.2 M (40 FT)	
Wall Reflectance (per cent)			70	30	70	30	70	30
	Room Width							
	M	FT						
	6.1	20	.0147	.0112	.0102	.0088	.0081	.0071
MAX	9.1	30	.0141	.0112	.0098	.0088	.0077	.0070
	12.2	40	.0137	.0112	.0093	.0086	.0072	.0069
	6.1	20	.0128	.0090	.0094	.0071	.0073	.0060
MID	9.1	30	.0083	.0057	.0062	.0048	.0050	.0041
	12.2	40	.0055	.0037	.0044	.0033	.0042	.0026
	6.1	20	.0106	.0071	.0082	.0054	.0067	.0044
MIN	9.1	30	.0051	.0026	.0041	.0023	.0033	.0021
	12.2	40	.0029	.0018	.0026	.0012	.0022	.0011

K

Ceiling Height			2.4 M (8 FT)		3 M (10 FT)		3.7 M (12 FT)		4.3 M (14 FT)	
Wall Reflectance (per cent)			70	30	70	30	70	30	70	30
	Room Width									
	M	FT								
	6.1	20	.124	.206	.140	.135	.111	.111	.0909	.0859
MAX	9.1	30	.182	.188	.140	.143	.111	.111	.0918	.0878
	12.2	40	.124	.182	.140	.142	.111	.111	.0936	.0879
	6.1	20	.123	.145	.122	.129	.111	.111	.100	.0945
MID	9.1	30	.0966	.104	.107	.112	.111	.111	.110	.105
	12.2	40	.0790	.0786	.0999	.106	.111	.111	.118	.118
	6.1	20	.0994	.108	.110	.114	.111	.111	.107	.104
MIN	9.1	30	.0816	.0822	.0984	.105	.111	.111	.121	.116
	12.2	40	.0700	.0656	.0946	.0986	.111	.111	.125	.132

work-plane, has little effect on the distribution of daylight within such rooms.) For illumination from the sky, the window height is from the sill up. In computing the gross transmission area of the windows, the length of the window is taken as the length of the equivalent room.

The coefficient of utilization for the equivalent room is obtained from Figs. 7–40 and 7–41, and the work-plane illuminance calculated as though the windows were at the outer edge of the overhang. Draw a graph of the work-plane illuminance (at three points) on a section through the equivalent room. To equate these illuminance figures to the actual room, pick the points on the illuminance curve which correspond to the three work-plane points in the actual room.

If direct sunlight reaches the ground under an overhang, an additional factor is applied to account for the light reflected into the room (see Fig. 7–42). First, determine the width of the sunlighted area [width of the overhang (A), less the width of the shaded area (B) as shown in Fig. 7–44]. Then determine the distance from each of the work surface prediction points to the edge of the shaded area (C), and add 6 when distances are in meters and 20 when in feet. Then, by dividing the former by the latter a factor X_e is derived.

$$\frac{A - B}{C + 6} = X_e \text{ or } \frac{A - B}{C + 20} = X_e$$

Find, on each of the three graphs in Fig. 7–42, the value of the X_f factor corresponding to the ceiling height and the variable X_e. Multiply the light on the vertical from the ground by these X_f values.

Example: A 6.1-meter wide (20-foot) by 9.1-meter long (30-foot) by 3.3-meter high (11-foot) room has a transparent window along the 9.1 meter length. The room has an 80 per cent ceiling reflectance, 50 per cent wall reflectance, and 30 per cent floor reflectance. The window has a transmittance of 90 per cent, a light loss factor of 70 per cent, and an 80 per cent ratio of net to gross area. The outside ground reflectance is 20 per cent. What is the interior illumination at 10:00 a.m. solar time on December 21 for an overcast day at a site of 34 degrees north latitude?

Step 1. From Fig. 7–20, the vertical illuminance on the window from the sky is 4000 lux. The horizontal illuminance on the ground (Fig. 7–17) is 9000 lux. Thus the vertical illuminance on the window from the ground is 9000 × 0.2 ground reflectance × 0.5 = 900 lux.

Step 2. The window from sill to ceiling is 2.4 meters (3.3 − 0.9 meters) high by 9.1 meters wide. Thus the gross area is 21.8 square meters.

Step 3. The net transmittance of the window is 0.9 glazing transmittance times 0.7 light loss factor times 0.8 net to gross ratio = 0.5.

Step 4. From Fig. 7–40, the "C" and "K" factors interpolated for 50 per cent wall reflectance are:

Overcast sky.

	C	*K*
MAX	0.0182	0.1165
MID	0.0094	0.111
MIN	0.0053	0.1078

Fig. 7–41. Coefficients of Utilization: "C" and "K" factors, and "V" factors for venetian blind angle[1]
(Ceiling reflectance = 80 per cent; Floor reflectance = 30 per cent)

A. Illuminance from the sky, with venetian blinds

C

Wall Reflectance (per cent)	Room Width M	Room Width FT	6.1M (20FT) 70	6.1M (20FT) 30	9.1M (30FT) 70	9.1M (30FT) 30	12.2M (40FT) 70	12.2M (40FT) 30
MAX	6.1	20	.0556	.0556	.0392	.0397	.0298	.0317
MAX	9.1	30	.0522	.0533	.0367	.0389	.0278	.0311
MAX	12.2	40	.0506	.0528	.0359	.0381	.0270	.0306
MID	6.1	20	.0556	.0556	.0418	.0411	.0320	.0364
MID	9.1	30	.0372	.0339	.0278	.0286	.0220	.0256
MID	12.2	40	.0217	.0211	.0192	.0186	.0139	.0164
MIN	6.1	20	.0556	.0556	.0422	.0456	.0320	.0409
MIN	9.1	30	.0294	.0233	.0222	.0203	.0189	.0194
MIN	12.2	40	.0139	.0110	.0133	.0108	.0120	.0100

K

Wall Reflectance (per cent)	Room Width M	Room Width FT	2.4M (8FT) 70	2.4M (8FT) 30	3M (10FT) 70	3M (10FT) 30	3.7M (12FT) 70	3.7M (12FT) 30	4.3M (14FT) 70	4.3M (14FT) 30
MAX			.154	.170	.129	.131	.107	.112	.091	.091
MID	6.1	20	.100	.106	.101	.106	.099	.102	.091	.091
MID	9.1	30	.074	.080	.086	.090	.091	.093	.091	.091
MID	12.2	40	.070	.074	.079	.084	.088	.091	.091	.091
MIN	6.1	20	.080	.080	.091	.091	.093	.093	.091	.091
MIN	9.1	30	.068	.068	.079	.079	.087	.087	.091	.091
MIN	12.2	40	.064	.064	.076	.076	.084	.084	.091	.091

V

Venetian Blind Setting	Wall Reflectance (per cent)	30° 70	30° 30	45° 70	45° 30	60° 70	60° 30
15° SUN ALT.	MAX	.0687	.0554	.0426	.0346	.0218	.0162
	MID	.0488	.0341	.0371	.0218	.0195	.0110
	MIN	.0376	.0228	.0276	.0156	.0142	.0078
30° SUN ALT	MAX	.0630	.050	.0394	.0312	.0208	.0156
	MID	.0462	.0324	.0337	.0216	.0176	.0110
	MIN	.0342	.0204	.0250	.0143	.0130	.0071
45° SUN ALT.	MAX	.0553	.0434	.0345	.0274	.0198	.0141
	MID	.0416	.0301	.0304	.0211	.0158	.0105
	MIN	.0308	.0182	.0225	.0127	.0117	.0064
60° SUN ALT.	MAX	.0464	.0362	.0313	.0236	.0190	.0135
	MID	.0370	.0264	.0270	.0185	.0140	.0092
	MIN	.0274	.0159	.0199	.0111	.0104	.0056

B. Illuminance from the ground, with venetian blinds

C

Wall Reflectance (per cent)	Room Width M	Room Width FT	6.1M (20FT) 70	6.1M (20FT) 30	9.1M (30FT) 70	9.1M (30FT) 30	12.2M (40FT) 70	12.2M (40FT) 30
MAX	6.1	20	.0556	.0556	.0392	.0426	.0303	.0348
MAX	9.1	30	.0528	.0539	.0370	.0433	.0289	.0337
MAX	12.2	40	.0506	.0544	.0359	.0426	.0278	.0344
MID	6.1	20	.0556	.0556	.0414	.0459	.0320	.0381
MID	9.1	30	.0367	.0356	.0274	.0308	.0217	.0270
MID	12.2	40	.0239	.0233	.0192	.0222	.0153	.0181
MIN	6.1	20	.0556	.0556	.0430	.0486	.0328	.0398
MIN	9.1	30	.0261	.0228	.0214	.0211	.0170	.0192
MIN	12.2	40	.0128	.0108	.0119	.0107	.0098	.0097

K

Wall Reflectance (per cent)	Room Width M	Room Width FT	2.4M (8FT) 70	2.4M (8FT) 30	3M (10FT) 70	3M (10FT) 30	3.7M (12FT) 70	3.7M (12FT) 30	4.3M (14FT) 70	4.3M (14FT) 30
MAX			.174	.200	.142	.157	.117	.123	.091	.091
MID	6.1	20	.104	.116	.110	.121	.106	.112	.091	.091
MID	9.1	30	.074	.082	.092	.099	.099	.106	.091	.091
MID	12.2	40	.058	.062	.079	.083	.092	.096	.091	.091
MIN	6.1	20	.078	.082	.093	.097	.099	.102	.091	.091
MIN	9.1	30	.058	.060	.074	.076	.090	.092	.091	.091
MIN	12.2	40	.052	.056	.070	.071	.086	.087	.091	.091

V

Wall Reflectance	30° 70	30° 30	45° 70	45° 30	60° 70	60° 30
MAX	.150	.108	.141	.102	.087	.063
MID	.141	.094	.118	.077	.067	.043
MIN	.124	.072	.096	.056	.049	.028

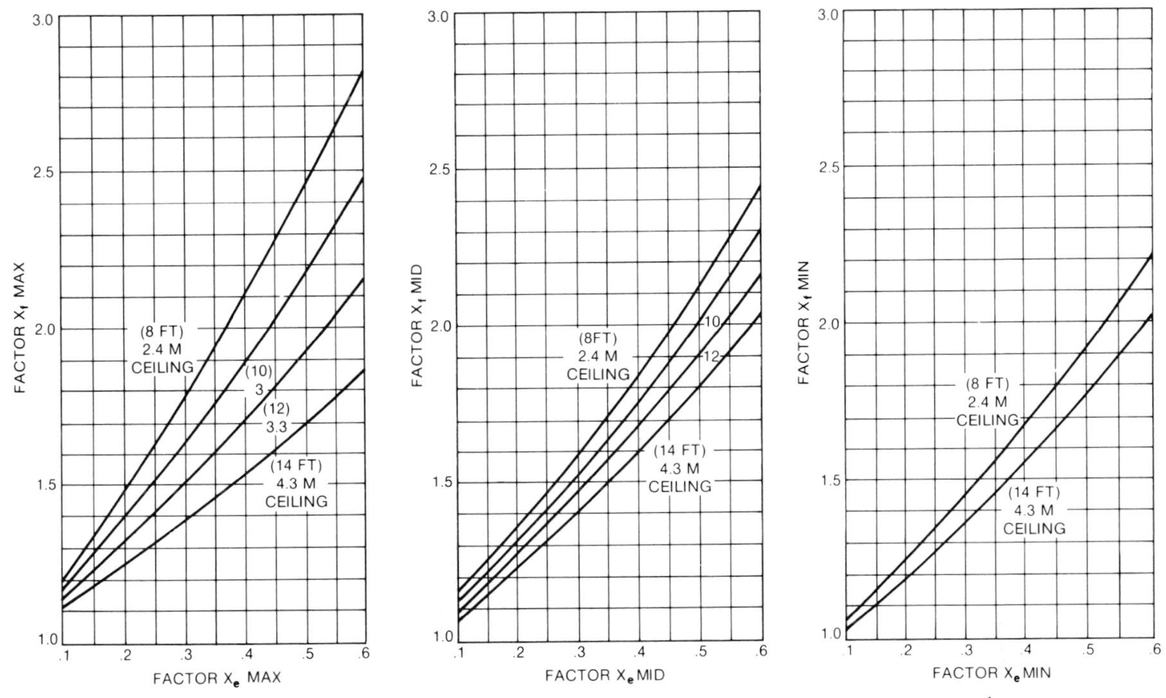

Fig. 7-42. Factors for determining illuminance from sunlighted areas under an overhang.[1]

Ground.

	C	K
MAX	0.0095	0.1243
MID	0.0083	0.1183
MIN	0.0068	0.1115

Interior illuminances from the sky:

MAX $E_n = 4000 \times 21.8 \times 0.5 \times 10 \times 0.0182 \times 0.1165 = 920$ lux
MID $E_n = 4000 \times 21.8 \times 0.5 \times 10 \times 0.0094 \times 0.111 = 455$ lux
MIN $E_n = 4000 \times 21.8 \times 0.5 \times 10 \times 0.0053 \times 0.1078 = 250$ lux

Interior illuminances from the ground:

MAX $E_n = 900 \times 21.8 \times 0.5 \times 10 \times 0.0095 \times 0.1243 = 115$ lux
MID $E_n = 900 \times 21.8 \times 0.5 \times 10 \times 0.0083 \times 0.1183 = 95$ lux
MIN $E_n = 900 \times 21.8 \times 0.5 \times 10 \times 0.0068 \times 0.1115 = 75$ lux

Thus the illuminance from the ground has little effect for this case. The total illuminance at the three points in the room are:

MAX $E_n = 1035$ lux
MID $E_n = 550$ lux
MIN $E_n = 325$ lux

Fig. 7-43. Coefficient of Utilization (CU) Table Selection Guide[13]

CU Table Description	Vertical/Horizontal Illuminance
Clear Sky	1.3
Uniform Sky	1.0
Overcast Sky	0.8

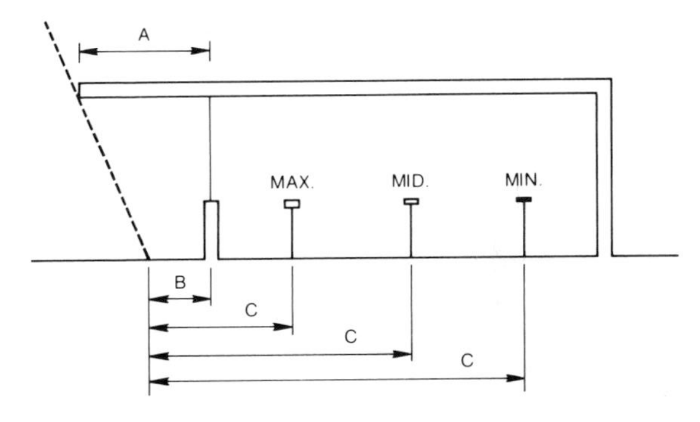

Fig. 7-44. Distances used in determining factor X_e.

Overhangs

The amount of window shaded from a long horizontal overhang mounted above the window (Fig. 7-45) can be calculated from the formula:

$$h = D_n \sec (\psi_z - \psi_n) \tan \theta$$

where

h = height of shadow below overhang
D_n = depth of overhang from the window surface
ψ_z = solar azimuth
ψ_n = angle between a line drawn (outward) normal to the window and true south
θ = solar altitude above the horizon.

The amount of shading from a vertical projection mounted at the side of a window can be obtained from the formula:

$$W = D_v \tan (\psi_z - \psi_n)$$

where

W = width of the shadow from the side projection

D_v = depth of vertical projection.

In planning fixed horizontal and vertical projections for sunlight control, it is advisable to calculate the window shading for several different months and times.

Daylight Factor Method

The Daylight Factor Method, which has been adopted by CIE[6], treats the illuminance from daylight that occurs at a point inside a room as a percentage of the simultaneous illuminance on a horizontal plane from an unobstructed sky outdoors. This ratio is called the Daylight Factor. The Daylight Factor is composed of three different contributions of daylight to the room:

1. Sky Component, or light received at the design point directly from the sky.
2. Externally Reflected Component, or light reflected from surfaces external to the room to the design point.
3. Internally Reflected Component, which is light reflected from interior room surfaces to the design point.

For example, for a 10 kilolux [1000 footcandle] overcast sky (a completely overcast sky producing 10 kilolux [1000 footcandles] on the horizontal plane) a simultaneous interior measurement of 200 lux [20 footcandles] at a specific point is expressed as a Daylight Factor of 2 per cent.

If sky luminance changes, the amount of illuminance measured at a point in the room will also change, but the Daylight Factor will normally remain approximately the same. Different quantities of light will occur at different points in a room depending on its geometry and the placement of the openings; yet the relative pattern will remain the same as sky luminance changes as long as the Overcast Sky distribution is maintained. The pattern of light distribution is calculated once and then can be multiplied by data on daylight availability to obtain actual levels to be expected for the minimum, average, or maximum daylighting conditions in a particular locale. The advantage of this concept is that the pattern of light distribution in a room can be computed once; that *pattern* will not change. The disadvantage is that it only works simply for the totally overcast sky, where the luminance distribution is symmetrical about an axis through the zenith, therefore, independent of orientation. Conditions of partly cloudy skies are so variable in distribution and occurrence as to make it pointless to try and predict exact effects. This process can also be followed for the clear sky since its distribution has been standardized[3], but it is a more complicated process due to the multiple conditions of sky luminance distribution as the sun changes position in the sky, and due to the effects of direct solar illumination in addition to the diffuse luminance of the sky vault. Effects of direct sunlight are not included in the Daylight Factor.

A minimum Daylight Factor is specified as a requirement for building in some countries. The CIE Overcast Sky is then used as the standard

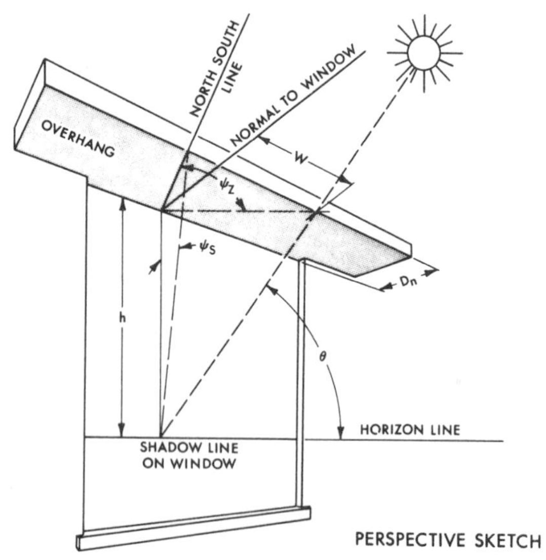

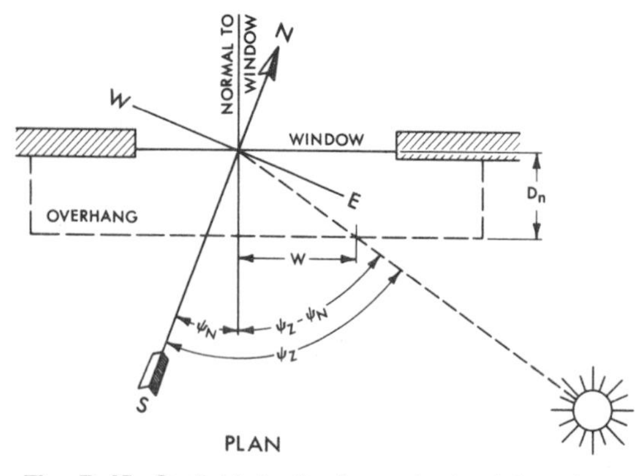

Fig. 7-45. Sunlight shading from a horizontal overhang.

"design sky" since it represents the conditions when minimum light will be available. The argument has been made[6] that it is a generally applicable design sky even for climates where clear skies prevail. Higher illuminance values are expected from clear skies due to higher horizon luminance and reflected sunlight, and since these higher levels are offset by shading devices required in these climates, use of the Overcast Sky as a standard provides a margin of safety.

Many methods of predicting the Daylight Factor in buildings have been developed since World War II at the Building Research Establishment in England, by research groups in other countries, and by commercial concerns. These methods include formulas, tables, charts, graphs and nomograms for quick approximations of minimum or average Daylight Factor for windows or skylights. Graphical projection techniques such as the Waldram Diagram[16] and Pleijel Diagram[17] have also been developed which, although time consuming to use, provide a visual yet precise evaluation of the Sky Component and Externally Reflected Component at a particular design point. A Daylight Factor Slide-Rule Calculator, available from the Building Research Establishment, allows speedy estimation of total Daylight Factor as well as the separate components for rooms with windows in one wall.

Two systems of daylighting prediction are of particular interest. The first is the Daylight Factor Protractor developed at the Building Research Establishment.[18] This calculation method is noteworthy for its application to almost all daylighting designs under the Uniform Sky or Overcast Sky: vertical glazing, sloped glazing at 30 and 60 degrees, horizontal glazing and unglazed openings, and combinations of the above. Use of the Uniform Sky and unglazed openings allows the consideration of other conditions, such as sky luminance from a particular clear sky distribution or special glazing. The suitable protractor is used with plan and section drawings of the room to obtain the Sky Component and externally Reflected Component, the Internally Reflected Component being obtained from tables. Several points can be figured in a room to obtain the distribution pattern, or the point where the minimum will occur can be calculated.

Another graphical method for predicting daylight illumination is explained in Reference 19 which has an accompanying set of overlays. The overlays can be used in several ways. One practical use is as a horizontal sundial which indicates date and time for sunlight studies at a certain latitude by a gnomon casting shadows onto the date and hour lines of the sundial. Sun penetration and shadowing effects can be studied quickly throughout the course of a year. The major use of the overlays is for daylighting analysis of proposed designs with interior elevation drawings. The overlays are designed so that they show the varying paths of the sun throughout the year across a building's facade (and thereby across fenestration) as seen from a chosen design point inside the room. Three types of information can be gathered from these analyses: visual data concerning sun penetration; quantitative data concerning amount of light from the overcast sky which can be expected at certain times of the year; and quantitative data concerning the amount of insolation to be expected at specific times. In addition, a "pepper pot" diagram can be used to predict illuminance from the Overcast Sky giving data for both the Sky Component and Externally Reflected Component of the Daylight Factor.[20] Different points, at the same distance from the window wall in a room, can be considered by moving the overlay horizontally. Points at different distances from the window can be considered by drawing the window elevation different sizes according to the selected scales (¼ scale and ⅛ scale). The effects of shading devices can also be analyzed.

REFERENCES

1. Daylighting Committee of the IES: "Recommended Practice of Daylighting," *Light. Des. Appl.*, Vol. 9, p. 25, February, 1979.
2. *Proposed Standardized C.I.E. Overcast Sky*, Proceedings 13th Session CIE, Vol. 11, Part 3-2, Zurich, 1955.
3. *Standardization of Luminance Distribution on Clear Skies*, CIE Publication No. 22, 1973.
4. IES Calculation Procedures Committee, "Recommended Practice for the Calculation of Daylight Availability," *J. Illum. Eng. Soc.*, Vol. 13, July, 1984.
5. Subcommittee on Measurement and Reporting on Daylighting of Interiors of the Daylighting Committee of the IES: "IES Guide for Measuring and Reporting Daylight Illumination," *Illum. Eng.*, Vol. LIII, p. 213, April, 1958.
6. CIE Committee E-3.2: *Daylight—International Recommendations for the Calculation of Natural Daylight*, CIE Publication No. 16, 1970.
7. Moon, P. and Spencer, D.: "Illumination from a Nonuniform Sky," *Illum. Eng.* Vol. 37, p. 707, December, 1942.
8. Murdoch, J. B.: "A Procedure for Calculating the Potential Savings in Lighting Energy from the Use of Skylights," *J. Illum. Eng. Soc.*, Vol. 6, p. 237, July, 1977.
9. Linforth, E.: "Efficiency of Domed Acrylic Skylights," *Illum. Eng.*, Vol. LIII, p. 544, October, 1958.
10. Pierson, O. L.: *Acrylics for the Architectural Control of Solar Energy*, Rohm and Haas, 1962.
11. Murdoch, J. B. and Nettleton, J. E., *A Systematic Procedure for Calculating the Average Illuminance on a Work-Plane from Skylights Located in a Pitched Roof*, Naval Civil Engineering Laboratory, Report CR 82.022, May, 1982.
12. Griffith, J. W., Arner, W. J. and Conover, E. W.: "A Modified Lumen Method of Daylighting Design," *Illum. Eng.*, Vol. L, p. 103, March, 1955.

13. Pierpoint, W.: "A Simple Sky Model for Daylighting Calculations," *Gen. Proc.*, ed. Vonier, T., 1983 International Daylighting Conference, Phoenix, Arizona, p. 47, February 16–18, 1983.
14. Bryan, H.: "A Simplified Procedure for Calculating the Effects of Daylight from Clear Skies," *J. Illum. Eng. Soc.*, Vol. 9, p. 142, April, 1980.
15. Gillette, G., Pierpoint, W. and Treado, S.: "A General Illuminance Model for Daylight Availability," *J. Illum. Eng. Soc.*, Vol. 13, July, 1984.
16. Waldram, P and Waldram, J.: Window Design and the Measurement and Predetermination of Daylight Illumination," *The Illum. Eng.* (London), Vol XIV, p. 90, April/May, 1923.
17. Pleijel, G.: "The Computation of Natural Radiation in Architecture and Town Planning," *Meddelande*, Bulletin No. 25, *Statens Nämnd För Byggnadsforskuing*, Stockholm, p. 92, 1954.
18. Longmore, J.: *BRS Daylight Protractors*, HMSO, London, 1968.
19. Turner, D. P., ed.: *Windows and Environment*, Pilkington Brothers Ltd., 1969, Architectural Press, 1971.
20. Lynes, J.: *Principles of Natural Lighting*, Amsterdam/New York, Elsevier Publishing Co., 1968.

Light Sources

SECTION 8

This section is devoted almost exclusively to a detailed description of the various light sources now available.* Fundamental information concerning the generation of light and the operating principles of electric light sources is given in Section 2 (see also Reference 1). The sun and sky as light sources are covered in Section 7.

INCANDESCENT FILAMENT LAMPS

The primary consideration of lamp design is that the lamp will produce the spectral radiation desired (light, infrared, ultraviolet) most economically for the application intended, or, in other words, obtain the best balance of over-all lighting cost in terms of the lighting results. Realization of this objective in an incandescent filament lamp requires the specification of the following: filament material, length, diameter, form, coil spacing and mandrel size (the mandrel is the form on which the filament is wound); lead-in wires; number of filament supports; method of mounting filament; vacuum or filling gas; gas pressure; gas composition; and bulb size, shape, glass composition and finish.

The manufacture of good quality lamps requires adherence to these specifications and necessitates careful process control.[2]

The Filament

The efficacy of light production by incandescent filament lamps depends on the temperature of the filament. An iron rod heated in a furnace will first glow a dull red, becoming brighter and whiter as its temperature increases. Iron, however, melts at about 1800 K and, with such a low melting point, is not an efficient source of light. The higher the temperature of the filament, the greater will be the portion of the radiated energy that falls in the visible region; for this reason it is important in the design of a lamp that the filament temperature be kept as high as is consistent with satisfactory life. Considerable research and investigation has been carried out on a variety of materials and metals in search of a suitable filament. The desirable properties of filament materials are high melting point, low vapor pressure, high strength, high ductility, and suitable radiation and electrical resistance characteristics.

Tungsten for Filaments. Tungsten has many desirable properties for use as an incandescent light source. (Early incandescent lamps utilized carbon, osmium and tantalum filaments.) Its low vapor pressure and high melting point (3655 K) permit high operating temperatures and consequently high efficacies. Drawn tungsten wire has high strength and ductility, allowing the drawing of wire of the high degree of uniformity necessary for the exacting specifications of present-day lamps. Tungsten has been alloyed with metals such as rhenium and used in lamp designs which can utilize the characteristics of the alloy. Thoriated tungsten wire is used in filaments for rough service applications.

Radiating Characteristics of Tungsten.[3,4] The emissivity of a blackbody is 1.0 for all wavelengths (see Section 2). Since the emissivity of tungsten is a function of the wavelength, tungsten is a selective radiator. Fig. 8–1 illustrates

* For history and references, see end of this section.

the radiation characteristics of tungsten and a blackbody. Curves A and B show the spectral radiant intensity for an area of one square centimeter of a blackbody and tungsten respectively at 3000 K. Curve B′ is for 2.27 square centimeters of tungsten at 3000 K and has the same amount of radiation in the visible region as the blackbody curve A. This illustrates that for the same amount of visible radiation, tungsten radiates only 76 per cent as much as the total radiation from a blackbody at the same temperature.

Only a small percentage of the total radiation from an incandescent source is in the visible region of the spectrum. As the temperature of a tungsten filament is raised, the radiation at shorter wavelengths increases more rapidly than that at longer wavelengths. Thus the radiation in the visible region (see Fig. 8–2) increases more rapidly than radiation in the infrared region. Therefore, the luminous efficacy increases. The luminous efficacy of an uncoiled tungsten wire at its melting point (3655 K) is approximately 53 lumens per watt. In order to obtain life, it is necessary to operate a filament at a temperature well below the melting point.

Resistance Characteristics of Tungsten. Tungsten has a positive resistance characteristic so that the resistance at operating temperature is much greater than its cold resistance. In general service lamps, the hot resistance is 12 to 16 times the cold resistance. Fig. 8–3 illustrates the change in resistance of the tungsten filament with temperature for various lamps. The low cold resistance of tungsten filaments results in an initial inrush of current which, because of the reactive impedance characteristic of the circuit, does not reach the theoretical value indicated by the ratio of the hot to cold resistance. Fig. 8–82 (page 8–72) gives the effect of the change in resistance on the current in incandescent filament lamps. The inrush current due to incandescent-filament-lamp loads is important in the design and adjustment of circuit breakers, in circuit fusing, and in the design of lighting-circuit switch contacts.

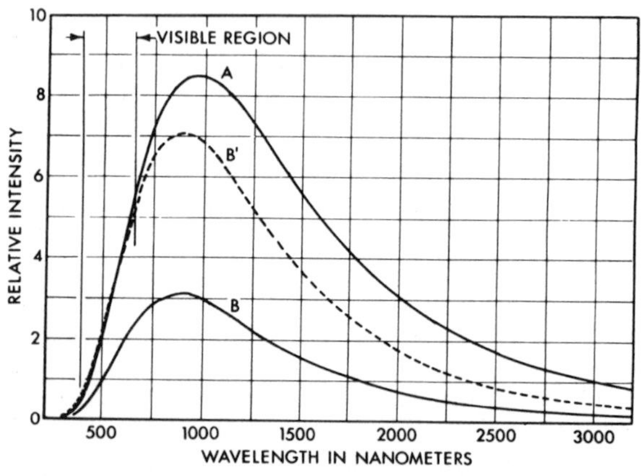

Fig. 8–1. Radiating characteristics of tungsten. Curve A: radiant flux from one square centimeter of a blackbody at 3000 K. Curve B: radiant flux from one square centimeter of tungsten at 3000 K. Curve B′: radiant flux from 2.27 square centimeters of tungsten at 3000 K (equal to curve A in visible region). (The 500-watt 120-volt general service lamp operates at about 3000 K.)

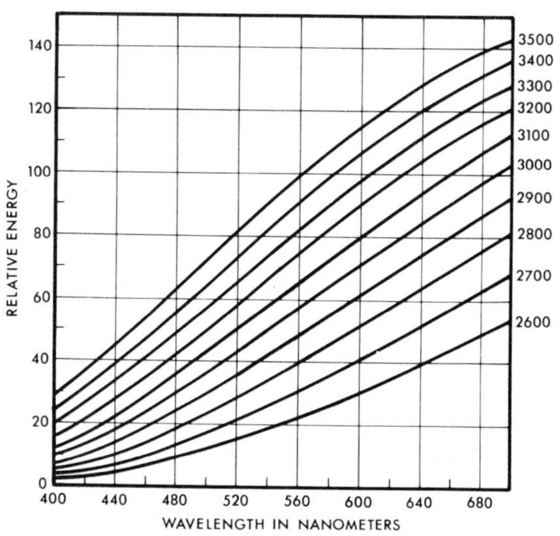

Fig. 8–2. Spectral power distribution in the visible region from tungsten filaments of equal wattage but different temperatures.

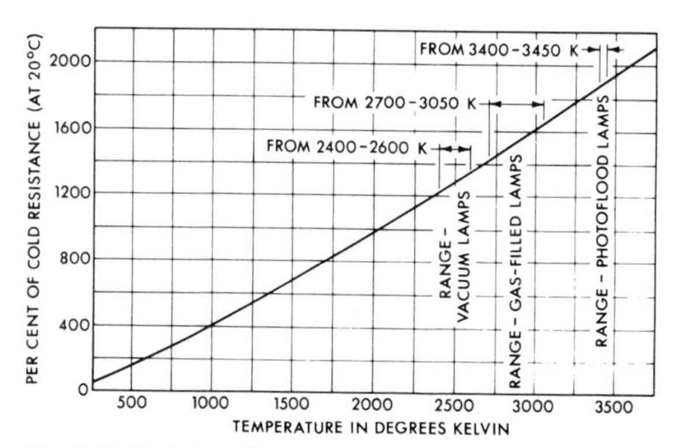

Fig. 8–3. Variation of tungsten filament hot resistance with temperature.

Color Temperature. In many applications, it is useful to know the apparent color temperature of an incandescent lamp. Fig. 8–4 expresses the approximate relationship between color temperature and luminous efficacy for a fairly wide range of wattages of gas filled lamps. Efficacy is frequently published, or it may be calculated from published lumen and wattage data. From this value it is possible to estimate the average color temperature of the filament.

Construction and Assembly. Fig. 8-5 shows the basic parts and steps in the assembly of a typical incandescent filament lamp. In miniature lamps three methods of construction are generally used: flange seal, butt seal and pinch seal (see Fig. 8-6).

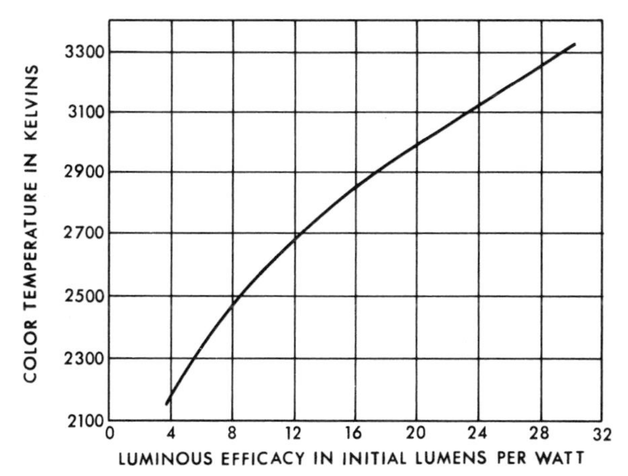

Fig. 8-4. Variation of color temperature with lamp efficacy.

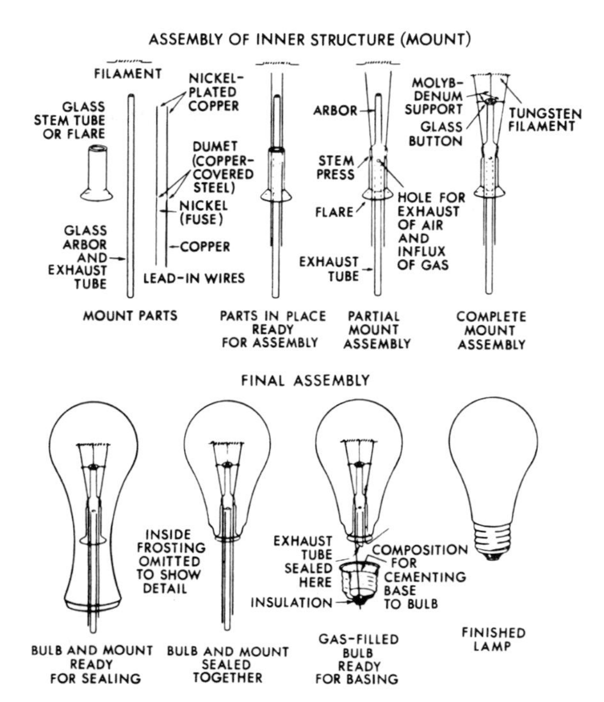

Fig. 8-5. Steps in the manufacture of a typical incandescent filament lamp.

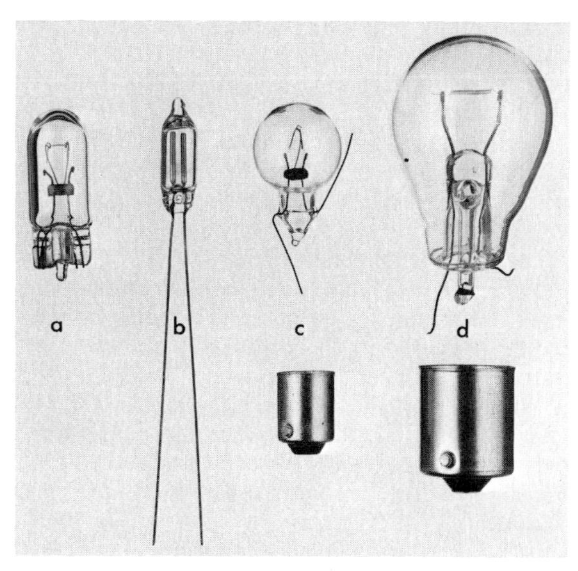

Fig. 8-6. Primary type of bulb construction: a. pinch seal with wedge base, b. pinch seal with lead-in wire terminals, c. butt seal, and d. flange seal.

The *flange seal* is used generally with lamps 20 millimeters (3/4 inch) and larger in envelope diameter. This construction features a glass stem with a flange at the bottom which is sealed to the neck of the bulb. When used with bayonet bases the plane of the filament and lead wires are normally at right angles to the plane of the base pins. However, a tolerance of ±15 degrees is generally permitted. The advantages of this construction are: (1) sufficiently heavy lead-in wires can be used for lamp currents up to 12 amperes; (2) filament can be accurately positioned; and (3) sturdy stem construction resists filament displacement and damage from shock and vibration.

Butt seal refers to the method of construction. A mount consisting of lead-in wires, bead and filament is dropped into the open end of the bulb. The lead-in wires are bent to locate the filament at approximately the desired distance from the bulb end. An exhaust tube is then dropped down and butted against the lead-in wire and glass bulb just prior to sealing and exhausting. The base, applied later, together with the basing cement, must not only provide the lamp contacts but also protect the delicate seal. Because of seal limitations, butt seal lamps are restricted to relatively small wire sizes with a current limit of about 1.0 ampere. The filament position varies considerably more than in flange seal lamps since there is no definite relationship between the plane of the filament and base pins. Occasionally butt seal lamps are used without bases; when done so these lamps should be carefully handled. When used with a base the advantages of butt seal construction are: (1) low cost and (2) small size (usually 20 millimeters [3/4 inch] and below).

The *pinch seal* is so named because glass is pinched or formed around the lead-in wires. Two forms are used: wire terminals or wedge base construction. For the smaller types of glow lamps the bulb is exhausted and tipped off at the end opposite the lead-in wires. With the newer wedge base lamps the exhaust tip is at the bottom rather than the top. Pinch seal construction eliminates the need for a conventional base. The advantages are: (1) low cost; (2) small size; (3) with filament lamps, the elimination of solder and cement which allows operation up to 300 °C; and (4) small space required for the wedge base lamps.

Filament Forms and Designations. Filament design is a careful balance of light output and life. Filament forms, sizes and support constructions vary widely with different types of lamps. See Fig. 8–7. Their designs are determined largely by service requirements. Filament forms are designated by a letter or letters followed by an arbitrary number. Most commonly used letters are: S—straight, meaning the wire is uncoiled; C—coiled, the wire is wound into a helical coil; and CC—coiled coil, the wire is wound into a double helical coil. Coiling the filament increases its luminous efficacy; forming a coiled-coil further increases efficacy. More filament supports are required in lamps designed for rough service and vibration service than for general service lamps.

The Bulb

Shapes and Sizes. Common lamp-bulb shapes are shown in Fig. 8–8. Bulbs are designated by a letter referring to the shape classification and a number which is the maximum diameter in multiples of 3.2 millimeters (eights of an inch). Thus, "R-40" designates a bulb of the "R" shape which is 40/8 or 5 inches or 127 millimeters in diameter.

Types of Glass. Most bulbs are made of regular lead or lime "soft" glass; other types are of borosilicate "heat-resisting" "hard" glass. The latter withstand higher temperatures and are used for highly loaded lamps. Under most circumstances they will better withstand exposure to moisture or luminaire parts touching the bulb. (See Bulb and Socket Temperatures, page 8–7.) Three specialized forms of glass are also used as lamp envelopes: fused silica (quartz), high silica and aluminosilicate glass. These materials can withstand still higher temperatures.

Bulb Finishes and Colors. Inside chemical frosting is applied to many types and sizes of bulbs. It produces moderate diffusion of the light with scarcely measurable reduction in output and with little reduction in bulb strength. The extremely high filament luminance of clear lamps

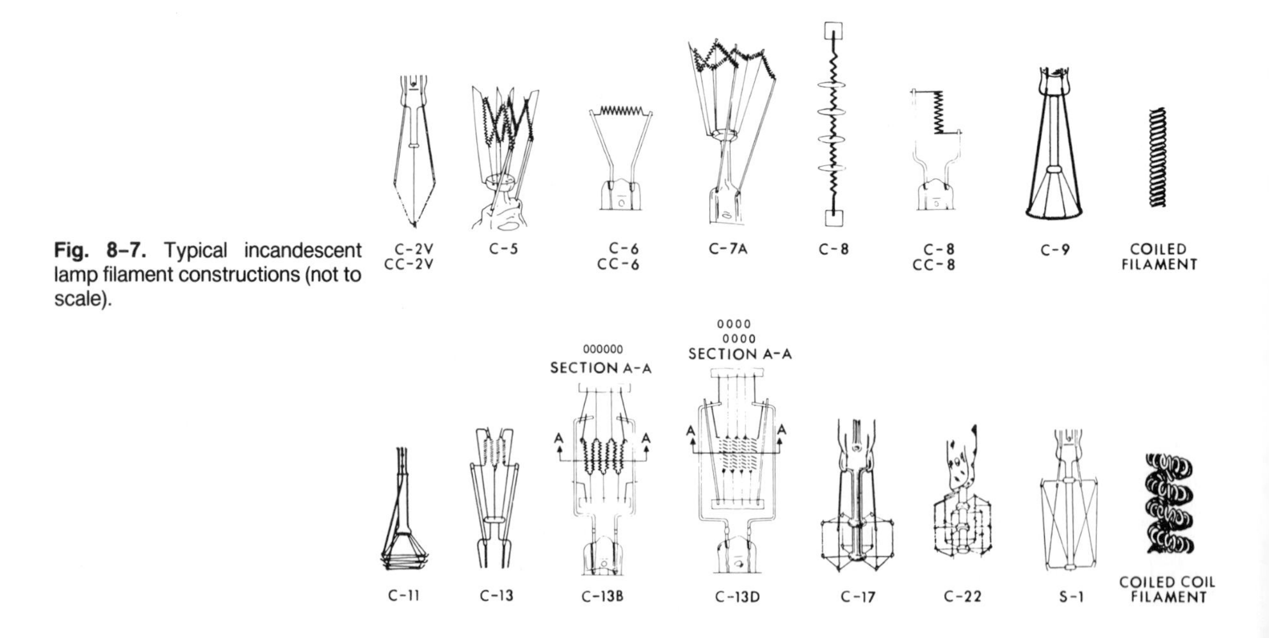

Fig. 8–7. Typical incandescent lamp filament constructions (not to scale).

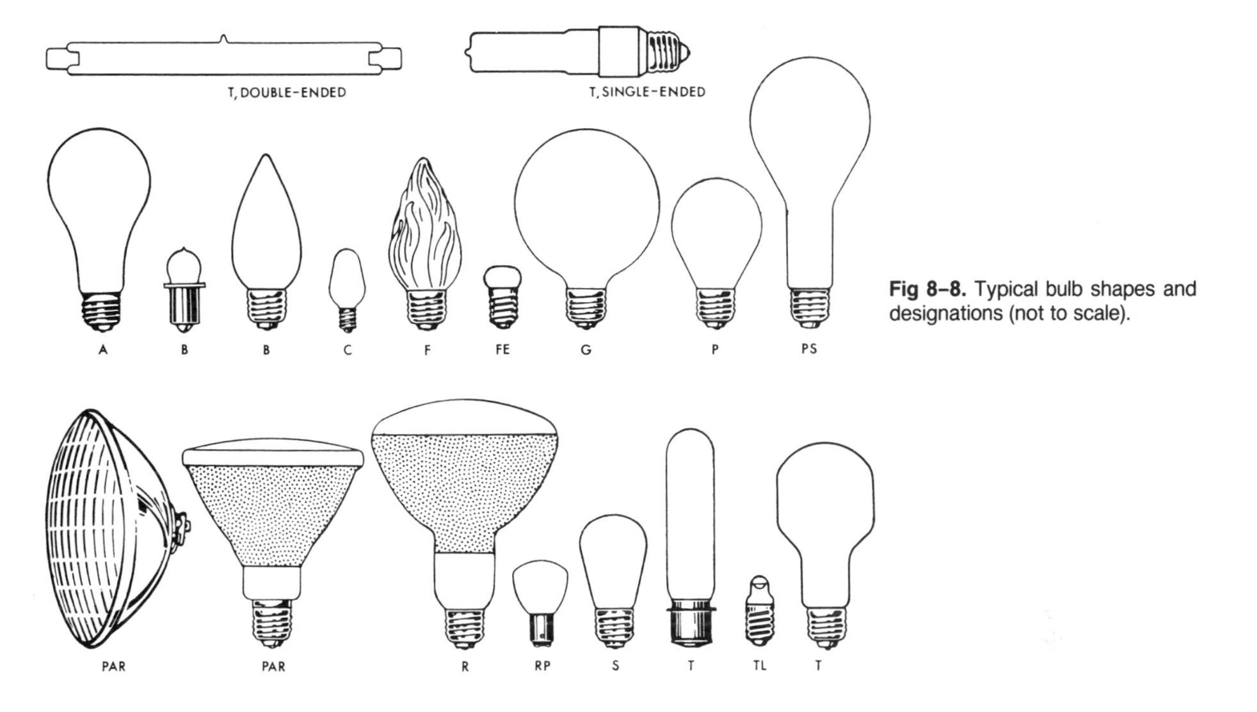

Fig 8-8. Typical bulb shapes and designations (not to scale).

is reduced and striations and shadows practically eliminated with the use of frosting. White lamps having an inside coating of finely powdered white silica provide a substantially greater degree of diffusion with very little absorption of light.

Daylight lamps have bluish glass bulbs which absorb some of the red and yellow light produced by the filament. The resulting light output is of a higher color temperature. This color, achieved at the expense of about 35 per cent reduction in light output through absorption, varies between 3500 and 4000 K. This is about midway between tungsten-filament light and natural daylight.

Colored lamps are available with inside and outside-spray-coated, outside-ceramic, transparent plastic coated, and natural-colored bulbs. Outside-spray-coated lamps are generally used for indoor use where not exposed to the weather. Their surfaces collect dirt readily and are not easily cleaned. Inside-coated bulbs have smooth outside surfaces which are easily cleaned; the pigments are not exposed to weather and, therefore, have the advantage in permanence of color. Ceramic-coated bulbs have the colored pigments fused into the glass, providing a permanent finish to the bulbs. They are suitable for indoor or outdoor use. Most transparent plastic coated bulbs also can be used both indoors and outside. The coating permits the filament to be observed directly. Natural-colored bulbs are made of colored glass. Reflectorized colored lamps utilize ceramic coated bulbs, stained bulbs, plastic coated bulbs, and dichroic interference filters to obtain the desired color characteristics.

The Base

Fig. 8-9 shows the most common types of lamp bases.

Most lamps for general-lighting purposes employ one of the various types of screw bases. Where a high degree of accuracy in positioning of light sources with relation to optical elements is important, as is the case in projection systems, bipost and prefocus bases insure proper filament location. The bipost construction eliminates the conventional stem seal and lead assembly and substitutes a supporting structure that is made of one piece of channel nickel. Such construction gives maximum strength to the long leads supporting the heavy filament and provides a greater exposed surface to dissipate heat. The metal prongs which hold the lamp in the socket carry the weight of the entire metal structure instead of putting the burden on the glass. Lamp wattage is often a factor in determining the base type.

Most bases are secured to the bulbs by cement and are cured by heat when the lamp is manufactured. Since this cement becomes weaker with age, particularly if exposed to excessive heat, some lamps intended for service that causes the base to run hot use a special heat-resisting basing cement or "mechanical" bases which employ no cement.

The Fill Gas

Up to about 1913, all lamps were of the vacuum type with uncoiled wire filaments. Attempts were made to reduce the rate of evaporation of the filament by the use of gas-filled bulbs which would allow higher filament temperatures and hence higher efficacies without a sacrifice in life. It was found that with uncoiled wire filaments the heat taken from the filament by the gas more than offset the advantage of the higher filament temperature. An incandescent filament operating in an inert gas is surrounded by a more or less stationary thin sheath of the gas and the percentage of the input energy that is lost becomes less as the filament diameter is increased. When the filament was coiled in a small helix it was found that the gas sheath surrounded the entire coil and greatly reduced the loss. The use of coiled filaments and gas-filled bulbs was a major improvement in incandescent lamp efficacies. A further improvement in efficacy has been obtained by the use of coiled-coil filaments in an increasing number of lamp types. General service 120-volt lamps below 40 watts are usually vacuum type lamps. In this wattage range gas filling does not improve the luminous efficacy.

Inert gases are necessary for filling incandescent filament lamps in order to prevent chemical action between the gas and the internal parts of the lamp. Nitrogen was first used for this purpose.[5] It was known that argon, due to its lower heat conductivity, was superior to nitrogen, but it was some years after the development of the gas-filled lamps before argon became available in sufficient quantity and purity, and at reasonable cost. Most present-day lamps are filled with argon with a small percentage of nitrogen, the nitrogen being necessary to suppress the tendency of arcing to take place between the lead-in wires.

The proportion of argon and nitrogen used depends on the voltage rating, the filament construction and temperature, and the lead-tip spacing. Typical amounts of argon in use are: 99.6 per cent for 6-volt lamps, 95 per cent for 120-volt general service coiled-coil lamps, 90 per cent for 230-volt lamps having fused lead wires, and 50 per cent or less for 230-volt lamps when no fuses are used in leads. Some projection lamps are 100 per cent nitrogen filled.

Krypton is used in some lamps where the increase in cost due to the use of krypton is justified by the increased efficacy or increased life. Krypton gas has lower heat conductivity than argon; also, the krypton molecule is larger than that of argon and therefore further retards the evaporation of the filament. Depending on the filament form, bulb size, and mixture of nitrogen and argon, conversion to krypton fill may increase efficacy by 7 to 20 per cent.[6] Krypton is used in some special lamps such as marine signal and miner's cap lamps where high efficacy

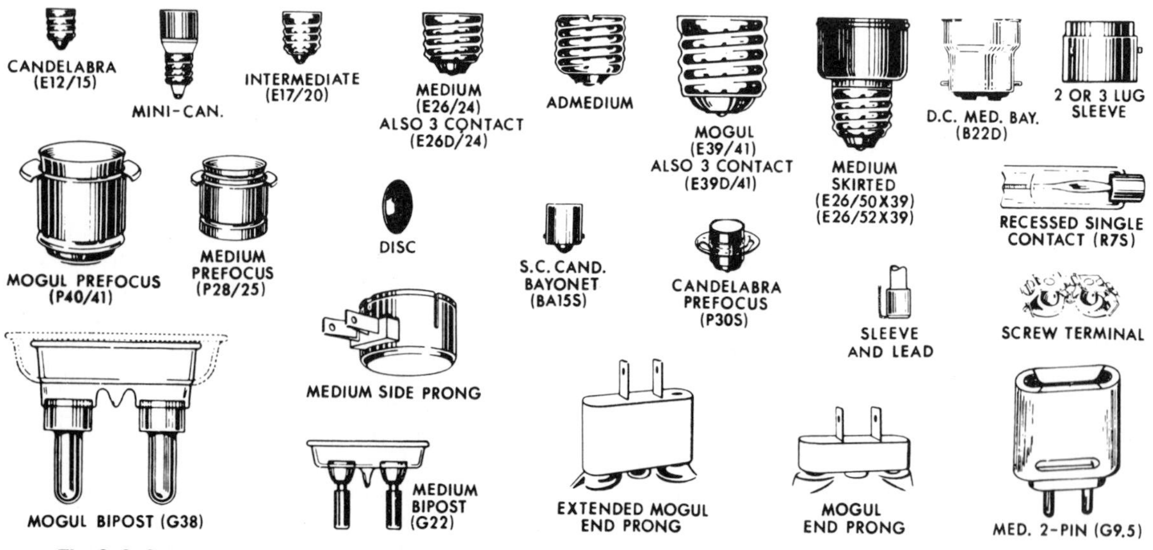

Fig. 8-9. Common lamp bases (not to scale). International Electrotechnical Commission (IEC) designations are shown, where available.

means less drain on the power source for a given amount of light produced.

Hydrogen gas has a high heat conductivity and is therefore useful for filling signaling lamps where quick flashing is desired.[7]

Tungsten-halogen lamps utilize a halogen regenerative cycle to provide excellent lumen maintenance, together with lamp compactness. They are a variation of incandescent filament lamps. The term halogen is the name given to a family of electronegative elements, *i.e.*, bromine, chlorine, fluorine and iodine.

Although the tungsten-halogen regenerative cycle has been known for many years, no practical method of using it was established until the use of small diameter fused quartz envelopes for filament lamps provided the proper temperature parameters. Iodine was the halogen used in the first tungsten-halogen lamp; today, other halogens are also being used.[8]

It is generally accepted that the regenerative cycle starts with the tungsten filament operating at incandescence evaporating tungsten off the filament. Normally the tungsten particles would collect on the bulb wall, resulting in bulb blackening, common with incandescent lamps, and most evident near end of life. However, in halogen lamps, the temperature of the bulb is high enough (controlled minimum 260 °C) so that the tungsten combines with the halogen within the lamp. This forms a gaseous tungsten-halogen compound. The gas continues to circulate inside the lamp until it comes in contact with the incandescent filament. Here, the heat is sufficient to break down the compound into tungsten, which is redeposited on the filament, and into the halogen, which is free to continue its role in the regenerative cycle. It should be noted, however, that since the tungsten does not necessarily redeposit exactly at the spot from where it came, the tungsten-halogen lamp has a finite life.

Energy Characteristics

The manner in which the energy input to a lamp is dissipated can be seen by reference to Fig. 8-83 (page 8-72) for typical general service lamps. The radiation in the visible spectrum (column 2) is the percentage of the input watts actually converted to visible radiation. The gas loss (column 4) indicates the amount of heat lost by the filament due to the conduction through and convection by the surrounding gas in gas-filled lamps. The end loss (column 6) is the heat lost from the filament by the lead-in wires and support hooks which conduct heat from the filament. Column 3 shows the total radiation beyond the bulb, which is less than the actual filament radiation due to absorption by the glass bulb and the lamp base.

Bulb and Socket Temperatures

Incandescent filament lamp operating temperatures are important for various reasons. Excessive lamp temperatures may affect the performance of the lamp itself. They may shorten the life of the luminaire or of the electrical supply circuit. Excessive lamp temperatures may also result in unsafe temperatures on combustible materials that form a part of the luminaire or of the material adjacent to the luminaire. Under certain atmosphere or dust conditions, high bulb temperatures (above 160 °C) may be hazardous. Bulb and socket temperatures for a 100-watt A-19 lamp and a 500-watt PS-35 lamp for different operating positions are shown in Fig. 8-10.

General service incandescent filament lamps are made of regular lead or lime "soft" glass, the maximum safe operating temperature of which is 370 °C. Some lamps for special applications, such as outdoor floodlighting lamps, have "hard" glass bulbs which have a safe temperature limit of 470 °C. Lamps with "still harder" glass bulbs can be operated up to 520 °C. For thermal shock conditions (liquid drops) safe operating temperatures would be substantially lower. There are many low wattage lamps which have inner parts that will not operate satisfactorily with such high bulb temperatures.

From a lamp performance standpoint, the maximum safe base temperature for general service lamps is 170 °C measured at the junction of the base and the bulb. In all cases excessive temperature may cause failure of the basing cement as well as softening of the solder used to connect the lead wires to the base. Silicone-type high-temperature basing cement, combined with a higher-melting-point solder, permits a base temperature up to 260 °C. Mechanical bases which are used in some lamps without basing cement generally should not be operated above 200 °C. There are some incandescent filament lamps which have internal parts which will withstand operation of the mechanical base up to 210 °C. In all cases care must be exercised to insure that allowable luminaire socket temperatures are not exceeded. Bipost bases carry considerable heat to the socket through the base pins, and the parts of the socket in contact with the base pins should be capable of withstanding a temperature of 290 °C. Tubular fused quartz infrared and tungsten-halogen cycle lamps have a maximum seal temperature limit of 350 °C to prevent oxidation. In addition, tungsten-halogen lamps have a *minimum* bulb temperature limit of 260 °C to insure proper functioning of the regenerative cycle.

Some of the factors affecting base temperature are filament type, light center length, use of or lack of a heat shield, bulb shape and size, and gas fill. Base temperatures may not parallel wattage ratings; that is, lamps of lower wattage do not necessarily have lower base temperatures. Medium base luminaires should be capable of accepting lamp base temperatures in the order of 135 to 150 °C. These limits relate to CSA* and IEC† standards for base temperature for A-type lamps. If the luminaire will accept R- and PAR-type lamps then the temperature capability of the luminaire should be for 170 to 185 °C if reasonable electrical insulation life is desired for the luminaire. Heat-transmitting dichroic-reflector lamps should be placed only in luminaires specifically designed for them; this is normally indicated on the luminaire.

For base up burning and where only slight enclosing of the lamp is provided by the luminaire, base temperature is the major factor affecting luminaire temperatures and lamp wattage is a minor consideration. As the luminaire provides more and more enclosing of the lamp, base temperatures have a lesser effect and wattage assumes more importance.

Measurements of the thermal effect of radiation from incandescent lamps indicates a variation with direction at right angles to the bulb axis to a degree which cannot be ignored if the life of the electrical insulation in luminaire components is not to be impaired. Fig. 8-11 illustrates the type and extent of variation that may occur.

* Canadian Standards Association.
† International Electrotechnical Commission.

Lamp Characteristics

Life, Efficacy, Color Temperature and Voltage Relationships. If the voltage applied to an incandescent filament lamp is varied, there is a resulting change in the filament resistance and temperature, current, watts, light output, efficacy, and life. These characteristics are interrelated and one cannot be changed without affecting the others. The following equations can be used to calculate the effect of a change from the design conditions on lamp performance (capital letters represent normal rated values, lower case letters represent changed values):

$$\frac{\text{life}}{\text{LIFE}} = \left(\frac{\text{VOLTS}}{\text{volts}}\right)^{d}$$

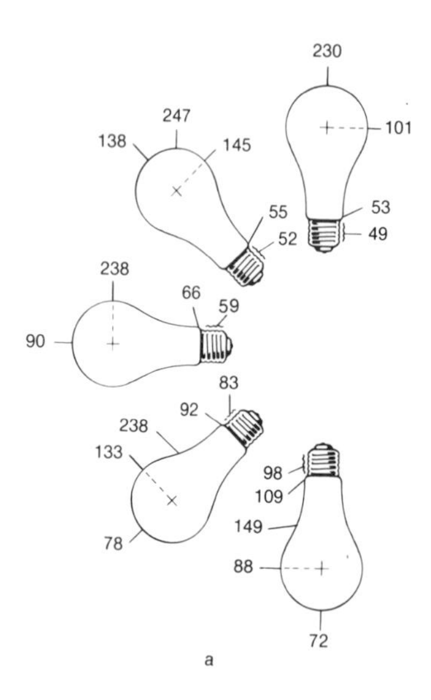

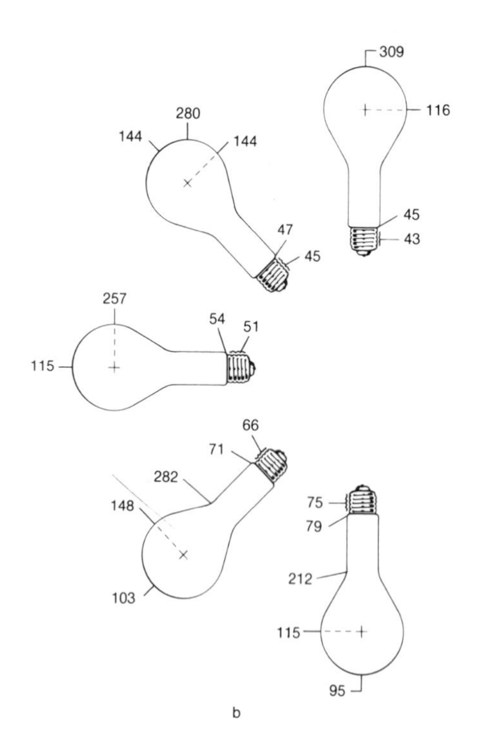

Fig. 8-10. Incandescent filament lamp operating temperatures in still air at 25°C ambient: (a) 100-watt CC-8, A-19 lamp; (b) 500-watt CC-8, PS-35 lamp. All temperatures shown are in degrees Celsius.

$$\frac{\text{lumens}}{\text{LUMENS}} = \left(\frac{\text{volts}}{\text{VOLTS}}\right)^k$$

$$\frac{\text{LPW}}{\text{lpw}} = \left(\frac{\text{VOLTS}}{\text{volts}}\right)^g$$

$$\frac{\text{watts}}{\text{WATTS}} = \left(\frac{\text{volts}}{\text{VOLTS}}\right)^n$$

$$\frac{\text{color temperature}}{\text{COLOR TEMPERATURE}} = \left(\frac{\text{volts}}{\text{VOLTS}}\right)^m$$

For *approximate* calculations, the following exponents may be used in the above equations: $d = 13$, $g = 1.9$, $k = 3.4$, $n = 1.6$ and $m = 0.42$. For more accuracy, the exponents must be determined by each lamp manufacturer from a comparison of normal and over or under voltage tests of many lamp groups. Exponents will vary for different lamp types, for different lamp wattages, and for various ranges of per cent voltage variation. The values given above are roughly applicable to vacuum lamps of about 10 lumens per watt and gas-filled lamps of about 16 lumens per watt in a voltage range of 90 to 110 per cent of rated volts. For information outside this range, refer to Fig. 8-12.

The curves of Fig. 8-12(a) show the effect of voltage variations[9] on the characteristics of lamps in general lighting (multiple) circuits while the curves of Fig. 8-12(b) show similar characteristics for series lamps for variations in current. (See also page 8-14 for the use of series and multiple circuits in street lighting).

Filament Notching. Ordinarily, for laboratory test operation, normal tungsten filament evaporation controls incandescent lamp life. Where normal filament evaporation is the dominant failure mechanism, lamps should reach their design-predicted life. In recent years another factor influencing filament life has become prominent. This is commonly referred to as "filament notching". Filament notching is the appearance of step-like or saw-tooth irregularities on all or part of the tungsten filament surface after substantial burning. These notches reduce the filament wire diameter at random points. In some cases, especially in fine-wire diameter filament lamps, the notching is so severe as to almost penetrate the entire wire diameter. Thus, faster spot evaporation due to this notching and reduced filament strength become the dominant factors influencing lamp life. Lamp life due to filament notching may be as much as one half of so-called ordinary or normal, predicted lamp life.

Filament notching is associated with at least three factors (primarily fine-wire filament lamps): (1) low temperature filament operation, as for long life lamp types such as 10,000- to 100,000-hour designs; (2) small filament wire sizes—less than 0.025 millimeter (0.001 inch) diameter in some cases; and (3) use of direct current operation.

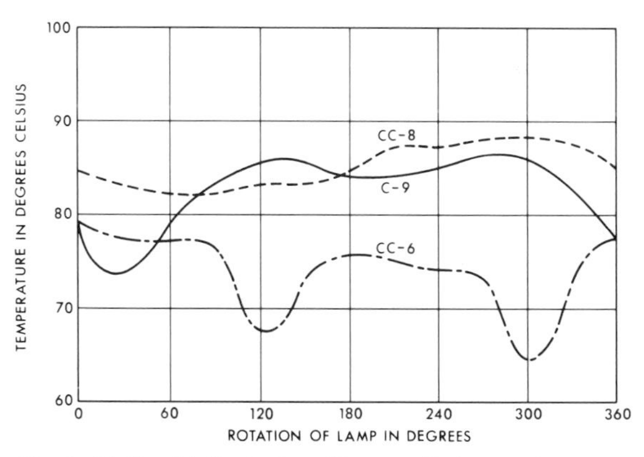

Fig. 8-11. Graphic illustration of how profile temperatures vary with the direction of radiation due to filament type.

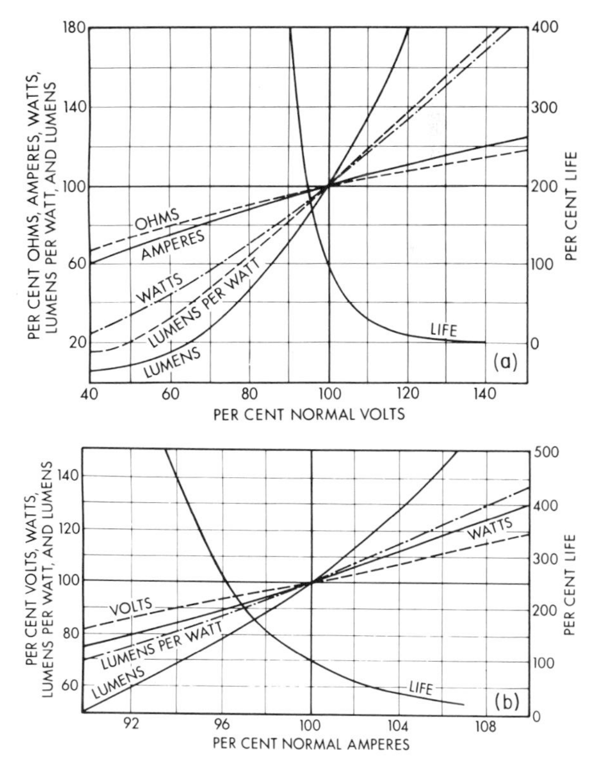

Fig. 8-12. Effect of voltage and current variation on the operating characteristics of: (a) incandescent filament lamps in general lighting (multiple) circuits and (b) incandescent filament lamps in series street lighting circuits.

Depreciation During Life

Over a period of time incandescent filaments evaporate and become smaller, which increases their resistance. In multiple circuits, the increase in filament resistance causes a reduction in amperes, watts and lumens. A further reduction in lumen output is caused by the absorption of light by the deposit of the evaporated tungsten particles on the bulb. Fig. 8–13(a) shows the change in watts, lumens and lumens per watt for a 200-watt lamp.

In series circuits having constant current regulators the increase in filament resistance during life causes an increase in the voltage across the lamp and a consequent increase in wattage and generated lumens. This increase in lumens is offset to varying degrees by the absorption of light by the tungsten deposit on the bulb. In low-current lamps the net depreciation in light output during life is very small, or in the smaller sizes there may be an actual increase in light output during life. In lamps of 15- and 20-ampere ratings, the bulb blackening is much greater and more than offsets throughout life the increase in lumens due to the increased wattage. Fig. 8–13(b) shows the change in watts, lumens and lumens per watt for high-current series lamps.

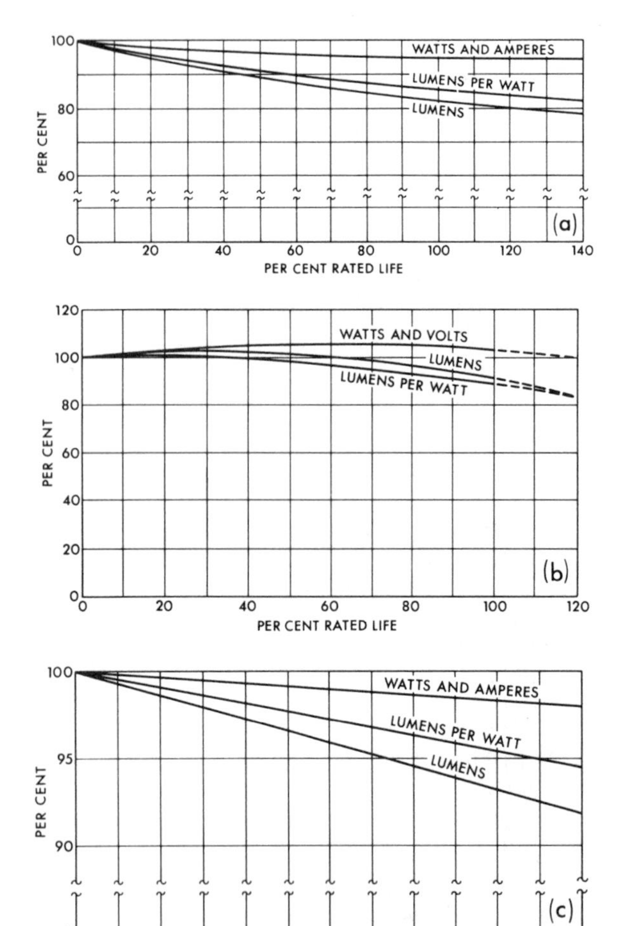

Fig. 8-13. Typical operating characteristics of lamps as a function of burning time: (a) general service lamps and (b) series lamps. Operating characteristics of a tungsten-halogen lamp are shown in (c). (*Note differences in scales.*)

The blackening in vacuum lamps is uniform over the bulb. In gas-filled lamps the evaporated tungsten particles are carried by convection currents to the upper part of the bulb. When gas-filled lamps are burned base-up most of the blackening occurs on the neck area where some of the light is normally intercepted by the base. Consequently, the lumen maintenance for base-up burning is better than for base-down or horizontal burning.

An appreciable gain in lumen maintenance can be obtained through the use of a coiled-coil filament located on or parallel to the bulb axis in a base-up burning lamp.

To reduce blackening from impurities in the gas filling, an active chemical, known as a "getter," is used inside the bulb. This "getter," during the first burning hours of the lamp, tends to combine with and absorb any traces of oxygen or water vapor remaining in the bulb. In certain other lamps where the blackening would not be reduced sufficiently by "getters" alone, various other means are employed. Some high-wattage lamps have a small amount of tungsten powder loose in the bulb, which, when shaken about, wipes off much of the blackening. Other lamps have a wire mesh "collector grid," mounted above the filament, to attract and condense the tungsten vapor and hold the tungsten particles, thereby reducing bulb blackening.

Tungsten-halogen cycle lamps have significantly less depreciation during life due to the regenerative cycle which removes the evaporated tungsten from the bulb and redeposits it on the filament. Fig. 8–13(c) shows the change in lumens, watts and lumens per watt for the 500-watt tungsten-halogen lamp.

Lamp Mortality

Many factors inherent in lamp manufacture and lamp materials make it impossible to have each individual lamp operate for exactly the life for which it was designed. For this reason lamp life is rated as the average of a large group. A

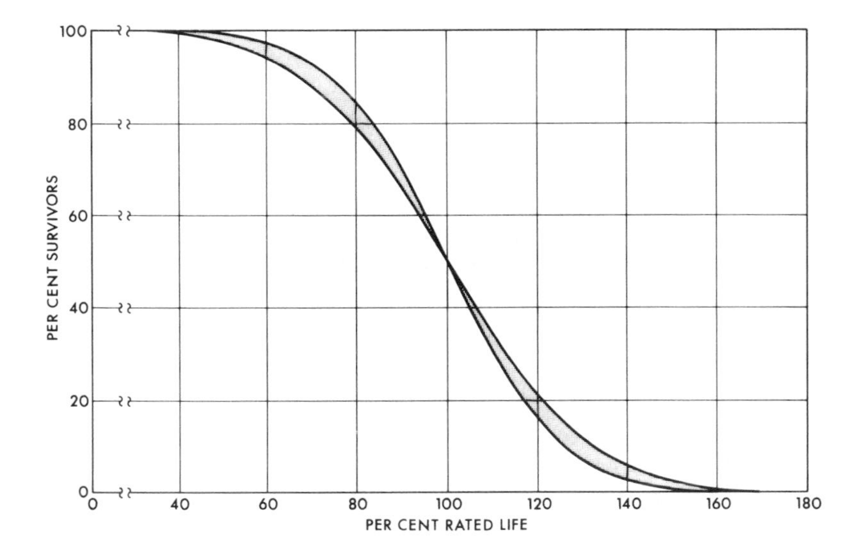

Fig. 8-14. Range of typical mortality curves (average for a statistically large group of good quality incandescent filament lamps).

typical mortality curve of a large group of lamps is illustrated in Fig. 8-14 and is representative of the performance of good quality lamps.

Classes of Lamps

Incandescent filament lamps are divided into three major groups: large lamps, miniature lamps and photographic lamps. These are cataloged separately by lamp manufacturers. There is no sharp dividing line between the groups. The large lamp classification generally refers to those with larger bulbs, with medium or mogul bases, and for operation in circuits of 30 volts or higher. The miniature classification generally includes lamps referred to as automotive, aircraft, flashlight, Christmas tree, radio panel, telephone switchboard, bicycle, toy train and many other small lamps generally operated from circuits of less than 50 volts. The photographic classification includes lamps designed for photographic or projection service. The following gives a brief description of a few of the many types of lamps that are regularly manufactured. More complete details are available in manufacturers' catalogs.

General Service. These are large lamps made for general lighting use on 120-volt circuits. General service lamps are made in sizes from 10 watts to 1500 watts and satisfy the majority of lighting applications. All sizes are made in both clear and inside frosted bulbs. Below 200 watts, inside white coated finishes are also available. Performance data are shown in Fig. 8-81.

High Voltage (220-300 Volts).* This voltage class refers to lamps designed to operate directly on circuits of 220 to 300 volts and represents a very small portion of the lamp demand.

High-voltage lamps have filaments of small diameter and longer length and the filaments require more supports than corresponding 120-volt lamps. They are, therefore, less rugged and produce 25 to 30 per cent fewer lumens per watt because of greater heat losses. The higher operating voltage causes these lamps to take less current for the same watts, permitting some wiring economy.[10] Operating characteristics and physical data are shown in Fig. 8-85.

Extended Service. Extended service lamps are intended for use in applications where a lamp failure causes great inconvenience, a nuisance, or a hazard, or where replacement labor cost is high or power cost unusually low. Therefore, for such applications where longer life is most important and a reduction of approximately 15 per cent in light output is acceptable, available extended service lamps with 2500 hours rated life are recommended. Where replacement of burned-out lamps is an easy, convenient operation, as in residential use, long life lamps are not usually recommended.[11] For most general use, incandescent lamps with the usual 750 or 1000 hours design life give a lower cost of light than extended service lamps.† See Fig. 8-87.

* It should be noted that lampholders should be Underwriters Laboratories or Canadian Standards Association "approved" for voltage level appropriate to the voltage rating of the lamps being used, *i.e.*, 250 volts for lamps up to 250 volts, 600 volts for lamps above 250 volts.

† Longer life is obtained by operating the lamp's filament at lower temperature than normal. This, however, lowers the lamp's luminous efficacy. In most general service use the cost of power used during the lamp life runs many times the lamp cost and, therefore, efficacy is important.

General Lighting Tungsten-Halogen. These lamps augment the regular incandescent sources. The advantages over regular incandescent lamps include excellent lumen maintenance and compactness. The source can also be described as providing a whiter light (higher color temperature) as well as longer life at a given light output. The lamps shown in Fig. 8–89 are tungsten-halogen lamps without an external bulb and are intended for use in general lighting applications. Tungsten-halogen lamps intended for specialty use and lamps with an external bulb are shown in the appropriate specialty tabulations.

There is relatively more ultraviolet radiation generated from tungsten-halogen lamps than from regular incandescent lamps due to higher filament temperature. The amount of ultraviolet radiation emitted is determined by the envelope material. Fused quartz and high silica glass transmit most of the ultraviolet radiated by the filament, while special high silica glass and aluminosilicate glasses absorb ultraviolet radiation. Precaution against hazard to both people and objects are recommended in applications where tungsten-halogen lamps are operated at color temperatures above 3100 K because both the visible light output and the relative ultraviolet output are sharply increased.[12]

Tungsten-halogen lamps are very temperature sensitive. Operating lamps at voltages above or below manufacturer's recommendations may have adverse effects on the internal chemical processes, because the temperature will differ from the design value. It is also important to follow manufacturer's recommendations as to burning position, bulb handling and luminaire temperatures.

The halogen cycle allows tungsten-halogen lamps to be designed for higher efficacy and/or longer life than normal incandescent lamps of the same wattage. For example, the 500-watt non-regenerative cycle lamp is rated at 10,600 lumens, 1000 hours, while the 500-watt T-3 tungsten-halogen lamp has 10,950 lumens for 2000 hours rated life.

Multilayer interference film (dichroic) coating technology has been applied to linear tungsten-halogen lamps to dramatically increase their efficacy 30 to 50 per cent. The transparent, slightly iridescent external coating transmits visible light but reflects infrared energy back to the filament, thereby reducing the input power required to reach a given filament temperature. Luminous efficacies of 28.8 to 35.6 lumens per watt are obtained with this technique without reducing life.

Train and Locomotive. Lamps designated as train and locomotive service are designed for several classes of low-voltage (30, 60, 75, etc.) service usually provided by generators, with a battery floated across the line. Data for these lamps are shown in Fig. 8–93. Low-voltage lamps have shorter and heavier filaments than 120-volt lamps of the same wattage and, consequently, are more rugged and generally produce more lumens per watt.

DC Series. Transit system voltages and some railway shop and yard voltages range from 525 to 625 volts. Lamps for this service (see Fig. 8–98) are operated five to twenty in series on these voltages. The design of voltages of individual lamps operated five in series are nominally 115, 120, and 125 volts. To identify dc series lamps, they are rated in odd wattages (36, 56, 94, 101, etc.).

Thirty-volt gas-filled lamps also are used for car lighting. The trolley voltage divided by 30 determines the number of lamps connected in series across the line. These lamps are equipped with short-circuiting cut-outs which short-circuit the lamps on burnout, thus preventing arcing and leaving the remainder of the lamps in a given circuit operating. These 30-volt lamps are rated in amperes, instead of the usual watt rating.

Spotlight and Floodlight. Lamps used in spotlights, floodlights, and other specialized luminaires for lighting theater stages, motion picture studios, and television studios have concentrated filaments accurately positioned with respect to the base. When the filament is placed at the focal point of a reflector or lens system, a precisely controlled beam is obtained. These lamps, listed in Fig. 8–94, are intended for use with external reflector systems. Because of their construction, these lamps must be burned in positions for which they are designed, to avoid premature failures.

Reflectorized. These lamps include those made in standard and special bulb shapes and which have a reflecting coating directly applied to part of the bulb surface. Both silver and aluminum coatings are used.[13] Silver coatings may be applied internally or externally, and in the latter case the silver coating is protected by an electrolytically-applied copper coating and sprayed aluminum finish. Aluminum coatings are applied internally by condensation of vaporized aluminum on the bulb surface. The following reflectorized lamps are regularly available: *bowl* *reflector lamps*, data for which are given in Fig. 8–95; *neck reflector lamps*; *reflector spot and reflector flood lamps* in R-type bulbs (certain sizes of reflectorized lamps are available with heat resisting glass bulbs; performance data are given in Fig. 8–96); *ellipsoidal reflector lamps* (ER-types) which allow substantially improved energy utilization in deep, well shielded downlights.[14] *PAR spot and PAR flood lamps* (PAR bulbs typically are constructed from two molded glass parts, the reflector and the lens, which are fused together; see Fig. 8–96 for performance data and Fig. 8–15 for typical candlepower distribution curves); *reflector showcase lamps* in T-type bulbs (see Fig. 8–90 for performance data). Colored R and PAR lamps are available. PAR lamps with heat-transmitting dichroic-reflectors are available for applications where it is desirable to reduce the infrared energy in the beam. General lighting PAR lamps utilizing tungsten-halogen cycle filament tubes are included in Fig. 8–96. Multi-faceted pressed glass ellipsoidal reflector lamps (MR-16) using 12-volt quartz tungsten-halogen filament tubes and infrared transmitting dichroic reflectors have been adapted from projection lamp designs (see page 8–18) for display lighting applications. See Fig. 8–91.

High-Temperature Appliance. These lamps are specially constructed for high temperature service. The most common types are clear, medium base, 40-watt A-15 appliance and oven; 50-watt, A-19; and 100-watt, A-23 bake oven lamps. Range oven lamps are designed to withstand oven temperatures up to 245 °C (475 °F) and bake oven up to 315 °C (600 °F).

Rough Service. To provide the resistance to filament breakage as required for portable extension cords, etc., rough service lamps employ special, multiple-filament-support construction. See Fig. 8–7, C-22. Because of the number of supports, the heat loss is higher and the efficacy lower than for general service lamps. Performance data are given in Fig. 8–86.

In using miniature lamps where rough service conditions are encountered, bayonet and wedge

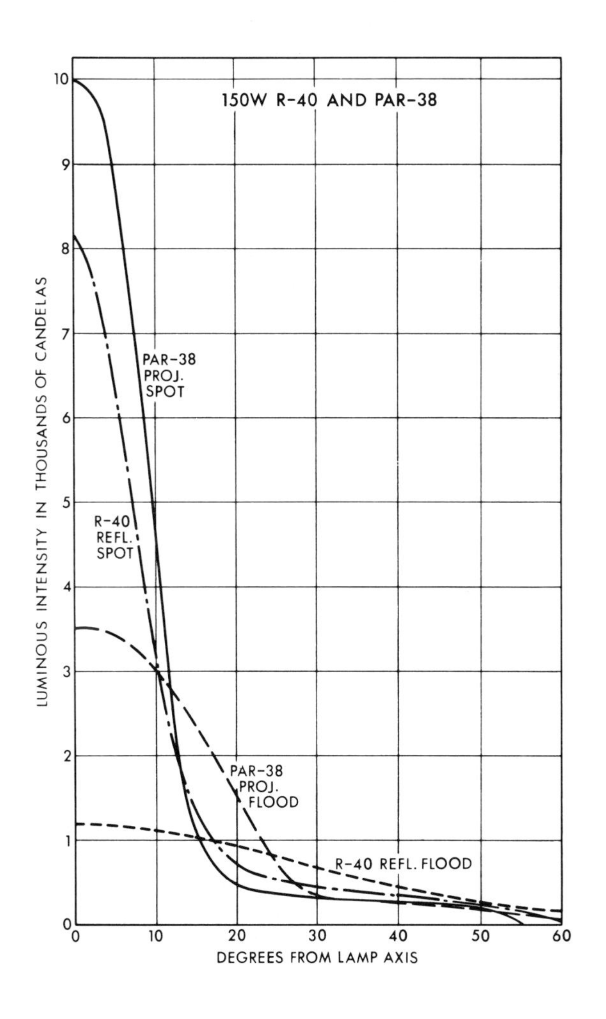

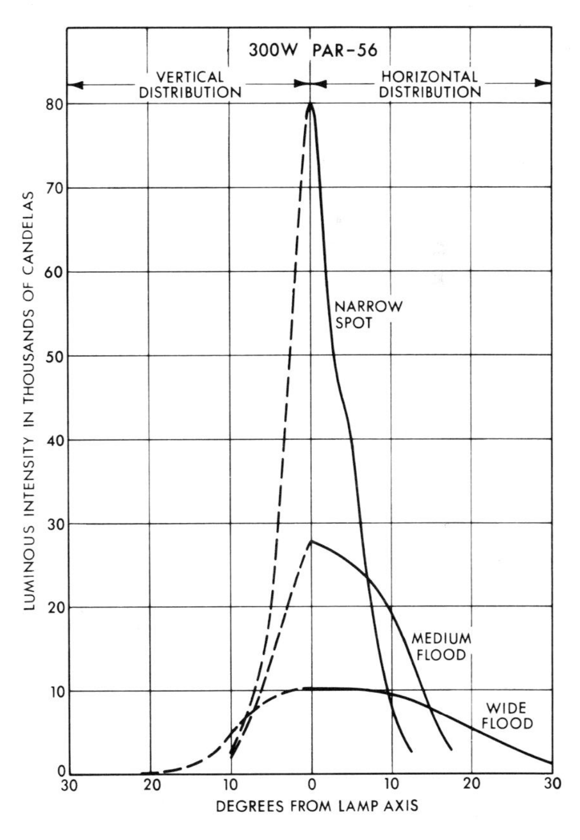

Fig. 8-15. Typical candlepower distribution characteristics of representative types of reflector and PAR lamps.

base lamps should be chosen instead of screw base lamps. Bayonet and wedge base lamps lock in the socket whereas screw base lamps tend to work loose.

Vibration Service. Most lamps have coiled filaments made of tungsten having high sag resistance. Vibration lamps, designed for use where high-frequency vibrations would cause early failure of general service lamps, are made with a more flexible tungsten filament. The sagging characteristics of the wire used allow the coils to open up under vibration, thus preventing short circuits between coils. See Fig. 8-88 for data.

Miniature radio panel lamps of 6.3 volts and under, incorporate mounts whose resonant frequency has been synchronized with that of the coiled filament to withstand shock and vibration.

Vibration and shock frequently accompany each other and sometimes only experiment will determine the best lamp for the purpose. Vibration-resisting sockets or equipment, utilizing a coiled spring or other flexible material to deaden vibration, have been employed where general service lamps are used under conditions of severe vibration.

Lumiline. The lumiline lamp has a tubular bulb (26 millimeters or 1 inch in diameter) and two metal disk bases: one at each end of the lamp, with the filament connected between them. The filament, in the form of a stretched coil, is supported on glass insulating beads along a small metal channel within the bulb. The 30- and 60-watt sizes are available in the 450-millimeter (17¾-inch) length. The 40-watt lamp is made in a 300-millimeter (11¾-inch) length. All sizes are available in either clear or inside-frosted tubes as well as white and various color coatings. See Fig. 8-92 for data.

Showcase. These use tubular bulbs and conventional screw bases. The longer lamps have elongated filaments with filament supports similar to lumiline lamps. The common sizes are 25- and 40-watt, but sizes up to 75 watts are available. See Fig. 8-90.

Three-Lite. These lamps employ two filaments, operated separately or in combination to provide three illuminance levels. The common lead-in wire is connected to the shell of the base; the other end of one filament is connected to a ring contact, and the end of the other filament to a center contact. Three-lite lamps are available in different wattage ratings. See Fig. 8-84.

Sign. While large numbers of gas-filled lamps are used in enclosed and other types of electric signs, those designated particularly as "sign" lamps are mostly of the vacuum type. Lamps of this type are best adapted for exposed sign and festoon service because the lower bulb temperature of vacuum lamps minimizes the occurrence of thermal cracks resulting from rain and snow. Some low-wattage lamps, however, are gas-filled for use in flashing signs. Bulb temperatures of these low-wattage, gas-filled lamps are sufficiently low to permit exposed outdoor use on high speed flashing circuits.

Decorative. A wide variety of lamps for decorative applications is available from lamp manufacturers. Different bulb shapes, together with numerous colors and finishes, are used to achieve the desired appearance. Lamp manufacturers' catalogs should be consulted for information on the various decorative types.

Series Street Lighting. Street series lamps are designed to operate in series on constant current circuits. The most common circuit uses 6.6 amperes, automatically regulated to maintain this value regardless of the number and size of the lamps in the circuit. Lamps for 15- and 20-ampere ratings usually are connected in the secondary of a transformer, the primary being in the 6.6-ampere circuit.

These lamps are designated by their initial lumen output and ampere rating; for example, the 2500-lumen, 6.6-ampere lamps, or the 6000-lumen, 20-ampere lamps. Watt and volt ratings, as used to designate multiple lamps, are not commonly employed. Because the lumen output of series lamps is generally specified in street lighting contracts, improvements in light output are reflected by reductions in watts and volts. This usually results in odd numbers and fractions for volts and watts; for example, the present 2500-lumen, 6.6-ampere lamp has an average rating of 21.6 volts and 142 watts. See Fig. 8-100.

Multiple Street Lighting. Multiple street lighting lamps are designed so that their mean lumens correspond approximately to the mean lumens of series lamps of the same initial lumen rating. Since the lamp's voltage is established by a multiple circuit, the watts come out in odd values in order to obtain the desired nominal lumen ratings. See Fig. 8-99 for performance data.

Traffic Signal Lamps. Lamps used in traffic signals are subjected to more severe service requirements than most applications of incandescent lamps. In order to provide uniformity of application, lamps must be compatible with the design requirements of optical systems of standard traffic signals. Data on typical traffic signal lamps are shown in Fig. 8–97.

Aviation. Lighting for aviation is divided into two classes: lighting on and around airports, and lighting on aircraft.

In airport lighting, both multiple and series type lamps are used. Most systems being installed use series type lamps of 6.6-ampere and 20-ampere designs for airport approach, runway and taxiway lighting, whereas multiple lamps are used for obstruction, hazard beacon and airport identification beacon lighting. On aircraft, small and miniature lamps are used exclusively for both interior and exterior lighting.

Most of the lamps used in airport lighting are designed to be used in precise projection type equipment to produce a controlled beam of light complying with required standards.

Hazard beacons and airport identification beacons signaling the presence of high obstructions or the whereabouts of the airport, use lamps of wattages ranging from 500 to 1200 watts. Lamps used on the airport proper range in size from 10 to 500 watts. Lamps used for aircraft lighting are in the miniature classification, although landing lamps as large as 1000 watts are used.

Lamps widely used for aviation service are listed in Figs 8–101 and 8–102. These tabulations include those lamps currently recommended for use in aircraft and airport lighting. In addition, tungsten-halogen lamps can be provided in place of many regular incandescent types. Tungsten-halogen lamps have the advantages of better lumen maintenance and longer life. Many of the types of lamps available but seldom used are not shown. Refer to the Application Volume for information on the application of lamps in the airport and aircraft lighting fields.

Ribbon Filament Lamps. Incandescent lamps made with ribbons or strips of tungsten for the filaments have been used in special applications where it is desirable to have a substantial area of fairly uniform luminance.[15] Ribbon dimensions vary from 0.7 to 4 millimeters in width and up to 50 millimeters in length. The 5- to 20-ampere ribbon filament lamps are usually employed in recorders, instruments, oscillographs and microscope illuminators. The 30- to 75-ampere lamps are used for pyrometer calibration standards and for spectrographic work. Fig. 8–16 shows typical lamps.

Miniature Lamps

The term "miniature" applied to light sources is really a lamp manufacturer's designation determined by the trade channels through which the lamps so identified are distributed, rather than by the size or characteristics of the lamps. In general, however, it *is* true that most miniature lamps *are* small, and consume relatively little power. The most notable exceptions to this generalization are the sealed-beam type lamps, such as automotive headlamps and aircraft landing lamps, some of which are classed as miniature lamps, even though they may be as large as 200 millimeters (eight inches) in diameter and consume up to 1000 watts.

The great majority of all miniature lamps are either incandescent filament lamps or glow lamps. (Glow lamps are covered under Miscellaneous Discharge Lamps, see page 8–54.) Some low wattage and relatively low voltage fluorescent lamps are listed as miniature types by at least one lamp manufacturer. Also, electroluminescent lamps and light emitting diodes (see pages 8–59 and 60) are included in the miniature lamp family. Incandescent miniature lamps[16] and glow lamps[17] are designated completely by numbers standardized and issued by the American National Standards Institute.

With the notable exception of multiple type Christmas lamps, miniature incandescent lamps are designed to operate under 50 volts. These

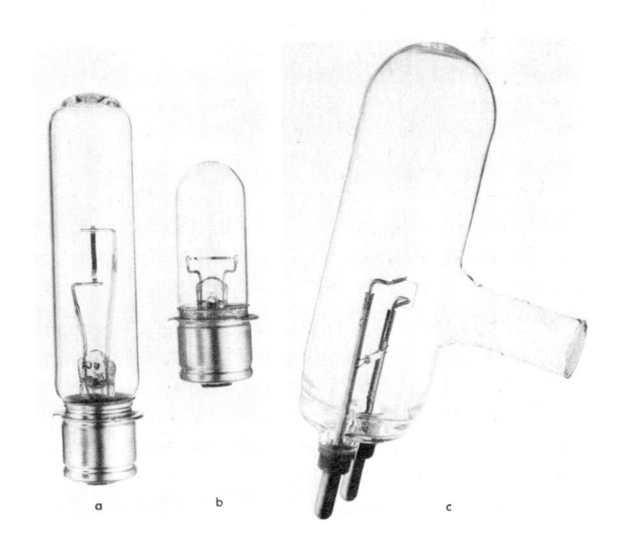

Fig. 8–16. Typical ribbon filament lamps: (a) 6-volt, 18-ampere, T-10, 2 mm., 3000 K microscope illuminator; (b) 6-volt, 9-ampere, T-8½, 1 mm., 3000 K optical source; and (c) 3.5-volt, 30-ampere, T-24 with quartz window, 3 mm., U-shaped filament 2300 K pyrometer and spectroscope source.

voltages may be obtained from batteries, generators or circuits with low voltage transformers.

Miniature lamps are used chiefly when conditions require a light source to be of a small size or consume little power. They have many uses, such as in the following principal fields: automotive, aircraft and decorative. Glow lamps are also used as components in electronic circuitry.

Sub-miniature lamps have increased in popularity. They range in size from T-2 down to T-1/8. Since early in World War II the T-1¾ has been used extensively for instruments and indicators. The T-1 size has become popular for aircraft instruments and indicators. The T-5/8 and smaller sizes down to T-1/8 are used chiefly in novelty applications such as tiny flashlights, jewelry, medical instruments, etc. Examples of variations in range and specifications are given in Fig. 8-110.

Power Sources. Most miniature incandescent filament lamps operate at voltages under 50 volts. However, miniature lamps may be used on 120-volt circuits when transformers, rectifiers or resistors are used to reduce the voltage.

The mean effective voltage delivered by the battery or circuit is generally higher than average voltage and should be the design voltage of the lamp. Design voltages for flashlight lamps have been determined by extensive tests. Proper lamp and battery combinations are shown in manufacturers' catalogs.

With transformers or resistors, delivered rather than rated voltage must be precisely known in order to obtain proper lamp life and output. On resistor operation, because the voltage increases as the filament evaporates, lamp life will generally be one-half of transformer operation.

Where space permits, larger rather than smaller dry cells should be chosen to reduce power costs. Fig. 8–17 gives sizes of batteries used with miniature lamps.

Flashlight, Handlantern and Bicycle. These lamps are commonly operated from dry cell batteries having an open circuit voltage of 1.5 volts per cell for new batteries and dropping to approximately 0.9 volts per cell at the end of battery life. This results in a big difference in light output, depending upon whether the batteries are new or old. Typical lamps of this group are listed in Fig. 8–105.

Automotive. Lamps for most passenger vehicles, trucks and coaches presently operate at 6 to 12 volts. The power source is a storage battery-rectified alternator system. Performance data of typical 6- and 12-volt lamps are given in Figs. 8–106 and 8–107 respectively.

Fig. 8–17. American Standard Sizes of Dry Cells*

Designation	Nominal Cell Dimensions			
	Diameter		Height	
	Millimeters	Inches	Millimeters	Inches
AAA	9.9	25/64	42.8	1-11/16
AA	13.5	17/32	47.6	1-7/8
C	23.8	15/16	46.0	1-13/16
D	31.8	1-1/4	57.2	2-1/4
F	31.8	1-1/4	87.3	3-7/16
No. 6	63.5	2-1/2	152.4	6

* American National Standard Specifications for Dry Cells and Batteries, C18.1-1979 (International Electrotechnical Commission (IEC) 86-1, 86-2, and 86-3) American National Standards Institute, 1430 Broadway, New York, N. Y. 10018.

Sealed Beam and Tungsten-Halogen Sealed Beam/Bonded. These lamps contain filaments, lens and reflectors in a precise rugged optical package available in a wide variety of sizes in voltages ranging from 6 to 120 volts. Sealed beam lamp lenses are made of borosilicate "hard" glass. The reflector is vaporized aluminum on glass, hermetically sealed to the lens cover. The advantages are: (1) accurate reflector contour for accurate beam control; (2) precise filament positioning on rugged filament supports; and (3) high efficacy and excellent luminous intensity maintenance. Vaporized aluminum on glass is an excellent reflector, does not deteriorate, and the normal bulb blackening has little effect on output throughout lamp life. The sealed beam lamp is particularly suitable where a large amount of concentrated light at low voltage is required. Tungsten-halogen sealed beam/bonded lamps contain (as the light source) a very small bulb of quartz, 96 per cent silica, or aluminosilicate glass surrounding the filament. The bulb also contains a high-pressure rare gas atmosphere with small additions of halogen compounds as required for operation of the tungsten-halogen cycle. Fig. 8–108 gives data on typical sealed beam lamps for automotive use.

Indicator, Radio and Television. Lamps for indicator, radio and television service are usually operated from low voltage transformers. Performance data on typical lamps are given in Fig. 8–103.

Flasher Lamps. Incandescent lamps which flash automatically (see Fig. 8–18), because of a built-in bi-metal strip similar to those used in thermostats, are available in several sizes. When the lamp lights, heat from the filament causes the bi-metal strip to bend away from the lead-in wire. This breaks the circuit and the lamp goes out. As the bi-metal strip cools it bends back to its original position against the lead-in wire and lights the lamp. This alternating cooling and heating keeps the lamp flashing. An exception to this is found in certain miniature screw base foreign made lamps. Some operate as described above, however, a few are of the "shorting" type. The bi-metal in the shorting type is so mounted that when it heats up it shorts across the lead-in wire. If these lamps, which are difficult to distinguish from the "opening" type, are inserted in sockets intended for the normal flasher lamps, they may cause run-down batteries, blown fuses, over-heated wires, or burned out transformers.

Some of the lamps listed in Fig. 8-104 (120-volt D26 and D27) are available also in transparent colors. The latter are designed particularly for Christmas and decorative effects.

Typical Uses. Indicator lamps provide a visual indication of existing circuit conditions. They are widely used in fire and police signaling systems, power plant switchboards, production machinery, motor switches, furnaces, and innumerable other devices requiring warning or pilot lights.

Miniature lamps may be wired in various ways with motor or heating elements and are used in many appliances to indicate that the current is flowing to the appliance or that it is functioning properly. They are used in many ways in instrumentation and in connection with photocells and relays, and play a vital part in computers and automation in general. Flashlights, radios, clocks, bicycles and toys account for many more uses.

Other applications include the use of miniature lamps for Christmas and other festive occasions, and for colorful patio and garden lighting. For garden lighting low voltage miniature equipment is available.

Incandescent Filament Lamps for Photography

The application of lamps to photography is covered in the Application Volume. The design of lamps specifically for photographic service is concerned with actinic quality—that is, providing sources which are best adapted to the response or sensitivity of several classes of film emulsions. Some lamps are specified in terms of color temperature, which serves as a basic rating for film exposure data. Thus, several lines of lamps are made available for the requirements of commercial studios as well as for home movies and still hobby photography. Photographic efficiency and unvarying spectral quality are of major concern to the photographer. Comparatively, life is of less importance. Lamps of various sizes are often matched for color temperature, and rated life varies as necessary with wattage to achieve the specified color temperature.

Lamps with Color Temperature Ratings. Lamps rated at 3200 K color temperature are used primarily in professional photography, with both black-and-white films and many types of color films. Another group of lamps, rated at 3350 K, is used primarily in professional color motion picture work, sometimes in conjunction with blue filters to simulate daylight color quality. Lamps rated at 3400 K are widely used with color film by hobbyists and some professional photographers. So-called "daylight blue" photoflood lamps provide light of approximately 5000 K. Their blue bulbs act as filters, absorbing red and yellow rays, producing a whiter light. The "photographic blue" lamps produce a spectral quality that approximates sunlight at approximately 5500 K color temperature.

Maintenance of color temperature throughout lamp life is important in color photography. Typical color-temperature rated lamps of conventional construction may drop about 100 K through life. There is negligible change in the color temperature of tungsten-halogen lamps during their life.

Fig. 8-18. Typical lamps for flashing: a. D27, b. 405, c. 407 (incandescent lamps with integral flasher); d. B6A (NE-21) (a glow lamp). (Shown here at nearly full size).

Photoflood. These are high efficacy sources of the same character as other incandescent filament lamps for picture taking, with color temperatures ranging from 3200 to 3400 K. Because of their high filament temperature, these lamps generally produce about twice the lumens and three times the photographic effectiveness of similar wattages of general service lamps. Relatively small bulb sizes are employed (*e.g.*, the 250-watt No. 1 photoflood has the size bulb formerly used for the 60-watt general service lamp) so that these lamps may be conveniently used in less bulky reflecting equipment or for certain effects in ordinary residential or commercial luminaires.

The photoflood family includes reflector (R) and projector (PAR) lamps with various beam spreads. Some of these have tungsten-halogen light sources; some have built-in 5000 K daylight filters. In addition, tungsten-halogen lamps in several sizes and color temperatures are classed as photofloods and used in especially-designed reflectors. See Fig. 8–94.

Photoflash. These are physically patterned after standard incandescent filament lamps but are actually "combustion" sources. The lamp bulb is simply a container for the flammable material (usually metallic aluminum or zirconium or a metallic compound), a tungsten filament with a small amount of chemical applied or a pressure sensitive chemical igniter, and oxygen. The bulbs are coated with lacquer to safeguard against shattering the glass. These lamps are designed to function over specified voltage ranges or within fixed force and energy requirements (pressure sensitive igniters).

Photoflash lamps have a burning life of only a few hundredths of a second and the design is predicated on the service, type of camera and the necessary synchronism of flash and shutter opening.

The use of more efficient flammable material such as zirconium has made possible the popular reflector lamp cube and the "flashbar". These lamps may be ignited either electrically or mechanically.

Photoflash lamps are rated in lumen-seconds while reflectored units are rated in beam-candela-seconds, which is a measure of their photographic effectiveness. Color-corrected (blue-bulb) photoflash lamps and cubes are used to simulate daylight at approximately 5500 K by filtering out red and yellow. This correction reduces the light output by about one-fourth. Deep red purple bulbs are used to filter out all visible light for taking pictures with infrared film. See Fig. 8–113 for lamp data.

Photo-Enlarger. Most equipment used in the process of enlarging requires a highly diffused light source. For this service incandescent filament enlarger lamps in diffuse white glass bulbs are available. These lamps are designed for high efficacy at a short life. See Fig. 8–114.

Projection. Lamps for projectors have carefully-positioned filaments and prefocus-type bases so the filament will be properly aligned with the optical system. The filaments are very compact and operate at relatively high temperatures. The efficacy, therefore, is high with consequent short life. Forced-air cooling is frequently required because of the high lamp temperatures.

The "4-pin" base is widely used, because it provides accurate light source positioning, allows minimum bulb height in "low profile" projectors, and facilitates the use of internal reflectors. Internal "proximity" reflectors, usually of metal, eliminate the need for the spherical mirror formerly used behind the lamp in conventional projectors. Internal ellipsoidal or parabolic reflectors—metal mirrors or glass with dichroic heat-transmitting mirror coating—focus light through the film aperture into the projection lens, replac ing both the conventional external heat-transmitting dichroic mirror and condenser lenses. For projection lamp data, see Fig. 8–109.

Tungsten-halogen lamps have nearly replaced normal incandescent types for use in the projection area. Halogen lamps are available for direct replacement of many existing incandescent projection types. In addition, special lamps and sockets have been devised to take advantage of the halogen lamps' compact size and greater efficacy.

One of the principal developments has been the adoption of halogen lamp types with integrated external dichroic mirrors. By carefully positioning the lamp filament in ellipsoidal or

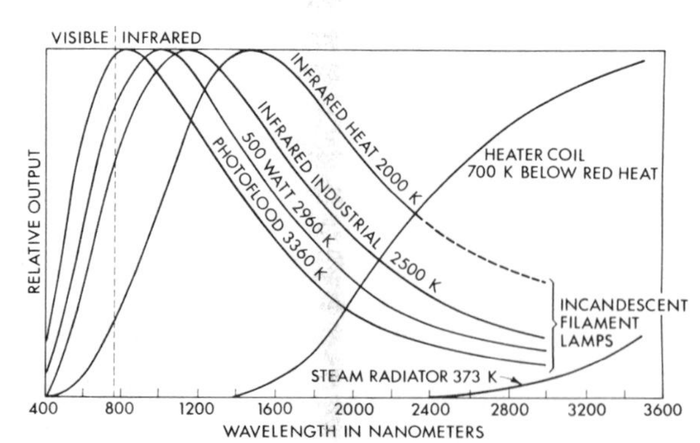

Fig. 8-19. Spectral power distribution from various infrared sources.

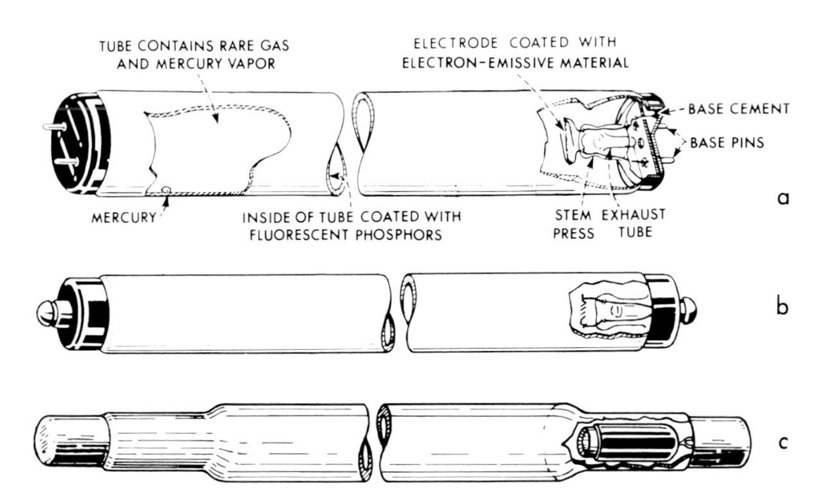

Fig. 8-20. Cutaway view of fluorescent lamps showing typical electrodes: (a) hot-cathode (filamentary) rapid-starting or preheat-starting; (b) hot-cathode (filamentary) instant-starting; (c) cold-cathode (cylindrical) instant-starting.

parabolic dichroic reflectors, precise beam control is possible. This eliminates the need for bulky and expensive external condensers and reflectors. The dichroic mirror is constructed to transmit nearly all of the infrared radiation and reflect light through the film plane. This results in a lower film gate temperature and longer film life.

Infrared Lamps

Any incandescent filament lamp is a very effective and efficient generator of infrared radiation. From 75 to 80 per cent of the wattage input to an incandescent filament lamp is radiated as infrared energy, and the greater portion of this energy is emitted in the invisible wavelength range from 770 to 5000 nanometers. Wavelengths longer than 5000 nanometers are absorbed to a large extent by the glass or fused quartz envelope. Lamps for heating applications are specially designed for low light output and long life.

Tungsten filament infrared lamps are available with ratings up to 5000 watts. Generally speaking, tungsten filament lamps for industrial, commercial and residential service operate at a filament color temperature of 2500 K. At this relatively low operating temperature compared to lighting lamp filament temperatures, the service life is well in excess of 5000 hours. Frequently, lamps using tungsten filaments have provided many years of operation because the service life is generally determined by mechanical breakage or rupture of the filament due to vibration or handling, rather than due to the rate of evaporation of tungsten as is the case with lighting lamps. Lamps having heat-resisting glass bulbs or tubular fused quartz envelopes are recommended where liquids may drop or be splashed on the bulb surface.

The distribution of power radiated by various infrared sources is shown in Fig. 8-19, while Fig. 8-111 tabulates the most popular types of infrared lamps and metal sheathed heaters. For application information, see the Application Volume.

FLUORESCENT LAMPS

The fluorescent lamp is a low-pressure gas discharge source, in which light is produced predominantly by fluorescent powders activated by ultraviolet energy generated by a mercury arc. The lamp, usually in the form of a long tubular bulb with an electrode sealed into each end, contains mercury vapor at low pressure with a small amount of inert gas for starting. The inner walls of the bulb are coated with fluorescent powders commonly called phosphors (see Fig. 8-20). When the proper voltage is applied an "arc" is produced by current flowing between the electrodes through the mercury vapor. This discharge generates some visible radiation, or light, but mostly invisible ultraviolet radiation. The ultraviolet in turn excites the phosphors to emit light. See Fluorescence, page 2-8.

The phosphors in general use have been selected and blended to respond most efficiently to ultraviolet energy at a wavelength of 253.7 nanometers[18, 19] the primary wavelength generated in a mercury low-pressure discharge.

Like most gas discharge lamps, fluorescent lamps must be operated in series with a current limiting device. This auxiliary, commonly called a *ballast*, limits the current to the value for which each lamp is designed. It also provides the required starting and operating lamp voltages.

Lamp Construction

Bulbs. Fluorescent lamps are most commonly made with tubular bulbs varying in diameter from approximately 16 millimeters (5/8 inches T-5) to 54 millimeters (2 1/8 inches T-17) and in over-all length from a nominal 150 to 2440 millimeters (6 to 96 inches). The bulb is historically designated by a letter indicating the shape followed by a number indicating the maximum diameter in eighths of an inch. Hence "T-17" indicates a Tubular bulb 17/8 inches, or 2 1/8 inches (54 millimeters) in diameter. The nominal length of the lamp includes the thickness of the standard lampholders. It is the back-to-back dimension of the lampholders with a seated lamp.

All fluorescent lamps are basically tubular bulbs of small cross sectional diameter relative to length. Shapes other than straight tubes are available. These include: a circle (circline), where the tube has been uniformly bent in a circle with the two ends adjacent to each other; a helicoid, in which the tube has a spiral groove running the length of the bulb; an intermittently grooved tube; a letter U, where the tube has been bent back upon itself; small diameter twin-tube lamps with a smaller tubular connection near the closed ends which is designed to control mercury pressure under reasonably heavy loaded conditions as are U-shaped lamps molded with squared corners (a special base includes the starter); and a U-shaped lamp bent on itself into another U-shape and enclosed in a plastic housing which includes a ballast and starter and a medium screw base.

Electrodes. Two electrodes are hermetically sealed into the bulb, one at each end. These electrodes are designed for operation as either "cold" or "hot" cathodes or electrodes, more correctly called glow or arc modes of discharge operation.

Electrodes for glow or *cold cathode* operation may consist of closed-end metal cylinders, generally coated on the inside with an emissive material. Cold cathode lamps operate at a current of the order of a few hundred milliamperes, with a high cathode fall, something in excess of 50 volts.

The arc mode or *hot cathode* electrode is generally constructed from a tungsten wire or a tungsten wire around which another very fine tungsten wire has been uniformly wound. The larger tungsten wire is coiled producing a triple coil electrode. When the fine wire is absent, the electrode is referred to as a coiled-coil electrode. This coiled-coil or triple-coiled tungsten wire is coated with a mixture of alkaline earth oxides to enhance electron emission. During lamp operation the coil and coating reach temperatures of about 1100 °C where the coil/coating combination thermally emits large quantities of electrons at a low cathode fall of the order of 10 to 12 volts. The normal operating current of hot cathode lamps presently range upwards to 1.5 amperes. As a consequence of the lower cathode fall associated with the "hot" cathode, more efficient lamp operation is obtained and, therefore, most fluorescent lamps are designed for "hot" cathode operation.

Gas Fill. The operation of the fluorescent lamp depends upon the development of a discharge between the two electrodes sealed at the extremities of the lamp bulb. This discharge is developed by ionization of mercury gas contained in the bulb. The mercury gas is maintained at a pressure of about 1.07 pascals (0.008 torr) which corresponds to a condensed or liquid mercury temperature of 40 °C (104 °F), the optimum temperature of operation for which most lamps are designed. In addition to the mercury, a rare gas or a combination of gases at low pressure, of the order of 100 to 400 pascals (1 to 3 torr), is added to the lamp to facilitate ignition of the discharge. Standard lamps employ argon gas; energy savings types, a mixture of krypton and argon, while still others employ a combination of neon and argon or neon, xenon and argon. As a consequence of ionization and mercury atom excitation, ultraviolet radiation is generated, particularly at a wavelength of 253.7 nanometers.

Phosphors. The color produced by a fluorescent lamp depends upon the blend of fluorescent chemicals (phosphors) used to coat the wall of the tube. (See Fig. 2-11 for a list of important phosphors.) There are different "white" and colored fluorescent lamps available, each having its own characteristic spectral power distribution. The most used lamps are in the "Cool White," "Cool White Deluxe," "Warm White" and "Warm White Deluxe" families; their spectral power distributions are shown in Fig. 8-21. All these types have a combination of continuous and line spectra.

There are other "white" lamps with good color rendering properties and with relatively high luminous efficacy available. They employ phosphors which radiate energy in several discrete wavelength bands, for example, blue-violet, green and red-orange. Such lamp designs are based on a concept that the human visual system responds most favorably to particular color regions.[20]

Another type of lamp has a continuous spec-

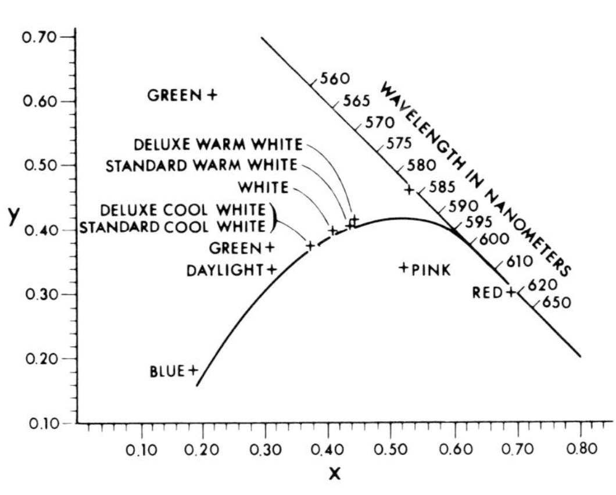

Fig. 8-21. (Left) CIE chromaticity diagram showing some white and colored fluorescent lamps in relation to the blackbody curve. (Below) Spectral power curves of light from typical fluorescent lamps. (Next page) Spectral power curves of high-intensity discharge and incandescent lamps (colors shown are approximate representations). Consult Section 5 for details on the significance of these diagrams.

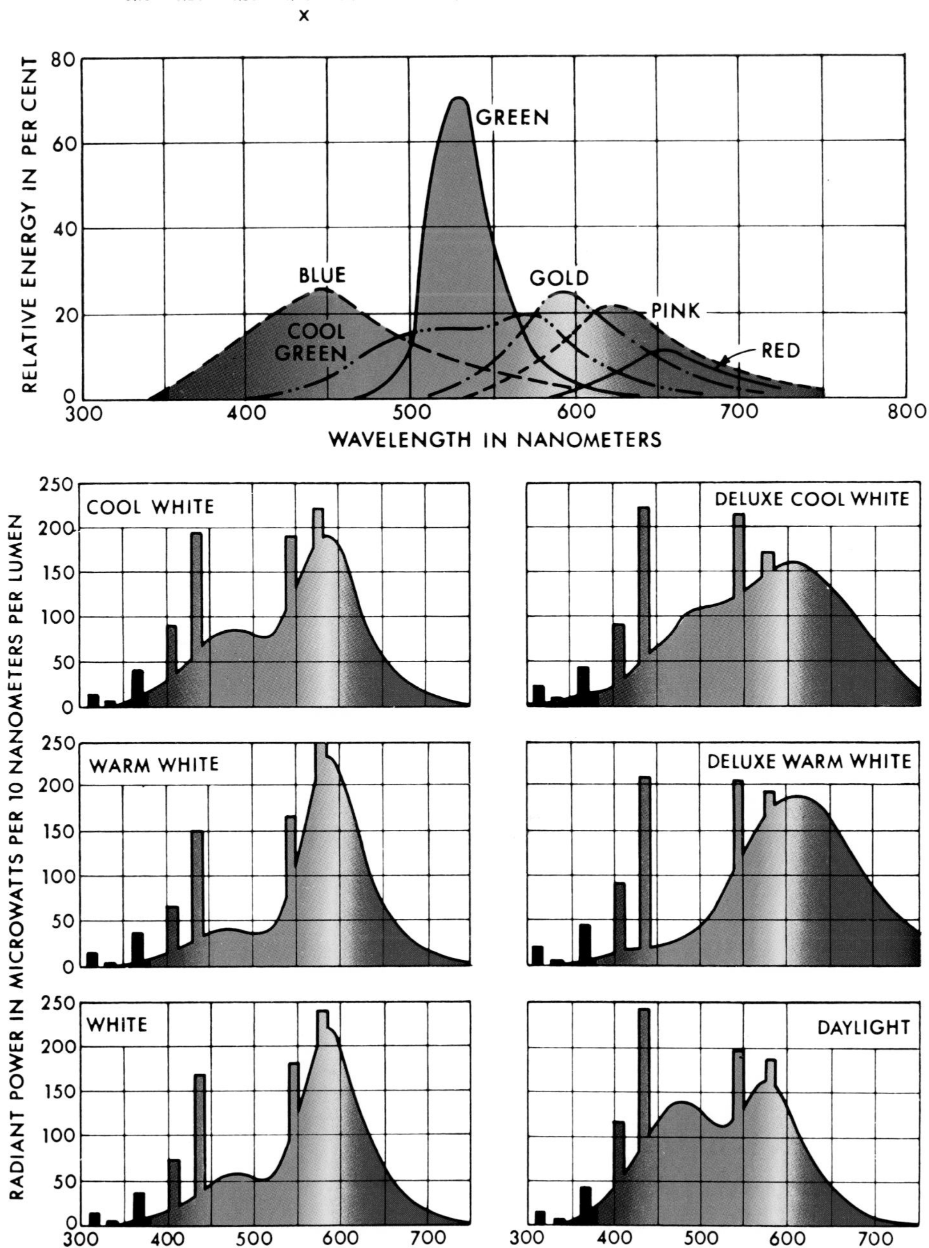

Fig. 8-21. *Continued*

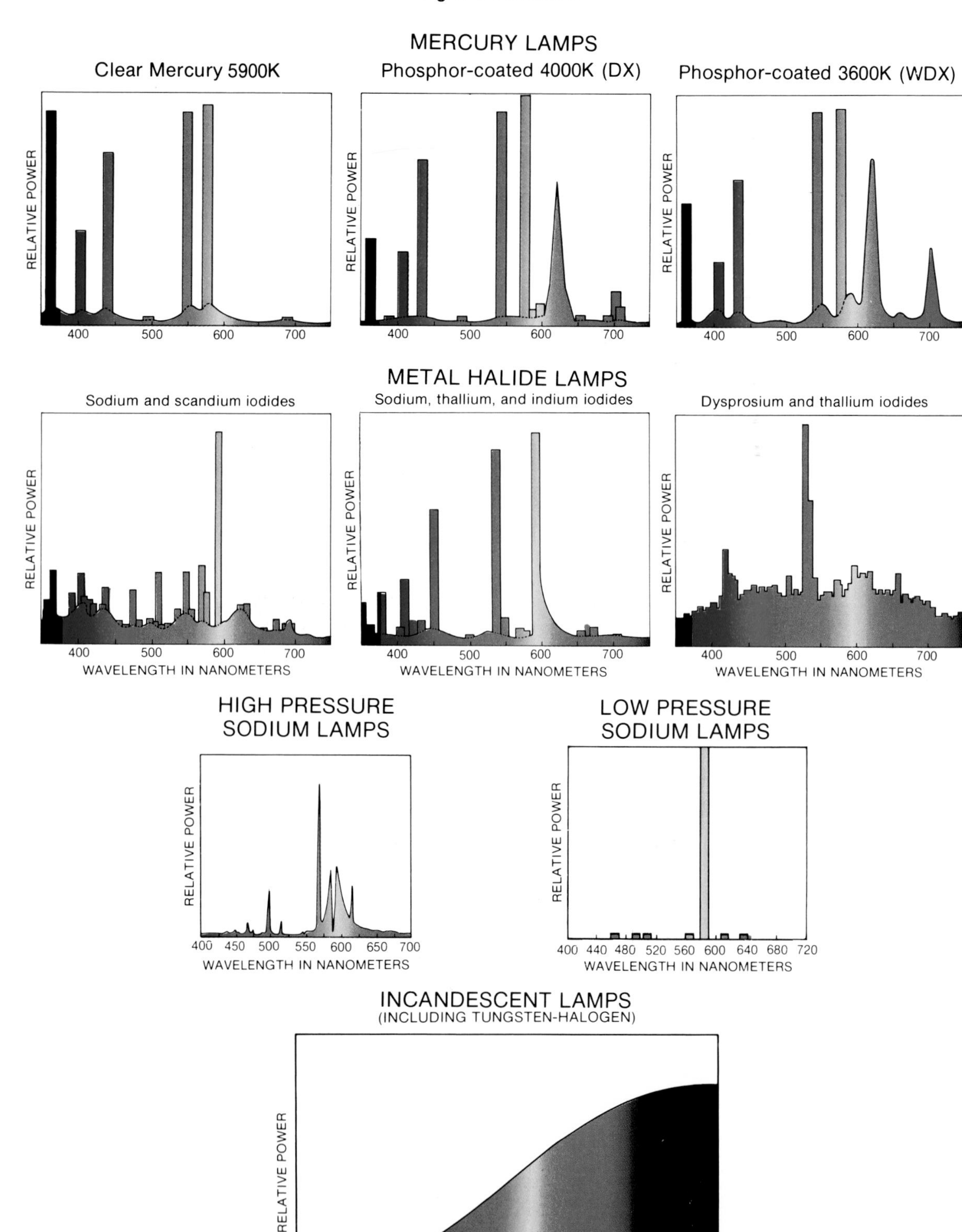

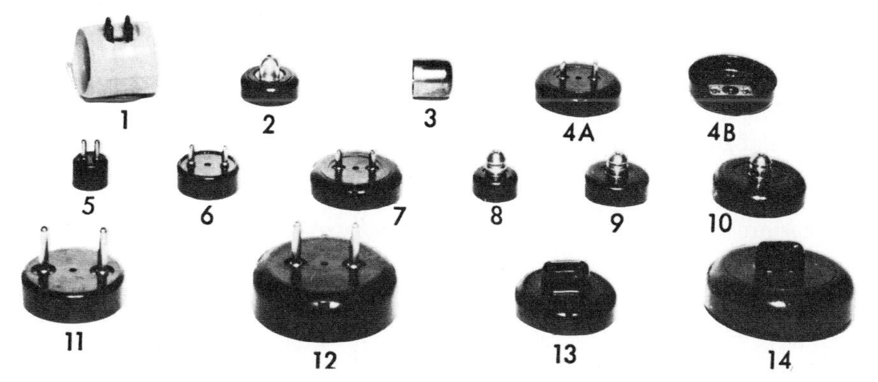

Fig. 8-22. Typical bases for fluorescent lamps: (a) Circular lamp; (2) Fluted single pin (K or clover leaf) cold cathode; (3) Ferrule type, cold cathode; (4) A. Medium bipin for T-12 lamps, showing B. internal connection between pins for instant-start lamp; (5) Miniature bipin for T-5 lamps; (6) Medium bipin for T-8 lamp; (7) Medium bipin for T-12 lamp; (8) Single pin for T-6 lamp; (9) Single pin for T-8 lamp; (10) Single pin for T-12 lamp; (11) Mogul bipin for T-12 lamp; (12) Mogul bipin for T-17 lamp; (13) Released double contact for T-12 lamp; (14) Recessed double contact for T-17 lamp.

tral power distribution which includes a controlled quantity of ultraviolet radiation. These lamps are designed with the objective of duplicating the daylight spectrum including its content of "near" ultraviolet radiation.[21]

Also, a variety of lamp types is available which generate radiation in particular wavelength regions for specific purposes, for example, for plant growth, merchandise enhancement and medical therapy.

When not lighted, most fluorescent powders appear as a matte white, translucent coating; however various colored lamps, such as red, green, gold, etc., are obtained by phosphor selection and/or filtration through pigments.

Bases. For satisfactory performance, each fluorescent lamp must be connected to a ballasted electrical circuit with proper voltage and current characteristics for its type. A number of different fluorescent lamp base designs are used. The bases physically support the lamp in most cases and provide a means of electrical connection (see Fig. 8-20).

The design of the base is dependent upon the lamp type. Lamps designed for instant start operation (see page 8-33) generally have a single connection at each end. As a consequence a single pin base is satisfactory. Such bases are shown in Fig. 8-22, numbers 8 through 10.

Preheat and rapid start lamps (see page 8-34) have four electrical connections; two at each end of the tube and, therefore, require dual contact bases. In the case of the circline lamp, a single four-pin connector is required. Examples of such bases are shown in Fig. 8-22. The base shown as number 1 is designed for use in a circline lamp, the bipin bases from 4 through 7 are used in low current applications (less than about 0.5 ampere), 11 and 12 in higher current applications, and numbers 13 and 14 for high current (0.8 to 1.5 amperes) rapid start lamp applications.

The 40-watt instant start lamp is manufactured with the medium bipin base; however, there is an internal connection between the pins, in effect producing a single contact for each cathode. Because of this construction these lamps should not be operated in preheat or rapid start circuits as the auxiliary equipment will be damaged.

Energy Saving Retrofit Lamps

With the advent of rising energy costs and the desire to conserve energy, a line of fluorescent lamps was developed which will operate on standard fluorescent ballasts,* but at reduced wattage, which may result in reduced light output. The wattage reduction is 10 to 20 per cent. This is accomplished by changing the composition of the rare gas in the lamp. Most energy saving lamps use a krypton-argon gas mixture to decrease the lamp operating voltage, thereby reducing the consumed wattage. This change in rare gas composition causes an increase in the voltage required to start the lamp. In the case of 40-watt and 30-watt types, a conductive coating of tin oxide is deposited on the bulb wall surface to achieve starting of those types on standard ballasts. Such lamps should not be operated at reduced current, on ballasts providing reduced

* Also any of the energy saving ballasts which operate standard lamps at rated current can be used, provided the ballast is listed for use with the lamps and this is stated on the ballast label. These lamps may not be used with any ballast which provides reduced wattage and thus reduced light output to a standard lamp nor with any ballast which does not list the lamp on its label.

lamp watts or at ambient temperatures below 16 °C (60 °F).

Additional energy savings (approximately 3 watts per lamp) is achieved by disconnecting the 3.5-volt electrode heating circuit of typical rapid start ballasts. This is accomplished inside the lamp with a bi-metal switch in series with each electrode. If lamps are turned off for several seconds there may be a delay of up to 60 seconds before relighting. Instant relight should occur after a momentary power interruption. Another form of energy saving fluorescent retrofit lamp is one which includes a reactive impedance built into the end of the lamp. These lamps are used to replace one of the two lamps operating on a two-lamp series rapid start type circuit. They reduce the wattage and output of the total system by approximately 33 or 50 per cent depending upon the lamp type chosen.

Other energy saving systems are available; one of them involves replacing one of the two lamps in a series circuit with a non-light-producing device which is in effect an impedance replacing the lamp impedance. Using this device reduces the system wattage by approximately 65 per cent and light output by approximately 70 per cent.

Caution must be exercised in using any lamp on a ballast other than those specifically listed on the ballast label since this may be in violation of the National Electric Code.

Special Fluorescent Lamps

In addition to the lamps described above, there are special lamps that utilize internal light control devices and those designed to produce radiation in the ultraviolet region of the spectrum. Lamps in the latter category use special phosphors or special glass without phosphor coatings. A third group of special lamps is designed for optimum light generation at temperatures other than 40 °C.

There are two types of lamps with light control from within the lamp: the reflector fluorescent lamp and the aperture fluorescent lamp. The reflector fluorescent lamp has an internal white powder layer between the phosphor and the envelope glass. This coating, which covers a major angular portion of the envelope wall, reflects a high percentage of the visible radiation striking it. The major portion of the light is emitted through the strip coated with just the fluorescent phosphor. A cross-sectional diagram and relative candlepower distribution for a 235-degree reflector lamp are shown in Fig. 8–23a. Reflector lamps with other angular width reflectors are available.

The aperture lamp is constructed similar to the reflector lamp except a clear window (no phosphor) exists for an angular aperture along the length of the tube. Elsewhere the phosphor coat is underlayed with a reflective layer. This results in a linear window of high luminance, up to ten times that of standard fluorescent lamps. The cross sectional diagram and candlepower distribution curve for a 60-degree aperture lamp are shown in Fig. 8–23b.

As a consequence of the reflector layer, absorption of generated light is somewhat higher in both types of lamps than in standard fluorescent lamps producing a somewhat reduced total lumen output. However, they are designed for applications which can effectively utilize the resulting light output distribution pattern.

Other special lamps are available for unusual ambient temperature applications. One family designed for low temperatures incorporates a jacket to conserve heat. Another group incorporates a mercury amalgam to optimize mercury vapor pressure at elevated temperatures. In both cases, these lamps are designed for optimization of the mercury vapor pressures at unusual temperatures.

Performance Parameters

To fully describe a light source, the performance parameters of the device need to be defined.

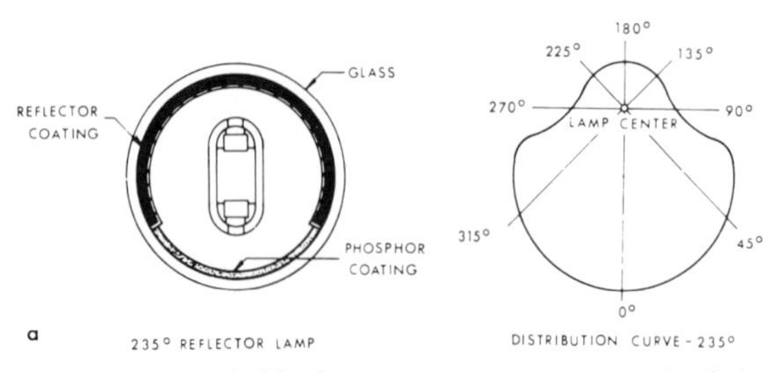

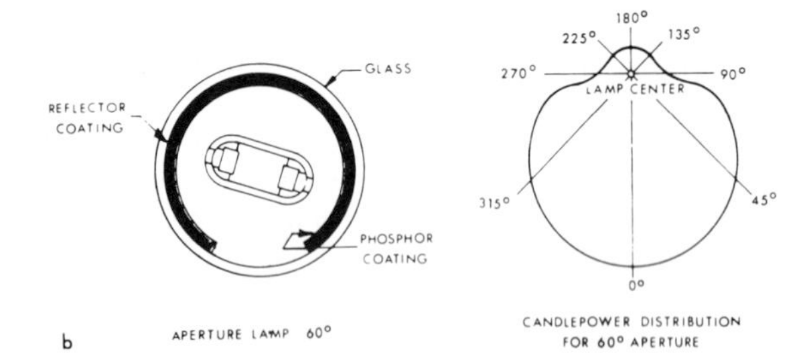

Fig. 8–23. Cross section diagrams and relative candlepower distribution curves for (a) 235-degree reflector lamp and (b) 60-degree aperture lamp.

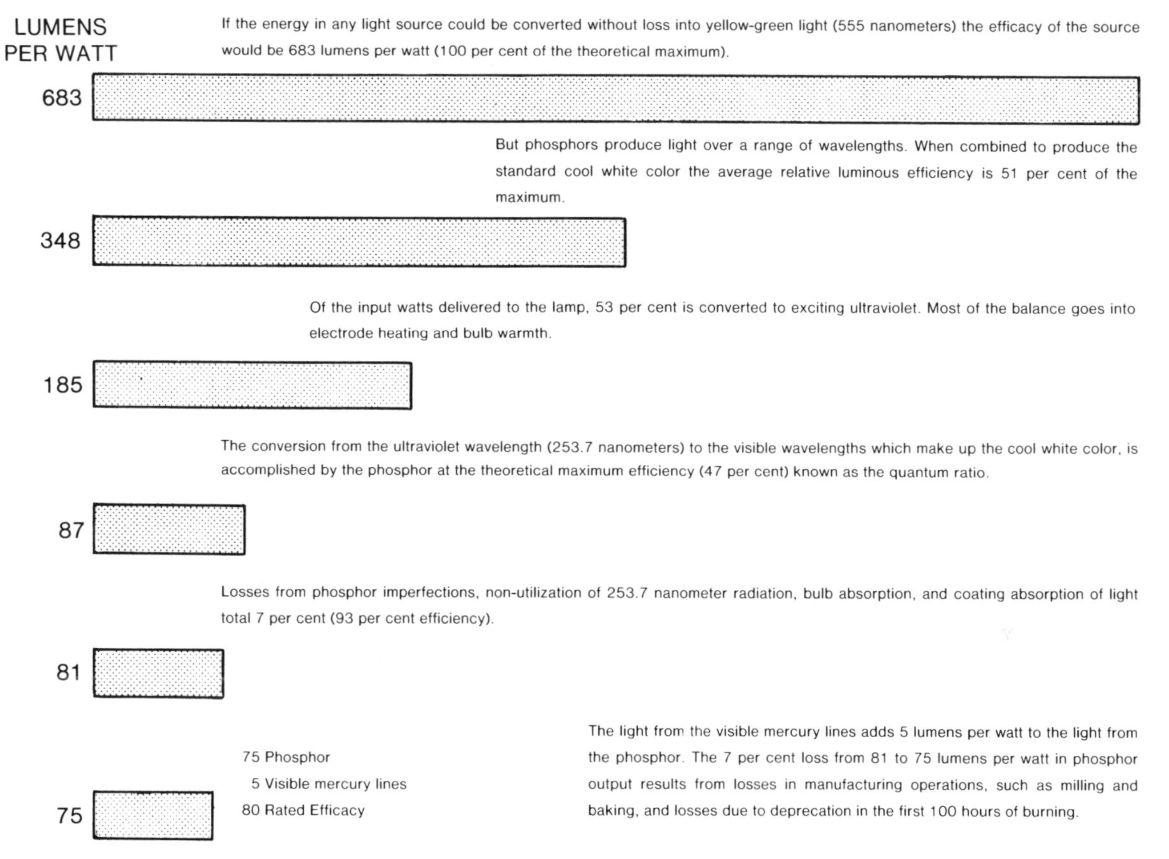

Fig. 8-24. Power conversion efficacy in a typical cool white fluorescent lamp.

Some of the more important parameters are:

1. Luminous efficacy.
2. Lamp life.
3. Light output maintenance with life.
4. Color (spectral power distribution and chromaticity).
5. Energy distribution.
6. Temperature effect on operation.
7. Flicker and stroboscopic effect.
8. Radio interference.

Luminous Efficacy: Light Output. Three main energy conversions occur during the process of generating light by a fluorescent lamp. Initially, electrical energy is converted into kinetic energy by accelerating elemental particles. These in turn yield their energy during particle collision to electromagnetic radiation, particularly ultraviolet. This ultraviolet energy in turn is converted to visible energy or light by the lamp phosphor. During each of these conversions some energy is lost so that only a small percentage of the input is converted into visible radiation light as shown in Fig. 8-24.

The geometric design and operating conditions of a lamp influence the efficacy with which energy conversions take place. Figs. 8-25 through 8-29 present data depicting the effect of bulb design and operating current.

Figs. 8-25 and 8-26 show the effect of the bulb design on lamp operation. As seen in Fig. 8-25, as lamp diameter increases efficacy increases, passes through a maximum, and decreases. There are at least two important reasons for this phenomenon. In lamps of small diameter, an

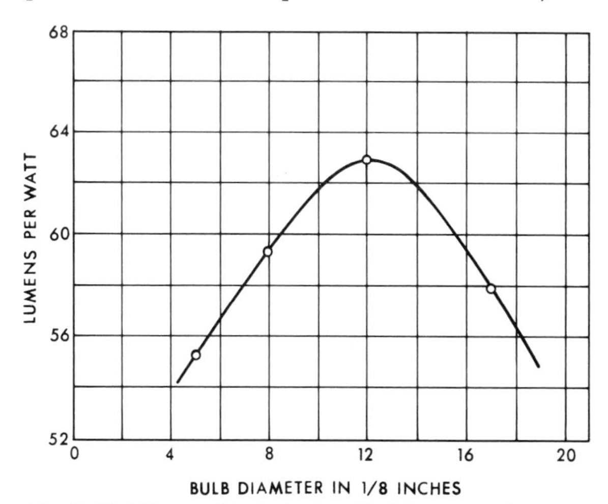

Fig. 8-25. Efficacy of typical argon filled fluorescent lamps as a function of bulb diameter in 3-millimeter (⅛-inch) units.

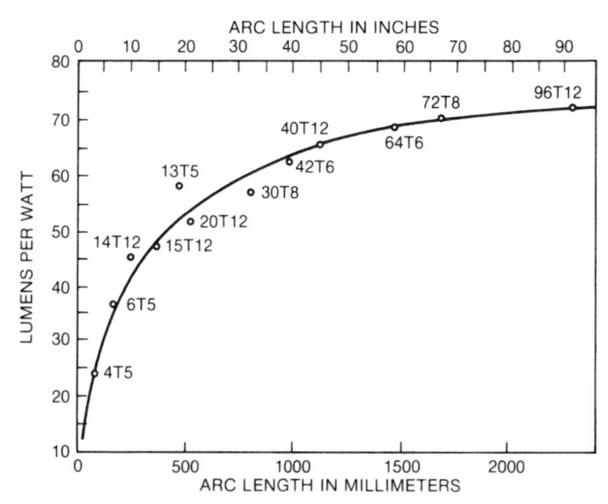

Fig. 8-26. Efficacy of typical fluorescent lamp as function of lamp length.[23]

excessive amount of energy is lost by recombinations of electrons with ions at the bulb wall. As the bulb diameter is increased, this loss becomes progressively smaller, but losses due to "imprisonment of radiation" become correspondingly larger.

As shown in Figs. 8-26 and 8-27, the length of a lamp influences its efficacy, *i.e.*, the longer the length the higher the efficacy. This is based on two separate energy losses within the lamp; the energy absorbed by the electrodes, which do not generate any appreciable light, and the energy loss directly associated with the generation of light. The electrode loss is essentially a constant, whereas the loss associated with light generation is a function of lamp length. As the lamp length increases, the effect of the electrode loss becomes less in comparison to the total losses.

Lamp operating voltage, similar to efficacy, is a function of lamp length. This effect is shown in Figs. 8-28 and 8-29. Fig. 8-28 presents data for hot cathode T-12 operation at four different current levels. Fig. 8-29 presents similar data for both hot and cold cathode T-8 lamp operation. The characteristic electrode drop for the two types of cathodes is indicated by the intersection of the curves with the ordinate corresponding to zero length.[23]

Figs. 8-115 through 8-117 and Figs. 8-119 and 8-120 show the light output data on typical fluorescent lamps. So far as possible, data presented on types produced by several manufacturers represent industry averages. Since these data are likely to differ slightly with specific figures for one manufacturer's product, it is advisable to check the manufacturers' sheets for detailed information on current production.

Lamp Life. Lamp life of hot cathode lamps is determined by the rate of loss of the electron emissive coating on the electrodes. Some of the coating is eroded from the filaments each time the lamp is started. Also, during lamp operation there is evaporation of emissive material. Although electrodes are designed to minimize both of these effects, the end of lamp life is reached when either the coating is completely removed from one or both electrodes or the remaining coating becomes non-emissive.

Fig. 8-27. Relationship of Arc Length and Lumens per Watt for Typical Cool White Fluorescent Lamps

Approximate Arc Length		Hot Cathode				Low Pressure T-8 Cold Cathode
		Approximate Efficacy (lumens per watt)	Approximate Lamp and Auxiliary Efficacy (lumens per watt)			Approximate Efficacy (lumens per watt)
(millimeters)	(inches)		Preheat Start	Instant Start	Rapid Start	
130	5	33	22			
250	10	45	34			
380	15	53	42			
500	20	59	48			
630	25	63	53			
760	30	66	57			
890	35	68	59			
1010	40	69	61	52	60	38
1140	45	70	62			
1270	50	71	63	54	63	41
1400	55	72				
1520	60	73		56	65	44
1650	65	73				
1780	70	74		59	67	47
1900	75	75				
2030	80	76		61	69	50
2160	85	77				
2290	90	77		63	71	52

Because some of the emissive coating is lost from the electrodes during each start, the frequency of starting hot cathode lamps influences lamp life. The rated average life of fluorescent lamps is usually based on three hours of operation per start. The effect of starting frequency on life is presented in Fig. 8-30. These data have been normalized to 100 per cent for life at three hours per start. Cold cathode lamps are not appreciably affected by starting frequency because of the type of electrode used.

There are many other variable conditions that affect lamp life in actual use. Ballast characteristics and starter design are key factors in the case of preheat circuits. Ballasts which do not provide specified starting requirements, or which do not operate lamps at proper voltage levels, can greatly affect lamp life. For preheat circuits, starters must also be designed to meet specified characteristics.

Proper electrode (cathode) heating current in rapid start lamps is critical and is affected not only by ballasts, but also by poor lamp to lampholder contact or improper circuit wiring. Improper seating of a lamp in a lampholder may result in no electrode (cathode) heating. Lamps operating in this mode will fail within a few hours—50 to 500 hours, depending on lamp type. Another factor in lamp life is line voltage. If line voltage is too high, it may cause instant starting of lamps in preheat and rapid start circuits. If it is low, slow starting of rapid start or instant start type lamps, or recycling of starters in preheat circuits may result. All of these conditions adversely affect lamp life. A typical mortality curve for a large group of good quality fluorescent lamps is given in Fig. 8-31.

Ballasts are available for low temperature starting of rapid start lamps. With higher temperatures encountered during the summer months, lamps operating on these ballasts will start before the electrodes are properly heated. This has an adverse effect on lamp life. Time delay relays are available to insure proper electrode heating prior to application of high ignition voltage to the lamp.

Light Output Maintenance with Life. The lumen output of fluorescent lamps decreases with accumulated burning time. Although the exact nature of the change in the phosphor which causes the phenomenon is not fully understood, it is known that at least during the first 4000 hours of operation the reduction in efficacy is related to arc-power to phosphor-area ratios. This relationship in several typical lamps is shown in Fig. 8-32. As depicted in Fig. 8-32a, lamps with different arc power to phosphor-area ratios have different lumen maintenance curves. This effect of lamp loading on lumen maintenance is presented in Fig. 8-32b.[24,25]

Another important effect that reduces light output is end darkening, resulting from emission coating deposition on the phosphor.

Color, Spectral Power Distribution and Chromaticity. Spectral power distribution curves for typical colors of fluorescent lamps are shown in Fig. 8-21. The CIE chromaticity diagram on which the position of various lamp colors are noted in conjunction with the blackbody curve is also shown in this figure. For a discussion of color and color rendering index, see Section 5.

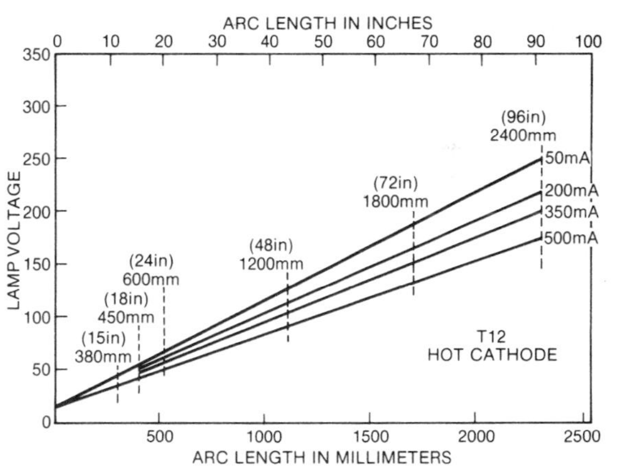

Fig. 8-28. Operating voltage of typical 38-millimeter (T-12) hot-cathode fluorescent lamps as a function of arc length.

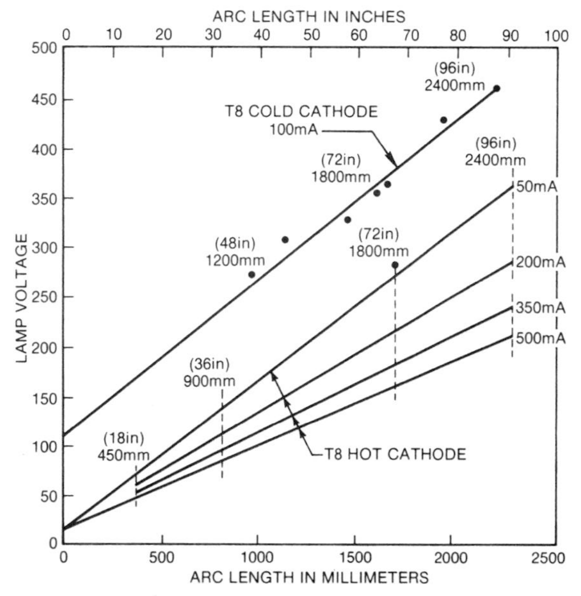

Fig. 8-29. Operating voltage of typical hot and cold-cathode 25-millimeter (T-8) fluorescent lamps as a function of arc length.

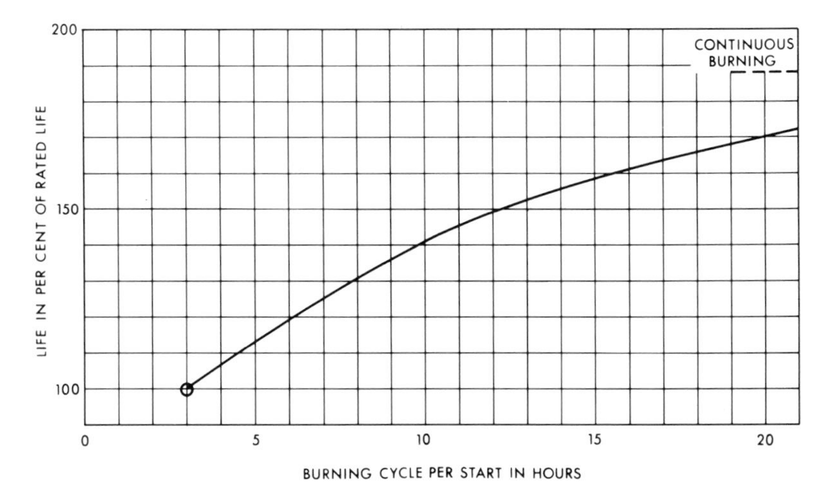

Fig. 8-30. Life of typical preheat or instant-start fluorescent lamp as a function of burning cycle. Variations from this curve can be expected with lamp loading.

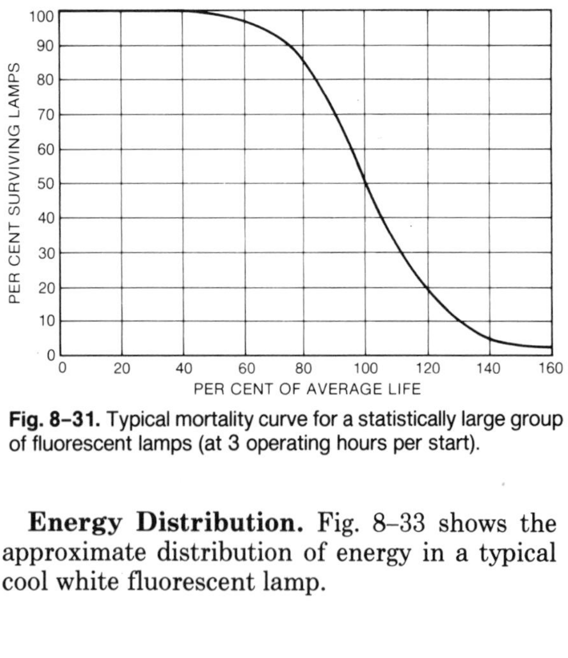

Fig. 8-31. Typical mortality curve for a statistically large group of fluorescent lamps (at 3 operating hours per start).

Energy Distribution. Fig. 8-33 shows the approximate distribution of energy in a typical cool white fluorescent lamp.

Temperature Effect On Operation

The characteristics of a fluorescent lamp are very much dependent upon the concentration of mercury vapor, or more specifically, the mercury vapor pressure within the lamp, which depends upon temperature.

A fluorescent lamp contains a larger quantity of liquid mercury than will become vaporized at any one time. This excess liquid mercury tends to condense at the coolest point or points on the lamp. If any particular location is significantly cooler than others in the lamp all the liquid mercury will collect at this point. The mercury pressure within the lamp will depend upon the temperature of this coolest point or points, the temperature depending upon lamp construction, lamp wattage, ambient temperature, luminaire design and wind or draft conditions.

The effects of temperature on mercury vapor pressure manifests itself as variations in light output and color.

Lamps using mercury amalgams are available for extending the usable ambient temperatures to higher values. The amalgam functions to stabilize and control the mercury pressure.

Effect of Temperature on Light Output. Light output depends upon mercury vapor pressure which in turn depends upon the temperature of the coolest point on the fluorescent lamp. Electrical characteristics are also affected. Typical characteristics of a lamp operated at constant current are shown in Fig. 8-34. The exact shape of the curves depends upon the lamp cross section, loading, and type of ballasting. However, all fluorescent lamps have essentially the same relation between light output and minimum bulb wall temperature, since this relation depends primarily upon mercury vapor pressure, which is a common function of temperature.

Light output and luminous efficacy reach optimum values at about 38 °C (100 °F). Fluorescent lamps intended primarily for indoor use are designed so that their coolest bulb-wall temperature will be near 38 °C (100 °F) when the lamp is operated in typical well-designed luminaires under usual indoor temperatures.

Curves for an 800-milliampere high output fluorescent lamp are shown in Fig. 8-35 (left). As these curves indicate, light output falls to very low values at temperatures below freezing. Lamps intended for indoor operation will produce poor low-temperature performance unless protected by suitable enclosures.

Fig. 8–35 (right) shows the relationship between ambient temperature and light output for a typical outdoor floodlight using 800-milliampere high output lamps. While considerable variation occurs with temperature change, satisfactory illumination is obtained for temperatures commonly encountered in most areas of the United States and Canada.

Each particular lamp-luminaire combination has its own distinctive characteristic of light output vs. ambient temperature. In general, the shape of the curve will be quite similar for all luminaires, but the temperature at which the highest light output will be reached may be quite different.

In areas with fluorescent lighting systems, high ambient temperatures will cause losses in the illuminance level and reductions in wattage. The lighting designer should take this into account in making his calculations.

Fig. 8–35 shows that the loss in light as the lamp is heated beyond the optimum temperature

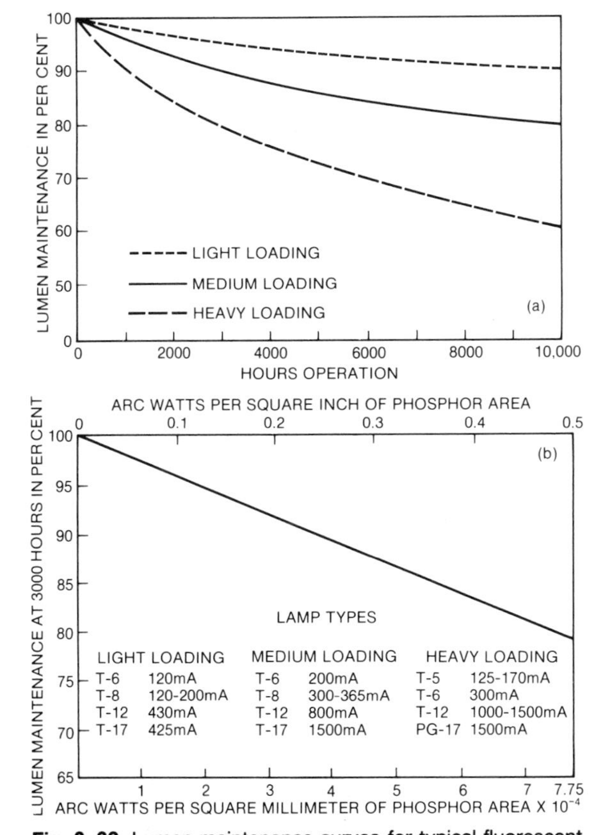

Fig. 8–32. Lumen maintenance curves for typical fluorescent lamps as a function of: (a) hours operation and (b) incident radiation power density on the phosphor surface.

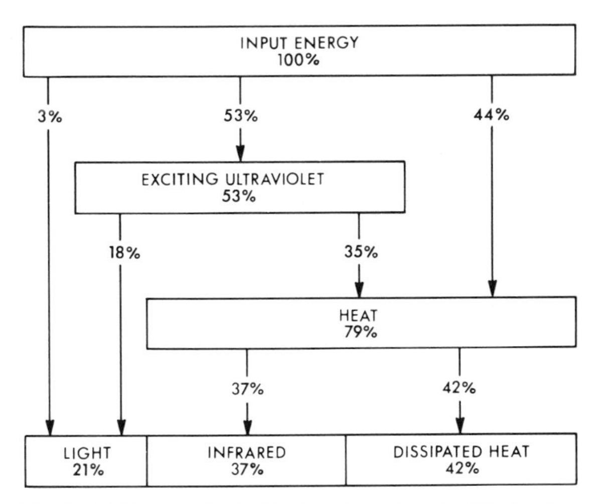

Fig. 8–33. Energy distribution in a typical cool white fluorescent lamp.

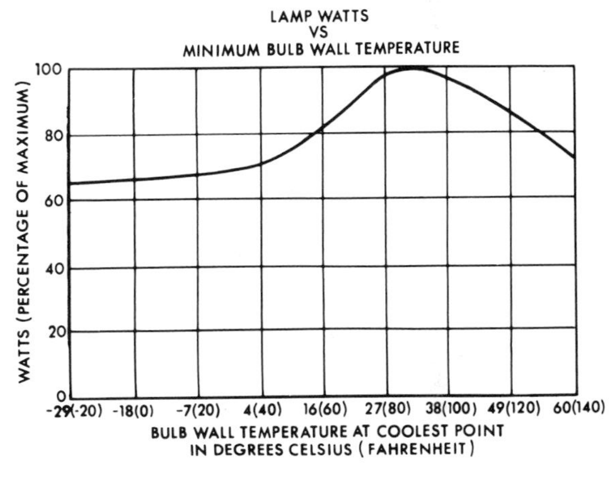

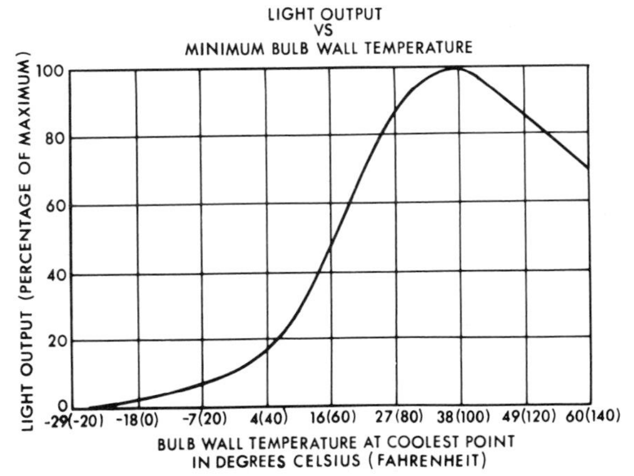

Fig. 8–34. Typical fluorescent lamp temperature characteristics. Exact shape of curves will depend upon lamp and ballast type; however, all fluorescent lamps have curves of the same general shape, since this depends upon mercury vapor pressure.

is nearly linear. From this fact an estimating rule can be derived which enables the designer to compensate for high ambient temperature condition: there will be a 1 per cent loss in light for every 1.1 °C (2 °F) that the ambient temperature around the lamp exceeds 25 °C (77 °F).

Effects of Temperature on Color. Just as temperature affects the light output of a lamp, temperature also affects the color of the light produced. The color of light from a fluorescent lamp has two components: that from the phosphor coating on the lamp and that from the mercury arc discharge within the lamp. Each of these components reacts independently to temperature changes. In general, the lamps shift toward the blue-green with increasing temperature due to the increasing contribution from the mercury arc visible spectrum.

Fig. 8-36 shows a typical color shift characteristic of a fluorescent lamp. In addition, the MacAdam four step ovals are noted for each color. These ovals depict the color tolerance limits for lamps operated at 25 °C (77 °F). These variations are due to manufacturing, color changes through life, etc. In a new interior lighting installation, color is much more uniform than this since all lamps are of the same manufacture, age, etc. Color shift becomes a concern to the lighting designer chiefly in cases where substantial differences in internal temperature may exist between adjacent luminaires. This may arise from the proximity of certain luminaires to air diffusers, open windows, etc.; differences in ceiling cavity conditions or ceiling material with surface and recessed equipment; differences in the tightness of enclosures with enclosed equipment; differences in lamp loading or number of lamps in identical luminaires; and use of some of the luminaires as air diffusers in conjunction with the air-conditioning system.

Flicker and Stroboscopic Effect

The ultraviolet energy generated by an arc discharge is a function of the instantaneous power input. As a consequence, the generated ultraviolet energy shows cyclic changes similar to the power input. The frequency of this variation is twice the input frequency. The cyclic variation of the ultraviolet energy is transferred to the emitted light output where the phosphors show both fluorescence and phosphoresence. As a consequence, some smoothing of the instantaneous light output occurs. Even so, cyclic variation in instantaneous light output results. This variation in light output is called *flicker.*

With a 60-hertz input frequency, the resulting 120-hertz variation is too fast to be noticed by the eye. However, when rapidly moving objects are viewed under these fluorescent systems, blurred "ghost" images may be observed. This effect is known as *stroboscopic effect.* The greater the flicker, the more noticeable is the stroboscopic effect.

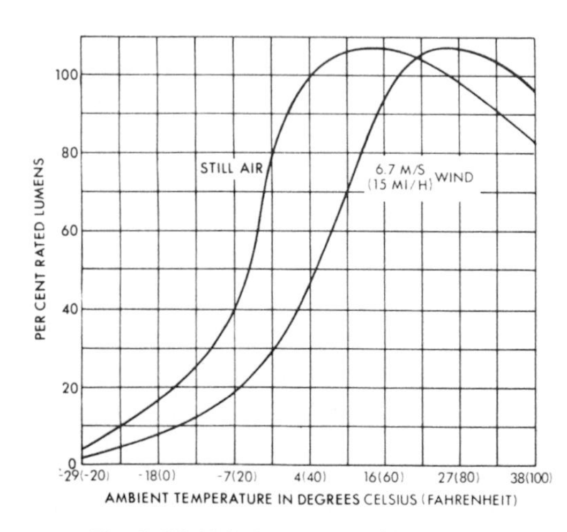

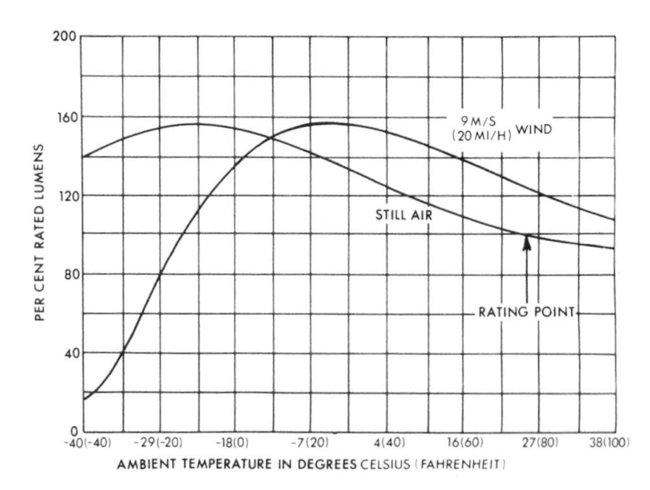

Fig. 8-35. Light output vs ambient temperature. (*Left*) F96T12/HO fluorescent lamp. Light output falls to low values at temperatures below freezing. Loss in light at high ambient temperatures is much less. (*right*) Typical outdoor floodlight with two F72T12/HO. By a suitable enclosure, the lamp is warmed to a high enough temperature to produce good light output under cold windy conditions.

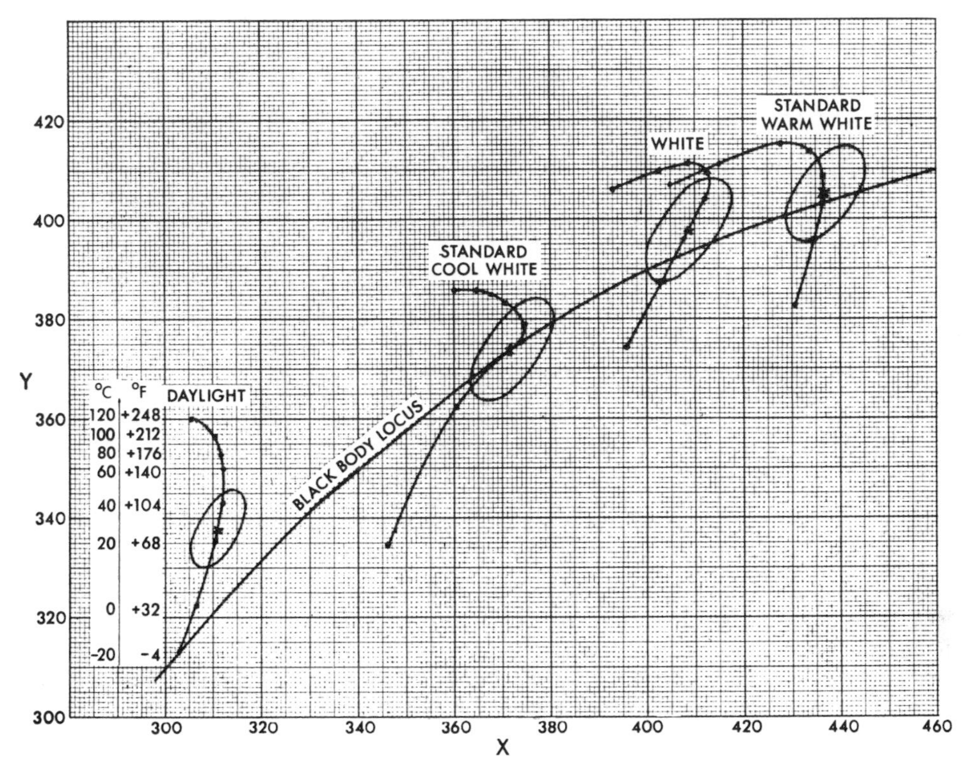

Fig. 8-36. Typical color shift characteristic of a fluorescent lamp with changes in ambient temperature. The MacAdam 4-step ovals illustrate the tolerance limits for the particular white. Color shift over a normal temperature range is in the same order of magnitude as that which may be experienced between lamps of the same nominal color due to manufacturing variations, depreciation through life, etc. (The lowest point on each curve is at −20°C (−4°F), the top point at 120°C (248°F), intermediate points are 20°C (36°F) apart; see daylight lamp.)

Flicker Index[26] has been established as a reliable, relative measure of the cyclic variation in output of various sources at a given power frequency. Previously this cyclic variation has been defined by *Per Cent Flicker.* However, Flicker Index is now considered to be a more reliable measure since it takes into account the wave form of the light output. It is calculated by dividing the area above the line of average light output by the total area under the light output curve for a single cycle (see Fig. 8–37). In Fig. 8–37, Per Cent Flicker is equal to $100(A - B)/(A + B)$, (per cent modulation).

The Flicker Index gives ratings from 0 to 1.0 with 0 for steady light output. The higher values of Flicker Index indicate an increased possibility of noticeable stroboscopic effect, as well as lamp flicker. In Fig. 8–38 the Flicker Index and Per Cent Flicker are listed for six "white" lamps when operated from typical circuits.

Some flicker may be seen from the ends of a lamp when viewed in the periphery of the retina (out of the "corner of one's eye"). This is due to the arc being initiated at alternate electrodes.

Radio Interference

The mercury arc in a fluorescent lamp emits electromagnetic radiation. This radiation may be picked up by nearby radios causing an audible sound. Because of the frequencies generated by the fluorescent lamp, interference is ordinarily limited to the AM broadcast band and nearby amateur and communications bands. FM, television, and higher frequencies are very rarely affected by radiated interference but may be affected by conducted interference. Most instant start ballasts and starters for preheat circuits have capacitors for reduction of radio interference.

Radio noise reaches the radio either by radiation to the antenna or by conduction over the power lines. Radiated interference may be eliminated by moving the antenna farther from the lamp. Three meters (ten feet) is usually sufficient. Where this is not practical, shielding media, such as electrically conducting glass or certain louver materials, will suppress the noise below the interference level. Conducted interference may be suppressed by an electric filter in the line at the luminaire. Fig. 8-39 shows a typical design. Luminaires with this type of filtering and appropriate shielding material have been qualified under pertinent military specifications for sensitive areas.

Lamp Circuits and Auxiliary Equipment

Like most arc discharge lamps fluorescent lamps have a negative volt-ampere characteristic and therefore, they require an auxiliary device to limit current flow. This device, normally called a ballast, may incorporate an added function which provides a voltage sufficient to insure ignition of the arc discharge. This voltage may vary between 1.5 to 4 times the normal lamp operating voltage.

The life and light output ratings of fluorescent lamps are based on their use with ballasts providing proper operating characteristics. The required operating characteristics have been established in the American National Standards for Dimensional and Electrical Characteristics of Fluorescent Lamps (C78 group). Ballasts that do not provide proper electrical values may reduce either lamp life or light output, or both. This auxiliary equipment consumes power and therefore reduces the over-all lumens-per-watt rating below that based on the power consumed by the lamp. Typical data are presented in Fig. 8-121.

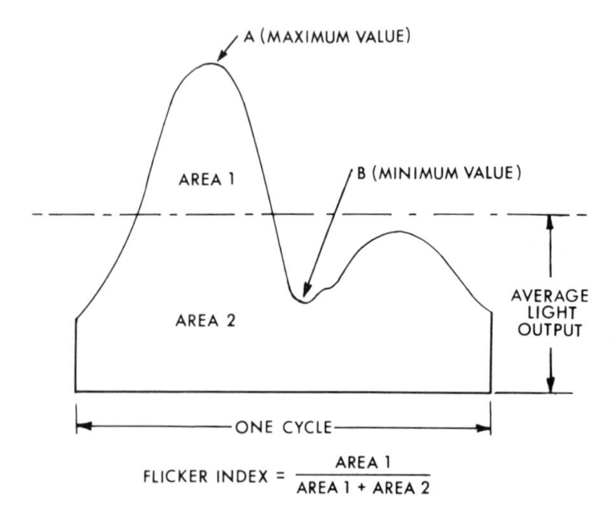

Fig. 8-37. Curve of the light output variation from a fluorescent lamp during each cycle, showing the method of calculating the Flicker Index.

Fig. 8-38. Flicker Index and Per Cent Flicker Values for "White" Fluorescent Lamps

	Single Lamp		2-Lamp Lead-Lag Instant Start		2-Lamp Lead-Lag Preheat Switch Start	
	Flicker Index	Per Cent Flicker	Flicker Index	Per Cent Flicker	Flicker Index	Per Cent Flicker
Cool white	.079	34	.071	26	.056	16
Deluxe cool white	.078	34	.075	27	.046	14
Warm white	.048	20	.044	16	.029	10
Deluxe warm white	.049	20	.043	16	.030	10
Daylight	.119	50	.107	41	.075	24
White	.058	25	.054	20	.042	12

Lamp Starting

The starting of a fluorescent lamp occurs in two stages. First, a sufficient voltage must exist between one lamp electrode and a nearby grounded conductor such as the metal surface of the luminaire. This will initiate ionization of the gas in the lamp. Secondly a sufficient voltage must exist across the lamp to extend the ionization throughout the lamp and develop an arc.

As ambient temperature is reduced, starting of all fluorescent lamps becomes more difficult. For reliable starting at low temperatures higher available output voltages are required. For more efficient ballast/lamp operation ballasts are available for each of the following temperature ranges:

Above 10 °C (50 °F) for indoor applications.
Above −18 °C (0 °F) for outdoor temperature applications.
Above −29 °C (−20 °F) for outdoor temperature applications.

A number of different means of lamp starting have been developed since the advent of the fluorescent lamp. The first was preheat starting which required an automatic or manual starting switch. Then came instant starting which required higher voltage. The most recent and probably the most important development was rapid starting where the use of continuously heated electrodes resulted in lamp starting without high voltage or starting switches. Several circuits for operating lamps are shown in Fig. 8-40.

Preheat Lamp/Ballast Operation. Early fluorescent lamp systems were all of the preheat type. As the name implies, the lamp electrodes are heated before application of the high voltage across the lamp. Lamps designed for such operation have bipin bases to facilitate electrode heating.

The preheating requires a few seconds and this is usually accomplished by an automatic switch which places the lamp electrodes in series across the output of the ballast. Current flows through both electrode filaments causing a temperature rise in the filaments. Subsequently, the switch opens applying the voltage across the lamp. Due to the opening of the switch under load, a transient voltage (an inductive kick) is developed in the circuit which aids in ignition of the lamp. If the lamp does not ignite, the switch will reclose and reheat the filaments. In some applications this preheating is accomplished by a manual switch.

The automatic switch is commonly called a *starter*. It may incorporate a small capacitor (0.006 microfarads) across the switch contacts to shunt high frequency oscillations which may cause radio interference.

Ballasts are available to operate some preheat lamps without the use of starters. These ballasts use the rapid start principle of lamp starting and operation and are popularly called *trigger start* ballasts.

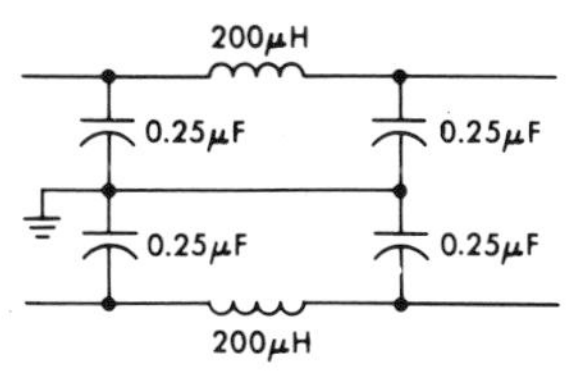

Fig. 8-39. Typical radio interference filter used in critical applications.

Instant Start Lamp/Ballast Operation. Arc initiation in instant start lamps depends solely on the application of a high voltage across the lamp. This high voltage (400 to 1000 volts) ejects electrons from the electrodes by field emission. These electrons flow through the tube, ionizing the gas and initiating an arc discharge. Thereafter the arc current provides electrode heating. Because no preheating of electrodes is required, instant start lamps need only a single contact at each end of the lamp. Thus, the single pin lamp is used on most instant start lamps.

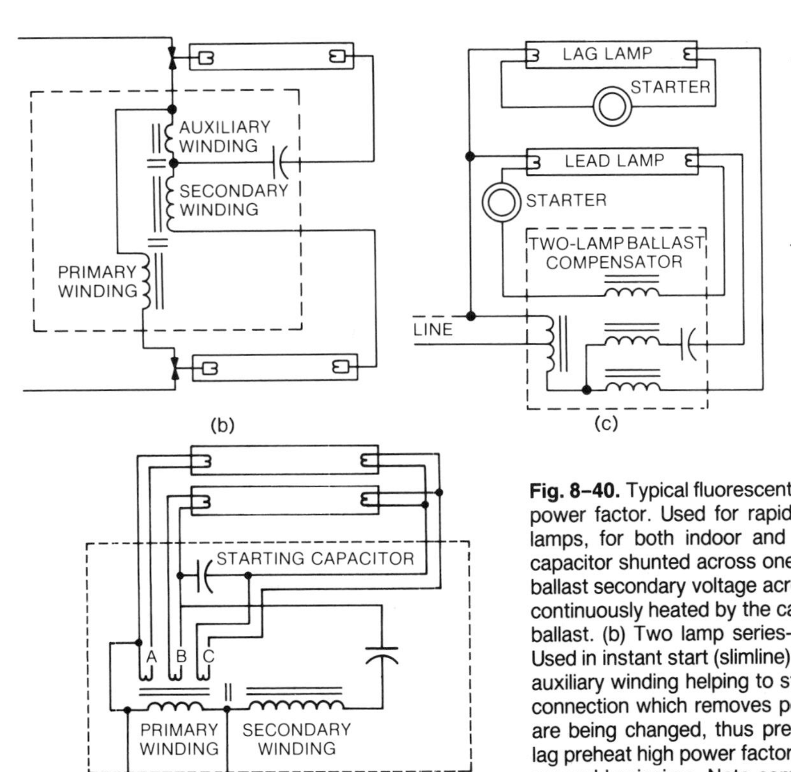

Fig. 8-40. Typical fluorescent lamp circuits: (a) Two lamp rapid start high power factor. Used for rapid start, high output and extra high output lamps, for both indoor and low temperature applications. The small capacitor shunted across one lamp momentarily applies nearly all of the ballast secondary voltage across the other lamp. The lamp cathodes are continuously heated by the cathode heating windings (A, B and C) in the ballast. (b) Two lamp series-sequence instant start high power factor. Used in instant start (slimline) indoor units. Lamps start in sequence with auxiliary winding helping to start first lamp. Note disconnect lampholder connection which removes power from the ballast primary when lamps are being changed, thus preventing electric shock. (c) Two-lamp lead-lag preheat high power factor. Used in 40-watt and 90-watt preheat type general luminaires. Note compensator winding which is needed to produce sufficient preheat current in the lead circuit. (d) Simple preheat circuit low power factor. Used for appliances, desk lamps, etc.

These are commonly called "slimline" lamps. A few instant start lamps use bipin bases with the pins connected internally. In the case of lamps designed for instant starting at 400 to 1000 volts open circuit, it is necessary to provide some means of counteracting the effect of humidity on the capacitive lamp-ground current which initiates the necessary glow discharge. Most manufacturers coat the outside of the bulb of this type of lamp with a transparent, non-wetting material; others apply a narrow conducting strip along the bulb. A grounded conducting plate, such as a metal reflector near the lamp, commonly known as a "starting aid," is necessary to obtain the lowest lamp starting.[27]

Rapid Start Lamp/Ballast Operation. The rapid start principle has been used in most recent fluorescent lamp developments. It can use low or high resistance electrodes which are heated continuously. Heating is accomplished through low voltage windings built into the ballast or through separate low voltage transformers designed for this purpose. This results in a starting voltage requirement similar to that of preheat lamps. Lamps usually start in 1 second, the time required to bring the filaments up to proper temperature.

A starting aid, consisting of a grounded conducting plate extending the length and adjacent to the lamp is a prerequisite to reliable starting. For lamps operating at 500 milliamperes or less the nominal distance between the lamp and a 25-millimeter (1-inch) wide conducting plate is 13 millimeters (½ inch); for lamps operating at currents greater than 500 milliamperes the nominal distance to the conducting strip is 25 millimeters (1 inch). Peak voltage recommendations for lamp ignition using starting aids other than the nominal are listed in Fig. 8-41.

Rapid start lamps are coated with a transparent non-wetting material to counteract the adverse effect of humidity in lamp starting. All 800-milliampere and most 1500-milliampere lamp types operate on the rapid start principle. Forty-watt and circline lamps designed for rapid start service can also be used in comparable preheat circuits.

Ballast Construction. The construction of a typical thermally protected rapid start ballast is shown in Fig. 8-42. The components consist of a transformer type core and coil. Depending upon the circuit, a capacitor may be part of the ballast. These components are the heart of the ballast, providing sufficient voltage for lamp ignition and lamp current regulation through their reactance.

The core and core assembly is made of laminated transformer steel wound with copper or aluminum magnet wire. The assembly is impregnated with a non-electrical conducting material that provides electrical insulation while aiding in heat dissipation, and with leads attached, is placed into a case. The case is filled with a potting material (hot asphalt for example) containing a filler such as silica. This compound completely fills the case encapsulating the core and coil and capacitor. The base is then attached.

Average ballast life at a 50 per cent duty cycle

Fig. 8-41. Effect of Starting Aid Dimensions on Peak Voltage Requirements for Reliable Starting of T-12 Argon-Filled Rapid Start Lamps

Starting Aid Dimensions				Per Cent Change in Peak Voltage	
Distance to Lamp		Width of Aid			
millimeters	inches	millimeters	inches	Lamp Operating Current Equal to or Less Than 500 mA	Lamp Operating Current Greater Than 500 mA
	Nominal*	200	8 or greater	−8	−8
	"	50	2	−4	−4
	"	25	1	−0	0
	"	13	½	+4	+4
	"	6	¼	+8	+8
	"	3	⅛	+12	+12
	"	1.5	1/16	+20	+20
75	3	25	1	+32	+25
50	2	25	1	+22	+15
38	1-½	25	1	+15	+8
25	1	25	1	+7	0
13	½	25	1	0	−7
6	¼	25	1	−8	−15
3	⅛ or less	25	1	−12	−20

* Nominal distance from starting aid to lamp is 13 millimeters (½ inch) for operating lamp current of 500 mA or less and 25 millimeters (1 inch) for operating lamp current of greater than 500 mA.

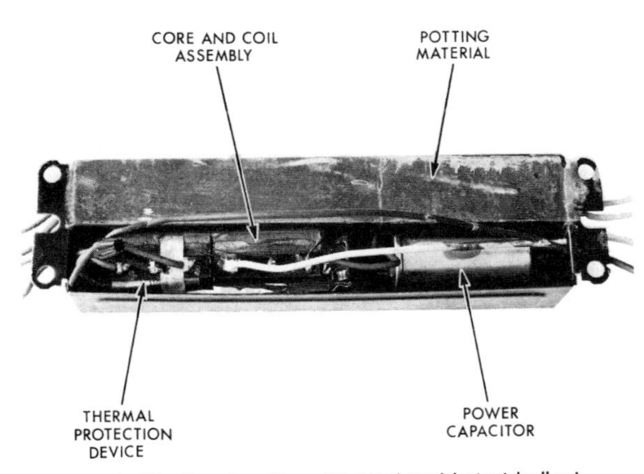

Fig. 8-42. Construction of typical rapid start ballast.

and a proper ballast operating temperature is normally estimated at about twelve years. Shorter ballast life will result at higher ballast temperature or longer duty cycle.

In the United States and Canada, it is now mandatory that all fluorescent lamp ballasts used indoors be internally thermally protected. This was done to prevent misapplication of the ballast as well as to protect against undesirable failure and conditions which can occur at end of ballast life. In the United States the thermally protected Underwriters Laboratories approval ballast is known and marked or labeled as "Class P."

Because of the magnetic elements in a ballast, vibrations are set up in the luminaire based on the input power frequency. This may produce an audible hum which is undesirable. The sound level produced will depend upon the ballast and luminaire construction and mounting. The acoustical characteristics of the space and the number of luminaires will have significant effect on the degree of audibility. Ballast manufacturers publish "sound ratings" which indicate the relative sound producing potential of their different models. However, no industry standards have as yet evolved that make possible the comparison of different brands. Some luminaire manufacturers also publish sound ratings for their units. (See Section 6.)

Reduced Wattage Ballasts. Ballasts are available which operate standard lamps at 50 to 80 per cent of their rated wattage. Energy saving lamps should not be used in combination with these ballasts since the arc will tend to striate.

Energy Saving Ballasts. Energy saving ballasts have lower losses than standard ballasts. These may be CBM* rated ballasts and may be used either with standard lamps or with reduced wattage lamps. For example the two-lamp 40-watt rapid start ballast losses have been reduced by 4 or 5 watts per lamp over standard ballasts. A typical two-lamp 40-watt unit with a low loss, energy saving ballast will consume approximately 86 watts compared with approximately 95 watts consumption for normal ballasts.

* Certified Ballast Manufacturers Association.

Energy Saving Systems. Specialized systems (lamp/ballast combinations) are available to achieve even greater energy savings. These include a 32-watt T-8 (4-foot) lamp with high efficiency ballast, and a 28-watt T-12 lamp also with a high efficiency ballast having internal solid state switches which turn off the typical rapid start cathode heater voltage. The latter ballast will also operate a 34-watt reduced wattage lamp.

Ballasts Power Factor. Characteristics of ballasts may result in low power factor. The measured watts of a low-power-factor ballast are approximately the same as the measured watts of the high-power-factor type when connected to the same load. The low-power-factor type draws more current from the power supply, and therefore, larger supply conductors may be necessary. The use of high-power-factor ballasts permits greater loads to be carried by existing wiring systems. Some public utilities have established penalty clauses in their rate schedules for low-power-factor installations. In some localities utilities require the use of equipment providing a high power factor. "High power factor" is defined as being above 90 per cent.

Starters for Preheat Circuits

The operation of a preheat circuit requires heating of the electrodes prior to application of voltage across the lamp. Preheating can be ef-

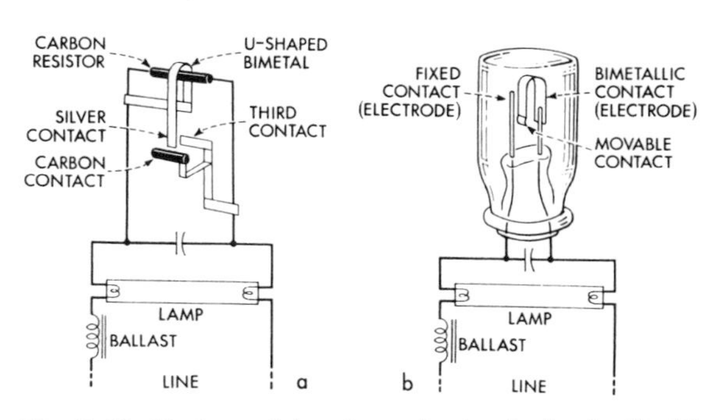

Fig. 8-43. Starter switches for preheat cathode circuits: (a) thermal type; (b) glow switch type.

fected by use of a manual switch or one which is activated by application of voltage to the ballast circuit. A number of designs of automatic switches are commercially available. Circuit diagrams of designs are presented in Fig. 8-43.

Thermal Switch Starter. The circuit diagramming this type starter is presented in Fig. 8-43. Initially the silver-carbon contact of the thermal starter is closed, placing the electrodes in series with the parallel combination of the bimetal and the carbon resistor. Upon closing the ballast supply circuit, output voltage of the ballast is applied to this series/parallel combination. The current heats the bimetallic strip in the starter, causing it to expand differentially, resulting in opening the silver/carbon contact. The time of opening is sufficient to raise the temperature of the electrodes to approximately their normal operating value. Upon opening the circuit, the ballast output voltage in series with an inductive spike (kick) voltage is applied to the lamp. If the lamp ignites, its normal operating voltage maintains a low current through the carbon resistor, developing and transferring sufficient heat to the bimetal to hold its contact open thereafter.

Should the lamp fail to start on the first attempt, the ballast open circuit voltage applied to the carbon resistor heats the bimetal sufficiently to cause the silver contact to move over against the third contact. This then short-circuits the carbon resistor, permitting preheating current to flow through the electrodes. As the bimetal cools, the circuit through the third contact is opened resulting in the application of the circuit voltage to the lamp again. This making and breaking of the circuit through the third contact continues until the lamp ignites. The bimetal circuit is held open thereafter as noted above. The carbon contact circuit functions only when the line voltage is initially applied to the ballast.

Thermal-switch starters consume some power (½ to 1½ watts) during lamp operation, but their design insures positive starting by providing: an adequate preheating period; a high induced starting voltage; and characteristics inherently less susceptible to line voltage variations. For these reasons they give good all-around performance of lamps under adverse conditions, such as direct-current operation, low ambient temperature and varying voltage.

Glow-Switch Starter. The circuit for this starter is presented in Fig. 8-43. The glass bulb shown is filled with an inert gas chosen for the voltage characteristics desired. On starting, the line switch is closed. There is practically no voltage drop in the ballast and the voltage at the starter is sufficient to produce a glow discharge between the contacts. The heat from the glow distorts the bimetallic strip, the contacts close, and electrode preheating begins. This short-circuits the glow discharge so that the bimetal cools and in a short time the contacts open. The open circuit voltage in series with an inductive spike voltage is applied to the lamp. If the lamp fails to ignite, the ballast open circuit voltage again develops a glow in the bulb and the sequence of events are repeated. This continues until the lamp ignites. During normal operation, there is not enough voltage across the lamp to produce further starter glow, so the contacts remain open and the starter consumes no power.

Cutout Starter. These starters may be made to reset either manually or automatically. They are designed to prevent the repeated blinking or attempts to start a deactivated lamp. This type of starter should be good for at least ten or more renewals.

Lamp Failure in Preheat Circuit. Starters which provide no means for deactivation when a lamp fails, will continue to function and attempt to start the lamp. A blinking of the lamp may result. This may lead to either ballast failure or starter failure. Thus it is important to remove a failed preheat lamp without significant delay.

Fluorescent Lampholders

Lampholders are designed for each base style. Several types are available for various spacings and mounting methods in luminaires. See Fig. 8-44. It is important that proper spacing be maintained between lampholders in luminaires

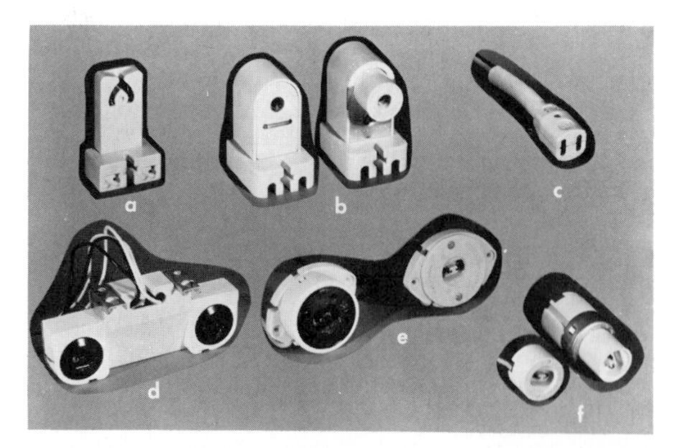

Fig. 8-44. Typical lampholder designs: (a) standard medium bipin lampholder; (b) standard single pin lampholder; (c) circline connector; (d) turret lampholder; (e) flange-mounted lampholder; (f) end-mounted lampholder.

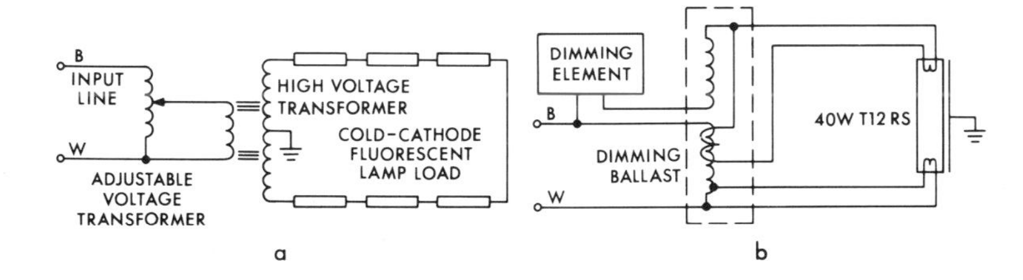

Fig. 8-45. Typical dimming circuits for: (a) series-connected cold-cathode lamps; (b) hot-cathode rapid start lamps.

to insure satisfactory electrical contact. Manufacturers' catalogs should be consulted for dimension and spacing information on any particular lampholder type.

When fluorescent lamps are used in circuits which may provide an open circuit voltage in excess of 300 volts, or in circuits which may permit a lamp to ionize and conduct current with only one end inserted in the lampholder, electrical codes usually require some automatic means for de-energizing the circuit when the lamp is removed. This is usually accomplished by the lampholder so that when the lamp is removed, the ballast primary circuit is opened. The use of recessed contact bases on 800- and 1500-milliampere fluorescent lamps has eliminated the need for this disconnect feature in lampholders for these lamps.

Dimming of Fluorescent Lamps[28]

Certain types of fluorescent lamps can be controlled in luminous intensity with commercially available dimming equipment. Dimming is achieved by reducing the effective current through the lamp. This is accomplished by either: lowering the primary voltage to the transformer or the ballast; adding impedance in the arc circuit; or shortening the time that the arc current flows each half cycle by gating action. Best dimming results when the open circuit voltage is high with respect to the lamp operating voltage. This permits adequate starting and restrike voltage at dimmer settings for low luminous output.

Restrike voltage can also be maintained by use of a peaking transformer incorporated into the dimming ballast. This permits dimming to low intensities with arc voltage maintained and starting qualities unimpaired, without requiring excessively high open circuit voltages.

The arrangement most readily adaptable for dimming the cold-cathode fluorescent lamp is the series circuit. A diagram showing typical arrangements for this circuit is shown in Fig. 8-45a (some manufacturers produce a solid-state device which replaces the autotransformer shown). When using this type of equipment, it is recommended that the lamp loading be about 20 per cent less than that recommended for general lighting requirements. With this arrangement, it is possible to reduce luminous intensity to a value of 10 per cent of the rated light output. It must be remembered that the constant wattage type transformer cannot be employed in dimming. High-power-factor cold-cathode-lamp transformers and ballasts cannot be used with saturable reactor magnetic amplifier dimmers or most solid-state thyristor dimmers.

The vast majority of all fluorescent lamp dimmers presently available are of the electronic solid-state variety and are designed for use with rapid start lamps only. These newer solid-state types are generally more efficient and less bulky than their autotransformer predecessors. The usable dimming range for solid-state types is also comparable or wider than that of the autotransformer type. Solid-state dimmers typically operate on a lamp current gating principle as mentioned above in conjunction with a special dimming ballast (see Fig. 8-45b). This type of ballast is unlike the standard ballast in that the secondary circuit is opened and brought outside the ballast as an extra conductor. It is this extra conductor (third wire) which provides a means of externally controlling lamp current. Solid-state fluorescent dimmers utilize thyristors to gate the current flow in this conductor. The gating method employed is generally similar to the variable phase angle scheme used in solid-state incandescent dimmers. Most fluorescent dimmers presently available operate on the "three-wire" scheme as described above, however, some manufacturers produce "two-wire" systems having a limited dimming range suitable for energy management applications.

Several installation factors will influence the performance of any fluorescent dimming system. The most important factor of all is to assure proper matching of the electronic dimming equipment with the ballasts and lamps. The ordinary non-dimming ballast generally cannot be used for dimming purposes. Energy saving lamps should not be used in dimming systems unless so specified by the dimmer manufacturer. Manufacturers should be consulted to assure proper coordination of the three components: the electronic dimming equipment, ballasts and lamps.

Stable flicker-free performance is most difficult to maintain at low lamp intensities and it is in this mode that the lamp cathode heaters become very important. It is necessary to maintain full cathode heating so as to maintain full emission capability at all intensities. To guarantee this performance, it is advisable to select premium quality sockets having knife-edge rather than leaf-spring contacts. This will assure the reliable contact needed to supply full heater efficiency. It is also advisable to avoid air handling luminaires or low ambient temperature applications, especially where maximum dimmer control range is desired. Since lamp efficacy is a strong function of bulb wall temperature, any non-uniform cooling in the luminaires will cause great variations in light output from luminaire to luminaire. This temperature effect is most pronounced at intensities less than 10 per cent light output. Finally, a ground plane, or luminaire starting aid spaced not more than 13 millimeters (0.5 inch) away from the lamp should be connected to an earth ground to assure reliable lamp starting.

Fluorescent dimmers do not provide continuous dimming down to *zero light* as do incandescent dimmers. The minimum dimmer intensity available in a standard product is generally 0.5 to 1.0 per cent of full output. Therefore, it is not advisable to use fluorescent dimmers for such applications as theatrical blackout effects.

Because of the high efficiency associated with solid-state dimmer control of fluorescent lamp intensity, most solid-state dimmers afford energy savings equal to the per cent of light reduction (See Fig. 8-46). This is especially true for dimmer settings above 50 per cent of full light output. This one-for-one relationship of power versus light, coupled with the flexibility of electronic control systems, makes it possible to provide for lighting energy management.

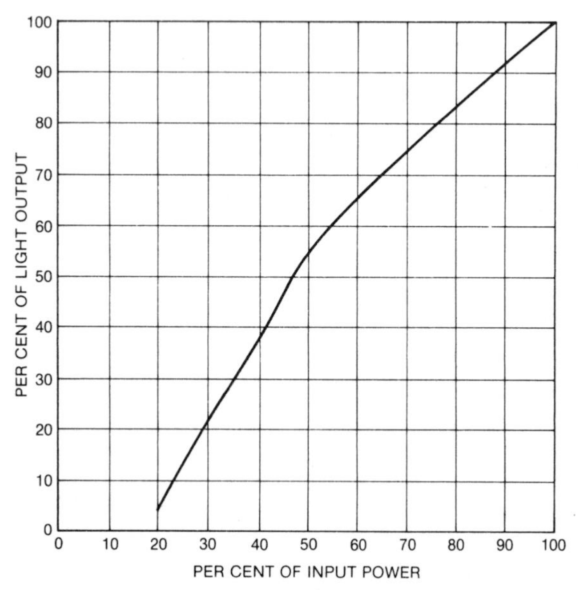

Fig. 8-46. Per cent light output vs input power for a typical 40-watt 120-volt rapid start fluorescent dimming system.

Flashing of Fluorescent Lamps[29]

Cold cathode and rapid start or preheat start hot cathode fluorescent lamps can be flashed with good performance. Cold cathode lamps are flashed through control of either transformer primary or secondary voltage. Hot cathode lamps can be flashed by means of special single-lamp or two-lamp ballasts designed to control lamp arc current while retaining cathode circuit voltage. An external flashing device is required with either system. This unit must be rated for the voltage and current involved and it is recommended that separate contacts be used for each ballast to prevent circulating currents between ballasts. The single-lamp circuit is similar to the single-lamp dimming circuit, shown in Fig. 8-45b except that a flashing device is required in place of the dimming element. Flashing of fluorescent lamps has had application in electrical advertising.

High Frequency Operation of Fluorescent Lamps

Operation of fluorescent lamps at frequencies above 60 hertz yields increased lamp efficacy (refer to Fig. 8-47). In addition to increased lamp efficacy, the ballast can often be reduced in size and weight.

At low frequencies, a reactor or a combination of reactor and capacitor is required for satisfactory lamp operation. At higher frequencies a low-loss capacitor can be used without introducing significant distortion of the wave shape of lamp electrical parameters. In addition, a reactor can be constructed with a low loss powdered iron (ferrite) core.

The high frequency voltage for operation of

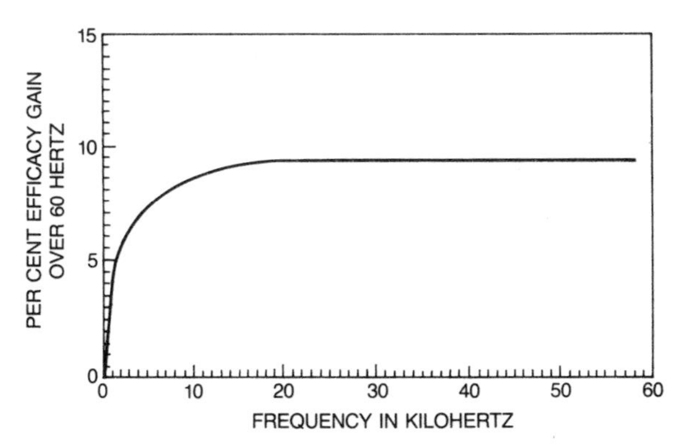

Fig. 8-47. Lamp efficacy gain at constant lumen output vs operating frequency for a 40-watt T-12 rapid start lamp.

the lamps can be generated by means of rotary converters, thyratron inverters or, more recently, by means of solid state inverters incorporating transistors or silicon controlled rectifiers. A transistor inverter changes low voltage direct current to high frequency alternating current. A basic circuit is shown in Fig. 8-48. Transistors T_1 and T_2 of Fig. 8-48 act as switches alternately connecting the direct current supply (B) across the primary winding P_1 and P_2. Feedback windings F_3 and F_4 are arranged in such a way that the base of the conducting transistor is negative whereas the base of the non-conducting transistor is positive. The collector current through the primary winding initially increases with time, thereby inducing voltage in the secondary windings. As soon as this current becomes constant (mainly determined by the inductance and resistance of the transformer) the induced voltage will be reduced to zero and the polarity of the feedback windings reversed. The first transistor will now be blocked and the second one will conduct. The ac voltage generated is transformed into the lamp circuit by transformer (T). This voltage supplies power for heating the electrodes and operating a discharge within the lamp. A small reactor (R) is used to limit the current flow in the lamp circuit. The efficiency of such a simple inverter is rather low but can be improved to nearly 90 per cent by reducing the commutation loss in the transistors and by capacitive tuning of the frequency.

The circuit of Fig. 8-48 is used when the inverter is contained within the luminaire. Other possibilities are a common inverter operating several luminaires with individual reactive or capacitive components to limit current in each lamp circuit.

The operating frequency should be high in order to increase the efficacy and to shift ballast noise to the inaudible range of the noise spectrum. High frequency presents problems in the distribution of power in the case of central inverters. Direct current to alternating current inverters are used extensively for illumination in buses, trains, boats trailers and planes, and for portable advertising signs on cars and for battery operated emergency lights.

With improvements in solid-state devices and the availability of sophisticated integrated circuit functions, there now are commercially available *electronic ballasts* that use 50 to 60 hertz alternating current for input to the ballast and operate the lamps at about 25 kilohertz with resulting improvement in both ballast efficiency and lamp efficacy. It is also noteworthy that the effectiveness of electronic control affords a new level of light output regulation which was previously unavailable in totally passive ballast design. With electronic lamp current regulation the lighting design variables of line voltage, ambient temperature and individual ballast differences can be minimized.

HIGH INTENSITY DISCHARGE LAMPS

High intensity discharge (HID) lamps include the groups of lamps commonly known as mercury, metal halide and high pressure sodium. The light producing element of these lamp types is a wall stabilized arc discharge contained within a refractory envelope (arc tube) with wall loading in excess of 3 watts per square centimeter.

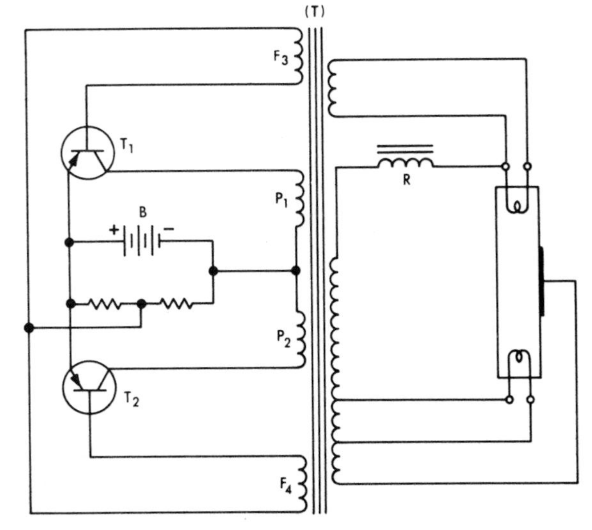

Fig. 8-48. Transistor inverter for high frequency operation of a fluorescent lamp.

Lamp Construction and Operation

Mercury Lamps.[30-34] In mercury lamps, light is produced by the passage of an electric current through mercury vapor. Since mercury has a low vapor pressure at room temperature, and even lower at cold temperatures, a small amount of more readily ionized argon gas is introduced to facilitate starting. The original arc is struck through the ionization of this argon gas. Once the arc strikes, its heat begins to vaporize the mercury, and this process continues until all of the mercury is evaporated. The amount of mercury in the lamp essentially determines the final operating pressure, which is usually 2 to 4 atmospheres (200 to 400 kilopascals) in the majority of lamps.

The operating electrodes used in mercury lamps are usually of the metal-oxide type, in which the emission material, composed of several metallic oxides, is embedded within the turns of a tungsten coil protected by an outer tungsten coil. The electrodes are heated to the proper electron-emissive temperature by bombardment energy received from the arc.

Most mercury lamps are constructed with two envelopes, an inner envelope (arc tube) which contains the arc, and an outer envelope which (a) shields the arc tube from outside drafts and changes in temperature, (b) usually contains an inert gas (generally nitrogen) which prevents oxidation of internal parts, and also maintains a relatively high breakdown voltage across the outer bulb parts, (c) provides an inner surface for coating of phosphors, and (d) normally acts as a filter to remove certain wavelengths of arc radiation (UV-B and UV-C). In most cases the arc tube is made of fused silica with thin molybdenum ribbons sealed into the ends as current conductors. The outer bulb is made of "hard" (borosilicate) glass in most cases, but may also be of other glasses for special transmission or where thermal shock or pollution attack is not a problem.

The essential construction details shown in Fig. 8-49 are typical of lamps with fused silica inner arc tubes within an outer envelope. (Other lamps such as those for special photochemical application, and self-ballasted types have different constructions.)

The pressure at which a mercury lamp operates accounts in a large measure for its characteristic spectral power distribution. In general, higher operating pressure tends to shift a larger proportion of emitted radiation into longer wavelengths. At extremely high pressure there is also a tendency to spread the line spectrum into wider bands. Within the visible region the mercury spectrum consists of five principal lines (404.7, 435.8, 546.1, 577, and 579 nanometers) which result in the greenish-blue light at efficacies of 30 to 65 lumens per watt without ballast losses. While the light source itself appears to be bluish-white, there is an absence of red radiation, especially in the low and medium pressure lamps, and most colored objects appear distorted in color rendition. Blue, green and yellow colors in objects are emphasized; orange and red appear brownish.

A significant portion of the energy radiated by the mercury arc is in the ultraviolet region. Through the use of phosphor coatings on the inside surface of the outer envelope some of this ultraviolet energy is converted to visible light by the same mechanism employed in fluorescent lamps. See Section 2. The most widely used lamps of this type are coated with a vanadate

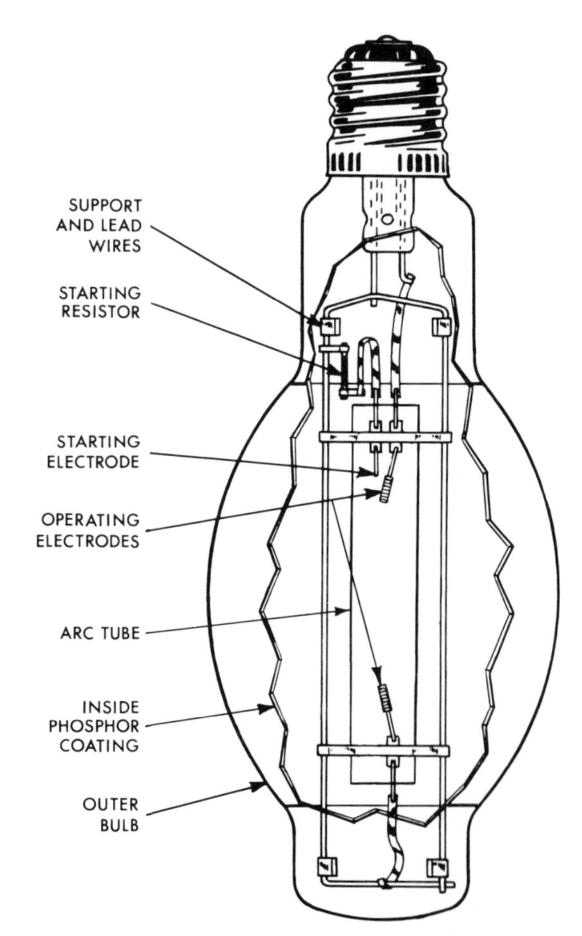

Fig. 8-49. A 400-watt phosphor-coated mercury lamp. Lamps of other sizes are constructed similarly.

phosphor (4000 K, designation "/DX") which emits orange-red radiation, improves efficacy and color rendition, and renders skin tones reasonably well. This phosphor is also blended with others to produce "cooler" or "warmer" colors. Fig. 8–21 shows the spectral distribution of a clear lamp and lamps using these phosphors. Special lamp types are also available for those applications where it is desirable to extinguish the arc if the outer bulb is punctured or broken. Thereby a potential exposure to ultraviolet energy is minimized.[35]

Metal Halide Lamps.[36–41] Metal halide lamps are very similar in construction to the mercury lamp, the major difference being that the metal halide arc tube contains various metal halides in addition to mercury and argon. When the lamp has attained full operating temperature, the metal halides in the arc tube are partially vaporized. When the halide vapors approach the high temperature central core of the discharge they are dissociated into the halogen and the metals, with the metals radiating their appropriate spectrum. As the halogen and metal atoms move near the cooler arc tube wall by diffusion and convection, they recombine, and the cycle starts over again. This cycle provides two desirable advantages. First, some metals, which in their metallic form cannot be vaporized at the temperatures which a fused silica arc tube can withstand, can be introduced into the discharge by dissociation of halides which vaporize at much lower temperatures. Typical metals introduced in this manner into metal halide lamps are thallium, indium, scandium and dysprosium. Secondly, some other metals, which react chemically with the arc tube, can be used in the form of a halogen which does not readily react with fused silica, since the metal is in the halogen form at the wall temperature, and the metal plus halogen form at the center of the discharge. A common example of a fused silica reactive metal which can be used in the halide form is sodium.

Compared with mercury lamps the efficacy of metal halide lamps is greatly improved. Commercially available metal halide lamps have efficacies of 75 to 125 lumens per watt (without ballast losses). Almost all varieties of available "white"-light metal halide lamps produce color rendering which is equal to or superior to the presently available phosphor coated mercury lamps.

Three typical combinations of halides used in metal halide lamps are: (1) sodium, thallium and indium iodides, (2) sodium and scandium iodides and (3) dysprosium and thallium iodides. Their spectral power distributions are shown in Fig. 8–21. Some halides, such as, sodium (589 nm), thallium (535 nm) and indium (435 nm) principally produce line spectra, while others, such as those of scandium, thorium, dysprosium and other rare earths produce multiline spectra across the full visible region. Other halides, such as those of tin, produce continuous spectra across the visible region.

Improved color balance can be produced by combining the spectra of elements which radiate in various regions of the spectrum as in the sodium-thallium-indium lamp; while it can also be achieved by use of the multiline spectra of scandium or dysprosium. High efficacy is produced by the use of halides such as thallium and sodium iodide.

High intensity discharge lamps of selected colors or for producing near ultraviolet can also be made by using the metal halide technique; for example, sodium for orange, thallium for green, indium for blue, and lead for ultraviolet.

Metal halide lamps are also available with phosphors applied to the outer envelopes to further modify the color and generally to lower the color temperature of the lamp.

A close look at a metal halide lamp will reveal several construction features which differ from mercury lamps. (1) The arc tubes are usually smaller than in mercury lamps for equivalent wattage, may or may not be specially shaped, with a coating or reflector at one or both ends of the arc tube. The end reflector(s) and small arc tube size serves to increase the temperature at the end of the arc tube to assure vaporization of the halides. (2) Some lamps include a system for either shorting the starting electrode to the operating electrode or opening the starting electrode circuit. This is required to prevent electrolysis in the fused silica between the starting and operating electrodes especially when a halide such as sodium iodide is used in the lamp. The type and location of the bimetal switch may, or may not, restrict lamp burning position. (3) In some lamps the electrical connection to the electrode at the dome of the lamp is made by means of a small nonmagnetic wire remote from the arc tube. This is to prevent diffusion of sodium through the arc tube by electrolysis caused by a photoelectric effect when the current lead is close to the arc tube. Most metal halide lamps require a higher open-circuit voltage to start than corresponding wattage mercury lamps. Therefore, they require specifically designed ballasts. Certain metal halide lamp designs, however, can be operated on some types of mercury ballasts in retrofit situations.

In the metal halide lamp family special types are also available where the arc is automatically extinguished if the outer envelope is broken or punctured. They may be required in locations where exposure to ultraviolet energy should be avoided.[35]

High Pressure Sodium Lamps. In the high pressure sodium lamp light is produced by electric current passing through sodium vapor. These lamps are constructed with two envelopes, the inner being polycrystalline alumina which has the properties of resistance to sodium attack at high temperatures as well as a high melting point, and good light transmission (more than 90 per cent) even though this material is translucent. The construction of a typical high pressure sodium lamp is shown in Fig. 8-50. Polycrystalline alumina cannot be fused to metal by melting the alumina without causing the material to crack. Therefore, the seal to metal (typically niobium) is made by means of an intermediate solder glass or metal between lead wire or tube, or ceramic plugs and the alumina. The arc tube contains xenon as a starting gas, and a small quantity of sodium-mercury amalgam which is partially vaporized when the lamp attains operating temperature. The mercury acts as a buffer gas to raise the gas pressure and operating voltage of the lamp to a practical level.

The outer borosilicate glass envelope is evacuated and serves to prevent chemical attack of the arc tube metal parts as well as maintaining the arc tube temperature by isolating it from ambient temperature effects and drafts.

Most high pressure sodium lamps can operate in any position. The burning position has no significant effect on light output. Lamp types are also available with diffuse coatings on the inside of the outer bulb to increase source luminous size or reduce source luminance, if required.

High pressure sodium lamps radiate energy across the visible spectrum. This is in contrast to the low pressure sodium lamp (see page 8-54) which radiates principally the doublet "D" lines of sodium at 589 nanometers. At higher sodium pressures, about 27 kilopascals (200 torr), in the high pressure sodium lamp, sodium radiation of the "D" line is self absorbed by the gas and is radiated as continuum on both sides of the "D" line frequency. This results in the "dark" region at 589 nanometers as shown in the typical spectrum in Fig. 8-21. The light produced by this lamp is consequently golden-white in color with all visible frequencies present. These lamps have efficacies of 60 to 140 lumens per watt depending on size. Increasing sodium pressure increases the percentage of red radiation and thus improves color rendition; however, life and efficacy are reduced.

Because of the small diameter of a high pressure sodium lamp arc tube, no starting electrode is built into the arc tube as in the mercury lamp. A high voltage high frequency pulse is required to start these lamps. Special high pressure sodium lamps, using a Penning starting gas mixture of argon and neon, and a starting aid inside the outer bulb, will start and operate on many mercury lamp ballasts. Consult the lamp manufacturers' literature for the suitable ballast types. These lamps are useful retrofit devices to upgrade mercury lamp systems, but are not as efficient as standard high pressure sodium lamp and ballast systems.

Lamp Designations

The current identifying designations of high intensity discharge lamps generally follow a system which is authorized and administered by the American National Standards Institute (ANSI). All designations start with a letter ("H" for

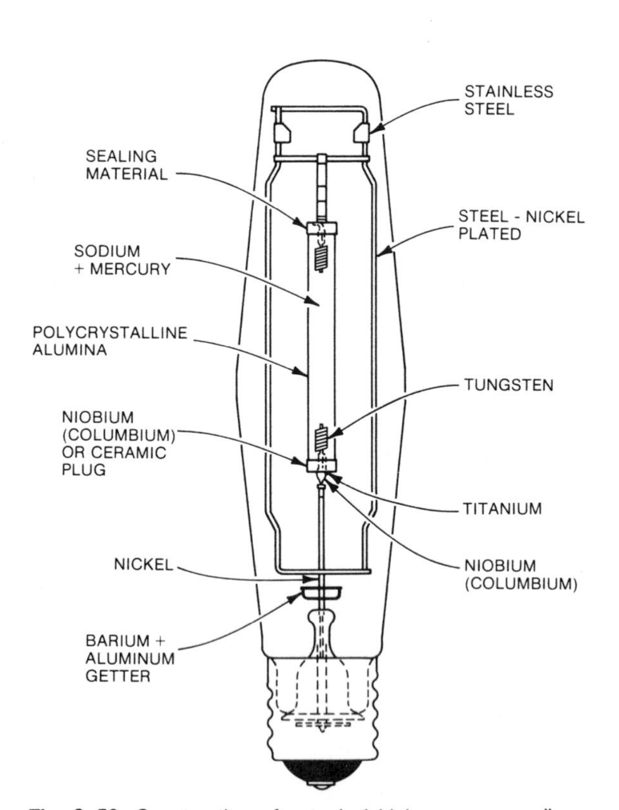

Fig. 8-50. Construction of a typical high pressure sodium lamp.

mercury, "M" for metal halide, "S" for high pressure sodium). This is followed by an ANSI assigned number which identifies the electrical characteristics of the lamp and consequently the ballast. After the number there are two arbitrary letters which identify the bulb size, shape, finish, etc., excluding color. After this section, the individual manufacturers may add, at their discretion any additional letters or numbers which they desire to indicate information not covered by the standard section of the designation, such as lamp wattage or color. Other systems being used start with one or two letters followed by lamp wattage, then an arbitrary letter identifying the bulb and finish, and finally the numbers defining electrical characteristics.

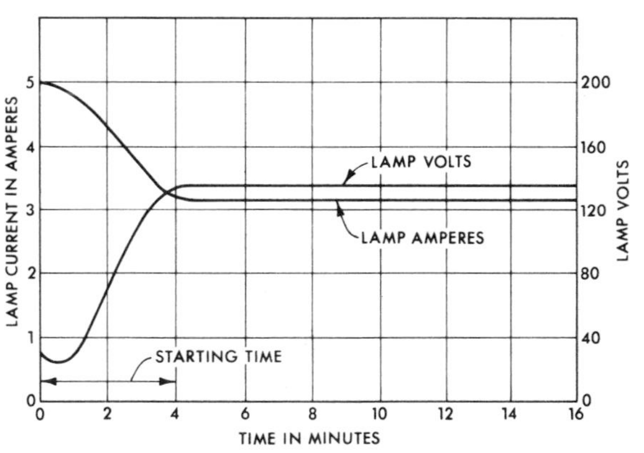

Fig. 8-51. Warm-up characteristics for 400-watt mercury lamp on a lag type ballast. This is typical of most mercury lamp types.

Lamp Starting

Mercury Lamps. Some special two-electrode types of mercury lamps and many photochemical types require a high open circuit voltage to ionize the argon gas and permit the arc to strike. In the more common three-electrode type lamps an auxiliary starting electrode placed close to one of the main electrodes makes it possible to start the lamp at a lower voltage. Here an electric field is first set up between the starting electrode (which is connected to the opposite main electrode through a current limiting resistor) and the adjacent main electrode. This causes an emission of electrons which develops a local glow discharge and ionizes the starting gas. The arc then starts between the main electrodes. The mercury gradually becomes vaporized from the heat of the arc and carries an increasing portion of the current. During this process the arc stream changes from the diffuse bluish glow characteristic of the argon arc, to the blue-green of mercury, increasing greatly in luminance and becoming concentrated in the center of the tube. At the instant the arc strikes, the lamp voltage is low. Normal operating values are reached after a warm-up period of several minutes during which the voltage rises until the arc attains a point of stabilization in vapor pressure (see Fig. 8-51 for a 400-watt lamp operated from a lag type ballast); the mercury is then entirely evaporated.

If the arc is extinguished, the lamp will not relight until it is cooled sufficiently to lower the vapor pressure to a point where the arc will restrike with the voltage available. The time from initial starting to full light output at ordinary room temperatures, with no enclosing lighting unit, and also the restriking time or cooling time until the lamp will restart, vary between 3 and 7 minutes for the various types of lamps.

Metal Halide Lamps. The method of starting metal halide lamps is the same as mercury lamps. However, because of the presence of the halide, the starting voltage required for some types is higher than that of mercury lamps. As the lamp warms up it may go through several color changes as the various halides begin to vaporize, until it reaches its equilibrium color and electrical characteristics after several minutes.

Since a metal halide arc tube operates at higher temperatures than a mercury lamp arc tube the time to cool and lower the vapor pressure is generally longer than that of the mercury lamp, consequently the restrike time may be as long as 15 minutes.

High Pressure Sodium Lamps. Since the high pressure sodium lamp does not contain a starting electrode, a high-voltage high-frequency pulse is used to ionize the xenon starting gas to facilitate starting. Once started, the lamp warms up to full light output in approximately 10 minutes, during which time the color changes until operating conditions are achieved.

Because the operating pressure of a high pressure sodium lamp is lower than that of a mercury lamp the restrike time is shorter. It will usually restrike in less than 1 minute and warm up in 3 to 4 minutes.

Lamp Operating Position

Mercury Lamps. When a mercury lamp is operated horizontally, the arc tends to bow upward because of the convection currents in the gas. This bowing of the arc will generally cause a small change in the electrical characteristics of the lamp as well as a small reduction of lamp output caused by a reduction in lamp wattage and efficacy. Many ballasts designed for horizontally operating lamps compensate for this wattage decrease by increasing the current through the lamp, but do not compensate for the loss in efficacy.

Metal Halide Lamps. When a metal halide lamp is operated horizontally the arc also bows upward, but the effect of change of operating position can be much greater than in the case of the mercury lamp especially with regard to color. Because a portion of the halides in metal halide lamps is not vaporized during lamp operation, any change in the cold spot temperature of the lamp will change the pressure of the halide and therefore the lamp color. Generally, the color temperature of these lamps may vary with operating position. When the burning position of a metal halide lamp is changed, as much as six hours may be required before the lamp characteristics, color, electrical characteristics and efficacy are stabilized. Some lamps have restricted burning positions. These should be observed if optimum performance is to be obtained.

Special shapes of arc tubes are used in certain lamps to increase initial lumens in specific burning positions. These include an arc tube designed for horizontal operation and curved to compensate for the tendency of the arc to rise toward the tube wall in the center, reducing watts and light output.

High Pressure Sodium Lamps. High pressure sodium lamps having small arc tube diameters permitting a very small arc bow, exhibit very little change in efficacy or electrical characteristics when operated in a horizontal compared to a vertical position.

Lamp Life and Lumen Maintenance

Mercury Lamps. General service mercury lamps have a long average life. These lamps usually employ an electrode with a mixture of metal oxides embedded in the turns of tungsten coils from which the electrode is assembled. During the life of the lamp, this emission material is very slowly evaporated or sputtered from the electrode and is deposited on the inner surface of the arc tube. This process results first in a white deposit on the inner surface of the arc tube, eventually in a blackening of the arc tube, and ultimately in the exhaustion of the emission material in the electrodes and the end of lamp life, when the starting voltage exceeds the open circuit voltage.

Mercury lamps are usually rated for "initial" lumens after 100 hours burning. Initial lumens for the various lamp types are given in Fig. 8-123. Fig. 8-52 illustrates a typical lumen maintenance curve. Neither life nor maintenance is materially affected by reasonable changes in burning cycle (5 hours or more per start).

Metal Halide Lamps. Chemical reaction between the iodine in a metal halide lamp and the emission materials included in mercury lamp electrodes prevents the use of those electrodes. Because the electrodes used evaporate more rapidly than corresponding mercury lamp electrodes, metal halide lamp life ratings are generally shorter. As with mercury lamps, rated "initial" lumens are based on 100 hours operation.

High Pressure Sodium. High pressure sodium lamps employ electrodes very similar to those used in mercury lamps. This fact plus the smaller diameter arc tube combine to give high pressure sodium lamps excellent lumen maintenance. Initial ratings also apply at 100 hours of operation.

The life of a high pressure sodium lamp is limited by a slow rise in operating voltage which occurs over the life of the lamp. This rise is

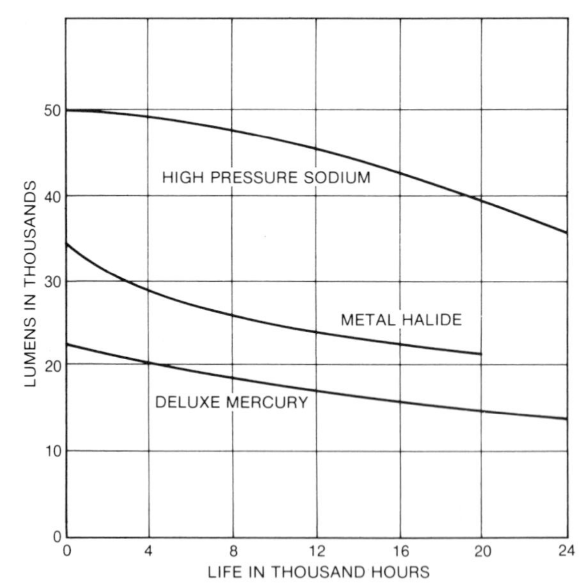

Fig. 8-52. Typical lumen maintenance curves for 400-watt high intensity discharge lamps.

principally caused by arc tube end blackening from electrode sputtering. The blackening absorbs radiation which heats up the arc tube ends and vaporizes additional sodium amalgam. This increases arc tube pressure and consequently arc voltage. Other reasons for arc tube voltage rise are the diffusion of sodium through the arc tube end seals and the removal of sodium from the arc stream by combination with impurities in the arc tube.

When the ballast can no longer supply enough voltage to reignite the arc during each electrical half-cycle, the lamp will extinguish. When it cools down, the lamp will again ignite and warm up until the arc voltage rises so that the ballast can not support the arc. This cycling process will occur until the lamp is removed.

Life ratings for high pressure sodium lamps are up to 24,000 hours, depending upon lamp design.

Effect of Ambient Temperature

The light output of the enclosed arc-tube type of high intensity discharge lamp is little affected by the ambient temperature. Experience has shown that these lamps are satisfactory for temperatures down to −29 °C (−20 °F) or lower. On the other hand, the single-envelope lamps, intended primarily for use as ultraviolet sources, are critically affected by low temperatures, particularly if the surrounding air is moving. They are not considered suitable for use below 0 °C (32 °F) without special protection since they may not warm up to full output. Ambient temperature affects the striking voltage of all discharge lamps and in some cases higher starting voltages than those listed in Fig. 8-123 for indoor use are recommended for roadway and floodlighting installations in cold climates. Ballasts for roadway lighting service and other low temperature applications are designed to provide the necessary voltage to start and operate each particular lamp at temperatures as low as −29 °C (−20 °F). Recommendations for starting voltages have been developed by the American National Standards Institute.

Lamp Operating Temperature

Because most high intensity discharge lamps are long-lived, operating temperatures are particularly important. The effect of heat on the lamp is partly a function of time, and the longer the rated life of the lamp the greater the possibility of damage from high temperatures. Excessive envelope and base temperatures may cause failures or unsatisfactory performance due to softening of the glass, damage to the arc tube by moisture driven out of the outer envelope, softening of the basing cement or solder, or corrosion of the base, socket or lead-in wires. Maximum bulb and base temperatures are prescribed by various standards associations. The use of reflecting equipment which concentrates heat and energy on either the inner arc tube or the outer envelope should be avoided.

In the case of metal halide and high pressure sodium lamps in which all the material is not vaporized, concentrated heat on the arc tube can materially effect their color as well as their electrical characteristics and shorten lamp life.

Auxiliary Equipment

High intensity discharge lamps have negative volt-ampere characteristics and, therefore, some current limiting device, usually in the form of a transformer and reactor ballast, must be provided to prevent excessive lamp and line currents. Lamps are operated on either multiple or series circuits. Fig. 8-53 illustrates schematic diagrams of several typical ballast types.

Basically, a distinction must be made between "lag circuit" and "lead circuit" ballasts. The lamp current control element of a lag type ballast consists of an inductive reactance in series with the lamp. The current control element in lead type ballasts consists of both inductive and capacitive reactances in series with the lamp; however, the net reactance of such a circuit is capacitive in mercury and metal halide ballasts, and inductive in high pressure sodium ballasts.

Mercury Lamps. There are a number of lamp ballasts in use for operating mercury lamps. Wattage losses in ballasts are usually in the order of 5 to 15 per cent of lamp wattage depending upon ballast and lamp type.

Normal Power Factor Reactor. The simplest lag circuit ballast is the reactor consisting of a single coil wound on an iron core placed in series with the lamp. The only function of the reactor is to limit the current delivered to the lamp. The reactor can only be used when the line voltage is within the specified lamp starting voltage requirement. Inherently the power factor of this circuit is about 50 per cent lagging and is

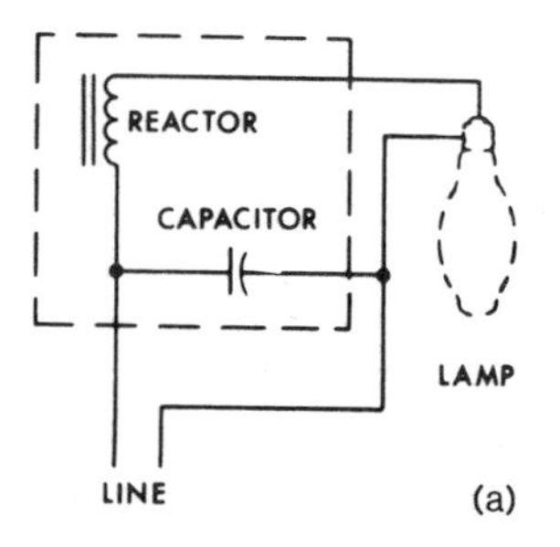

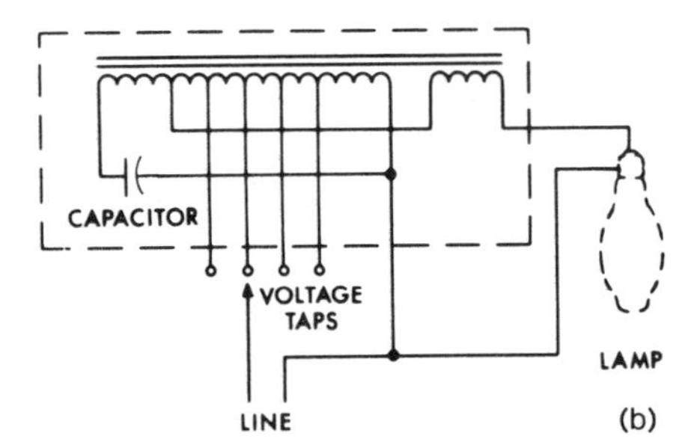

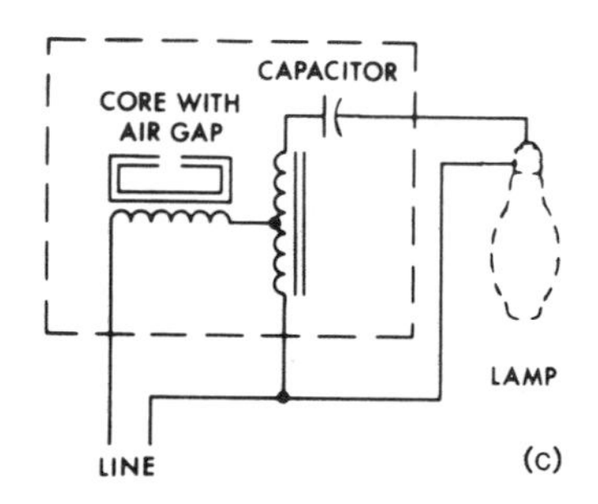

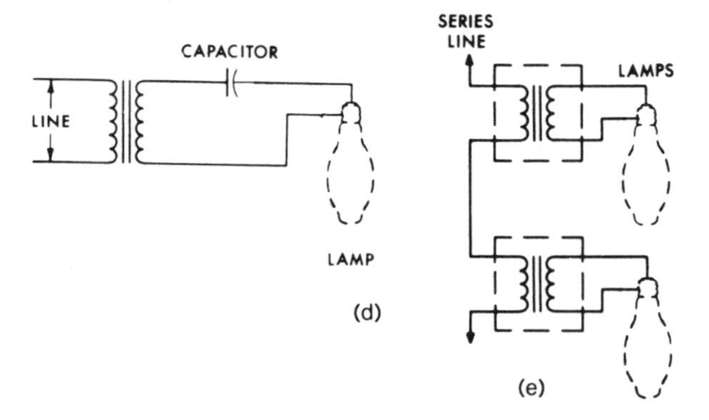

Fig. 8-53. Typical circuits for operating high intensity discharge lamps: (a) high power factor reactor mercury lamp ballast; (b) high power factor autotransformer mercury lamp ballast; (c) constant wattage autotransformer ballast for mercury lamps or peak-lead ballast for metal halide lamps; (d) constant wattage (isolated circuit) ballast for mercury lamps; (e) constant current series regulator ballast for mercury lamps.

commonly referred to as being normal or low power factor.

Since it only performs the function of current control, this reactor is the most economical, smallest and most efficient ballast. However, there are shortcomings which should be considered in application. The reactor provides little regulation for fluctuations in line voltage; for example, a 3 per cent change in line voltage would cause a 6 per cent change in lamp wattage. Therefore, the reactor is not recommended where line fluctuations exceed 5 per cent. The line current under starting conditions is approximately 50 per cent higher than normal operating current and, therefore, it is recommended that supply wiring be sized for approximately twice the normal operating current.

High Power Factor Reactor. This lag circuit ballast consists of a reactor in series with the lamp and a capacitor connected across the line. Both lamp current and regulation are essentially the same as described for the normal power factor reactor. The capacitor connected across the line does not affect the lamp circuit, but increases the power factor of the system to better than 90 per cent. It also reduces the input current under starting and operating conditions almost 50 per cent over the normal power factor system, allowing the use of a larger number of ballasts on a given circuit.

Normal Power Factor High Reactance Autotransformer. Where the line voltage is below or above the specified lamp starting voltage range, a transformer is used in conjunction with the reactor to provide proper starting voltage. This is normally accomplished with the combination of primary and secondary coils forming a one-piece single high reactance autotransformer. The power factor of this circuit is about 50 per cent lagging and has the same features and shortcomings as the normal power factor reactor lag circuit.

High Power Factor High Reactance Autotransformer. This lag circuit ballast is the same as the normal power factor high reactance autotransformer, except a capacitor is added to the primary circuit. In order to provide a more economical system, an extra capacitor winding is normally added to the autotransformer primary. This economical combination of extended windings with the capacitor increases the system power factor to above 90 per cent. The effect on input current is the same as in the high power factor reactor. Regulation and lamp performance are unchanged from those stated above.

Constant Wattage Autotransformer (CWA). This type of lead circuit ballast, commonly referred to as "CWA" type, is the most widely used in modern mercury lighting systems.

It consists of a high reactance autotransformer with a capacitor in series with the lamp. Use of the capacitor allows the lamp to operate with better wattage stability when the voltage on the branch circuit fluctuates. This ballast is used when line voltage is expected to vary more than 5 per cent. For example, a 10 per cent change in line voltage would result in only a 5 per cent change in lamp watts. Other advantages with the CWA ballast are high power factor, low line extinguishing voltage, and line starting currents that are lower than normal line currents. The CWA features allow maximum loading on branch circuits and provide an economical and efficient mercury lighting system.

The capacitor used with the CWA ballast performs an important ballasting function as in all lead-type circuits. The capacitor used in lag-type high power factor reactor and high power factor autotransformer ballasts is purely a power factor correction component and has no ballasting property.

Constant Wattage (CW). This type of ballast, also referred to as regulated or stabilized type, has operating characteristics similar to the CWA. Light output and wattage will vary by ± 2 per cent with a ± 13 per cent change in line voltage. The CW ballast, like the CWA ballast, uses a lead circuit; but unlike the CWA, the lamp circuit is completely isolated from the primary winding. It also has the same advantages as the CWA ballast, such as high power factor, low line extinguishing voltage and low line starting currents.

Two Lamp Lead-Lag Circuit. The lead-lag ballast design approach is commonly used to operate two 250-, 400-, or 1000-watt mercury or metal halide lamps in two independent circuits. A current limiting reactor operates one lamp, and a combination reactor and a capacitor connected in series operates the second. The lamps operate independently so that a failure of one has no effect on the other. The input current of the combination of capacitors and reactors is lower than the sum of the two individually operating currents. These elements provide a high power factor and reduce stroboscopic effect. This circuit can only be used when the line voltage is within the specified lamp starting voltage. It is the most economical two lamp system with regulation similar to the normal power factor reactor and autotransformer ballasts. A lag reactor may be used in one luminaire and a lead reactor in the next luminaire. An equal number of each in a branch circuit will result in a high branch circuit power factor.

Two Lamp Series (Isolated) Constant Wattage. This circuit is essentially the same as single lamp constant wattage, except it operates two lamps in series. The most effective use of this circuit is in applications where the ambient temperature is −18 °C (0 °F) or above. It is most popular with indoor 400-watt applications.

Constant Current Series Regulators. Mercury lamps are also operated in series on constant current series regulators. The most commonly used method employs a current transformer for each lamp (Fig. 8–53e). It differs in design from the more common multiple type of ballast. The usual design is a two-winding transformer as illustrated in Fig. 8–53e. Since the series circuit regulator reactance limits current in the circuit, the individual lamp current transformer is not designed to limit current but rather to transform it from the regulator secondary current to the rated lamp current. In addition, the transformer is made to limit secondary open circuit voltage so that no cutout is necessary in case of a lamp failure. Series transformers are available for the more popular lamps to operate from either 6.6-, 7.5-, or 20-ampere series circuits and can be operated on all types of constant current transformers. These circuits will normally be satisfactory for metal halide lamps and high pressure sodium lamps designed for operation on reactor-type mercury ballasts.

Two Level Mercury Ballasts. Two level operation of mercury lamps can be accomplished by switching capacitors on lead circuit ballasts. Such ballasts are available which will operate 125-, 250- and 400-watt mercury lamps at two levels. For example, a 400-watt mercury lamp may be operated at either 400 watts or 300 watts by switching leads at the ballast. These two level mercury ballasts are used for energy saving. Both lamp manufacturer and ballast manufacturer should be contacted for specific information.

This control technique is presently limited to horizontal operation above 10 °C (50 °F) and warm-up time is 50 per cent longer on low level.

Metal Halide Lamps. To insure proper starting and operating, a lead-peaked autotransformer ballast circuit specifically designed for all commercially available metal halide lamps is normally used. The regulation of such a ballast is reasonably good, providing a change in lamp watts of 7 to 10 per cent for a line voltage change of 10 per cent. Aside from regulation, the lead-peaked autotransformer ballast performs much like the CWA with similar operating features. Standard mercury lamps may be operated on this ballast. Certain metal halide lamp types can be operated on mercury lamp type ballasts. For operation under these conditions manufacturers' instructions should be consulted.

High Pressure Sodium Lamps. Unlike mercury and metal halide lamps which exhibit relatively constant lamp voltage with changes in lamp wattage, the high pressure sodium lamp voltage increases and decreases as the lamp wattage is varied. Because of this characteristic, trapezoids have been established which define maximum and minimum permissible lamp wattage vs. lamp voltage for purposes of ballast design. Fig. 8-54 shows the trapezoid for the 400-watt lamp as typical example. Four basic circuits have evolved which are designed to operate the lamps within the trapezoid limitations. In order to initiate the arc, an electronic starting circuit provides a pulse voltage which generates approximately 4000 volts peak for the 1000-watt type lamp, and 2500 volts peak for the other types, at least once per cycle until the lamp is started.

Lag or Reactor Ballast. This ballast type is similar to the mercry lamp reactor ballast. It is a simple reactor in series with the lamp specifically designed to keep the operating characteristics within the trapezoid limits. The starting circuit is incorporated to provide the starting pulse. Step-up or step-down transformers are provided where necessary to match the line voltage. In most cases, a power factor correcting capacitor is placed across the line or across a capacitor winding on the ballast primary. This type ballast usually provides good wattage regulation for changes in lamp voltage, but rather poor regulation for variations in line voltage. It is the least costly ballast with the lowest power loss of any high pressure sodium ballast type.

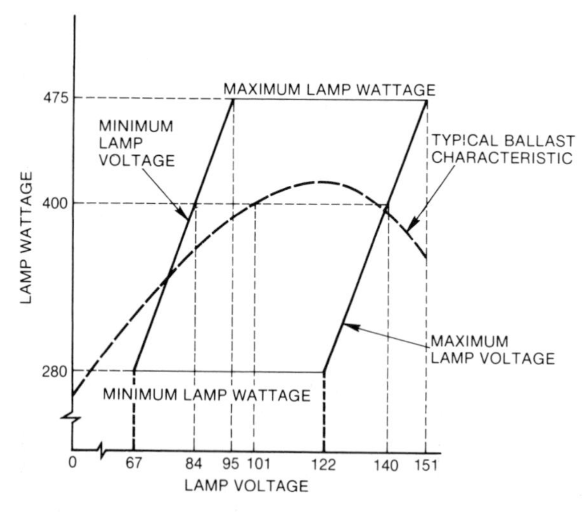

Fig. 8-54. Lamp wattage-voltage limits trapezoid for 400-watt high pressure sodium lamps.

Magnetic Regulator or Constant Wattage Ballast. This ballast consists essentially of a voltage regulating section which feeds a current limiting reactor and the pulse starting circuit. It provides good wattage regulation for changes in line voltage, as a result of the voltage regulating section, and good regulation for changes in lamp voltage which is the main characteristic of the reactor ballast.

It is a high cost ballast with the greatest losses, but generally provides good wattage regulation under all conditions of line and lamp voltage. A power factor correcting capacitor is usually included. It should be noted that this circuit differs from constant wattage mercury ballast circuits.

Lead Circuit Ballast. This circuit is similar in circuit configuration to the CWA mercury ballast. It operates with a combination of inductance and capacitance in series with the lamp. It differs in design from the CWA mercury ballast in that rather than maintaining a constant current through the lamp, the current decreases with increasing lamp voltage to keep the lamp operating wattage within the trapezoid. This ballast type provides wattage regulation for changes both in line voltage and lamp wattage. For a ± 10 per cent change in line voltage it maintains the lamp wattage within the trapezoid. It is intermediate in terms of cost and power losses.

Electronically Regulated Reactor Ballast. This is essentially a reactor ballast with an electronic circuit added to sense the lamp wattage. Correction is then made by varying the reactor impedance in order to control the lamp wattage to the desired value. This ballast type can give almost perfect wattage regulation for both line and lamp voltage changes. It is the most costly ballast, but has relatively low power losses.

Dimming of High Intensity Discharge Lamps

Advances in solid-state electronics have expanded the field of light source control to include the high intensity discharge (HID) group. Im-

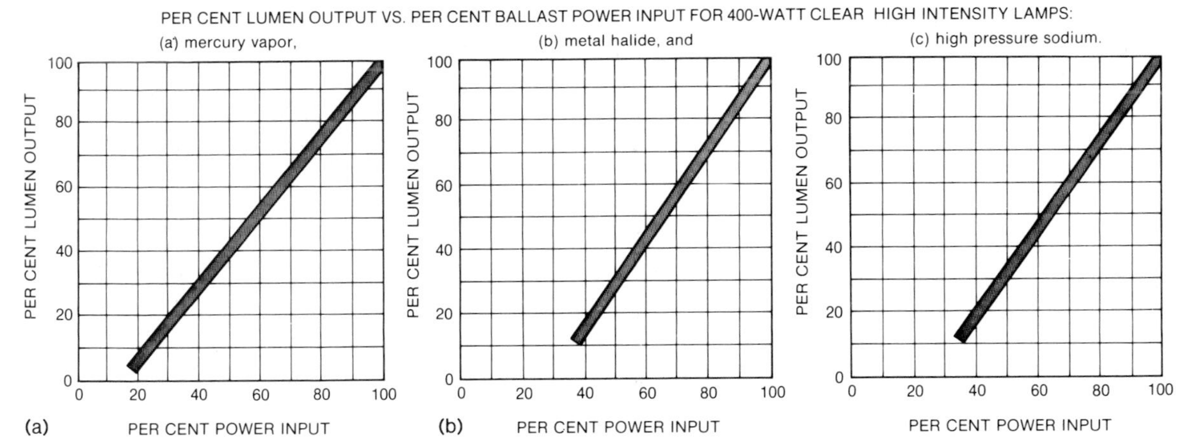

Fig. 8-55. Per cent lumen output vs per cent ballast input for 400-watt clear high intensity discharge lamps: (a) mercury vapor (b) metal halide, and (c) high pressure sodium.

provements in power semiconductors have provided a cost effective and reliable means to control lamp current and/or voltage. Coupled with this, similar advances in integrated circuits have provided the sophistication necessary to accomplish consistent and reliable dimming control of HID lamp output.

Lumen output versus ballast power characteristics and minimum controlled lumen output are shown in Figs. 8-55 and 8-56.

In dimming lamp color changes, especially at low lumen output. The following information describes general trends; however, the dimmer system manufacturer should be consulted for more specific detail.

Clear mercury lamps change very little in terms of color from 100 per cent to 25 per cent lumen output; the blue-green color which is characteristic of clear mercury sources is present at all dimmer settings. Color improved lamps will generally perform well down to about 30 per cent lumen output.

The color of clear metal halide lamps begins to fade at about 60 per cent lumen output where a blue-green color (characteristic of mercury vapor) starts to appear. This blue-green color is quite apparent at less than 40 per cent lumen output. The effect will be somewhat less with phosphor coated lamps.

High pressure sodium lamps generally provide their full output color quality down to about 50 to 60 per cent lumen output. Below 50 per cent lumen output, a strong yellow color (characteristic of low pressure sodium) begins to prevail.

Lamp life for dimmed HID lamps is generally as good as that of undimmed lamps; again individual system manufacturers should be consulted.

The slow warm-up and hot restrike delay which is characteristic of HID sources also applies in dimming applications. Some systems, however, may provide shorter warm-up and hot restrike times than conventional ballasts. HID lamps respond to dimmer settings much more slowly than incandescent or fluorescent sources and typical times required to go from minimum to maximum light output and vice versa are about 3 to 10 minutes. However, instantaneous response ranges for warm lamps may be as much as 50 to 80 per cent of full range depending on HID source type, lamp wattage and to some extent luminaire design.

The range and speed of response in general is not comparable to that of incandescent or fluorescent dimming; however, in all cases, the lamp

Fig. 8-56. HID Dimming System Performance for Clear Lamps

Light Source	Lamp Watts	Minimum Lumen Output (Per Cent of Full Lumen Output)
Mercury Vapor	100	2
	250	2
	400	2
	1000	3
Metal Halide	250	20
	400	10
	1000	12
High Pressure Sodium	250	12
	400	10
	1000	20

Fig. 8-57. Flicker Index for HID Lamps When Operated on Different Ballast Types

Lamp Type	Watts	Ballast Type	Per Cent Flicker (± 3 Per Cent)	Flicker Index (± 10 Per Cent)
Mercury	250	Reactor	88	.30
Mercury DX	400	Reactor	73	.25
Mercury	250	Lead-lag	25	.06
Mercury	400	Lead-lag	20	.05
Mercury	400	Regulator-Reactor	34	.10
Metal Halide (Na, Sc)	400	Reactor	38	.11
Metal Halide (Na, Sc)	1000	Reactor	34	.10
Metal Halide (Na, In, Tl)	1500	Reactor	28	.08
High Pressure Sodium	250	Reactor	95+	.29
High Pressure Sodium	400	Reactor	95+	.29
High Pressure Sodium	1000	Reactor	95+	.29
High Pressure Sodium	250	Lead-lag	60	.16

efficacy and color are reasonably good to at least 50 per cent lumen output. These characteristics are not well suited to dramatic lighting or theatrical effects, but are quite satisfactory for energy management applications. In energy management applications, the slow response time of HID lamps provides additional system stability and minimal occupant distraction.

A 50 per cent range of control can be employed to perform electronic lumen maintenance by means of a photoelectric servo control system. Automatic regulation of the lumen output in this manner can provide energy savings over a typical relamp cycle. Savings of 50 per cent or more can be obtained where available daylight is photoelectrically compensated for by a dimmer control system. Strong daylight contribution also will tend to compensate for color changes of the HID lamp at low lumen output.

Simple manual control of an HID dimming system can provide maximum uniformity, efficiency and flexibility in multi-purpose room applications. "Tuning" the light output for a specific task can provide energy savings. Additional controls can be employed to incorporate HID lighting reduction with time-of-day and demand reduction programs as administered by computer or simple time clock.

Self-Ballasted Lamps

Self-ballasted mercury lamps are available in various wattages. These lamps have a mercury vapor arc tube in series with a current limiting tungsten filament. In some types, phosphors coated on the outer envelope are used to provide additional color improvement. The overall efficacy is lower than that of other mercury lamps because of the lower efficacy tungsten filament. As the title denotes, these lamps do not require an auxiliary ballast as do standard mercury

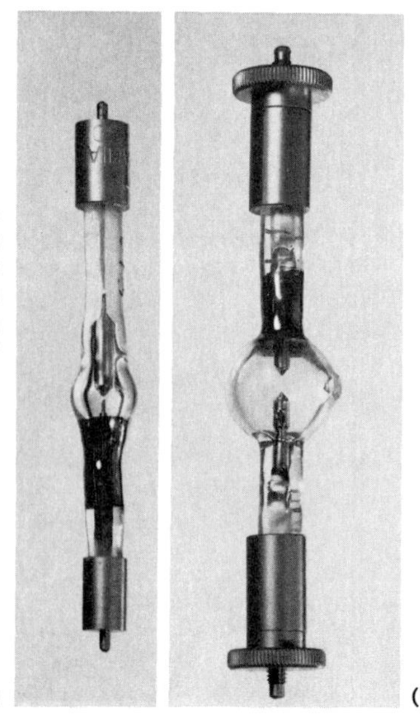

(a)

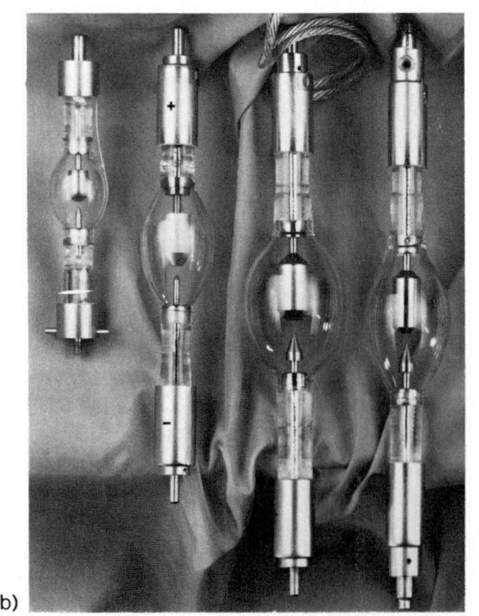

(b)

Fig. 8-58. Typical short-arc lamps. (a) Low wattage mercury-argon lamps (100-watt at left, 200-watt at right). (b) Medium wattage xenon lamps. From left: 1.6; 2.2; 4.2; 3.0-kilowatt. (For lamp dimensions see Fig. 8-130.)

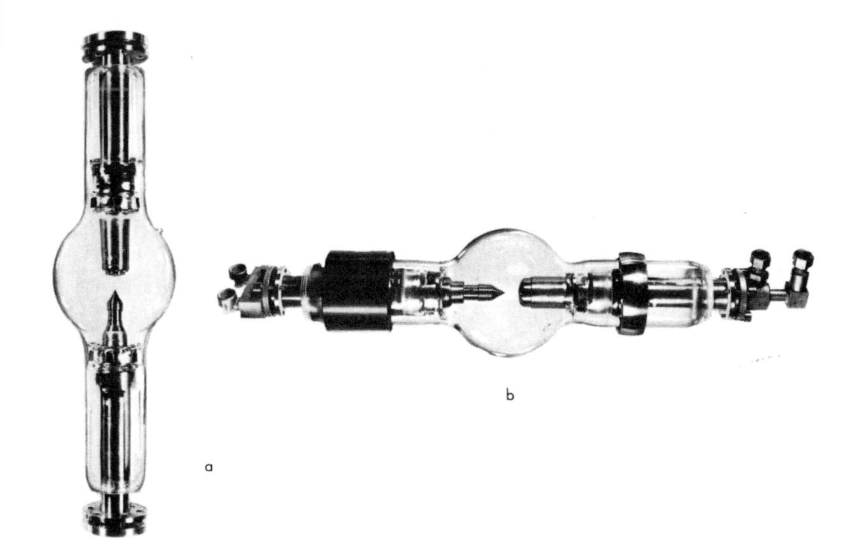

Fig. 8-59. Typical high power xenon compact arc lamps with liquid-cooled electrodes. (a) 30-kilowatt lamp for solar simulators (principal operating position is vertical) and (b) 20-kilowatt lamp for military searchlights (principal operating position is horizontal).

lamps. Typical self-ballasted mercury lamps are shown in Fig. 8-124.

Flicker and Stroboscopic Effect

The light output of all high intensity discharge lamps varies to some degree with the cyclic changes of the line voltage. This variation, or flicker, is a function of both the lamp type and ballast circuit. Fig. 8-57 illustrates the variation in both Per Cent Flicker and Flicker Index for mercury, metal halide and high pressure sodium lamps for several ballast types.[43] (See page 8-31 for definitions of Flicker Index and Per Cent Flicker.)

In most lighting applications stroboscopic effect from high intensity discharge sources is not a problem. However, it can be annoying to spectators in some types of ball games (*i.e.*, tennis, ping pong) and occasionally a nuisance or distraction around rotating machinery. To minimize stroboscopic effect in susceptible applications, systems with a flicker index of 0.1 or less are suggested or luminaires should be alternately wired on three-phase systems.

SHORT-ARC LAMPS

Short-arc or compact-arc lamps combine the high luminance of carbon arcs with the maintenance free and clean operation of regular gas discharge or incandescent lamps. They are primarily used in searchlights, projectors, display systems, optical instruments such as spectrophotometers, recording instruments, and for simulation of solar radiation in space. They are also specially suited as sources of modulated light, generated through current modulation.

Short-arc lamps are high pressure gas discharge lamps characterized by an electrode-stabilized arc which is short when compared with the diameter of the envelope. Depending on rated wattage and intended application, their arc length may vary from about a third of a millimeter to about 12 millimeters. These arcs have the highest luminance and radiance of any continuously operating light source and are the closest approach to a true "point" source of high luminance[44-57].

Some typical compact arc lamps are shown in Figs. 8-58 and 8-59. These lamps have optically clear fused silica (quartz) bulbs of spherical or ellipsoidal shape with two diametrically opposite seals. Four types of seals are used in compact-arc lamps. The graded seal and the molybdenum foil seal are current carrying seals while the molybdenum cup seal and the elastomer (mechanical) seal are separated from the current conductor by a cup or a flange.

Most short-arc lamps are designed for dc operation. Better arc stability and substantially longer life of the dc lamps have limited the use of ac compact arc lamps to special applications.

For applications requiring ozone free operation, the lamp envelopes are fabricated from quartz which does not transmit wavelengths below 210 nanometers.

Mercury and Mercury-Xenon Lamps[44, 45, 47, 51]

To facilitate lamp starting, short-arc mercury lamps contain argon or another rare gas at a pressure of several kilopascals (10 to 30 torrs), the same as standard mercury lamps. After the initial arc is struck, the lamp gradually warms up, the voltage increases and stabilizes as the mercury is completely vaporized. A mercury lamp requires several minutes to warm up to full operating pressure and light output. This warm-up time is reduced by about 50 per cent if xenon at a pressure exceeding one atmosphere is added to the mercury. Lamps with this type of filling are known as mercury-xenon short-arc or compact-arc lamps. The spectral power distribution in the visible region is essentially the same for both types, consisting mainly of the four mercury lines and some continuum, due to the high operating pressure. The luminous efficacy for these lamps is approximately 50 lumens per watt at a rated wattage of 1000 watts and about 55 lumens per watt at 5000 watts.

Mercury and mercury-xenon lamps are available for wattages from 30 to 5000 watts and are to be operated in a vertical position. For technical data see Fig. 8–130. A spectrum of a typical mercury-xenon lamp is shown in Fig. 8–60.

Xenon Lamps[46, 49–57]

Xenon short-arc lamps are filled with several atmospheres of xenon gas. They reach 80 per cent of the final output immediately after the start. The arc color very closely approximates daylight (color temperature approximately 6000 K). The spectrum is continuous from the ultraviolet through the visible into the infrared. Xenon lamps also exhibit strong lines in the near infrared between 800 and 1000 nanometers and some weak lines in the blue. A spectral power distribution curve is shown in Fig. 8–61.

Xenon short-arc lamps are made with rated wattages from 5 watts to 32,000 watts and are available for operation in vertical or horizontal positions. Lamps designed to be operated at wattages above 10 kilowatts require liquid cooling of the electrodes.

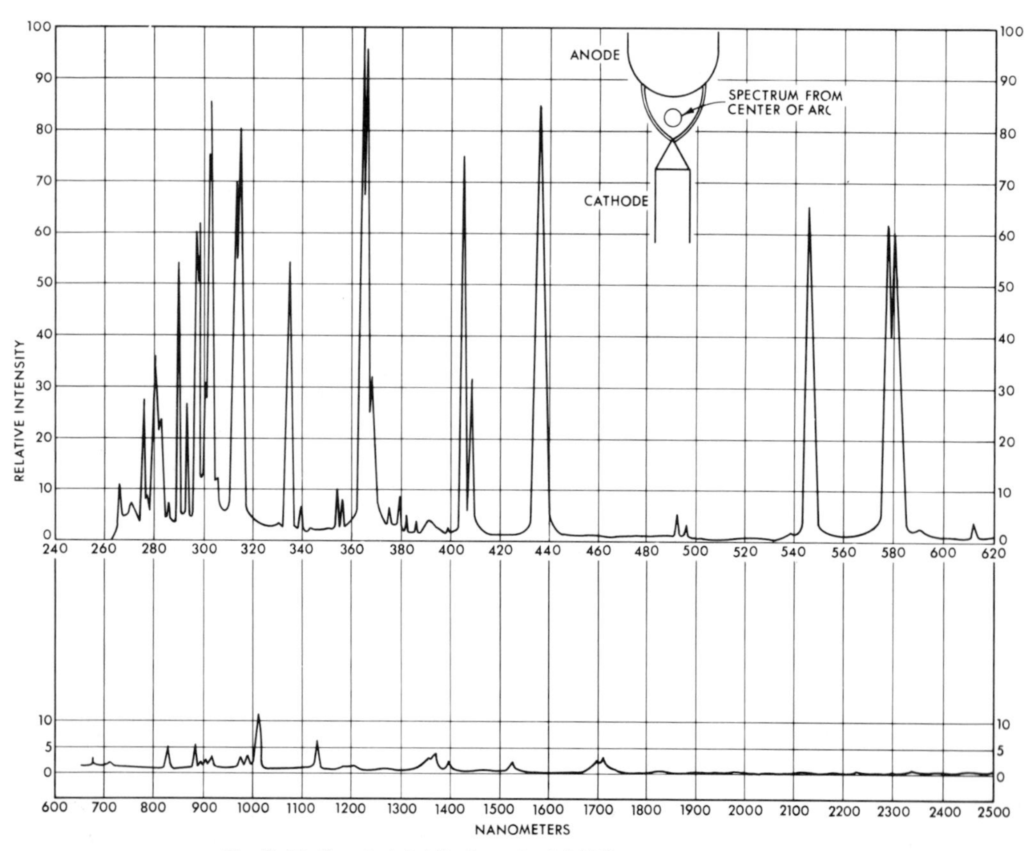

Fig. 8–60. Spectral distribution of a 2.5 kW mercury-xenon lamp.

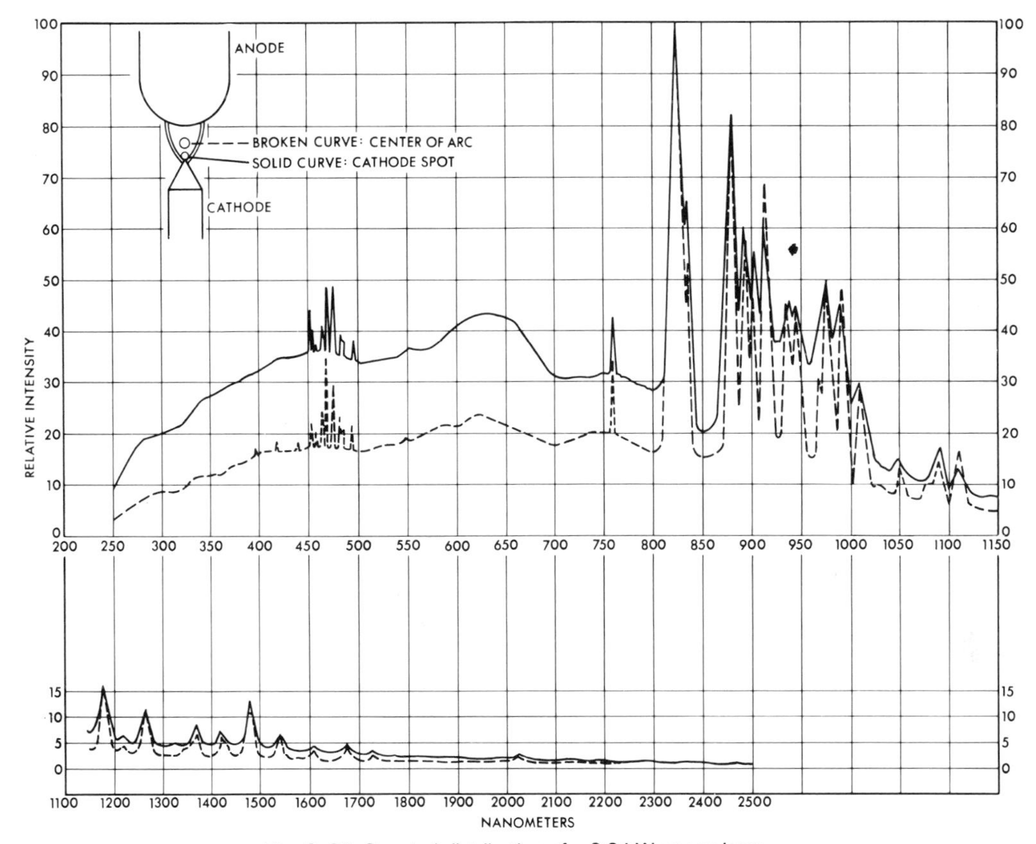

Fig. 8-61. Spectral distribution of a 2.2 kW xenon lamp.

The luminous efficacy of the xenon short-arc is approximately 30 lumens per watt at 1000 watts, 45 lumens per watt at 5000 watts, and over 50 lumens per watt at 20 kilowatts and higher input.

Short-Arc Illuminators

Short-arc illuminators combine the basic technology of short-arc lamps and an internal reflector that focuses the arc's energy through a sapphire window in the ceramic-to-metal housing. The window transmits energy in both the infrared and ultraviolet regions. These devices have the advantages of increased output, safety and ease of handling and installation; they eliminate the need for some peripheral equipment in many applications where short arc lamps are used. Short arc illuminator units are available with inputs from 150 to 1000 watts. Their spectral power distributions can be varied for special applications from the basic mercury-xenon or xenon spectrum by the material from which the reflector is made or by its coating.

Lamp Operating Enclosure

Short-arc mercury, mercury-xenon, and xenon lamps are under considerable pressure during operation (up to 50 atmospheres for small lamps and about 10 atmospheres for large units) and therefore must be operated in an enclosure at all times. In addition, precaution must be taken to insure protection from the powerful ultraviolet radiation emitted from the lamps.

In general, short-arc lamps up to about 2.5 kilowatts are designed to operate with natural air draft cooling. No special ventilation is required unless critical components of the lamps are subjected to excessive temperatures, caused by closely confined enclosures, excessive ambient temperatures or infrared radiation.

In order to eliminate a possible hazard during shipment, storage, or handling of xenon or mercury-xenon lamps, special protection cases are provided. These cases are made of metal or plastic and are so arranged around the bulb that the lamp can be electrically connected without removing the case. The case should not be removed until immediately before the lamp is energized.

Auxiliary Equipment

Like most vapor discharge lamps, short-arc lamps require auxiliary devices to start the arc and limit the current. For ac lamps either resistive or inductive ballasts may be used. Direct current lamps are best operated from specifically designed power systems which provide, with good efficiency, the high voltage pulses (up to 50,000 volts) required to break down the gap between the electrodes, ionize the gas and heat the cathode tip to thermionic emitting temperatures. Further, they provide enough open circuit dc voltage to assure the transition from the low-current high-voltage spark discharge initiated by the starter to the high-current low-voltage arc. With a properly designed system, a short-arc lamp will start within a fraction of a second. Many power supplies are regulated so that lamp operation is independent of line voltage fluctuation. Four basic types of sensors for power regulation are presently in use: current, voltage, power, and optical regulators. The type of power system used depends on the specifics of the application.

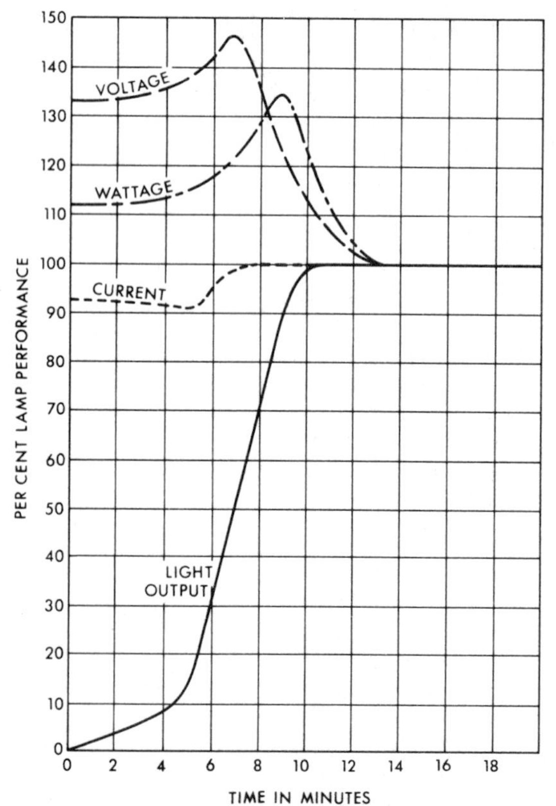

Fig. 8-62. Low pressure sodium lamp performance during starting.

Compact Source Metal Halide Lamps

Compact source or medium arc-metal halide lamps[58, 59] are based on a combination of the short-arc lamp and the metal halide lamp technology. Their arc discharge is predominantly electrode-stabilized and operates between tungsten electrodes spaced 9 to 35 millimeters apart in ellipsoidal or almost spherical quartz bulbs. They are filled with mercury and argon as basic elements for starting the arc and, as in some standard metal halide lamps, rare earth metal iodides and bromides are added in order to obtain a full spectrum. Both a high luminous efficacy and excellent color rendering with a correlated color temperature close to that of daylight are achieved, together with a high degree of source concentration.

These lamps, the result of European development work, are available under the designations HMI, CSI, DMI and CID in wattages from 200 to 12,000 watts; they operate on alternating current and require special ballasting equipment with high-voltage starting devices. They can be restarted in hot condition with suitably designed ignitors. Their main fields of application are: motion picture and television studio lighting, outdoor location lighting, theatrical lighting, sports lighting and solar simulation.

Fig. 8-131 gives the operating data of some commercially available compact source metal halide lamps.

MISCELLANEOUS DISCHARGE LAMPS

Low Pressure Sodium Lamps

In low pressure sodium discharge lamps, the arc is carried through vaporized sodium. The light produced by the low pressure sodium arc is almost monochromatic, consisting of a double line in the yellow region of the spectrum at 589.0 and 589.6 nanometers as shown in Fig. 8-21. The starting gas is neon with small additions of argon, xenon or helium. In order to obtain the maximum efficacy of the conversion of the electrical input to the arc discharge into light, the vapor pressure of the sodium must be in the order of 0.7 pascals (5×10^{-3} torr), which corresponds to an arc tube bulb wall temperature of approximately 260 °C. Any appreciable deviation from this pressure results in an undesirable loss of lamp efficacy. To maintain this proper oper-

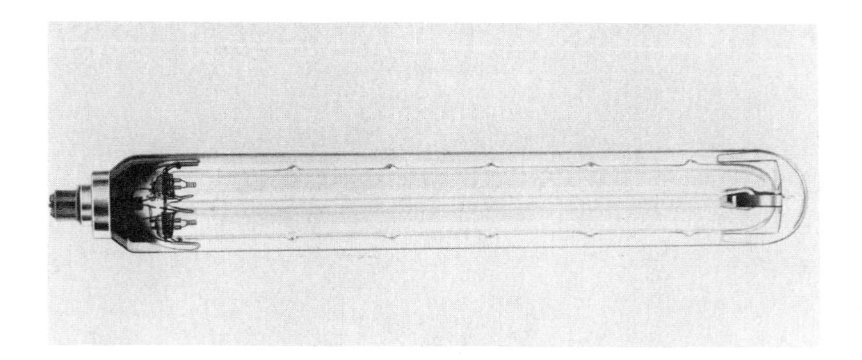

Fig. 8-63. Construction of low pressure sodium lamps (U-tube or hairpin type).

ating temperature, the arc tube is normally enclosed in a vacuum flask or in an outer bulb at high vacuum.

The run up time to full light output is 7 to 15 minutes. When first started, the light output is the characteristic red of the neon discharge and this gradually gives way to the characteristic yellow as the sodium is vaporized. Hot reignition is good and most low pressure sodium lamps will restart immediately after interruption of the power supply. Lamp performance during starting is shown in Fig. 8-62.

Efficacy. Low current density is vital to efficient generation of resonance radiation, as high densities result in excitation of atoms to higher energy levels and loss of resonance radiation. It is in the field of thermal insulation that greatest strides have been made in recent years, resulting in present day efficacies in excess of 180 lumens per watt* for the 180-watt "U" type low pressure sodium lamp. This thermal insulation consists of a light transparent infrared reflecting layer on the inside of the outer envelope. In current designs, this is an indium oxide layer, replacing the formerly-used tin oxide layer and even earlier internal glass sleeves.

Construction. There are two types of low pressure sodium lamps—the linear and the hairpin or "U" tube. The linear lamp has a double-ended arc tube, similar to a fluorescent lamp, with preheat electrodes sealed into each end. The arc tube, made of a special sodium resistant glass, is sealed in turn into an outer vacuum jacket and completed by a medium bipin base at each end. The hairpin type has the arc tube doubled back on itself, with its limbs very close together.

Two versions of the hairpin lamp are available based on differing approaches to maintaining even distribution of sodium in the arc tube throughout life. Since low pressure sodium lamps contain excess amounts of metallic sodium, the metal will tend to condense at the coolest part of the lamp. This occurs generally at the bend of the arc tube. If not controlled, sodium migration to the cool point will eventually result in the lamp "burning bare," that is, a return to a neon-argon arc in sections of the lamp.

One form of control is to provide dimples in the outer surface of the arc tube to present alternative cool points or reservoirs for the metallic sodium. The dimples also inhibit migration of sodium due to vibration or gravitational effects. This design ensures even distribution of sodium throughout the arc tube at the time of manufacture and control of the vapor pressure.

The alternative version uses a graded heat reflecting film along the inside of the outer envelope progressively increasing the amount of reflected heat as the natural cool point at the bend is appoached. In this version no dimples are used. The electrodes are sealed in at the pinches of the arc tube. The lamp is completed by a two-pin bayonet base. The construction of a low pressure sodium lamp with dimpled arc tube is shown in Fig. 8-63. The electrodes in this lamp are of a metal oxide type and are heated to electron-emissive temperature by ion bombardment. Electrical and light output values are given in Fig. 8-128.

Auxiliary Equipment. The low pressure sodium arc, in common with all discharge lamps, has a negative volt-ampere characteristic and a current limiting device, usually a transformer and reactor ballast, must be provided to prevent excessive lamp and line currents.

High power factor autotransformer ballasts are most commonly used, the required lamp starting voltages ranging between 400 and 550 volts. A capacitor wired in parallel on the primary side increases the power factor to 90 per cent or better. On this type of ballast, lamp

* Approximately 150 lumens per watt if ballast losses are added to lamp wattage.

Fig. 8-64. Typical glow lamps with American National Standards Institute numbers (old trade numbers).

regulation is excellent, lamp watts and lumen output remain within ±5 per cent from a varying line voltage range of ±10 per cent. Recently, constant wattage designs have been introduced.

Glow Lamps

These are low wattage, long life lamps designed primarily for use as indicator or pilot lamps, night-lights, location markers and circuit elements. They range from 1/17 to 3 watts and have an efficacy of approximately 0.3 lumens per watt. A group of typical glow lamps is shown in Fig. 8-64. These emit light having the spectral character of the gas with which they are filled. The most commonly used gas is neon, having a characteristic orange light output. The glow is confined to the negative electrode.

Glow lamps have a critical starting voltage, below which they are, in effect, an open circuit. The starting voltages for several glow lamps are shown in Fig. 8-132.

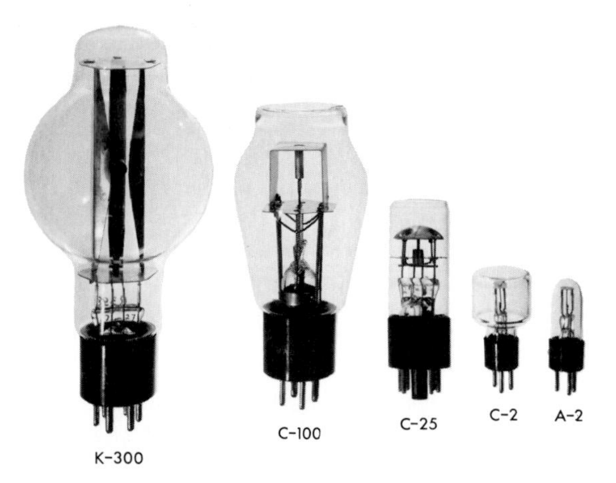

Fig. 8-65. Side and end-emission concentrated arc lamps.

Like other discharge lamps, glow lamps require a current limiting resistance in series. Glow lamps with screw bases have this resistor built into the base, while for unbased lamps or lamps with bayonet bases a resistor of the proper value must be employed external to the lamps.

Glow lamps filled with an argon mixture rather than neon radiate chiefly in the near ultraviolet region around 360 nanometers and are therefore used mainly to excite fluorescence in minerals and other materials as well as for some photographic applications.

Zirconium Concentrated Arc Lamps, Enclosed Type

These lamps are direct current arcs typified by a small concentrated point source of light of high intrinsic luminance, up to 4500 candelas per square centimeter. They are made with permanently fixed electrodes sealed into an argon-filled glass bulb. The light source is a small spot, 0.13 to 2.8 millimeters (0.005 to 0.11 inches) in diameter (depending on lamp wattage) which forms on the end of a zirconium oxide-filled tantalum tube which serves as the cathode. The spectral power distribution is similar to that of a blackbody with a color temperature of 3200 K. The spatial distribution is a cosine type.

These lamps require special circuits which generate a high voltage pulse for starting and a well filtered and ballasted operating current. Suitable power supplies are recommended by the manufacturer.

Fig. 8-65 illustrates various examples of side and end emission lamps. Fig. 8-127 gives essential characteristics of representative lamps.

Pulsed Xenon Arc (PXA) Lamps

These are ac nonpolarized xenon lamps with two active electrodes. A switching reactor in series with the low pressure lamp forces 50 to 100 peak amperes (pulsed 120 per second) through the lamp. The reactor also supplies a continuous current of 2 to 3 amperes to keep the lamp operating between pulses. The daylight spectrum produced is characteristic of xenon, *i.e.*, typically 6000 K. PXA lamps are available in linear to helical types.

The efficacy of these sources is about 35 to 40 lumens per watt. Available lamp wattages range from 300 to 8000 watts with forced air cooling a requirement during operation.

PXA lamps are used in the graphic arts industry where instant start, high stable light output, and daylight quality color temperature are required. Fig. 8-126 gives the main data of the principal types.

Flashtubes

These light sources are designed to produce high intensity flashes of extremely short duration. They are primarily used in photography applications; viewing and timing of reciprocating and rotating machinery; airport approach lighting systems, navigation aids, obstruction marking, warning and emergency lights; laser pumping; and other applications where a pulsing light is required or desired.

A conventional flashtube consists of a transparent tubular envelope of glass or fused silica (quartz) which has its main discharge electrodes internally located near the extremities and usually has an external electrode of wrapped wire. It generally contains very pure xenon gas at below atmospheric pressure usually in the range of 25 to 80 kilopascals (200-600 torr). Sometimes other gases such as argon, hydrogen and krypton are added to the xenon to obtain different spectral power distributions and/or different electrical, thermal and deionization characteristics. With a voltage applied across its main electrodes, the tube appears as a high impedance or open circuit until a trigger pulse ionizes the gas within the tube. The trigger pulse, usually applied to an external electrode, induces ionization and thereby causes the xenon gas to become conductive. A discharge then occurs between the main electrodes, whereupon the gas becomes highly luminescent. In some cases the trigger pulse is added to the voltage across the main electrodes and an external electrode is not required.

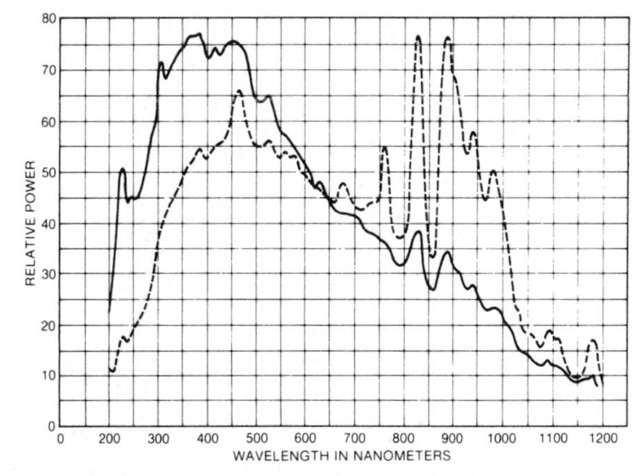

Fig. 8-66. Spectral power distribution of a typical xenon-filled flashtube for two different discharge conditions: (a) high voltage, low capacitance (solid line) and (b) low voltage, high capacitance (dashed line).

A xenon flashtube converts upward of 35 per cent of the input energy to light. The luminous efficacy ranges from 30 to 60 lumen-seconds per joule. Spectral quality is close to that of daylight having a correlated color temperature of approximately 6000 K, as the radiation encompasses the entire visible spectrum and extends into the ultraviolet and near infrared. See Fig. 8-66. Flashtubes are available in all sizes and shapes to suit the user and the type of optical system employed. The most common types are straight (linear), wound (helix) and "U" shape. Other configurations are available for special applications. Fig. 8-67 shows some typical, commercially available flashtubes.

Energy and Life. For single flash operation the limit to the amount of energy which can be consumed depends upon the desired tube life measured in useful flashes. This life is affected by the rate of envelope wall blackening and destruction of the tube or its parts. Flashtubes designed for very high loading have envelopes made of fused quartz which can withstand high thermal shock. The peak power encountered during a discharge produces a thermal shock which may be of sufficient magnitude to shatter the envelope, hence to maximize the energy per flash the thermal shock must be limited. This can be done by reducing the peak current which also lengthens the flash duration. To limit peak current and thermal shock as well as control the pulse duration, inductance may be added in series within the discharge loop.

Normally, the life expectancy of a flashtube can be approximated by the relationship of the "explosion energy" for that particular flashtube with respect to the energy per flash in a particular application. The "explosion energy" is defined by manufacturers as the energy level at a given flash duration that will cause the tube to fail within ten flashes, usually by disintegration of the envelope. Life can be approximated by the following:

Flash Energy (per cent "explosion energy")	Flashtube Life (number of flashes)
100	0–10
70	10–100
50	100–1000
40	1,000–10,000
30	10,000–100,000
20	100,000–1,000,000
5	over 1 million

Limits of Power Input. The average power input is a product of the energy per flash and the flash rate. The maximum power that any flashtube can dissipate is determined by the envelope area, type of envelope material and the method of cooling, such as free air convection, forced air convection or use of a liquid coolant. For fused quartz envelopes the maximum input power can be approximated as follows: 5 watts per square centimeter free air convection, 40 watts per square centimeter for forced air and 300 watts per square centimeter for liquid cooling.

Energy Storage Banks. The electrical energy which subsequently is discharged through the flashtube to produce light is stored in a capacitor bank. This bank must be capable of rapid discharge into very low impedance load. Therefore, it must have a relatively low inductance as well as a very low equivalent series resistance. It must also be capable of storing energy at a high voltage without significant leakage. Typical voltages vary from about 300 to 4000 volts. Modern banks use aluminum electrolytic, paper-oil or metalized paper capacitors designed specifically for energy storage applications. All these types are highly efficient in delivering energy to the flashtube. The type selected depends upon the voltage, temperature and life, as well as size and weight limitations.

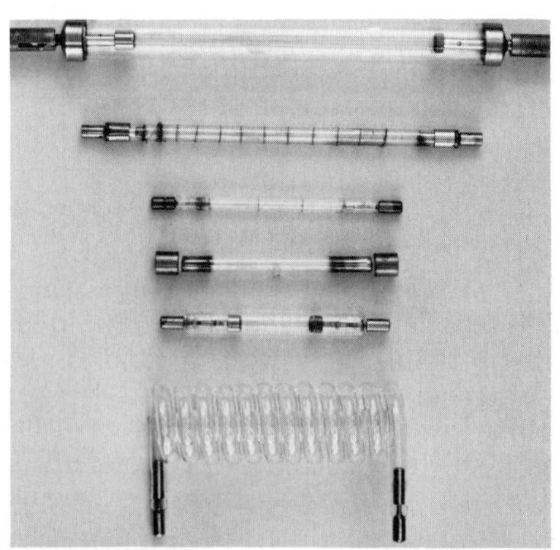

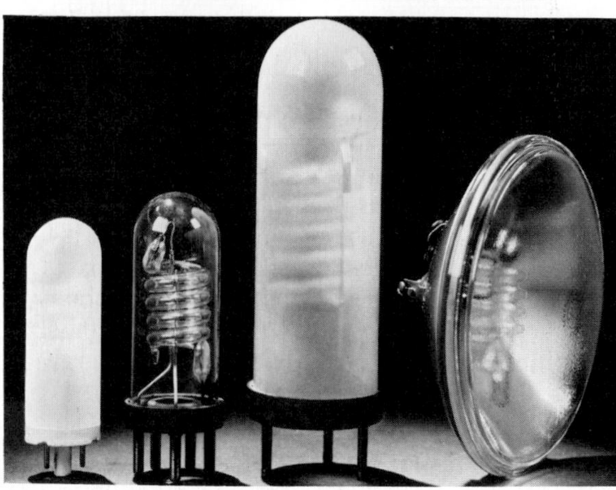

Fig. 8-67. Typical flashtubes.

Electronic Circuitry. In addition to discharge circuitry there are two other basic sections to the circuitry for a conventional xenon flash system—a charging circuit and a trigger circuit. See Fig. 8-68.

The charging circuit accepts primary electrical power at low voltage, transforms and rectifies it to higher voltage and applies it to the capacitor bank where it is stored as potential energy. The amount of light generated by the flashtube depends upon its loading which in turn depends upon the value of the energy storage capacitor and the voltage across it, in accordance with the formula:

$$\text{Loading in joules} = \frac{CV^2}{2}$$

where C is the capacitance in microfarads and V is the voltage across the tube (and capacitor) in volts.

The trigger circuit used for producing the high voltage ionizing pulse consists of a low-energy capacitor discharge system driving a pulse transformer. The output of the pulse transformer sets up an electric field which starts the ionization process and causes the gas to conduct. This pulse is usually applied to the external trigger wire (external electrode), but in some applications it may be applied across the main discharge electrodes by means of a pulse transformer with a very low secondary impedance connected in series with the flashtube discharge circuit.

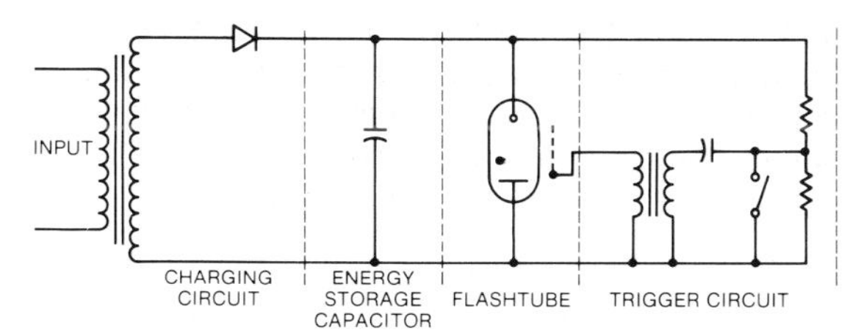

Fig. 8-68. Basic elements of a typical flashtube power supply.

By varying the voltage to which the capacitor bank is charged as well as its capacitance and by the insertion of inductance in the discharge circuit, it is possible to vary both the light output and the flash duration of the system. The flash duration is dependent upon the value of the capacitor, the inductance of the discharge circuit and the effective impedance of the flashtube. Although the flashtube is a non-linear circuit element, its effective impedance can be approximated according to the formula:

$$\text{Flashtube impedance} = \frac{\rho L}{A}$$

where ρ is the plasma impedance in ohm-centimeters, L is the arc length in centimeters and A is the cross section area in square centimeters. At current densities encountered in usual practice, ρ has a value of approximately 0.02 ohm-centimeter.

Flashtubes with their associated circuitry can be designed to operate with flash energies of fractional watt-seconds to 10,000 watt-seconds, and durations from approximately one microsecond to many milliseconds and with repetition rates from a single flash up to 1000 flashes per second. Higher frequencies can be attained with special circuitry and flashtube design.

Linear Arc Lamps

Linear arc quartz envelope lamps are available for both continuous wave and pulsed operation. Lamps operated in the pulsed mode are discussed above under flashtubes. Forced-air cooled long-arc xenon lamps are made with arc lengths up to 1.2 meters (4 feet), bore diameters up to 12 millimeters (0.47 inch) and wattages to 6 kilowatts. These lamps are used for special illumination requirements and solar simulation. Because of their low efficacy, 30 lumens per watt, they cannot compete with high intensity discharge lamps for general illumination usage.

Water cooled long-arc xenon and krypton lamps are made with arc lengths up to 0.3 meter (1 foot), bore diameters up to 10 millimeters (0.39 inch) and wattages up to 12 kilowatts. Their main application is for laser pumping; krypton arc lamps are especially suitable for pumping Nd:YAG lasers.

Forced-air cooled mercury and halide doped long-arc lamps are available in lengths up to 1.2 meters (4 feet), bore diameters up to 10 millimeters (0.39 inch) and wattages up to 5 kilowatts. They are used for ultraviolet photochemical applications, including the curing of paints, varnishes and coatings.

Mercury capillary lamps are made with arc lengths up to 150 millimeters (6 inches), bore diameters from 2 millimeters (.08 inch) and wattages up to 6 kilowatts. They are used for ultraviolet photo-exposure applications in the semiconductor and other industries. They are also finding use in the rapid thermal processing of silicon wafers.

All linear arc lamps use special ballasts and high voltage starting devices. Manufacturers recommendations for operation should be carefully followed.

ELECTROLUMINESCENT LAMPS

An electroluminescent lamp is a thin area source in which light is produced by a phosphor excited by a pulsating electrical field. In essence, the lamp is a plate capacitor with a phosphor embedded in its dielectric and with one or both of its plates translucent or transparent. Green, blue, yellow or white light may be produced by choice of phosphor. The green phosphor has the highest luminance. These lamps are available in ceramic and plastic form, flexible or with stiff backing, and are easily fabricated into simple or complex shapes. They have been used in decorative lighting, night lights, switchplates, instrument panels, clock faces, telephone dials, thermometers and signs. Their application is limited to locations where the general illuminance is low.

Luminance varies with applied voltage, frequency and temperature, as well as with the type of phosphor. See Fig. 8-69. At 120 volts, 60 hertz, the luminance of the ceramic form with the green phosphor is approximately 3.5 candelas per square meter [0.4 candelas per square foot]; the luminance of the plastic form may be as high as 27 candelas per square meter [2.7 candelas per square foot] under these conditions or up to 100

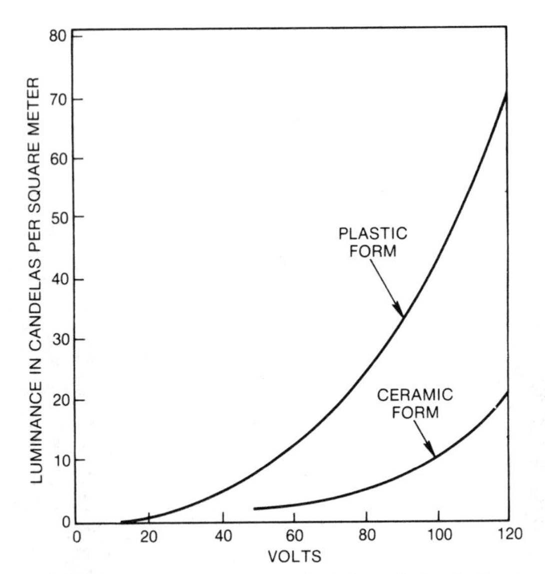

Fig. 8-69. Luminance of green ceramic and plastic forms of electroluminescent lamps operated at 400 hertz is dependent upon voltage.

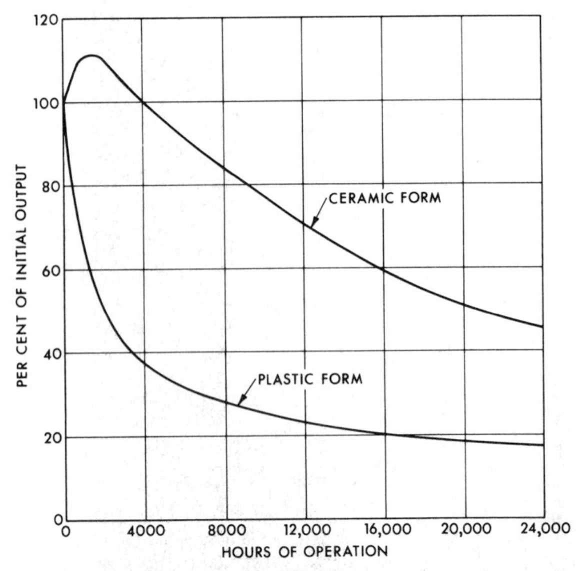

Fig. 8-70. Light output versus burning hours for green ceramic and plastic forms of electroluminescent lamps.

candelas per square meter [10 candelas per square foot] at 120 volts, 400 hertz. With the ceramic form at 600 volts, 400 hertz, a luminance of 70 candelas per square meter [7 candelas per square foot] has been achieved. These higher luminances are at the expense of useful lamp life.

Life is long and power consumption low. There is no abrupt point at which the lamp fails; the time at which the luminance has fallen to 50 per cent of initial is sometimes used as a measure of useful life. For the ceramic form, this is approximately 20,000 hours at 120 volts, 60 hertz. See Fig. 8–70. Approximate initial current and wattage values per tenth of a square meter (square foot) of lamp under these operating conditions is 60 milliamperes and 3.5 watts.

LIGHT EMITTING DIODES

The light emitting diode (LED) is a p-n junction semiconductor lamp which emits radiation when biased in a forward direction. The emitted radiation can be either invisible (infrared) or in the visible spectrum. Semiconductor light sources are available in a wide range of wavelengths, extending from the green region of the visible spectrum to the far-infrared region.

Visible solid-state lamps are used for long life indicator service. Infrared lamps have spectral outputs closely matched to the response of silicon photoreceivers. They are used in conjunction with these receivers for counting, sorting, sensing and positioning in applications as diverse as computer equipment, optical radar and burglar alarms.

The light producing material in a LED is a specially prepared semiconductor material of high purity, having small amounts of other elements added as controlled "impurities." Two classes of impurities are added: one to produce material having an excess of electrons and is called n-type material, the other to produce material having a shortage of electrons or "holes" which act as positive charges and is called p-type material. The impurities of the two classes are diffused into the same piece of semiconductor material so that an interface or "junction" is created between the p-type and the n-type materials. See Fig. 8–71a.

When a dc voltage is applied to the semiconductor material with polarity so that the n-type is negative and the p-type is positive the "holes" and the electrons are forced to meet at the junction where they combine to produce photons of light. The special characteristic of the energy or light is dependent upon the semiconductor material and the controlled impurities.

A typical LED construction using a metal header similar to that used for a transistor is shown in Fig. 8–71b. The lens is used to distribute the radiated energy.

The spectral output characteristics of four solid state lamp semiconductor materials are

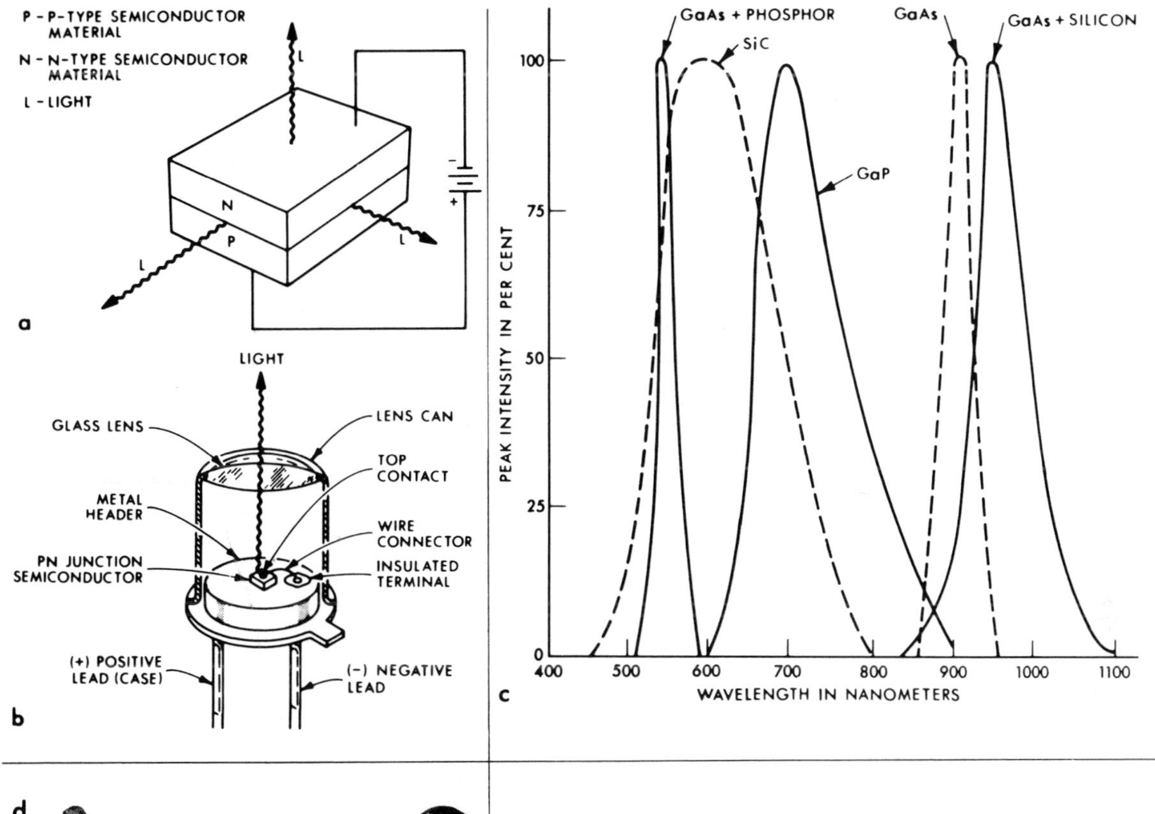

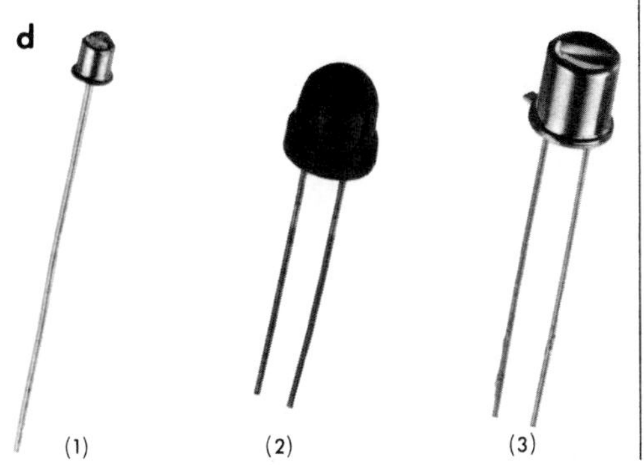

Fig. 8-71. a. A semiconductor junction. b. A cross-section view of a typical light emitting diode. c. Spectral output of several light emitting diode semiconductor materials (peak intensities are not equal for the different materials shown). d. Typical light emitting diodes: (1) coaxial, (2) plastic encapsulated, (3) header and lens cap.

shown in Fig. 8–71c: gallium arsenide and gallium arsenide with silicon producing infrared radiation, gallium phosphide producing red light, and silicon carbide producing yellow light. A phosphor, which absorbs infrared radiation at about 900 to 1000 nanometers and reradiates at about 540 nanometers, can be directly applied to gallium arsenide (with silicon) semiconductor material to produce green light.

The size of the "chip" (the piece of semiconductor material used in a solid state lamp) is generally 0.25 to 1.0 millimeter square (0.01 to 0.04 inch square) so lamps can be made very small in size.

When used in an indicator application, the LED semiconductor device (or "chip") will typically be encapsulated in a plastic package. Packages are typically molded out of epoxy or a similar translucent or transparent compound, thus providing strain relief for the leads connected to the semiconductor device itself. Lenses and diffusers are easily incorporated into the package design to focus or disperse the emitted light as required. The most common LED packages are the standard T-1 and T-1¾ (see Fig. 8–72), but a variety of other shapes and sizes are available. See Fig. 8–71d.

Certain specialized LED's are also available. One type combines two semiconductor devices in a common package. The devices are reversed biased so that when a positive current flow is reversed, the first device will turn off and the

Fig. 8-72. Light Produced by Various Types of Light Emitting Diodes (LED)

Type of LED	Typical Light Output (in millicandelas)*
T-1 Package:	
Red	0.7–2.5
Green	1.8–6.0
Yellow	1.8–4.0
Red (high efficiency)	1.5–2.5
T-1-3/4 Package:	
Red	0.5–2.0
Green	1.0–16.0
Yellow	1.0–16.0
Red (high efficiency)	2.0–24.0

* Light output as measured at design center of lens for recommended operating currents.

second device will begin emitting light, usually in a different color.

LED's generally operate in the range of 1 to 3 volts at currents of 10 to 100 milliamperes continuous. Output of visible solid-state lamps is of the order of 0.015 candela. Infrared lamps, operating continuously, emit up to 5.5 milliwatts at 100 milliamperes input current. Some lamps may be pulsed at currents exceeding 70 amperes to produce extremely short bursts of high infrared output with peak energies up to 0.5 watt.

The light output of a given LED is dependent on the size, color, current and specific LED semiconductor materials used. Light output increases linearly with current; however, lifetime will be shortened significantly as current is increased above recommended design limits.

NUCLEAR LIGHT SOURCES

Nuclear light sources are self-powered, self-contained light sources requiring no power supply. The light output is such that they can be easily seen by a person whose eyes are not dark adapted; some forms are visible at considerable distances and, therefore, provide a ready means for illuminating instrument panels, controls, locks and other devices.

These sources consist of a sealed glass tube or bulb internally coated with a phosphor and filled with tritium gas. Low energy beta particles (electrons) for the tritium, an isotope of hydrogen, strike the phosphor, which in turn emits light of a color characteristic of the type of phosphor used. Thus, the mechanism of light production is very similar to that in a conventional television tube where electrons strike a phosphor and cause it to emit visible light. With tritium having a half life of 12.3 years, the resulting light intensity should follow this decay. In reality, the time when half-intensity is reached is about six to seven years, and the useful life of the present lamp is approximately fifteen years.

The glass wall is impervious to tritium and completely absorbs any beta radiation not already absorbed by the phosphor. The unit is thus a completely sealed source and does not present any radiation hazard. Glass capsules can be produced in a wide variety of shapes and sizes and are usually made to normal glassworking tolerances.

The apparent luminance of a nuclear light source is determined by the beta-flux incident on the phosphor surface and by the color. The higher the tritium-phosphor area ratio, the greater the luminance. Luminance can range up to seven candelas per square meter, with a typical average of 1.7 candelas per square meter (this level is approximately that of an illuminated car instrument panel) and the sources can be supplied in a variety of colors. Highest apparent luminance is obtained in the green/yellow range and green is supplied as a standard color. Light sources can be supplied in very small sizes, down to 5 millimeters in diameter by 2 millimeters in length.

Use of these lamps in all applications is monitored by the Nuclear Regulatory Commission.

CARBON ARC LAMPS

Carbon arc lamps were the first commercial practical electric light sources. They now are used where extremely high luminance is necessary, where large amounts of radiant energy are required, and where their radiant power spectrum is advantageous. The distinct differences between the three basic types of carbon arcs (the low intensity, the high intensity, and the flame arc) are discussed below.

Types of Carbon Arcs

Low Intensity Arcs. Of the three principal types of carbon arcs in commercial use, the low intensity arc is the simplest. In this arc, the light source is the white-hot tip of the positive carbon. This tip is heated to a temperature near its sublimation point (3700 °C) by the concentration of a large part of the electrical energy of the discharge in a narrow region close to the anode surface. See Fig. 8–73.

The gas in the main part of the arc stream is extremely hot (in the neighborhood of 6000 °C) and so has a relatively high ion density and good electrical conductivity. The current is carried through this region largely by the electrons since, because of their small mass, move much more rapidly than the positive ions. However, equal numbers of positive ions and negative electrons are interspersed throughout the arc stream, so no net space charge exists, and the only resistance to the motion of the electrons is that supplied by collisions with inert atoms, molecules and other charged particles. Near the anode surface, the conditions are not as favorable for the conduction of current. The electrode tip is about 2000 °C cooler than the arc stream, and the gas immediately adjacent consists largely of carbon vapor in temperature equilibrium with the surface. At 3700 °C, this carbon vapor is a very poor conductor of electricity. Therefore, it requires a high voltage to force the current-carrying electrons through this vapor layer and into the anode. In a pure carbon arc this anode drop is about 35 volts. Most of the heat so developed is transferred to the surface of the positive carbon electrode, part by the impact of the highly accelerated electrons and part by thermal conduction. Finally, as the electrons reach the anode surface, they release their heat of condensation, contributing further to the high temperature of the electrode tip.

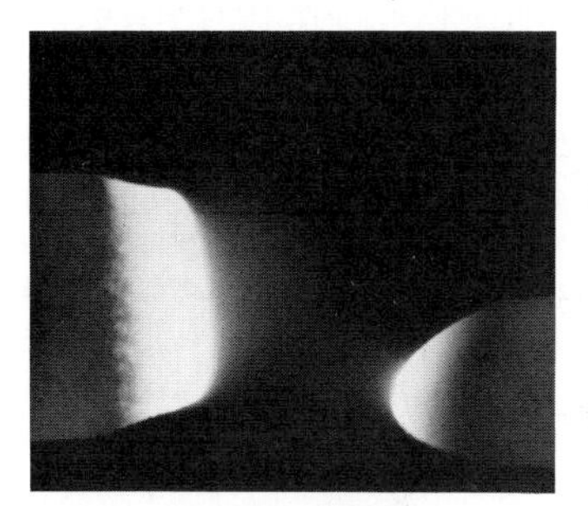

LOW-INTENSITY ARC

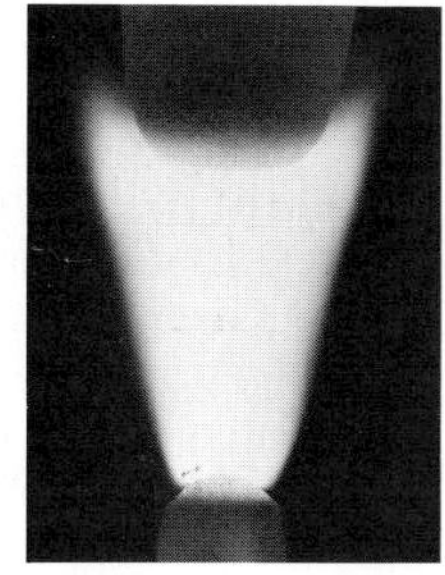

FLAME TYPE CARBON ARC

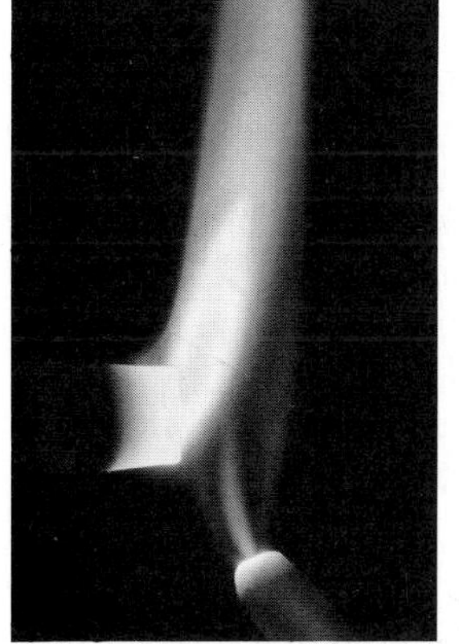

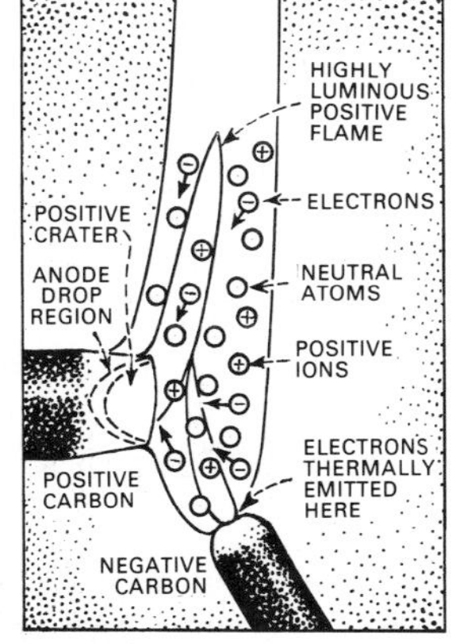

HIGH-INTENSITY ARC

Fig. 8–73. Low-intensity arc, 30 amperes, 55 volts, direct current. Flame arc, 60 amperes, 50 volts, alternating current. (Direct current flame arcs are very similar.) High-intensity arc, 125 amperes, 70 volts, direct current (rotating positive carbon).

The positive electrode of the low intensity arc may contain a core consisting of a mixture of soft carbon and a potassium salt. The potassium does not contribute to the light, but does increase the steadiness of the arc by lowering the effective ionization potential of the arc gas.

Flame Arcs. A flame arc (see Fig. 8–73) is obtained by enlarging the core in the electrodes of a low intensity arc and replacing part of the carbon with chemical compounds known as flame materials, capable of radiating efficiently in a highly heated gaseous form. These compounds are vaporized along with the carbon and diffuse throughout the arc stream, producing a flame of a color determined by the compounds used. Typical flame materials are iron for the ultraviolet, rare earths of the cerium group for white light, calcium compounds for yellow, and strontium for red.

Such flame materials have a considerably lower ionization potential than carbon. This greater ease of ionization reduces the temperature of the anode layer necessary for the conduction of current into the anode and results in a lower anode voltage drop (about 15 volts). The lower anode power input reduces the area and luminance of the anode spot so that its contribution to the total light becomes unimportant. The radiation from the flame arc consists chiefly of the characteristic line spectra of the elements in the flame material and the band spectra of the compounds formed. The excitation of the line and band spectra is thermal in nature, caused by the high temperature of the arc stream gas. The concentration of flame materials in the arc stream is not very high, so that the flame arc is considerably less bright than either the low or the high intensity arc. Since the whole arc flame is made luminous, the light source is one of large area and has radiating efficacies ranging up to 55 lumens per watt.

High Intensity Arcs. The high intensity arc is obtained from the flame arc by increasing the size and the flame material content of the core of the anode, and at the same time greatly increasing the current density, to a point where the anode spot spreads over the entire tip of the carbon. This results in a rapid evaporation of flame material and carbon from the core so that a crater is formed. The principal source of light is the crater surface and the gaseous region immediately in front of it. See Fig. 8–73. Since the flame material is more easily ionized than the carbon, a lower anode drop exists at the core area than at the shell of the carbon. This tends to concentrate the current at the core surface, and so encourages the formation of the crater.

The increased luminance of the high intensity arc is produced by radiation resulting from the combination of the heavy concentration of flame materials and the high current density within the confines of the crater. The spectrum of the radiation from the crater consists of a continuous portion plus line and band spectra characteristic of the gaseous components in the crater. This line and band radiation is greatly broadened, indicative of the many collisions, high temperature, and other processes and conditions existing in the crater. When compounds of the cerium family of rare earths are used as core materials, the combined effects produce radiation that is effectively continuous through the visual region of the spectrum and yield a luminance over ten times that of the low intensity arc.

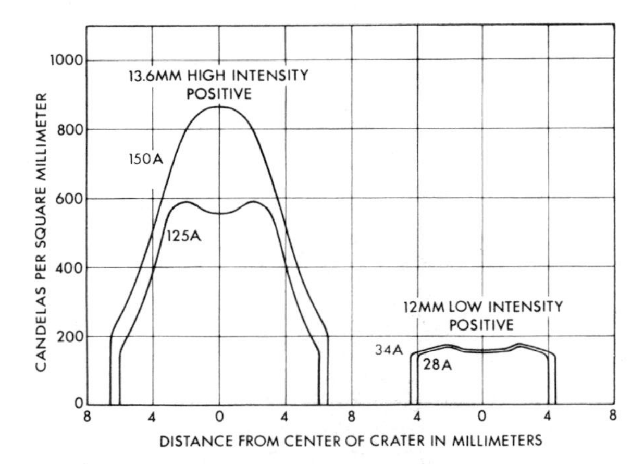

Fig. 8-74. Crater luminance distribution in forward direction for typical low and high intensity carbons.

Carbon Arc Lamps and Power Sources

Carbon arcs are operated in lamphouses which shield the outside from stray radiation. These lamphouses may incorporate optical components, such as lenses, reflectors and filters for eliminating undesired parts of the spectrum. They provide means to conduct the electrical current to the carbon electrodes and to feed the electrodes together to compensate for the portions consumed. They also provide for removal of the products of combustion. Some lamps employ directed streams of air to control the position of the flame from the positive electrode and to remove the combustion products.

Arcs for projection of motion pictures generally operate on direct current to prevent disturbing stroboscopic effects of the projector shutter. Both motor generators and rectifiers are employed. The flame arcs are widely used both on direct and alternating current. In some cases, alternating current arcs are operated directly from power lines and in others, from special transformers. Low current arcs have a negative volt-ampere characteristic and therefore must be operated from circuits which include ballast resistances or reactances (either in the generating or rectifying equipment or as separate units in the arc circuits). High intensity carbon arcs have positive volt-ampere characteristics and may be operated without ballasts. Suitable power supply equipment is available for all principal types of carbon arc lamps.

Carbon arcs may be started with a third (starting) electrode or the electrodes may be brought together momentarily after which they are separated to the proper distance to maintain the correct arc voltage and current. These conditions can be maintained and the carbons fed manually but in modern carbon arc lamps, automatic mechanisms feed the carbons as they are consumed and regulate the arc length and position of the light source. Some carbon arc lamps, such as those used to illuminate space environmental simulation chambers, are designed for continuous operation, without the necessity of shutdown for recarboning. Carbons are automatically dispensed from a magazine and joined to form a continuously burning electrode. Some electrodes are designed with a tapered joint to allow for complete consumption in the arc and thus eliminate the need for unburned carbon "stubs."

Low Intensity Arcs. Fig. 8–134 shows the characteristics of a typical low intensity carbon arc used for microscope illumination and projection. The light from a low intensity carbon arc has a luminance (150 to 180 candelas per square centimeter) and a color temperature (3600 to 3800 K) which exceed those from incandescent tungsten. When operated under prescribed conditions,[60] the low intensity carbon arc produces radiation which closely approximates that from a blackbody at 3800 K and is widely used as a radiation standard.

High Intensity Arcs. Fig. 8-134 shows the characteristics of a variety of typical high intensity carbon arcs ranging in power input from 2 to 30 kilowatts, in crater luminous intensity from 10,500 to 185,000 candelas, and in crater luminance from 55,000 to 145,000 candelas per square centimeter. The color temperature of the light ranges from as low as 2900 K to as high as 6500 K. These values will be further modified by the characteristics of the optical system used (see the Application Volume).

Fig. 8-74 shows the luminance distribution across the crater for typical low and high intensity carbons and illustrates two basic differences in the luminance characteristics: first, that the luminance is lower for the low intensity than the high intensity arcs; and secondly, that the luminance depends on the current, very markedly with the high intensity, but very little with the low intensity arc.

Flame Arcs. Fig. 8-135 gives characteristics of typical flame arcs. This compilation covers arcs with power input ranging from 1 to 4 kilowatts and shows the effect of different currents, voltages and flame materials. All arcs shown operate on alternating current with one exception. The wavelength intervals shown have been chosen to coincide with those important to various applications. The radiation in the indicated wavelength intervals is shown in two ways: first, as radiant intensity, expressed as microwatts incident on one square centimeter of area one meter distance from the arc; and secondly, as per cent of power input to the arc radiated over the entire sphere.

Rare earth flame materials exhibit high luminous efficacy, some exceeding 50 lumens per watt, and result in 15 to 20 per cent of the arc input energy radiated in the band of visible wavelengths (400 to 700 nanometers). The polymetallic flame materials with 3.6 to 7.2 per cent of the input power radiated below 320 nanometers are important to photochemical and therapeutic applications. Both rare earth flame arcs and the enclosed arc radiate efficiently in the 320- to 450-nanometer band so useful to various photoreproduction processes. Strontium concentrates the radiated energy in the red and near infrared regions of the spectrum.

Fig. 8-75 shows the spectral power distribution for flame arcs used for graphic arts applications. All of these show a concentration of radiation in the region 320 to 450 nanometers.

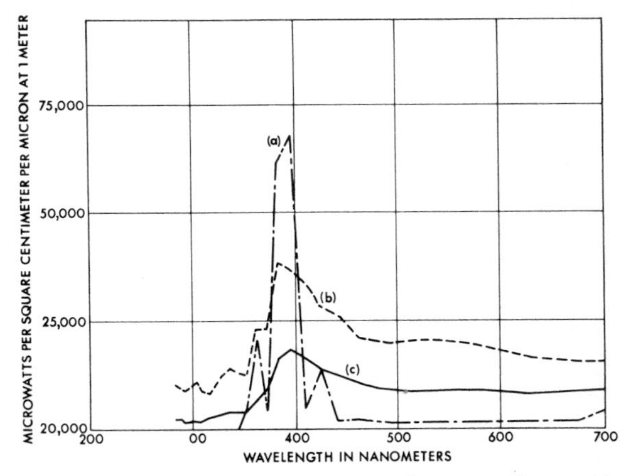

Fig. 8-75. Spectral power distribution of arcs used for graphic arts. (a) Half-inch enclosed arc carbons, 16 amperes, 138 volts. (b) Nine millimeter high-intensity photo carbons, 95 amperes, 30 volts. (c) Half-inch photographic white flame carbons, 38 amperes, 50 volts.

Fig. 8-76. Portable high intensity gaslight.

GASLIGHTS

Gaslights are devices which use gaseous fuels for lighting and decorative purposes. They use open gas flames or incandescent mantles of the upright and inverted types (see Fig. 8-76).

Mantle Construction

Mantles consist of a fabric such as rayon, silk, cotton ramie, or viscose woven into fabric tubing of the desired stitch and impregnated with a mixed solution of the nitrates of cerium, thorium, aluminum and magnesium. Rayon is the primary

Fig. 8-77. Light Output and Efficacies of Various Mantle Arrangements

Arrangement and Type of Mantle(s)	Total Input (watts)	Total Light Output (lumens)	Luminous Efficacy (lumens per watt)	Candelas Per 19 Watts*
One A† upright	645	434	0.67	1.0
Two No. 222‡ inverted	645	851	1.32	2.0
Three No. 222‡ inverted	967	1090	1.31	1.7

* 1.0 is minimum, per ANSI Z21.42-1971.
† Has a coated single wire top support and a single stitch weave.
‡ Representative of most inverted mantles.

material in use today. After impregnation, the knitted tubing is denitrated and cut into short lengths and attached to individual rings and mountings.

Mantles contain about 1 per cent ceric oxide (CeO_2) to 99 per cent thorium dioxide (ThO_2). They radiate selectively in the portion of the spectrum that includes wavelengths well suited to illumination. CeO_2, when heated alone to a high temperature, has high emissivity in the visible part of the spectrum, but it also has high emissivity in the infrared portion. A mantle made of pure CeO_2 cannot approach the temperature required to give high output in the visible spectrum. Due to its high emissivity in general, it keeps itself relatively cool. Because of this, a mantle made of CeO_2 is a poor light generator. ThO_2 has an emissive characteristic that is in a sense complementary to that of CeO_2. It is a poor heat radiator and of low emissivity in general.

If a mantle composed of CeO_2 and ThO_2 in about the proportion of 1 part to 99 is heated in a Bunsen flame, the preponderance of the ThO_2 will give the structure a very high temperature. CeO_2 is a drag on this temperature, but its mass is so small that it cannot prevent the attainment of a very high temperature in the combined mass. At the temperature of the combined oxides, CeO_2 gains high emissivity in the visible spectrum, which it could not attain without the aid of ThO_2. Any gain in temperature at about 2000 K results in a remarkable increase in emissive energy in the visible spectrum. This increase varies with the 11th or 12th power of the temperature for the visible but only with the fourth power for the total radiation. Luminous radiation of a very small mass of CeO_2 at high temperature is far greater than that of a large mass at lower temperature. Accordingly, mantle material so combined produces the maximum light output. In departing from the proportions of CeO_2 to ThO_2 either way, there is a loss of light output of the mantle. Lessening the proportion of CeO_2 beyond a certain point results in insufficient radiating surface. Increasing the CeO_2 beyond a certain amount results in lowered temperature, because of the large radiating power of CeO_2.

Mantles are available in two forms: soft and hard. Soft mantles are sold "unburned" and must be shaped on the burner, a familiar task for those who own gasoline lanterns.

Hard mantles are pre-shaped and pre-shrunk by burning out the fabric during manufacture. The remaining ash consists of oxides of the impregnating metals. A collodion coating is then applied to strengthen the burned mantle for handling and shipping; the coating, in turn, burns out when the mantle is first put to use.

It would be desirable, from the structural point of view, to make mantles much heavier, but if this were done, there would be a marked falling off in light output. Heavier strands in the finished mantle may be manufactured by giving heavier saturation to the knitting. These strands will not only be heavier, but of larger diameter and surface area according to the saturation, and will thus have larger capacity for absorbing and radiating energy in the form of light and heat. Because the energy in a given flame is fixed, it is evident that a larger radiation area would result in lower mantle temperature and, conse-

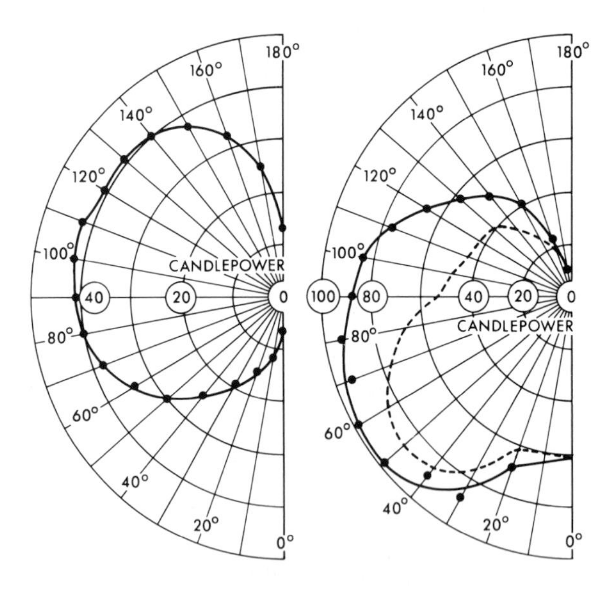

Fig. 8-78. Vertical candlepower distribution, in candelas, for mantles operated with optimum adjustment at total gas inputs of 645 watts: (*left*) a single type A upright mantle—view unobstructed by mantle support bracket; (*right*) a two-cluster No. 222 inverted mantle (solid line—both mantles foremost, dotted line—one mantle foremost). Note difference in scale: 180 degrees shown—the other 180 degrees are symmetrical.

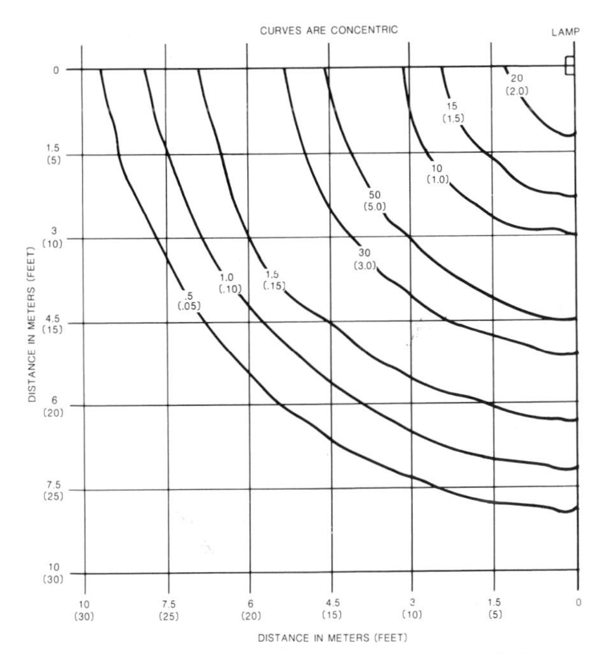

Fig. 8-79. Isofootcandle diagram (calculated) for 14 kilopascals (2 pounds per square inch) gaslight consisting of 4 inverted mantles 2.5 meters (8 feet 4 inches) above ground.

quently, in a marked loss of luminous efficiency. If, on the other hand, saturation were made lighter, smaller fiber surface and a consequent smaller radiating area would result in loss of total luminous and heat radiation, even though the luminous efficiency per unit area might be high. Furthermore, such mantles would be structurally weak.

Mantle Performance

Incandescence of a gas mantle is mainly dependent on an *exacting* cerium content, and on flame temperature and flame velocity. The type of gas, injection pressure, and burner design may contribute to the candlepower obtained. The weave of the mantle is well established for each type of gas and is not a factor in mantle life. The single stitch weave gives the same light output as the double stitch; however, light from a double stitch mantle may appear more yellow to the eye.

Light Output and Life. The light output, within design limits, is a direct function of the energy consumed in the mantle, and the corresponding mantle temperature. Gaslights are built so that a change in orifice size or the use of a properly adjusted pressure regulator is all that is required to accommodate the available gas pressures.

According to an ANSI standard which covers requirements for gas-fired illuminating appliances,[66] units incorporating a mantle should produce an average light output of at least one candela for each 19 watts (65 Btu per hour) of input rating. Fig. 8-77 gives performance data for three mantle arrangements. As it can be seen, the light output of a gaslight with a type A upright mantle (434 lumens) is about the same as that of a 40-watt incandescent lamp; the arrangement of two No. 222 inversed mantles has about the same light output (851 lumens) as a 60-watt lamp. Fig. 8-78 gives the candlepower distribution for a single upright mantle and a two-cluster inverted mantle. In the span of normal operating temperatures, light output varies as the tenth power of mantle temperature in degrees F. The luminous efficacy of a gas mantle is the equivalent of 0.67 to 1.32 lumens per watt.

Inverted mantles are generally more efficient than upright mantles because of superheating of the gas-air mixture in the mantle. The flame burns back on itself, concentrating more heat on the mantle fabric, and thus more light is generated.

A properly designed clear glass chimney (about 75-millimeter or 3-inch diameter) surrounding an upright mantle (not used with inverted mantles) will aid in directing the hot gases through the chimney and improve efficacy as much as 35 per cent above that of a bare mantle with no chimney. Opal domes may be used to produce a low-intensity upward light.

Unless mechanically abused, mantles have an indefinite life.

High Intensity Gaslights

Recent advances in gaslight technology make it possible to get brighter light from inverted gas mantles for a small increase in natural gas consumed in the mantle. The innovation is higher natural gas pressure, 14 to 135 kilopascals or 2 to 5 pounds per square inch gauge (psig), rather than the usual pressure of a few hundred pascals (inches of water column). Proper design as well as increased pressure provide significant increases in light output. Typical photometric data for a 14 kilopascals (2 psig) gaslight are shown in Fig. 8-79.

Gas, at increased pressures, aspirates more primary air and combines more thoroughly with that primary air. The resultant higher intensity flame, if matched to a suitable mantle, results in an appreciable increase in light output. Liquefied petroleum gas (propane and butane) at medium pressure (up to 100 kPa or 15 psig) has been used for many years in floodlights which use incandescent mantles.

Automatic Controls

When desired, gaslights may be turned down or off by clock or photocell actuated devices. However, most gaslights in use burn continuously at full light output.

Open Flame Gaslights

Open flame gaslights have no mantles and use non-primary aerated burners. They are used mainly for ornamental effects rather than for their lighting ability.

Fishtail Burners. These non-primary aerated burners yield a fan-shaped or fishtail-shaped flame. The head can either be slotted or have two ports angled so that their respective gas streams impinge. Such burners may be installed in gaslight assemblies if the maximum height of the flame is controlled and flame impingement is avoided (to minimize carbon deposits).

Gas Torches and Torch Lights. Also known as Luau torches, these units are often shaped like frustrums of right circular cones. Targets or flame spreaders, vanes to redirect drafts, and wire screens or porous plugs to coalesce and harden flames are used to improve the flame stability and wind resistance of these torches.

Portable Candle Lights. Another type of contemporary yellow flame burner is designed to operate with small portable cans of liquefied petroleum fuel and is meant to simulate a candle flame. A yellow flame about 25 millimeters (1 inch) high usually is adequate to create the desired effect. The light is normally operated in small lamp assemblies, which are basically miniature models of outdoor gaslight assemblies, and is therefore afforded some protection from drafts.

LIGHT SOURCE TABLES

The light source tables that follow on pages 8-137 to 8-141 represent an updating of and addition to the tables in the 1981 Reference Volume. The changes and additions are based on material received from light source manufacturers as of June 1984.

These tables are intended to provide the reader with a comprehensive listing of most light sources available in North America. They should be useful in preliminary design by helping determine appropriate lamps from both the physical and photometric aspects. Often, where there are more than one manufacturer for a particular lamp type, average photometric values are provided. For this reason and for latest data for a particular lamp, it is recommended that the lamp manufacturers be consulted.

The following is an abbreviated index for the tables.

Fig. 8-80. Approximate Luminance of Various Light Sources

Light Source			Approximate Average Luminance (cd/m^2)
Natural Light Sources:			
Sun (at its surface)			2.3×10^9
Sun (as observed from earth's surface)		At meridian	1.6×10^9
Sun (as observed from earth's surface)		Near horizon	6×10^6
Moon (as observed from earth's surface)		Bright spot	2.5×10^3
Clear sky		Average brightness	8×10^3
Overcast sky			2×10^3
Lightning flash			8×10^{10}
Combustion Sources:			
Candle flame (sperm)		Bright spot	1×10^4
Kerosene flame (flat wick)		Bright spot	1.2×10^4
Illuminating gas flame		Fish-tail burner	4×10^3
Welsbach mantle		Bright spot	6.2×10^4
Acetylene flame		Mees burner	1.1×10^5
Photoflash lamps			1.6×10^8 to 4×10^8 Peak
Nuclear Sources:			
Atomic fission bomb		0.1 millisecond after firing—30-m dia. ball	2×10^{12}
Self-luminous paints			0.2 to 0.3
Incandescent Lamps:			
Carbon filament		3.15 lumens per watt	5.2×10^5
Tantalum filament		6.3 lumens per watt	7×10^5
Tungsten filament		Vacuum lamp 10 lumens per watt	2×10^6
Tungsten filament		Gas filled lamp 20 lumens per watt	1.2×10^7
Tungsten filament		750-watt projection lamp 26 lumens per watt	2.4×10^7
Tungsten filament		1200-watt projection lamp 31.5 lumens per watt	3.3×10^7
RF (radio frequency) lamp		24-millimeter diameter disk	6.2×10^7
Blackbody at 6500 K			3×10^9
Blackbody at 4000 K			2.5×10^8
Blackbody at 2042 K			6×10^5
60-watt inside frosted bulb			1.2×10^5
25-watt inside frosted bulb			5×10^4
15-watt inside frosted bulb			3×10^4
10-watt inside frosted bulb			2×10^4
60-watt "white" bulb			3×10^4
Fluorescent Lamps:			
T-17 bulb	cool white	420 mA low loading	4.3×10^3
T-12 bulb	cool white	430 mA medium loading	8.2×10^3
T-12 bulb	cool white	800 mA high loading	1.1×10^4
T-17 grooved bulb	cool white	1500 mA extra high loading	1.5×10^4
T-12 bulb	cool white	1500 mA extra high loading	1.7×10^4
Electroluminescent Lamps:			
Green color at 120 volts 60 hertz			27
Green color at 600 volts 400 hertz			68
Electric Arcs:			
Plain carbon arc		Positive crater	1.3×10^8 to 1.6×10^8
High intensity carbon arc		13.6 mm rotating positive carbon	7×10^8 to 1.5×10^9
Electric Arc Lamps:			
High pressure mercury arc		Type H33, 2.5 atmospheres	1.5×10^6
High pressure mercury arc		Type H38, 10 atmospheres	1.8×10^6
High intensity mercury short arc		30 atmospheres	2.4×10^8 (4.3×10^9 peak)
Xenon short arc		900 W dc	1.8×10^8
Electronic flash tubes			1×10^9 to 3×10^9
Clear glass neon tube		15 mm 60 mA	1.6×10^3
Clear glass blue tube		15 mm 60 mA	8×10^2
Clear glass fluorescent			
daylight and white		15 mm 60 mA	5×10^3
green		15 mm 60 mA	9.5×10^3
blue and gold		15 mm 60 mA	3×10^3
pink and coral		15 mm 60 mA	2×10^3

Fig. 8–81. General Service Lamps for 115-, 120-, and 125-Volt Circuits (Will Operate in Any Position but Lumen Maintenance is Best for 40 to 1500 Watts When Burned Vertically Base-Up)

Watts	Bulb and Other Description	Base (see Fig. 8-9)	Filament (see Fig. 8-7)	Rated Average Life (hours)	Maximum Over-All Length		Average Light Center Length		Approximate Initial Filament Temperature (K)	Maximum Bare Bulb Temperature*		Base Temperature†		Approximate Initial Lumens	Rated Initial Lumens Per Watt‡	Lamp Lumen Depreciation** (per cent)
					(millimeters)	(inches)	(millimeters)	(inches)		(°C)	(°F)	(°C)	(°F)			
10	S-14 inside frosted or clear	Med.	C-9	1500	89	3 1/2	64	2 1/2	2420	41	106	41	106	80	8.0	89
15	A-15 inside frosted	Med.	C-9	2500	89	3 1/2	60	2 3/8	—	—	—	—	—	126	8.4	83
25	A-19 inside frosted	Med.	C-9	2500	98	3 7/8	64	2 1/2	2550	43	110	42	108	230	9.2	79
34[a,b]	A-19 inside frosted	Med.	CC-6 or CC-8	1500	113	4 7/16	76	3	—	—	—	—	—	410	12.1	
40	A-19 inside frosted and white¶	Med.	C-9	1500	108	4 1/4	75	2 15/16	2650	127	260	105	221	455	11.4	87.5
50	A-19 inside frosted	Med.	CC-6	1000	113	4 7/16	79	3 1/8	—	—	—	—	—	680	13.6	—
52[a,c]	A-19 clear or inside frosted	Med.	CC-8	1000	113	4 7/16	79	3 1/8	—	—	—	—	—	800	15.4	
55[a]	A-19 clear or white	Med.	CC-8	1000	113	4 7/16	79	3 1/8	—	—	—	—	—	870	15.8	
60	A-19 inside frosted and white¶	Med.	CC-6	1000	113	4 7/16	79	3 1/8	2790	124	25	93	200	860	14.3	93
67[a,d]	A-19 clear or inside frosted	Med.	CC-8	750	113	4 7/16	79	3 1/8	—	—	—	—	—	1130	16.9	
70[a]	A-19 clear or white	Med.	CC-8	750	113	4 7/16	79	3 1/8	—	—	—	—	—	1190	17.0	
75	A-19 inside frosted and white¶	Med.	CC-6	750	113	4 7/16	79	3 1/8	2840	135	275	96	205	1180	15.7	92
90[a]	A-19 clear or inside frosted	Med.	CC-8	750	113	4 7/16	79	3 1/8	—	—	—	—	—	1620	18.0	
95[a]	A-19 clear or white	Med.	CC-8	750	113	4 7/16	79	3 1/8	—	—	—	—	—	1710	18.0	
100	A-19 inside frosted and white¶	Med.	CC-8	750	113	4 7/16	79	3 1/8	2905	149	300	98	208	1740	17.4	90.5
100§	A-19 inside frosted and white	Med.	CC-8	1000	113	4 7/16	79	3 1/8	—	—	—	—	—	1680	16.8	—
100	A-21 inside frosted	Med.	CC-6	750	133	5 1/4	98	3 7/8	2880	127	260	90	194	1690	16.9	90
135[a]	A-21 clear or inside frosted	Med.	CC-8	750	139	5 1/2	101	4	—	—	—	—	—	2580	19.1	
150	A-21 inside frosted	Med.	CC-8	750	139	5 1/2	101	4	2960	—	—	—	—	2880	19.2	89
150	A-21 white	Med.	CC-8	750	139	5 1/2	101	4	2930	—	—	—	—	2790	18.6	89
150	A-23 inside frosted or clear or white	Med.	CC-6	750	157	6 3/16	117	4 5/8	2925	138	280	99	210	2780	18.5	89
150	PS-25 clear or inside frosted	Med.	C-9	750	176	6 15/16	133	5 1/4	2910	143	290	99	210	2660	17.7	87.5
200	A-23 inside frosted white or clear	Med.	CC-8	750	160	6 5/16	117	4 5/8	2980	174	345	107	225	4000	20.0	89.5
200	PS-25 clear or inside frosted	Med.	CC-6	750	176	6 15/16	133	5 1/4	—	—	—	—	—	3800	19.0	—

200	PS-30 clear or inside frosted	Med.	C-9	750	204	8 1/16	152	6	2925	152	305	99	210	3700	18.5	85
300	PS-25 clear or inside frosted	Med.	CC-8	750	176	6 15/16	131	5 3/16	3015	205	401	112	234	6360	21.2	87.5
300	PS-30 clear or inside frosted	Med.	C-9	750	204	8 1/16	152	6	3000	135	275	79	175	6100	20.3	82.5
300	PS-30 clear or inside frosted	Mog.	CC-8	1000	219	8 5/8	177	7	—	—	—	—	—	5960	19.8	—
300	PS-35 clear or inside frosted	Mog.	C-9	1000	238	9 3/8	177	7	2980	166	330	102	215	5860	19.6	86
500	PS-35 clear or inside frosted	Mog.	CC-8	1000	238	9 3/8	177	7	3050	213	415	79	175	10600	21.2	89
500	PS-40 clear or inside frosted	Mog.	C-9	1000	247	9 3/4	177	7	2945	199	390	102	215	10140	20.3	—
750	PS-52 clear or inside frosted	Mog.	C-7A	1000	331	13 1/16	241	9 1/2	2990	—	—	—	—	15660	20.9	—
750	PS-52 clear or inside frosted	Mog.	CC-8 or 2CC-8	1000	331	13 1/16	241	9 1/2	3090	—	—	—	—	17000	22.6	89
1000	PS-52 clear or inside frosted	Mog.	C-7A	1000	331	13 1/16	241	9 1/2	2995	249	480	113	235	21800	21.8	—
1000	PS-52 clear or inside frosted	Mog.	CC-8 or 2CC-8	1000	331	13 1/16	241	9 1/2	3110	—	—	—	—	23600	23.6	89
1500	PS-52 clear or inside frosted	Mog.	C-7A	1000	331	13 1/16	241	9 1/2	3095	266	510	129	265	34000	22.6	78

* Lamp burning base up in ambient temperature of 25°C (77°F).
† At junction of base and bulb.
‡ For 120-volt lamps.
§ Used mainly in Canada.
¶ Lumen and lumen per watt value of white lamps are generally lower than for inside frosted.
** Per cent initial light output at 70 per cent of rated life.

[a] Reduced wattage, high efficacy, 120 and 130 volt.
[b] Substitute for 40-watt lamp.
[c] Substitute for 60-watt lamp.
[d] Substitute for 75-watt lamp.

Fig. 8-82. Effect of Hot-Cold Resistance on Current in an Incandescent Filament (Laboratory Conditions)

Lamp Wattage	Voltage	Normal Current (amperes)	Theoretical Inrush: Basis Hot-to-Cold Resistance (amperes)*	Time for Current to Fall to Normal Value (seconds)
General Service Incandescent				
15	120	0.125	2.30	0.05
25	120	0.208	3.98	0.06
40	120	0.333	7.00	0.07
50	120	0.417	8.34	0.07
60	120	0.500	10.2	0.08
75	120	0.625	13.1	0.09
100	120	0.835	17.9	0.10
150	120	1.25	26.1	0.12
200	120	1.67	39.5	0.13
300	120	2.50	53.0	0.13
500	120	4.17	89.5	0.15
750	120	6.25	113	0.17
1000	120	8.3	195	0.18
1500	120	12.5	290	0.20
2000	120	16.7	378	0.23
Tungsten-halogen Lamps (C-8 Filament)				
300	120	2.50	62	**
500	120	4.17	102	**
1000	240	4.17	100	**
1500	240	6.24	147	**
1500	277	5.42	129	**

* The current will reach the peak value within the first peak of the supplied voltage. Thus the time approaches zero if the instantaneous supplied voltage is at peak, or it could be as much as 0.006 second.
** Not established. Estimated time is 5 to 20 cycles.

Fig. 8-83. Luminous and Thermal Characteristics of Typical Vacuum and Gas-Filled Incandescent Filament Lamps

1	2	3	4	5	6	7	8	9	10	
Watts	Radiated in Visible Spectrum (per cent of input wattage)	Total Filament Radiation Beyond Bulb (per cent of input wattage)	Gas Loss (per cent of input wattage)	Base and Bulb Loss (per cent of input wattage)	End Loss (Loss by conduction at filament ends) (per cent of input wattage)	Filament Heat Content (joules)	Heating Time to 90 Per Cent Lumens (seconds)	Cooling Time to 10 Per Cent Lumens (seconds)	Per Cent Flicker (see page 8-31)	
									60 hertz	25 hertz
6*	6.0	93.0	—	5.5	1.5	0.25	0.04	0.01	29	69
10*	7.1	93.5	—	5.0	1.5	0.62	.06	.02	17	40
25*	8.7	94.0	—	4.5	1.5	2.8	.10	.03	10	28
40†	7.4	71.3	20.0	7.1	1.6	2.5	.07	.03	13	29
60†‡	7.5	80.8	13.5	4.5	1.2	5.5	.10	.04	8	19
100†‡	10.0	82.0	11.5	5.2	1.3	14.1	.13	.06	5	14
200†	10.2	77.4	13.7	7.2	1.7	39.5	.22	.09	4	11
300†	11.1	79.8	11.6	6.8	1.8	80.0	.27	.13	3	8
500†	12.0	82.3	8.8	7.1	1.8	182.0	.38	.19	2	6
1000†	12.1	87.4	6.0	4.7	1.9	568.0	.67	.30	1	4

* Vacuum.
† Gas filled.
‡ Coiled-coil filament.

Fig. 8–84. Three-Lite Lamps for 115-, 120-, and 125-Volt Circuits (For Base-Down Burning Only)

Watts	Bulb and Other Description	Base	Filament (see Fig. 8–7)	Rated Average Life (hours)	Maximum Over-all Length (millimeters)	Maximum Over-all Length (inches)	Light Center Length (millimeters)	Light Center Length (inches)	Approximate Initial Lumens	Approximate Initial Lumens Per Watt	Lamp Lumen Depreciation* (per cent)
15	A-21	3 contact Med.	C-2R/CC-8	1500	135	5 5/16	98	3 7/8	85	5.7	
135				1200					2300	17.0	
150				1200					2385	15.9	
30	A-21 or T-19 white	3 contact Med.	C-8 or 2CC-6 or 2CC-8 or C-2R/CC-8	1200	135	5 5/16	95	3 3/4	290	9.7	
70									1035	14.8	
100									1315	13.2	86
50	A-23 or A-21 white	3 contact Med.	2CC-6 or 2CC-8 or C-2R/CC-8	1200	151	5 15/16	98	3 7/8	600	12.0	
100									1600	16.0	
150									2200	14.7	85
50	PS-25 or T-19 white	3 contact Med.	2C-2R or 2CC-6 or 2CC-8	1200	151	5 15/16	98	3 7/8	575	11.5	
100									1450	14.5	
150									2025	13.5	—
50	PS-25 inside frosted	3 contact Mog.	2C-2R or 2CC-8	1200	173	6 13/16	127	5	640	12.8	
100									1630	16.3	
150									2270	15.1	—
50	A-21	3 contact Med.	C-2R/CC-8	1500	112	4 7/16	79	3 1/8	580	11.6	
135				1200					2330	17.3	
185				1200					2910	15.7	
100	PS-25 white	3 contact Mog.	2CC-6 or 2CC-8 or C-2R/CC-8	1200	173	6 13/16	105	4 1/8	1360	13.6	
200									3400	17.0	
300									4760	15.9	72
50	PS-25, A-23 or T-21 white	3 contact Med.	2C-2R or 2CC-6 or 2CC-8 or C-2R/CC-8	1200	151	5 15/16	98	3 7/8	600	12.0	
200									3660	18.3	
250									4210	16.8	72

* Per cent initial light output at 70 per cent of rated life.

Fig. 8-85. Lamps for High Voltage Service (May be Burned in Any Position)

Watts	Bulb and Other Description	Base (see Fig. 8-9)	Filament (see Fig. 8-7)	Rated Average Life (hours)	Maximum Over-All Length		Average Light Center Length		230–250 Volts			277 Volts		
					(milli-meters)	(inches)	(milli-meters)	(inches)	Approx-imate Initial Lu-mens**	Rated Initial Lumens Per Watt**	Lamp Lumen Depre-cia-tion*** (per cent)	Approx-imate Initial Lumens	Rated Initial Lumens Per Watt	Lamp Lumen Depre-cia-tion*** (per cent)
25	A-19 inside frosted	Med.	C-17 or C-17A	1000	98	$3\frac{7}{8}$	65	$2\frac{9}{16}$	220	8.8	86	210	8.4	—
50	A-19 inside frosted	Med.	C-17 or C-17A	1000	98	$3\frac{7}{8}$	65	$2\frac{9}{16}$	490	9.8	79	490	9.8	—
100	A-21 inside frosted	Med.	C-7A or C-9	1000	133	$5\frac{1}{4}$	97	$3\frac{13}{16}$	1280	12.8	90	1280	12.8	—
150	PS-25 clear or inside frosted	Med.	C-7A or C-9	1000	176	$6\frac{15}{16}$	132	$5\frac{3}{16}$	2080	13.9	—	1950	13.0	—
200	PS-30 clear or inside frosted*	Med.	C-9	1000	205	$8\frac{1}{16}$	153	6	3040	15.2	90	2890	14.5	—
300	PS-30 clear	Med.	C-7A	1000	205	$8\frac{1}{16}$	153	6	4735	15.8	—	4450	14.8	—
300	PS-35 clear or inside frosted*	Mog.	C-7A	1000	238	$9\frac{3}{8}$	178	7	4890	16.3	89	4890	16.3	—
500	PS-40 clear or inside frosted*	Mog.	C-7A	1000	248	$9\frac{3}{4}$	178	7	9270	18.5	87	8900	17.8	—
750	PS-52 clear or inside frosted*	Mog.	C-7A	2000†	330	13	241	$9\frac{1}{2}$	13600	18.1	—	13500	18.0	—
1000	PS-52 clear or inside frosted*	Mog.	C-7A	2000†	330	13	241	$9\frac{1}{2}$	18000	18.0	82	18000	18.0	—
1500	PS-52 clear or inside frosted*	Mog.	C-7A	2000†	330	13	241	$9\frac{1}{2}$	27000	18.0	—	26000	17.3	—

* Clear only for 277-volt lamps.
** For 230-volt lamps.
*** Per cent initial light output at 70 per cent of rated life.
† 1000 hour life for 277-volt lamps.

Fig. 8-86. Rough Service Lamps for 115, 120, and 130-Volt Circuits (May Be Burned in Any Position)

Watts	Bulb and Other Description	Base (see Fig. 8–9)	Filament (see Fig. 8–7)	Rated Aver-age Life (hours)	Maximum Over-All Length		Light Center Length		Approxi-mate Ini-tial Lu-mens	Approxi-mate Ini-tial Lu-mens per Watt	Lamp Lumen Depre-ciation† (per cent)
					(millime-ters)	(inches)	(millime-ters)	(inches)			
25*	A-19 inside frosted	Med.	C-17	1000	100	$3\frac{15}{16}$	64	$2\frac{1}{2}$	230	9.2	—
50	A-19 inside frosted	Med.	C-22	1000	100	$3\frac{15}{16}$	64	$2\frac{1}{2}$	480	9.6	76
75*	A-21 inside frosted	Med.	C-22 or C-17	1000	113	$4\frac{7}{16}$	73	$2\frac{7}{8}$	750	10.0	
100	A-21 inside frosted	Med.	C-17	1000	135	$5\frac{5}{16}$	98	$3\frac{7}{8}$	1250	12.5	79
150*	A-23 inside frosted	Med.	C-17	1000	154	$6\frac{1}{16}$	111	$4\frac{3}{8}$	2130	14.2	—
150	PS-25 inside frosted	Med.	C-17	1000	176	$6\frac{15}{16}$	133	$5\frac{1}{4}$	2130	14.2	80
200*	PS-30 inside frosted or clear	Med.	C-9	1000	205	$8\frac{1}{16}$	152	6	3380	16.9	82.5
200*	A-23 inside frosted or clear	Med.	C-17	1000	154	$6\frac{1}{16}$	111	$4\frac{3}{8}$	3380	16.9	—

* 115-volt not available.
† Per cent initial light output at 70 per cent of rated life.

Fig. 8-87. Extended Service (2500 Hours Rated Life) Incandescent Filament Lamps†

Watts	Bulb and Finish	Base (see Fig. 8-9)	Filament (see Fig. 8-7)	Maximum Overall Length (millimeters)	Maximum Overall Length (inches)	Average Light Center Length (millimeters)	Average Light Center Length (inches)	Approximate Initial Lumens	Initial Lumens Per Watt	Lamp Lumen Depreciation (per cent)
For 120, 125, and 130 Volts										
15	A-15 inside frosted	Med.	C-9	89	3½	60	2⅜	125	8.3	83
25	A-19 inside frosted	Med.	C-9	100	3 15/16	64	2½	235	9.4	79
34‡	A-19 inside frosted	Med.	CC-6 or CC-8	113	4 7/16	76	3	390	11.5	
40	A-19 inside frosted	Med.	C-9	108	4¼	73	2⅞	420	10.5	87.5
52‡	A-19 inside frosted	Med.	CC-8	113	4 7/16	79	3⅛	740	14.2	
60	A-19 inside frosted	Med.	CC-6	113	4 7/16	79	3⅛	775	12.9	91.5
67‡	A-19 inside frosted	Med.	CC-8	113	4 7/16	79	3⅛	950	14.2	
90‡	A-19 inside frosted	Med.	CC-8	113	4 7/16	79	3⅛	1425	15.8	
100	A-19 inside frosted	Med.	CC-8	113	4 7/16	79	3⅛	1490	14.9	92.5
100	A-21 inside frosted	Med.	CC-6	135	5 5/16	98	3⅞	1450	14.5	—
135‡	A-19 inside frosted and clear	Med.	CC-8	137	5⅜	102	4	2200	16.3	
150	A-23 inside frosted and clear	Med.	CC-6	160	6 5/16	117	4⅝	2350	15.7	89
150	PS-25 inside frosted and clear	Med.	C-9	176	6 15/16	133	5¼	2300	15.3	85.5
200	A-23 inside frosted and clear	Med.	CC-8	157	6 3/16	117	4⅝	3410	17.0	87.5
200	A-25 inside frosted and clear	Med.	CC-6	176	6 15/16	133	5¼	3250	16.2	—
200	PS-25 inside frosted and clear	Med.	CC-6	176	6 15/16	133	5¼	3220	16.1	—
200	PS-30 inside frosted and clear	Med.	C-9	205	8 1/16	152	6	3260	16.3	81
300	PS-30 inside frosted and clear	Med.	C-9	205	8 1/16	152	6	5190	17.3	79
300	PS-35 inside frosted and clear	Mog.	C-9	238	9⅜	178	7	5190	17.3	84
500	PS-40 inside frosted and clear	Mog.	C-9	248	9¾	178	7	9070	18.1	80
750	PS-52 clear	Mog.	C-7A	332	13 1/16	241	9½	14200	18.9	83
1000	PS-52 clear	Mog.	C-7A	332	13 1/16	241	9½	19800	19.8	78
1500	PS-52 clear	Mog.	C-7A	332	13 1/16	241	9½	30000	20.0	—
For 230, 240, and 250 Volts										
100	A-21 inside frosted	Med.	C-7A or C-9	133	5¼	98	3⅞	1070	10.7	—
150	PS-25 inside frosted and clear	Med.	C-7A or C-9	176	6 15/16	133	5¼	1840	12.3	—
200	PS-30 inside frosted and clear	Med.	C-7A or C-9	205	8 1/16	152	6	2590	13.0	—
300	PS-35 inside frosted and clear	Mog.	C-7A	238	9⅜	178	7	4290	14.3	—
500	PS-40 inside frosted and clear	Mog.	C-7A	248	9¾	178	7	7930	15.9	—
750*	PS-52 clear	Mog.	C-7A	332	13 1/16	241	9½	13600	18.2	—
1000*	PS-52 clear	Mog.	C-7A	332	13 1/16	241	9½	18900	18.9	78
1500*	PS-52 clear	Mog.	C-7A	332	13 1/16	241	9½	27000	18.0	—

* Life is 2000 hours.
† Lamps of 3000-3500 hours rated life for 120, 125 and 130 volts also are available in some wattages.
‡ Reduced wattage, high efficacy.

Fig. 8-88. Vibration Service Lamps for 120- and 130-Volt Circuits

Watts	Bulb and Other Description	Burning Position	Base (see Fig. 8-9)	Filament (see Fig. 8-7)	Rated Average Life (hours)	Maximum Overall Length Millimeters	Maximum Overall Length Inches	Light Center Length Millimeters	Light Center Length Inches	Approximate Initial Lumens	Approximate Initial Lumens per Watt	Lamp Lumen Depreciation* (per cent)
25	A-19 inside frosted or clear	Any	Med.	C-9	1000	100	3 15/16	64	2½	260	10.4	—
50	A-19 inside frosted or clear	Not horizontal	Med.	C-9	1000	100	3 15/16	64	2½	550	11.0	72
75	A-21 inside frosted	Not horizontal	Med.	C-9	1000	133	5¼	98	3⅞	935	12.5	—
100	A-23 inside frosted or clear	Any	Med.	C-9	1000	151	5 15/16	113	4 7/16	1340	13.4	83
100	A-21 inside frosted or clear	Not horizontal	Med.	C-9	1000	135	5 5/16	98	3⅞	1400	14.0	—
150	PS-25 inside frosted	Any	Med.	C-9	1000	176	6 15/16	133	5¼	2390	15.9	—

* Per cent initial light output at 70 per cent of rated life.

Fig. 8-89. Tungsten-Halogen Lamps for General Lighting (Burning Position—Any, Except as Noted.)

Watts	Volts	Bulb and Finish	Base*	Filament	Maximum Overall Length (millimeters)	Maximum Overall Length (inches)	Average Light Center Length (millimeters)	Average Light Center Length (inches)	Rated Life (hours)	Approximate Initial Lumens	Approximate Initial Lumens Per Watt	Lamp Lumen Depreciation† (per cent)	Approximate Color Temperature (K)
Double-Ended Types													
200	120	T-3 clear	RSC	CC-8	79	3 1/8	—	—	1500	3460	17.3	96	2900
300	120	T-3 clear	RSC	C-8	119	4 11/16	—	—	2000	5950	19.9	96	2950
300	120	T-3 frosted	RSC	C-8	119	4 11/16	—	—	2000	5900	19.3	96	2950
300	120	T-3 clear	RSC	CC-8	79	3 1/8	—	—	2000	5650	18.5	—	2900
300	120	T-4 clear	RSC	CC-8	79	3 1/8	—	—	2000	5650	18.9	96	3000
300	130	T-4 clear	RSC	CC-8	79	3 1/8	—	—	2000	5650	18.9	—	3000
300	120	T-4 frosted	RSC	CC-8	79	3 1/8	—	—	2000	5300	17.7	—	3000
350	**	T-3 IR refl.	RSC	C-8	119	4 11/16	—	—	2000	10000	28.6	95	—
400	**	T-3 clear	RSC	CC-8	119	4 11/16	—	—	2000	9000	22.5	—	—
400	120	T-4 clear	RSC	CC-8	79	3 1/8	—	—	2000	7750	19.4	96	2950
400	125	T-4 clear	RSC	CC-8	79	3 1/8	—	—	2000	7500	18.7	—	3000
400	130	T-4 clear	RSC	CC-8	79	3 1/8	—	—	2000	7800	19.5	—	3000
500	120	T-3 clear	RSC	C-8§	119	4 11/16	—	—	1500	10500	20.8	96	—
500	120	T-3 clear	RSC	C-8	119	4 11/16	—	—	2000	10950	21.9	96	3000
500	120	T-3 frosted	RSC	C-8	119	4 11/16	—	—	2000	10700	21.4	96	3000
500	125	T-3 clear	RSC	C-8	119	4 11/16	—	—	2000	10500	21.0	—	3000
500	130	T-3 clear	RSC	C-8	119	4 11/16	—	—	2000	10750	21.5	96	3000
900	***	T-3 IR refl.	RSC	C-8	256	10 1/16	—	—	2000	32000	35.6	95	—
900	277	T-3 IR refl.	RSC	C-8	256	10 1/16	—	—	2000	31000	34.4	95	—
1000	120	T-6 clear	RSC or RSC (Rect)	CC-8	143	5 5/8	—	—	2000	23400	23.4	96	3000
1000	120	T-6 clear	RSC or RSC (Rect)	CC-8	143	5 5/8	—	—	4000	19800	19.8	93	2950
1000	120	T-6 frosted	RSC or RSC (Rect)	CC-8	143	5 5/8	—	—	2000	22700	22.7	96	3000
1000	120	T-6 clear	RSC or RSC (Rect)	CC-8	143	5 5/8	—	—	500	24500	24.5	—	3100
1000	220	T-3 clear	RSC	C-8	256	10 1/16	—	—	2000	21400	21.4	96	3000
1000	240	T-3 clear	RSC	C-8	256	10 1/16	—	—	2000	21400	21.4	96	3000
1200	***	T-3 clear	RSC	C-8	256	10 1/16	—	—	2000	29000	24.2	—	—
1250	208	T-3 clear	RSC	C-8	256	10 1/16	—	—	2000	28000	22.4	96	3050
1500	208	T-3 clear	RSC	C-8	256	10 1/16	—	—	2000	35800	23.0	96	3050
1500	220	T-3 clear	RSC	C-8	256	10 1/16	—	—	2000	35800	23.2	96	3050
1500	240	T-3 clear	RSC	C-8	256	10 1/16	—	—	2000	35800	23.2	96	3050
1500	240	T-3 frosted	RSC	C-8	256	10 1/16	—	—	2000	32000	21.3	—	3050
1500	277	T-3 clear	RSC	C-8	256	10 1/16	—	—	2000	33700	22.5	96	3050
1500	277	T-3 frosted	RSC	C-8	256	10 1/16	—	—	2000	31600	21.0	—	3050
Single-Ended Types													
5	12	T-3 clear	G4	C-6	31	1 7/32	—	—	2000	60	12	—	3000
10	6	T-3 clear	G4	C-6	31	1 7/32	—	—	2000	120	12	—	3000
10	12	T-3 clear	G4	C-6	31	1 7/32	—	—	2000	—	—	—	3000
20	††	T-3 clear	G4	C-6	31	1 7/32	—	—	2000	350	17.5	—	3000
50	12	T-4 clear	GY 6.35	C-6	44	1 23/32	—	—	2000	950	19	—	3000
50	24	T-4 clear	GY 6.35	C-6	44	1 23/32	—	—	2000	900	18	—	3000
75	28	T-3 clear	Min.	CC-6	60	2 3/8	30	1 3/16	2000	1400	21.4	—	3000
100	12	T-3 1/2 clear	Minican.	—	57	2 1/4	—	—	1000	2500	25.0	—	—
100	12	T-4 clear	GY 6.35	C-6	44	1 23/32	—	—	2000	2500	25	—	3000
100	24	T-4 clear	GY 6.35	C-6	44	1 23/32	—	—	2000	2000	20	—	3000
100	120	T-4 clear	Minican.	—	70	2 3/4	—	—	1000	1800	18.0	—	—
100	120	T-4 frosted	Minican.	—	70	2 3/4	—	—	1000	1750	17.5	—	—
100	120	T-4 clear	D.C. Bay.	—	62	2 7/16	—	—	1000	1800	18.0	—	—
100	120	T-4 frosted	D.C. Bay.	—	62	2 7/16	—	—	1000	1750	17.5	—	—
150	120	T-4 clear	Minican.	CC-2V	70	2 3/4	—	—	1500	2900	19.3	—	3000
150	120	T-4 clear	D.C. Bay	CC-2V	62	2 7/16	—	—	1500	2900	19.3	—	3000
150	120	T-4 frosted	D.C. Bay	—	62	2 1/16	—	—	1000	2700	18.0	—	—
150	120	T-10 clear or frosted	Med.	C-6	105	4 1/8	75	2 15/16	2000	2500	16.7	—	3000

See footnotes on page 8-77.

Continued on next page.

Fig. 8-89 *Continued*

Watts	Volts	Bulb and Finish	Base*	Filament	Maximum Overall Length (millimeters)	Maximum Overall Length (inches)	Average Light Center Length (millimeters)	Average Light Center Length (inches)	Rated Life (hours)	Approximate Initial Lumens	Approximate Initial Lumens Per Watt	Lamp Lumen Depreciation† (per cent)	Approximate Color Temperature (K)
150	120	T-5 clear or frosted	D.C. Bay.	C-6	95	3¾	—	—	2000	2000	16.7	—	3000
250	120	T-10 clear or frosted	Med.	C-6	105	4⅛	75	2¹⁵⁄₁₆	2000	4200	16.8	—	3000
250	120	T-5 clear or frosted	D.C. Bay.	C-6	95	3¾	—	—	2000	4200	16.8	—	3000
250	120	T-4 clear	Minican.	CC-8	79	3⅛	41	1⅝	2000	4850	19.4	96	2950
250	120	T-4 frosted	Minican.	CC-8	79	3⅛	41	1⅝	2000	4700	18.8	96	2950
250	120	T-4 clear	D.C. Bay.	CC-8	71	2 13/16	38	1½	2000	4850	19.4	—	2950
250	120	T-4 clear	D.C. Bay.	CC-8	71	2 13/16	41	1⅝	2000	4850	19.4	96	2950
250	120	T-4 frosted	D.C. Bay.	CC-8	71	2 13/16	41	1⅝	2000	4700	18.8	96	2950
250	130	T-4 clear	Minican.	CC-8	79	3⅛	41	1⅝	2000	4850	19.4	—	2950
400	120	T-4 frosted	Minican.	CC-8	92	3⅝	51	2	2000	8550	21.4	96	3050
400	120	T-4 clear	Minican.	CC-8	92	3⅝	51	2	2000	8800	22.0	96	3050
500	120	T-4 clear	Minican.	CC-8	95	3¾	51	2	2000	11500	23.0	—	3000
500	120	T-4 frosted	Minican.	CC-8	95	3¾	51	2	2000	10000	20.0	—	3000
500	120	T-4 frosted	D.C. Bay.	CC-8	87	3 7/16	54	2⅛	2000	10100	20.2	96	3000
500	120	T-4 clear	D.C. Bay.	CC-8	87	3 7/16	54	2⅛	2000	10450	20.9	96	3000
500	130	T-4 clear	Minican.	CC-8	95	3¾	51	2	2000	11500	23.0	—	3000
1000	120	T-24 clear	Med. Bipost	CC-8	233	9½	102	4	3000	22400	22.4	93	3050

* RSC = recessed single contact, RSC (Rect) = rectangular recessed single contact.
† Per cent initial light output at 70 per cent of rated life.
§ Lamp provides maximum filament straightness under severe operating conditions.
** 120, 130 volts
*** 208, 220, 240 volts
†† 12 and 24 volts

Fig. 8-90. Showcase Lamps for 120- and 130-Volt Circuits

Watts	Bulb and Other Description	Base (see Fig. 8-9)	Filament (see Fig. 8-7)	Rated Average Life (hours)	Maximum Over-All Length (millimeters)	Maximum Over-All Length (inches)	Approximate Initial Lumens	Lamp Lumen Depreciation‡ (per cent)
25*	T-6½ clear or inside frosted	Intermed.	C-8	1000	140	5½	250	76
25*	T-10 clear or inside frosted	Med.	C-8	1000	143	5⅝	255	79
25*	T-10 light inside frosted and side reflectorized	Med.	CC-8	1000	143**	5⅝**	230	80
40*	T-8 clear or inside frosted	Med.	C-23 or C-8	1000	302	11⅞	420	77
40*	T-10 clear or inside frosted	Med.	C-8	1000	143	5⅝	445	77
40*	T-10 light inside frosted and side reflectorized	Med.	CC-8	1000	143**	5⅝**	430	80
60†	T-10 clear	Med.	C-8	1000	143	5⅝	745	—
75*	T-10 clear	Med.	C-23 or C-8	1000	302	11⅞	800	—

* May be burned in any position.
** Exclusive of spring contact.
† Must be burned from base down to horizontal.
‡ Per cent initial light output at 70 per cent of rated life.

Fig. 8-91. Tungsten-Halogen 12-Volt* Lamps with External Ellipsoidal Reflector (MR-16)§

Watts	ANSI Code	Manufacturer's Ordering Code‡	Base	Average Life (hours)	Nominal Center Candlepower (candelas)	Approximate Color Temperature (kelvins)	Beam Spread (degrees)
10**		41960SP[c]	G4	2000	1500	3000	10
20	BAB[b]	Q20MR16/FL[a]	RM2P	2000–3000	460	2925	36
20	ESX[b]	Q20MR16/NSP[a]	RM2P	2000–3000	3300	2925	12
20		41900SP[c]	G4	2000	3800	3000	10
20		41900FL[c]	G4	2000	1000	3000	15
20		41905SP[c]	G4	2000	3100	2600	10
20		41905FL[c]	G4	2000	850	2600	15
20†		41930SP[c]	G4	2000	3000	3000	10
20		41970SP[c]	DC Bay.	2000	5000	3000	10
20		41970FL[c]	DC Bay.	2000	600	3000	30
20		41860SP[c]	RM2P	2000	10000	3100	5
20		41860FL[c]	RM2P	2000	1000	3100	20
35**		41830SSP[c]	—	2000	45000	3000	3
42	EYP[b]		RM2P	2500	990	3050	36
42	EYR[b]		RM2P	2500	7075	3025	1.5
42	EYS[b]	Q42MR16/NFL[a]	RM2P	2500–3000	2400	3075	22
50	EXN[b]	Q50MR16/FL[a]	RM2P	3000	1500	3050	37
50	EXT[b]	Q50MR16/NSP[a]	RM2P	3000	9150	3035	13
50	EXZ[b]	Q50MR16/NFL[a]	RM2P	3000	3000	3075	25
50		41990SP[c]	DC Bay.	2000	10000	3000	10
50		41990FL[c]	DC Bay.	2000	1100	3000	30
50		41995SP[c]	DC Bay.	2000	9000	2600	10
50		41995FL[c]	DC Bay.	2000	950	2600	30
50		41870SP[c]	RM2P	2000	15000	3100	7
50		41870FL[c]	RM2P	2000	2800	3100	20
75		41880SP[c]	RM2P	2000	17500	3100	10
75		41880FL[c]	RM2P	2000	4000	3100	20
75	EYC[b]	Q75MR16/FL[a]	RM2P	3500	2000	3050	39
75	EYF[b]	Q75MR16/NSP[a]	RM2P	3500	11500	3050	14
100	EXV[b]		RM2P	50	3300	3350	38
150*	FMG[b]		RM2P	25	5800	3300	38
250*	ENH[b]		RM2P	175	15000	3250	25
250*	EXX[b]		RM2P	25	7000	3300	40

§ Reflector may be dichroic or silvered or gold aluminum. Check manufacturer.
* 120 volts.
** 6 volts.
† 24 volts.
‡ If other than ANSI Code.
[a] General Electric.
[b] GTE Sylvania.
[c] Osram.

Fig. 8-92. Double Ended Linear Lamps* for 115- and 125-Volt Circuits (May Be Burned in Any Position)

Watts	Bulb and Other Description	Base (see Fig. 8-9)	Filament (see Fig. 8-7)	Rated Average Life (hours)	Over-All Length		Approximate Initial Lumens	Approximate Initial Lumens per Watt
					Millimeters	Inches		
30	T-8 clear	Disk	C-8	1500	451	17¾	255	8.5
30	T-8 inside frosted	Disk	C-8	1500	451	17¾	255	8.5
35	T-9½ inside white	Special	C-8	1000	300	11⅞	360	10.3
40	T-8 clear	Disk	C-8	1500	298	11¾	350	8.8
40	T-8 inside frosted	Disk	C-8	1500	298	11¾	350	8.8
60	T-8 clear	Disk	C-8	1500	451	17¾	540	9.0
60	T-8 inside frosted	Disk	C-8	1500	451	17¾	540	9.0
60	T-9½ inside white	Special	C-8	1000	500	19 11/16	730	12.2
120	T-8 inside white	Special	C-8	1000	1000	39⅜	1450	12.1

* Trade designations: Lumiline, Philinea and Tungsralin.

Fig. 8-93. Lamps for Train and Locomotive Service*

Watts	Volts	Bulb and Other Description	Base (see Fig. 8-9)	Filament (see Fig. 8-7)	Rated Average Life (hours)	Maximum Overall Length (millimeters)	Maximum Overall Length (inches)	Average Light Center Length (millimeters)	Average Light Center Length (inches)	Approximate Initial Lumens	Rated Initial Lumens Per Watt	Lamp Lumen Depreciation† (per cent)
						Train						
6	30	S-6 Indicator (A)	Cand.	C-2V	1500	47	$1\frac{7}{8}$	—	—	50	8.3	—
	60		Sc.	C-7A				—	—	48	8.0	—
6	30	S-6 Blue-Night Light (A)	Cand.	C-2V	1500	47	$1\frac{7}{8}$	—	—	—	—	—
	60		Sc.	C-7A				—	—	—	—	—
15	30	A-17 inside frosted (A)	Med.	C-9	1000	92	$3\frac{5}{8}$	60	$2\frac{3}{8}$	179	11.9	89
	60				—		—	—	—	147	9.8	—
25	30	A-19 inside frosted	Med.	C-9	1000	100	$3\frac{15}{16}$	63	$2\frac{1}{2}$	340	13.6	89
	60				—		—	—	—	282	11.3	90
25	30	T-$8\frac{1}{2}$ inside frosted	Med.	C-8	1000	139	$5\frac{1}{2}$	—	—	325	13.0	—
	60							—	—	280	11.2	—
40	30	A-19 inside frosted	Med.	C-9	1000	107	$4\frac{1}{4}$	73	$2\frac{7}{8}$	590	14.8	—
	60									525	13.1	—
50	30	A-21 inside frosted	Med.	C-9	1000	125	$4\frac{15}{16}$	85	$3\frac{3}{8}$	810	16.2	86
	60									700	14.0	—
75	30	A-23	Med.	C-9	1000	153	$6\frac{1}{16}$	111	$4\frac{3}{8}$	1230	16.4	—
	60									1140	15.2	—
100	30	A-23	Med.	C-9	1000	153	$6\frac{1}{16}$	111	$4\frac{3}{8}$	1800	18.0	—
	60									1650	16.5	—
						Locomotive						
6	60	S-6 Indicator (A)	S.C. Bay.	C-7A	1500	46	$1\frac{13}{16}$	26	$1\frac{1}{16}$	48	8.0	—
15	75	S-11 Marker (A)	D.C. Bay.	C-1	1000	60	$2\frac{3}{8}$	31	$1\frac{1}{4}$	142	9.5	—
15	34	S-14 inside frosted Cab (A)	Med.	C-9	1000	88	$3\frac{1}{2}$	63	$2\frac{1}{2}$	142	9.5	72
25	60	A-19 inside frosted	Med.	C-9	1000	98	$3\frac{7}{8}$	65	$2\frac{9}{16}$	265	10.6	—
	75	A-17 RS Cab (A)	Med.	C-9	1000	92	$3\frac{5}{8}$	63	$2\frac{1}{2}$	245	9.8	75
30	64	S-11 Marker	Bay.	C-7A	500	60	$2\frac{3}{8}$	31	$1\frac{1}{4}$	350	11.7	—
	75		S.C.-D.C.							336	11.2	—
50	60	A-21 inside frosted	Med.	C-9	1000	123	$4\frac{7}{8}$	87	$3\frac{7}{16}$	535	10.7	—
	75	A-19 RS Cab-Engine (A)	Med.	C-9	1000	98	$3\frac{7}{8}$	63	$2\frac{1}{2}$	545	10.9	57
75	75	PAR-36 RS Train Warning	Sc. Term.	CC-6	500	69	$2\frac{3}{4}$	—	—	—	—	—
100	32	A-21 Headlight	Med.	C-5	500	112	$4\frac{7}{16}$	76	3	1600	16.0	87
200	12	PAR-56 Headlight	Sc.	CC-8	500	114	$4\frac{1}{2}$	—	—	(B)	—	—
	30		Term.	CC-8	500	114	$4\frac{1}{2}$	—	—	(C)	—	—
250	32	P-25 Headlight	Med. Pref.	C-5A	500	127	5	55	$2\frac{3}{16}$	4550	18.2	89
			Med.		500	120	$4\frac{3}{4}$	76	3	4550	18.2	89

* Lamps may be burned in any position except the Headlight Lamps. They can be burned in any position except within 45 degrees of vertically base-up.

† Per cent initial light output at 70 per cent of rated life.

(A) Vacuum. (B) 180,000 max. candlepower. (C) 260,000 max. candlepower.

Fig. 8-94. Frequently Used Incandescent Filament and Tungsten-Halogen Lamps for Theatre Stages, Motion Picture and Television Production, and Still Photography (All Data for 120 Volt Lamps Except as Noted)

I. Tungsten-Halogen Double-Ended Lamps

Watts	Manufacturer's Ordering Code	ANSI Code	Color Temperature (K)	Average Life (hours)	Approximate Initial Lumens	Lighted Length		Bulb Finish
						(millimeters)	(inches)	
(A) T-3 or T-4 Bulb; Compact CC-8 Filament[b]; Recessed Single Contact Bases; Maximum Over-All Length: 79 millimeters (3 1/8 inches); Burning Position: Any								
120	FGY (20V)	FGY	2900	2500	1900	16	5/8	Clear
300	EHP	EHP	3000	2500	5600	16	5/8	Clear
300	EHX	EHX	3000	2500	4900	—	—	Frosted
375	DWZ (30V)	DWZ	2950	1500	6500	12	15/32	Clear
400	FDA	FDA	3200	250	10400	17	21/32	Clear
400	EHR	EHR	3000	2000	7800	16	9/16	Clear
650	FCA	FCA	3400	25	19000	—	—	Frosted
650	FAD	FAD	3200	100	16500	16	5/8	Clear
650	FBX	FBX	3200	100	16500	16	5/8	Frosted
650	DWY	DWY	3400	25	20000	16	5/8	Clear[a]
800	DXX (230, 240V)	DXX	3200	75	20500	25	1	Clear[a]
(B) T-4 or T-5 Bulb; Compact CC-8 Filament[b]; Recessed Single Contact Bases; Maximum Over-All Length: 95 millimeters (3 3/4 inches); Burning Position: Any								
600	FCB	FCB	3200	75	16500	15	19/32	Clear
1000	BRH	BRH	3200	75	30550	19	3/4	Clear
1000	FBZ	FBZ	3400	30	31500	—	—	Frosted
1000	DXW	DXW	3200	150	28000	22	7/8	Clear
1000	FBY	FBY	3200	150	26000	22	7/8	Frosted
1000	DXN	DXN	3400	50	33000	19	3/4	Clear[a]
(C) T-5 Bulb; Compact CC-8 Filament[b]; Recessed Single Contact Bases; Maximum Over-All Length: 111 millimeters (4 3/8 inches); Burning Position: Any								
1000	DYA	DYA	3200	150	28000	22	7/8	Clear
		DYN	3200	150	28000			Frosted
(D) T-3 or T-5 Bulb; Linear C-8 Filament[b]; Recessed Single Contact Bases; Maximum Over-All Length: 119 millimeters (4 11/16 inches); Burning Position: ± 4°								
300	EHM	EHM	3000	2000	6000	51	2	Clear
300	EHZ	EMZ	3000	2000	5700	—	—	Frosted
500	FCL	FCL	3000	2600	10000	51	2	Clear
500	FCZ	FCZ	3000	2600	9300	—	—	Frosted
500	Q500T3/4CL	FDF	3200	400	13250	60	2 3/8	Clear
500	Q500T3/4	FDN	3200	400	12800			Frosted
650	Q1000/650T3/4WM	FWM	3200	400	25500	60	2 3/8	Clear[n]
750	Q750T3/4CL	EJG	3200	400	20600	62	2 7/16	Clear
750	Q750T3/4	EMD	3200	400	19500			Frosted
800	Q800T3/4CL (240V)	EME	3200	250	22000	70	2 3/4	Clear
800	Q800T3/4 (240V)	EMF	3200	250	21400			Frosted
1000	Q1000T3/3CL (185V)	EJD	3350	100	33600	67	2 11/16	Clear
1000	Q1000T3/4CL	FCM	3200	500	28000	65	2 9/16	Clear
1000	Q1000T3/4	FHM	3200	500	28000			Frosted
1050	ELJ	ELJ	3200	75	30000	22	7/8[m]	Clear
(E) T-6 or T-8 Bulb; Compact CC-8 Filament[b]; Recessed Single Contact Bases; Maximum Over-All Length: 143 millimeters (5 5/8 inches); Burning Position: Any								
1000	Q1000T6/4CL	FER	3200	500	27500	19	3/4	Clear
1000	EHS	EHS	3100	500	28000	29	1 1/8	Clear
1000	DWT	DWT	3000	2000	22000	29	1 1/8	Clear
2000	FEX (240V)	FEX	3200	300	50000	35	1 3/8	Clear
2000	Q2000T8/4CL	FEY	3200	400	57000	43	1 11/16	Clear

See footnotes on pages 84 and 85.

Continued on next page.

Fig. 8-94 *Continued*

Watts	Manufacturer's Ordering Code	ANSI Code	Color Tempera-ture (K)	Average Life (hours)	Approxi-mate Initial Lumens	Lighted Length		Bulb Finish
						(millime-ters)	(inches)	
(F) T-3 or T-4 Bulb; Linear C-8 Filament[b]; Recessed Single Contact Bases; Maximum Over-All Length: 167 millimeters (6 9/16 inches): Burning Position: ± 4°								
1000	Q1000T3/1CL	FFT	3200	400	26400	65	2 9/16	Clear
1500	Q1500T4/4CL	FDB	3200	400	41200	57	2 1/4	Clear
1500	Q1500T4/4	FGT	3200	400	40200		[i]	Frosted
2000	FFW	FFW	3200	125	57000	79	3 1/8	Clear

II. Tungsten-Halogen Single-Ended Lamps

Watts	Bulb	Manufacturer's Order-ing Code	ANSI Code	Color Tempera-ture (K)	Average Life (hours)	Approx. Initial Lu-mens	Filament Form	Maximum Over-All Length		Bulb Finish
								(millime-ters)	(inches)	
(A) Medium 2-Pin Base; Light Center Length: 60 millimeters (2 3/8 inches); Burning Position: Any										
500	T4	Q500/5CL	EHC	3150	300	12700	CC-8[b]	102	4	Clear
500	T-4	Q500CL/TP	EHD	3000	2000	10600	CC-8[b]	102	4	Clear
750	T-6	Q750/4CL	EHF	3200	300	20400	CC-8[b]	105	4 1/8	Clear
	T-6	Q750CL/TP	EHG	3000	2000	15400	CC-8[b]	105	4 1/8	Clear
1000	T-6	Q1000/4CL	FEL	3200	300	27500	CC-8[b]	105	4 1/8	Clear
1000	T-6	Q1000/4	FCV	3200	300	26500	CC-8[b]	105	4 1/8	Frosted
1000	T-6	FEP (240V)	FEP	3200	150	25000	2CC-8	102	4	Clear
(B) Miniature 2-Pin Base; Light Center Length: 43 millimeters (1 11/16 inches); Burning Position: Base Down to Horizontal										
600	G-7	DYH	DYH	3200	75	17000	CC-6[b]	64	2 1/2	Clear
650	G-6	DVY	DVY	3400	25	20000	CC-6[b]	60	2 3/8	Clear[a]
650	G-7	DYJ (230V)	DYJ	3400	20	20000	CC-2P[b]	64	2 1/2	Clear[a]
(C) 2-Pin Prefocus Base; Light Center Length 37 millimeters (1 7/16 Inches); Burning Position: Burn with Coil Horizontal										
250	G-6	DYG (30V)	DYG	3400	15	7000	CC-6[b]	60	2 3/8	Clear[a]
420	G-7	EKB	EKB	3200	75	11000	CC-6[b]	64	2 1/2	Clear
600	G-7	DYS/DYV/BHC	DYS	3200	75	17000	CC-6[b]	64	2 1/2	Clear
650	G-6	EKD	EKD	3400	25	20000	CC-6[b]	64	2 1/2	Clear[a]
650	G-7	DYR (220, 240V)	DYR	3200	50	16500	2CC-8[b]	64	2 1/2	Clear[a]

III. Tungsten-Halogen and Incandescent Lamps

Watts	Manufacturer's Order-ing Code	ANSI Code	Color Tempera-ture (K)	Average Life (hours)	Beam Shape	Field Angle (degrees)[d]	Beam Angle (degrees)[c]	Approxi-mate Can-dle-Power[e]	Approxi-mate "Beam Lumens"[f]
(A) PAR-36 Bulb; Bases: Ferrule Contacts or Screw Terminals[h] or 2-Prong[j]: "Q" After "WATTS" Indicates Tungsten-Halogen Types, Burning Position: Any									
250	FGS/DWA	FGS	3400	4	MFL	—	—	18000	—
650Q	DWE[h]	DWE	3200	100	MFL	—	30 × 40	24000	—
650Q	DXK	DXK	3400	30	MFL	—	—	35000	—
650Q	FAY	FAY	5000	30	SP	—	15 × 22	35000	—
650Q	FBE[h]	FBE	5000	30	SP	—	15 × 22	35000	—
650Q	FBJ	FBJ	3400	30	SP	—	15 × 22	75000	—
250	FGR/FBM[j]	FGR	3400	4	MFL	—	—	18000	—
650Q	FBO[h]	FBO	3400	30	SP	—	15 × 22	75000	—
650Q	FCW	FCW	3200	100	FL	—	70 × 70	9000	—
650Q	FCX	FCX	3200	100	MFL	—	30 × 40	24000	—
650Q	FGK[h]	FGK	5000	30	MFL	—	30 × 40	24000	—

See footnotes on pages 84 and 85.

Continued on next page.

Fig. 8-94 *Continued*

Watts	Manufacturer's Ordering Code	ANSI Code	Color Temperature (K)	Average Life (hours)	Beam Shape	Field Angle (degrees)[d]	Beam Angle (degrees)[c]	Approximate Candle-Power[e]	Approximate "Beam Lumens"[f]
(B) PAR-64 Bulb; Bases: Extended Mogul End Prong or Screw Terminals[h]; Burning Position: Any Except Where Noted									
1000	Q1000PAR64/1	FFN	3200	600	VNSP	10 × 24	6 × 12	400000	11000
1000	Q1000PAR64/2	FFP	3200	600	NSP	14 × 26	7 × 14	330000	12000
1000	Q1000PAR64/5	FFR	3200	600	MFL	21 × 44	12 × 28	125000	14000
1000	Q1000PAR64/6	FFS	3200	600	WFL	45 × 71	24 × 48	40000	19000
1000	Q1000PAR64/3D	FGM	5200	200	NSP	12 × 24	6 × 13	200000	7000
1000	Q1000PAR64/7D	FGN	5200	200	MFL	20 × 43	11 × 27	70000	8200
(B-1) R-14 Bulb, D.C. Bayonet Base, Burn Horizontal									
100	BDK	BDK	3400	4	MFL	—	30 × 30	3200	—
(C) R-30 Bulb; Base: Medium Screw; Burning Position: Any									
300	BEP	BEP	3400	4					
375	EBR	EBR	3400	4					
(D) R-40 Bulb; Medium Screw Base; Burning Position: Any									
375	BFA	BFA	3400	4	MFL	—	—	12000	—
500	EAL	EAL	3200	15	MFL	—	60	8200	—
500	DXB	DXB	3400	6	SP	—	20	38000	—
500	DXC	DXC	3400	6	FL	—	90	5500	—
500	EAH (220, 240V)	EAH	3400	6	FL	—	90	5000	—

IV. Bipost-Based Lamps—Tungsten-Halogen and Incandescent

Watts	Bulb	Manufacturer's Ordering Code	ANSI Code	Color Temperature (K)	Average Life (hours)	Approximate Initial Lumens	Filament Form	Maximum Over-All Length (millimeters)	Maximum Over-All Length (inches)	Recommended Operating Position
(A) Medium Bipost Base; Light Center Length: 64 millimeters (2½ inches)										
500	T-20	500T20/63	DVG	3200	50	14500	C-13[b]	165	6½	BD45
750	T-24	750T24/16	DVH	3200	50	22000	C-13[b]	165	6½	BD45
750	T-24	750T24/13	EDH	3350	12	25000	C-13[b]	165	6½	BD45
1000	T-24	1M/T24/13	EBB	3350	12	33500	C-13[b]	165	6½	BD45
500	T-6	EGN	EGN	3200	100	13000	C-13D[b]	117	4⅝	BD45
750	T-7	Q750T7/4CL	EGR	3200	200	21000	C-13D[b]	127	5	BD45
1000	T-7	Q1000T7/4CL	EGT	3200	250	28500	C-13D[b]	127	5	BD45
(B) Mogul Bipost Base; Light Center Length 127 millimeters (5 Inches)										
2000	G-48	2M/G48/18	DVF	3200	100	61000	C-13[b]	238	9⅜	BD45
2000	G-48	2M/G48/14	ECK	3350	25	65000	C-13[b]	238	9⅜	BD45
2000	T-48	2M/T48/4	EDL	3350	25[g]	65000	C-13[b]	264	10⅜	BD45
1000	T-7	Q1000T7/4CL/BP	CYV	3200	200	28500	C-13D[b]	203	8	BDTH
1500	T-11	Q1500T11/4CL	CXZ	3200	300	44500	C-13[b]	216	8½	BDTH
2000	T-10	Q2000T10/4CL	CYX	3200	300	59000	C-13[b]	216	8½	BDTH
2000	T-8	BWA	BWA	3200	400	57000	CC-8	203	8	
2000	T-11	DCT	DCT	3050	900	47000	C-13D	216	8½	
(C) Mogul Bipost Base; Light Center Length: 165 millimeters (6½ inches)										
5000	G-64	5M/G64/7	ECN	3200	150[g]	145000	C-13[b]	302	11⅞	BD45
5000	T-64	5M/T64/1	EDN	3350	75[g]	161500	C-13[b]	340	13⅜	BD45
5000	T-20	Q5000T20/4CL	DPY	3200	500	143000	C-13[b]	279	11	BDTH
(D) Mogul Bipost Base; Light Center Length: 254 millimeters (10 inches)										
10000	G-96	10K/G96/1	EBA	3200	150[g]	295000	C-13	443	17 7/16	BD45
10000	G-96	10K/G96	ECP	3350	75[g]	335000	C-13	443	17 7/16	BD45
10000	T-24	Q10M/T24/4CL	DTY	3200	300	290500	C-13	391	15⅜	BD45

See footnotes on pages 84 and 85.

Continued on next page.

Fig. 8-94 *Continued*

V. Incandescent Prefocus-Base Lamps

(A) Medium Prefocus Base; Light Center Length: 56 millimeters (2 9/16 inches)

Watts	Bulb	Manufacturer's Ordering Code	ANSI Code	Color Temperature (K)	Average Life (hours)	Approx. Initial Lumens	Filament Form	Maximum Over-All Length (millimeters)	Maximum Over-All Length (inches)	Bulb Finish
250	T-20	250T20/47	—	2900	200	4600	C-13	146	5 3/4	BDTH
300	T-6	EEX	EEX	3200	50	7200	C-13	121	4 3/4	
500	T-20	500T20/64	—	2900	500	10000	C-13	146	5 3/4	BDTH
500	T-20	500T20/48	—	3000	200	11000	C-13	146	5 3/4	BDTH
500	T-20	DMX	DMX	3200	50	13200	C-13	146	5 3/4	BD30
500	T-6	BTL	BTL	3050	500	11000	C-13D	114	4 1/2	BDTH
500	T-6	BTM	BTM	3200	100	13000	C-13D	114	4 1/2	BDTH
750	T-20	750T20P/SP	BFE	3000	200	17000[l]	C-13	146	5 3/4	BD45
750	T-20	BFL/BFK	BFL	3050	200	13500	C-13	146	5 3/4	BD30
750	T-20	DPJ	DPJ	3250	25	19000	C-13D	146	5 3/4	BD30
750	T-7	BTN	BTN	3050	500	17000	C-13D	111	4 3/8	BDTH
750	T-7	BTP	BTP	3200	200	20000	C-13D	111	4 3/8	BDTH
1000	T-20	1M/T20P/SP	—	3050	200	23400	C-13D	146	5 3/4	BD45
1000	T-20	DRC	DRC	3250	50	30000	C-13	146	5 3/4	BD30
1000	T-20	DRS	DRS	3325	25	28500	C-13D	146	5 3/4	BD30
1000	T-20	DRB	DRB	3350	25	32000	C-13	146	5 3/4	BD30
1000	T-20	DWK(230V)	DWK	3100	50	25000	CC-13	146	5 3/4	BD30
1000	T-7	BTR	BTR	3200	200	27500	C-13D	111	4 3/8	BDTH

VI. Prefocus and Screw-Base Lamps—Tungsten-Halogen and Incandescent

(A) Mogul Prefocus Base; Light Center Length 100 millimeters (3 15/16 inches)

Watts	Bulb	Manufacturer's Ordering Code	ANSI Code	Color Temperature (K)	Average Life (hours)	Approximate Initial Lumens	Filament Form	Maximum Over-All Length (millimeters)	Maximum Over-All Length (inches)	Recommended Operating Position
1000	G-40	1M/G40/23	—	3050	200	24000	C-13	214	8 7/16	BDTH
1000	G-40	1M/G40PSP	—	3050	200	25000	C-5	214	8 7/16	BDTH
1000	T-7	BVT	BVT	3050	500	23000	C-13D	179	7	Any
1000	T-7	BVV	BVV	3200	200	27500	C-13D	179	7	Any
1500	G-40	1500G40/21	—	3000	200	35500	C-13D	214	8 7/16	BD30
1500	T-8	CWZ	CWZ	3200	325	38500	C-13D	191	7 1/2	Any
2000	T-11	BVW	BVW	3200	200	57500	C-13	191	7 1/2	Any

(B) Medium Screw Base; Light Center Length: 76 millimeters (3 inches)

Watts	Bulb	Manufacturer's Ordering Code	ANSI Code	Color Temperature (K)	Average Life (hours)	Approximate Initial Lumens	Filament Form	Maximum Over-All Length (millimeters)	Maximum Over-All Length (inches)	Recommended Operating Position
100	A-21	100A21/SP	—	2750	200	1340	C-5	111	4 3/8	Any
250	G-30	250G/FL	—	2800	800	3650	C-5	130	5 1/8	BDTH
250	G-30	250G/SP	—	2900	200	4500	C-5	130	5 1/8	BDTH
400	G-30	400G/FL	—	2850	800	6800	C-5	130	5 1/8	BDTH
400	G-30	400G/SP	—	2950	200	8400	C-5	130	5 1/8	BDTH
500	T-20	DMS	DMS	3200	50	13200	C-13	140	5 1/2	BD30

(C) Mogul Screw Base; Light Center Length: 108 millimeters (4 1/4 inches)

Watts	Bulb	Manufacturer's Ordering Code	ANSI Code	Color Temperature (K)	Average Life (hours)	Approximate Initial Lumens	Filament Form	Maximum Over-All Length (millimeters)	Maximum Over-All Length (inches)	Recommended Operating Position
500	G-40	500G/FL	—	2850	800	9300	C-5	179	7 1/16	BDTH

(D) Mogul Screw Base; Light Center Length: 133 millimeters (5 1/4 inches)

Watts	Bulb	Manufacturer's Ordering Code	ANSI Code	Color Temperature (K)	Average Life (hours)	Approximate Initial Lumens	Filament Form	Maximum Over-All Length (millimeters)	Maximum Over-All Length (inches)	Recommended Operating Position
1000	G-40	1M/G40FL	—	2950	800	20000	C-5	200	7 7/8	BDTH
1500	G-48	1500G48/6	—	2950	800	32200	C-5	217	8 9/16	BDTH
2000	T-8	Q2000/4CL	BWF	3200	500	59000	CC-8[b]	191	7 1/2	Any
2000	T-8	BWG	BWG	3200	400	55000		191	7 1/2	

See footnotes on pages 84 and 85.

Continued on next page.

Fig. 8-94 *Continued*

VII. Lamps for Ellipsoidal Reflector Spotlights—Tungsten-Halogen and Incandescent

Watts	Bulb	Manufacturer's Ordering Code	ANSI Code	Color Temperature (K)	Average Life (hours)	Approximate Initial Lumens	Filament Form	Maximum Over-All Length (millimeters)	Maximum Over-All Length (inches)	Recommended Operating Position
(A) Medium Prefocus Base; Light Center Length: 89 millimeters (3 1/2 inches)										
500	T-12	500T12/8	DEB	2850	800	9000	C-13D	156	6 1/8	BU30
500	T-12	500T12/9	DNS	2950	200	11000	C-13D	156	6 1/8	BU30
500	T-4	Q500CL/P	EGE	3000	2000	10450	CC-8[b]	152	6	Any
500	T-4	Q500/5CL/P	EGC	3150	500	12700	CC-8[b]	152	6	Any
750	T-12	750T12/9	DNT	3000	200	17000	C-13D	156	6 1/8	BU30
750	T-6	Q750CL/P	EGG	3000	2000	15750	CC-8[b]	152	6	Any
750	T-6	Q750/4CL/P	EGF	3200	500	20400	CC-8[b]	152	6	Any
1000	T-12	1M/T12/2	DNV	3050	200	25000	C-13D	156	6 1/8	BU30
1000	T-6	Q1000/4CL/P	EGJ	3200	500	27000	CC-8[b]	152	6	Any
1000	T-6	Q1000/4/P (frosted)	EGK	3200	500	26500	CC-8[b]	152	6	Any
1000	T-6	Q1000CL/P	EGM	3000	2000	21500	CC-8	152	6	Any
(B) Medium Bipost Base; Light Center Length: 102 millimeters (4 inches)										
500	T-14	500T14/7	—	2850	800	9000	C-13D[b]	164	6 7/16	BU30
500	T-5	EFX	EFX	3000	2000	10000	CC-8	165	6 1/2	Any
750	T-14	750T14	—	3000	200	17000	C-13D[b]	164	6 7/16	BU30
750	T-5	BSC	BSC	3200	250	20000	CC-8	165	6 1/2	Any
750	T-5	BSD	BSD	3000	2000	15000	CC-8	165	6 1/2	Any
1000	T-5	DZD	DZD	3200	400	25500	CC-8	165	6 1/2	Any
(C) Mogul Bipost Base; Light Center Length: 165 millimeters (6 1/2 inches)										
1500	T-24	1500T24/6	—	3000	200	33700	C-13D	254	10	BU30
2000	T-30	2M/T30/1	—	3050	200	48000	C-13D[b]	254	10	BU30
2100	T-24	2100T24/9 (60V)	—	3100	50	54100	C-13D	267	10 1/2	BU30
(D) Mogul Bipost Base; Light Center Length: 191 millimeters (7 1/2 inches)										
3000	T-32	3M/T32/2	—	3150	100	81000	C-13D[b]	292	11 1/2	BU30

VIII. Low-Wattage Lamps for Spotlights and Projectors

Watts	Bulb	Manufacturer's Ordering Code	Color Temperature (K)	Average Life (hours)	Approximate Initial Lumens	Filament Form	Maximum Over-All Length (millimeters)	Maximum Over-All Length (inches)	Recommended Operating Position
(A) Single-Contact Bayonet Candelabra Base; Light Center Length: 35 millimeters (1 3/8 inches)									
100	G16 1/2	100G16 1/2/29SC	2850	200	1660	CC-13	76	3	BDTH
(B) Double-Contact Bayonet Candelabra Base; Light Center Length: 35 millimeters (1 3/8 inches)									
50	T-8	CHY (230V)	2550	50	650	CC-13	79	3 1/8	BD30
100	G-16 1/2	100G16 1/2/29DC	2850	200	1660	CC-13	76	3	BDTH
100	T-4	Q100CL/DC (ESR)	2900	1000	1800	CC-2V	62	2 7/16	Any
100	T-8	CJX (230V)	2850	50	1650	CC-13	79	3 1/8	BD30
100	T-4	ETD	3000	1000	1750	CC-2V	62	2 7/16	
150	T-4	Q150CL/DC 2V (ESP)	2900	1000	2800	CC-2V	62	2 7/16	Any
200	B-12	BDJ	3200	20	5000	2CC-8	67	2 5/8	BD30
200	T-4	FEV	3200	50	5500	CC-2V	62	2 7/16	

[a] Ultraviolet-absorbing bulb.
[b] Low noise construction.
[c] Beam spread to 50 per cent peak candlepower.
[d] Beam spread to 10 per cent peak candlepower.
[e] Candlepower average in central 5° cone for "spot" beam lamps; average in central 10° cone for "flood" beam lamps.
[f] Measured within field angle.
[g] Tungsten powder cleaner in bulb. Useful lamp life and maintenance of output depend upon periodic removal from socket and scouring of bulb wall with tungsten powder to remove dark film that normally accumulates.
[h] Screw terminals.

NOTE—Footnotes and text continued on next page.

Fig. 8-94 *Continued*

IX. Screw-Base Lamps for Floodlights and Miscellaneous Special Effects—Tungsten-Halogen and Incandescent

Watts	Bulb	Manufacturer's Ordering Code	ANSI Code	Color Temperature (K)	Average Life (hours)	Approximate Initial Lumens	Bulb Finish	Filament Form	Maximum Over-All Length		Recommended Operating Position
									(millimeters)	(inches)	
(A) Medium Screw Base											
250	A-21	BBA	BBA	3400	3	8,500	I.F.	C-9	125	4 15/16	—
		BCA	BCA	4800 (21)	3	5,000	I.F.	C-9	125	4 15/16	—
	A-23	ECA	ECA	3200	20	6,500	I.F.	C-9	152	6	—
500	PS-25	EBV	EBV	3400	6	17,000	I.F.	C-9	176	6 15/16	—
		EBW	EBW	4800 (21)	6	10,500	I.F.	C-9	176	6 15/16	—
		ECT	ECT	3200	60	13,650	I.F.	C-9	176	6 15/16	—
(B) Mogul Screw Base; Light Center Length: 241 millimeters (9½ inches)											
1000	PS-52	Q1000PS52/4	DKZ/DSE	3200	750	28000	I.F.	CC-8[b]	330	13	Any
1500	PS-52	Q1500PS52/4	DKX/DSF	3200	1000	41000	I.F.	CC-8[b]	330	13	Any
2000	T-8	Q2000/4/95	BWL	3200	5	57200	Frost	CC-8[b]	298	11¾	Any
(C) Mogul Screw Base:											
1000	PS-40	ECV	ECV	3200	60	26500	I.F.	C-7A	248	9¾	Any
1000	PS-35	DXR/DXS	DXR	3400	10	31000	I.F.	C-9	238	9⅜	Any
1000	PS-35	DXT	DXT	4800[k]	10	19200	I.F.	C-9	238	9⅜	Any

X. Incandescent Lamps for Follow Spots, Effects Projectors

Watts	Bulb	Manufacturer's Ordering Code	ANSI Code	Color Temperature (K)	Average Life (hours)	Approximate Initial Lumens	Bulb Finish	Filament Form	Maximum Over-All Length		Recommended Operating Position
									(millimeters)	(inches)	
(A) Mogul Prefocus Base; Light Center Length: 87 millimeters (3 7/16 inches)											
1000	T-20	DPW	DPW	3200	50	28000	Clear	C-13	241	9½	BD30
1000	T-20	DSB	DSB	3225	25	28500	Clear	C-13D	241	9½	BD30
1500	T-20	DTJ	DTJ	3225	25	42500	Clear	C-13D	241	9½	BD30
1500	T-8	DTA	DTA	3200	100	39000	Clear	C-13D	168	6⅝	
(B) Mogul Bipost Base; Light Center Length: 102 millimeters (4 inches)											
2100	T-24	2100T24/8 (60V)	—	3100	50[9]	54100	Clear	C-13D	267	10½	BD30

[i] Apparent lighted length slightly longer than corresponding clear lamp.
[j] 2 prong base.
[k] Blue glass bulb. Apparent color temperature may vary among lamps.
[l] Top end of bulb is opaque-coated.
[m] CC-8 filament.
[n] Infrared reflecting coating.

Fig. 8-95. Silver-Bowl Lamps for 115, 120, and 125-Volt Circuits

Watts	Bulb	Base (see Fig. 8-9)	Approximate Initial Lumens**	Rated Average Life (hours)	Maximum Overall Length	
					(millimeters)	(inches)
60	A-19	Med. Sc.	750	1000	113	4 7/16
100*	A-21	Med. Sc.	1430	1000	133	5¼
150*	PS-25	Med. Sc.	2320	1000	176	6 15/16
200*	PS-30	Med. Sc.	3320	1000	205	8 1/16
300*	PS-35	Mog. Sc.	5325	1000	238	9⅜
500*	PS-40	Mog. Sc.	9420	1000	248	9¾
750*	PS-52	Mog. Sc.	14400	1000	330	13
1000*	PS-52	Mog. Sc.	20200	1000	330	13

* Base-up burning only. Use only porcelain sockets. For light center length see Fig. 8-81.
** Lumens based on 120 volts.

Fig. 8-96. Reflectorized R, ER and PAR Lamps for 120-Volt Circuits

Watts	Bulb	Description	Base (see Fig. 8-9)	Maximum Over-All Length (millimeters)	(inches)	Approximate Beam Spread (degrees)[c]	Approximate Beam Lumens	Approximate Total Lumens	Approximate Average Candle-power in Central 10 Degree Cone[f]	Rated Averaged Life (hours)
			R Lamps for Spotlighting and Floodlighting (Parabolic Reflector)							
30	R-20	Flood	Med.	100	$3\frac{15}{16}$	85	150	205	290	2000
50	R-20	Flood	Med.	100	$3\frac{15}{16}$	90	430	435	550	2000
75	R-30	Spot	Med.	136	$5\frac{3}{8}$	50	400	850	1730	2000
75	R-30	Flood	Med.	136	$5\frac{3}{8}$	130	610	850	430	2000
150	R-40	Spot	Med.	165	$6\frac{1}{2}$	37	835	1825	7000	2000
150	R-40	Flood	Med.	165	$6\frac{1}{2}$	110	1550	1825	1200	2000
300	R-40	Spot	Med.	165	$6\frac{1}{2}$	35	1800	3600	13500	2000
300	R-40	Flood	Med.	165	$6\frac{1}{2}$	115	3000	3600	2500	2000
300	R-40[a]	Spot	Med.	173	$6\frac{7}{8}$	35	1800	3600	13500	2000
300	R-40[a]	Flood	Med.	173	$6\frac{7}{8}$	115	3000	3600	2500	2000
300	R-40[a]	Spot	Mog.	184	$7\frac{1}{4}$	35	1800	3600	14000	2000
300	R-40[a]	Flood	Mog.	184	$7\frac{1}{4}$	115	3000	3600	2500	2000
500	R-40[a]	Spot	Mog.	184	$7\frac{1}{4}$	60	3300	6500	22000	2000
500	R-40[a]	Flood	Mog.	184	$7\frac{1}{4}$	120	5700	6500	4750	2000
1000	R-60[a,g]	Spot	Mog.	257	$10\frac{1}{8}$	32	11500	18300	135000	3000
1000	R-60[a,g]	Flood	Mog.	257	$10\frac{1}{8}$	110	15500	18300	15500	3000
			ER Lamps for Spotlighting and Floodlighting (Ellipsoidal Reflector)							
44	ER-30[b]	Light Inside Frost	Med.	162	$6\frac{3}{8}$	—	—	420	440	5000
50	ER-30	Light Inside Frost	Med.	162	$6\frac{3}{8}$	—	—	525	—	2000
67	ER-30[b]	Light Inside Frost	Med.	162	$6\frac{3}{8}$	—	—	700	1030	5000
75	ER-40	Light Inside Frost	Med.	162	$6\frac{3}{8}$	—	—	850	—	2000
90	ER-30[b]	Light Inside Frost	Med.	162	$6\frac{3}{8}$	—	—	950	1450	5000
90	ER-40	—	Med.	187	$7\frac{3}{8}$	—	—	—	1200	4000
120	ER-40	Light Inside Frost	Med.	187	$7\frac{3}{8}$	—	—	1475	—	2000
135	ER-40	—	Med.	187	$7\frac{3}{8}$	—	—	—	2050	4000
			R Lamps for General Lighting[d] (Parabolic Reflector)							
500	R-52	Wide Beam	Mog.	300	$11\frac{3}{4}$	90	—	7750	—	2000
750	R-52	Wide Beam	Mog.	300	$11\frac{3}{4}$	110	—	13000	—	2000
1000	R-52[a]	Wide Beam	Mog.	300	$11\frac{3}{4}$	110	—	16300	—	2000
1000	RB-52	Wide Beam	Mog.	322	$12\frac{11}{16}$	130	—	18900	—	2000
			PAR Lamps for Spotlighting and Floodlighting[e] (Parabolic Reflector)							
55*	PAR-38	Spot	Med. Skt.	135	$5\frac{5}{16}$	30 × 30	—	520	—	2000
55*	PAR-38	Flood	Med. Skt.	135	$5\frac{5}{16}$	60 × 60	—	520	—	2000
65*	PAR-38	Spot	Med. Skt.	135	$5\frac{5}{16}$	30 × 30	—	—	5100	2000
65*	PAR-38	Flood	Med. Skt.	135	$5\frac{5}{16}$	60 × 60	—	—	1850	2000
65*	PAR-38	Spot	Med. Side Prong	110	$4\frac{5}{16}$	30 × 30	—	—	—	2000
65*	PAR-38	Flood	Med. Side Prong	110	$4\frac{5}{16}$	60 × 60	—	—	—	2000
75	PAR-38	Spot	Med. Skt.	135	$5\frac{5}{16}$	30 × 30	465	750	3800	2000
75	PAR-38	Flood	Med. Skt.	135	$5\frac{5}{16}$	60 × 60	570	750	1500	2000
80*	PAR-38	Spot	Med. Skt.	135	$5\frac{5}{16}$	—	—	—	—	2000
80*	PAR-38	Flood	Med. Skt.	135	$5\frac{5}{16}$	—	—	—	—	2000
85*	PAR-38	Spot	Med. Skt.	135	$5\frac{5}{16}$	30 × 30	—	—	—	2000
85*	PAR-38	Flood	Med. Skt.	135	$5\frac{5}{16}$	60 × 60	—	—	—	2000
90*	PAR-38	Spot	Med. Skt.	135	$5\frac{5}{16}$	—	—	—	—	2000
90*	PAR-38	Flood	Med. Skt.	135	$5\frac{5}{16}$	—	—	—	—	2000
90[g]	PAR-38	Flood	Med. Skt.	135	$5\frac{5}{16}$	30 × 30	735	1450	3580	2000
100	PAR-38	Spot	Med. Skt.	135	$5\frac{5}{16}$	30 × 30	—	1250	—	2000
100	PAR-38	Flood	Med. Skt.	135	$5\frac{5}{16}$	60 × 60	—	1250	—	2000
120*	PAR-38	Spot	Med. Skt.	135	$5\frac{5}{16}$	30 × 30	—	1420	12000	2000

See footnotes on page 8-87

Continued on next page.

Fig. 8-96 *Continued*

Watts	Bulb	Description	Base (see Fig. 8-9)	Maximum Over-All Length		Approximate Beam Spread (degrees)[c]	Approximate Beam Lumens	Approximate Total Lumens	Approximate Average Candle-power in Central 10 Degree Cone[f]	Rated Averaged Life (hours)
				(millimeters)	(inches)					
PAR Lamps for Spotlighting and Floodlighting[e] (Parabolic Reflector)										
120*	PAR-38	Flood	Med. Skt.	135	$5\frac{5}{16}$	60 × 60	—	1420	4250	2000
120*	PAR-38	Spot	Med. Side Prong	110	$4\frac{5}{16}$	30 × 30	—	1420	—	2000
120*	PAR-38	Flood	Med. Side Prong	110	$4\frac{5}{16}$	60 × 60	—	1420	—	2000
150	PAR-38[i]	Spot	Med. Skt.	135	$5\frac{5}{16}$	30 × 30	1100	1735	11000	2000
150	PAR-38[i]	Flood	Med. Skt.	135	$5\frac{5}{16}$	60 × 60	1350	1735	3700	2000
150	PAR-38	Spot	Med. Side Prong	110	$4\frac{5}{16}$	30 × 30	1100	1735	11000	2000
150	PAR-38	Flood	Med. Side Prong	110	$4\frac{5}{16}$	60 × 60	1350	1735	3700	2000
200	PAR-46	Narrow Spot	Med. Side Prong	102	4	17 × 23	1200	2325	32500[h]	2000
200	PAR-46	Med. Flood	Med. Side Prong	102	4	20 × 40	1300	2325	11200[h]	2000
250	PAR-38[g]	Spot	Med. Skt.	135	$5\frac{5}{16}$	26 × 26	1600	3180	25000	4000
250	PAR-38[g]	Flood	Med. Skt.	135	$5\frac{5}{16}$	60 × 60	2400	3180	6500	4000
300	PAR-56[i]	Narrow Spot	Mog. End Prong	127	5	15 × 20	1800	3750	70000[h]	2000
300	PAR-56[i]	Med. Flood	Mog. End Prong	127	5	20 × 35	2000	3750	24000[h]	2000
300	PAR-56[i]	Wide Flood	Mog. End Prong	127	5	30 × 60	2100	3750	10000[h]	2000
500	PAR-64	Narrow Spot	Extended	153	6	13 × 20	3000	6000	110000[h]	2000
500	PAR-64	Med. Flood	Mog. End	153	6	20 × 35	3400	6000	35000[h]	2000
500	PAR-64	Wide Flood	Prong	153	6	35 × 65	3500	6000	12000[h]	2000
500	PAR-56[g]	Narrow Spot	Mog. End. Prong	127	5	15 × 32	4900	7650	96000	4000
500	PAR-56[g]	Med. Flood	Mog. End. Prong	127	5	20 × 42	5700	7650	43000	4000
500	PAR-56[g]	Wide Flood	Mog. End. Prong	127	5	34 × 66	5725	7650	19000	4000
1000	PAR-64[g]	Narrow Spot	Extended	153	6	14 × 31	8500	19400	180000	4000
1000	PAR-64[g]	Med. Flood	Mog. End	153	6	22 × 45	10000	19400	80000	4000
1000	PAR-64[g]	Wide Flood	Prong	153	6	45 × 72	13500	19400	33000	4000

* High efficacy, reduced wattage.
[a] Heat-Resistant glass bulb.
[b] Krypton filled.
[c] To 10 per cent of maximum candlepower.
[d] Some of these types are also available for 230 to 260-volt circuits.
[e] All PAR lamps have bulbs of molded heat-resistant glass.
[f] Central cone defined as 5-degree cone for all spots and 10-degree cone for all floods.
[g] Halogen cycle lamps.
[h] Horizontal operation. May be slightly lower for vertical operation.
[i] Also available with an interference filter reflector.

Fig. 8-97. Traffic Signal Lamps (Medium Screw Base—Voltage as Specified by User)

Nominal Traffic Signal Diameter		Wattage Group	Approximate Watts	Bulb	Light Center Length		Rated Average Life* (hours)	Average Initial Lumens
(millimeters)	(inches)				(millimeters)	(inches)		
200	8	40	40	A-21	62	$2\frac{7}{16}$	2000	370
			40	A-19	62	$2\frac{7}{16}$	6000	320
200	8	60	54*	A-19	62	$2\frac{7}{16}$	8000	530
			60	A-19	62	$2\frac{7}{16}$	6000	550
			60*	A-19	62	$2\frac{7}{16}$	8000	610
			60	A-21	62	$2\frac{7}{16}$	3000	675
			64	A-21	62	$2\frac{7}{16}$	3000	675
			67	A-21	62	$2\frac{7}{16}$	8000	625
			67	A-19	62	$2\frac{7}{16}$	8000	610
			69	A-19	62	$2\frac{7}{16}$	8000	630
200	8	100	90*	A-19	62	$2\frac{7}{16}$	8000	1040
			100	A-21	62	$2\frac{7}{16}$	6000	1080
			100	A-21	62	$2\frac{7}{16}$	3000	1280
			107	A-21	62	$2\frac{7}{16}$	3000	1280
			116	A-21	62	$2\frac{7}{16}$	8000	1280
300	12	150	135*	A-21	76	3	6000	1750
			150	A-21	76	3	6000	1750
			165	A-21	76	3	8000	1950

* Krypton filled lamps.

Fig. 8-98. Incandescent Filament Lamps for Series Operation on 600-Volt DC Circuits

Amperes or Watts	Volts	Bulb and Other Description	Base (see Fig. 8-9)	Filament (see Fig. 8-7)	Rated Average Life (hours)	Maximum Overall Length (millimeters)	Maximum Overall Length (inches)	Average Light Center Length (millimeters)	Average Light Center Length (inches)	Approximate Initial Lumens	Rated Initial Lumens Per Watt	Lamp Lumen Depreciation§ (per cent)
These Lamps Are Used Chiefly in Trolley Car, Trolley Bus, and Subway Car Lighting Operated 20 in Series												
Amperes												
1.0	30	A-19 inside frosted	Med.	C-2R	2000	87	3⅞	65	2 9/16	390	13.0	91.5
1.6	30	A-21 inside frosted*	Med.	C-2R	2000	111	4⅜	75	2 15/16	720	15.0	91.5
2.5	30	A-21 inside frosted and white*	Med.	C-2R	2000	113	4 7/16	73	2⅞	1120	14.9	—
These Lamps Are Used Chiefly in Trolley Car, Trolley Bus, and Subway Car Lighting Operated 5 in Series												
Watts***												
36‡	120**	A-21 inside frosted†	Med.	C-9	2000	111	4⅜	75	2 15/16	390	9.5	—
56#	120**	A-21 inside frosted†	Med.	C-9	2000	111	4⅜	75	2 15/16	630	10.1	—
94##	120**	P-25 St. Rwy. Headlight†	Med.	C-5	1000	121	4¾	53	2 1/16	985	9.5	—
101	120**	A-23 inside frosted	Med.	C-9	1500	151	5 15/16	113	4 7/16	1150	11.9	—

* Cut-out in lamp.
** Also available in 115- and 125-volt designs.
*** Nominal watts.
† Vacuum.
‡ Design current 0.342 amps.
§ Per cent initial light output at 70 per cent of rated life.
# Design current 0.519 amps.
Design current 0.863 amps.

Fig. 8-99. Incandescent Multiple Street Lighting Lamps* (Any Burning Position)

Watts	Nominal Lumens	Clear Bulb	Screw Base	Maximum Over-All Length (millimeters)	Maximum Over-All Length (inches)	Average Light Center Length (millimeters)	Average Light Center Length (inches)	Filament Form (see Fig. 8-7)	Approximate Initial Lumens**	Lamp Lumen Depreciation† (per cent)
A. 1500 Hour Life										
85	1000	A-23	Med.	153	6	113	4 7/16	C-9	1140	—
175	2500	PS-25	Med.	176	6 15/16	133	5¼	C-9	2800	—
268	4000	PS-35	Mog.	238	9⅜	178	7	C-9	4700	—
370	6000	PS-40	Mog.	248	9¾	178	7	C-9	6700	—
575	10000	PS-40	Mog.	248	9¾	178	7	C-7A	11000	—
B. 3000 Hour Life										
58	600	A-19	Med.	108	4¼	73	2⅞	C-9	665	—
92	1000	A-23	Med.	152	6	113	4 7/16	C-9	1185	89
189	2500	PS-25	Mog.	181	7⅛	137	5⅜	C-9	2900	86.5
189	2500	PS-25	Med.	176	6 15/16	133	5¼	C-9	2900	82
295	4000	PS-35	Mog.	238	9⅜	178	7	C-9	4840	82
405	6000	PS-40	Mog.	248	9¾	178	7	C-9	6850	80.5
620	10000	PS-40	Mog.	248	9¾	178	7	C-7A	11000	79.5
860	15000	PS-52	Mog.	332	13 1/16	241	9½	C-7A	15700	79.5
C. 6000 Hour Life										
103	1000	A-23	Med.	154	6 1/16	111	4⅜	C-9	1150	—
202	2500	PS-25	Med.	176	6 15/16	133	5¼	C-9	2800	—
202	2500	PS-25	Mog.	181	7⅛	137	5⅜	C-9	2800	—
327	4000	PS-35	Mog.	238	9⅜	178	7	C-9	4850	—
448	6000	PS-40	Mog.	248	9¾	178	7	C-9	6820	—
690	10000	PS-40	Mog.	248	9¾	178	7	C-7A	11000	—

* 120-volt operation. For HID lamps see Fig. 8-123.
† Per cent initial light output at 70 per cent of rated life.
** Values apply to vertical base-up burning only.

Fig. 8-100. Incandescent Series Street Lighting Lamps (Mogul Screw Base)

Rated Initial Lumens	Amperes	Bulb	Burning Position*	Filament (see Fig. 8-7)	Maximum Over-All Length (millimeters)	Maximum Over-All Length (inches)	Average Light Center Length (millimeters)	Average Light Center Length (inches)	Approximate Initial Volts	Approximate Initial Watts	Approximate Mean Lumens Throughout Rated Life
A. 2000 Hours Life											
600	6.6	PS-25	Any	C-8	181	7⅛	137	5⅜	6.4	42	600
1000	6.6	PS-25	Any	C-8	181	7⅛	137	5⅜	9.6	63	1000
1000	7.5	PS-25	Any	C-8	181	7⅛	137	5⅜	8.3	62	1000
2500	6.6	PS-25	B.U.	C-2V	181	7⅛	137	5⅜	21.2	140	2500
2500	6.6	PS-35	Any	C-2V	238	9⅜	178	7	21.6	142	2500
2500	7.5	PS-35	Any	C-2V	238	9⅜	178	7	19.0	142	2500
4000	6.6	PS-35	Any	C-2V	238	9⅜	178	7	32.8	215	4000
4000	7.5	PS-35	Any	C-2V	238	9⅜	178	7	28.6	214	4000
4000	15	PS-35	B.U.	C-2V	238	9⅜	178	7	13.8	205	3900
4000	15	PS-35	B.D.	C-2V	238	9⅜	159	6¼	13.8	205	3600
6000	6.6	PS-40	Any	C-2V	248	9¾	178	7	47.8	316	6000
6000	20	PS-40	B.U.	C-2V	248	9¾	178	7	15	300	5900
6000	20	PS-40	B.D.	C-2V	248	9¾	159	6¼	15.2	300	5400
10000	6.6	PS-40	Any	C-7A	248	9¾	178	7	79.7	525	9200
10000	20	PS-40	B.U.	C-7	248	9¾	178	7	24.8	496	9400
10000	20	PS-40	B.D.	C-7	248	9¾	159	6¼	24.8	496	8600
15000	20	PS-40	B.U.	C-7	248	9¾	178	7	35.9	712	13800
15000	20	PS-40	B.D.	C-7	248	9¾	159	6¼	36.4	729	12900
B. 3000 Hours Life											
600	6.6	PS-25	Any	C-8	181	7⅛	137	5⅜	6.7	44	600
800	6.6	PS-25	Any	C-8	181	7⅛	137	5⅜	8.2	54	—
1000	6.6	PS-25	Any	C-8	181	7⅛	137	5⅜	10.0	66	1000
1500	6.6	PS-25	Any	C-8	181	7⅛	137	5⅜	14.2	94	—
2500	6.6	PS-25	B.U.	C-2V	181	7⅛	137	5⅜	22.2	146	2500
2500	6.6	PS-35	Any	C-2V	238	9⅜	178	7	22.4	147	2500
2500	7.5	PS-25	Any	C-2V	238	9⅜	178	7	19.6	147	2500
4000	6.6	PS-35	Any	C-2V	238	9⅜	178	7	34.2	225	4000
4000	15	PS-35	B.U.	C-2V	238	9⅜	178	7	14.6	219	3900
4000	15	PS-35	B.D.	C-2V	238	9⅜	159	6¼	14.6	220	3600
6000	6.6	PS-40	Any	C-2V	248	9¾	178	7	50.1	330	6000
6000	20	PS-40	B.U.	C-2V	248	9¾	178	7	15.7	315	5900
6000	20	PS-40	B.D.	C-2V	248	9¾	159	6¼	16.0	319	5400
10000	6.6	PS-40	Any	C-7A	248	9¾	178	7	86.2	570	9200
10000	20	PS-40	B.U.	C-7	248	9¾	178	7	26.1	522	9400
10000	20	PS-40	B.D.	C-7	248	9¾	159	6¼	26.5	530	8600
15000	20	PS-40	B.U.	C-7	248	9¾	178	7	37.5	750	13500
C. 6000 Hours Life											
1000	6.6	PS-25	Any	C-8	181	7⅛	137	5⅜	11.1	73	
1500	6.6	PS-25	Any	C-8	181	7⅛	137	5⅜	15.7	104	
2500	6.6	PS-25	B.U.	C-2V	181	7⅛	137	5⅜	24.2	159	2500
2500	6.6	PS-35	Any	C-2V	238	9⅜	178	7	24.6	162	2500
4000	6.6	PS-35	Any	C-2V	238	9⅜	178	7	37.0	244	4000
4000	15	PS-35	B.U.	C-2V	238	9⅜	178	7	15.7	235	4000
6000	6.6	PS-40	Any	C-2V	248	9¾	178	7	54.3	358	6000
6000	20	PS-40	B.U.	C-2V	248	9¾	178	7	17.3	346	6000
10000	6.6	PS-40	Any	C-7A	248	9¾	178	7	92.0	608	10000

* B.U.—Base-up. B.D.—Base-down.

Fig. 8-101. Lamps for Airport and Airway Lighting
A. Incandescent Filament Types

Typical Service	Watts, Lumens, Amperes or Candelas	Volts or Amperes	Bulb[a]	Base	Filament	Light Center Length		Maximum Over-All Length		Rated[b] Average Lab Life (hours)	Approximate Initial Lumens	Approximate Initial Maximum Luminous Intensity (candelas)	Beamspread to 10 Per Cent Maximum Luminous Intensity (degrees)		Lamp Specifications		Application Specifications		Users
						(millimeters)	(inches)	(millimeters)	(inches)				Hor.	Vert.	FIIN 6240-	MIL Standards	FAA	MIL Specifications	
Flush Appr*, Thres, Rnwy	100W	20A	PAR-36	Sc. Term.	C-6			70	2¾	100		80,000	9	7	926-4342	21999		L-26202	A,C
Appr	300W	20A	PAR-56[c]	Mog. End Prong	C-6[k]			127	5	500		27,000	50	20	929-8003		L-982 E-2048	L-26764	A,C
Flush Appr, Thres	300W	20A	PAR-56[i]	Sc. Term	C-6[k]			114	4½	500		200,000	18	10	823-3179		CAN-1199 L-838	L-26202	A,C
Flush Appr, Thres, Rnwy	499W	20A	PAR-56	Sc. Term	CC-6[k]			114	4½	500		330,000	16	11	869-5077	24488-7		L-26202	A,N,C
Appr	500W	20A	PAR-56[c]	Mog. End Prong	CC-6[k]			127	5	500		48,000	50	25	869-5079	24348-3		L-26764	A,N,C
Thres, Rnwy	503W	20A	T-20[j]	Med. Bipost	C-13	67	2-21/32	165	6½	500	11,300				914-2549	24321-4		L-22252 L-26990	A,N
Med Int Rnwy, Txwy	30W	6.6A	T-10	Med. Pref.	C-2V	38	1-½	100	3 15/16	1000	400				196-4470	25012-1		L-7082	A,N,C
Rnwy Dist Mkr	45W	6.6A	PAR-38[c]	Med. Skt.	C-6			135	5 5/16	800		1,340	60	60	889-1777	24479-1			A
Lo Int Rnwy, Txwy	45W	6.6A	PAR-56	Sc. Term.	C-8			114	4½	1000		65,000	8	8	914-2546	24488-4		L-26202	A,N
Ctrln Rnwy, Txwy	45W	6.6A	T-2½	RSC	C-8			52	2 1/16	1000	710				889-1801				A,N,C
Ctrln Rnwy, Txwy	45W	6.6A	T-2½	Leads	C-8			44	1¾	1000	710						L-842		A,N,C
Txwy	45W	6.6A	T-4	2 Pin Pref.	C-6	39	1.54	64	2½	500	835								
Med Int Rnwy, Txwy	45W	6.6A	T-10	Med. Pref.	C-2V	38	1-½	100	3 15/16	1000	675				196-4472	25012-2	L-802 L-820, L-822	L-7082	A,N,C
Ctrln Rnwy	100W	6.6A	T-3	RSC	C-8			59	2 5/16	1000	2,080				196-6408				N
Ctrln Txwy	65W	6.6A	T-2½	Leads	C-8			44	1¾	1000	1,060						L-852		C
Rnwy, Txwy	100W	6.6A	T-4	2 Pin Pref.	C-6	39	1.54	64	2½	500	1,620								
Rnwy	115W	6.6A	T-4	2 Pin Pref.	C-6 Flat	39	1.54	64	2½	500	2,760								
Rnwy	150W	6.6A	T-4	2 Pin Pref.	C-6 Flat	39	1.54	64	2½	500	3,600								
Rnwy	200W	6.6A	T-4	2 Pin Pref.	CC-6	39	1.54	64	2½	500	5,150								
Ctrln Rnwy	200W	6.6A	T-4	D.C. Ring	CC-6	27	1-1/16	60	2⅜	500	4,900						L-850		C
Ship Fldltg	200W	6.6A	PAR-46	Sc. Term.	CC-6			95	3¾	500	(Filament behind reflector focus)				752-2423	17994-1		F-23528	N
Rnwy, Thres	200W	6.6A	PAR-56[i]	Sc. Term.	CC-6			114	4½	1000		200,000	12	8	935-6994	24488-6	L-838, L-845	L-26202	A,N,C
Appr	200W	6.6A	PAR-56[c]	Mod End Prong	CC-6			127	5	1000		16,000	50	20	538-8853	24348-1		L-26764	A,N
Rnwy Ctrln	200W	6.6A	T-4	RSC	CC-8			65	2 9/16	500	4,500				892-1580		L-850		A,N,C
Rnwy	200W	6.6A	T-4	Leads	CC-8			54	2⅛	500	4,500						L-843		A,N,C
Rnwy	210W	6.6A	T-14	Med. Pref.	C-13	56	2-3/16	146	5¾	500	4,500					25013-3	L-819	L-5004	C
Rnwy	204W	6.6A	T-14	Med. Pref.	C-13	56	2-3/16	146	5¾	500	4,200				299-6753	25013-2		L-5004	A,N

Fldltg	500W	20A	PAR-56	Sc. Term.	CC-6[k]			114	4½	500	(Filament behind reflector focus)								N
VASI	200W	6.6A	PAR-64	Mog. End Prong	CC-6[k]			114	4½	2000	(Filament behind reflector focus)						E-1328 E-2351		C
VASI	300W	6.6A	PAR-64	Mog. End Prong	CC-6[k]			114	4½	2000	(Filament behind reflector focus)				901–8612		E-1328 E-2351	L-27504	A,N,C
Deck Guide	3.04A	12.5V	RP-11	D.C. Bay.	C-2V	32	1¼	57	2¼	300	50				155–7940	15564–6			N
Wheels Watch Sig	5.3A	26V	PAR-46	Sc. Term.	4CC-8[d]			95	3¾	50		130000	10	10	155–7780	25240			N
Rnwy, Txwy	15W	115–125V	T-7	Inter.	C-7A			57	2¼	1000	115				617–1720		L-840		C
Thres, Rnwy, Txwy	40W	120V	T-10	Med. Pref.	CC-2V	38	1½	100	3 15/16	1000	425				295–2862		L-802, 822		C
Rnwy Dist Mkr	75W	75V	PAR-38[c]	Med. Sc.	CC-6			135	5 5/16	1000		1700	60	60					N
Rnwy, Thres	100W	90, 120V	T-10	Med. Pref.	CC-2V	38	1½	100	3 15/16	1000	1170							L-7802	N,C
Obstr*	100W	120V	A-21	Med. Sc.	C-9	62	2 7/16	113	4 7/16	2000	1260				143–7437		L-810		C
Obstr*	107W	120V	A-21	Med. Sc.	C-9	62	2 7/16	111	4⅜	3000	1260				617–1824		L-810		C
Obstr	116W	120V	A-21	Med. Sc.	C-9	62	2 7/16	111	4⅜	6000	1260				842–2887		L-810	L-7830	N,C
T'missomtr	120W	6V	PAR-64	Sc. Term.	C-6[d]			102	4	3000		180000	9	5	299–6769			T-4663	A,N,C
Airpt Ltg	399W	115V	PAR-56[c]	Mog. End Prong	CC-13			127	5	100		30000		20	299–4740	24348–4			N
Rotating Bm* Ceilometer	420W	12V	T-20[j]	Med. Bipost	C-8	64	2½	165	6½	1000	7560					28933–1			A,N,C
Flshg Obstr*	500W	120V	PS-40	Mog. Pref.	C-9	144	5 11/16	255	10 1/16	1000	9900				269–0948			L-6273	A,N,C
Airpt Bcn	500W	120V	T-20[j]	Med. Bipost	C-13B	76	3	191	7½	500	9250				244–5364				C
Flshg Obstr*	620W	120V	PS-40	Mog. Pref.	C-7A	144	5 11/16	255	10 1/16	3000	11200				295–0901	27269	L-606	L-6273	A,N,C
Flshg Obstr	700W	120V	PS-40[j]	Mog. Pref.	C-7A	144	5 11/16	255	10 1/16	6000	11200					27269–1		L-6273	A,N,C
Airpt Bcn*	1000W	120V	T-20[j]	Mog. Bipost	C-13	102	4	241	9½	500	20500				250–6435	25015–1	L-291		C
Airpt Bcn	1200W	115V	T-20[j]	Mog. Bipost	CC-8	102	4	241	9½	750	27500				556–8012	25015–2		L-7158	A,N

B. Gaseous Discharge Types

Typical Service	Watts or Lumens	Volts or Amperes	Bulb[a]	Base	Lamp Type	Light Center Length		Maximum Over-All Length		Approximate Initial Lumens	Rated[b] Average Lab Life (hours)	Lamp Specifications		Application Specifications		Users
						(millimeters)	(inches)	(millimeters)	(inches)			FIIN 6240-	MIL Standards	FAA	MIL Specifications	
Clg. Proj.*	900W	[e]	T-2	Sleeve	Mercury	—	—	83	3¼[g]	65000	100	—	—	—	—	A,N,C
Seq. Flshr. Appr.	60W-sec/fl	2000[e]	Helix	Giant 5-pin	Xenon	102	4	152	6	—	500[h]	725–1938	—	L-1106 L-1107 L-847	L-26311	A,N,C,
Seq. Flshr. Appr.	40W-sec/fl	450	Helix	Octal	Xenon	41	1⅝	64	2½	—	500	—	—	L-859		C
Seq. Flshr. Appr.	60W-sec/fl	2000[e]	PAR-56	Sc. Term	Xenon	—	—	114	4½	13700[l]	500[h]	—	—	L-847		—

[a] Clear bulb unless otherwise specified.
[b] Under specified test conditions.
[c] Lens cover, prismatic.
[d] Shielded filament.
[e] Lamp to be used with auxiliary of proper design.
[f] Life based on three hours per start.
[g] Nominal length.
[h] Life based on two flashes per second.
[i] Stippled cover.
[j] Heat resistant glass.
[k] Tungsten-halogen lamp within an outer bulb.
[l] Effective candlepower (candelas).
* Lamps now in use but not recommended for new designs.

Users:
A—United States Air Force or Army Air Bases
N—United States Navy Air Bases
C—Civil Airports

Fig. 8–102. Lamps for Aircraft

Service	Design Volts	Trade Number	Amperes or Watts	Mean Spherical Candelas	Bulb	Base	Filament	Light Center Length (millimeters)	Light Center Length (inches)	Maximum Over-All Length (millimeters)	Maximum Over-All Length (inches)	Amperes[a] at Design Volts	Rated Average Lab Life (hours)	Approx. Weight (grams)	Approx. Weight (pounds)	National Stock Number 6240-00	Military Standard
Landing	13.0	4509[b]	100W		PAR-36	Sc. Term.	C-6			69.8	$2\frac{3}{4}$	7.69	25	236	0.52	237-7867	MS25243-4509
	13.0	4537[b]	100W		PAR-46	Sc. Term.	C-6			79.4	$3\frac{1}{8}$	7.69	25	417	0.92	946-9636	
	13.0	4313[b]	250W		PAR-36	Sc. Term.	C-6			69.8	$2\frac{3}{4}$	19.23	25	236	0.52	946-4807	
	13.0	4522[b]	250W		PAR-46	Sc. Term.	C-2			79.4	$3\frac{1}{8}$	19.23	25	417	0.92	155-7920	MS25241-4522
	28.0	4553[b]	250W		PAR-46	Sc. Term.	CC-8			79.4	$3\frac{1}{8}$	8.92	25	47	0.92	725-4683	MS25241-4553
	28.0	Q4559[b]	600W		PAR-64	Sc. Term.	CC-8			95.2	$3\frac{3}{4}$	21.41	25	907	2.00	145-1110	MS25242-Q4559
	28.0	4581[b,c]	450W		PAR-46	Sc. Term.	CC-8			79.4	$3\frac{1}{8}$	16.08	10	417	0.92	557-3065	MS25241-4581
Taxiing	28.0	4551[b]	250W		PAR-46	Sc. Term.	CC-6			95.2	$3\frac{3}{4}$	8.92	25	417	0.92	583-3334	MS24517-4551
	28.0	4570[b]	150W		PAR-46	Sc. Term.	CC-6			95.2	$3\frac{3}{4}$	5.36	30	417	0.92	132-5328	MS28926-4570
Navigation	6.0	1680	4.1A	32	S-8	S.C. Bay.	C-6	31.8	$1\frac{1}{4}$	50.8	2	4.1	300	10	0.021	870-7778	MS35478-1680
	6.2	600[d]	26W		GG-10	S.C. Index	C-2R			55.6	$2\frac{3}{16}$	4.19	300	9	0.019		MS25309-600
	6.2	1163[d]	40W		GG-12	S.C. Index	C-6			63.5	$2\frac{1}{2}$	6.45	300	11	0.025		MS24513-1163
	6.2	1687[d]	40W		GG-12	S.C. Index	C-6			13.5	$2\frac{1}{2}$	6.45	150	11	0.025	870-0799	MS25338-1687
	10	1978X	100W	130	T-3	Spec	C-8			51.43	2.025	10.0	200	7	0.016		
	12.8	1777	1.52A	26	S-8	S.C. Bay.	C-2R	28.57	$1\frac{1}{8}$	50.8	2	1.52	400	10	0.021	941-2701	
	13.0	7512-12[d]	26W		GG-10	S.C. Index	C-2R			55.6	$2\frac{3}{16}$	2.0	300	9	0.019		
	14.0	4174-12[d]	40W		GG-12	S.C. Index	C-6			63.5	$2\frac{1}{2}$	3.08	150	11	0.025		
	14	7079B-12[d]	40W		GG-12	S.C. Index	C-6			63.5	$2\frac{1}{2}$	3.08	150	11	0.025		
	28	7079[d]	40W		GG-12	S.C. Index	CC-6			63.5	$2\frac{3}{16}$	1.43	150	11	0.025	789-2260	MS25338-7079
	28	7512[d]	26W		GG-10	S.C. Index	CC-6			55.6	$2\frac{3}{16}$	.93	300	9	0.019	504-2090	MS25309-7512
	28	4174[d]	40W		GG-12	S.C. Index	CC-6			63.5	$2\frac{1}{2}$	1.43	300	11	0.025	519-0854	MS24513-417A
	28	1683	1.02A	32	S-8	S.C. Bay.	2C-6	31.8	$1\frac{1}{4}$	50.8	2	1.02	500	10	0.021	044-6914	MS35478-1683
	28	1970X	100W	140	T-3	Spec.	CC-8			57.2	$2\frac{1}{4}$	3.57	1000	7	0.016		
	28	1967	150W	210	T-3	Spec.	CC-8			57.2	$2\frac{1}{4}$	5.36	1000	7	0.016	045-7173	
Wing Inspection	13	Q4631[b]	250W		PAR-36	Scr. Term.	C-6			69.8	$2\frac{3}{4}$	19.23	500	236	0.52		
	28	4502[b]	50W		PAR-36	Scr. Term.	CC-6			69.8	$2\frac{3}{4}$	1.78	400	236	0.52	196-4518	MS25243-4502
Interior	12.8	1141	1.44A	21	S-8	S.C. Bay.	C-6	31.8	$1\frac{1}{4}$	50.8	2	1.44	500	10	0.021	155-7799	MS35178-1141
	13.0	89	.58A	6	G-6	S.C. Bay.	C-2R	19.0	$\frac{3}{4}$	36.5	$1\frac{7}{16}$	.58	750	6	0.012	143-3159	MS15570-89
	28	456	.17A	2	G-$4\frac{1}{2}$	Min. Bay.	C-2F	14.3	$\frac{9}{16}$	27.0	$1\frac{1}{16}$	.17	500	6	0.012	941-8479	
	28	301	.17A	3	G-5	S.C. Bay.	C-2F	17.3	$\frac{11}{16}$	34.9	$1\frac{3}{8}$	.17	500	6	0.012	155-7947	MS25238-301
	28	303	.30A	6	G-6	S.C. Bay.	C-2F	19.0	$\frac{3}{4}$	365	$1\frac{7}{16}$	.30	500	6	0.012	155-7848	MS15570-308
	28	1309	.52A	15	B-6	S.C. Bay.	C-2F	27.0	$1\frac{1}{16}$	44.5	$1\frac{3}{4}$	.52	500	7	0.016	060-4707	
	28	305	.51A	15	S-8	S.C. Bay.	C-2V	28.6	$1\frac{1}{8}$	50.8	2	.51	300	10	0.021	155-7791	MS35478-305
	28	1691	.61A	15	S-8	S.C. Bay.	2C-2R	28.6	$1\frac{1}{8}$	50.8	2	.61	1000	10	0.021	295-2668	MS35478-1691
	28	3051F[e]	.51A	15	S-8IF	S.C. Bay.	C-2V			50.8	2	.51	300	10	0.021	295-1680	
	28	16911F[e]	.61A	15	S-8IF	S.C. Bay.	2C-2R			50.8	2	.61	1000	10	0.021	441-2708	
	28	2232	.643A	18	S-8	S.C. Bay.	CC-8	30.6	$1\frac{3}{16}$	50.8	2	.643	2000	10	0.021		MS35478-2232
	28	307	.67A	21	S-8	S.C. Bay.	C-2V	28.6	$1\frac{1}{8}$	50.8	2	.67	300	10	0.021	155-7784	MS35478-307
	28	3071F[e]	.67A	21	S-8IF	S.C. Bay.	C-2V			50.8	2	.67	300	10	0.021	122-0264	
	28	1665	.80A	21	S-8	S.C. Bay.	C-2V	28.6	$1\frac{1}{8}$	50.8	2	.80	1000	10	0.021	241-9703	
	28	16651F[e]	.80A	21	S-8IF	S.C. Bay.	C-2V			50.8	2	.80	1000	10	0.021	941-2709	MS35478-16651F
	28	2233	.766A	21	S-8	S.C. Bay.	CC-8	30.0	$1\frac{3}{16}$	50.8	2	.766	2000	10	0.021		MS35478-2233
	28	3011	1.29A	44	S-11	S.C. Bay.	C-2V	31.8	$1\frac{1}{4}$	60.3	$2\frac{3}{8}$	1.29	1000	10	0.021	353-6753	MS25235-3011
	[g]	5004WW[g]	4W		T-5	Min. Pinless				152	6[m]		7500[l]	20	0.045	053-8273	

Fig. 8–102 *Continued*

Service	Design Volts	Trade Number	Amperes or Watts	Mean Spherical Candelas	Bulb	Base	Filament	Light Center Length (millimeters)	Light Center Length (inches)	Maximum Over-All Length (millimeters)	Maximum Over-All Length (inches)	Amperes[a] at Design Volts	Rated Average Lab Life (hours)	Approx. Weight (grams)	Approx. Weight (pounds)	National Stock Number 6240-00	Military Standard
	[g]	5008WW[g]	8W		T-5	Min. Pinless				305	12[m]		7500[l]	34	0.075	955-9174	
	[g]	5013WW[g]	13W		T-5	Min. Pinless				533	21[m]		7500[l]	57	0.125	880-7800	
	[g]	5104WW[g]	4W		T-5	Min. Bi Pin				152	6[m]		7500[l]	20	0.045	916-8196	
	[g]	5106WW[g]	6W		T-5	Min. Bi Pin				228	9[m]		7500[l]	27	0.06	691-1397	
	[g]	5108WW[g]	8W		T-5	Min. Bi Pin				305	12[m]		7500[l]	34	0.075	955-9173	
	[g]	5113WW[g]	13W		T-5	Min. Bi Pin				533	21[m]		7500[l]	57	0.125	955-9164	
Indicator or Instrument	5.0	680	.06A	.03	T-1	Wire Term.	C-2R			6.35	1/4	.06	(j)	.2	0.0004	878-1965	MS24367-680
	5.0	682	.06A	.03	T-1	Sub-mid. Flange	C-2R	4.7	3/16	9.14	3/8	.06	(j)	.3	0.0006	879-4980	MS24515-682
	5.0	683	.06A	.05	T-1	Wire Term.	C-2R			6.35	1/4	.06	(k)	.2	0.0004	060-2941	MS24367-683
	5.0	685	.06A	.05	T-1	Sub-mid. Flange	C-2R	4.7	3/16	9.14	3/8	.06	(k)	.3	0.0006	752-2581	MS24515-685
	5.0	715	.115A	.15	T-1	Wire Term.	C-2R			6.35	1/4	.115	(k)	.2	0.0004	080-4508	MS24367-715
	5.0	718	.115A	.15	T-1	Sub-mid. Flange	C-2R	4.7	3/16	9.14	3/8	.115	(k)	.3	0.0006	764-8237	MS24515-718
	6	316	.7A	3.4	T-3 1/4	Min. Bay.	C-2R	15.8	5/8	30.0	1 3/16	.7	500	3	0.006	817-9803	MS25231-316
	6	328	.2A	.34[h]	T-1 3/4	Mid. Flang	C-6	9.5	3/8	15.8	5/8	.2	1000[n]	.5	0.0012	155-7857	MS25237-328
	13	89	.58A	6	G-6	S.C. Bay.	C-2R	19.0	3/4	36.5	1 7/16	.58	750	5	0.012	143-3159	MS15570-89
	13	1816	.33A	3	T-3 1/4	Min. Bay.	C-2V	15.8	5/8	30.0	1 3/16	.33	1000	3	0.006	155-7949	
	14	330	.08A	0.5	T-1 3/4	Mid. Flange	C-2F	9.5	3/8	15.8	5/8	.08	750	.5	0.0012	196-4491	
	28	301	.17A	3	G-5	S.C. Bay.	C-2F	17.3	11/16	34.9	1 3/8	.17	500	5	0.012	155-7947	MS25238-301
	28	303	.30A	6	G-6	S.C. Bay.	C-2F	19.0	3/4	36.5	1 7/16	.30	500	5	0.012	155-7848	MS15570-303
	28	313	.17A	3.5	T-3 1/4	Min. Bay.	C-2F	15.8	5/8	30.0	1 3/16	.17	500	3	0.006	155-8714	MS25231-313
	28	1864	.17A	3.0	T-3 1/4	Min. Bay.	C-2F	15.8	5/8	30.0	1 3/16	.17	1500	3	0.006	765-8443	
	28	387	.04A	0.3	T-1 3/4	Mid. flange	C-2F	9.5	3/8	15.8	5/8	.04	7000	.5	0.0012	763-7744	MS25237-387
	28	327	.04A	.34	T-1 3/4	Mid. Flange	C-2F	9.5	3/8	15.8	5/8	.04	4000	.5	0.0012	155-7836	MS25237-327
	28	1495	.30A	6	T-4 1/2	Min. Bay.	C-2F	15.8	5/8	34.9	10 3/8	.30	500	3	0.006	299-4742	MS25069-1495

[a] For purposes of wiring design maximum amperes at design volts may be approximately 10 per cent greater than design amperes.
[b] Beam data for PAR lamps:

Lamp No.	Type Cover Glass	Approximate Initial Maximum Candelas	Approximate Spread to 10 Per Cent Horizontal	Maximum Degrees Vertical
4313	Clear	150,000	15	8
4502	Lens	10,000	40	7
4509	Clear	110,000	11	6
4522	Clear	290,000	12	10
4537	Clear	200,000	11	6
4551	Lens	75,000	50	10
4553	Clear	300,000	11	12
Q4559	Clear	600,000	12	9
4570	Lens	32,000	50	9
4581	Clear	400,000	13	14
Q4631	Clear	90,000	13	11

[c] Consult lamp manufacturer before using.
[d] Reflectorized bulb.
[e] Inside frosted.
[g] Fluorescent lamp to be used with auxiliary of proper design.
[h] At 5.0 volts.
[j] Actual life depends on use and environment, theoretical design average life is 100,000 + hours.
[k] Actual life depends on use and environment, theoretical design average life is 40,000 + hours.
[l] Based on three burning hours per start.
[m] Nominal length.
[n] When operated at 5 volts, average laboratory life will be approximately 8,000 hours.
Reproduced by permission of the Society of Automotive Engineers. (Publication No. ARP-881A.)

Fig. 8–103. Lamps for Indicator, Radio, TV, Toy Train, and Other Services

Lamp Number	Bulb	Base	Design Volts	Amperes	Filament (see Fig. 8–7)	Light Center Length (millimeters)	Light Center Length (inches)	Maximum Over-All Length (millimeters)	Maximum Over-All Length (inches)	Rated Average Life (hours)
49	T-3¼	Min. Bay.	2.0	0.06	S-2	—	—	30	1 3/16	1000
253	TL-1¾	Midget Grooved	2.5	0.35	C-2R	—	—	17	11/16	10000
41	T-3¼	Min. Sc.	2.5	0.50	C-2R	25	31/32	30	1 3/16	3000
2158-2158D	T-1¾	Wire Term.	3.0	0.013–0.017	C-6	—	—	13	0.520	20000
1490	T-3¼	Min. Bay.	3.2	0.16	C-2R	20	25/32	30	1 3/16	3000
680	T-1	Wire Term.	5.0	0.06	C-2R	—	—	6	¼	60000
682	T-1	Sub-Midget Flanged	5.0	0.06	C-2R	5	3/16	10	⅜	60000
683	T-1	Wire Term.	5.0	0.06	C-2R	—	—	6	¼	50000
685	T-1	Sub-Midget Flanged	5.0	0.06	C-2R	5	3/16	10	⅜	50000
1850	T-3¼	Min. Bay.	5.0	0.09	C-2R	16	⅝	30	1 3/16	1500
715	T-1	Wire Term.	5.0	0.115	C-2R	—	—	6	¼	40000
718	T-1	Sub-Midget Flanged	5.0	0.115	C-2R	5	3/16	10	⅜	40000
328	T-1¾	S.C. Midget Flanged	6.0	0.20	C-2R	10	⅜	16	⅝	1000
1784-1784D	T-1¾	Wire Term.	6.0	0.20	C-2R	—	—	13	0.520	1000
1847	T-3¼	Min. Bay.	6.3	0.15	C-2R	20	25/32	30	1 3/16	5000†
1866	T-3¼	Min. Bay	6.3	0.25	C-2R	20	25/32	30	1 3/16	5000†
159	T-3¼	Wedge	6.3	0.15	C-2R	13	½	27	1 1/16	5000†
259	T-3¼	Wedge	6.3	0.25	C-2R	17	11/16	27	1 1/16	5000†
755	T-3¼	Min. Bay.	6.3	0.15	C-2R	20	25/32	30	1 3/16	20000
40	T-3¼	Min. Sc.	6.3	0.15	C-2R	25	31/32	30	1 3/16	3000
47	T-3¼	Min. Bay.	6.3	0.15	C-2R	20	25/32	30	1 3/16	3000
44	T-3¼	Min. Bay	6.3	0.25	C-2R	20	25/32	30	1 3/16	3000
46	T-3¼	Min. Sc.	6.3	0.25	C-2R	25	31/32	30	1 3/16	3000
380	T-1¾	S.C. Midget Flanged	6.3	0.04	C-2V	10	⅜	16	⅝	20000
2180-2180D	T-1¾	Wire Term.	6.3	0.04	C-2V	—	—	13	0.520	20000
381	T-1¾	S.C. Midget Flanged	6.3	0.20	C-2F	10	⅜	16	⅝	20000
2181-2181D	T-1¾	Wire Term.	6.3	0.20	C-2F	—	—	13	0.520	20000
455*	G-4½	Min. Bay.	6.5	0.50	C-2R	14	9/16	27	1 1/16	500
50	G-3½	Min. Sc.	7.5	0.22	C-2R	18	23/32	24	15/16	1000
344	T-1¾	S.C. Midget Flanged	10.0	0.0126–0.0154	C-2F	10	⅜	16	⅝	50000†
1869-1869D	T-1¾	Wire Term.	10.0	0.0126–0.0154	C-2F	—	—	13	0.520	50000†
428	G-4½	Min. Sc.	12.5	0.25	C-2V	18	23/32	27	1 1/16	250
382	T-1¾	S.C. Midget Flanged	14.0	0.08	C-2F	10	⅜	16	⅝	15000
2182-2182D	T-1¾	Wire Term.	14.0	0.08	C-2F	—	—	13	0.520	15000
430	G-4½	Min. Sc.	14.0	0.25	C-2V	18	23/32	27	1 1/16	250
1891	T-3¼	Min. Bay.	14.0	0.24	C-2F	16	⅝	30	1 3/16	500
1893	T-3¼	Min. Bay.	14.0	0.33	C-2F	16	⅝	30	1 3/16	7500
257*	G-4½	Min. Bay.	14.0	0.27	C-2R	14	9/16	27	1 1/16	500
256*	T-3¼	Min. Bay.	14.0	0.27	C-2R	16	⅝	30	1 3/16	500
1892	T-3¼	Min. Bay.	14.4	0.12	C-2F	16	⅝	30	1 3/16	1000
1458	G-5	Min. Bay.	20.0	0.25	C-2V	16	⅝	30	1 3/16	250
327	T-1¾	S.C. Midget Flanged	28.0	0.04	C-2F	10	⅜	16	⅝	4000
1762–1762D–1762U	T-1¾	Wire Term.	28.0	0.04	C-2F	—	—	13	0.520	4000
387	T-1¾	S.C. Midget Flanged	28.0	0.04	C-2F	10	⅜	16	⅝	7000
1829	T-3¼	Min. Bay.	28.0	0.07	C-2F	16	⅝	30	1 3/16	1000
757	T-3¼	Min. Bay.	28.0	0.08	C-2F	16	⅝	30	1 3/16	7500

* Flasher Lamp.
† At 6.6 volts.

Fig. 8–104. Specifications of Flasher-Filament Lamps

Lamp No.	Design Volts	Design Amperes or Watts	Approximate Flashes Per Minute	Average Useful Life (hours)	Bulb	Bulb Diameter (millimeters)	Bulb Diameter (inches)	Base	Maximum Over-All Length (millimeters)	Maximum Over-All Length (inches)
405	6.5	0.50	40–160	500	G-4½	14	9/16	Min. S.C.	27	1 1/16
406	2.6	0.30	40–160	50	G-4½	14	9/16	Min. S.C.	27	1 1/16
407	4.9	0.30	40–160	50	G-4½	14	9/16	Min. S.C.	27	1 1/16
D26	120	7W	45	750	C-7½	25	31/32	Cand. S.C.	54	2⅛
D27	120	7W	45	750	C-9½	30	1 3/16	Inter. S.C.	79	3⅛

Fig. 8-105. Lamps for Flashlight, Handlantern, Bicycle, and Other Services

Lamp Number	Bulb (see Fig. 8-8)	Base (see Fig. 8-9)	Design Volts	Amperes	Filament (see Fig. 8-7)	Light Center Length		Maximum Overall Length		Rated† Average Life (hours)	Service—Use with Cell	Luminous Intensity (candelas)
						(millimeters)	(inches)	(millimeters)	(inches)			
112	TL-3	Min. Sc.***	1.20	0.22	S-2	—	—	23	$\frac{59}{64}$	5	1-AA	—
401**	G-4½	Min. Sc.	1.25	0.22	S-2	22	$\frac{7}{8}$	27	$1\frac{1}{16}$	15	1-D	—
131	G-3½	Min. Sc.	1.30	0.10	S-2	18	$\frac{23}{32}$	24	$\frac{15}{16}$	50	1-D	—
222	TL-3	Min. Sc.***	2.25	0.25	C-6	—	—	23	$\frac{59}{64}$	5	2-AA	—
223	FE-3¾	Min. Sc.	2.25	0.25	C-6	17	$\frac{21}{32}$	21	$\frac{13}{16}$	5	2-AA	—
233	G-3½	Min. Sc.	2.33	0.27	C-2R	18	$\frac{23}{32}$	24	$\frac{15}{16}$	10	2-C	0.42
PR4	B-3½	S.C. Min. Fl.	2.33	0.27	C-2R	6	$\frac{1}{4}$	32	$1\frac{1}{4}$	10	2-C	0.4
PR2	B-3½	S.C. Min. Fl.	2.38	0.50	C-2R	6	$\frac{1}{4}$	32	$1\frac{1}{4}$	15	2-D	0.8
PR6	B-3½	S.C. Min. F1	2.47	0.30	C-2R	6	$\frac{1}{4}$	32	$1\frac{1}{4}$	30	2-D	0.45
14	G-3½	Min. Sc.	2.47	0.30	C-2R	18	$\frac{23}{32}$	24	$\frac{15}{16}$	15	2-D	0.5
406**	G-4½	Min. Sc.	2.60	0.30	C-2R	22	$\frac{7}{8}$	27	$1\frac{1}{16}$	50	2-D	—
PR3	B-3½	S.C. Min. Fl.	3.57	0.50	C-2R	6	$\frac{1}{4}$	32	$1\frac{1}{4}$	15	3-D	1.5
PR7	B-3½	S.C. Min. Fl.	3.70	0.30	C-2R	6	$\frac{1}{4}$	32	$1\frac{1}{4}$	30	3-D	0.9
13	G-3½	Min. Sc.	3.70	0.30	C-2R	18	$\frac{23}{32}$	24	$\frac{15}{16}$	15	3-D	0.98
4546	PAR-36	Sc. Term.	4.70	0.50	C-2R	—	—	70	$2\frac{3}{4}$	100	4-F	—
PR13	B-3½	S.C. Min. Fl.	4.75	0.50	C-2R	6	$\frac{1}{4}$	32	$1\frac{1}{4}$	15	4-F	2.2
PR15	B-3½	S.C. Min. Fl.	4.82	0.50	C-2R	6	$\frac{1}{4}$	32	$1\frac{1}{4}$	30	8-D*	1.9
27	G-4½	Min. Sc.	4.90	0.30	C-2R	18	$\frac{23}{32}$	27	$1\frac{1}{16}$	30	4-F	1.4
407**	G-4½	Min. Sc.	4.90	0.30	C-2R	21	$\frac{13}{16}$	27	$1\frac{1}{16}$	50	4-F	—
502	G-4½	Min. Sc.	5.10	0.15	C-2R	18	$\frac{23}{32}$	27	$1\frac{1}{16}$	100	4-F	0.6
PR12	B-3½	S.C. Min. Fl.	5.95	0.50	C-2R	6	$\frac{1}{4}$	32	$1\frac{1}{4}$	15	5-D	3.1
605	G-4½	Min. Sc.	6.15	0.50	C-2R	18	$\frac{23}{32}$	27	$1\frac{1}{16}$	15	5-D	3.2
PR18	B-3½	S.C. Min. Fl.	7.2	0.50	C-2R	6	$\frac{1}{4}$	32	$1\frac{1}{4}$	3	6-D	5.5
PR20	B-3½	S.C. Min. Fl.	8.63	0.50	C-2R	6	$\frac{1}{4}$	32	$1\frac{1}{4}$	15	7-D	5.5

* Two–4 cell groups in parallel.
** Flasher lamp.
† Laboratory.
*** Lens end lamp.

Fig. 8-106. Lamps for 6-Volt Automotive Service

Lamp Number	Luminous Intensity (candelas)	Bulb (see Fig. 8-8)	Base (see Fig. 8-9)	Design Volts	Design Current (amperes)	Filament (see Fig. 8-7)	Light Center Length		Maximum Over-All Length		Service*
							(millimeters)	(inches)	(millimeters)	(inches)	
51	1	G-3½	Min. Bay.	7.5	.22	C-2R	13	$\frac{1}{2}$	24	$\frac{15}{16}$	A, C
55	2	G-4½	Min. Bay.	7.0	.41	C-2R	14	$\frac{9}{16}$	27	$1\frac{1}{16}$	A, C
147	2	T-3¼	Wedge	7.0	.43	C-2R	14	$\frac{9}{16}$	27	$1\frac{1}{16}$	A, C, E, M
63	3	G-6	S.C. Bay.	7.0	.63	C-2R	19	$\frac{3}{4}$	37	$1\frac{7}{16}$	L, M, P, T
64	3	G-6	D.C. Bay.	7.0	.63	C-2R	19	$\frac{3}{4}$	37	$1\frac{7}{16}$	A, C, E
81	6	G-6	S.C. Bay.	6.5	1.02	C-2R	19	$\frac{3}{4}$	37	$1\frac{7}{16}$	E
82	6	G-6	D.C. Bay.	6.5	1.02	C-2R	19	$\frac{3}{4}$	37	$1\frac{7}{16}$	E
87	15	S-8	S.C. Bay.	6.8	1.91	C-2R	26	$1\frac{1}{8}$	51	2	E
88	15	S-8	D.C. Bay.	6.8	1.91	C-2R	26	$1\frac{1}{8}$	51	2	E
209	15	B-6	S.C. Bay.	6.5	1.78	C-6	27	$1\frac{1}{16}$	44	$1\frac{3}{4}$	E
210	15	B-6	D.C. Bay.	6.5	1.78	C-6	27	$1\frac{1}{16}$	44	$1\frac{3}{4}$	E
1129	21	S-8	S.C. Bay.	6.4	2.63	C-6	32	$1\frac{1}{4}$	51	2	E, D, S
1133	32	RP-11	S.C. Bay.	6.2	3.91	C-2R	32	$1\frac{1}{4}$	57	$2\frac{1}{4}$	B
1154	21/3	S-8	D.C. Index	6.4/7.0	2.63/.75	C-6/C-6	32	$1\frac{1}{4}$	51	2	D, L, S, T, P
1158	21/3	S-8	D.C. Bay.	6.4/7.0	2.63/.75	C-6/C-6	32	$1\frac{1}{4}$	51	2	D, L, S, T, P

* Letter designations are defined as follows:
A—Instrument
B—Back-up
C—Indicator
D—Turn signal
E—Interior
L—License
M—Marker, clearance, and identification
P—Parking
S—Stop
T—Tail

Fig. 8-107. Lamps for 12-Volt Automotive and Heavy Duty Truck Service

Lamp Number	Luminous Intensity (candelas)	Bulb (see Fig. 8-8)	Base (see Fig. 8-9)	Design Volts	Design Amperes	Filament (see Fig. 8-7)	Light Center Length		Maximum Over-All Length		Service*
							(millimeters)	(inches)	(millimeters)	(inches)	
1445	0.7	G-3½	Min. Bay.	14.4	.135	C-2V	13	½	24	15/16	A, C, H
53	1	G-3½	Min. Bay.	14.4	.12	C-2V	13	½	24	15/16	A, C
182	1	G-3½	Min. Bay.	14.4	.18	C-2F	13	½	24	15/16	A, C, H
53X	1	G-3½	Min. Bay.	14.4	.12	C-2V	13	½	24	15/16	A, C
161	1	T-3¼	Wedge	14.0	.19	C-2F	13	½	27	1 1/16	A, C, H
184	1	T-3¼	Wedge	14.4	.24	C-2F	14	9/16	27	1 1/16	A, C, H
257	1.6	G-4½	Min. Bay.	14	.27	C-2R	14	9/16	27	1 1/16	C (Flasher)
57	2	G-4½	Min. Bay.	14.0	.24	C-2V	14	9/16	27	1 1/16	A, C
57X	2	G-4½	Min. Bay.	14.0	.24	C-2F	14	9/16	27	1 1/16	A, C, H
158	2	T-3¼	Wedge	14.0	.24	C-2V	13	½	27	1 1/16	A, C
194	2	T-3¼	Wedge	14.0	.27	C-2F	14	9/16	27	1 1/16	A, C, H
193	2	T-3¼	Wedge	14.0	.33	C-2F	14	9/16	28	1 1/12	A, C, H
1895	2	G-4½	Min. Bay.	14.0	.27	C-2F	14	9/16	27	1 1/16	A, C, H
293	2	G-4½	Min. Bay.	14.0	.33	C-2F	14	9/16	27	1 1/16	A, C, H
557	2.5	T-3¼	Wedge	14	.42	C-2R	14	9/16	27	1 1/16	C (Flasher)
168	3	T-3¼	Wedge	14	.36	C-2F	14	9/16	27	1 1/16	A, C, H, M
1247	3	G-6	S.C. Bay.	14.0	.43	2C-2R	21	13/16	37	1 7/16	A, L, M, T, H
97	4	G-6	S.C. Bay.	13.5	.69	C-2V	21	13/16	37	1 7/16	A, C, H, L, M, T, P
67	4	G-6	S.C. Bay.	13.5	.59	C-2R	21	13/16	37	1 7/16	A, M, T, L, P
97	4	G-6	S.C. Bay.	13.5	.69	C-2V	21	13/16	37	1 7/16	A, M, T, L, H
68	4	G-6	D.C. Bay.	13.5	.59	C-2R	21	13/16	37	1 7/16	A, C, M, E
96	4	G-6	D.C. Bay.	13.5	.69	C-2V	21	13/16	37	1 7/16	A, C, M, E, H
1155	4	G-6	S.C. Bay.	13.5	.59	2C-2R	21	13/16	37	1 7/16	A, H, M, T, L
89	6	G-6	S.C. Bay.	13.0	.58	C-2R	19	¾	37	1 7/16	E, M
631	6	G-6	S.C. Bay.	14.0	.63	2C-2R	19	¾	37	1 7/16	A, E, M, T, H
98	6	G-6	S.C. Bay.	13.0	.62	C-2V	19	¾	37	1 7/16	E, M, H
90	6	G-6	D.C. Bay.	13.0	.58	C-2R	19	¾	37	1 7/16	A, E, M
99	6	G-6	D.C. Bay.	13.0	.62	C-2V	19	¾	37	1 7/16	A, E, M, H
105	12	B-6	S.C. Bay.	12.8	1.0	C-6	27	1 1/16	44	1¾	E, F, H
104	12	B-6	D.C. Bay.	12.8	1.0	C-6	27	1 1/16	44	1¾	E, F, H
93	15	S-8	S.C. Bay.	12.8	1.04	C-2R	29	1⅛	51	2	D, E
1093	15	S-8	S.C. Bay.	12.8	1.19	C-2R	29	1⅛	51	2	D, E, H
94	15	S-8	D.C. Bay.	12.8	1.04	C-2R	29	1⅛	51	2	E
1003	15	B-6	S.C. Bay.	12.8	.94	C-6	27	1 1/16	44	1¾	E
1004	15	B-6	D.C. Bay.	12.8	.94	C-6	27	1 1/16	44	1¾	E
1034	32/3	S-8	D.C. Index	12.8/14.0	1.80/.59	C-6/C-6	32	1¼	51	2	D, L, S, T, P
1073	32	S-8	S.C. Bay.	12.8	1.80	C-6	32	1¼	51	2	D, S, B
1141	21	S-8	S.C. Bay.	12.8	1.44	C-6	32	1¼	51	2	B, D, S
1159	21	S-8	S.C. Bay.	12.8	1.6	C-6	32	1¼	51	2	D, S, H
1142	21	S-8	D.C. Bay.	12.8	1.44	C-6	32	1¼	51	2	B, D, S
1176	21/6	S-8	D.C. Bay.	12.8/14.0	1.26/.57	C-6	32	1¼	51	2	E
1376	21/6	S-8	D.C. Bay.	12.8/14.0	1.6/.64	C-6/C-6	32	1¼	51	2	E, H
1156	32	S-8	S.C. Bay.	12.8	2.10	C-6	32	1¼	51	2	B, D, H, S
199	32	S-8	S.C. Bay.	12.8	2.25	C-6	32	1¼	51	2	B, D, S, H
1157	32/3	S-8	D.C. Index	12.8/14.0	2.10/.59	C-6/C-6	32	1¼	51	2	D, H, L, P, S, T
198	32/3	S-8	D.C. Index	12.8/14.0	2.25/.68	C-6/C-6	32	1¼	51	2	D, L, P, S, T, H
1293	50	RP-11	S.C. Bay.	12.5	3.00	C-2R	32	1¼	57	2¼	B, G
1195	50	RP-11	S.C. Bay.	12.5	3.00	C-2V	32	1¼	57	2¼	G
1295	50	S-8	S.C. Bay.	12.5	3.00	C-2R	32	1¼	51	2	B, G

* Letter designations are defined as follows:
A—Instrument
B—Back-up
C—Indicator or pilot
D—Turn-signal
E—Interior
G—Auxiliary service
H—Heavy duty
L—License
M—Marker, clearance, and identification
P—Parking
S—Stop
T—Tail

Fig. 8–108. Sealed Beam and Bonded Beam Lamps for Land Vehicles

Type of Service	Trade Number	Design Watts	Design Volts	Rated Average Laboratory Life (hours)	Filament	Bulb			Maximum Over-All Length		Base (see Fig. 8–9)		
						Type	Maximum Diameter		(millimeters)	(inches)	S.A.E. Type	Terminal	Contacts
							(millimeters)	(inches)					
Headlight	6006	50 40	6.1 6.2	75‡ 150‡	C-6 C-6	PAR-56	179	7.031	134	5¼	H-3	Lugs	3
Fog*	4012	35	6.2	80‡	C-6	PAR-46	145	5.70	95	3¾	G-2	Screw	2
Spotlight	4535	30	6.4	35‡	C-6	PAR-46	145	5.70	95	3¾	G-2	Screw	2
Spotlight	4515	30	6.4	35‡	C-6	PAR-36	113	4.46	70	2¾	G-2	Screw	2
Farm Tractor	4013	25	6.4	100‡	C-6	PAR-46	145	5.70	95	3¾	G-2	Screw	2
Farm Tractor	4019	30	6.2	80‡	C-6	PAR-46	145	5.70	95	3¾	G-2	Screw	2
Fog*	4015	35	6.2	80‡	C-6	PAR-36	113	4.46	70	2¾	G-2	Screw	2
Cycle Headlight	4568	25	12.8	300	C-6	PAR-36	113	4.46	70	2¾	—	Lugs	Slip-on 2
Farm Tractor	4419	35	12.8	300†	C-6	PAR-46	145	5.70	95	3¾	G-2	Screw	2
Farm Tractor	4413	35	12.8	300†	C-6	PAR-46	145	5.70	95	3¾	G-2	Screw	2
Headlight**	4001	37.5	12.8	300†	C-6	PAR-46	145	5.70	102	4	H-2	Lugs	2
Headlight	4101	55	12.8	200	C-6	PAR-46	145	5.7	102	4	—	Lugs	2
Headlight**	4002	50 37.5	12.8 12.8	500† 300†	C-6 C-6	PAR-46	145	5.70	102	4	H-3	Lugs	3
Headlight**	4000	60 37.5	12.8 12.8	320† 200†	C-6 C-6	PAR-46	145	5.70	102	4	H-3	Lugs	3
Headlight**	4005	50 37.5	12.8 12.8	500† 300†	C-6 C-6	PAR-46	145	5.70	102	4	H-3	Lugs	3
Headlight**	4040	60 37.5	12.8 12.8	320† 200†	C-6 C-6	PAR-46	145	5.70	102	4	H-3	Lugs	3
Headlight**	4006	37.5	12.8	300†	C-6	PAR-46	145	5.70	102	4	H-2	Lugs	2
Headlight*	6014	60 50	12.8 12.8	200† 320†	C-6 C-6	PAR-56	179	7.031	134	5¼	H-3	Lugs	3
Headlight	6015	60 50	12.8 12.8	250† 400†	C-6 C-6	PAR-56	179	7.031	134	5¼	H-3	Lugs	3
Headlight	6016	60 50	12.8 12.8	300 500	C-6 C-6	PAR-56	179	7.031	134	5¼	—	Lugs	3
Spotlight	4404	30	12.8	300	C-6	PAR-36	113	4.46	70	2¾	G-2	Screw	2
Spotlight	4405	30	12.8	100†	C-6	PAR-36	113	4.46	70	2¾	G-2	Screw	2
Cycle Headlight	4420	30 30	12.8 12.8	300 300	C-6 C-6	PAR-46	145	5.70	102	4	H-3	Lugs	3
Fog*	4412	35	12.8	100†	C-6	PAR-46	145	5.70	99	3⅞	G-2	Screw	2
Fog*	4415	35	12.8	100†	C-6	PAR-36	113	4.46	70	2¾	G-2	Screw	2
Aux. Headlight	Q4051	50	12.8	200†	C-6	PAR-46	145	5.70	76	3	—	Lugs	Slip-on 2
Headlight	4651	50	12.8	200†	C-6	Rectangular	100 × 165	—	114	—	—	Lugs	2
Headlight	4652	60 40	12.8 12.8	320 200	C-6 C-6	Rectangular	100 × 165	—	114	—	—	Lugs	3
Headlight	6052	65 55	12.8 12.8	150 320	C-6 C-6	Rectangular	142 × 200	—	121	—	—	Lugs	3
Headlight-Halogen	H4651	50	12.8	200	C-6	Rectangular	100 × 165	—	121	—	—	Lugs	2
Headlight-Halogen	H4656	35 35	12.8 12.8	200 320	C-6 C-6	Rectangular	100 × 165	—	121	—	—	Lugs	3

* Available with clear or amber lenses.
** Dual headlamp system (one unit of each type installed and used as a pair at each side on the front of vehicle).
† Rated life at 14.0 volts in these types.
‡ Rated life at 7.0 volts in these types.

Fig. 8–109. Commonly Used Projection Lamps

ANSI Code	Watts	Volts	Bulb	Base (see Fig. 8–9)	Filament (see Fig. 8–7)	Average Life (hours)	Light Center Length		Maximum Over-All Length		Approx-mate Initial Lumens	Approxi-mate Color Tempera-ture (kel-vins)	Footnotes
							(millime-ters)	(inches)	(millime-ters)	(inches)			
BLC	30	115–125	S-11	D.C. Bay.	CC-2V	50	35	1 3/8	60	2 3/8	400	2775	h, c
BLX	50	115–125	S-11	D.C. Bay.	CC-2V	50	35	1 3/8	60	2 3/8	780	2850	h, d
CAX	50	115–125, 130	T-8	D.C. Bay.	CC-13	50	35	1 3/8	79	3 1/8	775	2875	e, h
DFE	80	30	T-12	4-Pin	CC-8	15	40	1 9/16	78	3 1/16	—	—	k, t
DGB/DMD	80	30	T-12	4-Pin	CC-6	15	40	1 9/16	86	3 3/8	—	—	k, e
DLD/DFZ	80	30	T-14	4-Pin	CC-6	15	40	1 9/16	89	3 1/2	—	—	l, d
CAR	150	120	T-10	4-Pin	2CC-8	15	33	1 5/16	79	3 1/8	—	3100	b, e, i
DCH/DJA/DFP	150	120	T-12	4-Pin	CC-6	15	40	1 9/16	90	3 9/16	—	3150	k, e
DFN/DFC	150	125	T-12	4-Pin	CC-8	15	40	1 9/16	81	3 3/16	—	3150	k, t
DJL	150	120	T-14	4-Pin	CC-8	15	40	1 9/16	89	3 1/2	—	3150	j, t
DNE	150	120	MR-16	2-Pin Vented	CC-8	12	16	1 5/8	43	1 45/64	—	3350	a, s, p, f
ELD/EJN	150	21	MR-16	2-Pin	CC-6	40	—	—	44	1 3/4	—	3350	a, d, m, s
FCS	150	24	T-4	Glass 2-Pin	C-6 Oval	50	30	1 3/16	51	2	—	3300	a, s, d, g
EJL	200	24	MR-16	2-Pin	CC-6	50	—	—	44	1 3/4	—	3400	a, s, n, d
EMM/EKS	250	24	MR-14	2-Pin Vented	CC-6	50	16	5/8	43	1 45/64	—	3400	a, s, u, d
ENH	250	120	MR-16	Oval 2-Pin	CC-8	175	—	—	44	1 3/4	—	3250	a, s, q, d
CAL	300	120	T-10	4-Pin	C-13	25	40	1 9/16	102	4	—	3200	b, i, e
CLS/CLG	300	120–125	T-8 1/2	S.C. Bay.	C-13	25	35	1 3/8	105	4 1/8	7600	3200	h, b, e
ELH	300	120	MR-16	Oval 2-Pin	CC-8	35	—	—	44	1 3/4	—	3350	s, a, q, d
FAL	420	120	T-4	Recessed Single Contact	CC-8	75	—	—	67	2 5/8	11000	3200	s, a, c, g
CBA	500	120	T-6	4-Pin	C-13D	50	44	1 3/4	92	3 5/8	—	3200	b, i, d, g
CZA/CZB	500	120, 125	T-10	4-Pin	C-13D	25	40	1 9/16	102	4	—	3300	b, i, e
CZX/DAB	500	115–120, 125	T-10	Medium Prefocus	C-13D	25	56	2 3/16	146	5 3/4	12500	3200	b, e, g
DAY/DAK	500	120	T-10	4-Pin	C-13D	30	40	1 9/16	102	4	12500	3200	b, e, g
DEK/DFW	500	120	T-12	4-Pin	C-13D	25	44	1 3/4	92	3 5/8	—	3250	b, i, f
EHA	500	120	T-6	2-Pin	C-13D	50	37	1 7/16	76	3	—	3300	s, a, i, d, g
DYS/DYV/BHC	600	120	G-7	Prefocus 2-Pin	CC-6	75	37	1 7/16	64	2 1/2	17000	3200	s, a, f, g
CWA	750	115–125	T-12	Prefocus 4-Pin	C-13D	25	40	1 9/16	117		—	3250	b, i, e, g
DDB	750	115–120, 125	T-12	Medium Prefocus	C-13D	25	56	2 3/16	146	5 3/4	19500	3250	b, e, g
CTT/DAX	1000	115–125	T-12	4-Pin	C-13D	25	40	1 9/16	117	4 5/8	—	3300	b, i, e, g

[a] Tungsten-halogen lamp.
[b] Opaque top.
[c] Burning position—any.
[d] Burning position—base down to horizontal.
[e] Burning position—base down.
[f] Burning position—horizontal.
[g] Heat-resistant bulb.
[h] Base pins of lamp are approximately parallel to plane of lead wires.
[i] Proximity reflector.
[j] Internal reflector—focal distance, 44.5 millimeters (1 3/4 inches).
[k] Internal reflector—focal distance 57 millimeters (2 1/4 inches).
[l] Internal dichroic reflector—focal distance 44.5 millimeters (1 3/4 inches).

[m] Mounted in dichroic reflector 50 millimeters (2 inches) diameter.
[n] Mounted in dichroic reflector 50 millimeters (2 inches) diameter. Mounting distance 32 millimeters (1¼ inches).
[o] Mounted in dichroic reflector 50 millimeters (2 inches) diameter. Mounting distance 65.9 millimeters (2 19/32 inches).
[p] Mounted in dichroic reflector 50 millimeters (2 inches) diameter. Mounting distance, 69 millimeters (2 23/32 inches).
[q] Mounted in dichroic reflector, 50 millimeters (2 inches) diameter. Mounting distance 142 millimeters (5 19/32 inches).
[r] On these "non-jacketed" tungsten halogen lamps, screening techniques should be used where appropriate to protect people and surroundings in case of shattering.
[s] Lamps should be protected from scratches and abrasions and should not be used at over 110% of rated voltage.
[t] Burning position—base down or horizontal.
[u] Mounted in dichroic reflector 44 millimeters (1¾ inches) diameter. Mounting distance 65.9 millimeters (2 19/32 inches).

Fig. 8-110. Examples of Variations in Range and Specifications of Subminiature Lamps

Lamp No.	Luminous Intensity (candelas)	Volts Rated/ Design	Amperes Design	Watts	Battery	Bulb Shape	Base	Length: Light Center (millimeters)	Length: Light Center (inches)	Length: Over-All (millimeters)	Length: Over-All (inches)	Filament or Electrode	Approximate Life (hours)	Maximum Safe Temperature (°C)	Maximum Safe Temperature (°F)	Pre-focused
A9A (NE-2E)	.003[a]	105–125	.0007	1/12	—	T-2	Wire Term	—	—	19	¾	W-11	25,000+	150	300	No
B7A (NE-45)	.01[a]	105–125	.002	¼	—	T-4½	Cand. S.C.	—	—	41	1⅝	P-3	7,500+	150	300	No
327	.34[a]	28	.04	—	—	T-1¾	S.C. Mid. Fls.	10	⅜	16	⅝	C-2F	4,000	150	300	No
14	.5[a]	2.47	.3	—	2-D Cells	G-3½	Min. S.C.	18	23/32	24	15/16	C-2R	15	175	350	No
PR2	.8[a]	2.38	.5	—	2-D Cells	B-3½	S.C. Min. Fl.	6	¼	32	1¼	C-2R	15	—	—	Yes
44	.9[a]	6.3	.25	—	—	T-3¼	Min. Bay.	20	25/32	30	1 3/16	C-2R	3,000	175	350	No
1157	32/3[a]	12.8/14.0	2.10/.59	—	—	S-8	D.C. Indexing	32	1¼	51	2	C-6/ C-6	1200/5000	175	350	No
51	1[a]	6/7.5	.22	—	—	G-3½	Min. Bay.	13	½	24	15/16	C-2R	1,000	175	350	No
4535[c]	95000[b]	6/6.4	—	30	—	PAR-46	S.C. Term.			95	3¾	C-6	100	175	350	Yes
222[d]	5[b]	2.25	.25	—	2-AA Cells	TL-3	Min. S.C.			23	59/64	C-6	5	175	350	No
194	2[a]	14.	.27	—	—	T-3¼	Wedge	14	.562	27	1 1/16	C-2F	2,500	315	600	No
259	.65[a]	6.3	.25	—	—	T-3¼	Wedge	17	11/16	27	1 1/16	C-2R	indf. long	315	600	No

[a] Mean spherical luminous intensity.
[b] Maximum luminous intensity in beam.
[c] PAR-type lamp-reflectorized.
[d] Bulb has built-in lens.

Fig. 8–111. Infrared Energy Sources

A. Sheathed Resistance Radiant Heaters* (1500–14,000 nm) (For Use with External Reflectors)

Watts	Watts per Length		Volts	Terminals	Description	Nominal Sheath Temperature		Heated Length	
	Millimeter	Inch				°C	°F	Millimeters	Inches
1000**	0.6	14	230	Bayonet plug-in	Hairpin	450	850	1780	70
2000**	1.7	44	230	Bayonet plug-in	Hairpin	815	1500	1170	46
600†	1.2	30	115 or 230	Screw	Straight Rod	700	1300	430	17
800†	1.6	40	115 or 230	Screw	Straight Rod	790	1450	430	17
1000†	1.5	37	230	Screw	Straight Rod	760	1400	580	23
1500†	1.3	33	230	Screw	Straight Rod	730	1340	990	39
2500†	1.6	40	230 or 460	Screw	Straight Rod	790	1450	1350	53
3000†	1.6	40	230 or 460	Screw	Straight Rod	790	1450	1650	65
3600†	1.6	40	230 or 460	Screw	Straight Rod	790	1450	1950	77

* Rated life in excess of 7000 hours; may be burned in any position.
** General Electric, North American Philips Lighting, Tuttle & Kift.
† Edwin L. Wiegand Company.

B. Tungsten Filament Lamps* (500–4000 nm)

Watts	Bulb and Other Description	Base (see Fig. 8–9)	Service	Filament (see Fig. 8–7)	Maximum Over-all Length		Light Center Length		Voltage Range
					Millimeters	Inches	Millimeters	Inches	
125	G-30 clear	Med. Skt.	Industrial	C-7A C-11	181 186	7 1/8 7 5/16	127 124	5 4 7/8	115–125
125	R-40 light inside frosted reflector	Med. Skt.	Industrial	C-9 C-11	187 187	7 3/8 7 3/8	—	—	115–125
250	G-30 clear	Med. Skt.	Industrial	C-7A C-11	183 186	7 3/16 7 5/16	127 124	5 4 7/8	115–125
250	R-40 light inside frosted reflector	Med. Skt.	Industrial	C-9 C-11	187 187	7 3/8 7 3/8		—	115–125
250	R-40 light inside frosted reflector, heat resistant bulb	Med. Skt.	Industrial	C-9 C-11	191 184	7 1/2 7 1/4	—	—	115–125
250	R-40 light inside frosted reflector	Med.	Home and farm	CC-6 C-9	165 167	6 1/2 6 9/16	—	—	115–125
250	R-40 reflector, red bowl, heat resistant bulb	Med.	Home and farm	C-9 C-11	171	6 3/4	—	—	115–125
250	PS-30 reflector	Med.	Chicken brooding	C-9	205	8 1/16	—	—	115–125
375	G-30 clear	Med. Skt.	Industrial	C-7A C-11	183 186	7 3/16 7 5/16	127 124	5 4 7/8	115–125
375	R-40 light inside frosted reflector	Med. Skt.	Industrial	C-9 C-7A C-11	187 187	7 3/8 7 3/8	—	—	115–125
375	R-40 light inside frosted reflector heat resistant bulb	Med. Skt.	Industrial	C-9 C-7A C-11	191 191	7 1/2 7 1/2	—	—	115–125
375	R-40 reflector, red bowl, heat resistant bulb	Med. Skt.	Industrial	C-9	191 194	7 1/2 7 5/8	—	—	115–125
500	G-30 clear	Med. Skt.	Industrial	C-7A C-11	183 183	7 3/16 7 3/16	127	5	115–125
1000	T-40 clear	Med. Bipin**	Industrial	Triang.	186	7 5/16	83	3 1/4	115–125

* Rated average life in excess of 5000 hours.
** With 6-inch flexible connectors.

* Also see Fig. 8–125A.
[a] This lamp has a tungsten-filament resistance and a thermal switch enclosed in its reflector-type outer envelope. It is operated without external ballast on 110 to 125 volts, 50 to 60 hertz alternating current only.
[b] These lamps contain their own internal reflector; output figures apply to the energy directed out the face of the lamp.
[c] The amount of radiant flux which will produce the same erythemal effect as 10 microwatts of radiant flux at wavelength 296.7 nanometers.

Fig. 8-111. *Continued*

C. Tubular Quartz Heat Lamps

(Burning Position—Horizontal, except as noted. All lamps listed are translucent quartz except as noted.)

Watts	Bulb and Other Description	Base (see Fig. 8-9)	Service	Filament (see Fig. 8-7)	Maximum Over-all Length		Lighted Length		Voltage Range
					Millimeters	Inches	Millimeters	Inches	
300	T-3[(1)]	Sleeve**	Industrial	C-8	217	8 9/16	107	4 7/32	115-125
375	T-3[(1)]	Sleeve**	Industrial	C-8	224	8 13/16	127	5	115-125
375	T-3[(1)]	R.S.C.***	Industrial	C-8	211	8 5/16	127	5	115-125
500	T-3[(1)]	Sleeve**	Industrial	C-8	224	8 13/16	127	5	100-110, 115-125
500	T-3 clear[(1)]	Sleeve**	Industrial	C-8	224	8 13/16	127	5	115-125
500	T-3 clear[(1, 2)]	Sleeve**	Special equip.	C-8	224	8 13/16	127	5	115-125
500	T-3[(1)]	R.S.C.***	Comfort	C-8	211	8 5/16	127	5	115-125
800	T-3	Sleeve**	Commercial	C-8	303	11 15/16	203	8	115-125
1000	T-3	Sleeve**	Industrial	C-8	351	13 13/16	254	10	200-220, 230-250
1000	T-3 clear	Sleeve**	Industrial	C-8	303	11 15/16	254	10	230-250
1000	T-3 clear[(2)]	Sleeve**	Special equip.	C-8	303	11 15/16	254	10	230-250
1000	T-3 clear[(2)]	Sleeve**	Special equip.	C-8	351	13 13/16	254	10	230-250
1200	T-3 clear	Sleeve**	Special equip.	C-8	224	8 13/16	152	6	144
1600	T-3	Sleeve**	Industrial	C-8	503	19 13/16	406	16	200-220, 230-250, 277
1600	T-3	R.S.C.***	Comfort	C-8	498	19 5/8	406	16	200-220, 230-250, 277
1600	T-3 clear	Sleeve**	Industrial	C-8	503	19 13/16	406	16	230-250
1600	T-3 vertical burning	Sleeve**	Industrial	C-8	503	19 13/16	406	16	230-250
1600	T-3 clear	Sleeve**	Special equip.	C-8	456	17 15/16	406	16	230-250
1600	T-3 clear[(2)]	Sleeve**	Special equip.	C-8	456	17 15/16	406	16	230-250
2000	T-3 clear	Sleeve**	Special equip.	C-8	303	11 15/16	248	9 3/4	230-250
2000	T-3 clear[(2)]	Sleeve**	Special equip.	C-8	303	11 15/16	248	9 3/4	230-250
2000	T-3 clear[(2)]	Sleeve**	Special equip.	C-8	351	13 13/16	254	10	230-250
2500	T-3	Sleeve**	Industrial	C-8	732	28 13/16	635	25	460-500, 575-625
2500	T-3	R.S.C.***	Comfort	C-8	727	28 5/8	635	25	460-500
2500	T-3 vertical burning	Sleeve**	Industrial	C-8	732	28 13/16	635	25	460-500
2500	T-3 clear	Sleeve**	Industrial	C-8	732	28 13/16	635	25	460-500
2500	T-3 clear[(2)]	Sleeve**	Special equip.	C-8	732	28 13/16	635	25	460-500
3800	T-3	Sleeve**	Industrial	C-8	1062	41 13/16	965	38	550-600
3800	T-3 clear	Sleeve**	Industrial	C-8	1062	41 13/16	965	38	550-600
3800	T-3 vertical burning	Sleeve**	Industrial	C-8	1062	41 13/16	965	38	550-600
5000	T-3 clear[(2)]	Sleeve**	Special equip.	C-8	732	28 13/16	635	25	575-625
5000	T-3 clear	Sleeve**	Industrial	C-8	1367	53 13/16	1270	50	920-1000
6000	T-3 clear	Sleeve**	Industrial	C-8	300	11 15/16	248	9 3/4	480

* Rated average life in excess of 5000 hours. ** With flexible connectors. *** Recessed single contact.
(1) Burning position: any. (2) High temperature construction.

Fig. 8-112. Sunlamps*

Designation	FS20	FS40	FS72T12	RSM
Type of lamp	Fluorescent Preheat Start	Fluorescent Rapid Start	Fluorescent Instant Start	Mercury Self Ballasted
Bulb	T-12	T-12	T-12	R-40
Diameter				
(millimeters)	38	38	38	127
(inches)	1.5	1.5	1.5	5
Over-all length (nominal)				
(millimeters)	600	1200	1800	186 (max.)
(inches)	24	48	72	7 5/16 (max.)
Base	Med. Bipin	Med. Bipin	Single Pin	Mog.
Rated power (watts)	20	40	57	275
Lamp potential drop (volts)	57	100	149	[a]
Arc current (amperes)	0.38	0.43	0.425	[a]
Number of applications at 10 minutes per application	5000	5000	5000	1200
Approximate output in erythemal units[c] (initial, bare lamp)	50000	155000	225000	35000[b]

Footnotes on left page.

Fig. 8-113. Typical Photoflash Lamps*

Class	Designations	Approximate Time to Peak (milliseconds)	Approximate Lumen Seconds or Beam Candela Seconds	Bulb or Shape	Maximum Over-All Length		Base
					(millimeters)	(inches)	
Clear Lacquer-Coated for Black and White Film (Approximate Color Temperature—3800 K)							
M	M3	17	16000[a]	T-6½	46	$1\frac{13}{16}$	Min. Bay. Pinless
M	Press 25	20	21000[a]	B-12	67	$2\frac{5}{8}$	S.C. Bay.
M	2	20	70000[a]	A-19	121	$4\frac{3}{4}$	Med.
S	3	30	100000[a]	A-23	151	$5\frac{15}{16}$	Med.
FP	FP26	—	18000[a]	B-12	67	$2\frac{5}{8}$	S.C. Bay.
Blue Lacquer-Coated for Daylight Color Film (Approximate Color Temperature—5500 to 6000 K)							
MF	AG-1B	12	5300[a]	T-3¾	33	$1\frac{5}{16}$	Glass Groove
M	M3B	17	10000[a]	T-6½	44	$1\frac{3}{4}$	Min. Bay. Pinless
M	5B, 25B	20	10000[a]	B-11, B-12	67	$2\frac{5}{8}$	S.C. Bay.
M	2B	20	33500[a]	A-19	121	$4\frac{3}{4}$	Med.
S	3B	30	53000[a]	A-23	151	$5\frac{15}{16}$	Med.
FP	FP26B	—	9800[a]	B-12	67	$2\frac{5}{8}$	S.C. Bay.
Multiple Flash (5500 K)							
MF	Flashcube	13	2000[b]	cube	36	$1\frac{13}{32}$	Plastic
MF	Hi-Power	13	4800[b]	cube	36	$1\frac{13}{32}$	Plastic
MF	Magicube	7	2000[b]	cube	40	$1\frac{19}{32}$	Plastic
MF	Flashbar	16	3600[b]	array	108	$4\frac{1}{4}$	Contact Tab
MF	Flipflash	14	2000[b]	array	138	$5\frac{7}{16}$	Special Tab

* Voltage ranges usually 3, 3–45, or 3–125 (see manufacturers' catalogs).
[a] Lumen seconds.
[b] Beam candela seconds.

Fig. 8-114. Typical Enlarger Lamps

Manufacturers' Designations	Nominal Voltage	Rated Watts	Average Rated Life (hours)	Approximate Initial Lumens	Bulb	Base	Footnotes
PH/111A	125	75	25	1120	S-11	S.C. Bay.	a, f
PH/113	115–125	50	50	700	S-11	S.C. Bay.	a, f
PH/140	120	75	50	—	S-14	Med.	a
PH/211	115–125	75	100	1200	A-21	Med.	a
PH/212	115–125	150	100	2850	A-21	Med.	a
PH/213	115–125	250	3	7000	A-21	Med.	a
PH/300	115–125	150	100	3300	PS-30	Med.	a
PH/302	115–125	500	100	11000	PS-30	Med.	a
BEV	20	150	50	4300	G-16½	S.C. Pf.	h
DLS/DLG/DHX	21.5	150	15	—	T-14	4-Pin	e, g
EJL	24	200	50	—	MR-16MM	2-Pin	b, c, f
EJV	21	150	40	—	MR-16	2-Pin	b, d, f
ELC	24	250	50	—	MR-16MM	2-Pin	b, c, f
ESD	120	150	12	—	MR-16MM	Oval 2-Pin	b, d, f

[a] White diffusing bulb.
[b] Reflector type tungsten-halogen lamp. Dichroic reflector, 50 millimeters (2 inches) diameter.
[c] Mounting distance, 31.8 millimeters (1¼ inches).
[d] Mounting distance, 44.5 millimeters (1¾ inches).
[e] Internal dichroic reflector with focal distance of 44.5 millimeters (1¾ inches).
[f] Burning position—base down to horizontal.
[g] Burning position—base down or horizontal.
[h] Burning position—base up.

Fig. 8–115. Energy Saving Fluorescent Lamps*

	Rapid Start						Preheat Start		Instant Start		
Nominal length											
Millimeters	900	1200	2400	2400	2400	1200	1200	1500	1200	2400	2400
Inches	36	48	96	96	96	48	48	60	48	96	96
Bulb	T-12 38-mm	T-12 38-mm	T-12 38-mm	T-12 38-mm	PG-17 54-mm	PG-17 54-mm	T-12 38-mm	T-17 54-mm	T-12 38-mm	T-8 25-mm	T-12 38-mm
Base	Med. Bipin	Med. Bipin	Recess D.C.	Recess D.C.	Recess D.C.	Recess D.C.	Med. Bipin	Mog. Bipin	Single Pin	Single Pin	Single Pin
Lamp amperes	0.453	0.45	0.81	1.58	1.57	1.53	0.45				0.44
Approx. lamp volts	64	84	134	137	144	64	84				153
Approx. lamp watts[a]	25	34–35[d]	95	185–195	185	95	34	82–84	30–32	40–41	60
Lamp watts replaced[a]	30	40	110	215	215	110	40	90	40	50	75
Rated life (hours)[b]	18000	20000	12000	10000–12000	12000	12000	15000	9000	9000	7500	12000
Initial lumens[c]											
Cool White	2000	2770	8300	13000	14000	6550	2800	6175	2550	3725	5500
Deluxe Cool White		1925	6000								4000
Warm White	2050	2820	8500		13000		2900				5700
Deluxe Warm White		1925	5720								3870
White		2820	8300					6400		3795	5600
Daylight		2300						5400			4800
Supermarket White		2050									4250
Royal White		2900	8800								5700
Lite-White		2925	8800	13800	14900				2700		5850
Lite White Deluxe		2925		8800					2700		5850
Warm Lite Deluxe		2925		8800					2700		5850
Regal White		2450									5100
Super Saver II		3050							2750		6000
3 K		2900	8800						2700		5800
Chroma 50											4100
Optima 50		2025									
Super White		2630									5220
Optima 32		2260									4200
Vita-Lite		2010									4015
Ultralume 83		2980									5860
Ultralume 84		2980									5860
Ultralume 85		2980									5860
Colortone 50											4100
Design 50		2250									
SP 30		2900									5950
SP 35		2900	8550								5800
SP 41		2850	8600								5800

* The life and light output ratings of fluorescent lamps are based on their use with ballasts that provide proper operating characteristics. Ballasts that do not provide proper electrical values may substantially reduce either lamp life or light output, or both.

[a] For rapid start lamps, includes watts for cathode heat.

[b] Rated life under specified test conditions with three hours per start. At longer burning intervals per start, longer life can be expected.

[c] At 100 hours. Where color is made by more than one manufacturer, lumens and footlamberts represent average of manufacturers.

[d] Also in 32-watt cathode-cutout, but with reduced life.

Note: All electrical and lumen values apply only under standard photometric conditions.

Fig. 8–116. Typical Hot-Cathode Fluorescent Lamps (Rapid Starting)[a]

	Circline				U-Shaped		Lightly Loaded Lamps						
Nominal length (millimeters) (inches) Bulb	165 dia. 6½ dia. T-9	210 dia. 8¼ dia. T-9	300 dia. 12 dia. T-10	400 dia. 16 dia. T-10	600 24 T-12	600 24 T-12	900 36 T-12	1200 48 T-12	1200 48 T-12	1200 48 T-10	900 36 T-8	1200 48 T-8	1500 60 T-8
Base	4-Pin	4-Pin	4-Pin	4-Pin	Med. Bipin	Med. Bipin	Med. Bipin	Med. Bipin	Med. Bipin	Med. Bipin	Med. Bipin	Med. Bipin	Med. Bipin
Leg spacing					92 mm (3⅝ in)	152 mm (6 in)							
Approx. lamp current (amperes)	0.38	0.37	0.425	0.415	0.42	0.43	0.43	0.43	0.43	0.42	0.265	0.265	0.265
Approx. lamp volts	48	61	81	108	103	100	81	101	101	104	100	135	172
Approx. lamp watts[f]	20	22.5	33	41.5	41	40.5	32.4	41	41	41	25	32	40
Rated life (hours)[b]	12000	12000	12000	12000	12000	12000	18000	20000	15000	24000	20000	20000	20000
Lamp lumen depreciation (LLD)[c]		72	82	77	84	84	81	84					
Initial lumens[d]													
Cool White	800	1065	1870	2580	2900	2935	2210	3150	3250	3200	k	k	k
Deluxe Cool White		875	1425	2000	2020	2065	1555	2200		2270			
Warm White	825	1065	1835	2550	2850	2965	2235	3175	3250	3250			
Deluxe Warm White	630	800	1375	1950	1980	2040	1505	2165					
White		1100	1870	2650	2850	2965	2255	3185	3250	3250			
Daylight		906	1550	2165			1900	2615	2650	2700			
Factors for Calculating Luminance[e]													
Candelas per Square Meter	11.5	8.4	4.6	3.3	2.65	2.65	3.58	2.65	2.65	3.18	5.42	4.03	3.18
Candelas per Square Foot	1.07	0.78	0.43	0.31	0.25	0.25	0.33	0.25	0.25	0.30	0.50	0.37	0.30

	Medium Loaded Lamps									
Nominal length (millimeters) (inches) Bulb	600 24 T-12	750 30 T-12	900 36 T-12	1050 42 T-12	1200 48 T-12	1500 60 T-12	1600 64 T-12	1800 72 T-12	2100 84 T-12	2400 96 T-12
Base	Recess D.C.	Recess D.C.	Recess D.C.	Recess D.C.	Recess D.C.	Recess D.C.	Recess D.C.	Recess D.C.	Recess D.C.	Recess D.C.
Approx. lamp current (amperes)	0.8	0.8	0.8	0.8	0.8	0.8	0.8	0.78	0.8	0.79
Approx. lamp volts	41		59		78	98		117	135	153
Approx lamp watts[f]	37		50		63	75.5		87	100	113
Rated life (hours)[b]	9000	9000	9000	9000	12000	12000	12000	12000	12000	12000
Lamp lumen depreciation (LLD)[c]	77		77	77	82	82	82	82	82	82
Initial lumens[d]										
Cool White	1700	2290	2885	3516	4300	5400	5800	6650	7800	9150
Deluxe Cool White					3050			4550		6533
Warm White	1700				4300	5500		6500		9200
Deluxe Warm White										6475
White								6475		9200
Daylight	1400		2476	3100	4300	4650	4900	5600	6867	7800
Factors for Calculating Luminance[e]										
Candelas per Square Meter	5.54	4.35	3.58	3.04	2.65	2.1	1.95	1.7	1.44	1.24
Candelas per Square Foot	0.51	0.40	0.33	0.28	0.25	0.20	0.18	0.16	0.13	0.12

	Highly Loaded Lamps									
Nominal length (millimeters) (inches)	1200 48	1800 72	2400 96	1200 48	1500 60	1800 72	2400 96	1200 48	1800 72	2400 96
Bulb	T-10[g]	T-10[g]	T-10[g]	T-12[i]	T-12	T-12[i]	T-12[i]	PG-17	PG-17	PG-17
Base	Recess D.C.	Recess D.C.	Recess D.C.	Recess D.C.	Recess D.C.	Recess D.C.	Recess D.C.	Recess D.C.	Recess D.C.	Recess D.C.
Approx. lamp current (amperes)	1.5	1.5	1.5	1.5	1.5	1.52	1.5	1.5	1.5	1.5
Approx. lamp volts	80	120	160	84		125	163	84	125	163
Approx. lamp watts[f]	105[h]	150[h]	205[h]	116		168	215	116	168	215
Rated life (hours)[b]	9000	9000	9000	9000	9000	9000	9000	9000	9000	9000
Lamp lumen depreciation (LLD)[c]	66	66	66	69		72	72	69	69	69
Initial lumens[d]										
Cool White	6700[h]	10000[h]	14000[h]	6900	8950	10640	15250	7450	11500	16000
Deluxe Cool White	4690[h]	7000[h]	9800[h]	4900		7400	10750	5200		11200
Warm White				6700		10500	14650	7000		15000
White						10500	15000			
Daylight				5700		9300	12650	6000	9300	13300
Factors for Calculating Luminance[e]										
Candelas per Square Meter	3.18	2.04	1.49	2.65	2.1	1.7	1.24	1.91[j]	1.2[j]	0.89[j]
Candelas per Square Foot	0.30	0.19	0.14	0.25	0.20	0.16	0.12	0.18[j]	0.11[j]	0.08[j]

[a] The life and light output ratings of fluorescent lamps are based on their use with ballasts that provide proper operating characteristics. Ballasts that do not provide proper electrical values may substantially reduce either lamp life or light output, or both.
[b] Rated life under specified test conditions at three hours per start. At longer burning intervals per start, longer life can be expected.
[c] Per cent of initial light output at 70 per cent rated life at three hours per start. Average for cool white lamps. Approximate values.
[d] At 100 hours. Where lamp is made by more than one manufacturer, light output is the average of manufacturers. For the lumen output of other colors of fluorescent lamps, multiply the cool white lumens by the relative light output value from the table below.
[e] To calculate approximate lamp luminance, multiply the lamp lumens of the lamp color desired by the appropriate factor. Factors derived using method by E. A. Linsday, "Brightness of Cylindrical Fluorescent Sources", Illuminating Engineering, Vol. XXXIX, January, 1944, p. 23.
[f] Includes watts for cathode heat.
[g] A jacketed T-10 design is also available for use in applications where lamps are directly exposed to cold temperatures.
[h] Peak value. At 25 °C (77 °F) lumen and wattage values are lower.
[i] These lamps available in several variations (outdoor, low temperature, jacketed) with the same, or slightly different, ratings.
[j] Average luminance for center section of lamp. Parts of surface of lamp will have higher luminance.
Note: All electrical and lumen values apply only under standard photometric conditions.
[k] Initial Lumens for 3100 K and 4100 K color lamps (GTE Sylvania) are: 25-W, 2150; 32-W, 2900; and 40-W, 3650.

Approximate Relative Light Output of Various Fluorescent Lamp Colors Compared to Cool White

Color	Relative output	Group
Blue	0.40	A
Green	1.38	A
Pink	0.37	A
Gold	0.72	A
Red	0.06	A
Natural*	0.67	B
Soft White (new-3000 K)	0.69	B
Cool Green	0.83	B

*Formerly, Soft White/Natural.

Color	Manufacturer	Relative output
Incandescent/Fluorescent	(GTE Sylvania)	0.55
Modern White Deluxe	(North American Philips Lamp)	0.77
Sign White	(GE)	0.76
Design White	(GTE Sylvania)	0.78
Supermarket White	(North American Philips Lamp)	0.74
Optima 32	(Duro-Test)	0.76
Super White	(Duro-Test)	0.93
Turquoise	(Duro-Test)	0.96
Super Deluxe 45	(Duro-Test)	0.79
Daylight 65	(Duro-Test)	0.65
Vita-Lite, Vita-Lux	(Duro-Test)	0.69
Ultralume 83, 84, 85	(North American Philips Lamp)	0.92

Color	Manufacturer	Relative output
Lite White Deluxe	(GTE Sylvania)	1.05
3 K	(GTE Sylvania)	1.05
Color 86	(Philips Electronics, Canada)	1.05
High CRI, 5000 K: Chroma 50 (GE), Colortone 50 (North American Philips) and Color-Matcher 50, Color Classer 50, Optima 50, and Magnalux (Duro-Test).		0.70
High CRI, 7500 K: Chroma 75 (GE), Colortone 75 (North American Philips Lamp), and Color Classer 75, and Color-Matcher 75 (Duro-Test).		0.67
Lite White† (GE, North American Philips, and GTE Sylvania)		1.08
Royal White	(GTE Sylvania)	1.03
Regal White	(GE)	0.90

† Currently available in energy saving fluorescent lamps only. Consult the lamp manufacturers for the availability and lumen output of colors in other lamp sizes.
The colors in groups A and B are made by most lamp manufacturers and in several lamp sizes. The balance of the lamp colors are generally made by just one lamp company and in just a few lamp sizes. Practically all colors are made in the 40-watt preheat start/rapid start size. Consult the lamp manufacturers for the availability and lumen output of colors in other lamp sizes.

Fig. 8-117. Compact Fluorescent Lamps

Manufacturer's Lamp Designation	Lamp Watts	Maximum Over-All Length		Bulb	Base	Approx. Lamp Current Amperes	Approx. Lamp Voltage Volts	Rated Life[a]	Initial Lumens[h]	Lamp Lumen Depreciation (LLD)	Color Rendering Index
		Millimeters	Inches								
Dulux[e]	5	105	4 1/8	T-4[k]	G23	0.18	40	10000	250	—	86
F7TT/SPX27[b], F7TT/27K[d], Dulux[e]	7	135	5 5/16	T-4[k]	G23	0.18	45	10000	400	83	84
PL7/82[c]	7	133	5 1/4	T-4[m]	G23	0.18	45	10000	400	—	82
F9TT/SPX27[b], F9TT/27K[d], Dulux[e]	9	167	6 9/16	T-4[k]	G23	0.23	59	10000	600	85	84
PL9/82[d]	9	165	6 1/2	T-4[m]	G23	0.18	59	10000	600	—	82
Dulux-D[e]	9	116	4 9/16	T-4[k]	G24	0.185	66	10000	600	—	85
PL13/82[c]	13	188	7 7/16	T-4[m]	GX23	0.30	60	10000	900	—	82
F13TT/27K[d]	13	188	7 7/16	T-4[k]	GX23	0.30	60	10000	900	—	82
Dulux[e]	13	177	6 31/32	T-4[k]	GX23	0.30	60	10000	900	—	86
Dulux-D[e]	13	146	5 3/4	T-4[k]	G24	0.175	92	10000	900	—	85
2D[g]	16	135	5 5/16	T-4[q]	[r]	0.20	97	10000	1050	—	82
Dulux-D[e]	17	170	6 11/16	T-4[k]	G24	0.24	106	10000	1250	—	85
SL-18[c]	18	183	7 3/16	[n]	Med.	0.24	—	7500	1100	—	81
Dulux-T5[e]	18	251	9 7/8	T-5	TBA	0.38	57	10000	1250	—	85
FBU19T8[f]	19	222	8 3/4	T-8[o]	Med.	—	—	7500	[p]	—	—
Dulux T5[e]	24	362	14 1/4	T-5	TBA	0.34	91	10000	2000	—	85
Dulux-D[e]	25	190	7 1/2	T-4[k]	G24	0.30	114	10000	1800	—	85
Dulux T5[e]	35	443	17 7/16	T-5	TBA	0.43	108	10000	2900	—	85

[a] At three hours per start.
[b] General Electric.
[c] North American Philips Lighting.
[d] GTE Sylvania.
[e] Osram.
[f] Interlectric.
[g] Thorn EMI.
[h] 2700K color.
[k] U-shaped molded with square corners.
[m] Twin-tube with small connection near top.
[n] U-shaped bent into another U-shape and with enclosing plastic housing.
[o] U-shaped band.
[p] Bent to appear as 2 joined D shapes.
[q] Special recessed 2-pin base.
[r] Cool white 665, Warm white 675 and Tru-lite 520.

Fig. 8-118. Typical Cold-Cathode Instant-Starting Fluorescent Lamps

Manufacturers' designation	48T8			72T8			96T8			U6-96T8 (Hairpin)[d]		
ANSI designation	25-millimeter 45-Inch			25-Millimeter 69-Inch			25-Millimeter 93-Inch			——		
Lamp length[a]												
Millimeters	1125			1725			2325			1125[d]		
Inches	45			69			93			45[d]		
Lamp base	Ferrule			Ferrule			Ferrule			Ferrule		
Bulb	25-mm, T-8			25-mm, T-8			25-mm, T-8			25-mm, T-8		
Open-circuit volts (rms) for starting												
type L.P.	450			600			750			750		
type H.P.	600			750			835			835		
Lamp current (milliamperes)[b]	120	150	200	120	150	200	120	150	200	120	150	200
Lamp voltage—lamp type L.P.	250	240	—	330	310	—	420	400	—	420	400	—
Lamp voltage—lamp type H.P.	270	250	240	350	330	310	450	425	400	450	425	400
Lamp power (watts)												
type L.P.	26	30	—	34	40	—	42	49	—	42	49	—
type H.P.	28	33	40	37	43	52	46	54	65	46	54	65
Initial Lumens												
Warm white[c]	1100	1300	1600	1700	2000	2350	2300	2700	3400	2300	2700	3400
White (3500 K)[c]	1050	1250	1550	1650	1900	2300	2250	2650	3300	2225	2650	3300
White (4500 K)[c]	1000	1200	1500	1600	1850	2200	2200	2600	3200	2200	2600	3200
Daylight[c]	950	1150	1450	1550	1800	2150	2150	2550	3100	2150	2550	3100
Luminance (candelas per square meter)												
Warm white	4040	4800	5860	4350	5140	6100	4490	5240	6610	4490	5240	6610
White (3500 K)	3940	4590	5690	4250	5040	5930	4385	5140	6440	4385	5140	6440
White (4500 K)	3770	4385	5480	4145	4930	5650	4280	5070	5820	4280	5070	5820
Daylight	3730	4210	5310	4080	4830	5480	4210	4970	5650	4210	4970	5650
Rated lamp life (thousands of hours)[e]												
type L.P.	15	12.5	—	15	12.5	—	18	16	—	18	16	—
type H.P.	25	20	15	25	20	15	30	25	20	30	25	20

[a] Length of lamp without lamp holders. [b] Lamps can be operated up to 200 mA.
[c] Initial rating after 100 hours for types LP and HP. Other color lamps are available.
[d] Extended lamp length 2325 millimeters (93 inches) formed to U shape with 180-degree, 150-millimeter (6-inch) arc.
[e] Life not affected by number of starts.

Fig. 8-119. Typical Hot Cathode

Nominal lamp watts Nominal length (millimeters) (inches)	4 150 6	6 225 9	8 300 12	13 525 21	14 375 15	15 450 18
Bulb	T-5 (16 mm)	T-5 (16 mm)	T-5 (16 mm)	T-5 (16 mm)	T-12 (38 mm)	T-8 (25 mm)
Base (bipin)	min.	min.	min.	min.	med.	med.
Approx. lamp current (amperes)	0.17	0.16	0.145	0.165	0.38	0.305
Approx. lamp volts	29	42	57	95	40	55
Approx. lamp watts	4.5	6.0	7.2	13	14	14
Preheat curent-max. amperes	0.25	0.25	0.25	0.27	0.65	0.65
Preheat current-min. amperes	0.16	0.16	0.16	0.18	0.44	0.44
Rated life (hours)[b]	6000	7500	7500	7500	9000	7500
Lamp lumen depreciation (LLD)[c]	67	67	75	72	82	79
Initial lumens[d]						
Cool white	138	293	400	833	687	873
Deluxe Cool White	105	205	280	—	—	613
Warm White	127	290	400	880	710	880
Deluxe Warm White	85	180	275	580	476	610
White	138	275	405	880	710	880
Daylight	115	230	333	750	593	750
Factors for Calculating Luminance[e]						
Candelas per Square Meter	78.9	44.4	31.4	15.2	10.3	12.0

[a] The life and light output ratings of fluorescent lamps are based on their use with ballasts that provide operating characteristics. Ballasts that do not provide proper electrical values may substantially reduce either lamp life or light output, or both.

[b] Rated life under specified test conditions at three hours per start. At longer burning intervals per start, longer life can be expected.

[c] Per cent of initial light output at 70 per cent rated life at three hours per start. Average for cool white lamps. Approximate values.

[d] At 100 hours. Where lamp is made by more than one manufacturer, light output is the average of manufacturers. For the lumen output of other colors of fluorescent lamps, multiply the cool white lumens by the relative light output value from the table below Fig. 8-116.

Fig. 8-120. Typical Hot-Cathode

Nominal length (millimeters) (inches) Bulb	1050 42 T-6	1600 64 T-6	1800 72 T-8	2400 96 T-8	1200 48 T-12	1500 60 T-17
Base	Single Pin	Single Pin	Single Pin	Single Pin	Med. Bipin	Mog. Bipin
Approx. lamp current (amperes)	0.200[f]	0.200[f]	0.200[f]	0.200[f]	0.425	0.425
Approx. lamp volts	150	233	220	295	104	107
Approx. lamp watts	25.5	38.5	38	51	40.5	42
Rated life (hours)[b]	7500	7500	7500	7500	7500–12000	7500–9000
Lamp lumen depreciation (LLD)[c]	76	77	83	89	83	89
Initial lumens[d]						
Cool White	1835	3000	3030	4265	3100	2900
Deluxe Cool White	1265	2100	2100	2910		2020
Warm White	1875	3050	3015	4215	3150	2940
Deluxe Warm White	1275	2100				1990
White	1900	2945	3050	4225	3150	2940
Daylight	1605	2600	2650	3525	2565	2410
Factors for Calculating Luminance[e]						
Candelas per Square Meter	6.1	3.93	2.55	1.87	2.65	1.5

[a] The life and light output ratings of fluorescent lamps are based on their use with ballasts that provide proper operating characteristics. Ballasts that do not provide proper electrical values may substantially reduce either lamp life or light output, or both.

[b] Rated life under specified test conditions at three hours per start. At longer burning intervals per start, longer life can be expected.

[c] Per cent of initial light output at 70 per cent rated life at three hours per start. Average for cool white lamps. Approximate values.

[d] At 100 hours. Where lamp is made by more than one manufacturer, light output is the average of manufacturers. For the lumen output of other colors of fluorescent lamps, multiply the cool white lumens by the relative light output value from the table below Fig. 8-116.

Fluorescent Lamps (Preheat Starting)[a]

15 450 18	20 600 24	25 700 28	25 825 33	30 900 36	40 1200 48	90 1500 60
T-12 (38 mm) med.	T-12 (38 mm) med.	T-12 (38 mm) med.	T-12 (38 mm) med.	T-8 (25 mm) med.	T-12 (38 mm) med.	T-12 or T-17 mog.
0.325	0.38	0.46	0.46	0.355	0.43	1.5
47	57	63	61	99	101	65
14.5	20.5	25	25.5	30.5	40	90
0.65	0.65	0.95	0.95	0.65	0.75	2.2
0.44	0.44	0.41	0.41	0.40	0.55	1.45
9000	9000	7500	7500	7500	20000	9000
81	85	79	79	79	82	85
793	1270	1725	1915	2200	3150	6400
537	858	—	—	1555	2200	—
800	1310	—	1935	2235	3175	6350
532	848	—	—	1505	2165	5710
800	1290	—	1925	2255	3185	6350
650	1050	1450	1635	1900	2615	5525
8.0	5.5	4.68	3.93	5.4	2.65	2.1, 1.5[f]

[e] To calculate approximate lamp luminance, multiply the lamp lumens of the lamp color desired by the appropriate factor.
[f] First factor is for the T-12 lamp; second factor is for the T-17 lamp.
Note: All electrical and lumen values apply only under standard photometric conditions.

Fluorescent Lamps (Instant Starting)[a]

600 24 T-12	900 36 T-12	1050 42 T-12	1200 48 T-12	1500 60 T-12	1600 64 T-12	1800 72 T-12	2100 84 T-12	2400 96 T-12
Single Pin	Single Pin	Single Pin	Single Pin	Single Pin	Single Pin	Single Pin	Single Pin	Single Pin
0.425	0.425	0.425	0.425	0.425	0.425	0.425	0.425	0.425
53	77	88	100	123	131	149	172	197
21.5	30	34.5	39	48	50.5	57	66.5	75
7500–9000	7500–9000	7500–9000	7500–12000	7500–12000	7500–12000	7500–12000	7500–12000	12000
81	82	80	82	78	78	89	91	89
1065	2000	2350	3000	3585	3865	4650	5400	6300
			2065			3175		4465
1150	2050	2500	3000		3950	4675		6500
			2050			3200		4365
			3000			4700		6400
1010	1715	2015	2500	3135	3250	3850	4450	5425
5.54	3.58	3.04	2.65	2.1	1.95	1.7	1.44	1.24

[e] To calculate approximate lamp luminance, multiply the lamp lumens of the lamp color desired by the appropriate factor.
[f] These lamps can also be operated at 0.120 and 0.300 amperes.
Note: All electrical and lumen values apply only under standard photometric conditions.

Fig. 8-121. Fluorescent Lamp Power Requirements

Lamp Designation	Nominal Length (millimeters)	Nominal Length (inches)	Minimum Required rms Voltage across lamp for reliable starting[a]	Operating Current (milliperes)	Approximate Watts Consumed[i]: Lamp	Single Lamp Circuit: Ballast	Single Lamp Circuit: Total	Two Lamp Circuit: Ballast	Two Lamp Circuit: Total
Preheat Start									
4T5	150	6	[b]	170	4.5	2	6.5		
6T5	225	9	[b]	160	6	2	8		
8T5	300	12	[b]	145	7.2	2	9.2		
15T8	450	18	[b]	305	15	4.5	19.5	9	39
14T12	375	18	[b]	380	14	5	19	5	33
15T12	450	18	[b]	325	14.5	4.5	19	9	38
20T12	600	24	[b]	380	20.5	5	25.5	10	41
25T12	825	33	[b]	460	25.5	6	31.5		
13T5	525	21	180	165	13	5	18		
30T8	900	36	176	355	30.5	10.5	41	17	78
40T12	1200	48	176	430	40	12	52	16	96
90T17	1500	60	132	1500	90	20	110	24	204
Instant Start with Bipin Bases									
40T12	1200	48	385	425	40.5	23.5	64	25	106
40T12	1500	60	385	425	42	24	66		
40T17	1500	60	385	425	42	24	66	26	110
with Single Pin Bases (Slimline Lamps)									
42T6[d]	1050	42	405	200	25.5	13	38.5	21	72
64T6[d]	1600	64	540	200	38.5	16	54.5	22	99
72T8[d]	1800	72	540	200	38	16	54	31	107
96T8[d]	2400	96	675	200	51	16	67		
48T12 (lead-lag)	1200	48	385	425	39			26	104
(Series)	1200	48	385	425	39			17	95
72T12 (lead-lag)	1800	72	475	425	57			47	161
(Series)	1800	72	475	425	57			25	139
96T12 (lead-lag)	2400	96	565	425	75			40	190
(Series)	2400	96	565	425	75			22	172
96T12	2400	96	565	425	75				160
96T12	2400	96	565	440	60				128
Rapid Start[c]									
Lightly Loaded									
20T9[e]	165	6½	150	380	19	11.5	30.5		
22T9[e]	210	8¼	180	370	22.5	11.5	34		
25T8	900	36	170	265	25	5	30	8	58
28T12[j]	1200	48	231	330	28				60
32T8	1200	48	200	265	32	5	37	7	71
32T10[e]	305	12	200	425	33	12	45		
40T10[e]	406	16	205	415	41.5	12.5	54		
30T12	900	36	150	430	33.5	10.5	44		
"32"T12[j]	1200	48			32				76
40T8	1500	60	250	265	40	6	46	11	91
40T10	1200	48	200	420	41	13	54	13	95
40T12	1200	48	200	430	41	13	54	13	95
40W-U3[f]	600	24	200	420	41	13	54	13	95
40W-U6[f]	600	24	200	430	40.5	13	53.5	13	94
Medium Loaded									
24T12	600	24	[b]	800	37		70		115
48T12	1200	48	155	800	63		85		146
72T12	1800	72	260	800	87		106		200
84T12	2100	84	280	800	100		119		213
96T12	2400	96	296	800	113		140		252
Highly Loaded									
48T12, 48PG17	1200	48	160	1500	116		146		252
72T12, 72PG17	1800	72	225	1500	168		213		326
96T12, 96PG17	2400	96	300	1500	215		260		450

Continued on following page.

Fig. 8-121 *Continued*

Lamp Designation	Nominal Length (millimeters)	Nominal Length (inches)	Minimum Required rms Voltage across lamp for reliable starting[a]	Operating Current (milliperes)	Approximate Watts Consumed[i]: Lamp	Single Lamp Circuit: Ballast	Single Lamp Circuit: Total	Two Lamp Circuit: Ballast	Two Lamp Circuit: Total
Cold Cathode									
96T8LP	2400	96	750	120	42	9	51	17	101
				150	49				
				200	59	19	78	30	148
96T8HP	2400	96	835	120	46	12	58	24	115
				150	54			28	136
				200	65			30	160
Using Energy Saving Ballasts[g]									
Rapid Start[c]									
40T12	1200	48	200	430	40	7	47	6	86
40T12[h]	1200	48	200	460	34-35		41		74
Slimline									
96T12	2400	96	565	425	75				160
96T12[h]	2400	96	565	440	60				128
Rapid Start[c] Medium Loaded									
96T12	2400	96	296	800	113				237
96T12[h]	2400	96	296	840	95				207

[a] Between 10°C (50°F) and 43°C (110°F). The voltage shown is that required across one lamp. A ballast to operate two rapid start lamps in series will require an open circuit voltage roughly 40 per cent higher than shown here. Consult lamp manufacturer for information.
[b] Suitable for operation on 120-watt ac lines with series reactor as ballast.
[c] Requires starting aid.
[d] T-6 and T-8 slimline lamps also operate at 120 and 300 milliamperes.
[e] Circular lamp-dimension given is nominal outside diameter.
[f] U-shaped lamp.
[g] Energy saving CBM rated ballasts are designed to operate standard fluorescent lamps at CBM rated light output with lower ballast losses than the usual fluorescent lamp ballasts. The data shown here are typical for this type of ballast for the three most common fluorescent lamps and the energy saving fluorescent lamp that can be used as a replacement.
[h] Energy saving fluorescent lamp.
[i] Lamp watts and light output will vary with lamp temperature.
[j] Cathode-cutout type.

Fig. 8-122. Representative Fluorescent Lamp Ballast Factors

Lamp	Ballast: Standard	Ballast: Low Loss	Ballast: High Performance
4-Foor Rapid Start System (F40T12)			
Standard	0.95	0.95	0.97
Reduced Wattage	0.89	0.87	0.95
8-Foot Slimline Systems (F96T12)			
Standard	0.94	0.93	0.97
Reduced Wattage	0.87	0.85	0.96
8-Foot High Output Systems			
Standard	0.98	—	1.03
Reduced Wattage	0.93	—	0.98

Fig. 8–123. Typical High-Intensity Discharge Lamps*
A. Mercury Lamps

ANSI Lamp Designation[i]	Manufacturer's Lamp Designation[j]	ANSI Ballast Number	Lamp Watts	Approximate Ballast Watts	Approximate Lumens** (vertical)		Bulb[h]	Outer Bulb Finish	Maximum Over-All Length		Light Center Length		Arc Length		Base	Average Rated Life (hours)	Lamp Voltage and Current
					Initial (100 hours)	Mean			Millimeters	Inches	Millimeters	Inches	Millimeters	Inches			
H45AY-40/50 DX	H40/50 DX 45-46[a]	45	40	7–12	1140	910	E17, ED17	Phos. coat	130	5⅛	79	3⅛	—	—	Med.	16000	A
H45/46DL-40/50DX	H45/46 DL-40/50/DX[c]	46	50		1575	1260	E17	Phos. coat	130	5⅛	79	3⅛	—	—	Med.	16000	A
	H45AY-40/50DX[t]	45	40		1350	1070	B21	Phos. coat	165	6½	95	3¾	—	—	Med.	16000	A
		46	50		1680	1330	B21	Phos. coat	165	6½	95	3¾	—	—	Med.	16000	A
H46DL-40-50/DX	H46DL-40-50/DX[b]	45	40	7–12	1140	910	B17	Phos. coat	130	5⅛	79	3⅛	—	—	Med.	24000+	A
		46	50		1575	1260	B17	Phos. coat	130	5⅛	79	3⅛	—	—	Med.	24000+	A
H46DL-40-50/N	H46DL-40-50/N[b]	45	40		800	580	B17	Phos. coat	130	5⅛	79	3⅛	—	—	Med.	24000+	A
		46	50		1100	830	B17	Phos. coat	130	5⅛	79	3⅛	—	—	Med.	24000+	A
H43AZ-75	H43AZ-75[b]	43	75	8.0	2800	2400	B21	Clear	165	6½	95	3¾	27	1 1/16	Med.	24000	B
H43AV-75/DX	H75DX43[a] H43AV-75/DX[b,c]	43	75	8.0	3000	2430	B17,21 ED17	Phos. coat	165	6½	95	3¾	—	—	Med.	24000	B
H43AV-75/N	H43AV-75/N[b]	43	75	8.0	2800	2200	B21, ED17	Phos. coat	165	6½	95	3¾	—	—	Med.	24000+	B
H43/44-75/100/PFL	H75/100PFL43-44/4[a] H43/44KL-75-100[b] H43/44-75/100/PFL[c]	43	75	8.0	1780	1165	PAR38	Clear	110	4 11/32	—	—	—	—	Med. side prong	16000	B
		44	100	10–35	2585	1755	PAR38	Clear	110	4 11/33	—	—	—	—	Med. side prong	16000	B
H38AV-100/DX/N	H38AV-100/DX/N[c]	38	100	10–35	4400	3560	ED17	Phos. coat	165	6½	95	3¾		—	Med.	24000+	C
H38BM-100	H100RFL38-4[a] H38BM-100[b]	38–4	100	10–35	2850	2250	R40	I.F. Refl.	178–191	7-7½		—		—	Med.	24000+	C
H38BP-100/DX	H100RDXFL38-4[a] H38BP-100/DX[b]	38–4	100	10–35	2865	2300	R40	Phos. Refl.	178 191	7 7½		—		—	Med.	24000+	C
H38BP-100/N	H38BP-100/N[b]	38–4	100	10–35	2450	1950	R40	Phos. Refl.	191	7½		—		—	Med.	24000+	C

H38HT-100	H100A38-4[a] H38HT-100[b]	38–4	100	10–35	4040	3415	BT25 or E23½	Clear	191	7½	127	5	29	1⅛	Mog.	24000+	C
H38JA-100/C	H38JA-100/c[c]	38–4	100	10–35	4100	3230	BT25	Phos. coat	191	7½	127	5		—	Mog.	24000	C
H38JA-100/DX	H100DX38-4[a] H38JA-100/DX[b,c]	38–4	100	10–35	4425	3620	BT25 or E23½	Phos. coat	191	7½	127	5		—	Mog.	24000+	C
H38JA-100/N	H38JA-100/N[b,c]	38–4	100	10–35	4600	3700	BT25 ED23½	Phos. coat	191	7½	127	5		—	Mog. Med.	24000+	C
H38JA-100/WDX	H100WDX38-4[a] H38JA-100/WDX[d]	38–4	100	10–35	4000	3100	BT25 or E23½	Phos. coat	191	7½	127	5		—	Mog.	24000+	D
H38LL-100	H100A38-4/A23[a] H38LL-100[b]	38–4	100	10–35	3900	3225	A23	Clear	138	5 7/16	89	3½	29	1⅛	Med.	24000	C
H38MP-100/DX	H100DX38-4/A23[a] H38MP-100/DX[b,c]	38–4	100	10–35	4275	3500	A23	Phos. coat	138	5 7/16	89	3½		—	Med.	24000	C
H38MP-100/N	H38MP-100/N[b,c]	38–4	100	10–35	4195	3185	A23	Phos. coat	138	5 7/16	89	3½		—	Med.	24000+	C
H44GS100	H100PSP44-4[a] H44GS-100[b,c]	44–4	100	10–35	2585	1820	PAR38	Refl. spot	138	5 7/16		—		—	Admed.	16000	C
H44JM100	H100PFL44-4[a] H44JM-100[b,c]	44–4	100	10–35	2585	1820	PAR38	Refl. Flood	138	5 7/16		—		—	Admed.	16000	C
H44AB100	H100A4/T[a] H44AB100[b]	4	100	10–35	3725	2740	T10	Clear	143	5⅝	87	3 7/16		—	Admed.	12000	C
H39BM-175	H175RFL39-22[a] H39BM-175[b]	39–22	175	15–35	6030	5075	R40	I.F. Refl.	178–191	7–7½		—		—	Med.	24000+	D
H39BN-175/DX	H39BM-175/DX[c]	39–22	175	15–35	5800	4900	R40	Phos. Refl.	191	7½		—		—	Med.	24000+	D
H39BP-175/DX	H175RDXFL39-22[a] H39BP-175/DX[b,c]	39–22	175	15–35	5715	4685	R40	Phos. I.F. Refl.	178–191	7–7½		—		—	Med.	24000+	D
H39BP-175/N	H39BP-175/N[b]	39–22	175	15–35	4600	3650	R40	Phos. I.F. Refl.	191	7½		—		—	Med.	24000+	D
H39BS-175	H175RFL39-22/M[a] H39BS-175[b]	39–22	175	15–35	5830	5075	R40	I.F. Refl.	178–191	7–7½		—		—	Mog.	24000+	D
H39BV-175/DX	H39BV-175/DX[b]	39–22	175	15–35	5725	4800	R40	Phos. Refl.	191	7½				—	Mog.	24000+	D

Fig. 8–123—*Continued*

ANSI Lamp Designation[i]	Manufacturer's Lamp Designation[j]	ANSI Ballast Number	Lamp Watts	Approximate Ballast Watts	Approximate Lumens** (vertical)		Bulb[h]	Outer Bulb Finish	Maximum Over-All Length		Light Center Length		Arc Length		Base	Average Rated Life (hours)	Lamp Voltage and Current
					Initial (100 hours)	Mean			Millimeters	Inches	Millimeters	Inches	Millimeters	Inches			
H39KB-175	H175A39–22[a] H39KB-175[b,c,g]	39–22	175	15–35	7975	7430	BT28 or E28	Clear	211	8 5/16	127	5	51	2	Mog.	24000+	D
H39KC-175/C	H39KC-175/C[c,b]	39–22	175	15–35	7850	7140	BT28	Phos. coat	211	8 5/16	127	5		—	Mog.	24000	D
H39KC-175/DX	H175DX39–22[a] H39KC-175/DX[b,c,t,g]	39–22	175	15–35	8600	7640	BT28 or E28	Phos. coat		8 5/16	127	5		—	Mog.	24000+	D
H39KC175/N	H39KC-175/N[b,c]	39–22	175	15–35	8800	7830	BT28	Phos. coat	211	8 5/16	127	5		—	Mog.	24000+	D
H39KC175/WDX	H175WDX39-22[a]	39–22	175	15–35	7650	6600	BT28 or E28	Phos. coat	211	8 5/16	127	5		—	Mog.	24000+	D
H37FS-250/DX	H37FS-250/DX[b]	37–5	250	25–35	8000		R60	Phos. Refl.	257	10 1/8		—		—	Mog.	24000	E
H37KB-250	H250A37-5[a] H37KB-250[b,c,g]	37–5	250	25–35	11825	10625	BT28 or E28	Clear	211	8 5/16	127	5	57	2 1/4	Mog.	24000+	E
H37KC-250/C	H37KC-250/C[c,b]	37–5	250	25–35	11850	10540	BT28	Phos. coat	211	8 5/16	127	5		—	Mog.	24000+	E
H37KC-250/DX	H250DX37-5[a] H37KC-250/DX[b,c,t,g]	37–5	250	25–35	12775	10790	BT28 or E28	Phos. coat	211	8 5/16	127	5		—	Mog.	24000+	E
H37KC-250/N	H37KC-250/N[b,c]	37–5	250	25–35	12345	9335	BT28	Phos. coat	211	8 5/16	127	5		—	Mog.	24000+	E
H37KC-250/WDX	H37KC-250/WDX[a]	37–5	250	25–35	10750	8950	BT28 or E28	Phos. coat	211	8 5/16	127	5		—	Mog.	24000+	E
H33CD-300[b] H33GL-300/DX[b]	H33CD-300[b] H33GL-300/DX[b]	33 33	300 300	— —	14000 15700	— —	BT37 BT37	Clear Phos. coat	292 292	11 1/2 11 1/2	178 178	7 7		— —	Mog. Mog.	16000+ 16000+	— —
H33GL-300/N[b]	H33GL-300/N[b]	33	300	—	1300	—	BT37	Phos. coat								16000+	—
H33AR-400	H400A33-1/T16[a] H33AR-400[c]	33–1	400	20–55	19750	18200	T16	Clear	279	11	178	7	70	2 3/4	Mog.	12000	F
H33CD-400	H400A33-1[a] H33CD-400[b,c,g]	33–1	400	20–55	21000	19150	BT37 or E37	Clear	292	11 1/2	178	7	70	2 3/4	Mog.	24000+	F
H33DN400/C	H33DN400/C[c]	33–1	400	20–55	21000	18800	R57	Phos. Refl.	324	12 3/4		—			Mog.	24000	F

H33DN400/DX	H400RDX33-1[a] H33DN400/DX[b,c]	33–1	400	20–55	22670	19370	R52 or R57	Phos. Refl.	298	11¾ 12¾		—			Mog.	24000+	F
H33FP-400	H400RSP33-1[a]	33–1	400	20–55	15300	12200	R60	Clear Refl.	257	10⅛		—			Mog.	24000+	F
	H33FP-400[d]																
H33FS-400/DX	H400RDXFL33-1[a] H33FS-400/DX[b,c]	33–1	400	20–55	15775	12810	R60	Phos. Refl.	257	10⅛ 10⅞		—			Mog.	24000+	F
H33FY-400	H400R33–1[a] H33FY-400[b,c]	33–1	400	20–55	18270	16670	R52 or R57	I.F. Refl.	298	11¾ 12¾		—			Mog.	24000+	F
H33GL-400/C	H33GL-400/C[c]	33–1	400	20–55	20500	18570	BT37	Phos. coat	292	11½	178	7			Mog.	24000+	F
H33GL-400/DX	H400DX33-1[a] H33GL-400/DX[b,c,t,g]	33–1	400	20–55	23125	19840	BT37 or E37	Phos. coat	292	11½	178	7			Mog.	24000+	F
H33GL-400/DX/BT	H400DX33-1/BT[a]	33–1	400	20–55	22100	18800	BT37	Phos. coat	287	11 5/16	178	7			Mog.	24000+	F
H33GL-400/N	H33GL-400/N[b,c]	33–1	400	20–55	23000	20000	BT37	Phos. coat	292	11½	178	7			Mog.	24000+	F
H33GL-400/WDX	H400WDX33-1[a]	33–1	400	20–55	21500	18000	BT37 or E37	Phos. coat	292	11½	178	7			Mog.	24000+	F
H33HS-400 H33LN400	H33HS-400[b] H33LN-400[b,c]	33–1 33–1	400 400	20–55 20–55	17500 16660	15200 15300	R57 R60	I.F. Refl. Clear Refl.	324 276	12¾ 10⅞		— —			Mog. Mog.	24000+ 24000+	F F
H35NA-700	H700A35-18[a] H35NA-700[b,c]	35–18	700	35–65	40500	36250	BT46	Clear	368	14½	241	9½	127	5	Mog.	24000+	G
H35ND-700/C	H35ND-700/C[c]	35–18	700	35–65	41000	36490	BT46	Phos. coat	368	14½	241	9½		—	Mog.	24000+	G
H35ND-700/DX	H700DX35-18[a]	35–18	700	35–65	42750	36045	BT46	Phos. coat	368	14½	241	9½		—	Mog.	24000+	G
	H35ND-700/DX[b,c]																
H35ND-700/N	H35ND-700/N[b]	35–18	700	35–65	36400	29850	BT46	Phos. coat	368	14½	241	9½		—		24000+	G
H34GW-1000/C	H346W-1000/C[c]	34–12	1000	40–100	57500	49450	BT56	Phos. coat	391	15⅜	241	9½		—	Mog.	24000	H
H34GW-1000/DX	H1000DX34-12[a]	34–12	1000	40–100	61670	47670	BT56	Phos. coat	391	15⅜	241	9½		—	Mog.	24000+	H
	H34GW-1000/DX[b,c]																
H34GV-1000	H1000A34-12[a]	34–12	1000	40–100	55330	48480	BT56	Phos. coat	391	15⅜	241	9½		—	Mog.	24000+	H
	H34GV-1000[b,c,g]																
H36FA-1000/DX	H1000RDXFL36-15[a]	36–15	1000	40–90	48330	30100	R80	Phos. Refl.	352	13⅞		—		—	Mog.	24000+	I

Fig. 8-123—*Continued*

ANSI Lamp Designation[i]	Manufacturer's Lamp Designation[j]	ANSI Ballast Number	Lamp Watts	Approximate Ballast Watts	Approximate Lumens** (vertical)		Bulb[h]	Outer Bulb Finish	Maximum Over-All Length		Light Center Length		Arc Length		Base	Average Rated Life (hours)	Lamp Voltage and Current
					Initial (100 hours)	Mean			Millimeters	Inches	Millimeters	Inches	Millimeters	Inches			
	H36FA-1000/DX[b]																
H36FB-1000	H1000RFL36-15[a] H36FB-1000[b]	36–15	1000	40–90	43500	33130	R80	I.F. Refl.	352	13⅞		—		—	Mog.	24000+	I
H36HR-1000/DX	H1000RSDX36-15[a]	36–15	1000	40–90	43000	25200	R40	Semi Phos. Refl.	352	17⅞		—		—	Mog.	24000+	I
H36GV-1000	H1000A36-15[a] H36GV-1000[b,c,g]	36–15	1000	40–90	56150	48400	BT56	Clear	391	15⅜	241	9½	152	6	Mog.	24000+	I
H36GW-1000/C	H36GW-1000/C[c]	36–15	1000	40–90	55000	46200	BT56	Phos. goat	391	15⅜	241	9½		—	Mog.	24000+	I
H36GW-1000/DX	H1000DX36-15[a]	36–15	1000	40–90	63000	48380	BT56	Phos. coat	391	15⅜	241	9½		—	Mog.	24000+	I
	H36GW-1000/DX[b,c,g]																
H36GW-1000/N	H36GW-1000/N[b]	36–15	1000	40–90	53500	38000	BT56	Phos. coat	391	15⅜	241	9½		—	Mog.	24000+	I
H36GW-1000/R	H36GW-1000/R[b]	36–15	1000	40–90	63000	44700	BT56	Phos. coat	391	15⅜	241	9½		—	Mog.	24000+	I
H36GW-1000/WDX	H1000WDX36-15[a]	36–15	1000	40–90	58000	39440	BT56	Phos. coat	383	15$\frac{1}{16}$	241	9½		—	Mog.	24000+	I
H36KY-1000/DX	H1000RDX36-15[a]	36–15	1000	40–90	58250	38750	BT56	Semi. Phos. Refl.	391	15⅜	241	9½		—	Mog.	24000+	I
	H36KY-1000/D[b]																

* The life and light output ratings of mercury lamps are based on their use with ballasts that provide proper operating characteristics. Ballasts that do not provide proper electrical values may substantially reduce either lamp life or light output, or both. Unless otherwise noted, ratings apply to operation in ac circuits.

** Average of manufacturers' rating.

[a] General Electric. [b] North American Philips Lighting. [c] GTE Sylvania. [e] Osram. [f] Westron. [t] Action Tungsram.

[g] Duro-Test (T versions only).

[h] All bulbs are of heat resistant glass.

[i] The basic ANSI designation adds an R in front of the wattage designation for non-self extinguishing mercury lamps and a T for self extinguishing mercury lamps to comply with Federal Standard #21 CFR 1040.30 See page 8-41. The system does not distinguish between different bulb coatings for the same lamp type. Each manufacturer adds a suffix to identify the coating.

[j] The R or T designation is added after the H for manufacturer[a], and in front of the wattage for manufacturer [b,c,d,&g].

Lamp Voltage and Current for Mercury Lamps

	Lamp Voltage (volts)			Lamp Current (amperes)	
	Minimum for starting at −29°C (−20°F)		Operating	Starting	Operating
	rms	peak			
A	180	255	90	0.80	0.53
B	225	320	130	0.92	0.64
C	225	320	130	1.3	0.85
D	225	320	130	2.5	1.5
E	225	320	130	3.1	2.1
F	225	320	135	5.0	3.2
G	375	530	265	5.0	2.8
H	300	425	135	12.0	8.0
I	375	375	265	6.5	4.0

Fig. 8-123—*Continued*
B. Metal Halide Lamps*

ANSI Lamp Designation[p]	Manufacturer's Lamp Designation[p]	ANSI Ballast Number	Lamp Watts	Approximate Lumens				Bulb	Outer Bulb Finish	Maximum Over-All Length		Light Center Length		Arc Length		Base	Rated Life (hours)	Lamp Voltage and Current
				Initial (100 hours)		Mean				Millimeters	Inches	Millimeters	Inches	Millimeters	Inches			
				Vert.	Horiz.	Vert.	Horiz.											
M85PX-70	HQI-TS70W/WDL[e]	M85	75		5000		4000	T-6.3	Clear	114	$4\frac{1}{2}$	57	$2\frac{1}{4}$	7	$\frac{9}{32}$		15000	
M81PS-150	HQI-TS150W/NDL[e]	M81	150		11250		9000	T-7.2	Clear	138	$5\frac{7}{16}$	69	$2\frac{23}{32}$	18	$\frac{23}{32}$		15000	
M57PE-175/X	MXR175/BU,BD[a,g]	M57	175	16600		13300		E-$23\frac{1}{2}$	Clear	197	$7\frac{3}{4}$	127	5	—	—	Mog.	10000	A
M57PF-175/X	MXR175/C/BU,BD[a,g]	M57	175	15750		12600		E-$23\frac{1}{2}$	Diffuse	197	$7\frac{3}{4}$	127	5	—	—	Mog.	10000	A
—	MS175/3K/HOR[h,c]	M57	175		14000		10500	BT-28	Phos. Coat	211	$8\frac{5}{16}$	127	5	—	—	Mog.[q]	10000	A
M57PE-175	MV175/U[a,k] MH175/4[b,k,g] M175BU,BD-only[c,k,g]	M57	175	14000	12000	10800	8300	BT-28 or E-28	Clear	211	$8\frac{5}{16}$	127	5	25	1	Mog.	7500	A
M57PF-175	MV175/C/U[a,k] MH175/C[b,k,g] M175/C/BU,BD-only[c,k,g]	M57	175	14000		10200		BT-28 or E-28	Phos Coat	211	$8\frac{5}{16}$	127	5	—	—	Mog.	7500	A
—	MS175/HOR[c,h]	M57	175		15000		12000	BT-28	Clear	211	$8\frac{5}{16}$	127	5	—	—	Mog.[q]	10000	A
—	MS175/C/HOR[c,h]	M57	175		15000		11300	BT-28	Phos. Coat	211	$8\frac{5}{16}$	127	5	—	—	Mog.[q]	10000	
M58PG-250	MV250/U[a,k] MH250/U[b] M250U[c,k]	M58	250	20500	19500	17000	14000	BT-28 or E-28	Clear	211	$8\frac{5}{16}$	127	5	36	1.4	Mog.	10000	B
M58PH-250	MV250/C/U[a,k] MH250/C/U[b] M250/C/U[c,k]	M58	250	20500	19500	16000	13500	BT-28 or E-28	Phos. Coat	211	$8\frac{5}{16}$	127	5	—	—	Mog.	10000	B

Fig. 8-123—*Continued*
B. Metal Halide Lamps*

ANSI Lamp Designation[p]	Manufacturer's Lamp Designation[p]	ANSI Ballast Number	Lamp Watts	Approximate Lumens				Bulb	Outer Bulb Finish	Maximum Over-All Length		Light Center Length		Arc Length		Base	Rated Life (hours)	Lamp Voltage and Current
				Initial (100 hours)		Mean				Millimeters	Inches	Millimeters	Inches	Millimeters	Inches			
				Vert.	Horiz.	Vert.	Horiz.											
M80PR-250[e]	HQI-TS250W/D[e]	M80	250		19000		13300	T-8	Clear	165	6½	83	3¼	27	1 1/16	FC2[r,s]	15000	C
M88PN-250	HQI-T250W/D[e]	M88	250		19000		13300	T-14½	Clear	220	8 21/32	150	5 29/32	27	1 1/16	Mog.	15000	C
M80PR-250	HQI-TS250W/NDL[e]	M80	250		20000		16000	T-8	Clear	165	6½	83	3¼	27	1 1/16	FC2[r,s]	15000	C
—	913[d]	S50	250		19000			T-14½	Clear	222	8¾	146	5¾	—	—	Mog.	10000	
—	HGMI250/D[w]	—	250	20000				T-14	Clear	320	12 19/32	150	5 29/32	—	—	Mog.	12000	
—	HGMIL250/D[w]	—	250	14000				E-28	Phos. Coat	227	8 15/16	—	—	—	—	Mog.	12000	
—	MS250/3K/HOR[c,h]	M58	250		21500		16500	BT-28	Phos. Coat	211	8 5/16	127	5	36	1 13/32	Mog[q]	10000	
—	MS250/HOR[c,h]	M58	250		23000		18000	BT-28	Clear	211	8 5/16	127	5	36	1 11/32	Mog.[q]	10000	
—	MS250/C/HOR[c,h]	M58	250		23000		17000	BT-28	Phos. Coat	211	8 5/16	127	5	36	1 11/32	Mog.[q]	10000	
—	MH300BU,BD/4[b]	M59	300	24000				BT-37	Clear	292	11½	178	7	—	—	Mog.	10000	—
—	MH300BU,BD/C[b]	M59	300	24000				BT-37	Phos. Coat	292	11½	178	7	—	—	Mog.	10000	—
—	MV325/I/U/WM[a,l]	M33[i]	325	28000	26000	18200	16400	E-37	Clear	178	7	287	11 5/16	27	1 1/16	Mog.	15000	E
—	MV325/C/I/U/WM[a,l]	H33[i]	325	28000	26000	17600	15800	E-37	Phos. Coat	178	7	287	11 5/16	27	1 1/16	Mog.	15000	E
M86PZ-360	HQI-T400W/DH/DV[e]	M86	360	28000	25000	19600	17500	T-14½	Clear	285	11 7/32	175	6⅞	48	1 29/32	Mog.	15000	F
M86PY-360	HQI-TS400W/D[e]	M86	360		25000		17500	T-9½	Clear	206	8⅛	103	4 1/16	48	1 29/32	FC2[r,s]	15000	F
M59PJ-400	MV400/U[a,l] MH400/U[b,l] M400/U[c,n]	M59	400	34000	32000	25600	22600	BT-37 or E-37	Clear	178	7	292	11½	33	1.3	Mog.	15-20000	D
M59PK-400	MV400/C/U[a,l] MH400/C[b] M400/C/U[c,n]	M59	400	34000	32000	24600	22600	BT-37 or E-37	Phos. Coat	178	7	292	11½	—	—	Mog.	15-20000	D
M59PL-400	MH400/E[b]	M59	400	34000		25600		E-18	Clear	248	9¾	146	5¾	—	—	Mog.	10000	D
—	MS400/HOR-only[c,h]	M59	400	[h]	40000		32000	BT-37	Clear	292	11½	178	7	—	—	Mog.[q]	20000	D
—	MS400/C/HOR-only[c,h]	M59	400	[h]	40000		31000	BT-37	Phos. Coat	292	11½	178	7	—	—	Mog.[q]	20000	D
M59PJ-400	MS400/BD,BU-only[c]	M59	400	32000				BT-37 or E-37	Clear	292	11½	178	7	—	—	Mog.	15000	D
—	MS400/C/BD,BU-only[c]	M59	400	31000				BT-37 or E-37	Phos. Coat	292	11½	178	7	—	—	Mog.	15000	D
M59PJ-400/ VBD,VBU	MV400/VBD,VBU[a,g] MS400/BU/BD-only[c,t]	M59	400	40000		32000		BT-37 or E-37	Clear	287	11 5/16	178	7	33	1.3	Mog.	2000	D
M59PK-400/ VBD,VBU	MV400/C/VBD,VBU[a,g] MS400/C/BU/BD-only[c,t]	M59	400	40000		31000		BT-37 or E-37	Phos. Coat	287	11 5/16	178	7	—	—	Mog.	20000	D

M59PJ-400/I/U	MV400/I/U[a,l]	H33M59	400	34000	32000	20400	18400	BT-37 or E-37	Clear	287	11 5/16	178	7	33	1.3	Mog.	15000	D
M59PK-400/I/U	MV400/C/I/U[a,l]	H33M59	400	34000	32000	19600	17600	BT-37 or E-37	Phos. Coat	287	11 5/16	178	7	—	—	Mog.	15000	D
—	MV400/BD,BU/I[a,g]	H33[m]M59	400	34000		26500		BT-37 or E-37	Clear	287	11 5/16	178	7	—	—	Mog.	15000	D
	MS400/3K/HOR-only[c]	M59	400		37000		28000	BT-37	Phos. Coat	292	11½	178	7	—	—	Mog.[q]	20000	D
M59PK-400	MS400/3K/BU-only[c]	M59	400	37000		28000		BT-37	Phos. Coat	292	11½	178	7	—	—	Mog.[q]	20000	D
	914[d]	S51	400		28000			T-14½	Clear	250	9⅞	146	5¾	—	—	Mog.	10000	
	M750/BU-only[c,t]		750	83500		66800		BT-37	Clear	292	11½	178	7	—	—	Mog.	5000	
M47PA-1000	MV1000/U[a,f] MV1000/VBU,VBD[a,g] MH1000/U[b] M1000/U[c,n]	M47	1000	11000		88000		BT-56	Clear	391	15⅜	241	9½	89	3½	Mog.	10-12000	G
M47PB-1000	MV1000/C/U[a,l] MH1000/C/U[b] M1000/C/U[c,h]	M47	1000	10500		82000		BT-56	Phos. Coat	391	15⅜	241	9½			Mog.	10-12000	G
M47PA-1000/BD,BU/I	MV1000/BD,BU[a,l]	H36,M47	1000	115000		92000		BT-56	Clear	383	15 1/16	238	9⅜	89	3½	Mog.	10000	H
—	MS1000/BD,BU-only[c,g]	M47	1000	125000		100000		BT-56	Clear	391	15⅜	241	9½	—	—	Mog.	10000	H
—	MS1000/C/BD,BU-only[c,g]	M47	1000	125000		95800		BT-56	Phos. Coat	391	15⅜	241	9½	—	—	Mog.	10000	H
—	HQI-T1000W/D[e]		1000		90000		63000	T-25	Clear	340	13¼	220	8⅝	70	2¾	Mog.	10000	I
M48PC-1500/-	MV1500/HBU/E,HBD/E[a] MH1500/BD,BU[b] M1500/BD,BU-HOR[c,k]	M48	1500	15500		142000		BT-56	Clear	391	15⅜	214	9½	89	3½	Mog.	3000	J
—	MW1500T-7/7H[c,k]	—	1500		150000		130000	T-7	Clear	256	10 1/16			178	7	RSC[s]	3000	—
—	MB1L[f]	—	1500	—	140000			T-8	Frosted	256	10.1	128	5			RSC[s]	4000	K
M82PT-2000	HQI-T2000W/D[e]	M82	2000		170000		119000	T-32	Clear	430	19 5/16	260	10¼	105	4⅛	E40[r,s]	3000	L
M82PW-3500	HQI-T3500W/D[e]	M84	3500		300000		210000	T-32	Clear	430	19 5/16	260	10¼	150	5 15/16	E40[r,s]	2500	M

* The life and light output ratings of metal halide amps are based on their use with ballasts that provide proper operating characteristics. Ballasts that do not provide proper electrical values may substantially reduce either lamp life or light output, or both. Unless otherwise noted, ratings apply to operation in ac circuits.

[a] General Electric. [b] North American Philips Lighting. [c] GTE Sylvania. [d] Duro-Test. [e] Osram. [f] Thorn. [w] Action Tungsram.

[g] ±15° of vertical operation and enclosed luminaires only.

[h] ±15° horizontal operation and enclosed luminaires only.

[i] CW/CWA mercury ballasts only.

[j] Minimum starting volts at −18°C (0°F) and crest factor = 2.0.

[k] Operation restricted to enclosed luminaires only.

[l] Open of enclosed luminaires when operated within ±15° vertical, all other enclosed.

[m] Do not use on CW or CWA mercury ballasts, consult lamp manufacturer.

[n] Enclosed operation when operated horizontal or ±60° of horizontal.

[o] Base up to 15° below horizontal only (HBU), base down to 15° above horizontal only (HBD).

Fig. 8-123—*Continued*
B. Metal Halide Lamps*

[p] The basic ANSI designation adds R in front of the wattage designation for self-extinguishing metal halide lamps and a T for self-extinguishing metal halide lamps to comply with Federal standard #21 CFR 1040.30. See page 8-42.
[q] Position oriented mogul screw base.
[r] European base.
[s] Double ended construction.
[t] Operation restricted to ±15° of vertical only.

	Lamp Voltage (volts)			Lamp Current (amperes)	
	Minimum for starting at −29°C (−20°F)		Operating	Starting	Operating
	rms	peak			
A	382	540	130	1.8	1.4
B	382	540	130	2.8	2.1
C	198	3500	100	6.0	3.0
D	382	540	135	5.0	3.2
E	190[j]	380[j]	111	5.0	3.5
F	198	3500	120	7.0	3.5
G	530	750	250	6.0	4.3
H	440	622	250	6.0	4.3
I	230	4500	125	19.0	9.5
J	530	750	270	9.0	6.0
K	340	2000	250	—	6.7
L	380	4500	230	22.0	10.3
M	380	4500	220	38.0	18.0

Fig. 8-123—*Continued*
C. High Pressure Sodium Lamps

ANSI Lamp Designation	Manufacturer's Lamp Designation	ANSI Ballast Number	Lamp Watts*	Approximate Lumens		Bulb	Outer Bulb Finish	Maximum Over-All Length		Light Center Length		Base	Average Rated Life (hours)[j]	Lamp Voltage and Current
				Initial	Mean[j]			(millimeters)	(inches)	(millimeters)	(inches)			
S76HA-35	LU35/MED[a,c,t] C35S76/M[b]	S76	35	2250	2025	E-17, B-17 or ED-17	Clear	138	5 7/16	87	3 7/16	Med.	16000	A
S76HB-35	LU35/D/MED[a,c,t] C35S76/D/M[b]	S76	35	2150	1935	E-17, B-17 or ED-17	Diffuse	138	5 7/16	87	3 7/16	Med.	16000	A

S68MS-50	{LU50[a,c] C50S68[b]}	S68	50	4000	3600	BT-25 or E-23½	Clear	197	7¾	127	5	Mog.	24000	A
S68MT-50	{LU50/D[a] C50S68/D[b]}	S68	50	3800	3420	BT-25 or E-23½	Diffuse	197	7¾	127	5	Mog.	24000	A
—	{LU50/MED[a,c,t] C50S68/M[b]}	S68	50	4000	3600	E-17 or B-17	Clear	138	5 7/16	87	3 7/16	Med.	24000	A
—	{LU50/D/MED[a,c,t] C50S68/D/M[b]}	S68	50	3800	3420	E-17 or B-17	Diffuse	138	5 7/16	87	3 7/16	Med.	24000	A
S62ME-70	{LU70[a,c] C70S62[b]}	S62	70	6000	5500	BT-25 or E-23½	Clear	197	7¾	127	5	Mog.	24000	B
S62MF-70	{LU70/D[a,c] C70S62/D[b]}	S62	70	5800	5000	BT-25 or E-23½	Diffuse	197	7¾	127	5	Mog.	24000	B
—	{LU70/(MED)[c,t] C70S62/M[b]}	S62	70	6000	5500	E-17 or B-17	Clear	138	5 7/16	87	3 7/16	Med.	24000	B
—	C70S62/PAR[b]	S62	70	4500	—	PAR-38		110	4 5/16		—	Med. Side Prong	7500	B
—	C70S62/PAR/M[b]	S62	70	4500	—	PAR-38	—	135	5 5/16	—	—	Med.	7500	B
S62LF-70	{LU70/D/MED[c,t] C70S62/D/M[b]}	S62	70	5985	5390	B-17	Diffuse	140	5½	—	—	Med.	24000	B
S54SB-100	{LU100[a,c,t] C100S54[b]}	S54	100	9500	8550	BT-25 or E-23½	Clear	197	7¾	127	5	Mog.	24000	C
S54MC-100	{LU100/D[a,c,t] C100S54/D[b]}	S54	100	8800	7920	BT-25 or E-23½	Diffuse	197	7¾	127	5	Mog.	24000	C
—	{LU100/MED[c] C100S54/M[b]}	S54	100	9500	8550	B-17	Clear	140	5½	87	3 7/16	Med.	24000	C
—	{LU100/D/MED[c] C100S54/D/M[b]}	S54	100	8800	7920	B-17	Diffuse	140	5½	87	3 7/16	Med.	24000	C
S55SC-150	{LU150/55[a,c,t] C150S55[b]}	S55	150	16000	14400	BT-25 or E-23½	Clear	194	7⅝	127	5	Mog.	24000	D
S55MD-150	{LU150/55/D[a,t] C150S55/D[b]}	S55	150	15000	13500	BT-25 or E-23½	Diffuse	194	7⅝	127	5	Mog.	24000	D
S56SD-150	{C150S56[b] LU150/100[c]}	S56	150	16000	14400	BT-28 or E-28	Clear	211	8 5/16	127	5	Mog.	24000	E
S56SE-150	C150S56/D[b]	S56	150	15000	13500	BT-28	Diffuse	211	8 5/16	127	5	Mog.	24000	E
—	150 SONDL-E[e,f]	S56[l]	150	13000	11700	E-28	Diffuse	227	18 15/16	142	5⅝	Mog.	12000	E
—	150 SONDL-T[e,f]	S56[l]	150	13500	12150	T-15	Clear	210	8 3/16	127	5	Mog.	12000	E

Fig. 8–123—*Continued*

C. High Pressure Sodium Lamps

ANSI Lamp Designation	Manufacturer's Lamp Designation	ANSI Ballast Number	Lamp Watts*	Approximate Lumens		Bulb	Outer Bulb Finish	Maximum Over-All Length		Light Center Length		Base	Average Rated Life (hours)[j]	Lamp Voltage and Current
				Initial	Mean[j]			(millimeters)	(inches)	(millimeters)	(inches)			
S63MG-150	LUH150/BU,BD/EZ[a] ULX150[c] C150S63/EL[b]	H39**	150	13000	11700	BT-28 or E-28	Clear	211	8 5/16	127	5	Mog.	12000–16000	F
S63MH-150	LUH150/D/BU,BD/EZ[a] ULX150/D[c]	H39**	150	11330	10000	BT-28 or E-28	Diffuse	211	8 5/16	127	5	Mog.	12000–16000	F
—	LU150/55/MED[c] C150S55/M[b]	S55	150	16000	14400	B-17	Clear	140	5½	95	3¾	Med.	24000	D
—	LU150/55/D/MED[c] C150S55/D/M[b]	S55	150	15000	13500	B-17	Diffuse	140	5½	95	3¾	Med.	24000	D
S66MN-200	LU200[a,c] C200S66[b]	S66	200	22000	19800	E-18	Clear	248	9¾	146	5¾	Mog.	24000	G
S65ML-215	LUH215/BU,BD/EZ C215S65/EL[b] ULX215[c]	H37**	215	19750	17775	BT-28 or E-28	Clear	229	9	127	5	Mog.	12000	H
S65MM-215	LUH215/D/BU,BD/EZ LU215H/D[b]	H37**	215	18250	16350	E-28	Diffuse	229	9	127	5	Mog.	12000	H
S50VA-250	LU250[a,c,t] C250S50[b]	S50	250	27500	24750	E-18	Clear	248	9¾	146	5¾	Mog.	24000	I
S50VA-250/S	LU250S[a,t,c] C250S50/S[b]	S50	250	30000	27000	E-18	Clear	248	9¾	146	5¾	Mog.	24000	I
S50VB-250	250 SON-TD[e]	S50	250	27000[j]	24300	T-7	Clear	191	7½	95	3½	RSC[k]	24000	I
—	250 SONDL-E[e,f]	S50[l]	250	23000	20700	E-28	Diffuse	227	8 15/16	142	5⅝	Mog.	12000	I
—	250 SONDL-T[e,f]	S50[l]	250	23500	21150	T-15	Clear	257	10⅛	158	6¼	Mog.	12000	I
—	LU250/T7/RSC[c,s]	S50	250	25500	23400	T-7	Clear	256	10 1/16	130	5⅛	RSC[k]	24000	I
S50VC-250	LU250/D[a,c] C250S50/D/28[b]	S50	250	26000	23400	BT-28 or E-28	Diffuse	229	9	127	5	Mog.	24000	I
S67MR-310	LU310[a,c] C310S67[b]	S67	310	37000	33300	E-18	Clear	248	9¾	146	5¾	Mog.	24000	J
S64MJ-360	ULX360[c] C360S64/EL[b]	H33**	360	38000	35000	BT-37 or E-37	Clear	292	11½	178	7	Mog.	16000	—
S64MK-360	ULX360/D[c]	H33**	360	36000	32500	E-37	Diffuse	292	11½	178	7	Mog.	16000	—

S51WA-400	LU400[a,c,t] / C400S51[b]	S51	400	50000	45000	E-18	Clear	248	9¾	146	5¾	Mog.	24000	K
S51WB-400	LU400/D[a,c] / C400S51/D[b]	S51	400	47500	42750	BT-37 or E-37	Diffuse	292	11½	178	7	Mog.	24000	K
S51WC-400	400 SON-TD[e]	S51	400	50000	45000	T-7	Clear	256	10¹⁄₁₆	128	5	RSC[k]	24000	K
—	400 SONDL-E[e,f]	S51[l]	400	39000	35100	E-37	Diffuse	286	11¼	173	6¹³⁄₁₆	Mog.	12000	K
—	400 SONDL-T[e,f]	S51[l]	400	41000	36900	T-15	Clear	285	11¼	173	6¹³⁄₁₆	Mog.	12000	K
—	LU400/T7/RSC[c,s]	S50	400	46000	41400	T-7	Clear	256	10¹⁄₁₆	130	5⅛	RSC[k]	24000	I
—	C400S51/DE[b]	S51	400	45000	40500	T-6	Clear	256	10¹⁄₁₆	—	—	RSC[k]	24000	K
—	ULX880[c]	H36**	880	102000	91800	E-25	Clear	383	15¹⁄₁₆	241	9½	Mog.	16000	—
S52XB-1000	LU1000[a,c] / C1000S52[b]	S52	1000	140000	126000	E-25	Clear	383	15¹⁄₁₆	222	8¾	Mog.	24000	L
S52XE-1000	1000 SON-TD[e]	S52	1000	140000	126000	T-7	Clear	334	13⅛	167	6⁹⁄₁₆	RSC[k]	24000	L

* Actual lamp watts may vary depending on the ballast characteristic curve. Use only on ballasts that provide proper electrical values.
** Do not use on CW- or CWA-type mercury ballasts. Consult lamp manufacturer.
[a] General Electric. [b] North American Philips Lighting. [c] GTE Sylvania. [e] Thorn. [t] Action Tungsram.
[f] Deluxe color CRI ≥ 70.
[i] Horizontal operation ±20°.
[j] At 10 hours per start.
[k] Double ended construction.
[l] Ballasts are restricted to certain types; consult lamp manufacturer.

Lamp Voltage and Current for High Pressure Sodium Lamps

	Lamp Volts (Operating)	Ballast Open-Circuit Volts (minimum)†	Lamp Current (amperes)	
			Starting	Operating
A	52	110	1.85	1.18
B	52	110	2.4	1.6
C	55	110	3.2	2.1
D	55	110	5.0	3.3
E	100	—	—	1.8
F	130	—	—	1.5
G	100	195	3.5	2.3
H	130	—	2.1	—
I	100	195	4.5	3.0
J	100	190	5.5	3.6
K	100	195	7.0	4.7
L	250	456	7.0	4.7

† Requires a 2500- to 5000-volt pulse to start.

Fig. 8–124. Self-Ballasted Mercury Lamps Used for General Lighting

Watts	Bulb	Volts[h]	Bulb Finish	Approximate Lumens[i]		Lamp Current (amperes)[i]		Rated Life (hours)[j]	Base	Max. Over-All Length		Light Center Length	
				Initial (100 hours)	Mean	Starting	Operating			(millimeters)	(inches)	(millimeters)	(inches)
100	B-21[b,f,k]	Std.[l]	Phos. Coat	1650		1.1	.8	12000	Med.	159	6¼		
110	PAR-38[b,f,m]	Std.	I. M. Refl.			1.3	.9	12000	Med. Skt.	165	6½		
110	R-40[b,f,m]	Std.	I. M. Refl.			1.3	.9	12000	Med. Skt.	191	7½		
160	B-21[d]	Std.	Phos. Coat	2800		1.6	1.3	12000	Med.	178	7	122	4 13/16
	E-23[f,d]	Std.	Phos. Coat	2800–3125	1700	1.6	1.3	12000	Med.	173	6 13/16		
	R-40[b,f,m]	High	I. M. Refl.			.9	.6	12000	Med. Skt. or Mog.	178–191	7–7½		5½
	PS-30[a,b,f,g]	Std.	Phos. Coat	2100–3125		1.6	1.3	8–16000	Med.	184	7¼	140	4 13/16
	B-21[d]	High	Phos. Coat	3000		.9	.8	12000	Med.	178	7	122	
	E-23[b,f]	High	Phos. Coat	3520		.9	.8	12000	Med.	173	6 13/16		
	PAR-38[b,d,f,g]	Std.	Clear Flood	1970		1.6	1.3	12–14000	Med. Skt.	165	6½	100	3 15/16
	R-40[b,d,f,g]	Std.	I. F. Refl.	1850		1.6	1.3	12000	Med.	167	6 9/16		
250	E-28[a,b,d,f]	Std.	Phos. Coat	5000–5990		2.8	2.2	12000	Med. or Mog.	213	8⅜	146	5¾
	E-28[b,d,f]	High	Phos. Coat	5000–6450		1.3	1.1	12–14000	Med.	213	8⅜	146	5¾
	E-28[b,f,k]	High	Phos. Coat	6250		1.5	1.1	12000	Med. or Mog.	213	8⅜		
	PS-30[g]	Std.	Phos. Coat	4800		2.8	2.2	10–12000	Med.	205	8 1/16	152	6
	PS-35[b,f,g]	Std.	Phos. Coat	4700–5750		2.8	2.2	10–14000	Med. or Mog.	229–238	9–9⅜	165–178	6½–7
	PS-35[b,g]	High	Phos. Coat	5900		1.3	1.1	12–14000	Mog.	238	9⅜	178	7
	R-40[b,d,f,g]	Std.	I. F. Refl.	2650–3810		2.8	2.2	12–14000	Med. or Mog.	178	7		
300	E-28[b,f,m]	Std.	Phos. Coat	6750		3.5	2.9	14000	Med. Skt.	229	9		
	E-28[b,f,m]	Std.	Phos. Coat	6750		3.5	2.9	14000	Mog.	216	8½		
	R-40[b,f]	Std.	I. F. Refl.			3.5	2.9	14000	Med. Skt., or Mog.	178–191	7–7½		
	E-28[f]	High	Phos. Coat	7400		1.6	1.1	14000	Mog.	211	8 5/16		
450	BT-37[d]	Std.	Clear	9100	8280	6	4	16000	Mog.	292	11½	178	7
	BT-37[a,d]	Std.	Phos. Coat	8200–9500	6970–7125	6	4	16000	Mog.	290–292	11⅓–11½	178–187	7–7⅜
	BT-37[d,g]	High	Clear	9850–11800	9050–9700	3.5	2.3	14–16000	Mog.	292	11½	178	7
	BT-37[d,g]	High	Phos. Coat	8900–12200	7750–8500	3.5	2.3	14–16000	Mog.	292	11½	178	7
	E-37[b,f,m]	Std.	Phos. Coat	10400		5.5	3.8	16000	Mog.	260	10¼		
	E-37[b,f,m]	High	Phos. Coat	13000		3.4	2.2	16000	Mog.	260	10¼		
	PS-40[g]	Std.	Clear	9500		6	4	14–16000	Mog.	248	9¾	178	7
	PS-40[g]	Std.	Phos. Coat	10200		6	4	14–16000	Mog.	248	9¾	178	7
	PS-40[g]	High	Phos. Coat	12000–12200		3.5	2.3	14–16000	Mog.	248	9¾	178	7
	R-57[g]	Std.	Phos. Refl.			6	4	14–16000	Mog.	324	12¾	213	8⅜
	R-60[d,g]	Std.	I. F. Refl.	8170		6	4	14–16000	Mog.	260	10¼		
	R-60[g]	Std.	Phos. Refl.			6	4	14–16000	Mog.	260	10¼		
	R-60[g]	High	I. F. Refl.			3.5	2.3	14–16000	Mog.	260	10¼		
	R-60[g]	High	Phos. Refl.			3.5	2.3	14–16000	Mog.	260	10¼		
500	E-37[b,d,f]	Std.	Phos. Coat	10850–11780	8670	5.6	4.2	16000	Mog.	298	11¾	178	7
	E-37[b,d,f]	High	Phos. Coat	13590–14750	10880	3.3	2.2	16000	Mog.	286	11¼		
	R-57[d,f]	Std.	I. F. Refl.			5.6	4.2	16000	Mog.	324	12¾		
	R-57[b,f]	Std.	Phos. Refl.			5.6	4.2	16000	Mog.	324	12¾		

	R-57[b,f]	High	I. F. Refl.			3.3	2.2	16000	Mog.	257	10⅛		
	R-57[d]	High	Phos. Refl.			3.3	2.2	16000	Mog.	324	12¾		
	R-60[d]	Std.	I. F. Refl.	9160		5.6	4.2	16000	Mog.	257	10⅛		
	R-60[d]	Std.	Phos. Refl.	7800		5.6	4.2	16000	Mog.	257	10⅛		
750	BT-46[b,d,f]	Std.		18900	15120	8.8	6.6	16000	Mog.	368	14½		
	BT-46[b,f,m]	High	Phos. Coat	22000		4.8	3.4	16000	Mog.	368	14½		
	R-57[g]	Std.	Clear	18500		8.8	6.6	14–16000	Mog.	324	12¾	213	8⅜
	R-57[b,f,g]	Std.	I. F. Refl.	14000	11200	8.8	6.6	14–16000	Mog.	324	12¾		
	R-57[a,b,f,g]	Std.	Phos. Refl.	14000–19200	10500	8.8	6.3	14–16000	Mog.	324	12¾		
	R-57[g]	High	Clear	22200		5.1	3.6	18–20000	Mog.	324	12¾		
	R-57[b,f,g]	High	I. F. Refl.			5.1	3.6	18–20000	Mog.	324	12¾		
	R-57[a,b,f]	High	Phos. Refl.	14500–23000	12325	5.1	3.6	16–20000	Mog.	324	12¾		
	R-60[d]	Std.	I. F. Refl.	12120		8.8	6.6	16000	Mog.	257	10⅛		
	R-60[d]	Std.	Phos. Refl.	8160–11500		8.8	6.6	16000	Mog.	276	10⅞		
	R-60[d]	High	I. F. Refl.	12310–15000		5.1	3.6	16000	Mog.	257–276	10⅛–10⅞		
	R-60[d]	High	Phos. Refl.	8160–11250		5.1	3.6	16000	Mog.	257	10⅛		
1250	BT-56[d,g]	High	Clear	38000		10	6.4	14–16000	Mog.	381–391	15–15⅝	241	9½
	BT-56[b,d,f,g]	High	Phos. Coat	37780–41000	30230	10	6.4	14–16000	Mog.	381–391	15–15⅝	241	9½
	R-57[f]	High	I. F. Refl.			10	6.4	8000	Mog.	324	12¾		
	R-80[b,f]	High	I. F. Refl.			10	6.4	16000	Mog.	352	13⅞		

[a] General Electric.
[b] Public Service Lamp.
[d] North American Philips Lighting.
[f] Westron.
[g] Duro-Test.
[h] Std. = 120 volts ac nominal. High = 208, 220, 240 or 277 volts ac nominal.
[i] Values apply to lower voltage where two different voltage lamps were indicated.
[j] Life expectancy dependent on line voltage.
[k] For indoor use. Outdoors, use enclosed luminaire.
[l] Can be used in voltage range 115 volts to 130 volts.
[m] Hard glass. For indoor or outdoor use.
[o] Action Tungsram.

Fig. 8–125. Characteristics of Typical Low-Pressure Mercury-Vapor Sources of Ultraviolet Energy
A. Near Ultraviolet Output (320–420 Nanometers)

Designation	Nominal Lamp Watts	Rated Life[a] (hours)	Relative Black Light Energy[d] (100 hours)	Nominal Over-All Length[e] (millimeters)	Nominal Over-All Length[e] (inches)	Useful Arc Length (millimeters)	Useful Arc Length (inches)	Diameter (millimeters)	Diameter (inches)	Base	Minimum rms Starting Voltage[g]	Lamp Volts	Lamp Arc Current (amperes)
Preheat Start Lamps													
MF-3RP12/BL[h]	3	—	3	65	2 9/16[f]	—	—	38	1½	D.C. Bay.	10–16 dc	8	0.500 max.
MF-5000[h]	4	—	3	65	2 9/16[f]	—	—	38	1½	D.C. Bay.	24–28 dc	10	0.500 max.
F4T4/BL	7	3500	6	133	5¼	152	6	13	½[h]	Oval Small 4 Pin	120	51–55	0.09—0.16
F4T5/BL F4T5/BLB[b]	4	6000	4 3	152	6	64	2½	16	⅝	Min. Bipin	120	29	0.17
F6T5/BL[i] F6T5/BLB[b]	6	7500	7 6	229	9	140	5½	16	⅝	Min. Bipin	120	42	0.16
F8T5/BL F8T5/BLB[b]	8	7500	13 8	305	12	216	8½	16	⅝	Min. Bipin.	120	57	0.145
F14/T8/BL	14	7500	21	381	15	254	10	25	1	Med. Bipin	120	45	0.365
F15T8/BL F15T8/BLB[b]	15	7500	25 20	457	18	330	13	25	1	Med. Bipin	120	55	0.305
F20T12/BL F20T12/BLB[b]	20	9000	42 31	610	24	483	19	38	1½	Med. Bipin	120	56	0.38
F25T12/BL/28	25	7500	53	711	28	584	23	38	1½1	Med. Bipin	120	63	0.46
F25T12/BL/33	25	7500	57	838	33	711	28	38	1½	Med. Bipin	120	61	0.48
F30T8/BL F30T8/BLB[b]	30	7500	65 54	914	36	787	31	25	1	Med. Bipin	176	98	0.355
F40BL F40BLB[b]	40	20,000+	100 81	1219	48	1092	43	38	1½	Med. Bipin	176	102	0.43
U-Shaped Lamps (Preheat Starting)													
U3-15T8/BL U3-15T8/BLB[b]	15	7500	25 20	224	8 13/16	330	13	25	1	Med. Bipin	120	55	0.305
U3-25T8/BL U3-25T8/BLB[b]	25	7500	35 28	224	8 13/16	330	13	25	1	Med. Bipin	120	48	0.600
U3-30T8/BL U3-30T8/BLB[b]	30	7500	65 54	452	17 13/16	787	31	25	1	Med. Bipin	176	98	0.355
U3-40T8/BLB	40	7500	70	452	17 13/16	787	31	25	1	Med. Bipin	176	85	0.545
Rapid Start Lamps													
FC8T9/BL[c]	22.5	10,500	38	210	8¼	483	19	29	1.125	4-Pin	185	62	0.39
FC12T9/BL[c]	32	12,000	63	305	12	—	—	—	—	4-Pin	—	—	—
FC12T10/BL[c]	32	10,500	63	305	12	787	31	32	1.250	4-Pin	205	80	0.43
F24T12/BL/HO	35	9000	—	610	24	—	—	38	1.5	Recess D.C.	—	39.5	0.800
F36T12/BL/HO	45	9000	—	914	36	—	—	38	1.5	Recess D.C.	—	58	0.800
F40BL F40BLB[b]	40	12,000	100 81	1219	48	1092	43	38	1.5	Med. Bipin	200	100	0.43
F48T12/BL/HO	60	12,000	—	1219	48	—	—	38	1.5	Recess D.C.	—	76	0.800
F60T12/BL/HO	75	12,000	—	1524	60	—	—	38	1.5	Recess D.C.	—	94	0.800

F72T12/BL/HO	85	12,000	—	1829	72	—	—	38	1.5	Recess D.C.	—	112	0.800
F48T12/SDI/HO	60	12,000	—	1219	48	—	—	38	1.5	Recess D.C.	—	76	0.800
F71T12/BL/HO	84	12,000	—	[k]		1727	68	38	1.5	Med. Bipin	—	120	0.800
F72T12/SDI/HO	85	12,000	—	1829	72	—	—	38	1.5	Recess D.C.	—	112	0.800
F84T12/SDI/HO	100	12,000	—	2134	84	—	—	38	1.5	Recess D.C.	—	132	0.800
F96T12/SDI/HO	110	12,000	—	2438	96	—	—	38	1.5	Recess D.C.	—	149	0.800
F54T12/SDI/VHO	120	9000	—	[k]		1295	51	38	1.5	Med. Bipin	200	90	1.500
F59T12/SDI/140	140	9000	—	[k]		1425	56	38	1.5	Med. Bipin	225	100	1.750
F72T12/BL/VHO	160	9000	—	1829	72	1607	63	38	1.5	Recess D.C.	225	125	1.500

[a] At 3 hours per start. Useful life may be less, depending on the requirements of the application.
[b] Blue glass bulb to filter out much of the visible radiation.
[c] Circline lamp.
[d] Relative value of 100 equals 8100 fluorens.
[e] One lamp plus two standard lamp holders.
[f] Maximum over-all length.
[g] Minimum required rms voltage between lamp electrodes for reliable starting from 10°C (50°F) to 45°C (110°F).
[h] RP-12 bulb.
[i] Operation on same "universal" ballast.
[k] Nominal over-all length does not apply; not a standard general lighting size.

B. Germicidal Ultraviolet Output (253.7 Nanometers)

Lamp Designation	Glass Transmission below 200 nanometers Wavelength[b]	Special Characteristics	Type Cathode	Nominal Over-All Length[c] Millimeters (inches)	Useful Arc Length Millimeters (inches)	Nominal Diameter Millimeters (inches)	Base	Lamp Current (milliamperes)	Potential Drop between Electrodes (volts)	Minimum rms Starting Voltage[d]	Approximate Lamp Input Power (watts)	Approximate 254-nm Ultraviolet Output (watts)	Approximate 254-nm Ultraviolet Output (μW/cm² at 1 m)	Rated Life[e] (hours)
Preheat Starting														
G4T4/1[a, g]	Very low	U-Shaped	Hot	133 (5¼)	150 (6)	13 (½)	4 Pin	80 140	60 52	120 120	4 6.5	0.7 1.1	7.5 12	3500 3500
G4T5[a, v]	Very low		Hot	150 (6)	63 (2½)	16 (⅝)	Min. Bipin	170	29	120	4.5	0.5	5.4	—
G6T5[a, v]	Very low		Hot	225 (9)	140 (5½)	16 (⅝)	Min. Bipin	160	42	120	6	1	11	—
G8T5[a, g, s, v, p]	Very low		Hot	300 (12)	216 (8.5)	16 (⅝)	Min. Bipin	145	57	120	7.2	1.5	17	7500
GPH197T5-VH[v]	Very high		Hot	215 (8½)	117 (4⅝)	15 (9/16)	Min. Bipin	400	30	120	10	1.7	20	7500
GPH211T5-L[v]	Very low		Hot	225 (9)	130 (5⅛)	15 (9/16)	Min. Bipin	400	31	120	11	1.9	22	7500
GPH287T5-L[v] GPH287T5-H[v] GPH287T5-VH[v]	Very low High Very High	Quartz bulb	Hot	300 (12)	207 (8½)	15 (9/16)	Min. Bipin	400	42	120	14	3	39	7500
G15T8[a, g, p, s, v, p]	Very low		Hot	450 (18)	350 (14)	25 (1)	Med. Bipin	305	55	120	15	3.5	38	7500
G25T8[a, g, p, v, p]	Very low		Hot	450 (18)	350 (14)	25 (1)	Med. Bipin	600	48.5	120	25	5[k]	55[k]	7500
G30T8[a, g, p, s, v, p]	Very low		Hot	900 (36)	800 (32)	25 (1)	Med. Bipin	355	99	200	30	8.3	85	7500
Instant Starting														
G4S11[a, s]	High	S11 bulb	Hot	57 (2¼)	10 (⅜)	35 (1⅜)	Int. Screw	350	10	24	3.8	0.1	1.5	6000
782L10[a, v, p] 782H10[a, v, p]	Very low High		Cold	429 (16⅞)	250 (10)	16 (⅝)	Single Pin	60[i] 90[i]	240 235	950 900	12 20	2 2.8	20 28	17500 15000
GCC369N[a] GCC369H[a]	Very low High		Cold	435 (17⅛)	229 (9)	16 (⅝)	Single Pin	55[i] 90[i]			9 14.7	1.6 2.4	16 24	
782L20[a, v, p] 782H20[a, v, p]	Very low High		Cold	683 (26⅞)	500 (20)	16 (⅝)	Single Pin	50[i] 90[i]	325 315	950 900	14 24	3.9 6	35 54	17500 15000
782L30[a, v, p] 782H30[a, v, p]	Very low High		Cold	937 (36⅞)	750 (30)	16 (⅝)	Single Pin	50[i] 90[i]	410 390	950 900	17 29	5.2 8.3	46 73	17500 15000

Continued on page 8-128.

Fig. 8–125—*Continued*
B. Germicidal Ultraviolet Output (253.7 Nanometers)

Lamp Designation	Glass Transmission below 200 nanometers Wavelength[b]	Special Characteristics	Type Cathode	Nominal Over-All Length[c] Millimeters (inches)	Useful Arc Length Millimeters (inches)	Nominal Diameter Millimeters (inches)	Base	Lamp Current (milliamperes)	Potential Drop between Electrodes (volts)	Minimum rms Starting Voltage[d]	Approximate Lamp Input Power (watts)	Approximate 254-nm Ultraviolet Output (watts)	Approximate 254-nm Ultraviolet Output (μW/cm² at 1 m)	Ratio Life[e] (hours)
782VH29[a, v, p]	Very high	Quartz bulb	Cold	900 (36)	750 (30)	16 (5/8)	Single Pin	90	390	900	29	8.3	73	15000
93A-1[h, v]	Very high	Quartz bulb	Cold	412 (16½)	292 (11½)	13 (½)	Single Pin	30	300	3000	8	1.9	21	17500
84A-1[h, v]	Very high	Quartz bulb	Cold	746 (29⅜)	628 (24¾)	13 (½)	Single Pin	30	450	3000	14	4.1	46	17500
83A-1[h, v]	Very low		Cold	412 (16¼)	273 (10¾)	17.5 (11/16)	Single Pin	120	200	600	22	3.1	35	17500
94A-1[h, v]	Very low		Cold	768 (30¼)	628 (24¾)	17.5 (11/16)	Single Pin	120	300	600	32	7.2	80	17500
782L25½[a]	Very low		Cold	768 (30¼)	648 (25½)	15.8 (5/8)	Single Pin	120			34	7.3	75	
86A-45[h]	Very low	U-Shaped	Cold	133 (5¼)	114 (4½)	17.5 (11/16)	Bipin	120	200	600	8	1.4	16	17500
87A-45[h]	Very low	U-Shaped	Cold	206 (8⅛)	267 (10½)	17.5 (11/16)	Bipin	120	200	600	22	4.3	47	17500
88A-45[h, v]	Very low	U-Shaped	Cold	384 (15⅛)	622 (24½)	17.5 (11/16)	Bipin	120	300	600	32	10.4	113	17500
GSL312T5L[v]	Very low		Hot	384 (15⅛)	232 (9⅛)	15 (9/16)	Single pin	400	45	450	16	3.3	40	7500
G10T5½L[a, v, p]	Low													
G10T5½H[a, v, p]	High		Hot	429 (16⅞)	276 (10⅞)	15 (9/16)	Single Pin	400	55	450	16	5.3[k]	55[k]	7500
G10T5½VH[v]	Very high	Quartz bulb												
G36T6[a, g, v]	Very low							120[j]	180	660	17	8	70	7500
G36T6L[a, v, p]	Very low		Hot	914 (36)	760 (30)	15 (9/16)	Single Pin	200[j]	150	660	25	11.4	100	7500
G36T6H[a, v, p]	High							300[j]	130	660	32	13.1	115	7500
G36T6VH[v]	Very high	Quartz bulb						425[j]	120	660	39	13.8[k]	120[k]	7500
G37T6VH[a, v, p]	Very high	Quartz bulb	Hot	930 (37)	785 (31)	16 (5/8)	Single Pin	425	120	660	39	14.3[k]	130[k]	7500
G64T5L[a, v]	Very low		Hot	1625 (64)	1470 (58)	15 (9/16)	Single Pin	425	220	—	72	25	200	10000
G64T5VH[v]	Very high	Quartz bulb												
G64T6L[g, v]	Very low		Hot	1625 (64)	1470 (58)	19 (¾)	Single Pin	425	180	—	65	25	200	8500
G64T6L[a]	Very low													
GSL620T5L[v]	Very low		Hot	692 (27¼)	540 (21¼)	15 (9/16)	Single pin	425	86	660	30	9.5	100	7500

[a] American Ultraviolet. [g] General Electric. [h] Canrad-Hanovia. [p] North American Philips Lighting. [s] GTE Sylvania. [v] Voltarc.

[b] Low pressure mercury discharges generate some energy at 184.9 nanometers. In most lamps this is filtered out by the glass bulb. Germicidal lamps are made with several different glasses to allow transmission of more of less of this short wave ultraviolet energy. Ultraviolet energy of wavelengths below 200 nanometers can convert some of the oxygen (O_2) in the air into ozone (O_3). Ozone is a toxic gas, even in relatively low concentrations, so care should be exercised in using germicidal lamps that transmit high or very high amounts of this radiation.

[c] With double ended lamps, length includes two standard lampholders.

[d] Minimum required rms voltage between lamp electrodes for reliable starting from 10°C (50°F) to 43°C (110°F).

[e] At 8 hours per start except for the G4T4/1 which is at 3 hours per start.

[i] Ratings apply to either of the two lamps in this group.

[j] Any of the four G36T6 lamps can be operated at any of these currents with the output ratings shown.

[k] Approximate output in still air at 25°C (77°F). The output is increased in cool or moving air.

Fig. 8-126. Pulsed Xenon Arc (PXA) Lamps

Ordering Number	Description	Effective Source Envelope		Light Center Length		Maximum Over-All Length		Approximate Life (hours)	Maximum Watts at Design Voltage	Approximate Maximum Lumens	Application
		(millimeters)	(inches)	(millimeters)	(inches)	(millimeters)	(inches)				
PXA-41	Linear	9.5 × 178	3/8 × 7	—	—	267	10 1/2	150	700	—	Copyboard
PXA-43	Linear	9.5 × 76	3/8 × 3	—	—	165	6 1/2	50	300	—	Copyboard
PXA-44	Linear	9.5 × 152	3/8 × 6	—	—	241	9 1/2	150	600	—	Copyboard
PXA-45	Linear	9.5 × 305	3/8 × 12	—	—	394	15 1/2	150	1500	37 500	Copyboard
PXA-47	Linear	9.5 × 610	3/8 × 24	—	—	699	27 1/2	150	3000	84 000	Copyboard
PXA-50	Helical	48 × 60	1 7/8 × 2 3/8	83	3 1/4	118	4 5/8	75	4000	125 000	Printing
PXA-56	Helical	38 × 51	1 1/2 × 2	70	2 3/4	102	4	75	1500	38 000	Printing
PXA-80	Helical	89 × 57	3 1/2 × 2 1/4	83	3 1/4	118	4 5/8	75	8000	240 000	Printing
PH/1550/HT3	Helical	38 × 51	1 1/2 × 2	73	2 7/8	114	4 1/2	75	1550	60 000	Printing

Fig. 8-127. Enclosed Concentrated-Arc Lamps*

Watts	Volts	Amperes	Average Light Source Diameter		Average Luminance (candelas per square centimeter)	Average Candlepower (candelas)	Rated Life (hours)	Maximum Temperature			
								Bulb		Base	
			(millimeters)	(inches)				°C	°F	°C	°F
2**	27	0.099	0.18	0.007	1800	0.30	150	60	140	38	100
2	38	0.055	0.13	0.005	2400	0.30	150	60	140	38	100
10	20	0.50	0.38	0.015	4500	4.7	450	107	225	54	130
25	20	1.25	0.76	0.030	3500	16.0	325	179	355	63	145
100	16	6.25	1.83	0.072	3800	100	375	243	470	71	160
300	20	15.0	2.79	0.110	4500	275	250	271	520	82	180

* Nominal color temperature 3200 K on all lamps except 2**.
Various types of bulbs are available for all wattages of lamps.
** Tungsten arc 3085 K color temperature; all others—zirconium arcs.

Fig. 8-128. Typical Low Pressure Sodium Lamps†

Manufacturers' Designations	ANSI Designations	Type**	Rated Watts	Nominal Volts	Lamp Current (amperes)	Bulb Diameter (millimeters)	Length (millimeters)	Initial Lumens*	Rated Average Life (hours)
SOX10W	L77RG-10	"U" Tube	10	55	0.20	40	150	1000	10000
SOX18W	L69RA-18	"U" Tube	18	57	0.35	54	216	1800	14000
SOX35W	L70RB-35	"U" Tube	35	70	0.60	54	311	4800	18000
SOX55W	L71RC-55	"U" Tube	55	109	0.59	54	425	8000	18000
SOX90W	L72RD-90	"U" Tube	90	112	0.94	68	528	13500	18000
SOX135W	L73RE-135	"U" Tube	135	164	0.95	68	775	22500	18000
SOX180W	L74RF-180	"U" Tube	180	240	0.91	68	1120	33000	18000

* After 100 burning hours.
** Linear types are still manufactured for replacement purposes.
† All have BY22d bases—an IEC designation for a specific D.C. bayonet base.

Fig. 8-129. Medium and High Pressure Mercury

Lamp Designation	Lamp Watts (rated)	Outer Bulb[a]	Base	Rated Initial Lumens	Rated Average Life (hours at 5 hours per start)	Operating Volts	Starting Amperes
					Short Lamps (Less Than 25		
H43-AZ	75	B-21	Med.	2800	2400	130	0.95
H85A3[f] H85A3/uv[f]	85	T-10	Med.	3050	500	250	0.6
H100A4/T[i,g] H44AB H4AB100	100	T-10	Admed.	3700	9000	130	1.3
H38HT-100[i,g]	100	BT-25 or ED-23½	Mog.	4000	24000	130	1.3
H100PSP44-4[f] H100PFL44-4[f] H44GS-100[i,g] H44-JM[i,h]	100	PAR-38	Ad. Sk.	2450	12000	130	1.3
HPR125[g]	125	110	Med.	2800	2000	125	—
HPW125[g]	125	E24	Med.	—	2000	125	—
H175A39-22[f] H39KB-175[i,g]	175	BT-28 or E-28	Mog.	7900	24000+	130	2.15
H250A37-5[f] H37KB-250[i,g]	250	BT-28 or E-28	Mog.	11500	24000+	130	2.9
LA5.5-6MU7[s]	300	8 mm	Wire Term.	—	500	120	—
79A[h]	325	T-6½[b]	Wire Term.	—	1000	135	3.5
H33CD-400[i,g]	400	E-37 or BT-37	Mog.	21000	24000+	135	5.0
H33-IAR[i]	400	T-16	Mog.	19500	12000	135	5.0
H2T5½[f]	400	T-5½[b]	RSC[l]	18000	2000	130	5.0
64A[h]	400	—	—	—	1000	195	3.3
H33-FY[i]H33HS	400	R-57	Mog.	18500	16000	135	5.0
H40-17MA[i]	425	BT-37	Mog.	2500	16000	265	2.5
73A[h]	560	T-7½[b]	Wire Term.	—	1000	145	6.0
51B[h]	600	T-8	⅜ Sleeve	—	1000	500	1.9
74A[h]	700	T-7½[b]	Wire Term.	—	1000	155	7.5
H700A35-18[f] H35-18NA[i]	700	BT-46	Mog.	39000	24000+	265	4.5
H12T3[f]	750	T-3[l,n]	RSC	30000	750	500	2.65
4A[h]	750	—	—	—	1000	500	2.7
H750T3/12A,B,C[i]	750	T-3[b]	[m]	30000	1000	500	2.45
H3T7/OF2	750	T-7[b]	RSC	43000	2000	130	10.0
P-750[s]-E1 thru E14	750	21 mm	Wire	—	—	—	—
H850T3/14B,C[i]	850	T-3[b]	[m]	34000	1000	500	2.8
H1000A36-15[f] H36GV-1000[i,g]	1000	BT-56	Mog.	57000	24000+	265	6.0
H1000A34-12[f] H34GV-1000[i,g]	1000	BT-56	Mog.	56000	16000+	135	12.0
H34-12FC[i]	1000	R-80	Mog.	53500	16000	135	12.0
H36-15FC[i]	1000	R-80	Mog.	54000	16000	265	6.0
HTQ-4[g]	1000	12.5 mm	Wire Term.	—	1000	700	2.4
H1200T3/18 B,C[i]	1200	T-3[b]	[m]	48000	1000	1200	1.6
67A[h]	1200	—	—	—	1000	450	5.0
89A[h]	1200	T-6½	Wire Term.	—	1000	270	6.3
H13C/6A[e]	1300	25 mm	—	—	1000	285	—
P1300E1[s] thru E14	1300	21 mm	—	—	—	—	—
H25C/12A[e]	2500	25 mm	—	—	1000	420	—
P2500E1[s] Thru E14	2500	21 mm	—	—	—	—	—
6512A.431[h]	2500	T-6½	—	—	2000	420	8.3
H37C/18A[e]	3700	25 mm	—	—	1000	615	—
P3700E1[s] Thru E14	3700	21 mm	—	—	—	—	—
6538A.431	3750	T-7¾	—	—	2000	450	14

Lamps for the Production of Ultraviolet Energy

Operating Amperes	Warm-up Time[d] (minutes)	Maximum Over-All Length (millimeters)	Maximum Over-All Length (inches)	Light Center Length (millimeters)	Light Center Length (inches)	Lighted Length (millimeters)	Lighted Length (inches)	Blacklight	Blue & White Printing	Copy board Lighting	Vacuum Frame Printing	General Ultraviolet (300–400 nm)	General Ultraviolet (220–400 nm)	Therapeutic	Photochemical
Inches Over-All Length)															
0.66	3	165	6.5	95	3.75	27	1 1/16	×				×			×
0.4	6	137	5 3/8	76	3	17	11/16					×	×	×	×
0.9	3–6	137	5 5/8	87	3 7/16	29	1 1/8	×				×			×
0.9	3	191	7 1/2	127	5	27	1 1/16	×				×			×
0.9	3–8	138	5 7/16	—	—	25	1	×				×			×
1.15	3.5	222	—	—	—	—	—	×	×	×					
1.15	2.5	172	—	—	—	—	—	×				×			×
1.5	4–5	211	8 5/16	127	5	51	2	×				×			×
2.1	4	211	8 5/16	127	5	60	2 3/8	×				×			×
2.5	—	—	—	—	—	140	5.5		×			×			×
2.8	5	279	11	—	—	114	4 1/2					×	×	×	×
3.2	4–5	292	11 1/2	179	7	70	2 3/4	×				×			×
3.2	4	279	11	179	7	70	2 3/4	×				×			×
3.3	2–3	114	4	—	—	60	2 3/8	×		×	×	×	×		×
2.3	—	387	15 1/4	—	—	298	11 3/4		×	×	×	×	×		×
3.2	6	324	12 3/4	—	—	71	2 13/16	×				×			×
1.7	4	298	11 1/2	179	7	97	3 13/16	×			×				×
4.4	5	279	11	—	—	114	4 1/2				×	×	×		
1.34	5	448	17 5/8	—	—	305	12	×	×	×	×	×	×		×
5.3	5	343	13 1/2	—	—	186	7 1/4					×	×	×	×
2.8	4–5	368	14 1/2	241	9 1/2	149	5 7/8	×		×	×	×			×
1.65	2–3	360	14 3/16	—	—	305	12	×	×	×	×	×	×		×
1.7	5	457	18	—	—	343	13 1/2	×	×	×	×	×	×		×
1.65	4	360	14 3/16	—	—	305	12	×		×	×	×	×		×
6.0	2.3	140	5 1/2	—	—	68	2 11/16	×		×	×	×	×		×
4.2	—	165	6 1/2	—	—	97	—		×	×		×	×		×
1.9	4	411	16 3/16	—	—	356	14						×		×
4.0	4	383	15 1/16	238	9 3/8	149	5 7/8	×		×	×	×			×
8.0	4	383	15 1/16	238	9 3/8	143	5 5/8	×		×	×	×			×
8.0	6	352	13 7/8	—	—	143	5 5/8	×		×	×	×			×
4.0	6	352	13 7/8	—	—	143	5 7/8	×		×	×	×			×
1.7	<5	495	—	—	—	440	—						×		×
1.1	4	513	20 3/16	—	—	457	18						×		×
3.1	—	559	22	—	—	445	17 1/2		×	×	×	×	×		×
5.1	—	432	17	—	—	305	12		×	×	×	×	×		×
4.7	—	308	12 1/8	—	—	150	5 15/16						×		×
4.7	—	308	12 1/8	—	—	152	6.0		×	×		×	×		×
6.3	—	473	18 5/8	—	—	329	12 15/16						×		×
6.3	—	473	18 5/8	—	—	318	12.5		×	×		×	×		×
6.3	5	473	18 5/8	—	—	318	12.5					×	×		×
6.6	—	625	24 5/8	—	—	469	18 15/32						×		×
6.6	—	625	24 5/8	—	—	470	18.5		×	×		×	×		×
8.8	5	473	18 5/8	—	—	333	13 1/8					×	×		×

Fig. 8–129

Lamp Designation	Lamp Watts (rated)	Outer Bulb[a]	Base	Rated Initial Lumens	Rated Average Life (hours at 5 hours per start)	Operating Volts	Starting Amperes
						Long Lamps (Over 25	
71A[h]	1200	—	—	—	1000	1200	1.8
H24T3[f]	1440	T-3[b,n]	RSC	58000	1000	1200	2.2
H1440T3/24B,C[i]	1440	T-3[b]	[m]	58000	1000	1200	2.0
33D[h]	1500	—	—	—	1000	1500	1.8
70A[h]	1800	—	—	—	1000	1000	3.1
78A[h]	2000	T-7½[b]	Wire Term.	—	1000	550	6.0
30A[h]	2000	—	—	—	1000	550	6.0
34A[h]	2200	—	—	—	1000	1000	3.9
462A[h]	2200	—	—	—	1000	1000	3.9
H22C/24B	2200	T-6¼[c]	Wire Term.	—	1000	950	3.9
158C[h]	2750	—	—	—	1000	1700	2.8
24A[h]	3000	—	—	—	1000	550	9.0
153A[h]	3000	T-4½[b]	19/32 Sleeve	—	1000	1500	3.0
166B[h]	3500	—	—	—	1000	1500	3.9
47A[h]	3500	—	—	—	1000	925	6.0
H36C/12A[e]	3600	25 mm	—	—	1000	450	—
61A[h]	4000	—	—	—	1000	1500	4.5
H40C/20A[e]	4000	25 mm	—	—	1000	670	—
77A[h]	4500	—	—	—	1000	955	6.6
57A[h]	4800	T-7½[b]	Wire Term.	—	1000	1250	6.0
H50C/24A[e]	5000	25 mm	—	—	1000	850	—
H50C/25A[e]	5000	25 mm	—	—	1000	840	—
5000-E1[s] Thru E14	5000	21 mm	—	—	—	—	—
175B[h]	5000	—	—	—	1000	1500	5.8
59B[h]	5500	—	—	—	1000	1250	6.6
H60C/30A[e]	6000	—	—	—	1000	840	—
H64C/32A[e]	6400	—	—	—	1000	900	—
56B[h]	7500	—	—	—	1000	1650	7.3
H75C/25A[e]	7500	25 mm	—	—	1000	925	—
H76C/38A[e]	7600	—	—	—	1000	1225	—
H76C/38B[e]	7600	—	—	—	1000	935	—
6538A.431[h]	7600	T-6½	—	—	2000	935	15
H80C/40A[e]	8000	—	—	—	1000	950	—
H84C/42A[e] and B[e]	8400	—	—	—	1000	1300 1000	—
40B[h]	9000	—	—	—	1000	1650	8.8
H90C/30A[e]	9000	25 mm	—	—	1000	840	—
H97C/48A[e]	9750	—	—	—	1000	1570	—
H10K/50A[e]	10000	—	—	—	1000	1570	—
P10000[s]E/1 Thru E/14	10000	—	—	—	—	—	—
H11K/55A[e]	11000	—	—	—	1000	1870	—
H11K/38A[e]	11000	25 mm	—	—	1000	1300	—
6838A.431[h]	11400	T-7¾	—	—	2000	930	—
H12K/60A[e]	12000	—	—	—	1000	1870	—
H13K/65A[e]	13000	—	—	—	1000	1870	—
H15K/50A[e]	15000	25 mm	—	—	1000	1700	—
H15K/77A[e]	15000	—	—	—	1000	1870	—
6577A.431	15000	T-6½	—	—	15000	1870	—
H18K/60A[e]	18000	25 mm	—	—	1000	1950	—

[a] May be burned in any position. All outer bulbs have clear finish except H44-4GS, H100PSP44-4, H44-4JM and H100PFL44-4 which have reflector bulbs.
[b] Only single bulb used.
[c] Single bulb. Can be furnished with outer bulb and fittings.
[d] Approximate-depends upon ballast, luminaire, and ambient temperatures.

Continued

Operating Amperes	Warm-up Time[d] (minutes)	Maximum Over-All Length (millimeters)	Maximum Over-All Length (inches)	Light Center Length (millimeters)	Light Center Length (inches)	Lighted Length (millimeters)	Lighted Length (inches)	Blacklight	Blue & White Printing	Copy board Lighting	Vacuum Frame Printing	General Ultraviolet (300–400 nm)	General Ultraviolet (220–400 nm)	Therapeutic	Photochemical
Inches Over-All Length)															
1.2	—	978	38 1/2	—	—	889	35		×	×	×	×	×		×
1.35	2-3	665	26 3/16	—	—	610	24		×	×	×	×	×		×
1.35	4	665	26 3/16	—	—	610	24						×		×
1.2	—	1371	54	—	—	1219	48		×	×	×	×	×		×
2.2	—	699	27 1/2	—	—	559	22		×	×	×	×	×		×
4.0	5	651	25 5/8	—	—	546	21 1/2		×	×	×	×	×		×
4.2	—	1371	54	—	—	1291	48			×	×		×		
2.4	5	749	29 1/2	—	—	635	25	×	×	×	×	×	×		×
2.4	5	686	27	—	—	559	22	×	×	×	×	×	×		×
2.6	—	667	26 1/8	—	—	597	23 1/2	×	×	×	×	×	×		×
1.9	—	1372	54	—	—	1219	48		×	×	×	×	×		×
6.3	—	1372	54	—	—	1219	48		×	×	×	×	×		×
2.15	—	1332	52 7/16	—	—	1219	48		×	×	×	×	×		×
2.6	—	1372	54	—	—	1219	48		×	×	×	×	×		×
4.3	—	1372	54	—	—	1219	48		×	×	×	×	×		×
8.8	—	473	18 5/8	—	—	318	12.5					×	×		×
3.0	—	1372	54	—	—	1219	48		×	×	×	×	×		×
6.3	—	664	26 5/32	—	—	508	20					×	×		×
5.0	—	1232	48 1/2	—	—	1067	42		×	×	×	×	×		×
4.5	—	1372	54	—	—	1219	48		×	×	×	×	×		×
6.3	—	778	30 5/8	—	—	622	24 15/16					×	×		×
6.3	—	803	31 5/8	—	—	647	25 15/16					×	×		×
6.3	—	803	31 5/8	—	—	635	25		×			×	×		×
3.8	—	1372	54	—	—	1219	48		×	×	×	×	×		×
5.0	—	1626	64	—	—	1499	59		×	×	×	×	×		×
7.6	—	930	36 5/8	—	—	774	30 15/32					×	×		×
7.7	—	981	38 5/8	—	—	825	32 15/16					×	×		×
4.7	—	1372	54	—	—	1219	48		×	×	×	×	×		×
8.8	—	803	31 5/8	—	—	635	25					×	×		×
6.75	—	1133	44 5/8	—	—	977	38 15/32					×	×		×
8.8	—	1133	44 5/8	—	—	977	38 15/32					×	×		×
8.8	5	1133	44 5/8	—	—	978	38 1/2					×	×		×
9.0	—	1184	46 5/8	—	—	1028	40 15/32					×	×		×
6.8 9.0	—	1223	48 5/32	—	—	1067	42					×	×		×
5.7	—	1626	64	—	—	1499	59		×	×	×	×	×		×
9.1	—	930	36 5/8	—	—	774	30 15/32					×	×		×
6.75	—	1387	54 5/8	—	—	1231	48 15/32					×	×		×
7.0	—	1426	56 5/32	—	—	1270	50					×	×		×
7.0	—	1426	56 5/32	—	—	1270	50					×	×		×
6.3	—	1576	62 1/32	—	—	1419	55 7/8					×	×		×
9.3	—	1133	44 5/8	—	—	977	38 15/32					×	×		×
14.5	5	1133	44 5/8	—	—	994	39.125					×	×		×
7.0	—	1692	66 5/8	—	—	1536	60 15/32					×	×		×
7.5	—	1807	71 5/32	—	—	1651	65					×	×		×
9.2	—	1426	56 5/32	—	—	1270	50					×	×		×
8.8	—	2124	83 5/8	—	—	1968	77 15/32					×	×		×
8.8	—	2124	83 5/8	—	—	1968	77 15/32					×	×		×
10.0	—	1692	66 5/8	—	—	1536	60 15/32					×	×		×

[e] Voltarc Tubes.
[f] General Electric.
[g] North American Philips Lighting.
[h] Canrad-Hanovia.
[i] GTE Sylvania.
[s] ILC Technology.
[l] Recessed Single Contact-RSC.
[m] A-ceramic tubular, B-RSC, C-axial lead with ceramic support.
[n] Ozone free.
[x] See Fig. 8-125 for other ultraviolet producing lamps.

Fig. 8–130. Approximate Operating Characteristics of Mercury-Argon, Mercury-Xenon, and Xenon Short-Arc Lamps

Rated Power (watts)	Operating			Arc Length (millimeters)	Bulb Diameter[n] (millimeters)	Over-All Length		Gas	Initial Lumens	Average Luminance (candelas per square millimeter)	Envelope Material	Rated Life (hours)	Manufacturers and Remarks[m]
	Position	Volts	Amperes			(millimeters)	(inches)						
5	Horiz. or Vert.	12	0.5	0.3	4	40	1.570	Xe	5	15	i	1000	c
75	Vertical	14	5.4	0.5	10	90	3.54	Xe	1000	400	k	400	e,f
75	Vertical	15	5.0	0.9	12	110	4.33	Xe	1600	200	i	400	h
75	Horiz. or Vert.	14	5.3	0.5	6	44	1.75	Xe	1400	500	l		c
150	Vertical	20	7.5	2.2	20	150	5.91	Xe	3000	200	k	1200	1,f,e,g,h
150	Vertical	20	7.5	1.7	20	117	4.61	Xe	2200	180	i	1200	e
150	Vertical	17.5	8.5	1.9	20	150	5.91	Xe	2900	180	i	1200	e
150	Horiz. or Vert.	18	8.3	1.4	8	53	2.08	Xe	3200	150	l		c
250	Vertical	14	18	1.7	25	226	8.90	Xe	4800	260	i	1200	e
300	Horizontal	19	15	2.5	24	171	6.75	Xe	8000	200	j	1000	a
300	Vertical	20	15	2.3	25	175	6.89	Xe	7600	200	i	900	h
300	Horiz. or Vert.	18	16.1	2.2	10	53	2.08	Xe	7050	320	l		c
350	Horizontal	22	16	3.0	22	155	6.10	Xe	9000	440	i	600	h
450	Vertical	18	25	2.7	30	260	10.23	Xe	13000	350	j	2000	f,d,e,g
450	Vertical	18	25	2.7	30	177	6.96	Xe	13000	350	j	2000	g
450	Horizontal	18.5	25	2.2	29	260	10.24	Xe	13000	450	i	800	e
450	Vertical	18	25	3.2	29	260	10.24	Xe	15000	450	j	1500	a
500	Horizontal	20	28	1.9	29	260	10.24	Xe	15000	450	j	1000	a
500	Vertical	20	25	3.5	29	234	9.21	Xe	18000	420	i	1500	h
500	Horizontal	18	28	2.5	35	190	7.48	Xe	14500	400	j	2000	g
700	Horizontal	19	37	2.9	39	229	9.00	Xe	21000	400	j	1500	a,e
750	Horizontal	20	37	3.2	42	234	9.20	Xe	22000	450	j	2000	d
900	Vertical	20	45	3.3	40	318	12.50	Xe	30000	500	j	2000	a,f,e,g
900	Vertical	20	45	3.5	40	325	12.80	Xe	33000	620	i	2000	h
1000	Horizontal	20	50	3.5	40	229	9.00	Xe	32000	600	j	2000	a,d,e
1000	Horizontal	22	45	4.0	46	325	12.80	Xe	35000	450	j	2000	g
1000	Vertical	22	45	4.4	40	315	12.40	Xe	37000	600	i	1500	h
1600	Vertical	25	65	4.2	50	368	14.47	Xe	60000	650	j	2000	f,e,g
1600	Horizontal	22	65	3.7	47	235	9.25	Xe	60000	700	j	2000	d,e
2000	Horizontal	29	70	4.8	60	370	14.57	Xe	80000	750	j	2000	f,e,g
2000	Horizontal	25	80	5.5	72	368	14.50	Xe	75000	530	i	1250	b
2000	Horizontal	23	80	5.9	52	341	13.43	Xe	85000	750	j	2000	d
2000	Horizontal	25	80	6.0	57	370	14.56	Xe	70000	650	j	1500	a
2200	Horizontal	24	90	5.0	57	341	13.44	Xe	81000	750	j	1500	d
2500	Vertical	30	83	6.7	57	428	16.85	Xe	100000	620	j	1500	f,e,g
3000	Vertical	30	100	7.7	61	408	16.06	Xe	125000	650	j	1500	a
3000	Vertical	30	100	5.5	56	423	16.65	Xe	130000	850	j	1500	f
3000	Horizontal	30	100	6.7	72	381	15.00	Xe	125000	750	k	1500/1000	a,b
3000	Horizontal	30	100	5.5	56	360	14.17	Xe	130000	850	j	1200	f
3000	Horizontal	30	100	6.9	57	341	13.43	Xe	130000	800	j	1500	d
3000	Horizontal	30	100	5.7	66	370	14.57	Xe	130000	850	j	1500	e
4000	Vertical	32	125	8.0	57	432	17.00	Xe	145000	700	j	1000	a
4000	Vertical	33	120	7.5	54	432	17.00	Xe	190000	800	i	1000	e,f

4000	Horizontal	31	130	7.4	70	433	17.03	Xe	150000	800	j	1000	[a]
4000	Horizontal	31	130	6.0	61	423	16.66	Xe	180000	950	j	900	[f]
4000	Horizontal	30	135	6.0	70	410	16.14	Xe	155000	900	j	800	[e]
4000	Horizontal	33	120	7.5	60	432	17.00	Xe	180000	700	i	1000	[g]
4200	Vertical	30	140	8.0	61	421	16.59	Xe	160000	830	j	1000	[a]
4200	Horizontal	31	135	7.9	80	411	16.18	Xe	160000	830	j	1000	[a]
4500	Horizontal	31	145	7.4	94	540	21.25	Xe	200000	1050	i	1000	[i]
4500	Horizontal	32	140	7.9	80	412	16.22	Xe	160000	950	j	1000	[f]
5000	Vertical	34	147	10.0	71	435	17.13	Xe	210000	600	i	1000	
6500	Vertical	41	160	9.0	60	483	19.00	Xe	325000	950	i	500	[e,f]
7000	Horizontal	46	150	12.0	80	431	16.97	Xe	400000	700	j	600	[a]
7500	Vertical	38	200	10.0	70	397	15.63	Xe	400000	260	i	500	[a]
8000	Vertical	32	250	7.0	73	334	13.13	Xe	410000	300	i	500	[a]
10000	Horizontal	38	260	10.0	85	397	15.64	Xe	410000	3600*	i	500	[a]
12000	Horizontal	32	375	10.0	114	483	19.03	Xe	540000	3700*	i	600	[b]
20000	Horizontal	45	450	11.0	121	474	18.66	Xe	1000000	4800*	i	400	[a]
20000	Horizontal	44	455	13.0	137	487	19.16	Xe	900000	4400*	i	500	[b]
20000	Horizontal	43	465	14.0	137	613	24.13	Xe	880000	4400*	i	400	[b]
30000	Horizontal	51	600	12.0	121	474	18.66	Xe	1600000	7500*	i	300	[b]
32000	Horizontal	49	700	13.0	137	487	19.16	Xe	1200000	7500*	i	250	[b]
50	Vertical	40	1.4	1.0	9.5	53	2.09	Hg-Ar	2000	300	i	100	[e]ac
50	Vertical	22	2.3	0.35	9.5	53	2.09	Hg-Ar	1300	900	i	200	[e]
75	Vertical	51	1.6	0.7	30.0	122	4.80	Hg-Ar	2650	65	i	400	[e]ac
100	Vertical	20	5.0	0.25	10.0	90	3.54	Hg-Xe	2000	1700	i	100	[e]ac
100	Vertical	20	5.0	0.25	10.0	90	3.54	Hg-Xe	2200	1700	i	200	[e]
100	Vertical	20	5.0	2.0	10.0	85	3.35	Hg-Xe	2000	1700	i	200	[f]
100	Vertical	20	5.0	0.25	10.0	90	3.54	Hg-Ar	2200	1700	i	200	[h]
200	Vertical	57	3.5	2.2	18.0	128	5.04	Hg-Ar	10000	330	i	400	[h]
200	Vertical	57	3.9	2.2	18.0	108	4.25	Hg-Ar	9500	400	i	200	ac or dc[e]
200	Vertical	57	3.9	2.2	17.0	128	5.04	Hg-Xe	10000	400	i	200/400	ac or dc[e]
200	Vertical	57	3.5	2.2	17.0	122	4.80	Hg-Xe	10000	450	i	400	[f]
200	Vertical	57	3.4	2.2	18.0	122	4.80	Hg-Xe	9500	330	i	200	[f]
200	Vertical	22	9.0	1.9	18.0	114	4.50	Hg-Xe	9500	420	i	1000	[a]
200	Vertical	22	9.0	1.9	18.0	114	4.50	Hg-Xe	9500	420	j	1000	[a]
350	Vertical	67	5.3	2.7	20.0	128	5.04	Hg-Xe	19500	500	i	400	[e]
350	Vertical	60	5.8	2.5	21.0	124	4.88	Hg-Ar	21000	590	i	400	[h]
500	Vertical	60	8.3	4.1	30.0	190	7.48	Hg-Ar	25000	300	i	600	[h]
500	Vertical	74	6.6	4.1	27.0	170	6.69	Hg-Ar	28500	300	i	200	ac or dc[e]
500	Vertical	74	6.6	4.1	27.0	170	6.69	Hg-Ar	30000	300	i	400/200	ac or dc[e]
600	Vertical	22	2.6	3.3	25.0	191	7.50	Hg-Xe	23000	700	i	1000	[a]
600	Vertical	22	2.6	3.3	25.0	191	7.50	Hg-Xe	23000	700	j	1000	[a]
1000	Vertical	65	15.5	6.1	38	191	7.50	Hg-Xe	50000	250	j	1000	[a]
1000	Vertical	35	28.5	3.5	38	214	8.44	Hg-Xe	40000	360	i	1000	[a]
1000	Vertical	35	28.5	3.5	38	214	8.44	Hg-Xe	40000	360	j	1000	[a]
1000	Vertical	63	18	5.8	40	240	9.45	Hg-Ar	60000	140	i	700	[h]
1000	Vertical	38	28	2.6	40	240	9.45	Hg-Ar	45000	350	i	300	[h]
1000	Vertical	80	12.5	4.2	43	285	11.22	Hg-Xe	50000	350	i	400	[f]
1000	Vertical	80	13.3	4.2	43	285	11.22	Hg-Xe	50000	350	i	200	ac[f]
1250	Vertical	35	28.5	3.5	38	208	8.19	Hg-Xe	40000	360	j	1000	[d]

Continued on page 8–136.

Fig. 8–130. *Continued*

Rated Power (watts)	Operating: Position	Operating: Volts	Operating: Amperes	Arc Length (millimeters)	Bulb Diameter[n] (millimeters)	Over-All Length (millimeters)	Over-All Length (inches)	Gas	Initial Lumens	Average Luminance (candelas per square millimeter)	Envelope Material	Rated Life (hours)	Manufacturers and Remarks[m]
1250	Vertical	65	15.5	6.1	38	171	6.74	Hg-Xe	50000	250	j	1000	d
2500	Vertical[o]	50	50	5.0	62	273	10.75	Hg-Xe	122000	760	i	1000	d
2500	Vertical[o]	50	50	5.0	62	273	10.75	Hg-Xe	122000	760	j	1000	a
2500	Vertical	50	50	5.0	57	254	10.00	Hg-Xe	127000	790	j	1000	d
2500	Vertical	50	50	5.0	57	254	10.00	Hg-Xe	127000	790	j	1000	d
3500	Vertical[o]	56	62	6.6	72	330	13.00	Hg-Xe	175000	650	i	1000	a
5000	Vertical[o]	60	82	7.2	85	330.2	13.00	Hg-Xe	265000	860	i	1000	a
5000	Vertical	60	83	7.2	86	307.3	12.10	Hg-Xe	265000	860	j	1000	d
7000	Vertical	70	100	12.0	72	365.25	14.38	Hg-Xe	400000	260	i	500	a
7000	Vertical	70	100	12.0	86	327.66	12.90	Hg-Xe	400000	970	j	500	d

[a] Canrad-Hanovia.
[b] Duro-Test.
[c] ILC Technology.
[d] Optical Radiation.
[e] Osram.
[f] Philips.
[g] Action Tungsram.
[h] Ushio.
[i] Pure quartz envelope.
[j] Ozone free quartz envelope.
[k] Made with either pure quartz or ozone free quartz.
[l] Sapphire envelope.
[m] Lamps are for operation on direct current (dc) only unless otherwise noted here.
[n] Not including bulb tip. Approximately 6 millimeters (¼ inch) additional clearance required at tip location.
[o] Cathode up.

Fig. 8–131. Compact Source Metal Halide Lamps

Lamp Watts	200	200	200	400	500	500	575	575	1000	1000	1000	1000	1200	1200	2500	2500	4000	6000
Manufacturers' Designations	HMI[a] BA200[g]	CID[b]	BB200[g]	CSI[b]	CID[b]	CID[b]	HMI[a] DMI[c] BA575[g]	BB575[g]	CSI[b]	CSI[b]	CID[b]	CID[b]	HMI[a] DMI[c] BA1200[g]	BB1200[g]	HMI[a] DMI[c] BA2500[g]	CID[b]	HMI[a] DMI[c] BA4000[g]	DMI[c]
Luminous Flux-Initial (lumens)	16000	14000	14000	32000	35000	—[e]	49000	45000	90000	—[d]	70000	—[d]	110000	100000	240000	200000	410000	650000
Luminous Intensity (candelas)	—	—	—	—	—	370000	—	—	—	1,500,000	—	850,000	—	—	—	—	—	—
Luminous Efficacy (lumens per watt)	80	70	—	80	70	—	85	—	90	—	70	—	92	—	96	80	102	—
Average Luminance (mega candelas per square meter)	50	—	—	—	—	—	100	—	—	—	—	—	120	—	120	—	110	—
Correlated Color Temperature (kelvins)	5600	5500	5600	4200	5500	5500	5600	5600	4200	4200	5500	5500	5600	5600	5600	5500	5600	5600
Time to Full Output (minutes)	2	1	—	1	1	1	2	—	1	1	1	1	2	—	2	1	2	2
Rated Life (hours)	300	200	500	500	500	1000	750	700	500	3500[f]	500	2000	750	1000	500	500	500	300
Arc Length (millimeters)	10	5.5	10	9	9	9	11	12	14–15	14–15	14–15	14–15	13	13	20	16–18	34	—
Bulb Diameter (millimeters)	14	15	[i]	20	21	145.7	21	[e]	32	205[d]	32	205[d]	27	[d]	30	36	38	—
Maximum Over-All Length (millimeters)	75	57	69	55	105	145	145	145	115	175	115	175	220	175	355	185	405	—
Burning Position	Horiz. ±15°	any	Horiz. ±15°[h]	any	any	any	any	Horiz. ±15°[h]	any	any	any	any	any	Horiz. ±15°[h]	Horiz. ±15°	any	Horiz. ±15°	Horiz. ±15°
Operating Volts	80	65–80	80	100	80	80	95	90	70–85	70–85	70–85	70–85	100	100	115	95	200	135
Operating Current (amperes)	3.1	3.3	3.1	4	7	7	7	7	15	15	15	15	13.8	13.8	25.6	30	24	55

[a] Osram Corp., Newburgh, NY
[b] Thorn EMI Lighting
[c] ILC Technology, Sunnyvale, CA
[d] In a PAR-64 bulb.
[e] In a PAR-46 bulb
[f] Average life to 50% failure, 4 hrs. per start
[g] GTE Sylvania
[h] Refers to capsule position
[i] In a PAR-36 bulb

HMI,BA and DMI lamps are double ended
CSI,BB and CID lamps are single ended

Fig. 8-132. Typical Glow Lamps

Lamp Designation[a]	A9A (NE-2E)	C2A (NE-2H)	C9A (NE-2J)	A1B	A1C	B1A (NE-51)	B2A (NE-51H)
Gas	Neon	Neon	Neon	Neon	Neon	Neon	Neon
Brightness class	Standard	High	High	Standard	High	Standard	High
Nominal watts	1/12	1/4	1/4	1/25	1/7	1/25	1/7
Circuit volts	105–125 ac or dc	105–125 ac 150 dc	105–125 ac 150 dc	105–125 ac or dc	105–125 ac 150 dc	105–125 ac or dc	105–125 ac 150 dc
Bulb (clear)	T-2 (6 mm)	T-2 (6 mm)	T-2 (6mm)	T-2 (6 mm)	T-2 (6 mm)	T-3¼ (10 mm)	T-3¼ (10 mm)
Base[b] (see Fig. 8-9)	Wire Term.	Wire Term.	S.C. Mid. Fl.	Wire Term.	Wire Term.	Min. Bay.	Min. Bay.
Max. over-all length							
(millimeters)	19	19	24	13	13	30	30
(inches)	3/4	3/4[d]	15/16	1/2[d]	1/2[d]	1 3/16	1 3/16
Max. breakdown voltage (volts) initial							
ac	65	95	95	65	95	65	95
dc	90	135	135	90	135	90	135
Series resistance (ohms)[e]	100,000	30,000	30,000	220,000	47,000	220,000	47,000
Average life (hours)[c]	25,000	25,000	25,000	25,000	25,000	15,000	25,000

Lamp Designation[a]	B7A (NE-45)	B9A (NE-48)	J2A (AR-3)	J3A (AR-4)	F4A (NE-58)	J5A (NE-30)	L5A (NE-32)	R6A (NE-40)
Gas	Neon	Neon	Argon	Argon	Neon	Neon	Neon	Neon
Brightness class	Standard	Standard	Standard	Standard	Standard	Standard	Standard	Standard
Nominal watts	1/4	1/4	1/4	1/4	1/2	1	1	3
Circuit volts	105–125 ac or dc	105–125 ac or dc	105–125 ac 135 dc	105–125 ac 135 dc	210–250 ac or dc	105–125 ac or dc	105–125 ac or dc	105–125 ac or dc
Bulb (clear)	T-4½ (14 mm)	T-4½ (14 mm)	T-4½ (14 mm)	T-4½ (14 mm)	T-4½ (14 mm)	S-11 (35 mm)	G-10 (32 mm)	S-14 (44 mm)
Base[b] (see Fig. 8-9)	Cand. Sc.	D.C. Bay.	Cand. Sc.	D.C. Bay.	Cand. Sc.	Med. Sc.	D.C. Bay.	Med. Sc.
Max. over-all length								
(millimeters)	39	38	39	38	39	57	54	89
(inches)	1 17/32	1½	1 17/32	1½	1 17/32	2¼	2⅛	3½
Max. breakdown voltage (volts) initial								
ac	65	65	85	85	75	65	75	60
dc	90	90	120	120	100	90	100	85
Series resistance (ohms)[e]	None[e]	30,000	None[e]	15,000	None[e]	None[e]	7,500	None[e]
Average life (hours)[c]	7,500	7,500	150[f]	150[f]	7,500	10,000	10,000	10,000

[a] ANSI. Existing designation in parenthesis.

[b] Wire Term—Wire terminal (generally available with attached resistor, if desired); Min. Bay—Single contact, miniature bayonet; Cand. Sc.—Candelabra screw; D.C. Bay.—Double contact, bayonet candelabra; S.C. Mid. Fl.—Single contact midget flanged; Med.Sc.—Medium screw.

[c] Ac life ratings shown. For standard brightness lamps, life on dc is approx. 60 per cent of ac values. For high brightness lamps, where a minimum of 150 volts is recommended for dc operation, life will be somewhat lower than the 60 per cent figure and current and wattage will increase.

[d] Glass parts only.

[e] Screw base lamps have resistors built into base; all others must be added externally.

[f] Ultraviolet output drops to 50 per cent of initial during life shown in table. Visible light decreases to 50 percent of initial in 1000 hours.

Fig. 8-133. Typical Flashtubes

Part Number	Configuration	Effective Source Envelope. Dia. x Length. (millimeters)	Mounting	Voltages (Volts)			Typical Operation		Nominal Trigger Voltage (kilovolts)	Maximum Continuous Power (watts)			Explosion Energy (joules)	Typical Application
										Air Cooled				
				Minimum	Typical	Maximum	Energy (joules)	Life (flashes)		Convection	Blower	Fluid Cooled		
FT-34HP[i]	PAR-56		Screw Term	1500	2000	2500	60	4×10^6	20	120	—	—	—	Airport
R-4336[j]	Helical	58 × 45	Plug in	1800	2000	2200	60	3.6×10^6	9	120	—	—	—	Airport
N-106[k]	Helical	38 × 44	End Lugs	1200	1700	2500	100	10^6	30	200	1200	—	—	Photographic
L-833[m]	Helical	30 × 30	End Lugs	1200	1400	4000	120	7×10^6	16	430	860	4300	3250[a]	Visual Warning
FT-38[i]	Helical	29 × 38	Wire Term	325	400	450	13	3×10^6	5	24	—	—	—	Visual Warning
FX-46C[n]	Helical	29 × 32	End Lugs	1500	2000	2500	—	—	25	40	675	—	2600[a]	Laser Pumping
FT-506[i]	Helical	29 × 25	Plug in	600	900	1250	1000	10^4	9	35	—	—	—	General Purpose
FT-152A[i]	Helical	29 × 21	Plug in	325	400	450	200	10^4	5	15	—	—	—	Visual Warning
R-4339[j]	Helical	28 × 28	Plug in	1700	—	2300	20	7×10^6	5	40	—	—	—	Visual Warning
R-4337[j]	Helical	27 × 50	Plug in	850	—	1500	60	3.6×10^6	8	120	—	—	—	Airport
L-224[m]	Helical	25 × 25	Plug in	700	1400	3000	20	7.5×10^5	16	110	220	1100	1000[a]	High Speed Photography
R-4321C[j]	Helical	25 × 20	Plug in	250	—	400	20	2.5×10^6	0.2[b]	25	—	—	—	Visual Warning
R-4429[j]	Helical	24 × 30	Plug in	1500	—	2300	40	5×10^6	9.5	80	—	—	—	Visual Warning
FT-118G[i]	Helical	22 × 22	Wafer	400	500	550	200	10^4	6	25	—	—	—	Photographic
FT-217[i]	Helical	22 × 22	Plug in	600	900	1000	200	10^4	6	15	—	—	—	Photographic
FT-218[i]	Helical	22 × 22	Wafer	600	900	1000	200	10^4	6	15	—	—	—	Photographic
FT-120GC[i]	Helical	19 × 19	Plug in	350	500	550	200	10^4	6	20	—	—	—	Photographic
FT-151[i]	Helical	19 × 19	Plug in	375	450	500	125	10^4	9	15	—	—	—	Visual Warning
L-2102[m]	Helical	15 × 25	Wire Term	600	900	2500	10	10^8	14	60	120	600	550[a]	Stroboscope
FT-106[i]	U	16 × 16	Wafer	250	300	350	50	10^4	4	5	—	—	—	Photographic
FX-77C-9[n]	Linear	19 × 229	End Lugs	1200	2000	4000	—	—	25	150	1200	15000	11000[a]	Laser Pumping Reprographic
L-2810[m]	Linear	15 × 1120	Wire Term	5000	14000	16000	20000	10^4	14[c]	5200	10400	52000	58800[d]	Laser Pumping
FX-47C-12[n]	Linear	13 × 300	End Lugs	1300	2500	5000	—	—	25	125	1000	12500	11000[a]	Laser Pumping Reprographic
FT-61-17H[i]	Linear	7 × 432	End Lugs	3500	3900	4200	62	15×10^6	18	125	—	—	—	Reprographic
FT-61-12H[i]	Linear	7 × 305	End Lugs	2500	3100	3500	38	15×10^6	18	75	—	—	—	Reprographic
FXQ-293-3[n]	Linear	7 × 76	End Lugs	600	1500	2500	—	—	15	—	—	3000	2160[d]	Laser Pumping Reprographic
N-187C[k]	Linear	7 × 76	End Lugs	1000	1500	2000	200	10^6	20	60	480	4000	—	Laser Pumping
L-1988[m]	Linear	6 × 100	End Lugs	700	1200	3500	400[e]	3×10^6	18	275	550	3750	850[f]	Laser Pumping
L-484[m]	Linear	5 × 240	End Lugs	1200	3800	4000	250	2×10^6	18	550	1100	6500	3800[a]	Reprographic
R-4338[j]	Linear	5 × 76	Wire Term.	1800	2000	2200	10	3.6×10^6	5	20	—	—	—	Visual Warning
L-3525[m]	Linear	4 × 196	Wire Term.	1000	11000	17000	140	2×10^3	11[c]	370	740	5000	350[g]	UV Source
L-1832[m]	Linear	4 × 100	Wire Term.	500	8000	10000	15[e]	3×10^6	8[c]	200	400	2400	200[f]	Laser Pumping
FX-38C-3[n]	Linear	4 × 76	End Lugs	700	1500	2500	—	—	15	20	160	1000	840[a]	Laser Pumping Reprographic
FXQ-289-3[n]	Linear	4 × 76	End Lugs	600	1500	2500	—	—	15	150	300	—	1170[d]	Laser Pumping Reprographic
L-349[m]	Linear	4 × 54	Wire Term.	550	900	3000	10	10^7	16	100	200	1200	650[a]	Laser Pumping
FX-139C-3.5[n]	Linear	3.5 × 89	End Lugs	—	10000	—	—	—	10[c]	100	—	—	—	Laser Pumping

N-722C[k]	Linear	3.5 × 3.5	End Lugs	2500	10000	25000	5	10^5	2.5[c]	75	300	—	—	Photochemistry
N-701C[k]	Linear	3 × 89	End Lugs	3500	10000	25000	12	10^5	3.5[c]	50	100	1000	—	Laser Pumping
N-734C[k]	Linear	3 × 76	End Lugs	3500	10000	25000	12	10^5	3.5[c]	50	100	1000	—	Photochemistry
N-725C[k]	Linear	3 × 38	End Lugs	3000	10000	25000	5	10^4	3[c]	40	80	800	—	High Speed Photography
N-189C[k]	Linear	1.3 × 165	End Lugs	1100	1800	2500	500	10^7	25	100	800	10000	—	Laser Pumping
1CPN[o]	Linear	Point Source	End Lugs	120	—	1900	0.2	10^9	6	40	—	—	—	General Purpose
2CPN[o]	Linear	Point Source	End Lugs	120	—	1900	0.4	5×10^8	6	75	—	—	—	General Purpose
3CPN[o]	Linear	Point Source	End Lugs	120	—	1900	2.0	3×10^7	8	100	—	—	—	General Purpose
2UP1.5[o]	Bulb	Point Source	Plug in[h]	120	—	1500	0.02	2×10^9	—[h]	10	—	—	—	General Purpose
2UP3.0[o]	Bulb	Point Source	Plug in[h]	200	—	1500	0.1	5×10^8	—[h]	10	—	—	—	General Purpose
FX-198[n]	Bulb	5 × 5	Plug in[h]	300	800	1500	—	—	—[h]	10	15	—	20	High Speed Strobe Photocomposition
FX-200[n]	Bulb	5 × 5	Plug in[h]	300	800	1500	—	—	—[h]	15	20	—	30	High Speed Strobe Photocomposition
FX-193[n]	Bulb	5 × 5	Plug in[h]	300	800	1500	—	—	—[h]	40	80	—	40	High Speed Strobe Photocomposition
N114[k]	Wound in Flat Circle	7 × 76	End Lugs	1000	2000	3000	1000	10^4	25	200	1200	—	—	Special
N-787B[k]	End view source	1.6 × 1.6	Clamp	—	—	—	0.01	10^6	6[c]	1	—	—	—	High Speed Photography
N-789V[k]	End view source	.02 × .04	Clamp	—	—	—	0.02	10^5	6[c]	1	—	—	—	High Speed Photography

[a] At 500 microseconds duration.
[b] Input to self contained trigger transformer.
[c] Applied across main electrodes through a series switch.
[d] At 1 millisecond duration.
[e] Simmer pulse operation.
[f] At 100 microseconds duration.
[g] At 3 microseconds duration.
[h] Plugs into special trigger module.
[i] General Electric.
[j] GTE Sylvania.
[k] Xenon Corp.
[m] ILC Technology.
[n] E G & G.
[o] United States Scientific Instruments.

Fig. 8–134. Direct-Current Carbon Arc

	Low Intensity	Non-Rotating High Intensity		Rotating High Intensity				
	Application Number*							
	1	2	3	4	5	6	7	8
Type of carbon	Microscope	Projector	Projector	Projector	Projector	Projector ***	Searchlight	Studio "Yellow Flame"
Positive carbon								
Diameter, millimeters	5	7	8	10	11	13.6	16	16
Length, millimeters (inches)	200 (8)	300–355 (12–14)	300–355 (12–14)	500 (20)	500(20)	318 (12.5)	560 (22)	560 (22)
Negative carbon								
Diameter, millimeters (inches)	6	6	7	9 (11/32)	10 (3/8)	13 (1/2)	11	13.5 (17/32)
Length, millimeters (inches)	114 (4½)	230 (9)	230 (9)	230 (9)	230 (9)	230 (9)	300 (12)	230 (9)
Arc current, amperes	5	50	70	105	120	160	150	225
Arc volts, dc	59	40	42	59	57	66	78	70
Arc power, watts	295	2000	2940	6200	6840	10600	11700	15800
Burning rate, millimeters (inches) per hour								
Positive carbon	114 (4.5)	295 (11.6)	345 (13.6)	546 (21.5)	419 (16.5)	432 (17.0)	226 (8.9)	513 (20.2)
Negative carbon	53 (2.1)	109 (4.3)	109 (4.3)	74 (2.9)	61 (2.4)	56 (2.2)	99 (3.9)	56 (2.2)
Approximate crater diameter, millimeters (inches)	3 (0.12)	6 (0.23)	7 (0.28)	9 (0.36)	10 (0.39)	13 (0.50)	14 (0.55)	15 (0.59)
Maximum luminance of crater, candelas per square centimeter	15000	55000	83000	90000	85000	96000	65000	68000
Forward crater candlepower, candelas	975	10500	22000	36000	44000	63000	68000	99000
Crater lumens**	3100	36800	77000	126000	154000	221000	250000	347000
Total lumens†	3100	55000	115000	189000	231000	368000	374000	521000
Total lumens per Arc Watt	10.4	29.7	39.1	30.5	33.8	34.7	32.0	33.0
Color temperature, kelvins‡	3600	5950	5500–6500	5500–6500	5500–6500	5500–6500	5400	4100

* Typical applications: 1, microscope illumination and projection; 2, 3, 4, 5, and 6, motion picture projection; 7, searchlight projection; 8, motion picture set lighting and motion picture and television background projection.

** Includes light radiated in forward hemisphere.

† Includes light from crater and arc flame in forward hemisphere.

‡ Crater radiation only.

*** Jointed carbon.

Fig. 8-135. Flame-Type Carbon Arcs

	Application Number[a]				
	1	2	3	4[b]	5[c, e]
Type of carbons	"Sunshine"	"Sunshine"	Enclosed arc	Photo[b]	Photo[c]
Flame materials	Rare earth	Rare earth	None	Rare earth	Rare earth
Burning position[d]	Vertical	Vertical	Vertical	Vertical	Horizontal
Upper carbon[e]					
Diameter, millimeters (inches)	22 (7/8)	22 (7/8)	13 (1/2)	13 (1/2)	9 (3/8)
Length, millimeters (inches)	300 (12)	300 (12)	75 to 400 (3 to 16)	300 (12)	200 (8)
Lower carbon[e]					
Diameter, millimeters (inches)	13 (1/2)	13 (1/2)	13 (1/2)	13 (1/2)	9 (3/8)
Length, millimeters (inches)	300 (12)	300 (12)	75 to 400 (3 to 16)	300 (12)	200 (8)
Arc current, amperes	60	80	16	38	95
Arc voltage, ac[f]	50	50	138	50	30
Arc power, kilowatts	3.0	4.0	2.2	1.9	2.85
Candlepower[g] (candelas)	9100	10000	1170	6700	14200
Lumens	100000	110000	13000	74000	156000
Lumens per arc watt	33.3	27.5	5.9	39.8	54.8
Color temperature, kelvins	12800[h]	24000[h]	[h]	7420[h]	8150
Spectral intensity (microwatts per square centimeter one meter from arc axis[i])					
Below 270 nm	102	140	—	95	—
270–320 nm	186	244	—	76	100
320–400 nm	2046	2816	1700	684	1590
400–450 nm	1704	2306	177	722	844
450–700 nm	3210	3520	442	2223	3671
700–1125 nm	3032	3500	1681	1264	5632
Above 1125 nm	9820	11420	6600	5189	8763
Total	20100	24000	10600	10253	20600
Spectral radiation (per cent of input power)					
Below 270 nm	.34	.35	—	.5	—
270–320 nm	.62	.61	—	.4	.35
320–400 nm	6.82	7.04	7.7	3.6	5.59
400–450 nm	5.68	5.90	.8	3.8	2.96
450–700 nm	10.70	8.80	2.0	11.7	12.86
700–1125 nm	10.10	8.75	7.6	6.7	19.75
Above 1125 nm	32.70	28.55	29.9	27.3	30.69
Total	67.00	60.00	48.0	54.0	72.20

[a] Typical applications: 1 and 2, accelerated exposure testing, or accelerated plant growth; 3, 4, and 5 blueprinting diazo printing, photo copying, and graphic arts.
[b] Photographic white flame carbons.
[c] High intensity photo 98 carbons.
[d] All combinations shown are operated coaxially.
[e] Both carbons are same in horizontal, coaxial ac arcs.
[f] All operated on alternating current.
[g] Horizontal candlepower, transverse to arc axis.
[h] Deviate enough from blackbody colors to make color temperature of doubtful meaning.
[i] See 1981 Application Volume, Section 11, Fig. 11-18 for spectral power distribution curve.

REFERENCES

1. Elenbaas, W.: *Light Sources*, MacMillan, London and Crane, Russak & Co., New York, 1972.
2. Hall, J. D.: "The Manufacture of Incandescent Lamps," *Elect. Eng.*, Vol. 60, p. 575, December, 1941. Millar, P. S.: "Safeguarding the Quality of Incandescent Lamps," *Trans. Illum. Eng. Soc.*, Vol. XXVI, p. 948, November, 1931.
3. Coolidge, W. D.: "Ductile Tungsten," *Trans. Am. Inst. Elect. Engrs.*, Vol. XXIX, p. 961, May, 1910. Forsythe, W. E., and Adams, E. Q.: "The Tungsten Filament Incandescent Lamp," *J. Sci. Lab.*, Denison University, April, 1937.
4. Smithells, C. J.: *Tungsten*, Chemical Publishing Co., New York, 1953. Rieck, G. D.: *Tungsten and Its Compounds*, Pergamon Press Ltd., London and New York, 1967.
5. Langmuir, I.: "Convection and Conduction of Heat in Gases," *Phys. Review*, Vol. XXXIV, No. 6, p. 401, June, 1912; *Proc. Am. Inst. Elect. Engrs.*, Vol. XXXI, p. 1229, 1912.
6. Thouret, W. E., Anderson, H. A. and Kaufman, R.: "Krypton Filled Large Incandescent Lamps," *Illum. Eng.*, Vol. 65, p. 231, April, 1970. Thouret, W. E., Kaufman, R. and Orlando, J. W.: "Energy and Cost Saving Krypton Filled Incandescent Lamps," *J. Illum. Eng. Soc.*, Vol. 4, p. 188, April, 1975.
7. Morris, R. W.: "Consideration Affecting the Design of Flashing Signal Filament Lamps," *Illum. Eng.*, Vol. XLII, p. 625, June, 1947.
8. Zubler, E. G. and Mosby, F. A.: "An Iodine Incandescent Lamp with Virtually 100 Per Cent Lumen Maintenance," *Illum. Eng.*, Vol. LIV, p. 734, December, 1959.
9. Forsythe, W. E., Adams, E. Q., and Cargill, P. D.: "Some Factors Affecting the Operation of Incandescent Lamps," *J. Sci. Lab.*, Denison University, April, 1939. Merrill, G. S.: "Voltage and Incandescent Electric Lighting," *Proc. Intern. Illum. Congr.*, Vol. II, 1931.
10. The Industry Committee on Interior Wiring Design, *Handbook of Interior Wiring Design*, 420 Lexington Ave., New York, 1941.
11. Potter, W. M. and Reid, K. M.: "Incandescent Lamp Design Life for Residential Lighting," *Illum. Eng.* Vol. LIV, p. 751, December, 1959.
12. "Questions and Answers on Light Sources," *Illum. Eng.*, Vol. LXII, p. 139, March, 1967.
13. Whittaker, J. D.: "Applications of Silver Processed Incandescent Lamps with Technical Data," *Trans. Illum. Eng. Soc.*, Vol. XXVIII, p. 418, May, 1933.
14. Evans, M. W., LaGuisa, F. F. and Putz, J. M.: "An Evaluation of a New Ellipsoidal Incandescent Reflector Lamp," *Light. Des. Appl.*, p. 22, March 1977.
15. Leighton, L. G.: "Characteristics of Ribbon Filament Lamps," *Illum. Eng.*, Vol. LVII, p. 121, March, 1962.
16. *Method for the Designation of Miniature Incandescent Lamps*, C78.390-1983, American National Standards Institute, New York.
17. *Method for the Designation of Glow Lamps*, C78.381-1961, American National Standards Institute, New York.
18. Waymouth, J. F.: *Electric Discharge Lamps*, The M.I.T. Press, Cambridge, MA, 1971.
19. Townsend, M. A.: "Electronics of the Fluorescent Lamp," *Trans. Am. Inst. Elect. Engrs.*, Vol. 61, p. 607, August 1942.
20. Haft, H. H. and Thornton, W. A.: "High Performance Fluorescent Lamps," *J. Illum. Eng. Soc.*, Vol. 2, p. 29, October, 1972. Verstegen, J.M.P.J., Radielovic, D. and Vrenken, L. E.: "A New Generation Deluxe Fluorescent Lamp," *J. Illum. Eng. Soc.*, Vol. 4, p. 90, January, 1975.
21. Thorington, L., Schiazzano, G. and Parascandola, L.: "Spectral Design of the 40-Watt Fluorescent Lamp," *Illum. Eng.*, Vol. LXI, p. 381, May, 1966. Thorington, L., Parcscandola, L. and Cunningham, L.: "Visual and Biologic Aspects of an Artificial Sunlight Illuminant," *J. Illum. Eng. Soc.*, Vol. 1, p. 33, October, 1971.
22. Lowry, E. F., Gungle, W. C., and Jerome, C. W.: "Some Problems Involved In The Design of Fluorescent Lamps," *Illum. Eng.*, Vol. XLIX, p. 545, November, 1954.
23. Lowry, E. F., Frohock, W. S., and Meyers, G. A.: "Some Fluorescent Lamp Parameters and Their Effect on Lamp Performances," *Illum. Eng.*, Vol. XLI, p. 859, December, 1946. Lowry, E. F.: "The Physical Basis for Some Aspects of Fluorescent Lamp Behavior," *Illum. Eng.*, Vol. XLIII, p. 141, February, 1948
25. Lowry, E. F. and Mager, E. L.: "Some Factors Affecting the Life and Lumen Maintenance of Fluorescent Lamps," *Illum. Eng.*, Vol. XLIV, p. 98, February, 1949.
26. Eastman, A. A., and Campbell, J. H.: "Stroboscopic and Flicker Effects from Fluorescent Lamps," *Illum. Eng.*, Vol. XLVII, p. 27, January, 1952.
27. McFarland, R. H., and Sargent, T. C.: "Humidity Effect on Instant Starting of Fluorescent Lamps," *Illum. Eng.*, Vol. XLV, p. 423, July, 1950.
28. Carpenter, W. P.: "Application Data for Proper Dimming of Cold Cathode Fluorescent Lamp," *Illum. Eng.*, Vol. XLVI, p. 306, June, 1951. Campbell, J. H., Schultz, H. E., and Abbott, W. H.: "Dimming Hot Cathode Fluorescent Lamps," *Illum. Eng.*, Vol. XLIX, p. 7, January, 1954. Von Zastrow, E. E.: "Fluorescent Lamp Dimming with Semiconductors," *Illum. Eng.*, Vol. LVIII, p. 312, April, 1963.
29. Campbell, J. H., and Kershaw, D. C.: "Flashing Characteristics of Fluorescent Lamps," *Illum. Eng.*, Vol. LI, p. 755, November, 1956. Bunner, R. W., and Dorsey, R. T.: "Flashing Applications of Fluorescent Lamps," *Illum. Eng.*, Vol. LI, p. 761, November, 1956.
30. Elenbaas, W.: *The High Pressure Mercury Discharge*, Interscience Publishers, Inc., New York, 1951.
31. Elenbaas, W.: *High-Pressure Mercury-Vapor Lamps and Their Applications*, Philips Technical Library, Eindhoven, 1965.
32. Till, W. S., and Pisciotta, M.: "New Designations for Mercury Lamps," *Illum. Eng.*, Vol. LIV, p. 594, September, 1959.
33. Till, W. S., and Unglert, M. C.: "New Designs for Mercury Lamps Increase Their Usefulness," *Illum. Eng.*, Vol. LV, p. 269, May, 1960.
34. Jerome, C. W.: "Color of High Pressure Mercury Lamps," *Illum. Eng.*, Vol. LVI, p. 209, March, 1961.
35. U.S. Food and Drug Administration, Bureau of Radiological Health: "Radiation Safety Performance Standard for Mercury Vapor Lamps," *Federal Register*, Vol. 44, No. 175, September 7, 1979.
36. Larson, D. A., Fraser, H. D., Cushing, W. V., and Unglert, M. C.: "Higher Efficiency Light Source Through Use of Additives to Mercury Discharge," *Illum. Eng.*, Vol. LVIII, p. 434, June, 1963.
37. Martt, E. C., Simialek, L. J., and Green, A. C.: "Iodides in Mercury Arcs—For Improved Color and Efficacy," *Illum. Eng.*, Vol. LIX, p. 34, January, 1964.
38. Waymouth, J. F., Gungle, W. C., Harris, J. M., and Koury, F.: "A New Metal Halide Arc Lamp." *Illum. Eng.*, Vol. LX, p. 85, February, 1965.
39. Reiling, G. H.: "Characteristics of Mercury Vapor-Metallic Iodide Arc Lamps," *J. Opt. Soc. Am.*, Vol. LIV, p. 532, April, 1964.
40. Kühl, B.: "Mercury High Pressure Lamps with Iodide Additives," *Lichttechnik*, Vol. XVI, p. 68, January, 1964.
41. Fromm, O. C., Seehawer, J. and Wagner, W. J.: "A Metal Halide High Pressure Discharge Lamp with Warm White Colour and High Efficacy," *Light. Res. Technol.*, Vol. II, No. 1, p. 1, 1979.
42. *American National Standard Specification for 400-Watt High-Pressure Sodium Lamp S51*, ANSI C78. 1350-1976, American National Standards Institute, New York.
43. Frier, J. P. and Henderson, A. J.: "Stroboscopic Effect of High Intensity Discharge Lamps," *J. Illum. Eng. Soc.*, Vol. 3, p. 83, October, 1973.
44. Rompe, R. and Thouret, W. E.: "Mercury Vapor Lamps of High Brightness," *Zeitsch. Techn. Physik*, Vol. XIX, p. 352, 1938.
45. Rompe, R., Thouret, W. E. and Weizel, W.: "The Problem of Stabilisation of Free Burning Arcs," *Zeitsch. Physik*, Vol. CXXII, p. 1, 1944.
46. Schulz, P.: "Xenon Short Arc Lamps," *Ann. Phys.*, Vol. 1, p. 95, 1947.
47. Bourne, H. K.: *Discharge Lamps for Photography and Projection*, Chapman A. Hall, Ltd., London, 1948.
48. Thouret, W. E.: "New Designs of Quartz Lamps," *Lichttechnik*, Vol. II, p. 73, 1950.
49. Thouret, W. E. and Gerung, G. W.: "Xenon Short Arc Lamps and Their Application," *Illum. Eng.*, Vol. XLIX, p. 520, November, 1954.

50. Anderson, W. T.: "Xenon Short Arc Lamps," *J. Soc. Mot. Pict. Telev. Eng.*, Vol. LXIII, p. 96, 1954.
51. Thouret, W. E.: "Tensile and Thermal Stresses in the Envelope of High Brightness High Pressure Discharge Lamps," *Illum. Eng.*, Vol. LV, p. 295, May, 1960.
52. Retzer, T. C.: "Circuits for Short-Arc Lamps," *Illum. Eng.*, Vol. LIII, p. 606, November, 1958.
53. Thouret, W. E., and Strauss, H. S.: "New Designs Demonstrate Versatility of Xenon High-Pressure Lamps," *Illum. Eng.*, Vol. LVII, p. 150, March, 1962.
54. Lienhard, O. E., and McInally, J. A.: "New Compact-Arc Lamps of High Power and High Brightness," *Illum. Eng.*, Vol. LVII, p. 173, March, 1962.
55. Thouret, W. E., Strauss, H. S., Cortorillo, S. F. and Kee, H.: "High Brightness Xenon Lamps with Liquid-Cooled Electrodes," *Illum. Eng.*, Vol. LX, p. 339, May, 1965.
56. Lienhard, O. E.: "Xenon Compact-Arc Lamps with Liquid-Cooled Electrodes," *Illum. Eng.*, Vol. LX, p. 348, May, 1965.
57. Thouret, W. E., Strauss, H. S., Leyden, J., Kee, H. and Shaffer, G.: "20 to 30 KW Xenon Compact Arc Lamps for Searchlights and Solar Simulators," *J. Illum. Eng. Soc.*, Vol. 2, p. 8, October, 1972.
58. Lemons, T. M.: "HMI Lamps," *Light. Des. Appl.*, Vol. 8, p. 32, August, 1978.
59. Hall, R.: "Development of High Intensity Discharge Sources with Integral Fixtures for Floodlighting Applications," a paper presented at the IES Annual Technical Conference, Sept. 16–20, 1979, Atlantic City, N.J.
60. Null, M. R. and Lozier, W. W.: "Carbon Arc as a Radiation Standard," *J. Opt. Soc. Amer.*, Vol. 52, p. 1156, October, 1962.
61. Payne, E. C., Mager, E. L., and Jerome, C. W.: "Electroluminescence—A New Method of Producing Lighting," *Illum. Eng.*, Vol. XLV, p. 688, November, 1950.
62. Ivey, H. F.: "Problems and Progress In Electroluminescent Lamps," *Illum. Eng.*, Vol. LV, p. 13, January, 1960.
63. Blazek, R. J.: "High Brightness Electroluminescent Lamps of Improved Maintenance," *Illum. Eng.*, Vol. LVII, p. 726, November, 1962.
64. Weber, K. H.: "Electroluminescence—An Appraisal of Its Short-Term Potential," *Illum. Eng.*, Vol. LIX, p. 329, May, 1964.
65. Hall II, J. W.: "Solid State Lamps—How They Work and Some of Their Applications," *Illum. Eng.*, Vol. LXIV, p. 88, February, 1969.
66. *American National Standard for Gas-Fired Illuminating Appliances*, Z-21.42-1971, American National Standards Institute, New York.

History of Light Sources

67. Hammer, W. J.: "The William J. Hammer Historical Collection of Incandescent Electric Lamps," *Trans. New York Elec. Soc.*, New Series, No. 4, 1913.
68. Schroeder, H.: "History of Electric Light," Smithsonian Miscellaneous Collections, Vol. 76, No. 2, August 15, 1923.
69. Howell, J. W., and Schroeder, H.: "History of the Incandescent Lamp," The Maqua Company, 1927.
70. "The Development of the Incandescent Electric Lamp Up to 1879," *Trans. Illum. Eng.*, Vol. XXIV, p. 717, October, 1929.
71. "The Lamp Makers' Story," *Illum. Eng.*, Vol. LI, p. 1, January, 1956.

Lighting Calculations

SECTION 9

Lighting calculations are part of the lighting design process in that they generate information that is required to make design decisions. Even in their simplest form, lighting calculations are mathematical models of complex physical processes. It is usually true that more extensive mathematical models of lighting systems require more extensive calculations and provide the benefit of more accurate information. Less complicated models require less work but usually provide less accuracy. For the purpose of the techniques described in this Section, *accuracy* can be defined as the degree to which the calculated information agrees with physical reality. In many cases, great accuracy is not a necessary characteristic of the information provided by lighting calculations, and simple calculation techniques are perfectly adequate. Occasionally, the lighting design process requires higher accuracy of the information provided by lighting calculations. In this case, a more elaborate mathematical model is required, along with the resulting computational complexity. This Section presents techniques that offer a range of complexity and accuracy.

For calculation techniques not covered in this Section, such as for daylighting, roadways, searchlights and underwater application, see the Index.

Computers in Lighting Calculations

Calculation procedures should be chosen commensurate with the requirements of time and accuracy. In some cases, the more accurate models require calculations of such complexity that only the use of a computer makes their use practical.

Computer programs for lighting design and analysis are available for computers of sizes ranging from hand-held calculators to large in-house computer systems. For many simple lighting design analysis needs, it may be cost effective to write a program to solve a particular or general problem. Throughout this Section an effort has been made to provide algorithms for the more important computational methods which lend themselves to computer use, and with some knowledge of computer programming these algorithms can be turned into effective lighting analysis tools.

Calculated Quantities and Data Required

The information provided by lighting calculations almost always centers on the quantities illuminance and luminance. Other quantities used to evaluate lighting systems, such as visual comfort probability and the various measures of visibility, all use values of illuminance and/or luminance. Most of this Section deals with the calculation of these two quantities. The amount of illuminance or luminance is calculated by modeling the interaction of light and the environment. Thus, lighting calculations require two types of data: photometric and environmental.

The photometric data is a description of the light emitting characteristics of the lighting equipment. This is usually the luminous intensity (candela) distribution of the luminaire describing the spacial distribution of light from the luminaire, usually in spherical coordinates. It is important to note that this information is usually available only in a finite number of directions. Thus, interpolation is required when the

NOTE: References are listed at the end of each section.

intensity is needed in some direction other than those used during the photometric test (see Section 4). The interpolation technique used can affect the result of a calculation considerably. A technique should be chosen on the basis of the required accuracy for the computational result and the complexity of the data.

Environmental data describes the nature of the interaction of light and surfaces, usually in the form of reflectances. This interaction can be very complex and involve the directional and spectral dependence of reflectances. Except in some luminance calculation, the directional characteristics of reflectances are assumed to be uniform and, thus, the reflectance is assumed to be perfectly diffuse, as from a matte surface. In the absence of spectral information, the reflectance is assumed to be more or less independent of the spectral composition of incident flux (the gray assumption). These two assumptions, diffuseness and spectral flatness, greatly simplify lighting calculations. Care should be exercised in the specification of reflectances for surfaces that have a highly saturated color or that exhibit highly directional reflectances. The extensive data required to describe these conditions complicates calculations considerably.

Criteria Rating

Historically, only the simplest lighting calculations have been employed due to the considerable computational resources required of the more elaborate techniques. The simple techniques usually result in a single number, such as an average, used to characterize an entire room or lighting system. With the advent of programmable calculators and powerful microcomputers, computer programs can be employed to calculate any of the various quantities described in this Section (illuminance, luminance, visual comfort probability, visibility, etc.) at many points in the visual environment. As an alternative to a simple average, which hides or fails to use much of the information generated by point calculation techniques, criteria rating can be used.

Criteria rating is a technique that determines the probability that specific criteria will be met anywhere in a defined area. It can be used along with (or instead of) concepts such as average, minimum and maximum, and provides significantly more information. Lighting criteria to which this technique may be applied include luminance, illuminance, visual comfort probability (VCP), equivalent sphere illumination (ESI), contrast and relative visual performance (RVP). The name of the criteria rating assumes the name of the criteria being rated. For example, the criteria rating for illuminance becomes the illuminance rating; VCP becomes VCP rating; etc. Assuming, for example, that an illuminance of 500 lux [50 footcandles] has been established as the design criteria for a specific space, the *illuminance rating* defines the likelihood that any point on the work-plane in that space will have an illuminance equal to or greater than 500 lux [50 footcandles].

Criteria rating is determined by evaluating (by calculation or measurement) the appropriate quantity over the area in question. The distance between evaluation points must not exceed one-fifth the distance from any luminaire to the evaluation plane. The percentage of points in compliance with the criteria is the criteria rating.

Criteria rating may be expressed in shorthand notation by listing the rating in per cent followed by the criteria itself and separated by an "@." For example, a lighting system producing a luminance of 20 candelas per square foot over 60 per cent of the specified area could have its luminance rating expressed as 60%@20cd/ft^2. For dimensionless criteria, such as contrast and VCP, the shorthand form for the criteria rating would include the name and value of the criteria expressed as a decimal, not in per cent form; *i.e.*, 92%@.70VCP for 92 per cent of the area has a VCP of 70(%) or better.

Example. As an example of the use of the criteria rating technique (using the foot as the unit of length), assume a room 30 by 30 feet square, as shown in Fig. 9–1, with an 8-foot ceiling height, a 3-foot high work-plane and the use of recessed luminaires. The tabulated values in Fig. 9–1 are calculated illuminances in the 8- by 12-foot shaded area. The required illuminance is 50 footcandles in the shaded area. The distance

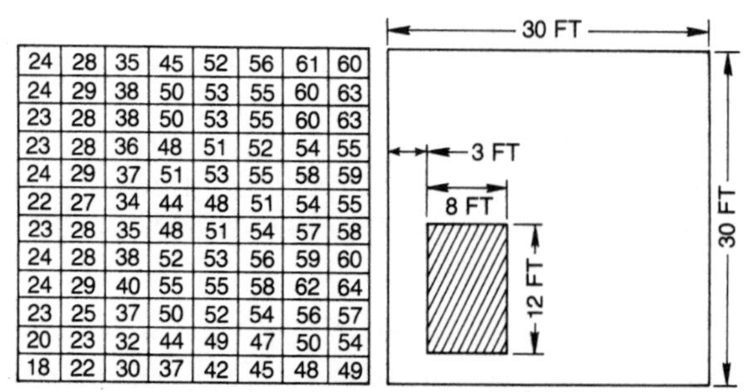

24	28	35	45	52	56	61	60
24	29	38	50	53	55	60	63
23	28	38	50	53	55	60	63
23	28	36	48	51	52	54	55
24	29	37	51	53	55	58	59
22	27	34	44	48	51	54	55
23	28	35	48	51	54	57	58
24	28	38	52	53	56	59	60
24	29	40	55	55	58	62	64
23	25	37	50	52	54	56	57
20	23	32	44	49	47	50	54
18	22	30	37	42	45	48	49

Fig. 9–1. Layout and data for an example of the use of the criteria rating technique. The numbers in the grid represent the calculated values of illuminance, in footcandles, in the center of each one-foot square of the shaded portion of the work-plane.

from the work-plane to the luminaires is 5 feet (8 feet − 3 feet). Thus, the distance between rows and columns of analysis points must be no greater than 1 foot (5 feet divided by 5 feet). The calculated illuminance values are then examined for criteria compliance. It is found that 47 of the 96 locations receive an illuminance of 50 footcandles or more. The illuminance rating of this lighting system for the shaded area is then 48.9% (47 ÷ 96 × 100). This can be expressed as 48.9%@50fc.

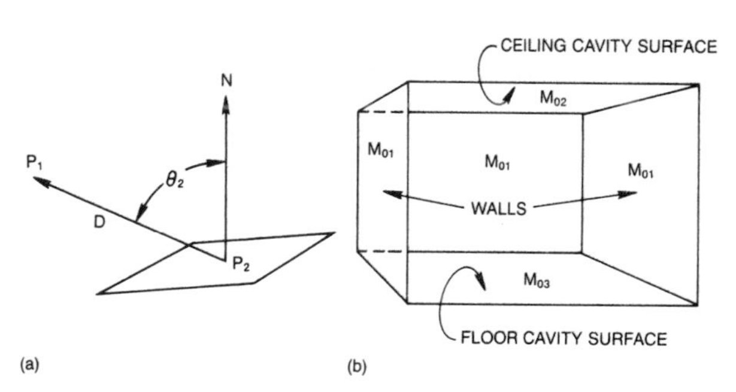

Fig. 9-2. Radiative transfer theory diagrams.

RADIATIVE TRANSFER THEORY

Radiative transfer theory describes the transport of luminous flux from one location to another. Only the simplest elements of the theory are presented here, describing those portions of the general theory commonly used in lighting calculations.

The simplest case is that of a point (small) source of luminous flux, P_1, illuminating a small area of a surface, located at point P_2 in Fig. 9-2a. The illuminance, E_{P_2}, produced at point P_2 is given by the inverse-square cosine law:

$$E_{P_2} = (I/D^2) \cos \theta_2$$

The angle θ_2 is that between the normal to the surface containing point P_2, N, and the direction-distance D. I is the luminous intensity of the point source P_1 in the direction of P_2.

If the source is now assumed to be an area (rather than a point) and the area is a uniform, perfectly diffuse emitter, then the illuminance at point P_2 is:

$$E_{P_2} = M_1 \cdot C_{2-1}$$

where M_1 is the exitance of A_1, the area located at P_1, and is equal to $L\pi$, and C_{2-1} is the *configuration factor* from surface 1 to point 2. It is defined by an area integral involving only the geometry relating the area A_1 to the point P_2. This integration has been performed analytically for many arrangements and explicit equations are available for many practical situations.[1]

If an area A_2 is now illuminated (rather than a point) by the area source A_1, the average illuminance E_2 is given by:

$$E_{A_2} = \frac{M_1 \cdot A_1 f_{1-2}}{A_2}$$

where f_{1-2} is the *form factor* of surface 2 with respect to surface 1. It is defined by a double area integral involving only the geometry relating A_1 to A_2. Explicit equations are available for form factors for many practical situations. It can be seen that an equivalent definition for f_{1-2} is the percentage of total flux that leaves A_1 which falls on A_2.

Note that both configuration factor and form factor assume that the source is a perfectly diffuse emitter. Equations for C_{2-1} and f_{1-2} are easy to evaluate, lend themselves to programmable calculator use and can help determine the illuminance from area sources quickly. Form factors are also used in more elaborate calculations, including the lumen method for average illuminance (see page 9-24).

To explain the use of form factors, consider a normal room cavity as shown in Fig. 9-2b. Assume all walls have the same reflectance and that the lighting system is symmetrical. This allows all four walls to be lumped together and treated as one surface. Assume that the three inside surfaces are completely diffuse and have initial exitances M_{01}, M_{02} and M_{03} respectively. These initial exitances might, for example, be due to the flux falling directly on them from a luminaire within the enclosure. Express exitance in lumens per unit area and illuminance in lumens per unit area.

Since there will be reflections, the final exitances M_1, M_2 and M_3 will be higher than the initial ones. The increase will be due to light falling on the surfaces from other surfaces in the room. The final exitance of surface 3, then, can be expressed as follows:

$$M_3 = M_{03} + \rho_3 E_3$$

where ρ_3 is the reflectance of surface 3, and E_3 is the illuminance on that surface due to the other surfaces in the room. What is now needed is to find E_3.

This is where the form factor comes in. From the preceding paragraphs, the flux falling on surface 3 from surface 2 will equal the total emission of surface 2 times the form factor $f_{2\to3}$. The total emission of surface 2 will equal $M_2 A_2$, where M_2 is the final exitance of surface 2 and A_2 is its area.

The illuminance on surface 3 due to surface 2, therefore, is

$$E_{3(2)} = \text{Flux/Area} = \frac{M_2 A_2 f_{2\to3}}{A_3}$$

This equation, and those to follow, can be simplified by making use of what is called the "Reciprocity Relationship," *i.e.*, $A_m f_{m\to n} = A_n f_{n\to m}$. When this is done, the equation reduces to:

$$E_{3(2)} = M_2 f_{3\to2}$$

Applying the same reasoning to the flux from surface 1, an equation results for M_3 in these terms:

$$M_3 = M_{03} + \rho_3(M_2 f_{3\to2} + M_1 f_{3\to1})$$

All of the above applies equally well to surface 2, giving a similar equation for it, which is:

$$M_2 = M_{02} + \rho_2(M_3 f_{2\to3} + M_1 f_{2\to1})$$

Surface 1, the walls, has one difference from those mentioned above, however. Not all of the flux from it goes to the other two surfaces. Some of it goes to other parts of itself. This flux interchange between the walls must be taken into account. There is an illuminance component on the walls due to the walls themselves, or

$$E_{1(1)} = M_1 f_{1\to1}$$

When this is added, the result is:

$$M_1 = M_{01} + \rho_1(M_1 f_{1\to1} + M_2 f_{1\to2} + M_3 f_{1\to3})$$

With these three equations and the form factors, any interior illuminance or exitance can be determined since M_1, M_2 and M_3 are the unknowns in a set of three simultaneous equations. The illuminance on surface 3 (the work-plane) is obtained by dividing M_3 by ρ_3. Because of the assumed lambertian surfaces, the luminances in candela per unit area are determined from the exitances by

$$L_n = (1/\pi)M_n$$

The above set of simultaneous equations is the basis of generating the tables of coefficients of utilization, the wall exitance coefficients and the ceiling cavity exitance coefficients. Form factors are a function of length to width to height ratios of rectangular rooms. However, they are relatively insensitive to the ratio of length to width for normal rooms.[2] In accordance with CIE practice, the standard lighting coefficient tables are based upon a length-to-width ratio of 1.6 and given as a function of room cavity ratio (RCR) only. Fig. 9-3 gives numerical values for the various form factors. Some useful relationships are:

Fig. 9-3. Form Factors for Zonal-Cavity System (Length/Width = 1.6)

RCR	$f_{1\to1}$	$f_{1\to2}$ $f_{1\to3}$	$f_{2\to1}$ $f_{3\to1}$	$f_{2\to3}$ $f_{3\to2}$
0	0.000	0.500	0.000	1.000
1	.133	.434	.173	.827
2	.224	.388	.311	.689
3	.298	.351	.421	.579
4	.361	.320	.511	.489
5	.415	.292	.585	.415
6	.463	.269	.645	.355
7	.504	.248	.694	.306
8	.540	.230	.735	.265
9	.573	.214	.769	.231
10	.601	.199	.798	.202

$$f_{2\to1} = f_{3\to1} = 1 - f_{2\to3}$$

$$f_{1\to2} = f_{1\to3} = \frac{A_2}{A_1}(1 - f_{2\to3}) = \frac{1 - f_{2\to3}}{0.4(RCR)}$$

$$f_{1\to1} = 1 - 2\frac{A_2}{A_1}(1 - f_{2\to3}) = 1 - \frac{1 - f_{2\to3}}{0.2(RCR)}$$

Thus, all of the form factors can be expressed in terms of $f_{2\to3}$. The analytical expression for $f_{2\to3}$ is complex; however, it can be approximated by the expression* (error < 0.4 per cent for $0 \le \text{RCR} \le 20$):

$$f_{2\to3} \cong 0.026 + 0.503 \exp(-0.270 \text{ RCR}) + 0.470 \exp(-0.119 \text{ RCR})$$

Although more complicated, exactly the same procedure can be used for models or radiative transfer in a room that uses more than three surfaces. The result is a more detailed knowledge of the room surface luminances.

LIGHT LOSS FACTORS[3-5]

Light loss factors adjust lighting calculations from controlled laboratory to actual field conditions. They account for differences in lamp lumen output, luminaire output and surface reflectances between the two conditions. Lighting calculations based on laboratory data alone are likely to be unrealistic if they are not adjusted by the light loss factors that follow.

* $\exp(x) \equiv e^x$

Light loss factors are divided into two groups: recoverable and nonrecoverable (see Fig. 9-4). Recoverable factors are those that can be changed by regular maintenance, such as cleaning and relamping luminaires and cleaning or painting room surfaces. Nonrecoverable factors are those attributed to equipment and site conditions and cannot be changed with normal maintenance.

Light loss factors are multiplicative. The Total Light Loss Factor (LLF) is the product of all the factors listed in Fig. 9-4. No factor should be ignored (set equal to 1.0) until investigations prove otherwise.

Lighting calculations should not be attempted until light loss factors are considered.

Nonrecoverable Factors

The "nonrecoverable" factors usually are not controlled by lighting maintenance procedures. Some will exist initially and continue through the life of the installation—either being of such little effect as to make correction impractical, or being too costly to correct. However, all should be studied, because they can diminish the planned output of the lighting system.

Temperature Factor. The effect of ambient temperature on the output of some luminaires is considerable. Variations in temperature, above or below those normally encountered in interiors, have little effect on the light output of incandescent and high intensity discharge lamp luminaires but have an effect on light output of fluorescent luminaires. Each particular lamp-luminaire combination has its own distinctive characteristic of light output *vs* ambient temperature. To apply a factor for light loss due to ambient temperature, the designer needs to know the highest or lowest temperature expected and to have data showing the variation in light output *vs* ambient temperature for the specific luminaire to be used, and the application (mounting) conditions and effect of heat transfer (if applicable). Test data for the effects of air temperature and flow rate are developed for air-handling luminaires.

Fig. 9-4. Light Loss Factors

Nonrecoverable	Recoverable
Temperature Factor	Lamp Lumen Depreciation Factor
Line Voltage Factor	Lamp Burnout Factor
Ballast Factor	Luminaire Dirt Depreciation Factor
Lamp Position (Tilt) Factor	Room Surface Dirt Depreciation Factor
Equipment Operating Factor	
Luminaire Surface Depreciation Factor	

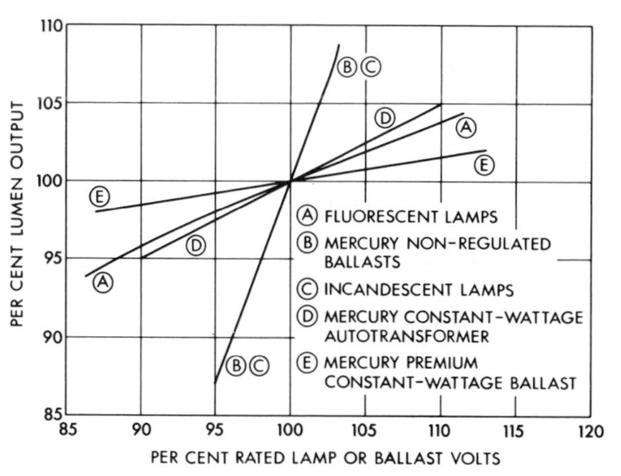

Fig. 9-5. Light output change due to voltage change.

Line Voltage Factor. In-service voltage is difficult to predict, but high or low voltage at the luminaire will affect the output of most luminaires. For incandescent units, small deviations from rated lamp voltage cause approximately three per cent change in lumens for each one per cent of voltage deviation. For mercury luminaires with high reactance ballasts there is a change of approximately three per cent in lamp lumens for each one per cent change in primary voltage deviation from rated ballast voltage. When regulated output ballasts are used, the lamp lumen output is relatively independent of primary voltage within the design range. Fluorescent luminaire output changes approximately one per cent for each two and a half per cent change in primary voltage. Fig. 9-5 shows these variations in graph form. See also Section 8.

Ballast Factor. The lumen output of fluorescent lamps depends on the ballast used to drive the lamps. The lumen output from lamps on commercial ballasts generally differs from the rated lumen output used for photometric test data. For this reason, a multiplicative ballast factor is required to correct photometric test data to actual luminaire performance. The ballast factor is the fractional flux of a fluorescent lamp(s) operated on a ballast compared to the flux when operated on the standard (reference) ballasting specified for rating lamp lumens. Ballast factors are determined in accordance with the *American National Standard Methods of Measurement of Fluorescent Lamp Ballasts.*[6] Manufacturers should be consulted for necessary factors. Some representative values are shown in

Table 8-122. (Note that when uncertified ballasts are used, there may be no reliable data available.)

Changes in lamp lumen output due to lamp temperature changes within the luminaire, from the 25 °C (77 °F) ambient conditions used in rating lamp lumens, are incorporated in the luminaire photometric data and coefficients of utilization. Fluorescent photometry is generally presented for standard lamp loadings on magnetic ballasts. Electronic ballasts and various combinations of "energy conserving" lamps and/or ballasts may change lamp temperature in the luminaire from that occurring during luminaire photometry. A correction for this change should be included in lighting analysis. Data for this correction are available in various forms as part of the luminaire photometric report or from the manufacturer.

Equipment Operating Factor. The lumen output of high intensity discharge (HID) lamps depends on the ballast, the lamp operating position and the effect of reflected power from the luminaire back onto the lamp. These effects are collectively incorporated in the equipment operating factor. The equipment operating factor is the fractional flux of an HID lamp-ballast-luminaire combination, in a given operating position, compared to the flux of the lamp-luminaire combination operating in the position for rating the lamp lumens and using the standard (reference) ballasting specified for rating lamp lumens. Equipment operating factors are determined in accordance with the "IES Approved Method for Determining Luminaire-Lamp-Ballast Combination Operating Factors for High Intensity Discharge Luminaires."[7]

Lamp Position (Tilt) Factor. For HID lamps, the lamp position factor (sometimes known as the tilt factor) is the fractional flux of an HID lamp in a given operating position compared to the flux when the lamp is operated in the position at which the lamp lumens are rated. This factor is determined at constant lamp wattage and comprises part of the equipment operating factor. The lamp position factor is reason-

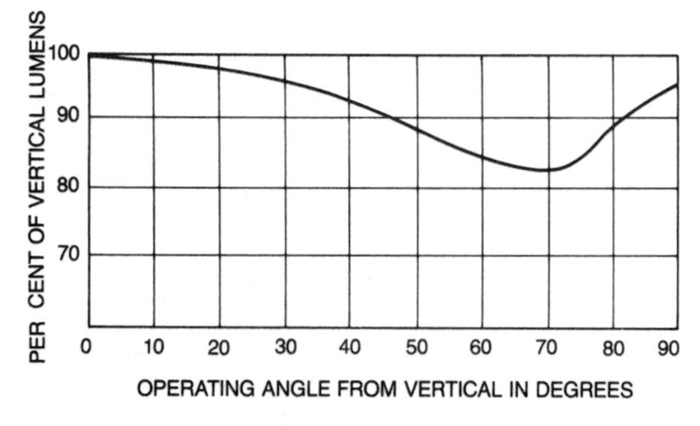

Fig. 9-6. Typical lumen output characteristics of a metal halide lamp as a function of tilt from its vertical position.

Fig. 9-7. Procedure for Determining Luminaire Maintenance Categories

To assist in determining Luminaire Dirt Depreciation (LDD) factors, luminaires are separated into six maintenance categories (I through VI). To arrive at categories, luminaires are arbitrarily divided into sections, a *Top Enclosure* and a *Bottom Enclosure*, by drawing a horizontal line through the light center of the lamp or lamps. The characteristics listed for the enclosures are then selected as best describing the luminaire. Only one characteristic for the top enclosure and one for the bottom enclosure should be used in determining the category of a luminaire. Percentage of uplight is based on 100 per cent for the luminaire.

The maintenance category is determined when there are characteristics in both enclosure columns. If a luminaire falls into more than one category, the lower numbered category is used.

Maintenance Category	Top Enclosure	Bottom Enclosure
I	1. None.	1. None
II	1. None 2. Transparent with 15 per cent or more uplight through apertures. 3. Translucent with 15 per cent or more uplight through apertures. 4. Opaque with 15 per cent or more uplight through apertures.	1. None 2. Louvers or baffles
III	1. Transparent with less than 15 per cent upward light through apertures. 2. Translucent with less than 15 per cent upward light through apertures. 3. Opaque with less than 15 per cent uplight through apertures.	1. None 2. Louvers or baffles
IV	1. Transparent unapertured. 2. Translucent unapertured. 3. Opaque unapertured.	1. None 2. Louvers
V	1. Transparent unapertured. 2. Translucent unapertured. 3. Opaque unapertured.	1. Transparent unapertured 2. Translucent unapertured
VI	1. None 2. Transparent unapertured. 3. Translucent unapertured. 4. Opaque unapertured.	1. Transparent unapertured 2. Translucent unapertured 3. Opaque unapertured

Fig. 9-8. Evaluation of Operating Atmosphere[4]
Factors for Use in Table Below

1 = Cleanest conditions imaginable
2 = Clean, but not the cleanest
3 = Average
4 = Dirty, but not the dirtiest
5 = Dirtiest conditions imaginable

Type of Dirt*	Area Adjacent to Task Area						Filter Factor (per cent of dirt passed)		Area Surrounding Task						Sub Total
	Intermittent Dirt		Constant Dirt		Total				From Adjacent		Intermittent Dirt		Constant Dirt		
Adhesive Dirt		+		=		×		=		+		+		=	
Attracted Dirt		+		=		×		=		+		+		=	
Inert Dirt		+		=		×		=		+		+		=	
											Total of Dirt Factors				

0–12 = Very Clean	13–24 = Clean	25–36 = Medium	37–48 = Dirty	49–60 = Very Dirty

* See step 2 under Luminaire Dirt Depreciation.

Fig. 9-9. Five Degrees of Dirt Conditions

	Very Clean	Clean	Medium	Dirty	Very Dirty
Generated Dirt	None	Very little	Noticeable but not heavy	Accumulates rapidly	Constant accumulation
Ambient Dirt	None (or none enters area)	Some (almost none enters)	Some enters area	Large amount enters area	Almost none excluded
Removal or Filtration	Excellent	Better than average	Poorer than average	Only fans or blowers if any	None
Adhesion	None	Slight	Enough to be visible after some months	High—probably due to oil, humidity or static	High
Examples	High grade offices, not near production; laboratories; clean rooms	Offices in older buildings or near production; light assembly; inspection	Mill offices; paper processing; light machining	Heat treating; high speed printing; rubber processing	Similar to Dirty but luminaires within immediate area of contamination

ably consistent for mercury lamp types; however, for metal halide lamps, it is variable from lamp to lamp and depends on past operating history, *i.e.*, the function is not repeatable for a given lamp.[8,9] Fig. 9-6 presents typical average data for the lamp position factor; manufacturers should be consulted regarding specific lamp types.

Luminaire Surface Depreciation Factor. Luminaire surface depreciation results from adverse changes in metal, paint and plastic components which result in reduced light output. Surfaces of glass, porcelain or processed aluminum have negligible depreciation and can be restored to original reflectance. Baked enamel and other painted surfaces have a permanent depreciation due to all paints being porous to some degree. For plastics, acrylic is least susceptible to change, but its transmittance may be reduced by usage over a period of 15 to 20 years in certain atmospheres. For the same usage, polystyrene will have lower transmittance than acrylic and will depreciate faster.

Because of the complex relationship between the light-controlling elements of luminaires using more than one type of material (such as a lensed troffer) it is difficult to predict losses due to deterioration of materials. Also, for luminaires with one type of surface the losses will be affected by the type of atmosphere in the installation. No factors are available at present.

Recoverable Factors

The "recoverable" factors that follow always need to be considered in determining the total light loss factor. The magnitude of each will depend on the maintenance procedures to be used in addition to the physical environment and the lamps and luminaires to be installed.

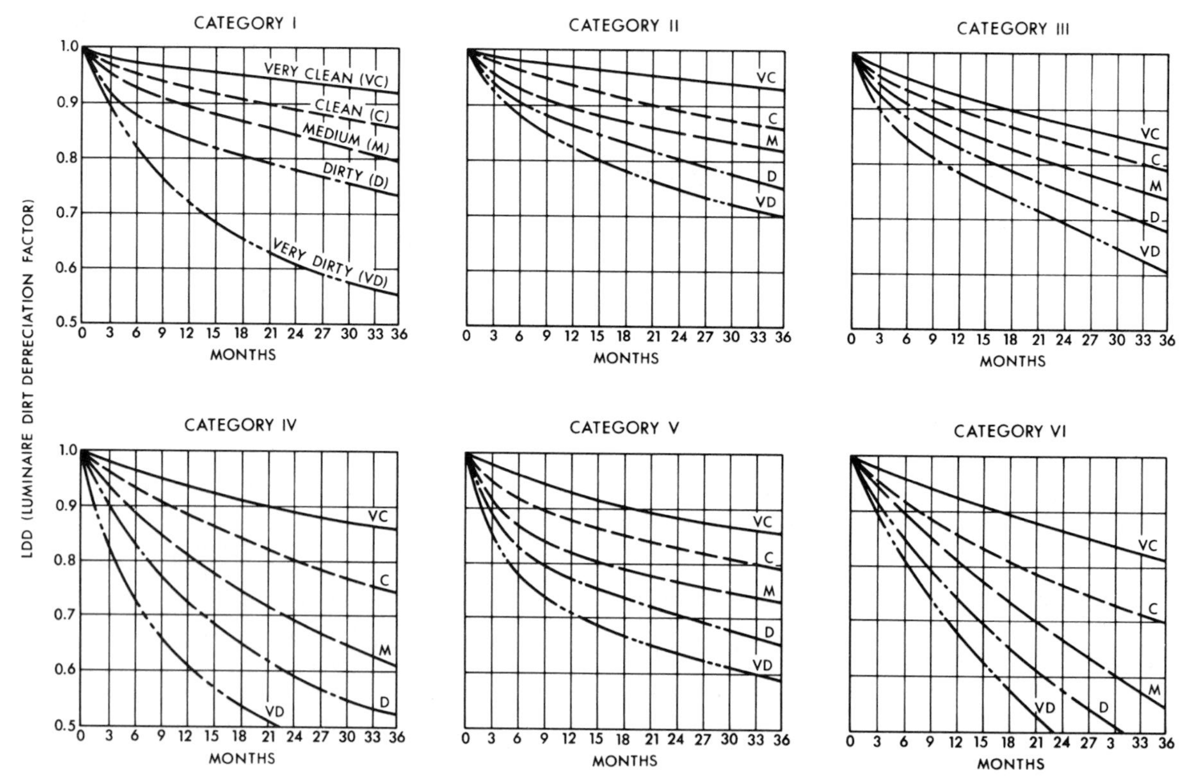

Fig. 9-10. Luminaire Dirt Depreciation factors (LDD) for six luminaire categories (I to VI) and for five degrees of dirtiness as determined from either Fig. 9-8 or 9-9.

Lamp Lumen Depreciation Factor. Information about lamp lumen depreciation is available from manufacturers' tables and graphs for lumen depreciation and mortality of the chosen lamp. Rated average life should be determined for the specific hours per start; it should be known when burnouts will begin in the lamp life cycle. From these facts, a practical group relamping cycle will be established and then, based on the hours elapsed to lamp removal, the specific Lamp Lumen Depreciation (LLD) factor can be determined. Consult the tables at the end of Section 8 or manufacturers' data for LLD factors. Seventy per cent of average rated life is the minimum reached in an installation where burnouts are promptly replaced, whether the planned relamping is to be a group or a random replacement program. It should be noted that some electronic ballasting systems compensate to varying degrees for change in lamp lumen output through life either by an average correction or by feedback control.

Luminaire Dirt Depreciation Factor. The accumulation of dirt on luminaires results in a loss in light output, and therefore a loss on the work-plane. This loss is known as the Luminaire Dirt Depreciation (LDD) Factor and is determined as follows:

1. The luminaire maintenance category is selected from manufacturers' data or by using Fig. 9-7.

2. The atmosphere (one of five degrees of dirt conditions) in which the luminaire will operate is found as follows:

Dirt in the atmosphere will have come from two sources: that passed from adjacent atmosphere(s) and that generated by work done in the surrounding atmosphere. Dirt may be classified as adhesive, attracted or inert and it may come from intermittent or constant sources. Adhesive dirt will cling to luminaire surfaces by its stickiness, while attracted dirt is held by electrostatic force. Inert dirt will vary in accumulation from practically nothing on vertical surfaces, to as much as a horizontal surface will hold before the dirt is dislodged by gravity or air circulation. Examples of adhesive dirt are: grease from cooking, particles from machine operation borne by oil vapor, particles borne by water vapor as in a laundry, fumes from metal-pouring operations or plating tanks. Examples of attracted dirt are: hair, lint, fibers or dry particles which are electrostatically charged from machine operations.

Inert dirt is represented by non-sticky, uncharged particles such as dry flour, sawdust, fine cinders and the like. Figs. 9-8 and 9-9 are useful for evaluating the atmosphere.

3. From the appropriate luminaire maintenance category curve of Fig. 9-10, the applicable dirt condition curve and the proper elapsed time in months of the planned cleaning cycle, the LDD factor is found. For example, if the Category is I, the atmosphere dirty and cleaning every 20 months, LDD is approximately .80.

An alternative procedure to Fig. 9-10 is to use the curve-fitted equation:

$$LDD = e^{-A(t^B)}$$

where constants A and B are found from Fig. 9-11, based on the luminaire maintenance category and the atmosphere condition involved, and t is time in decimal years.

Fig. 9-11. Luminaire Dirt Depreciation Constants (Used for Calculating the LDD for Six Luminaire Categories and for Five Degrees of Dirtiness.)

Luminaire Maintenance Category	B	A				
		Very Clean	Clean	Medium	Dirty	Very Dirty
I	.69	.038	.071	.111	.162	.301
II	.62	.033	.068	.102	.147	.188
III	.70	.079	.106	.143	.184	.236
IV	.72	.070	.131	.216	.314	.452
V	.53	.078	.128	.190	.249	.321
VI	.88	.076	.145	.218	.284	.396

Room Surface Dirt Depreciation Factor. The accumulation of dirt on room surfaces reduces the amount of luminous flux reflected and interreflected to the work-plane. To take this into account, Fig. 9-12 has been developed to provide Room Surface Dirt Depreciation (RSDD) Factors for use in calculating maintained average illuminance levels. These factors are determined as follows:

1. Find the expected dirt depreciation using the Area Atmosphere determined using Fig. 9-8 or 9-9 as guides, the time between cleaning and the curves in Fig. 9-12. For example, if the atmosphere is dirty and room surfaces are cleaned every 24 months, the expected dirt depreciation would be approximately 30 per cent.

2. Knowing the expected dirt depreciation, the type of luminaire distribution (see 1981 Application Volume, Section 1) and the room cavity ratio (see page 9-26), determine the RSDD factor from Fig. 9-12. For example, for a dirt depreciation of 30 per cent, a direct luminaire and a Room Cavity Ratio (RCR) of 4, the RSDD would be .92.

Lamp Burnout Factor. Unreplaced burned-out lamps will vary in quantity, depending on the kinds of lamps and the relamping program used. Manufacturers' mortality statistics should be consulted for the performance of each lamp type to determine the number expected to burn out before the time of planned replacement is

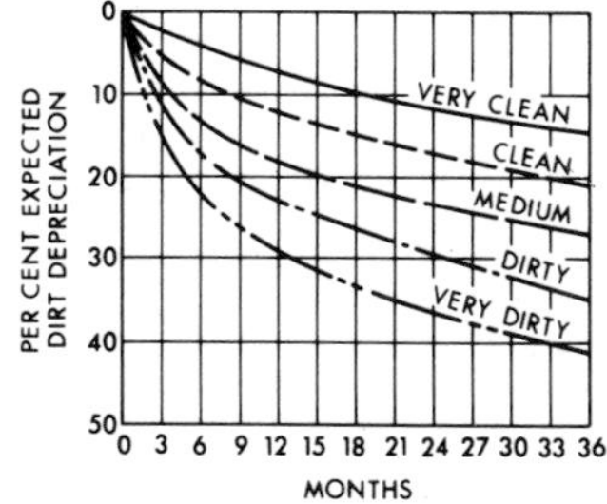

Fig. 9-12. Room Surface Dirt Depreciation Factors

	Luminaire Distribution Type																			
	Direct				Semi-Direct				Direct-Indirect				Semi-Indirect				Indirect			
Per Cent Expected Dirt Depreciation	10	20	30	40	10	20	30	40	10	20	30	40	10	20	30	40	10	20	30	40
Room Cavity Ratio																				
1	.98	.96	.94	.92	.97	.92	.89	.84	.94	.87	.80	.76	.94	.87	.80	.73	.90	.80	.70	.60
2	.98	.96	.94	.92	.96	.92	.88	.83	.94	.87	.80	.75	.94	.87	.79	.72	.90	.80	.69	.59
3	.98	.95	.93	.90	.96	.91	.87	.82	.94	.86	.79	.74	.94	.86	.78	.71	.90	.79	.68	.58
4	.97	.95	.92	.90	.95	.90	.85	.80	.94	.86	.79	.73	.94	.86	.78	.70	.89	.78	.67	.56
5	.97	.94	.91	.89	.94	.90	.84	.79	.93	.86	.78	.72	.93	.86	.77	.69	.89	.78	.66	.55
6	.97	.94	.91	.88	.94	.89	.83	.78	.93	.85	.78	.71	.93	.85	.76	.68	.89	.77	.66	.54
7	.97	.94	.90	.87	.93	.88	.82	.77	.93	.84	.77	.70	.93	.84	.76	.68	.89	.76	.65	.53
8	.96	.93	.89	.86	.93	.87	.81	.75	.93	.84	.76	.69	.93	.84	.76	.68	.88	.76	.64	.52
9	.96	.92	.88	.85	.93	.87	.80	.74	.93	.84	.76	.68	.93	.84	.75	.67	.88	.75	.63	.51
10	.96	.92	.87	.83	.93	.86	.79	.72	.93	.84	.75	.67	.92	.83	.75	.67	.88	.75	.62	.50

reached. Practically, quantity of lamp burnouts is determined by the quality of the lighting services program incorporated in the initial design procedure and by the quality of the physical performance of the program.

Lamp burnouts contribute to loss of light. If lamps are not replaced promptly after burnout, the average illuminance will be decreased proportionately. In some instances, more than just the faulty lamp may be lost. For example, when series sequence fluorescent ballasts are used and one lamp fails, both lamps go out. The Lamp Burnout (LBO) Factor is the ratio of the lamps remaining lighted to the total, for the maximum number of burnouts permitted.

Total Light Loss Factor. The Total Light Loss Factor (LLF) is simply the product of multiplying all the contributing factors described above. Where factors are not known, or applicable, they are assumed to be unity. At this point, if it is found that the total light loss factor is excessive it may be desirable to reselect the luminaire.

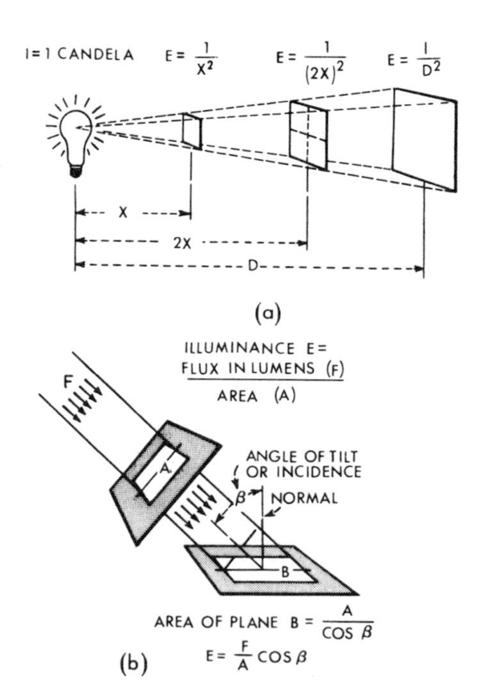

Fig. 9-14. Point calculations assume a point source and involve applications of the inverse-square and cosine laws.

ILLUMINANCE CALCULATIONS

An important aspect in evaluating lighting designs is the ability to calculate illuminance at a specific point in space, such as a visual task, and the average in a space on a specific plane.

Illuminance at a Point—Direct Component[10]

The calculation of the illuminance at a point, whether on a horizontal, a vertical or an inclined plane consists of two parts: the Direct Component and the Reflected Component. The total of these two components is the illuminance at the point in question.

Fig. 9-13 is a matrix of the most commonly used systems of computing the direct component. Although each of the methods can be used to make calculations under all the conditions if sufficient data is available, the × in the matrix indicated the conditions under which the method is easier to use. The following is a brief description of each method.

Inverse-Square Method. Variations in the formula involving the inverse-square law are used to determine the illuminance at definite points where the distance from the source is at least five times the maximum dimension of the source. In such situations the illuminance is proportional to the candlepower of the source in the

Fig. 9-13. Methods of Determining the Direct Illumination Component of Illumination at a Point

Source	Point			Linear			Area		
Plane	Horiz.	Vert.	Inclined	Horiz.	Vert.	Inclined	Horiz.	Vert.	Inclined
1. Inverse Square	×	×	×						
2. Plan-Scale Method	×	×							
3. Angular Coord-DIC Method				×	×				
4. IES-London-Aspect Factor Method				×	×	×			
5. Illumination Charts and Tables	×	×	×	×	×	×	×	×	×
6. Idealized Source Chart							×	×	
7. Configuration Factor							×	×	
8. Real Source Method	×	×	×	×	×	×	×	×	×

given direction, and inversely proportional to the square of the distance from the source:

Illuminance on Plane Normal to Light Ray

$$= \frac{\text{Candlepower of Source* in Direction of Ray}}{(\text{Distance from Source to Plane})^2}$$

$$= \frac{I^*}{D^2} \text{ (see Fig. 9–14a).}$$

If the surface on which the illuminance to be determined is tilted, instead of normal to the rays, the light will be spread over a greater area, reducing the illuminance in the ratio of the area of plane A to the area of plane B, as shown in Fig. 9–14b. This ratio is equal to the cosine of the angle of incidence or tilt, thus:

Illuminance on Plane B

$$= \frac{\text{Candlepower of Source* in Direction of Ray}}{(\text{Distance from Source to Point in Plane})^2} \times \cos\beta$$

$$= \frac{I^*}{D^2} \times \cos\beta$$

where β equals the angle between the light ray and a perpendicular to the plane at that point.

Fig. 9–15 illustrates the particular cases where the plane on which the illuminance is to be determined is either horizontal or vertical. In Fig. 9–15 H is the vertical mounting height of the light source above the plane of measurement; R is the horizontal distance from the light source to the point whose illuminance is being computed; D is the actual distance from the light source to the point; and I is the candlepower of the source in the direction of the point (from the distribution curve). For the horizontal plane, angle β equals angle θ. Furthermore, since $\cos\theta = \cos\beta = H/D$, the formula in Fig. 9–15a may be expressed as follows:

Illuminance on Horizontal Plane (E_h)

$$= \frac{I^* \times \cos\theta}{D^2} = \frac{I^* \times \cos\beta}{D^2}$$

$$= \frac{I^* \times H}{D^3} = \frac{I^* \times \cos^3\theta}{H^2}$$

For the vertical plane $\sin\theta = \cos\beta = R/D$, the formula in Fig. 9–15b may be expressed as follows:

Illuminance on Vertical Plane (E_v)

$$= \frac{I^* \times \sin\theta}{D^2} = \frac{I^* \times \cos\beta}{D^2}$$

$$= \frac{I^* \times R}{D^3} = \frac{I^* \times \cos^2\theta \sin\theta}{H^2}$$

It is often more convenient to obtain the illuminance from tables than by computation from formulas. Fig. 9–16 gives illuminances on horizontal and vertical planes calculated for various mounting heights and distances from a light source of 100 candelas. Illuminance for other values of luminous intensity is simply a matter of percentages.

Plan-Scale Method. The Plan-Scale Method provides a rapid means for finding the illuminance at a point due to several luminaires without tediously applying the inverse-square law many times. The following is one method, although others have been described:[11]

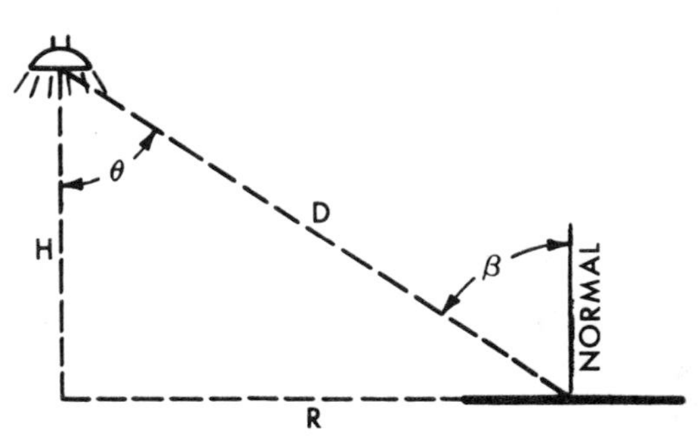

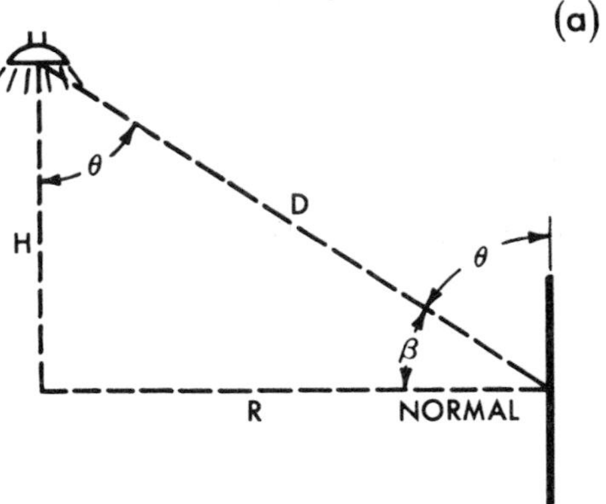

Fig. 9–15. Fundamental relationships for point calculations where the inverse-square law applies.

* Candlepower of the source at some specific time. The candlepower should be multiplied by appropriate light loss factors.

Fig. 9-16. Initial Illuminance Computed for Points at Various Locations on a Horizontal Plane in Terms of 100 Candela Sources* (When distances [horizontal and height] are in meters, the illuminances are in lux, when the distances are in feet, the illuminances are in footcandles.)

Height of Light Source Above Surface	Horizontal Distance From Axis of Light Source																										
	0	1	2	3	4	5	6	7	8	9	10	11	12	13	14	15	16	18	20	22	24	26	28	30	35	40	50
	Illuminances for Each 100 Candela																										
2	0°0′ 25.00	27° 17.85	45° 8.850	56° 4.275	63° 2.245	68° 1.298	71° .802	74° .528	76° .355	78° .255	79° .190	80° .142	81° .113	81° .090	82° .070	82° .058	83° .048	84° .038	84° .025	85° .020	85° .015	86° .013	86° .008	86° .007	87° .004	87° .000	87° .000
3	0°0′ 11.11	18° 9.500	34° 6.400	45° 3.933	53° 2.400	59° 1.522	63° 1.000	67° .680	69° .477	72° .356	73° .264	75° .205	76° .161	77° .126	78° .100	79° .084	80° .070	81° .050	81° .036	82° .027	83° .021	83° .016	84° .012	84° .011	85° .007	86° .004	87° .002
4	0°0′ 6.250	14° 5.707	27° 4.472	37° 3.200	45° 2.210	51° 1.524	56° 1.066	60° .764	63° .559	66° .419	68° .320	70° .249	72° .198	73° .159	74° .130	75° .107	76° .090	78° .064	79° .047	80° .037	81° .028	81° .022	82° .018	82° .015	84° .009	84° .006	86° .003
5	0°0′ 4.000	11° 3.771	22° 3.202	31° 2.522	39° 1.904	45° 1.414	50° 1.050	54° .785	58° .595	61° .458	63° .358	66° .283	67° .228	69° .185	70° .152	72° .126	73° .106	74° .077	76° .057	77° .044	78° .034	79° .027	80° .022	81° .017	82° .010	83° .008	84° .004
6	0°0′ 2.778	9° 2.673	18° 2.372	27° 1.987	34° 1.600	40° 1.260	45° .982	49° .766	53° .600	56° .474	59° .378	61° .305	63° .249	66° .205	67° .170	68° .142	69° .120	71° .088	73° .066	75° .051	76° .040	77° .032	78° .026	79° .021	80° .013	81° .009	83° .005
7	0°0′ 2.041	8 1.980	16° 1.814	23° 1.585	30° 1.336	36° 1.100	41° .893	45° .722	49° .583	52° .473	55° .385	58° .316	60° .261	62° .218	63° .183	65° .154	66° .131	69° .097	71° .074	72° .057	74° .045	75° .036	76° .029	77° .024	79° .016	80° .010	82° .006
8	0°0′ 1.563	7° 1.527	14° 1.427	21° 1.283	27° 1.118	32° .958	37° .800	41° .672	45° .552	48° .458	51° .381	54° .318	56° .267	58° .225	60° .191	62° .163	63° .140	66° .105	68° .080	70° .063	72° .050	73° .040	74° .032	75° .026	77° .018	79° .012	81° .007
9	0°0′ 1.235	6° 1.212	13° 1.148	18° 1.054	24° .943	29° .825	34° .711	38° .697	42° .515	45° .437	48° .370	51° .314	53° .267	55° .228	57° .196	59° .168	61° .146	63° .110	66° .085	68° .067	69° .053	71° .043	72° .035	73° .029	76° .019	77° .013	80° .008
10	0°0′ 1.000	5°43′ .985	11° .943	17° .879	22° .801	27° .716	31° .631	35° .550	39° .476	42° .411	45° .354	48° .305	50° .263	52° .227	54° .196	56° .171	58° .149	61° .115	63° .089	66° .071	67° .057	69° .046	70° .038	72° .032	74° .021	76° .014	79° .008
11	0°0′ .826	5°12′ .816	10° .787	15° .742	20° .686	24° .623	29° .559	32° .496	36° .437	39° .383	42° .335	45° .292	48° .255	50° .223	52° .195	54° .171	56° .150	59° .117	61° .092	63° .074	65° .060	67° .049	69° .040	70° .034	73° .023	75° .015	78° .009
12	0°0′ .694	4°46′ .687	9° .668	14° .634	18° .593	23° .546	27° .497	30° .448	34° .400	37° .356	40° .315	43° .278	45° .246	47° .217	49° .191	51° .169	53° .150	56° .119	59° .094	61° .076	63° .065	65° .051	67° .043	68° .036	71° .024	73° .017	77° .009
13	0°0′ .592	4°24′ .587	9° .571	13° .547	17° .517	21° .481	25° .447	28° .404	32° .366	35° .329	38° .295	40° .263	43° .235	45° .209	47° .187	49° .166	51° .148	54° .119	57° .096	59° .078	62° .064	63° .053	65° .044	67° .037	70° .025	72° .017	76° .010
14	0°0′ .510	4°5′ .506	8° .495	12° .477	16° .454	20° .426	23° .396	27° .365	30° .334	33° .304	36° .275	38° .248	41° .223	43° .201	45° .180	47° .162	49° .146	52° .118	55° .096	58° .079	60° .065	62° .054	63° .046	65° .039	68° .026	71° .018	75° .011
15	0°0′ .444	3°49′ .442	8° .433	11° .419	15° .401	18° .380	22° .356	25° .331	28° .305	31° .280	34° .256	36° .233	39° .212	41° .192	43° .174	45° .157	47° .142	50° .117	53° .096	56° .079	58° .066	60° .055	62° .047	63° .040	67° .027	69° .019	73° .011
16	0°0′ .391	3°35′ .388	7° .382	11° .371	14° .357	17° .339	21° .321	24° .300	27° .280	29° .259	32° .238	35° .219	37° .200	39° .183	41° .167	43° .152	45° .138	48° .115	51° .095	54° .080	56° .067	58° .056	60° .048	62° .041	66° .028	68° .020	72° .012
17	0°0′ .346	3°22′ .344	7° .339	10° .331	13° .319	16° .306	19° .290	22° .274	25° .256	28° .239	30° .222	33° .205	35° .189	37° .174	39° .159	41° .146	43° .134	47° .112	50° .094	52° .079	55° .069	57° .057	59° .048	60° .042	64° .029	67° .021	71° .012
18	0°0′ .309	3°11′ .307	6° .303	9° .297	13° .287	16° .276	18° .264	21° .250	24° .236	27° .221	29° .206	31° .192	34° .178	36° .165	38° .152	40° .140	42° .129	45° .109	48° .092	51° .079	53° .067	55° .057	57° .049	59° .042	63° .030	66° .021	70° .012
19	0°0′ .277	3°1′ .276	6° .273	9° .267	12° .260	15° .251	18° .240	20° .229	23° .217	25° .205	28° .192	30° .180	32° .167	34° .156	36° .145	38° .134	40° .124	43° .106	46° .090	49° .077	52° .066	54° .057	56° .049	58° .042	62° .030	65° .022	69° .013
20	0°0′ .250	2°51′ .249	5°43 .246	9° .242	11° .236	14° .228	17° .219	19° .210	22° .200	24° .190	27° .179	29° .168	31° .158	33° .147	35° .137	37° .128	39° .119	42° .103	45° .088	48° .076	50° .066	52° .057	54° .049	56° .043	60° .030	63° .022	68° .013
21	0°0′ .227	2°44′ .226	5°26′ .224	8° .220	11° .215	13° .210	16° .201	18° .194	21° .185	23° .176	25° .167	28° .158	30° .144	32° .139	34° .131	36° .122	37° .114	41° .099	44° .086	46° .075	49° .065	51° .056	53° .049	55° .043	59° .031	62° .023	67° .014
22	0°0′ .207	2°36′ .206	5°10′ .205	8° .201	10° .196	13° .192	15° .185	18° .179	20° .171	22° .164	25° .155	27° .148	29° .140	31° .132	33° .124	34° .114	36° .109	39° .096	42° .084	45° .073	47° .064	50° .056	52° .049	54° .043	58° .031	61° .023	66° .014

* Upper figures—angle between light ray and vertical. Lower figures—illuminance on a horizontal plane produced by source. Illuminances on the vertical surface—at a point that lies in a vertical plane which also includes the light source—may be determined by using the multiplying factor found when the table headings are reversed, *i.e.*, the height of the light source is read on the horizontal distance scale, etc.

Step 1: Prepare a plan drawing of the lighting installation to some convenient scale as illustrated in Fig. 9-17a.

Step 2: Compute the illuminance on the workplane at several horizontal distances from the luminaire for the luminaire mounting height involved. One of the most convenient methods is to apply the equation:

$$E_h = \frac{I^* \times \cos^3 \theta}{H^2}$$

The values of E_h thus computed are used to make a special scale, as in Fig. 9-17c. Since the calculated values will seldom be whole numbers and the scale will be easier to use if it has mostly whole numbers, an intermediate graph, as in Fig. 9-17b will be helpful. Compute and plot a few points; drawing a curve through them will permit the determination of illuminances at distances other than those calculated. The scale used to draw the illuminance marks should be the same as that used for the lighting plan of Step 1.

Example. Suppose that the illuminance at point P due to luminaire A in Fig. 9-17a is to be found. The zero distance point on the scale (marked L in Fig. 9-17c) could be placed at luminaire A and the illuminance read from the scale where it touches point P. However, the illuminance at P depends only on the distance between P and A. Consequently, the scale can be turned around as shown in Fig. 9-17d. In the example shown in Fig. 9-17 the illuminance at point P is 12.5 footcandles due to luminaire A. The contribution of each luminaire at point P can be found by rotating the scale about point P and reading the scale as it touches each luminaire.

The above procedure applies when the candlepower distribution of a luminaire is symmetrical about the vertical centerline. If the distribution is not symmetrical, the candlepower distribu-

* Candlepower of the source at some specific time. The candlepower should be multiplied by appropriate light loss factors.

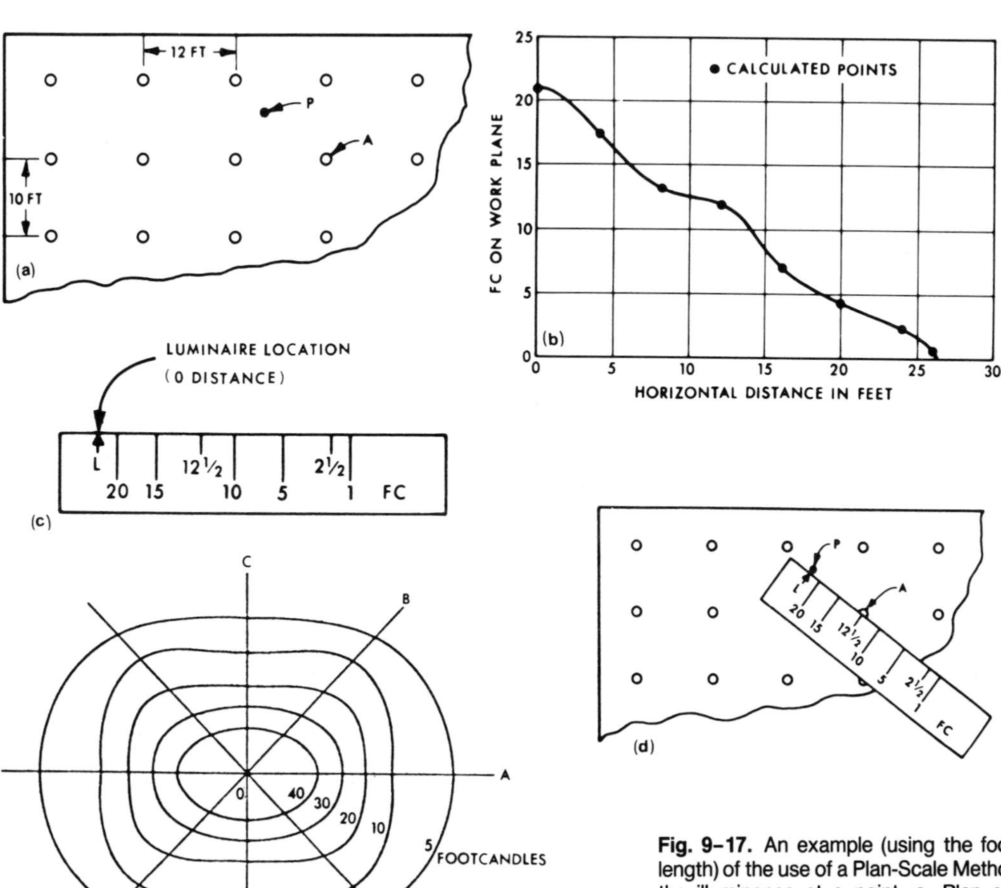

Fig. 9-17. An example (using the foot as the unit of length) of the use of a Plan-Scale Method for calculating the illuminance at a point. a. Plan of room used to illustrate the method. b. Intermediate graph. c. Scale made from data on graph (b). d. Applying the scale to drawing of (a). e. Isolux curves on tracing paper.

tions are given for at least three directions (parallel, 45-degree, and perpendicular planes) on the photometric report for the luminaire. Illuminance values on the work-plane are graphed for each of these directions in a manner similar to Fig. 9-17b.

Then a plot of lines indicating various illuminance levels is made on tracing paper as illustrated in Fig. 9-17e. Illuminance values along the three directions *OA*, *OB*, and *OC* are read from the three graphs, and lines are sketched in the upper right-hand quadrant. The other three quadrants are symmetric. The distance scale used to prepare this transparent overlay is the same as the scale used for the lighting plan. To determine illuminance at a point *Q*, place the transparent overlay on the lighting plan with *O* over the point *Q*.

The directions of the *OA* and *OC* should be parallel to the corresponding directions of the luminaires. Now, the illuminance at point *Q* due to each luminaire can be read from the intersection of the luminaire with the curves on the overlay.

Angular Coordinate—DIC* Method.[12] The Angular Coordinate Method is most applicable to continuous rows of fluorescent luminaires. Two angles are involved in this calculation, a longitudinal angle α and a lateral angle β. See Fig. 9-18. Angle α is the angle between a vertical line (perpendicular to the ceiling) passing through the seeing task (point *P*), and a line from the seeing task to the end of the rows of the luminaires. If the seeing task is not in the vertical plane of a row of luminaires, a parallel reference plane is created for the specification of angle α.

Angle α is easily determined graphically from a chart showing angles α and β for various combinations of *V*, the vertical distance from the seeing task to the plane of the luminaire, and *H*, the horizontal distance parallel to the luminaires from the seeing task to the end of the row of luminaires. See Fig. 9-18. Usually, all rows of luminaires have the same α coordinates, one coordinate for each end of the row.

Angle β is the angle between the vertical plane of the row of luminaires and a tilted plane containing both the seeing task and the luminaire or row of luminaires. This angle is determined from the same chart as angle α, again using *V*, the mounting height of the luminaires above the seeing task, and *H*, which in determining angle β, is the horizontal distance from the seeing task to the row of luminaires, measured perpendicular to the luminaires. Angle β is different for each row of luminaires. Each row has only one β coordinate.

The direct illuminance component for each luminaire or row of luminaires is determined by referring to a table of direct illuminance components for the specific luminaire. Tables such as shown in Fig. 9-19 list direct illuminance components for specific luminaires for determining the illuminance on the horizontal and on vertical planes either parallel to, or perpendicular to the luminaires. The direct illuminance components are based on the assumption that the luminaire is mounted a given height (1.8 meters or 6 feet) above the seeing task. If this mounting height above the task is other than that height, the direct illuminance component

* Direct Illuminance Component.

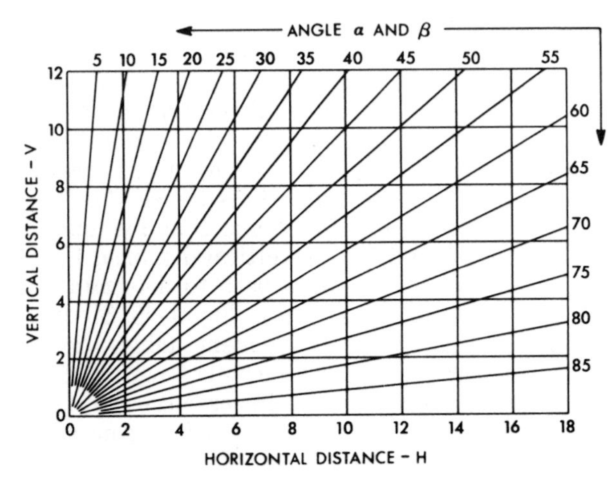

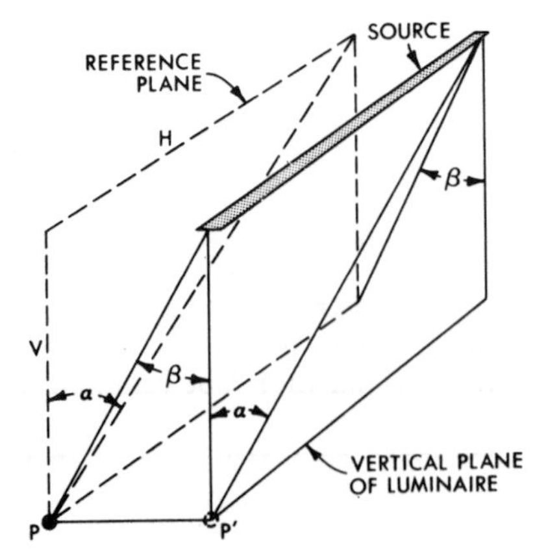

Fig. 9-18. Relationship of angles and distances used in the Angular Coordinate Method.

Fig. 9–19. Typical Footcandle Presentation of Angular Coordinate Method Data

Direct Illuminance Components																
β	5	15	25	35	45	55	65	75	5	15	25	35	45	55	65	75
α	Vertical Surface Illuminance, Footcandles at a Point On a Plane Parallel to Luminaires								Vertical Surface Illuminance, Footcandles at a Point On a Plane Perpendicular to Luminaires							
0–10	1.0	2.8	3.8	3.4	2.0	1.1	.4	...	1.0	.9	.7	.4	.2	.1	...	...
0–20	2.0	5.4	7.4	6.7	4.0	2.2	.9	...	4.0	3.5	2.7	1.6	.7	.3	.1	...
0–30	2.8	7.6	10.6	9.6	5.9	3.3	1.4	...	8.3	7.4	5.9	3.5	1.6	.7	.2	...
0–40	3.4	9.3	12.9	12.0	7.5	4.4	1.9	...	13.3	11.9	9.4	5.9	2.7	1.2	.4	...
0–50	3.8	10.4	14.4	13.6	8.7	5.4	2.5	...	17.7	15.9	12.7	8.2	3.9	1.9	.7	...
0–60	4.0	10.9	15.2	14.6	9.6	6.2	3.1	...	20.7	18.7	15.2	10.2	5.1	2.7	1.1	...
0–70	4.1	11.1	15.5	15.1	10.1	6.8	3.7	...	22.3	20.2	16.7	11.6	6.2	3.5	1.7	...
0–80	4.1	11.2	15.6	15.2	10.3	7.1	4.1	...	23.0	20.9	17.4	12.3	6.9	4.2	2.3	.1
0–90	4.1	11.2	15.6	15.2	10.3	7.1	4.2	.2	23.2	21.1	17.6	12.5	7.1	4.4	2.5	.3
	Footcandles at a Point on Work-Plane								Category V							
0–10	11.5	10.6	8.1	4.9	2.0	.8	.2	...								
0–20	22.7	20.4	15.7	9.6	4.0	1.6	.4	...								
0–30	31.9	28.7	22.5	13.7	5.9	2.4	.6	...								
0–40	39.1	35.1	27.5	17.1	7.5	3.2	.8	...								
0–50	43.5	39.1	30.8	19.4	8.7	3.9	1.1	...								
0–60	45.6	41.1	32.5	20.8	9.6	4.4	1.4	...								
0–70	46.3	41.8	33.2	21.5	10.1	4.8	1.7	...								
0–80	46.5	42.0	33.4	21.7	10.3	5.0	1.9	...	2 T-12 Lamps—430 mA							
0–90	46.5	42.0	33.4	21.7	10.3	5.0	1.9	...	1′ Wide Prismatic Wrap-Around							

Fig. 9–20. Layout and Data Obtained Using the Angular Coordinate Method

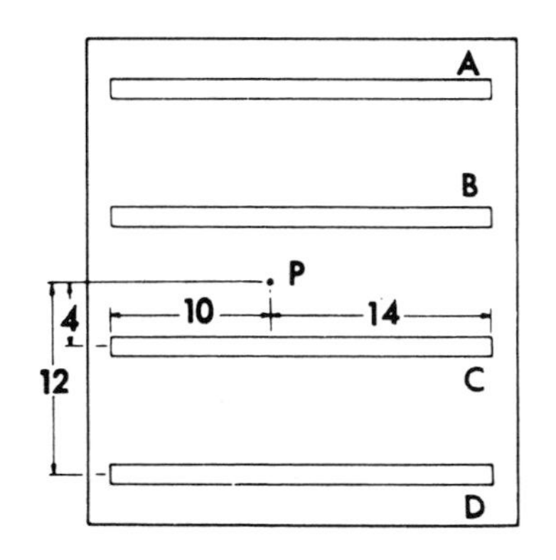

Row	α_1	α_2	β	Direct Illum. Component from Left End	Direct Illum. Component from Right End	Total
A	50	60*	55	3.9	4.4	8.3
B	50	60*	25	30.8	32.5	63.3
C	50	60*	25	30.8	32.5	63.3
D	50	60*	55	3.9	4.4	8.3
					Total	143.2

* Actually, α_2 is 59° but is rounded off to 60°.

shown in Fig. 9–19 must be multiplied by the height (1.8 meters or 6 feet)/V where V is the mounting height above the task. Thus, the total direct illuminance component would be the product of 1.8 or 6/V and the sum of the individual direct illuminance components for each row. If maintained illuminance is desired, the sum must be multiplied by the Total Light Loss Factor.

Example. As an example of the method (using the foot as the unit of length), assume four rows of six four-foot luminaires for which data are shown in Fig. 9–19. They are surface mounted on 8-foot centers in a room 28 by 30 feet. Assume the ceiling reflectance is 80 per cent and that of the walls is 50 per cent. Floor cavity reflectance is 20 per cent. The mounting height of the luminaires is 8½ feet above the work-plane. The initial illuminance on the horizontal work-plane at point P in Fig. 9–20 is desired.

It is first necessary to determine angle α for both ends of the rows of luminaires and angle β. For angle α, H is 10 feet for α_1; and 14 feet for α_2. The vertical distance, V, is 8½ feet. For angle β, H is 12 feet for rows A and D and is four feet for rows B and C. The vertical distance, V, still is 8½ feet. Fig. 9–18 is used to determine the angles and values shown in Fig. 9–20. Since the direct illuminance component tables are all based on an assumed mounting height of 6 feet above the point, and in this case the luminaires are actually 8½ feet above point P, it is necessary to multiply the total of 143.2 by 6/8½. The resultant direct component is 101 footcandles.

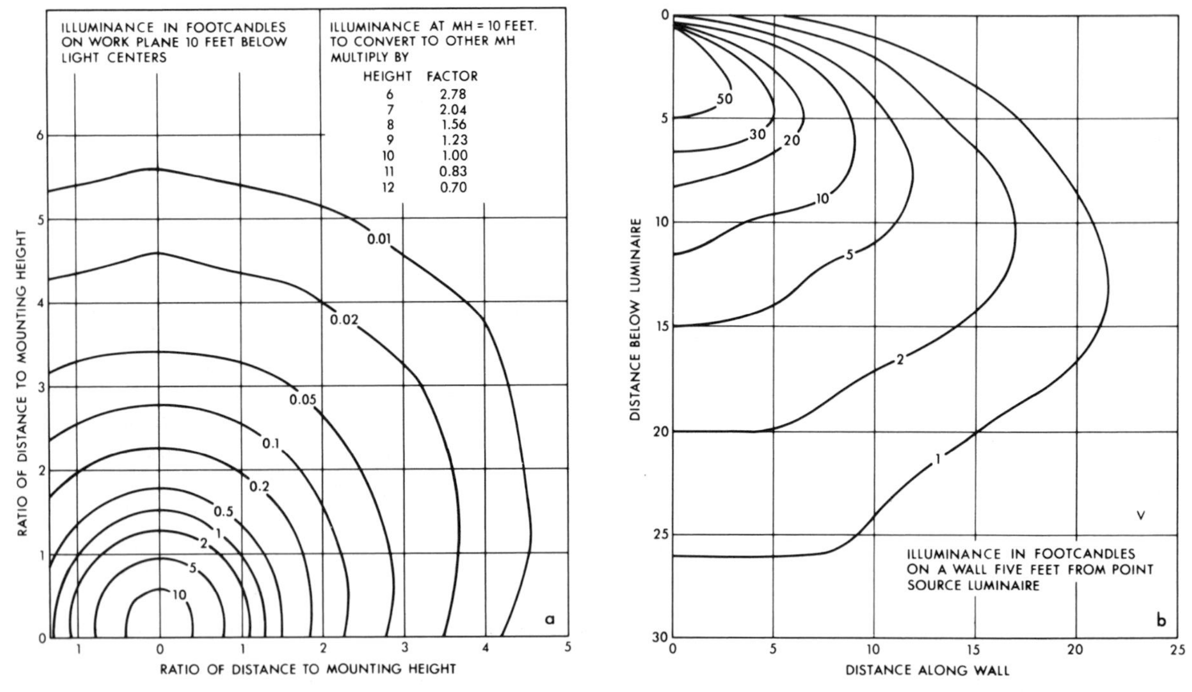

Fig. 9-21. Examples of isofootcandle charts useful in determining the direct component of illuminance. a. For horizontal surfaces (or plane perpendicular to nadir or axis of luminaire). b. For vertical surfaces.

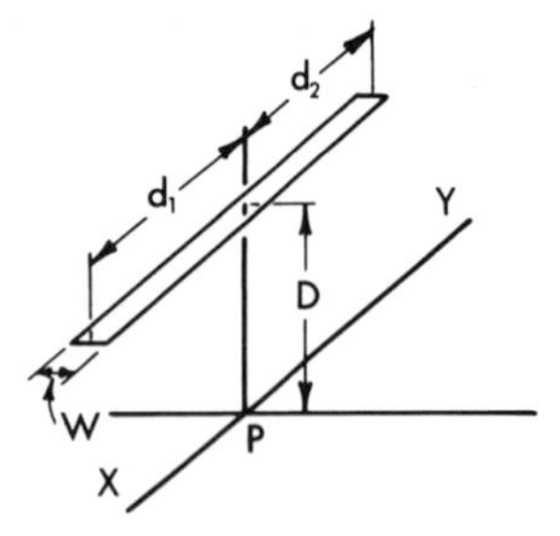

Fig. 9-22. Symbols used in computing illuminance at specific points with line sources.

Aspect-Factor Method—IES (London).[13] In this method there are published tables of multipliers known as aspect factors, based on angular coordinates, which permit rapid and simple calculation of illuminance on horizontal, vertical and inclined planes from linear rows of various luminaires.

Illuminance Charts and Tables. The isolux or isofootcandle methods of calculating illuminance at a point make use of predetermined plots of lines of equal illuminance on horizontal or vertical surfaces. By referring to a set of coordinates on a graph which coincide with the location of the point at which the illuminance is to be determined, the contribution of illuminance from one luminaire to that point may be determined.

The use of these charts is illustrated by reference to Fig. 9-21a, isofootcandle charts, where the locus of equal footcandles from a single fluorescent luminaire mounted 10 feet above the work-place is shown. By locating a point at certain distances with reference to the luminaire, the direct illuminance at the point can be read directly. If many luminaires are involved, all of the various contributions can be read and added together to obtain total direct illuminance at a given point. Charts such as these are drawn on the basis of ratios of distances to mounting heights and may be used at other mounting heights than those for which they are measured by applying appropriate conversion factors given on the chart.

Isofootcandle and isolux charts are also prepared to show vertical illuminance at a point (see Fig. 9-21b). These may show lines of equal illuminance on a wall produced by a luminaire mounted from the ceiling. This method is also advantageous in determining the spacing of luminaires for uniform illuminance or for a given degree of departure from uniformity. The data also can be displayed in tabular form.

Idealized Source Charts. Exact calculations of the direct illuminance component due to linear and area sources are possible when the luminance distribution of the luminaire is known. In general, the variation of luminance is too complicated to make such calculations practical. However, certain special cases are useful; for example, rectangular area sources with constant luminance (lambertian surface) and rectangular area sources with a typical luminance distribution of the overcast sky (see Daylight Calculations, page 7–18). The lambertian surface can be used to approximate luminaires with diffusing panels or to approximate diffuse reflecting areas such as walls. The overcast sky distribution is used to estimate the effect of daylighting through windows.

Calculations with Linear Sources.[14-16] For line sources whose intensity characteristics are either a lambertian or a toroidal distribution* the following simplified expression may be used for determining the illuminance at point P lying on line XY or its extension (see Fig. 9–22):

$$E_p = \frac{M \times W}{2D}$$

where

E_p = Illuminance at P in lux (footcandles)
M = exitance† of source in lumens per square meter (lumens per square foot)
W = width of source
D = distance from source to point P

and W and D are in the same linear units.

* Either of these distributions is approximated by the direct component of most linear source installations, with the exception of clear glass tube sources such as neon tubes, etc. The most common types of line sources have either a toroidal distribution (as approximated by a fluorescent lamp), or a lambertian (tangent sphere) distribution (as produced by a diffusing troffer).

† Exitance at some specific time. The exitance should be multiplied by appropriate Light Loss Factors.

This expression is exact only in the case of an infinitely long source, but it will be accurate to within 10 per cent if both d_1 and d_2 are greater than $1.5D$, and accurate to within 5 per cent if both d_1 and d_2 are greater than $2D$.

Note that the illuminance from a line source of infinite length varies inversely as the distance to the source, *not* inversely as the square of the distance as is the case with point sources.

The following equations provide a means of determining the illuminance, at point P not directly below the luminaire (see Fig. 9–23), produced by a source with either a toroidal (as approximated by a fluorescent lamp) or lambertian distribution. The formulas are applicable when P is on any of the three major planes provided line J is perpendicular to the axis of the source and passes through one end of the source. If the source extends beyond J, the illuminance may be obtained by considering the source to consist of two parts separated by line J, and adding the results of each computation. Conversely, if the source does not extend far enough to reach line J, the problem is solved by assuming a longer source and subtracting the illuminance that would be provided by the imaginary source that was added to reach line J.

For source and P locations shown in Fig. 9–23a:

Lambertian distribution

$$E = \frac{MWV^2}{2\pi J^2}\left(\frac{1}{J}\tan^{-1}\frac{d}{J} + \frac{d}{R^2}\right)$$

Toroidal distribution

$$E = \frac{MWV}{2\pi J}\left(\frac{1}{J}\tan^{-1}\frac{d}{J} + \frac{d}{R^2}\right)$$

In the above formulas, and other which follow,

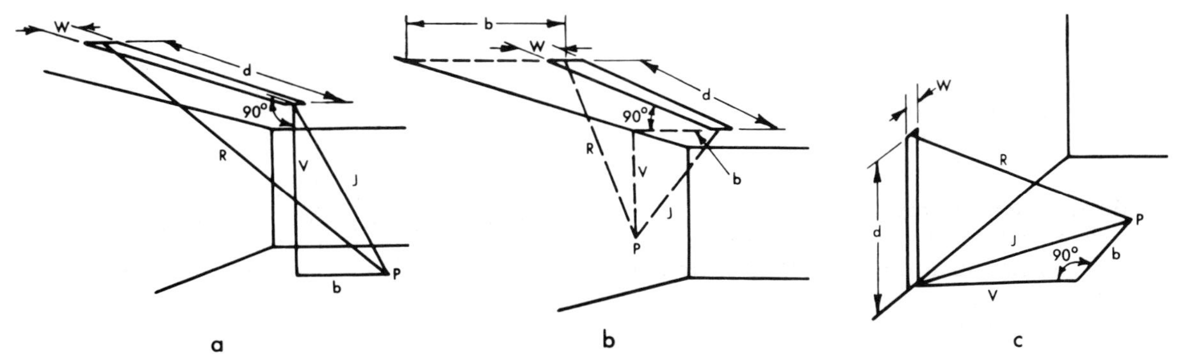

Fig. 9-23. Geometric relationships for calculating illuminance at point P from a linear source $d \times W$. ($J = \sqrt{V^2 + b^2}$, $R = \sqrt{V^2 + b^2 + d^2}$)

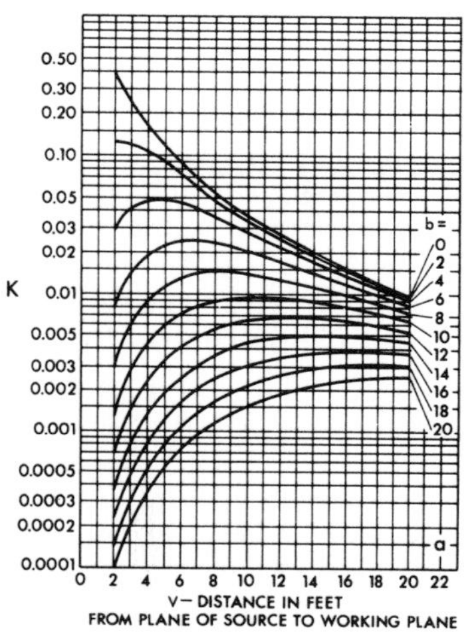

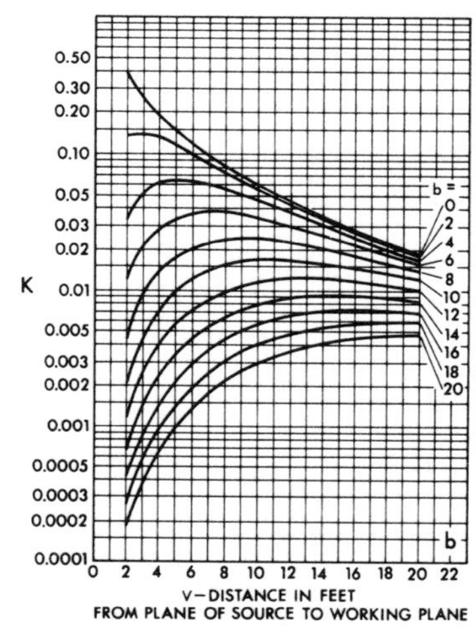

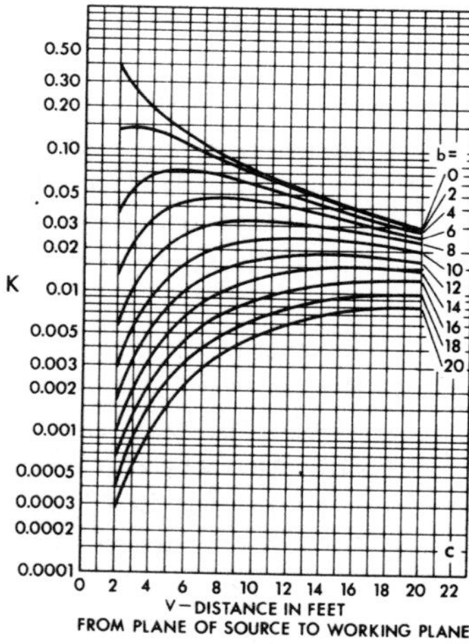

Fig. 9-24. Maximum candlepower per foot times K equals the illuminance in footcandles at a point *P* (Fig. 9-23a) which is *b* feet distant from the projection of the linear source on the work-plane (which is parallel to the source-plane):[17] a. 4-foot source, b. 8-foot source, c. 16-foot source. Note: These data hold only for sources terminating at the plane *VbPJ* in Fig. 9-23a. For sources extending on both sides of the plane determine the illuminance components from each side of the plane and add the two. For sources that do not extend to the plane calculate the contribution for the over-all length to the plane and subtract the hypothetical contribution of the portion missing.

R, W, b, V, d and J are in the same units (any linear unit may be chosen)

Angles are in radians

M = exitance* of source in lumens per square meter (lumens per square foot)

E = illuminance in lux (footcandles)

For source and P locations shown in Fig. 9-23b:

Lambertian sphere distribution

$$E = \frac{MWbV}{2\pi J^2}\left(\frac{1}{J}\tan^{-1}\frac{d}{J} + \frac{d}{R^2}\right)$$

Toroidal distribution

$$E = \frac{MWb}{2\pi J}\left(\frac{1}{J}\tan^{-1}\frac{d}{J} + \frac{d}{R^2}\right)$$

For source and P locations shown in Fig. 9-23c:

Lambertian sphere distribution

$$E = \frac{MWV}{2\pi J^2}\left(\frac{d^2}{R^2}\right)$$

Toroidal distribution

$$E = \frac{MW}{2\pi J}\left(\frac{d^2}{R^2}\right)$$

* Exitance at some specific time. The exitance should be multiplied by appropriate Light Loss Factors.

Fig. 9-24 gives illuminance values (in footcandles) for Lambertian distribution.

Calculations with Area Sources. A lambertian source is one for which the luminance is constant in all directions from the surface. As a consequence, the intensity of a plane lambertian element varies as the cosine of the angle from a normal to that element. From this, the relation between the luminance (L) and the exitance (M) for a lambertian surface is derived as $M = \pi L$, where M is in lumens per square meter and L in candela per square meter, or M in lumens per square foot and L in candela per square foot.†

An infinitely large plane lambertian source radiating light to a parallel work-plane produces illuminance as follows:

$$\text{Illuminance on Work-Plane} = \text{Exitance of Infinite Source}$$

Both are expressed in lumens per square meter, or both are expressed in lumens per square foot.

With such a source (large skylight or uniformly bright ceiling in large room), the illuminance is theoretically independent of the distance. For surface sources whose greatest dimension is less than one-fifth the distance from the source to the work-plane, point source formulas may be used with reasonable accuracy.

† If the deprecated luminance unit of the footlambert is applied to a lambertian surface, it is numerically equal to the exitance of that surface in lumens per square foot.

The illuminance from a rectangular source is usually computed at a point perpendicular to the corner of the source. The variables involved are the distance from the source D, the height of the source H, the width W, and the exitance M. (See Fig. 9-25.) The illuminance usually calculated for horizontal planes is E_h, and for a vertical plane, E_v.

The illuminance can be determined for any point on a plane, in relation to the source, by the principle of superposition. For instance, if it is desired to calculate the illuminance at point P, from source D, Fig. 9-26, the illuminance is first calculated from source $ABCD$ and then from sources A, AC, and AB. The illuminance from source D is then equal to the illuminance from source $ABCD$ minus the sum of the illuminance from source AB and AC, plus the illuminance from source A.

The illuminance from a *rectangular source of uniform illuminance*, assuming no contribution or reflection from any surrounding surface, can be obtained from Figs. 9-27 and 9-28. For example, to obtain the illuminance at distance 10 meters from a rectangular uniform source 20 meters wide and 10 meters high, which is perpendicular to the illuminated plane, obtain the values of source width to height (20/10) and the distance from the source to source height (10/10). From Fig. 9-28 follow the curve $W/H = 2$ until it strikes the ordinate $D/H = 1$; the ratio of illuminance to source exitance for this condition is 0.07. Assuming a uniform source luminance of 1000 candela per square meter, then the illuminance is $0.07 \times \pi \times 1000 = 219.9$ lumens per square meter.

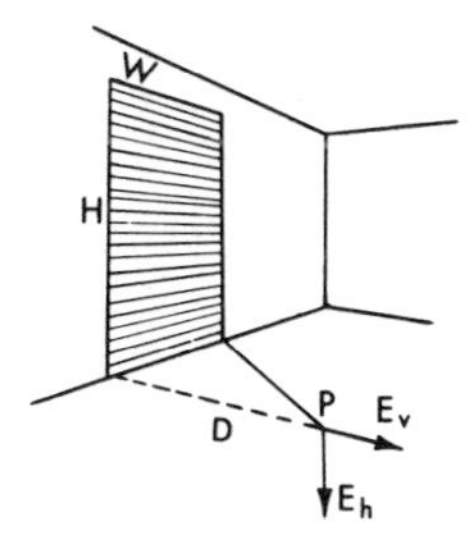

Fig. 9-25. Rectangular surface source perpendicular to illuminated surface.

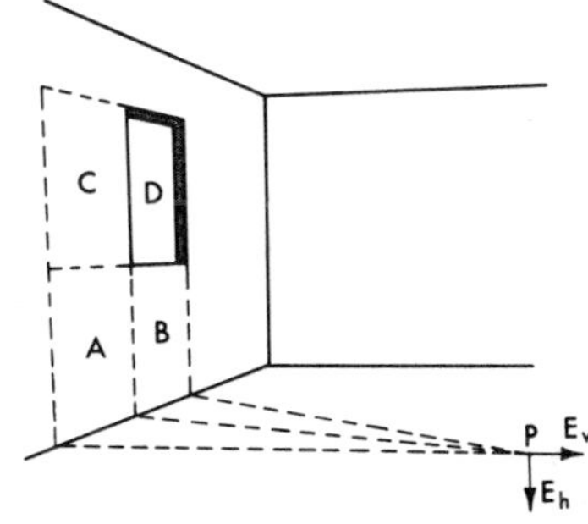

Fig. 9-26. Example of illuminance calculation from area source D.

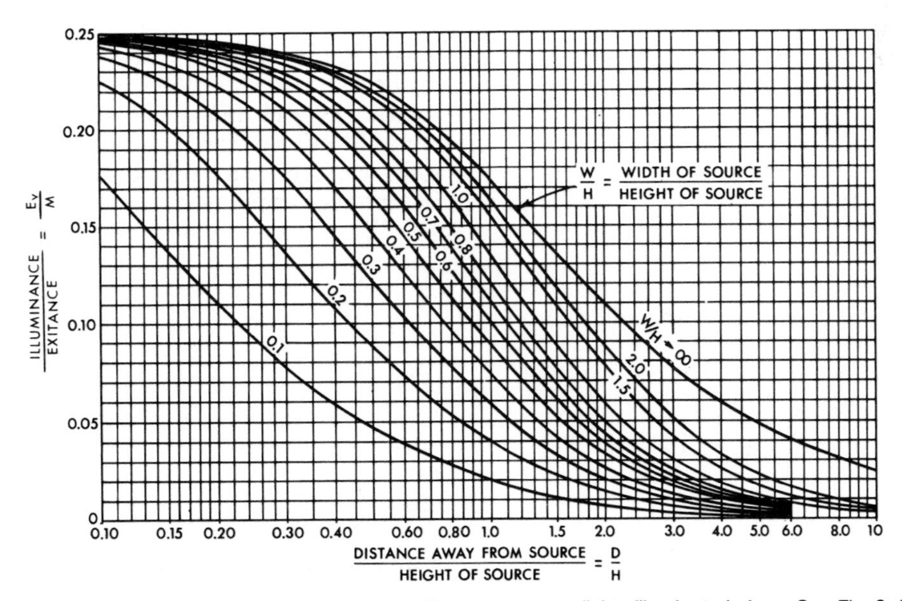

Fig. 9-27. Illuminance from a rectangular uniform source parallel to illuminated plane. See Fig. 9-25.

$$\frac{E_v}{M} = \frac{1}{2\pi}\left[\frac{H}{\sqrt{D^2 + H^2}} \sin^{-1}\frac{W}{R} + \frac{W}{\sqrt{D^2 + W^2}} \sin^{-1}\frac{H}{R}\right] \quad \text{where } R = \sqrt{D^2 + H^2 + W^2}$$

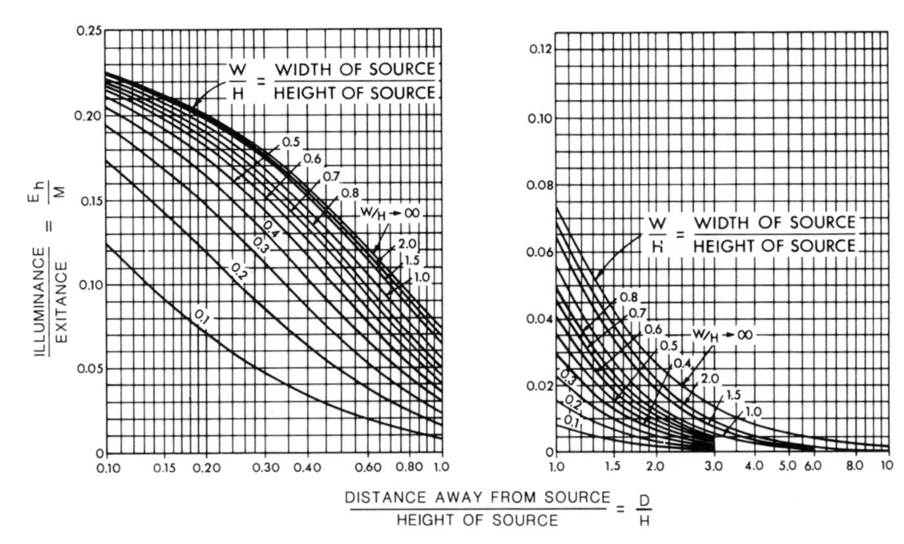

Fig. 9-28. Illuminance from a rectangular uniform source perpendicular to illuminated plane. See Fig. 9-25.

$$\frac{E_h}{M} = \frac{1}{2\pi}\left[\tan^{-1}\frac{W}{D} - \frac{D}{\sqrt{D^2 + H^2}}\sin^{-1}\frac{W}{R}\right] \quad \text{where } R = \sqrt{D^2 + H^2 + W^2}$$

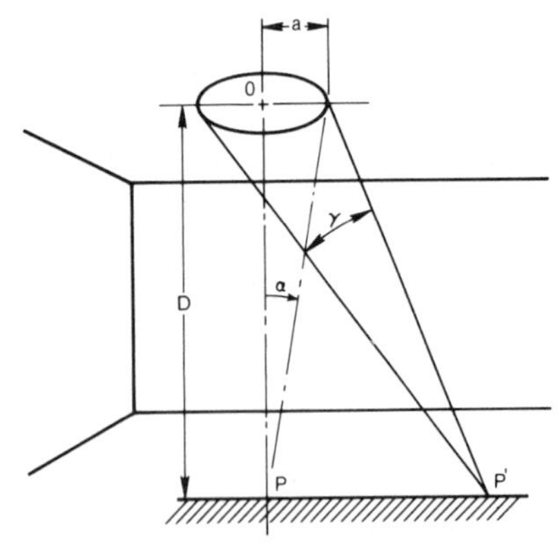

Fig. 9-29. Geometric relationships for calculating illuminance at point P from a circular surface source.

For uniform circular lambertian surface sources, the illuminance at points along the axis of the source can be determined from the following formulas (see Fig. 9-29):

Illuminance at point on axis:

$$E_p = M\frac{a^2}{a^2 + D^2}$$

or

$$E_p = \frac{\text{(Lumen Output of Circular Surface Source)}}{\pi\,(a^2 + D^2)}$$

or

$$E_p = M\sin^2\alpha$$

If the point lies off the axis, as P' of Fig. 9-29, the general formula may be employed:

$$E_{p'} = \tfrac{1}{2}M(1 - \cos\gamma)$$

where E and M are both in lumens per square meter or both in lumens per square foot; the units of a and D are meters or feet accordingly.

Configuration-Factor Method.* Use of configuration factors permits the calculation of illuminance at a point from a diffuse surface. It is necessary to know only the luminance of the surface and the spacial relationship of the surface to the point. This is an approximate method[18] bypassing computation of the exact configuration factor and the use of the superposition theorem.* Fig. 9-30 is a precalculated table of Room Position Multipliers (RPM) to be used for the illuminance at a point due to the reflected component of flux from the room surfaces (see Illuminance at a Point—Reflected Component below).

Providing that a light source is nearly lambertian and is a long narrow strip source, a similar approximation can be made[19] using the strip source room position multipliers given in Fig. 9-32.

Example. Fig. 9-33 shows a room with one row of 600 millimeter by 1200 millimeter lambertian distribution luminaires (I_o = 3000 candelas) mounted 2 meters above the work-plane.

* For a more precise method see Calculations with Area Sources, page 9-18.

Fig. 9-30. Room Position Multipliers
(For All Room Cavity Ratios and All Points Designated by a Number and a Letter as Shown in Fig. 9-31.)

Points Along Width of Room	RCR	Points Along Length of Room: A	B	C	D	E	F
0	1	.24	.42	.47	.48	.48	.48
	2	.24	.36	.42	.44	.46	.46
	3	.23	.32	.37	.40	.42	.42
	4	.22	.28	.32	.35	.37	.37
	5	.21	.25	.28	.31	.33	.33
	6	.20	.23	.26	.28	.29	.30
	7	.18	.21	.23	.25	.26	.27
	8	.17	.18	.21	.22	.22	.23
	9	.15	.17	.18	.19	.20	.20
	10	.14	.16	.16	.17	.18	.18
1	1	.42	.74	.81	.83	.84	.84
	2	.36	.51	.60	.63	.66	.68
	3	.32	.40	.48	.51	.53	.57
	4	.28	.33	.40	.42	.44	.48
	5	.25	.29	.33	.36	.38	.42
	6	.23	.26	.29	.31	.33	.36
	7	.21	.23	.26	.28	.29	.30
	8	.18	.20	.23	.25	.26	.26
	9	.17	.18	.20	.21	.22	.23
	10	.16	.17	.18	.19	.19	.20
2	1	.47	.81	.90	.92	.93	.93
	2	.42	.60	.68	.72	.78	.83
	3	.37	.48	.58	.61	.64	.67
	4	.32	.40	.48	.50	.52	.57
	5	.28	.33	.40	.42	.44	.48
	6	.26	.29	.35	.37	.38	.40
	7	.23	.26	.30	.32	.33	.34
	8	.21	.23	.26	.27	.28	.29
	9	.18	.20	.23	.24	.25	.25
	10	.16	.18	.19	.21	.22	.22
3	1	.48	.83	.92	.94	.95	.95
	2	.44	.63	.72	.77	.82	.85
	3	.40	.51	.61	.65	.69	.71
	4	.35	.42	.50	.54	.58	.61
	5	.31	.36	.42	.46	.49	.52
	6	.28	.31	.37	.39	.41	.43
	7	.25	.28	.32	.34	.35	.36
	8	.22	.25	.27	.29	.30	.30
	9	.19	.21	.24	.25	.26	.26
	10	.17	.19	.21	.22	.23	.23
4	1	.48	.84	.93	.95	.96	.97
	2	.46	.66	.78	.82	.85	.86
	3	.42	.53	.64	.69	.73	.75
	4	.37	.44	.52	.58	.62	.64
	5	.33	.38	.44	.49	.52	.54
	6	.29	.33	.38	.41	.43	.45
	7	.26	.29	.33	.35	.37	.37
	8	.22	.26	.28	.30	.31	.32
	9	.20	.22	.25	.26	.26	.27
	10	.18	.19	.22	.23	.23	.24
5	1	.48	.84	.93	.95	.97	.97
	2	.46	.68	.83	.85	.86	.87
	3	.42	.57	.67	.71	.75	.77
	4	.37	.48	.57	.61	.64	.66
	5	.33	.42	.48	.52	.54	.56
	6	.30	.36	.40	.43	.45	.47
	7	.27	.30	.34	.36	.37	.38
	8	.23	.26	.29	.30	.31	.32
	9	.20	.23	.25	.26	.27	.27
	10	.18	.20	.22	.23	.24	.25

The illuminance at P is desired. In the solution, the first step is to define a "room" whose width is that of the luminaire row length and whose length is twice that value, *i.e.*, 6 by 12 meters (GHJK). This room cavity is oriented so that its left wall is the centerline of the luminaires. The height of the cavity is the mounting height of the luminaires above the work-plane, and the RCR† of this fictitious room is:

$$5 \times 2\ (6 + 12)/(6 \times 12) = 2.5.$$

Next, a grid is determined for the 2 to 1 room ratio, dividing it into 10 by 20 divisions as in Fig. 9-34. The point P corresponds to F3 in Fig. 9-34.

From Fig. 9-32 for strip source room position multipliers, the factor is interpolated between 0.0142 and 0.0209 as 0.0175. The factors are based on a source whose width to length ratio is 0.1. Consequently, a correction must be made for the actual luminaire ratio:

$$0.0175 \times (0.8/6)/(0.1) = 0.0233.$$

The luminaire exitance is:

$$\pi I_o/A = (3000\pi)/(0.8 \times 1.2) = 9817\ \text{lm/m}^2.$$

Illuminance at point P is $9817 \times 0.0233 = 229$ lm/m^2 (lux). If the required illuminance point such as Q in Fig. 9-35 is beyond the end of the luminaire row T, the illuminance due to the row

† See page 9-26.

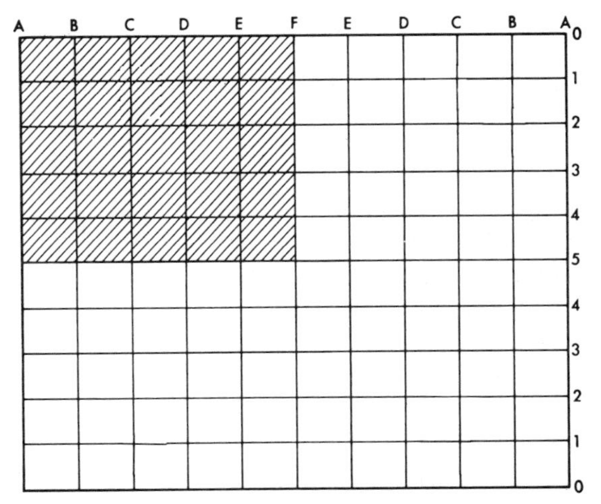

Fig. 9-31. Chart for locating points on the work-plane in a room. Letters are placed along length of room and numbers along width. Each block represents ten per cent of length or width. In a square room each position could have two designations. Position A0 is corner of a room and position F5 is the center.

Fig. 9-32. Strip Source Room Postion Multipliers (Strip Width to Length of 0.1).

RCR	Positions										
	A0	B0	C0	D0	E0	F0	G0	I0	K0	P0	U0
1	.1873	.0958	.0313	.0124	.0058	.0031	.0018	.0007	.0004	.0001	0
2	.0931	.0763	.0473	.0269	.0154	.0092	.0058	.0026	.0013	.0003	.0001
3	.0611	.0557	.0434	.0307	.0209	.0141	.0097	.0048	.0025	.0007	.0003
4	.0046	.0423	.0363	.0290	.0221	.0164	.0121	.0067	.0038	.0011	.0004
5	.0345	.0333	.0300	.0256	.0210	.0167	.0131	.0079	.0048	.0016	.0006
6	.0275	.0268	.0249	.0222	.0190	.0159	.0131	.0086	.0056	.0020	.0008
7	.0225	.0221	.0209	.0191	.0169	.0147	.0125	.0088	.0060	.0024	.0010
8	.0187	.0184	.0176	.0164	.0149	.0133	.0116	.0086	.0062	.0027	.0012
9	.0157	.0155	.0150	.0142	.0131	.0119	.0106	.0082	.0062	.0029	.0014
10	.0134	.0133	.0129	.0123	.0115	.0106	.0097	.0077	.0060	.0030	.0015
	A1	B1	C1	D1	1	F1	G1	I1	K1	P1	U1
1	.3124	.1557	.0470	.0169	.0075	.0038	.0021	.0008	.0004	.0001	0
2	.1338	.1079	.0644	.0350	.0194	.0112	.0069	.0029	.0014	.0004	.0001
3	.0797	.0722	.0553	.0383	.0254	.0168	.0113	.0054	.0028	.0007	.0003
4	.0549	.0519	.0442	.0348	.0261	.0191	.0139	.0074	.0042	.0012	.0004
5	.0408	.0393	.0352	.0298	.0242	.0191	.0148	.0088	.0053	.0017	.0006
6	.0316	.0308	.0285	.0252	.0215	.0179	.0146	.0094	.0060	.0021	.0009
7	.0252	.0247	.0233	.0212	.0188	.0162	.0137	.0095	.0064	.0025	.0011
8	.0206	.0203	.0194	.0180	.0163	.0144	.0126	.0092	.0066	.0028	.0013
9	.0171	.0169	.0163	.0153	.0141	.0128	.0114	.0088	.0065	.0030	.0014
10	.0144	.0143	.0138	.0132	.0123	.0113	.0103	.0082	.0063	.0031	.0016
	A2	B2	C2	D2	E2	F2	G2	I2	K2	P2	U2
1	.3595	.1792	.0558	.0202	.0088	.0044	.0025	.0009	.0004	.0001	0
2	.1595	.1287	.0768	.0414	.0226	.0129	.0078	.0032	.0016	.0004	.0001
3	.0943	.0852	.0649	.0446	.0293	.0191	.0126	.0059	.0030	.0008	.0003
4	.0636	.0600	.0508	.0397	.0296	.0214	.0154	.0081	.0045	.0013	.0005
5	.0462	.0444	.0398	.0335	.0270	.0211	.0162	.0095	.0056	.0017	.0007
6	.0351	.0342	.0316	.0278	.0236	.0195	.0158	.0101	.0064	.0022	.0009
7	.0276	.0270	.0255	.0231	.0204	.0175	.0147	.0101	.0068	.0026	.0011
8	.0222	.0219	.0209	.0193	.0175	.0154	.0134	.0098	.0069	.0029	.0013
9	.0183	.0180	.0174	.0163	.0150	.0136	.0121	.0092	.0068	.0031	.0015
10	.0152	.0151	.0146	.0139	.0130	.0119	.0108	.0086	.0066	.0032	.0016

R is determined and then the contribution of the fictitious section S is determined and subtracted.

Real Source Method. The most general means yet devised for calculating illuminance at

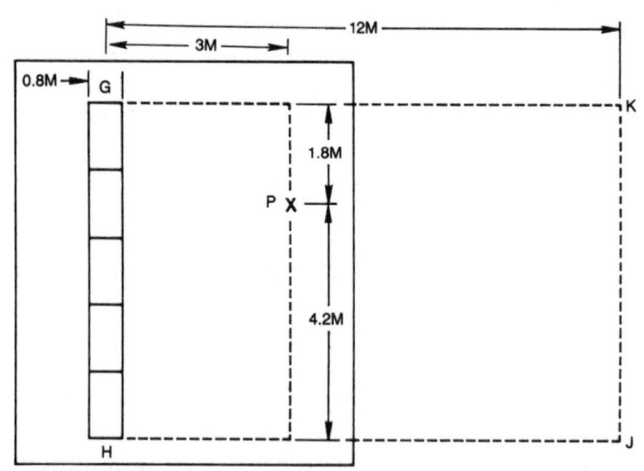

Fig. 9-33. Layout for example of use of the Configuration Factor Method. Luminaires are 0.8- × 1.2-meter units mounted 2 meters above the work-plane.

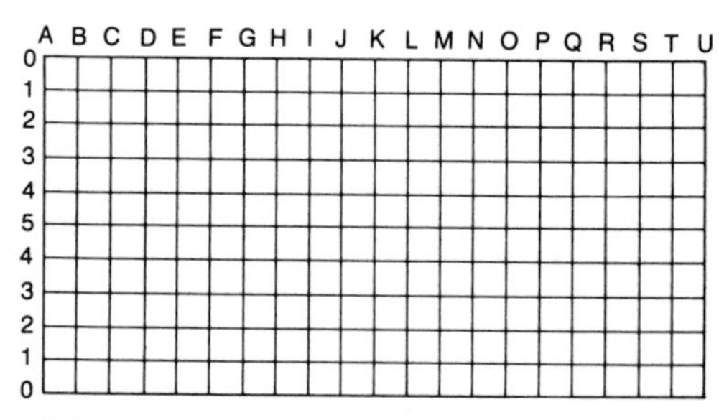

Fig. 9-34. Proportional divisions for strip source room position mutipliers.

a point permits the luminaire to have any arbitrary luminous intensity distribution, the calculation point to be arbitrarily close to the luminaire, and the luminaire's area to be taken into account.[20] The method assumes that the luminaire has a homogeneous distribution characteristic over its area.

The coordinate system used in this method is the perpendicular-plane angular (PPA) coordi-

See Fig. 9–34 for Proportional Dimensions for Strip Source Room Position Multipliers.

RCR	Positions										
	A3	B3	C3	D3	E3	F3	G3	I3	K3	P3	U3
1	.3639	.1864	.0597	.0220	.0097	.0048	.0027	.0010	.0005	.0001	0
2	.1719	.1393	.0839	.0456	.0249	.0142	.0085	.0035	.0016	.0004	.0001
3	.1037	.0938	.0714	.0490	.0321	.0209	.0137	.0063	.0032	.0008	.0003
4	.0699	.0659	.0557	.0434	.0322	.0232	.0166	.0086	.0047	.0013	.0005
5	.0503	.0484	.0432	.0363	.0291	.0227	.0174	.0100	.0059	.0018	.0007
6	.0378	.0368	.0339	.0298	.0253	.0209	.0168	.0106	.0067	.0023	.0009
7	.0294	.0288	.0271	.0246	.0216	.0185	.0155	.0106	.0071	.0026	.0011
8	.0235	.0231	.0220	.0204	.0184	.0162	.0140	.0102	.0072	.0030	.0013
9	.0191	.0189	.0182	.0171	.0157	.0142	.0126	.0095	.0070	.0032	.0015
10	.0159	.0157	.0152	.0145	.0135	.0124	.0112	.0088	.0068	.0033	.0016
	A4	B4	C4	D4	E4	F4	G4	I4	K4	P4	U4
1	.3716	.1888	.0612	.0229	.0101	.0051	.0028	.0010	.0005	.0001	0
2	.1771	.1440	.0873	.0477	.0262	.0149	.0089	.0036	.0017	.0004	.0001
3	.1086	.0983	.0751	.0515	.0338	.0219	.0144	.0066	.0033	.0008	.0003
4	.0735	.0693	.0586	.0457	.0338	.0243	.0174	.0089	.0048	.0013	.0005
5	.0528	.0508	.0453	.0380	.0304	.0237	.0181	.0104	.0061	.0018	.0007
6	.0395	.0384	.0354	.0311	.0263	.0216	.0174	.0109	.0068	.0023	.0009
7	.0306	.0299	.0281	.0255	.0224	.0191	.0160	.0109	.0072	.0027	.0011
8	.0243	.0239	.0228	.0210	.0190	.0167	.0144	.0104	.0073	.0030	.0013
9	.0197	.0194	.0187	.0176	.0161	.0145	.0129	.0098	.0072	.0032	.0015
10	.0163	.0161	.0156	.0148	.0138	.0126	.0114	.0090	.0069	.0033	.0016
	A5	B5	C5	D5	E5	F5	G5	I5	K5	P5	U5
1	.3722	.1893	.0617	.0232	.0103	.0052	.0029	.0011	.0005	.0001	0
2	.1786	.1453	.0883	.0484	.0266	.0152	.0091	.0037	.0017	.0004	.0001
3	.1101	.0997	.0762	.0524	.0343	.0223	.0146	.0067	.0033	.0008	.0003
4	.0747	.0704	.0596	.0464	.0343	.0247	.0176	.0091	.0049	.0013	.0005
5	.0536	.0516	.0460	.0386	.0309	.0240	.0183	.0105	.0061	.0018	.0007
6	.0401	.0390	.0359	.0315	.0266	.0219	.0176	.0111	.0069	.0023	.0009
7	.0310	.0303	.0285	.0258	.0226	.0193	.0162	.0110	.0073	.0027	.0011
8	.0245	.0241	.0230	.0213	.0191	.0169	.0146	.0105	.0074	.0030	.0013
9	.0199	.0196	.0189	.0177	.0163	.0147	.0130	.0098	.0072	.0032	.0015
10	.0164	.0162	.0157	.0149	.0139	.0127	.0115	.0091	.0069	.0034	.0016

nate system. The intensity distribution of the luminaire is first transformed out of the usual spherical coordinates and into PPA coordinates. These two dimensional data are then curve-fit with a finite Fourier series. The inverse-square cosine law is then used to derive an expression for the illuminance produced at a point by a differential area of the luminaire. This expression is then integrated over the luminaire's area to yield an expression for the illuminance produced by the entire luminaire. An extension to this method[21] treats indirect luminaires in a similar manner.

These techniques are suitable for computer program applications.

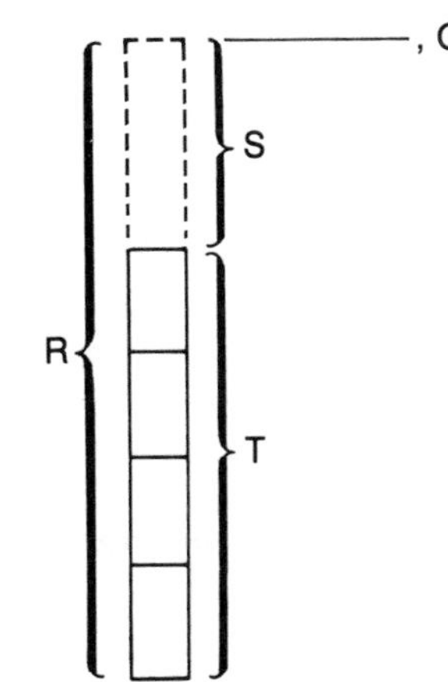

Fig. 9–35. Example of Configuration Factor Method using superposition.

Illuminance at a Point—Reflected Component

For Horizontal Surfaces. The reflected illuminance component at a point on a horizontal surface is calculated in exactly the same manner as the average illuminance is computed using the lumen method except that the RRC, the reflected radiation coefficient, is substituted for the CU, coefficient of utilization.

Reflected Illuminance (Horizontal)

$$= \frac{\text{Lamp Lumens per Luminaire} \times \text{RRC}}{\text{Area per Luminaire (on Work-Plane)}}$$

where

RRC $= WEC + \text{RPM}(CCEC - WEC)$
WEC = wall exitance coefficient (see Fig. 9-62)
$CCEC$ = ceiling cavity exitance coefficient (see Fig. 9-62)
RPM = room position multiplier (see Fig. 9-30)

For Vertical Surfaces. To determine the average illuminance reflected to vertical surfaces the approximate average value is determined using the same general formula, but substituting the WRRC, wall reflected radiation coefficient, for the CU, coefficient of utilization.

Reflected Illuminance (Vertical)

$$= \frac{\text{Lamp Lumens per Luminaire} \times \text{WRRC}}{\text{Area per Luminaire (on Work-Plane)}}$$

where

$$\text{WRRC} = \frac{WEC}{\rho_W} - \text{WDRC}$$

ρ_W = average wall reflectance
WDRC = the wall direct radiation coefficient (see Fig. 9-62)

Average Illuminance Calculations

Lumen Method. The lumen method is used in calculating the illuminance that represents the average of all points on the work-plane in an interior. It is based on the definition of illuminance as luminous flux per unit area, or:

$$\text{Illuminance} = \frac{\text{Luminous Flux}}{\text{Area}}$$

where luminous flux is expressed in lumens. If the area is in square feet, the illuminance is in footcandles (lumens per square foot): if the area is in square meters, the illuminance is in lux (lumens per square meter).

Because not all the lamp lumens will reach the work-plane due to losses in the luminaire and at the room surfaces, they must be multiplied by a coefficient of utilization which represents the portion that reaches the work-plane. Thus:

$$\text{Initial Illuminance} = \frac{\text{Lamp Lumens} \times \text{Coefficient of Utilization}}{\text{Area}}$$

Since the design objective is usually maintained illuminance, factors must be applied to account for the estimated depreciation in lamp lumens, the estimated losses from dirt collection on the luminaire surfaces (including lamps), etc. The formula thus becomes:

$$\text{Maintained Illuminance} = \frac{\text{Lamp Lumens} \times \text{CU} \times \text{LLF}}{\text{Area}}$$

where

CU = the Coefficient of Utilization
LLF = the Light Loss Factor (see page 9-4)

The lamp lumens in the formula are most conveniently taken as the total rated lamp lumens in the luminaire, and the area then becomes the area per luminaire. Thus:

$$\text{Maintained Illuminance} = \frac{\text{Lamp Lumens per Luminaire} \times \text{CU} \times \text{LLF}}{\text{Area per Luminaire}}$$

or, if the desired illuminance is known, the area per luminaire (and hence the spacing between luminaires) to produce this illuminance may be obtained by:

$$\text{Area per Luminaire} = \frac{\text{Lamp Lumens per Luminaire} \times \text{CU} \times \text{LLF}}{\text{Maintained Illuminance}}$$

A lighting system can be designed with spacings between units to approximate this area, but if the total number of luminaires is also desired, then:

$$\text{Total Number of Luminaires} = \frac{\text{Total Room Area}}{\text{Area per Luminaire}}$$

For a typical form for calculating illuminance, see Fig. 9-36.

Limitations. The illuminance computed by the lumen method is an average value that will be representative only if the luminaires are spaced to obtain reasonably uniform illuminance. The calculation of the coefficients of utilization is based on empty interiors having surfaces that exhibit perfectly diffuse reflectance.

The average illuminance determined by the lumen method is defined to be the total lumens

Fig. 9–36. Average Illuminance Calculation Sheet

GENERAL INFORMATION

Project identification: ______________________________
(Give name of area and/or building and room number)

Average maintained illuminance for design: ___ lux or
: ___ footcandles

Lamp data:
Type and color: ________
Number per luminaire: ________
Total lumens per luminaire: ________

Luminaire data:
Manufacturer: ________
Catalog number: ________

SELECTION OF COEFFICIENT OF UTILIZATION

Step 1: Fill in sketch at right.

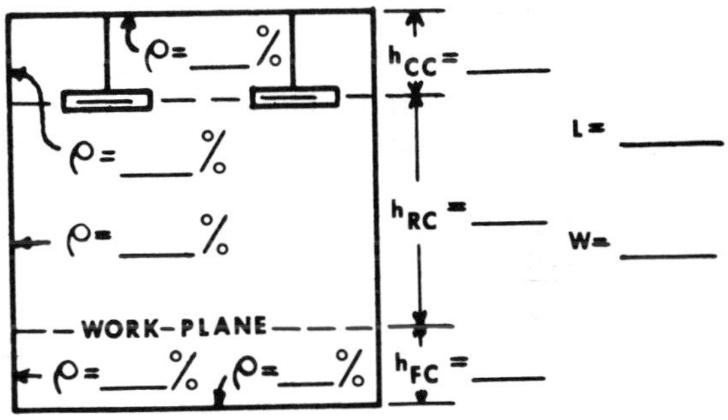

Step 2: Determine Cavity Ratios from Fig. 9–38, or by formulas.

Room Cavity Ratio, RCR = ________
Ceiling Cavity Ratio, CCR = ________
Floor Cavity Ratio, FCR = ________

ρ= ___% ρ= ___% ρ= ___%
WORK-PLANE
ρ= ___% ρ= ___%
h_{CC} = ____
h_{RC} = ____
h_{FC} = ____
L = ____
W = ____

Step 3: Obtain Effective Ceiling Cavity Reflectance (ρ_{CC}) from Fig. 9–39. ρ_{CC} = ____

Step 4: Obtain Effective Floor Cavity Reflectance (ρ_{FC}) from Fig. 9–39. ρ_{FC} = ____

Step 5: Obtain Coefficient of Utilization (CU) from Manufacturer's Data. CU = ____

SELECTION OF LIGHT LOSS FACTORS

Nonrecoverable
Luminaire ambient temperature (See page 9–5.) ____
Voltage to luminaire (See page 9–5.) ____
Ballast factor (See page 9–5.) ____
Luminaire surface depreciation (See page 9–7.) ____

Recoverable
Room surface dirt depreciation RSDD (See page 9–9.) ____
Lamp lumen depreciation LLD (See page 9–8.) ____
Lamp burnouts factor LBO (See page 9–9.) ____
Luminaire dirt depreciation LDD (See page 9–8.) ____

Total light loss factor, LLF (product of individual factors above) = ____

CALCULATIONS

(Average Maintained Illuminance)

$$\text{Number of Luminaires} = \frac{(\text{Illuminance}) \times (\text{Area})}{(\text{Lumens per Luminaire}) \times (\text{CU}) \times (\text{LLF})}$$

= ______________________ =

$$\text{Illuminance} = \frac{(\text{Number of Luminaires}) \times (\text{Lumens per Luminaire}) \times (\text{CU}) \times (\text{LLF})}{(\text{Area})}$$

= ______________________ =

Calculated by: ______________________ Date: ____________

reaching the work-plane, divided by the area of the work-plane. The average value determined this way may vary considerably from that obtained by averaging discrete values of illuminance at several points.

In addition to measurement uncertainties, calculated illuminance values may differ from measured values due to luminaire input data, assumed room and system parameters and mathematical modeling of the lighting system. For example: individual lamps may vary from nominal ratings; individual luminaires may differ from the nominal photometric data due to manufacturing and lamp positioning differences; and assumed values for room reflectances and ballast factor may vary from actual values. In addition, the mathematical model is not an exact representation of most real rooms. For a complete discussion of uncertainties, see reference 25.

Calculation Procedure. Fig. 9-36 provides a procedure for calculating average maintained illuminance using the *Zonal-Cavity Method.*[22-24] The paragraphs that follow discuss the calculation of *cavity ratios* and *effective cavity reflectances* and the selection of *luminaire coefficients of utilization* to be used in the Method.

Cavity Ratios. In the Zonal-Cavity Method, the effects of room proportions, luminaire suspension length, and work-plane height upon the coefficient of utilization are respectively accounted for by the *Room Cavity Ratio, Ceiling Cavity Ratio* and *Floor Cavity Ratio.* These ratios are determined by dividing the room into three cavities as shown by Fig. 9-37 and substituting dimensions (in feet or meters) in the following formula:

$$\text{Cavity Ratio} = \frac{5h(\text{Room Length} + \text{Room Width})}{(\text{Room Length}) \times (\text{Room Width})}$$

where

$h = h_{RC}$ for the Room Cavity Ratio, RCR
$= h_{CC}$ for the Ceiling Cavity Ratio, CCR
$= h_{FC}$ for the Floor Cavity Ratio, FCR

Note that

$$\text{CCR} = \text{RCR}\,\frac{h_{CC}}{h_{RC}}$$

and

$$\text{FCR} = \text{RCR}\,\frac{h_{FC}}{h_{RC}}$$

Cavity Ratios may also be obtained from Fig. 9-38.

The illuminance in rooms of irregular shape can be determined by calculating the Room Cavity Ratio using the following formula and solving the problem in the usual manner:

$$\text{Cavity Ratio} = \frac{2.5 \times (\text{Cavity Height}) \times (\text{Cavity Perimeter})}{(\text{Area of Cavity Base})}$$

Effective Cavity Reflectances. Fig. 9-39 provides a means of converting the combination of wall and ceiling or wall and floor reflectances into a single *Effective Ceiling Cavity Reflectance,* ρ_{CC}, and a single *Effective Floor Cavity Reflectance,* ρ_{FC}. In calculations, ceiling, wall and floor reflectances should be initial values. The RSDD factor (see page 9-9) compensates for the decrease of reflectance with time. Note that for surface-mounted and recessed luminaires, CCR = 0 and the ceiling reflectance may be used as ρ_{CC}.

A rectangular cavity consists of four walls, each having a reflectance of ρ_W, and a base of reflectance ρ_B (ceiling or floor). The effective reflectance, ρ_{eff}, of this cavity is the ratio of flux reflected out, divided by the flux entering the cavity through its opening. If the reflectances are assumed to be perfectly diffuse and the flux is assumed to enter the cavity in a perfectly diffuse way, it is possible to calculate the effective cavity reflectance using flux transfer theory (see page 9-3). The result is:

$$\rho_{eff} = \frac{\rho_B\rho_W f\left[\dfrac{2A_B}{A_W}(1-f) - f\right] + \rho_B f^2 + \rho_W \dfrac{A_B}{A_W}(1-f)^2}{1 - \rho_B\rho_W \dfrac{A_B}{A_W}(1-f)^2 - \rho_W\left[1 - 2\dfrac{A_B}{A_W}(1-f)\right]}$$

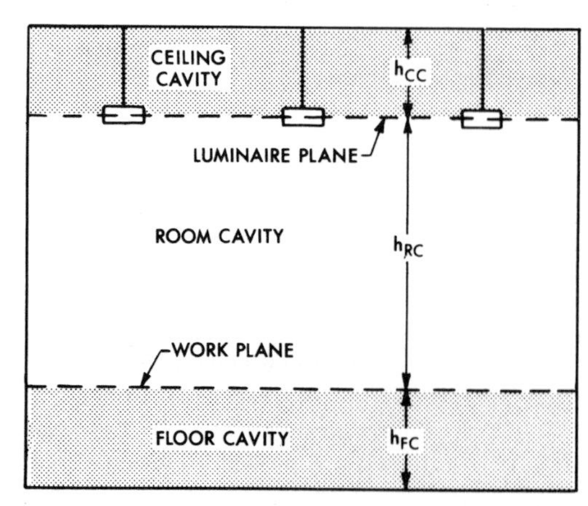

Fig. 9-37. Cavities used in the Zonal-Cavity Method.

Fig. 9-38. Cavity Ratios

For room and cavity dimensions other than those below, the cavity ratio can be calculated by the formulas on page 9-26. If smaller room and cavity dimensions are required divide width, length *and* cavity depth by 10.

Room Dimensions		Cavity Depth																			
Width	Length	1	1.5	2	2.5	3	3.5	4	5	6	7	8	9	10	11	12	14	16	20	25	30
8	8	1.2	1.9	2.5	3.1	3.7	4.4	5.0	6.2	7.5	8.8	10.0	11.2	12.5	—	—	—	—	—	—	—
	10	1.1	1.7	2.2	2.8	3.4	3.9	4.5	5.6	6.7	7.9	9.0	10.1	11.3	12.4	—	—	—	—	—	—
	14	1.0	1.5	2.0	2.5	3.0	3.4	3.9	4.9	5.9	6.9	7.8	8.8	9.7	10.7	11.7	—	—	—	—	—
	20	0.9	1.3	1.7	2.2	2.6	3.1	3.5	4.4	5.2	6.1	7.0	7.9	8.8	9.6	10.5	12.2	—	—	—	—
	30	0.8	1.2	1.6	2.0	2.4	2.8	3.2	4.0	4.7	5.5	6.3	7.1	7.9	8.7	9.5	11.0	—	—	—	—
	40	0.7	1.1	1.5	1.9	2.3	2.6	3.0	3.7	4.5	5.3	5.9	6.5	7.4	8.1	8.8	10.3	11.8	—	—	—
10	10	1.0	1.5	2.0	2.5	3.0	3.5	4.0	5.0	6.0	7.0	8.0	9.0	10.0	11.0	12.0	—	—	—	—	—
	14	0.9	1.3	1.7	2.1	2.6	3.0	3.4	4.3	5.1	6.0	6.9	7.8	8.6	9.5	10.4	12.0	—	—	—	—
	20	0.7	1.1	1.5	1.9	2.3	2.6	3.0	3.7	4.5	5.3	6.0	6.8	7.5	8.3	9.0	10.5	12.0	—	—	—
	30	0.7	1.0	1.3	1.7	2.0	2.3	2.7	3.3	4.0	4.7	5.3	6.0	6.6	7.3	8.0	9.4	10.6	—	—	—
	40	0.6	0.9	1.2	1.6	1.9	2.2	2.5	3.1	3.7	4.4	5.0	5.6	6.2	6.9	7.5	8.7	10.0	12.5	—	—
	60	0.6	0.9	1.2	1.5	1.7	2.0	2.3	2.9	3.5	4.1	4.7	5.3	5.9	6.5	7.1	8.2	9.4	11.7	—	—
12	12	0.8	1.2	1.7	2.1	2.5	2.9	3.3	4.2	5.0	5.8	6.7	7.5	8.4	9.2	10.0	11.7	—	—	—	—
	16	0.7	1.1	1.5	1.8	2.2	2.5	2.9	3.6	4.4	5.1	5.8	6.5	7.2	8.0	8.7	10.2	11.6	—	—	—
	24	0.6	0.9	1.2	1.6	1.9	2.2	2.5	3.1	3.7	4.4	5.0	5.6	6.2	6.9	7.5	8.7	10.0	12.5	—	—
	36	0.6	0.8	1.1	1.4	1.7	1.9	2.2	2.8	3.3	3.9	4.4	5.0	5.5	6.0	6.6	7.8	8.8	11.0	—	—
	50	0.5	0.8	1.0	1.3	1.5	1.8	2.1	2.6	3.1	3.6	4.1	4.6	5.1	5.6	6.2	7.2	8.2	10.2	—	—
	70	0.5	0.7	1.0	1.2	1.5	1.7	2.0	2.4	2.9	3.4	3.9	4.4	4.9	5.4	5.8	6.8	7.8	9.7	12.2	—
14	14	0.7	1.1	1.4	1.8	2.1	2.5	2.9	3.6	4.3	5.0	5.7	6.4	7.1	7.8	8.5	10.0	11.4	—	—	—
	20	0.6	0.9	1.2	1.5	1.8	2.1	2.4	3.0	3.6	4.2	4.9	5.5	6.1	6.7	7.3	8.6	9.8	12.3	—	—
	30	0.5	0.8	1.0	1.3	1.6	1.8	2.1	2.6	3.1	3.7	4.2	4.7	5.2	5.8	6.3	7.3	8.4	10.5	—	—
	42	0.5	0.7	1.0	1.2	1.4	1.7	1.9	2.4	2.9	3.3	3.8	4.3	4.7	5.2	5.7	6.7	7.6	9.5	11.9	—
	60	0.4	0.7	0.9	1.1	1.3	1.5	1.8	2.2	2.6	3.1	3.5	3.9	4.4	4.8	5.2	6.1	7.0	8.8	10.9	—
	90	0.4	0.6	0.8	1.0	1.2	1.4	1.6	2.0	2.5	2.9	3.3	3.7	4.1	4.5	5.0	5.8	6.6	8.3	10.3	12.4
17	17	0.6	0.9	1.2	1.5	1.8	2.1	2.3	2.9	3.5	4.1	4.7	5.3	5.9	6.5	7.0	8.2	9.4	11.7	—	—
	25	0.5	0.7	1.0	1.2	1.5	1.7	2.0	2.5	3.0	3.5	4.0	4.5	5.0	5.5	6.0	7.0	8.0	10.0	12.5	—
	35	0.4	0.7	0.9	1.1	1.3	1.5	1.7	2.2	2.6	3.1	3.5	3.9	4.4	4.8	5.2	6.1	7.0	8.7	10.9	—
	50	0.4	0.6	0.8	1.0	1.2	1.4	1.6	2.0	2.4	2.8	3.1	3.5	3.9	4.3	4.5	5.4	6.2	7.7	9.7	11.6
	80	0.4	0.5	0.7	0.9	1.1	1.2	1.4	1.8	2.1	2.5	2.9	3.3	3.6	4.0	4.3	5.1	5.8	7.2	9.0	10.9
	120	0.3	0.5	0.7	0.8	1.0	1.2	1.3	1.7	2.0	2.3	2.7	3.0	3.4	3.7	4.0	4.7	5.4	6.7	8.4	10.1
20	20	0.5	0.7	1.0	1.2	1.5	1.7	2.0	2.5	3.0	3.5	4.0	4.5	5.0	5.5	6.0	7.0	8.0	10.0	12.5	—
	30	0.4	0.6	0.8	1.0	1.2	1.5	1.7	2.1	2.5	2.9	3.3	3.7	4.1	4.5	4.9	5.8	6.6	8.2	10.3	12.4
	45	0.4	0.5	0.7	0.9	1.1	1.3	1.4	1.8	2.2	2.5	2.9	3.3	3.6	4.0	4.3	5.1	5.8	7.2	9.1	10.9
	60	0.3	0.5	0.7	0.8	1.0	1.2	1.3	1.7	2.0	2.3	2.7	3.0	3.4	3.7	4.0	4.7	5.4	6.7	8.4	10.1
	90	0.3	0.5	0.6	0.8	0.9	1.1	1.2	1.5	1.8	2.1	2.4	2.7	3.0	3.3	3.6	4.2	4.8	6.0	7.5	9.0
	150	0.3	0.4	0.6	0.7	0.8	1.0	1.1	1.4	1.7	2.0	2.3	2.6	2.9	3.2	3.4	4.0	4.6	5.7	7.2	8.6
24	24	0.4	0.6	0.8	1.0	1.2	1.5	1.7	2.1	2.5	2.9	3.3	3.7	4.1	4.5	5.0	5.8	6.7	8.2	10.3	12.4
	32	0.4	0.5	0.7	0.9	1.1	1.3	1.5	1.8	2.2	2.6	2.9	3.3	3.6	4.0	4.3	5.1	5.8	7.2	9.0	11.0
	50	0.3	0.5	0.6	0.8	0.9	1.1	1.2	1.5	1.8	2.2	2.5	2.8	3.1	3.4	3.7	4.4	5.0	6.2	7.8	9.4
	70	0.3	0.4	0.6	0.7	0.8	1.0	1.1	1.4	1.7	2.0	2.2	2.5	2.8	3.0	3.3	3.8	4.4	5.5	6.9	8.2
	100	0.3	0.4	0.5	0.6	0.8	0.9	1.0	1.3	1.6	1.8	2.1	2.4	2.6	2.9	3.1	3.7	4.2	5.2	6.5	7.9
	160	0.2	0.4	0.5	0.6	0.7	0.8	1.0	1.2	1.4	1.7	1.9	2.1	2.4	2.6	2.8	3.3	3.8	4.7	5.9	7.1
30	30	0.3	0.5	0.7	0.8	1.0	1.2	1.3	1.7	2.0	2.3	2.7	3.0	3.3	3.7	4.0	4.7	5.4	6.7	8.4	10.0
	45	0.3	0.4	0.6	0.7	0.8	1.0	1.1	1.4	1.7	1.9	2.2	2.5	2.7	3.0	3.3	3.8	4.4	5.5	6.9	8.2
	60	0.3	0.4	0.5	0.6	0.7	0.9	1.0	1.2	1.5	1.7	2.0	2.2	2.5	2.7	3.0	3.5	4.0	5.0	6.2	7.4
	90	0.2	0.3	0.4	0.6	0.7	0.8	0.9	1.1	1.3	1.6	1.8	2.0	2.2	2.5	2.7	3.1	3.6	4.5	5.6	6.7
	150	0.2	0.3	0.4	0.5	0.6	0.7	0.8	1.0	1.2	1.4	1.6	1.8	2.0	2.2	2.4	2.8	3.2	4.0	5.0	5.9
	200	0.2	0.3	0.4	0.5	0.6	0.7	0.8	1.0	1.1	1.3	1.5	1.7	1.9	2.0	2.2	2.6	3.0	3.7	4.7	5.6
36	36	0.3	0.4	0.6	0.7	0.8	1.0	1.1	1.4	1.7	1.9	2.2	2.5	2.8	3.0	3.3	3.9	4.4	5.5	6.9	8.3
	50	0.2	0.4	0.5	0.6	0.7	0.8	1.0	1.2	1.4	1.7	1.9	2.1	2.5	2.6	2.9	3.3	3.8	4.8	5.9	7.2
	75	0.2	0.3	0.4	0.5	0.6	0.7	0.8	1.0	1.2	1.4	1.6	1.8	2.0	2.3	2.5	2.9	3.3	4.1	5.1	6.1
	100	0.2	0.3	0.4	0.5	0.6	0.7	0.8	0.9	1.1	1.3	1.5	1.7	1.9	2.1	2.3	2.6	3.0	3.8	4.7	5.7
	150	0.2	0.3	0.3	0.4	0.5	0.6	0.7	0.9	1.0	1.2	1.4	1.6	1.7	1.9	2.1	2.4	2.8	3.5	4.3	5.2
	200	0.2	0.2	0.3	0.4	0.5	0.6	0.7	0.8	1.0	1.1	1.3	1.5	1.6	1.8	2.0	2.3	2.6	3.3	4.1	4.9
42	42	0.2	0.4	0.5	0.6	0.7	0.8	1.0	1.2	1.4	1.6	1.9	2.1	2.4	2.6	2.8	3.3	3.8	4.7	5.9	7.1
	60	0.2	0.3	0.4	0.5	0.6	0.7	0.8	1.0	1.2	1.4	1.6	1.8	2.0	2.2	2.4	2.8	3.2	4.0	5.0	6.0
	90	0.2	0.3	0.3	0.4	0.5	0.6	0.7	0.9	1.0	1.2	1.4	1.6	1.7	1.9	2.1	2.4	2.8	3.5	4.4	5.2
	140	0.2	0.2	0.3	0.4	0.5	0.5	0.6	0.8	0.9	1.1	1.2	1.4	1.5	1.7	1.9	2.2	2.5	3.1	3.9	4.6
	200	0.1	0.2	0.3	0.4	0.4	0.5	0.6	0.7	0.9	1.0	1.1	1.3	1.4	1.6	1.7	2.0	2.3	2.9	3.6	4.3
	300	0.1	0.2	0.3	0.3	0.4	0.5	0.5	0.7	0.8	0.9	1.1	1.3	1.4	1.5	1.7	1.9	2.2	2.8	3.5	4.2

Fig. 9-38. *Continued*

Room Dimensions		Cavity Depth																			
Width	Length	1	1.5	2	2.5	3	3.5	4	5	6	7	8	9	10	11	12	14	16	20	25	30
50	50	0.2	0.3	0.4	0.5	0.6	0.7	0.8	1.0	1.2	1.4	1.6	1.8	2.0	2.2	2.4	2.8	3.2	4.0	5.0	6.0
	70	0.2	0.3	0.3	0.4	0.5	0.6	0.7	0.9	1.0	1.2	1.4	1.5	1.7	1.9	2.0	2.4	2.7	3.4	4.3	5.1
	100	0.1	0.2	0.3	0.4	0.4	0.5	0.6	0.7	0.9	1.0	1.2	1.3	1.5	1.6	1.8	2.1	2.4	3.0	3.7	4.5
	150	0.1	0.2	0.3	0.3	0.4	0.5	0.5	0.7	0.8	0.9	1.1	1.2	1.3	1.5	1.6	1.9	2.1	2.7	3.3	4.0
	300	0.1	0.2	0.2	0.3	0.3	0.4	0.5	0.6	0.7	0.8	0.9	1.0	1.1	1.3	1.4	1.6	1.9	2.3	2.9	3.5
60	60	0.2	0.2	0.3	0.4	0.5	0.6	0.7	0.8	1.0	1.2	1.3	1.5	1.7	1.8	2.0	2.3	2.7	3.3	4.2	5.0
	100	0.1	0.2	0.3	0.3	0.4	0.5	0.5	0.7	0.8	0.9	1.1	1.2	1.3	1.5	1.6	1.9	2.1	2.7	3.3	4.0
	150	0.1	0.2	0.2	0.3	0.3	0.4	0.5	0.6	0.7	0.8	0.9	1.0	1.2	1.3	1.4	1.6	1.9	2.3	2.9	3.5
	300	0.1	0.1	0.2	0.2	0.3	0.3	0.4	0.5	0.6	0.7	0.8	0.9	1.0	1.1	1.2	1.4	1.6	2.0	2.5	3.0
75	75	0.1	0.2	0.3	0.3	0.4	0.5	0.5	0.7	0.8	0.9	1.1	1.2	1.3	1.5	1.6	1.9	2.1	2.7	3.3	4.0
	120	0.1	0.2	0.2	0.3	0.3	0.4	0.4	0.5	0.6	0.8	0.9	1.0	1.1	1.2	1.3	1.5	1.7	2.2	2.7	3.3
	200	0.1	0.1	0.2	0.2	0.3	0.3	0.4	0.5	0.5	0.6	0.7	0.8	0.9	1.0	1.1	1.3	1.5	1.8	2.3	2.7
	300	0.1	0.1	0.2	0.2	0.2	0.3	0.3	0.4	0.5	0.6	0.7	0.7	0.8	0.9	1.0	1.2	1.3	1.7	2.1	2.5
100	100	0.1	0.1	0.2	0.2	0.3	0.3	0.4	0.5	0.6	0.7	0.8	0.9	1.0	1.1	1.2	1.4	1.6	2.0	2.5	3.0
	200	0.1	0.1	0.1	0.2	0.2	0.3	0.3	0.4	0.4	0.5	0.6	0.7	0.7	0.8	0.9	1.0	1.2	1.5	1.9	2.2
	300	0.1	0.1	0.1	0.2	0.2	0.2	0.3	0.3	0.4	0.5	0.5	0.6	0.7	0.7	0.8	0.9	1.1	1.3	1.7	2.0
150	150	0.1	0.1	0.1	0.2	0.2	0.2	0.3	0.3	0.4	0.5	0.5	0.6	0.7	0.7	0.8	0.9	1.1	1.3	1.7	2.0
	300	—	0.1	0.1	0.1	0.1	0.2	0.2	0.2	0.3	0.3	0.4	0.5	0.5	0.6	0.6	0.7	0.8	1.0	1.2	1.5
200	200	—	0.1	0.1	0.1	0.1	0.2	0.2	0.2	0.3	0.3	0.4	0.5	0.5	0.6	0.6	0.7	0.8	1.0	1.2	1.5
	300	—	0.1	0.1	0.1	0.1	0.1	0.2	0.2	0.2	0.3	0.3	0.4	0.4	0.5	0.5	0.6	0.7	0.8	1.0	1.2
300	300	—	—	0.1	0.1	0.1	0.1	0.1	0.2	0.2	0.2	0.3	0.3	0.3	0.4	0.4	0.5	0.5	0.6	0.7	0.8
500	500	—	—	—	—	0.1	0.1	0.1	0.1	0.1	0.1	0.2	0.2	0.2	0.2	0.2	0.3	0.3	0.4	0.5	0.6

where

A_B, A_W = areas of the cavity base and walls, respectively

ρ_B, ρ_W = reflectance of cavity base and walls, respectively

f = form factor between the cavity opening and the cavity base; where

$$f = \frac{2}{\pi x y} \left\{ \ln \left[\frac{(1 + x^2)(1 + y^2)}{1 + x^2 + y^2} \right]^{1/2} + y(1 + x^2)^{1/2} \tan^{-1} \left[\frac{y}{(1 + x^2)^{1/2}} \right] + x(1 + y^2)^{1/2} \tan^{-1} \left[\frac{x}{(1 + y^2)^{1/2}} \right] - y \tan^{-1}(y) - x \tan^{-1}(x) \right\}$$

where

$$x = \frac{\text{Cavity Length}}{\text{Cavity Depth}}$$

$$y = \frac{\text{Cavity Width}}{\text{Cavity Depth}}$$

and the arc tangents are expressed in radians.

If it is assumed that the cavity is square, then

$$x = y = \frac{10}{\text{Cavity Ratio}}$$

The effective ceiling cavity reflectance of non-horizontal ceilings can be determined[26] by the following formula and solving the problem in the usual manner:

$$\rho_{CC} = \frac{\rho}{\dfrac{A_s}{A_o} - \rho \left(\dfrac{A_s}{A_o} - 1 \right)}$$

where

A_o = area of ceiling opening

A_s = area of ceiling surface

ρ = reflectance of ceiling surface

If the ceiling surface reflectance is not the same for all parts of the ceiling, use an area weighted average. Thus, if the ceiling has several sections 1, 2, 3, · · ·

$$\rho = \frac{\rho_1 A_1 + \rho_2 A_2 + \rho_3 A_3 + \cdots}{A_1 + A_2 + A_3 + \cdots}$$

This formula for ρ_{CC} applies to concave ceilings such as a hemispherical dome where all parts of the ceiling are exposed to all other parts of the ceiling.

Luminaire Coefficients of Utilization. Absorption of light in a luminaire is accounted for in the computation of coefficients of utiliza-

Fig. 9-39. Per Cent Effective Ceiling or Floor Cavity Reflectances for Various Reflectance Combinations
See Page 9-26 for Algorithms for Calculation of Effective Cavity Reflectance*

Per Cent Base† Reflectance	90									
Per Cent Wall Reflectance	90	80	70	60	50	40	30	20	10	0
Cavity Ratio										
0.2	89	88	88	87	86	85	85	84	84	82
0.4	88	87	86	85	84	83	81	80	79	76
0.6	87	86	84	82	80	79	77	76	74	73
0.8	87	85	82	80	77	75	73	71	69	67
1.0	86	83	80	77	75	72	69	66	64	62
1.2	85	82	78	75	72	69	66	63	60	57
1.4	85	80	77	73	69	65	62	59	57	52
1.6	84	79	75	71	67	63	59	56	53	50
1.8	83	78	73	69	64	60	56	53	50	48
2.0	83	77	72	67	62	56	53	50	47	43
2.2	82	76	70	65	59	54	50	47	44	40
2.4	82	75	69	64	58	53	48	45	41	37
2.6	81	74	67	62	56	51	46	42	38	35
2.8	81	73	66	60	54	49	44	40	36	34
3.0	80	72	64	58	52	47	42	38	34	30
3.2	79	71	63	56	50	45	40	36	32	28
3.4	79	70	62	54	48	43	38	34	30	27
3.6	78	69	61	53	47	42	36	32	28	25
3.8	78	69	60	51	45	40	35	31	27	23
4.0	77	69	58	51	44	39	33	29	25	22
4.2	77	62	57	50	43	37	32	28	24	21
4.4	76	61	56	49	42	36	31	27	23	20
4.6	76	60	55	47	40	35	30	26	22	19
4.8	75	59	54	46	39	34	28	25	21	18
5.0	75	59	53	45	38	33	28	24	20	16
6.0	73	61	49	41	34	29	24	20	16	11
7.0	70	58	45	38	30	27	21	18	14	08
8.0	68	55	42	35	27	23	18	15	12	06
9.0	66	52	38	31	25	21	16	14	11	05
10.0	65	51	36	29	22	19	15	11	09	04

Per Cent Base† Reflectance	80									
Per Cent Wall Reflectance	90	80	70	60	50	40	30	20	10	0
Cavity Ratio										
0.2	79	78	78	77	77	76	76	75	74	72
0.4	79	77	76	75	74	73	72	71	70	68
0.6	78	76	75	73	71	70	68	66	65	63
0.8	78	75	73	71	69	67	65	63	61	57
1.0	77	74	72	69	67	65	62	60	57	55
1.2	76	73	70	67	64	61	58	55	53	51
1.4	76	72	68	65	62	59	55	53	50	48
1.6	75	71	67	63	60	57	53	50	47	44
1.8	75	70	66	62	58	54	50	47	44	41
2.0	74	69	64	60	56	52	48	45	41	38
2.2	74	68	63	58	54	49	45	42	38	35
2.4	73	67	61	56	52	47	43	40	36	33
2.6	73	66	60	55	50	45	41	38	34	31
2.8	73	65	59	53	48	43	39	36	32	29
3.0	72	65	58	52	47	42	37	34	30	27
3.2	72	65	57	51	45	40	35	33	28	25
3.4	71	64	56	49	44	39	34	32	27	24
3.6	71	63	54	48	43	38	32	30	25	23
3.8	70	62	53	47	41	36	31	28	24	22
4.0	70	61	53	46	40	35	30	26	22	20
4.2	69	60	52	45	39	34	29	25	21	18
4.4	69	60	51	44	38	33	28	24	20	17
4.6	69	59	50	43	37	32	27	23	19	15
4.8	68	58	49	42	36	31	26	22	18	14
5.0	68	58	48	41	35	30	25	21	18	14
6.0	66	55	44	38	31	27	22	19	15	10
7.0	64	53	41	35	28	24	19	16	12	07
8.0	62	50	38	32	25	21	17	14	11	05
9.0	61	49	36	30	23	19	15	13	10	04
10.0	59	46	33	27	21	18	14	11	08	03

Per Cent Base† Reflectance	70									
Per Cent Wall Reflectance	90	80	70	60	50	40	30	20	10	0
Cavity Ratio										
0.2	70	69	68	68	67	67	66	66	65	64
0.4	69	68	67	66	65	64	63	62	61	58
0.6	69	67	65	64	63	61	59	58	57	54
0.8	68	66	64	62	60	58	56	55	53	50
1.0	68	65	62	60	58	55	53	52	50	47
1.2	67	64	61	59	57	54	50	48	46	44
1.4	67	63	60	58	55	51	47	45	44	41
1.6	67	62	59	56	53	47	45	43	41	38
1.8	66	61	58	54	51	46	42	40	38	35
2.0	66	60	56	52	49	45	40	38	36	33
2.2	66	60	55	51	48	43	38	36	34	32
2.4	65	60	54	50	46	41	37	35	32	30
2.6	65	59	54	49	45	40	35	33	30	28
2.8	65	59	53	48	43	38	33	30	28	26
3.0	64	58	52	47	42	37	32	29	27	24
3.2	64	58	51	46	40	36	31	28	25	23
3.4	64	57	50	45	39	35	29	27	24	22
3.6	63	56	49	44	38	33	28	25	22	20
3.8	63	56	49	43	37	32	27	24	21	19
4.0	63	55	48	42	36	31	26	23	20	17
4.2	62	55	47	41	35	30	25	22	19	16
4.4	62	54	46	40	34	29	24	21	18	15
4.6	62	53	45	39	33	28	24	21	17	14
4.8	62	53	45	38	32	27	23	20	16	13
5.0	61	52	44	36	31	26	22	19	16	12
6.0	60	51	41	35	28	24	19	16	13	09
7.0	58	48	38	32	26	22	17	14	11	06
8.0	57	46	35	29	23	19	15	13	10	05
9.0	56	45	33	27	21	18	14	12	09	04
10.0	55	43	31	25	19	16	12	10	08	03

Per Cent Base† Reflectance	60									
Per Cent Wall Reflectance	90	80	70	60	50	40	30	20	10	0
Cavity Ratio										
0.2	60	59	59	59	58	57	56	56	55	53
0.4	60	59	59	58	57	55	54	53	52	50
0.6	60	58	57	56	55	53	51	51	50	46
0.8	59	57	56	55	54	51	48	47	46	43
1.0	59	57	55	53	51	48	45	44	43	41
1.2	59	56	54	51	49	46	44	42	40	38
1.4	59	56	53	49	47	44	41	39	38	36
1.6	59	55	52	48	45	42	39	37	35	33
1.8	58	55	51	47	44	40	37	35	33	31
2.0	58	54	50	46	43	39	35	33	31	29
2.2	58	53	49	45	42	37	34	31	29	28
2.4	58	53	48	44	41	36	32	30	27	26
2.6	58	53	48	43	39	35	31	28	26	24
2.8	58	53	47	43	38	34	29	27	24	22
3.0	57	52	46	42	37	32	28	25	23	20
3.2	57	51	45	41	36	31	27	23	22	18
3.4	57	51	45	40	35	30	26	23	20	17
3.6	57	50	44	39	34	29	25	22	19	16
3.8	57	50	43	38	33	29	24	21	19	15
4.0	57	49	42	37	32	28	23	20	18	14
4.2	56	49	42	37	32	27	22	19	17	14
4.4	56	49	42	36	31	27	22	19	16	13
4.6	56	49	41	35	30	26	21	18	16	13
4.8	56	48	41	34	29	25	21	18	15	12
5.0	56	48	40	34	28	24	20	17	14	11
6.0	55	45	37	31	25	21	17	14	11	07
7.0	54	43	35	30	24	20	15	12	09	05
8.0	53	42	33	28	22	18	14	11	08	04
9.0	52	40	31	26	20	16	12	10	07	03
10.0	51	39	29	24	18	15	11	09	07	02

Per Cent Base† Reflectance	50									
Per Cent Wall Reflectance	90	80	70	60	50	40	30	20	10	0
Cavity Ratio										
0.2	50	50	49	49	48	48	47	46	46	44
0.4	50	49	48	48	47	46	45	45	44	42
0.6	50	48	47	46	45	44	43	42	41	38
0.8	50	48	47	45	44	42	40	39	38	36
1.0	50	48	46	44	43	41	38	37	36	34
1.2	50	47	45	43	41	39	36	35	34	29
1.4	50	47	45	42	40	38	35	34	32	27
1.6	50	47	44	41	39	36	33	32	30	26
1.8	50	46	43	40	38	35	31	30	28	25
2.0	50	46	43	40	37	34	30	28	26	24
2.2	50	46	42	38	36	33	29	27	24	22
2.4	50	46	42	37	35	31	27	25	23	21
2.6	50	46	41	37	34	30	26	23	21	20
2.8	50	46	41	36	33	29	25	22	20	19
3.0	50	45	40	36	32	28	24	21	19	17
3.2	50	44	39	35	31	27	23	20	18	16
3.4	50	44	39	35	30	26	22	19	17	15
3.6	50	44	39	34	29	25	21	18	16	14
3.8	50	44	38	34	29	25	21	17	15	13
4.0	50	44	38	33	28	24	20	17	15	12
4.2	50	43	37	32	28	24	20	17	14	12
4.4	50	43	37	32	27	23	19	16	13	11
4.6	50	43	36	31	26	22	18	15	13	10
4.8	50	43	36	31	26	22	18	15	12	09
5.0	50	42	35	30	25	21	17	14	12	09
6.0	50	42	34	29	23	19	15	13	10	06
7.0	49	41	32	27	21	18	14	11	08	05
8.0	49	40	30	25	19	16	12	10	07	03
9.0	48	39	29	24	18	15	11	09	07	03
10.0	47	37	27	22	17	14	10	08	06	02

* Values in this table are based on a length to width ratio of 1.6.
† Ceiling, floor or floor of cavity.

Fig. 9–39. *Continued**

Per Cent Base† Reflectance: 40

Cavity Ratio \ Per Cent Wall Reflectance	90	80	70	60	50	40	30	20	10	0
0.2	40	40	39	39	39	38	38	37	36	36
0.4	41	40	39	39	38	37	36	35	34	34
0.6	41	40	39	38	37	36	34	33	32	31
0.8	41	40	38	37	36	35	33	32	31	29
1.0	42	40	38	37	35	33	32	31	29	27
1.2	42	40	38	36	34	32	30	29	27	25
1.4	42	39	37	35	33	31	29	27	25	23
1.6	42	39	37	35	32	30	27	25	23	22
1.8	42	39	36	34	31	29	26	24	22	21
2.0	42	39	36	34	31	28	25	23	21	19
2.2	42	39	36	33	30	27	24	22	19	18
2.4	43	39	35	33	29	27	24	21	18	17
2.6	43	39	35	32	29	26	23	20	17	15
2.8	43	39	35	32	28	25	22	19	16	14
3.0	43	39	35	31	27	24	21	18	16	13
3.2	43	39	35	31	27	23	20	71	15	13
3.4	43	39	34	30	26	23	20	17	14	12
3.6	44	39	34	30	26	22	19	16	14	11
3.8	44	38	33	29	25	22	18	16	13	10
4.0	44	38	33	29	25	21	18	15	12	10
4.2	44	38	33	29	24	21	17	15	12	10
4.4	44	38	33	28	24	20	17	14	11	09
4.6	44	38	32	28	23	19	16	14	11	08
4.8	44	38	32	27	22	19	16	13	10	08
5.0	45	38	31	27	22	19	15	13	10	07
6.0	44	37	30	25	20	17	13	11	08	05
7.0	44	36	29	24	19	16	12	10	07	04
8.0	44	35	28	23	18	15	11	09	06	03
9.0	44	35	26	21	16	13	10	08	05	02
10.0	43	34	25	20	15	12	08	07	05	02

Per Cent Base† Reflectance: 30

Cavity Ratio \ Per Cent Wall Reflectance	90	80	70	60	50	40	30	20	10	0
0.2	31	31	30	30	29	29	29	28	28	27
0.4	31	31	30	30	29	28	28	27	26	25
0.6	32	31	30	29	28	27	26	26	25	23
0.8	32	31	30	29	28	26	25	25	23	22
1.0	33	32	30	29	27	25	24	23	22	20
1.2	33	32	30	28	27	25	23	22	21	19
1.4	34	32	30	28	26	24	22	21	19	18
1.6	34	33	29	27	25	23	22	20	18	17
1.8	35	33	29	27	25	23	21	19	17	16
2.0	35	33	29	26	24	22	20	18	16	14
2.2	36	32	29	26	24	22	19	17	15	13
2.4	36	32	29	26	24	22	19	16	14	12
2.6	36	32	29	25	23	21	18	16	14	12
2.8	37	33	29	25	23	21	17	15	13	11
3.0	37	33	29	25	22	20	17	15	12	10
3.2	37	33	29	25	22	19	16	14	12	10
3.4	37	33	29	25	22	19	16	14	11	09
3.6	38	33	29	24	21	18	15	13	10	09
3.8	38	33	28	24	21	18	15	13	10	08
4.0	38	33	28	24	21	18	14	12	09	07
4.2	38	33	28	24	20	17	14	12	09	07
4.4	39	33	28	24	20	17	14	11	09	06
4.6	39	33	28	24	20	17	13	10	08	06
4.8	39	33	28	24	20	17	13	10	08	05
5.0	39	33	28	24	19	16	13	10	08	05
6.0	39	33	27	23	18	15	11	09	06	04
7.0	40	33	26	22	17	14	10	08	05	03
8.0	40	33	26	21	16	13	09	07	04	02
9.0	40	33	25	20	15	12	09	07	04	02
10.0	40	32	24	19	14	11	08	06	03	01

Per Cent Base† Reflectance: 20

Cavity Ratio \ Per Cent Wall Reflectance	90	80	70	60	50	40	30	20	10	0
0.2	21	20	20	20	20	20	19	19	19	17
0.4	22	21	20	20	20	19	19	18	18	16
0.6	23	21	21	20	19	19	18	18	17	15
0.8	24	22	21	20	19	19	18	17	16	14
1.0	25	23	22	20	19	18	17	16	15	13
1.2	25	23	22	20	19	17	17	16	14	12
1.4	26	24	22	20	18	17	16	15	13	12
1.6	26	24	22	20	18	17	16	15	13	11
1.8	27	25	23	20	18	17	15	14	12	10
2.0	28	25	23	20	18	16	15	13	11	09
2.2	28	25	23	20	18	16	14	12	10	09
2.4	29	26	23	20	18	16	14	12	10	08
2.6	29	26	23	20	18	16	14	11	09	08
2.8	30	27	23	20	18	15	13	11	09	07
3.0	30	27	23	20	17	15	13	11	09	07
3.2	31	27	23	20	17	15	12	11	09	06
3.4	31	27	23	20	17	15	12	10	08	06
3.6	32	27	23	20	17	15	12	10	08	05
3.8	32	28	23	20	17	15	12	10	07	05
4.0	33	28	23	20	17	14	11	09	07	05
4.2	33	28	23	20	17	14	11	09	07	04
4.4	34	28	24	20	17	14	11	09	07	04
4.6	34	29	24	20	17	14	11	09	07	04
4.8	35	29	24	20	17	13	10	08	06	04
5.0	35	29	24	20	16	13	10	08	06	04
6.0	36	30	24	20	16	13	10	08	05	02
7.0	36	30	24	20	15	12	09	07	04	02
8.0	37	30	23	19	15	12	08	06	03	01
9.0	37	29	23	19	14	11	08	06	03	01
10.0	37	29	22	18	13	10	07	05	03	01

Per Cent Base† Reflectance: 10

Cavity Ratio \ Per Cent Wall Reflectance	90	80	70	60	50	40	30	20	10	0
0.2	11	11	11	10	10	10	10	09	09	09
0.4	12	11	11	11	11	10	10	09	09	08
0.6	13	13	12	11	11	10	10	09	08	08
0.8	15	14	13	12	11	10	10	09	08	07
1.0	16	14	13	12	12	11	10	09	08	07
1.2	17	15	14	13	12	11	10	09	07	06
1.4	18	16	14	13	12	11	10	09	07	06
1.6	19	17	15	14	12	11	09	08	07	06
1.8	19	17	15	14	13	11	09	08	06	05
2.0	20	18	16	14	13	11	09	08	06	05
2.2	21	19	16	14	13	11	09	07	06	05
2.4	22	19	17	15	13	11	09	07	06	05
2.6	23	20	17	15	13	11	09	07	06	04
2.8	23	20	18	16	13	11	09	07	05	03
3.0	24	21	18	16	13	11	09	07	05	03
3.2	25	21	18	16	13	11	09	07	05	03
3.4	26	22	18	16	13	11	09	07	05	03
3.6	26	22	19	16	13	11	09	06	04	03
3.8	27	23	19	17	14	11	09	06	04	02
4.0	27	23	20	17	14	11	09	06	04	02
4.2	28	24	20	17	14	11	09	06	04	02
4.4	28	24	20	17	14	11	08	06	04	02
4.6	29	25	20	17	14	11	08	06	04	02
4.8	29	25	20	17	14	11	08	06	04	02
5.0	30	25	20	17	14	11	08	06	04	02
6.0	31	26	21	18	14	11	08	06	03	01
7.0	32	27	21	17	13	11	08	06	03	01
8.0	33	27	21	17	13	10	07	05	03	01
9.0	34	28	21	17	13	10	07	05	02	01
10.0	34	28	21	17	12	10	07	05	02	01

Per Cent Base† Reflectance: 0

Cavity Ratio \ Per Cent Wall Reflectance	90	80	70	60	50	40	30	20	10	0
0.2	02	02	02	01	01	01	01	00	00	0
0.4	04	03	03	02	02	02	01	01	00	0
0.6	05	05	04	03	03	02	02	01	01	0
0.8	07	06	05	04	04	03	02	02	01	0
1.0	08	07	06	05	04	03	02	02	01	0
1.2	10	08	07	06	05	04	03	02	01	0
1.4	11	09	08	07	06	04	03	02	01	0
1.6	12	10	09	07	06	05	03	02	01	0
1.8	13	11	09	08	07	05	04	03	01	0
2.0	14	12	10	09	07	05	04	03	01	0
2.2	15	13	11	09	07	06	04	03	01	0
2.4	16	13	11	09	08	06	04	03	01	0
2.6	17	14	12	10	08	06	05	03	02	0
2.8	17	15	13	10	08	07	05	03	02	0
3.0	18	16	13	11	09	07	05	03	02	0
3.2	19	16	14	11	09	07	05	03	02	0
3.4	20	17	14	12	09	07	05	03	02	0
3.6	20	17	15	12	10	08	05	04	02	0
3.8	21	18	15	12	10	08	05	04	02	0
4.0	22	18	15	13	10	08	05	04	02	0
4.2	22	19	16	13	10	08	06	04	02	0
4.4	23	19	16	13	10	08	06	04	02	0
4.6	23	20	17	13	11	08	06	04	02	0
4.8	24	20	17	14	11	08	06	04	02	0
5.0	25	21	17	14	11	08	06	04	02	0
6.0	27	23	18	15	12	09	06	04	02	0
7.0	28	24	19	15	12	09	06	04	02	0
8.0	30	25	20	15	12	09	06	04	02	0
9.0	31	25	20	15	12	09	06	04	02	0
10.0	31	25	20	15	12	09	06	04	02	0

* Values in this table are based on a length to width ratio of 1.6.

† Ceiling, floor or floor of cavity.

Fig. 9-40. Multiplying Factors for Other than 20 Per Cent Effective Floor Cavity Reflectance

% Effective Ceiling Cavity Reflectance, ρ_{CC}	80				70				50			30			10		
% Wall Reflectance, ρ_W	70	50	30	10	70	50	30	10	50	30	10	50	30	10	50	30	10
For 30 Per Cent Effective Floor Cavity Reflectance (20 Per Cent = 1.00)																	
Room Cavity Ratio																	
1	1.092	1.082	1.075	1.068	1.077	1.070	1.064	1.059	1.049	1.044	1.040	1.028	1.026	1.023	1.012	1.010	1.008
2	1.079	1.066	1.055	1.047	1.068	1.057	1.048	1.039	1.041	1.033	1.027	1.026	1.021	1.017	1.013	1.010	1.006
3	1.070	1.054	1.042	1.033	1.061	1.048	1.037	1.028	1.034	1.027	1.020	1.024	1.017	1.012	1.014	1.009	1.005
4	1.062	1.045	1.033	1.024	1.055	1.040	1.029	1.021	1.030	1.022	1.015	1.022	1.015	1.010	1.014	1.009	1.004
5	1.056	1.038	1.026	1.018	1.050	1.034	1.024	1.015	1.027	1.018	1.012	1.020	1.013	1.008	1.014	1.009	1.004
6	1.052	1.033	1.021	1.014	1.047	1.030	1.020	1.012	1.024	1.015	1.009	1.019	1.012	1.006	1.014	1.008	1.003
7	1.047	1.029	1.018	1.011	1.043	1.026	1.017	1.009	1.022	1.013	1.007	1.018	1.010	1.005	1.014	1.008	1.003
8	1.044	1.026	1.015	1.009	1.040	1.024	1.015	1.007	1.020	1.012	1.006	1.017	1.009	1.004	1.013	1.007	1.003
9	1.040	1.024	1.014	1.007	1.037	1.022	1.014	1.006	1.019	1.011	1.005	1.016	1.009	1.004	1.013	1.007	1.002
10	1.037	1.022	1.012	1.006	1.034	1.020	1.012	1.005	1.017	1.010	1.004	1.015	1.009	1.003	1.013	1.007	1.002
For 10 Per Cent Effective Floor Cavity Reflectance (20 Per Cent = 1.00)																	
Room Cavity Ratio																	
1	.923	.929	.935	.940	.933	.939	.943	.948	.956	.960	.963	.973	.976	.979	.989	.991	.993
2	.931	.942	.950	.958	.940	.949	.957	.963	.962	.968	.974	.976	.980	.985	.988	.991	.995
3	.939	.951	.961	.969	.945	.957	.966	.973	.967	.975	.981	.978	.983	.988	.988	.992	.996
4	.944	.958	.969	.978	.950	.963	.973	.980	.972	.980	.986	.980	.986	.991	.987	.992	.996
5	.949	.964	.976	.983	.954	.968	.978	.985	.975	.983	.989	.981	.988	.993	.987	.992	.997
6	.953	.969	.980	.986	.958	.972	.982	.989	.977	.985	.992	.982	.989	.995	.987	.993	.997
7	.957	.973	.983	.991	.961	.975	.985	.991	.979	.987	.994	.983	.990	.996	.987	.993	.998
8	.960	.976	.986	.993	.963	.977	.987	.993	.981	.988	.995	.984	.991	.997	.987	.994	.998
9	.963	.978	.987	.994	.965	.979	.989	.994	.983	.990	.996	.985	.992	.998	.988	.994	.999
10	.965	.980	.989	.995	.967	.981	.990	.995	.984	.991	.997	.986	.993	.998	.988	.994	.999
For 0 Per Cent Effective Floor Cavity Reflectance (20 Per Cent = 1.00)																	
Room Cavity Ratio																	
1	.859	.870	.879	.886	.873	.884	.893	.901	.916	.923	.929	.948	.954	.960	.979	.983	.987
2	.871	.887	.903	.919	.886	.902	.916	.928	.926	.938	.949	.954	.963	.971	.978	.983	.991
3	.882	.904	.915	.942	.898	.918	.934	.947	.936	.950	.964	.958	.969	.979	.976	.984	.993
4	.893	.919	.941	.958	.908	.930	.948	.961	.945	.961	.974	.961	.974	.984	.975	.985	.994
5	.903	.931	.953	.969	.914	.939	.958	.970	.951	.967	.980	.964	.977	.988	.975	.985	.995
6	.911	.940	.961	.976	.920	.945	.965	.977	.955	.972	.985	.966	.979	.991	.975	.986	.996
7	.917	.947	.967	.981	.924	.950	.970	.982	.959	.975	.988	.968	.981	.993	.975	.987	.997
8	.922	.953	.971	.985	.929	.955	.975	.986	.963	.978	.991	.970	.983	.995	.976	.988	.998
9	.928	.958	.975	.988	.933	.959	.980	.989	.966	.980	.993	.971	.985	.996	.976	.988	.998
10	.933	.962	.979	.991	.937	.963	.983	.992	.969	.982	.995	.973	.987	.997	.977	.989	.999

tion (CU) for that particular luminaire. Fig. 9-62 (page 9-54) is a tabulation of coefficients of utilization calculated by the Zonal-Cavity Method for representative luminaire types. These coefficients are for an Effective Floor Cavity Reflectance of 20 per cent, but any CU obtained from the table may be corrected for a different value of ρ_{FC} by applying the appropriate multiplier from Fig. 9-40.

Fig. 9-62 is based on generic type luminaires that can be readily identified. As an example, there are many variations of the flat bottom fluorescent troffer using prismatic lenses, and since there are not well-known subgroups, a single set of entries for wide and narrow cutoffs (luminaire nos. 42 and 43) are given using data averaged for the entire class.

Luminaires in Fig. 9-62 are not to be considered as recommended luminaires. Some present useful data even though the the specific luminaires are no longer common in terms of practice. For example, luminaire nos. 2 and 33 are no longer in common use; however, the coefficients apply to any indirect luminaire of similar efficiency (see note 3 of Fig. 9-62) and direct component, since the coefficients do not depend on the shape of the upward intensity distribution.

Since the Light Loss Factor includes the effect of dirt depositing on wall surfaces, the selection of the proper column of wall reflectances, ρ_W, should be based upon the initial values expected. The wall reflectance should also represent the weighted average of the painted areas, glass fenestration or daylight controls, chalkboards, shelves, etc., in the area to be lighted. The weighting should be based on the relative areas of each type of surface.

In using Fig. 9-62, it will often be necessary to interpolate between Room Cavity Ratios and/or Effective Ceiling Cavity Reflectances. This is most easily accomplished by interpolating first between RCRs to obtain CUs for Effective Ceiling Cavity Reflectances that straddle the actual ρ_{CC} and then interpolating between these CUs.

CALCULATION OF LUMINANCE AND EXITANCE

Luminance Calculation

Methods for evaluating the visibility of visual work in an interior space require the prediction (or measurement) of luminance at a point. All methods require values of adaptation luminance and task contrast (see page 3-14). Both these quantities are derived from the task detail luminance, L_t, and the task background luminance, L_b, at some task location. These quantities vary from location to location not only because the illuminance varies, but because of the highly directional nature of the reflectance of the task detail and task background. The procedure described below is not limited to the luminances required for visibility calcuations, but applies whenever it is necessary to calculate the luminance of a surface that exhibits a very directionally sensitive reflectance.

A luminance predetermination method should be able to predict the luminance at any point in a room for a given surface with reasonable accuracy. While it is not required that any given method be able to take all possible ranges of the following nine parameters into account, any restrictions which the method imposes should be stated clearly:

1. Room size and shape.
2. Room surface reflectances.
3. Luminaire characteristics.
4. Number and location of luminaires.
5. Nature of the given surface.
6. Observer location, line of sight and viewing angle.
7. Nature and luminance of all other surfaces in the environment.
8. Body shadow effects.
9. Polarization effects.

For example, a method may be limited to calculation of luminance in a rectangularly shaped room with flat floors and ceilings.

In order to compute the luminance, it is necessary to generate a mathematical description, usually in the form of sets of equations or inequalities. In generating such a model one must be careful to avoid oversimplication, which results in erroneous or misleading results at best and meaningless results at worst.

The problem of computing the details of lighting system performance is relatively complex, because the fundamental physical phenomena involved interact in a complicated manner and a sizable number of parameters must be handled simultaneously. Under such circumstances, the use of a computer may be considered desirable. It is important to note that acceptable methods of luminance calculation may be developed in the future which do not require computer facilities.

In order to understand any luminance calculation method, it is necessary to review flux transfer theory, that is, the concepts which underlie the transfer of energy from a real or virtual source to the point of interest. See page 9-3. In the present context it is doubly important to do so since the mathematical model on which any luminance calculation is based must rely heavily on such a theory either explicitly or implicitly. First, an exact theory is developed and the reasons for approximations discussed. Then, the problem is divided into direct and reflected components of illumination, and the manner in which they can be approximated for ease of computation is discussed.

Fundamental Equations. To predict the luminance in a particular luminous environment at a particular location, it is necessary to solve the following equation:

$$L = \int dE(\theta,\psi) f_r(\theta,\psi)$$

where

L = luminance at a point on a surface
θ,ψ = spherical coordinates, declination (co-latitude) and azimuth (longitude), respectively
$dE(\theta,\psi)$ = differential amount of illuminance at the point in the plane of the surface from a direction indicated by (θ,ψ)
$f_r(\theta,\psi)$ = bidirectional reflectance-distribution function of surface material

Fig. 9-41 indicates the necessary coordinates.

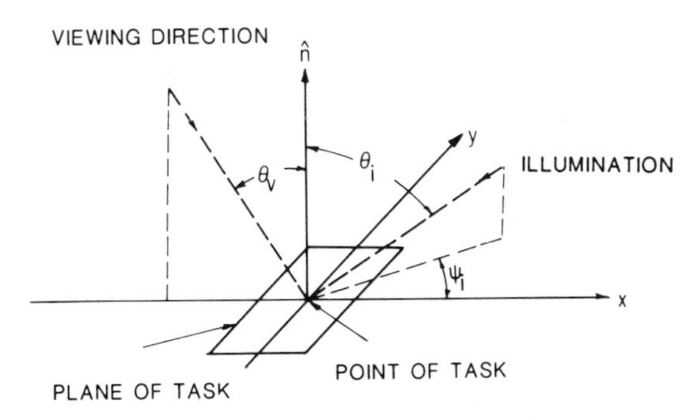

Fig. 9-41. Task and illumination coordinates.

This expression represents the total effect of all components of illuminance multiplied by the appropriate bidirectional reflectance-distribution function, to give luminance.

It should be noted that unlike perfectly diffuse reflectances in which the resulting differential luminances are independent of both the incident direction of the differential illuminance and the direction of view, the bidirectional reflectance-distribution functions are bidirectional, and may be rewritten:

$$f_r(\theta_v,\psi_v,\theta_i,\psi_i)$$

where the subscript i represents incident illumination, θ_v indicates viewing angle and ψ_v indicates direction of view. However, if the ψ_i-coordinate of the direction from which the illumination arrives is measured from the plane described by the viewing direction and the normal to the surface of the task, then the ψ_v-coordinate, indicating viewing direction, can be considered a constant, and the specification of bidirectional reflectance-distribution functions becomes:

$$f_r(\theta_v;\theta_i,\psi_i)$$

This gives,

$$L = \int dE(\theta_i,\psi_i)f_r(\theta_v;\theta_i,\psi_i)$$

θ_v, the viewing angle, normally will be held constant.

Some surfaces exhibit sensitivity to polarization and it is possible to separate two orthogonal components of the bidirectional reflectance-distribution functions associated with horizontal and vertical plane polarization components. The bidirectional reflectance-distribution functions are:

$$f_{rh}(\theta_v;\theta_i,\psi_i),\ f_{rv}(\theta_v;\theta_i,\psi_i)$$

where the subscripts h and v indicate horizontal and vertical plane polarization components, respectively. In a complementary fashion, two orthogonal components of the illuminance can be considered separately, and will be indicated by dE_h and dE_v. The above equation thus is expanded to:

$$L = \int \{dE_h(\theta_i,\psi_i)f_{rh}(\theta_v;\theta_i,\psi_i) + dE_v(\theta_i,\psi_i)f_{rv}(\theta_v;\theta_i,\psi_i)\}$$

The bidirectional reflectance-distribution functions f_{rh}, f_{rv} are properties dependent only upon the surface and viewing angle, while the illuminance values $dE_h(\theta_i,\psi_i)$ and $dE_v(\theta_i,\psi_i)$ are functions of the luminous environment only. The expression for L is perfectly general and applicable to all situations. Since the illuminance values and luminance factors must be expressed as analytic functions in order to attempt integration of these expressions there can be, in general, no closed form analytic expression for L applicable to all problems of interest. Numerical approximation by method of finite steps allows their solution by introducing the necessary flexibility for their application over a wide range of problems. The approximation is:

$$L \cong \Sigma\{\Delta E_h(\theta_i,\psi_i)f_{rh}(\theta_v;\theta_i,\psi_i) + \Delta E_v(\theta_i,\psi_i)f_{rv}(\theta_v;\theta_i,\psi_i)\}$$

An accurate determination of the defined value of L may be obtained by the mathematical technique of iteration. Although such a technique is unsuited for everyday use, it may be used to find defined values of luminance for a range of lighting conditions. These values then can be used as the basis for checking the accuracy of practical techniques.

The bidirectional reflectance - distribution functions now take the form of a finite number of values that approximate the continuous functions, the appropriate values being determined by the coordinates θ_i and ψ_i for a given viewing angle. A modification will be required to take body shadow into account. (Values of bidirectional reflectance-distribution functions for a reference pencil task, in increments of θ_i and ψ_i at the 25-degree viewing angle are given in reference 20.) Illuminance at a surface is now considered as an assemblage of discrete values, each having a unique incident direction.

The sum is taken over all the discrete values of illuminance. The number of discrete steps determines the accuracy of the approximation. It can be seen that although many values of f_{rh}

and f_{rv} may be needed to represent their rapidly changing values in a particular range of θ_i and ψ_i, only a few values of E may be necessary to represent the actual pattern found in the luminous environment in the particular range of θ_i and ψ_i at a particular location of the task. Thus, several values of f_{rh} and f_{rv} may be associated with the same value of $\Delta E_h(\theta_i,\psi_i)$, for example. The reverse is also true.

Each element of the sum represents an incremental luminance $\Delta L_t(\theta_i,\psi_i)$ of the surface. These incremental values are due to illumination from a particular incident direction when viewed from a specific viewing direction. They are not hemispherical or diffuse luminances.

It is convenient to separate the illuminance at a point into two components; the illuminance due to luminaires in the space, or the direct component, and the illuminance due to the surfaces of the enclosure made luminous by interreflections, or the reflected component.

Direct Component. The basic principle for illuminance calculations is the Inverse-Square Law (see page 9-10). In some cases is is sufficient to photometer a luminaire as if it were a point source, determine several values of intensity in the most important directions and use this information for calculating illuminance as if it pertained to a point source.

Depending upon the luminaire, it may be possible to determine the intensity distribution characteristics of a differential element of the large area source and integrate over the luminous surface to give the resulting incident illuminance at a point due to the large area source.

Reflected Component. The most practical theory for determining the luminance pattern of room surfaces due to interreflections, and consequently the illuminance at the point due to the luminous surfaces, is called the lumped parameter, network, or finite difference theory.[27] The enclosing surfaces are broken up or discretized into zones, each of which, for the purposes of this theory, is assumed to have a constant luminance over its area. This implies a constant reflectance, which is one of the lumped parameters. A flux balance equation is written for each zone, each containing the unknown luminances of all the zones. A set of linear, independent, simultaneous equations result. The coefficients of the unknowns in each equation are determined from the geometry of the problem and are seen to be radiative exchange form factors.[28]

Thus the walls, ceiling and floor are discretized into zones, each of which is assumed to have a different, but constant, luminance. Discretizing the enclosing surfaces into zones allows not only an approximation to the actual luminance pattern in the room, but allows the effect of doors, windows, bulletin boards, chalkboards, etc., on the radiative transfer within the space to be considered. Since these surfaces may exhibit a reflectance that is different from the walls, their final illuminances will be different and, consequently, their effect on illuminance will be different from that of the walls.

It should be noted that maintained values of luminance may differ slightly from initial values. Should computation of maintained values of luminance be required, surface reflectances used in the computation of the reflected component should be the reflectances expected under maintained conditions in the particular environment.

The finite zone or network technique assumes that all zones exhibit perfectly diffuse reflectance. The luminous exitance of a zone, therefore, will be numerically equal to its luminance in lumens per unit area. As will be seen, the accuracy of the approximation to the actual luminance pattern will depend upon the size of the zones, and be limited by the non-lambertian reflectance of real surfaces.

Each zone has a single value of luminance, and the degree to which the luminance pattern in the enclosure is represented by the assemblage of discrete zone luminances depends upon zone size. However, this process exhibits diminishing returns. As the zone size is decreased, the form factors will become less accurate, since the radiant exchange characteristics of two small zones separated by distances large compared to their dimensions are a strong function of the real directional reflectance characteristics. The lambertian approximation used here can be expected to yield useful results for zones that have dimensions of the same order of magnitude as their separating distance. Unfortunately, quantitative criteria are difficult to derive and apply, since the geometric characteristics will vary from enclosure to enclosure. It has been found that a maximum zone dimension of one-fifth the maximum room dimension produces satisfactory results. The resulting matrix equation has the usual form, as shown at the top of page 9-35,

where

m = number of zones in the system,
A_i = area of the ith zone,
M_i = exitance of the ith zone, due to direct and interreflected flux. The equation is solved for these values.
ρ_i = hemispherical, perfectly diffuse reflectance of ith zone,

$$
\begin{vmatrix} -E_{01} \\ -E_{02} \\ -E_{03} \\ \cdot \\ \cdot \\ \cdot \\ \cdot \\ -E_{0,m-1} \\ -E_{0,m} \end{vmatrix}
=
\begin{vmatrix}
f_{11}-1/\rho_1 & f_{1,2} & f_{1,2}\cdots\cdots f_{1,m-1} & f_{1,m} \\
\frac{A_1}{A_2}f_{2,1} & f_{2,2}-1/\rho_2 & f_{2,3}\cdots\cdots f_{2,m-1} & f_{2,m} \\
\frac{A_1}{A_3}f_{3,1} & \frac{A_2}{A_3}f_{3,2} & f_{3,3}-1/\rho_3\cdots f_{3,m-1} & f_{3,m} \\
\cdot & & & \\
\cdot & & & \\
\cdot & & & \\
\cdot & & & \\
\frac{A_1}{A_{m-1}}f_{m-1,1} & \frac{A_2}{A_{m-1}}f_{m-1,2}\cdots & f_{m-1,m-1}-1/\rho_{m-1} & f_{m-1,m} \\
\frac{A_1}{A_m}f_{m,1} & \frac{A_2}{A_m}f_{m,2}\cdots & \frac{A_{m-1}}{A_m}f_{m,m-1} & f_{m,m}-1/\rho_m
\end{vmatrix}
\begin{vmatrix} M_1 \\ M_2 \\ M_3 \\ \\ \\ \\ \\ M_{m-1} \\ M_m \end{vmatrix}
$$

E_{oi} = direct or initial illuminance at i^{th} zone, due only to luminaires in the enclosure,
f_{ij} = form factor from zone i to zone j. Note that for planar zones $f_{ii} = 0$.

f_{ij}, the form factor, describes the fraction of flux emitted by zone i that is directly intercepted by zone j. The assumption that each zone is a perfectly diffuse reflector allows the flux balance equations to be written in the above form and allows easy computation of the necessary form factor. Note that $f_{ij} = f_{ji}$, thus making it necessary to compute only half of the off-diagonal form factors.

If the system of equations is then solved for the unknown exitances $M_1, \ldots, M_m$, these values of exitance can be used to determine any zone's contribution to illuminance at the task. Each zone will produce an ΔE_v and ΔE_h. The ΔE_v and ΔE_h for each zone are calculated, and by using these in conjunction with the values of direct illuminance components, all sources of illuminance at the task will have been taken into account.

Exitance Calculations

The ability to predict the luminance[29] of the visual environment is needed in order to (1) design lighting that promotes both visual comfort and good visual performance, and (2) evaluate various criteria of the lighting system, such as Visual Comfort Probability and Veiling Reflections.

The ability to predict exitance is needed in order to predict illuminance at specific points within the environment (see page 9-10). For lambertian surfaces, luminance and exitance are related by a constant. Exitance calculations are greatly simplified through the use of exitance coefficients (EC). These coefficients, like coefficients of utilization, may be computed for any specific luminaire. The wall and ceiling cavity exitance coefficients for certain generic luminaires are found in Fig. 9-62.

Exitance coefficients are similar to coefficients of utilization. They may be substituted into a variation of the lumen method formula in place of the coefficient of utilization. The result obtained is either the average wall exitance or the average ceiling cavity exitance, rather than illuminance on the work-plane. Thus:

Average Initial Wall Exitance

$$= \frac{\text{Total Bare Lamp Lumens} \times \text{Ceiling Exitance Coefficient}}{\text{Floor Area}}$$

and

Average Initial Ceiling Cavity Exitance

$$= \frac{\text{Total Bare Lamp Lumens} \times \text{Ceiling Cavity Exitance Coefficient}}{\text{Floor Area}}$$

If the area is expressed in square feet, the exitance is in lumens per square foot; if the area is in square meters, the exitance is in lumens per square meter. Numerically, the exitance coefficients are identical to the formerly used luminance coefficients when luminance was expressed in the obsolete unit of footlambert.[30]

If the maintained average wall exitance or the maintained average ceiling cavity exitance is required, a Light Loss Factor is introduced into these equations in the same manner as it is used for maintained average illuminance.

For suspended luminaires the average ceiling cavity exitance obtained is the average exitance of the imaginary plane at the level of the luminaires. This exitance does not include the weighted average exitance of the luminaires as seen from below. It is, rather, the average exitance of the background against which the luminaires are seen. In the case of recessed or ceiling mounted luminaires the average ceiling cavity exitance obtained is the average exitance of the ceiling between luminaires.

Limitations. The limitations for exitance calculations are similar to those for average illuminance. In addition, the wall reflectance used to enter the tables is a weighted average of the reflectances for the various parts of the walls: the wall exitance found by the wall exitance coefficient is the value that would occur if the walls were of a uniform and perfectly diffuse reflectance equal to the average reflectance used. Thus, many parts of the wall may have exitance values that differ from the calculated average value. A correction can be applied to determine the approximate exitance of any part of the wall. For any area on the wall,

$$\text{Exitance} = \frac{\begin{pmatrix}\text{Average Wall}\\ \text{Exitance}\end{pmatrix} \times \begin{pmatrix}\text{Reflectance}\\ \text{of Area}\end{pmatrix}}{(\text{Average Wall Reflectance})}$$

SPECIFIC APPLICATION METHODS

Calculation of Equivalent Sphere Illumination (ESI)

Equivalent sphere illumination (ESI) is a tool that may be used in determining the effectiveness of controlling veiling reflections and as part of the evaluation of lighting systems. The ESI of a visual task at a specific location in a room, illuminated with a specific lighting system is defined as that level of perfectly diffuse (sphere) illuminance which makes the visual task as visible in the sphere as it is in the real lighting environment. Using that definition of ESI, the following formula for ESI is obtained:

$$\text{ESI} = \frac{1}{\beta_{bo}} \text{RCS}^{-1}(C)\left[\frac{C}{C_o}\text{RCS}(L_b)\right]$$

where

$\text{RCS}(L_b)$ is a value of the contrast sensitivity function for an adaptation luminance L_b.

$\text{RCS}^{-1}(C)$ (the inverse RCS function) is a value of adaptation luminance for a value of reference task contrast, C.

C is physical contrast of the practical task in the real lighting environment.

C_o is physical contrast of the practical task under sphere illumination.

β_{bo} is the luminance factor of the task's background, under sphere illumination.

and

$$\text{RCS}(L_b) = 10^{\alpha}$$

where

$$\alpha = \left[\frac{2.19572}{1 + \dfrac{1}{1.75885(L_b)^{0.2}}}\right]$$

(L_b) = Adaptation luminance in candelas per square meter

$$L(\text{RCS}) = \text{RCS}^{-1}(\text{RCS}) = \left[\frac{0.568558}{\dfrac{2.19572}{\text{Log(RCS)}} - 1}\right]^{5}$$

Calculation of Contrast Rendition (Rendering) Factor (CRF)

The ratio of contrasts, C/C_o, as defined above, is called the contrast rendition (rendering) factor, CRF. It is the heart of predicting ESI.

$$\text{CRF} = \frac{\dfrac{L_b - L_t}{L_b}\ \text{environment}}{\dfrac{L_b - L_t}{L_b}\ \text{sphere}}$$

The luminance in the above expression, L_t and L_b, are the task and background luminances of the visual task display for which ESI is being calculated. These are luminances at a point and

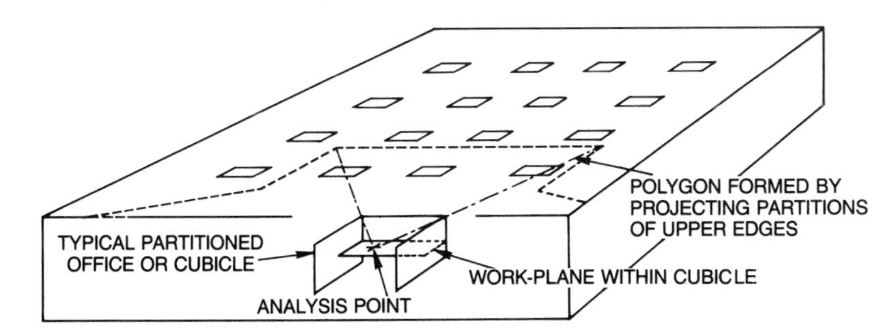

Fig. 9-42. Projection of lines from analysis point past partitions to the ceiling.

are calculated from the following equations, where the symbols are the same as those used on page 9-32 and subscripts t and b stand for task and background, respectively.

$$L_t = \int \{dE_h(\theta_i,\psi_i)f_{rth}(\theta_v;\theta_i,\psi_i) + dE_v(\theta_i,\psi_i)f_{rtv}(\theta_v;\theta_i,\psi_i)\}$$

$$L_b = \int \{dE_h(\theta_i,\psi_i)f_{rbh}(\theta_v;\theta_i,\psi_i) + dE_v(\theta_i,\psi_i)f_{rbv}(\theta_v;\theta_i,\psi_i)\}$$

Approximations are

$$L_t \cong \Sigma\{\Delta E_h(\theta_i,\psi_i)f_{rth}(\theta_v;\theta_i,\psi_i) + \Delta E_v(\theta_i,\psi_i)f_{rtv}(\theta_v;\theta_i,\psi_i)\}$$

$$L_b \cong \Sigma\{\Delta E_h(\theta_i,\psi_i)f_{rbh}(\theta_v;\theta_i,\psi_i) + \Delta E_v(\theta_i,\psi_i)f_{rbv}(\theta_v;\theta_i,\psi_i)\}$$

Illuminance at a Point in Partitioned Spaces

The calculation of illuminance in partitioned spaces is complicated by the shadowing effects of the partitions. As with other illuminance calculations, there is a direct and reflected component of illuminance, although in some cases the direct component will be zero.

The procedure begins with the selection of a specific point on the work-plane within the partitioned space and the illuminance calculation at that point.

Direct Component Determination. Each partition which is in the immediate vicinity of the work surface has the potential of obscuring luminaires as seen from the specific point; that is, luminaires that might otherwise produce a direct component of illuminance at the point are hidden from the point's "view" and do not contribute to the illuminance directly. To determine which are contributing luminaires, a projection of the upper edges of the neighboring partitions is made on the ceiling, using the point of interest on the work-plane as the point of projection. The partitions' upper edges will project a polygon (very often a rectangle) onto the ceiling, defining that portion of the ceiling seen by the point. See Fig. 9-42. Only luminaires within this polygon will generate a direct component of illuminance at the point. Any of the methods for determining the direct illuminance at a point from a luminaire can be used for those luminaires within the polygon.

Note that for some points on the work-plane and/or partition positions, none of the luminaires may be within the projected polygon and the direct component will be zero. In other cases the polygon will include only a portion of some luminaire and only that portion of the luminaire within the polygon should be used to determine that luminaire's contribution to the direct component. An indirect lighting system will produce no direct component.

Reflected Component. The reflected component of horizontal illuminance at the point on the work-plane is determined by a two step process:

1. Determine the reflected component of average illuminance on a plane at the top of the partitions, and
2. Determine the resulting reflected component that is, in turn, produced at the point on the work-plane down inside the partitioned space.

The first step is performed as follows. Using the lumen method and the zonal cavity coefficients of utilization for the luminaires being used, the total average illuminance for the entire room is calculated for a plane at the top of the partitions. The effective floor cavity reflectance for this calculation should be that of the partitioned space treated as a cavity. Note that all the rooms' luminaires are used since presumably they all contribute to the illuminance on a plane at the partition tops. This same calculation is now repeated, but using zero room surface re-

Fig. 9-43. Coefficients of Utilization for a Perfectly Diffuse Emitter

Floor Cavity Reflectance (ρcc = 20 Per Cent)					
Wall Reflectance ρw (Per Cent)	70	50	30	10	0
RCR					
0	1.00	1.00	1.00	1.00	1.00
1	.90	.88	.86	.84	.84
2	.81	.77	.74	.71	.70
3	.74	.69	.64	.61	.59
4	.67	.61	.56	.52	.50
5	.61	.54	.49	.44	.42
6	.56	.44	.43	.38	.36
7	.52	.44	.38	.33	.31
8	.48	.39	.33	.29	.27
9	.44	.35	.29	.25	.23
10	.41	.32	.26	.22	.20

flectances. This results in the direct average illuminance for the entire room on a plane at the top of the partitions. This direct average is subtracted from the total average, resulting in the reflected average illuminance at the partition tops. Note that zero reflectance coefficients of utilization are required.

For indirect lighting systems, the average direct component will be zero and there will only be an average reflected component on a plane at the partition tops.

The second step is performed as follows. The partitioned space is now treated as a separate room, lighted with a "virtual" luminaire occupying the entire upper opening of the partitioned space. Using the cubical dimensions and reflectances, a Room Cavity Ratio is calculated and a zonal cavity coefficient of utilization is determined, assuming that the candela distribution of this "virtual" luminaire is perfectly diffuse (cosine). A CU table for a perfectly diffuse emitter is given in Fig. 9-43. In many partitioning arrangements each cubical has only three sides formed by partitions, the other side being open. Thus, the reflectance of the wall surface used in determining this CU must be the area weighted averge reflectance of the three partition surfaces and the cubical opening. It is reasonable and conservative to assign the cubical opening a reflectance of zero.

Multiplying this CU by the average reflected illuminance at the partition tops (determined in step 1) gives the average reflected illuminance down on the work-plane in the cubical. This can be taken as the reflected illuminance at the point of interest.

This procedure assumes that the reflected component of illuminance is very uniform within the partitioned space; that is, only the direct component is assumed to vary with position inside the cubical.

The sum of the direct and reflected components calculated are added together to give the illuminance at the specific point being considered.

Coves and Valances

In many instances, illuminance at a point from coves and valances can be calculated using both direct and reflected components.

Direct Component. The direct component illuminance is calculated using the Inverse-Square Method or the Configuration Factor Method and the light emitting portion of the luminaire that can be seen from the point on the work-plane at which the illuminance is to be determined.

Reflected Component. It is practical to manually calculate only the first bounce of the reflected component. This is accomplished by

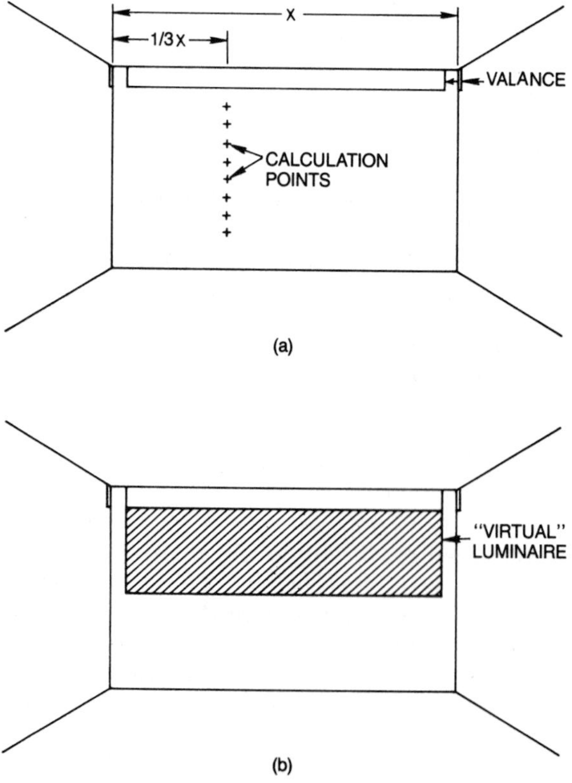

Fig. 9-44. Locations of (a) calculation points and (b) "virtual" luminaire for valance calculations. For coves, the calculation point and "virtual" image are on the ceiling.

first establishing a "virtual" luminaire on the wall or ceiling and then using the Configuration Factor Method to calculate the illuminance at a point on the work-plane.

The "virtual" luminaire is determined by calculating the illuminance at equally spaced points along the wall or ceiling (whichever is appropriate) along a line running normal to the luminaire and spaced no more than 0.1 meter (four inches) apart, starting 0.1 meter (four inches) from the luminaire. This line should be located at approximately one-third the luminaire's length, as shown in Fig. 9-44a. Then, the point at which the illuminance drops below one-fifth the maximum defines the "virtual" luminaire on that surface, rectangular in shape, as long as the cove or valance, and having a width that extends from the luminaire to the one-fifth maximum point as shown in Fig. 9-44b. The exitance of this "virtual" luminaire is equal to the average of the calculated illuminances (out to the one-fifth maximum point) multiplied by the reflectance of the surface.

The illuminance on the work-plane is the sum of the direct and reflected components.

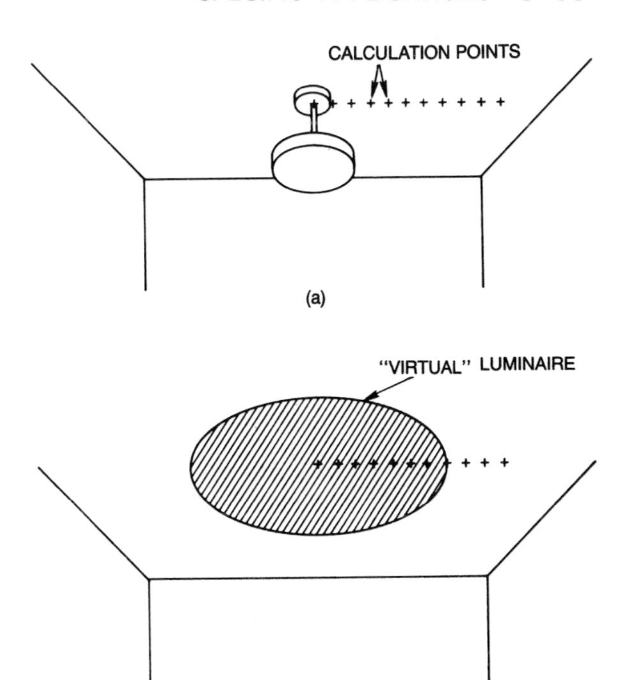

Fig. 9-45. Locations of (a) calculation points and (b) "virtual" luminaire for indirect lighting calculations.

Indirect Lighting

In calculating illuminance at a point from indirect luminaires, only the reflected component can be evaluated because the work-plane analysis point cannot see the source directly. Of this reflected component, only the first bounce from the ceiling can be manually calculated. This is done by first establishing a "virtual" luminaire on the ceiling and then calculating the illuminance from that "virtual" luminaire at the point on the work-plane using the Configuration Factor Method, see page 9-20 and Fig. 9-29.

The "virtual" luminaire is determined by first calculating direct component illuminance on the ceiling (using the Inverse-Square Method or the Configuration Factor Method) at equally spaced points starting directly above the luminaire and extending away in a straight line as shown in Fig. 9-45a. The recommended point spacing is one-fifth the distance from the luminaire to the ceiling. Using these points, the point at which the illuminance drops below one-fifth the maximum is located. The "virtual" luminaire then is a disc whose radius is the distance out to the one-fifth maximum illuminance (see Fig. 9-45b) and whose exitance is the average of the illuminances out to that point multiplied by the ceiling's reflectance.

Merchandising Applications

Empirical data are available to compute illuminance levels for a great many specialized applications. The data shown are for typical lighting equipment and will vary according to actual equipment design.

Shelf and Garment Case Illumination. To estimate the illuminance normal to a vertical surface from a continuous row of fluorescent lamps located in close proximity, the charts shown in Fig. 9-46 may be used. Such data, while particularly adapted to shelf and garment case illumination, can be used for any similar type of vertical surface lighting installations.

Show Window Illumination. For estimating average illuminances in show windows, two planes are generally used to represent the average display surfaces as shown in Fig. 9-47. These vary in size and position in different windows. They are divided into zones *A*, *B* and *C* (*B* being of equal width to *C*) to permit designing for either a variation in illuminance level between parts of the display or for a uniform level equally effective throughout. In selecting a zone for use in estimating average illuminance consider the nature of the trim and whether the back will be open or closed and proceed as follows:

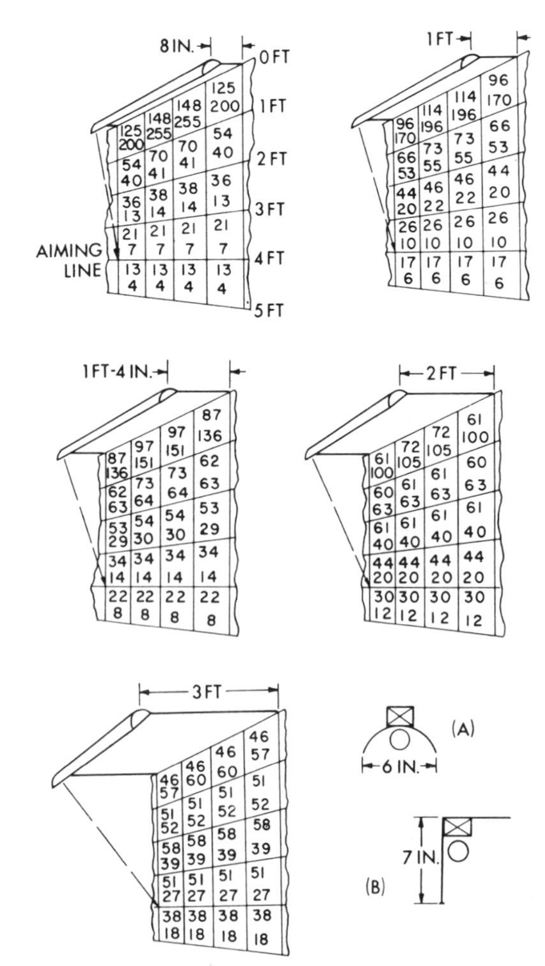

Fig. 9-46. Chart showing a four-foot wide section of a vertical area such as a wall, display background, garment rack, shelves, etc., with a height of five feet, lighted by a single continuous row of T-12 fluorescent lamps of 2500 lumens each. Figures shown are initial footcandles in the center of one-foot squares. Upper figures: typical polished parabolic trough reflector (A), aimed as shown in the diagram. Lower figures: typical white cornice (B), with a 75 per cent reflectance.

1. Illuminance can be approximately doubled using two rows of lamps.
2. Charts may be inverted for lamps mounted below the vertical surface.
3. Improved uniformity as well as increased illuminance can be obtained using lamps top and bottom. For areas up to nine feet high overlap the upright and inverted charts as necessary to obtain required height. Add together the overlapping footcandles.
4. Illuminance from other types of T-12 lamps may be determined by multiplying the footcandles shown by a factor of

$$\frac{\text{lamp lumens per foot}}{625}$$

This factor also applies for other diameter lamps in a white cornice or in a concentrating parabolic reflector having a width approximately four times the lamp diameter.

5. For illuminances in lux multiply values in the squares by 10.76, also converting the length figures to meters or millimeters.

A. For lamps in external reflectors (Fig. 9-48).

1. Determine how many lamp lumens per linear meter (foot) of window must be provided.

Lamp Lumens Needed per Meter (Foot)

$$= \text{Lux (Footcandles) Desired} \times \text{Height in Meters (Feet)} \times K \times P \times Q$$

where

K = a constant that takes into account the height-to-depth (H/D) ratio of the window, and the light distribution
P = length factor [introduces the effects of the length-to-height (L/H) ratio and of the kinds of ends the window has]
Q = shielding factor (adjusts for the effects of type and orientation of louvering used).

The K, P and Q factors for incandescent filament and fluorescent lamps can be found in Figs. 9-48A and 9-48B respectively. For each H/D ratio, the lowest K factor in the selected zone identifies the reflector distribution that will provide the desired illuminance most efficiently.

2. (*For incandescent filament lamps.*) Make tentative selection of lamp size. One way to do it is to choose the smallest lamp size that has a lumen output rating higher than the lamp lumens needed per ⅓ meter or per foot, found in step 1. Then:

Lamp Spacing in Millimeters

$$= 1000 \times \frac{\text{Rated Lumens per Lamp}}{\text{Lamp Lumens Needed per Meter}}$$

Lamp Spacing in Inches

$$= 12 \times \frac{\text{Rated Lumens per Lamp}}{\text{Lamp Lumens Needed per Foot}}$$

If spotlighting equipment is to be mounted between the general lighting units, the spacing found from this formula may not always be great enough to accommodate them. In this case, select

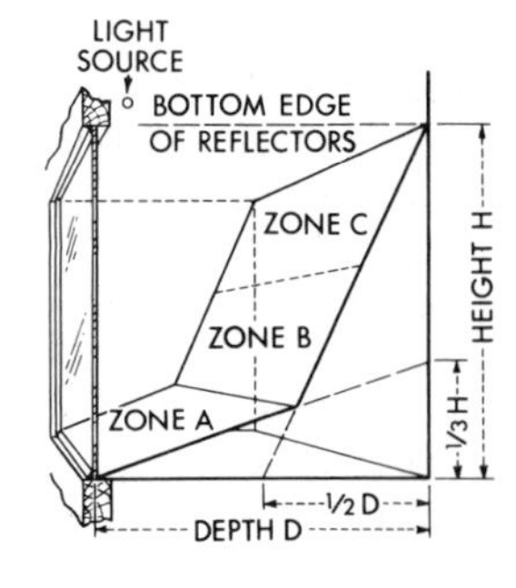

Fig. 9-47. Show windows are divided into three zones (*A*, *B*, and *C*) when it is desired to estimate the number of lamps necessary for a given illuminance level.

Fig. 9-48. Multiplying Factors for Lamps in External Reflectors
A. For Incandescent Filament Lamps

K Factors*

Cross Section	Zone A			Zone B			Zone C		
H/D	Wide	Semi-Conc.	Conc.	Wide	Semi-Conc.	Conc.	Wide	Semi-Conc.	Conc.
4.0	4.2	3.4	2.0	6.8	5.5	3.9	1.6	2.0	3.0
3.5	3.6	3.0	1.8	6.0	5.0	4.1	1.8	2.3	3.6
3.0	3.2	2.6	1.7	5.5	4.6	4.5	2.0	2.8	4.1
2.5	3.0	2.4	1.7	4.6	4.0	5.0	2.4	3.7	5.0
2.0	2.9	2.3	1.7	4.3	3.7	5.5	3.1	5.1	6.1
1.5	3.0	2.4	1.9	4.1	3.6	6.3	4.1	8.3	9.4
1.0	3.3	2.9	2.3	4.6	4.1	7.5	7.5	20.3	20.0

P (Length) Factors

Type of Equipment	Window Length Divided by Height (*L/H*)							
	0.5		1.0		1.5		2.0	
	1 Glass End	Solid Ends	1 Glass End	Solid Ends	1 Glass End	Solid Ends	1 Glass End	Solid Ends
Wide	1.40	1.25	1.10	1.05	1.00	1.00	1.00	0.95
Semi-conc.	1.30	1.20	1.05	1.00	1.00	1.00	1.00	0.95
Conc.	1.20	1.10	1.00	1.00	1.00	1.00	1.00	1.00

Q (Shielding Factors (Factor for Unlouvered Lamps = 1.00)

	Louvers at Right Angles to Plate Glass			Louvers Parallel to Plate Glass			Eccentric Ring Louvers		
H/D	Zone A	Zone B	Zone C	Zone A	Zone B	Zone C	Zone A	Zone B	Zone C
4.0	1.3	1.4	1.4	1.2	1.4	2.2		Not Usually Employed	
3.5	1.3	1.4	1.4	1.2	1.4	2.3		Not Usually Employed	
3.0	1.3	1.4	1.5	1.2	1.5	2.6		Not Usually Employed	
2.5	1.3	1.4	1.6	1.2	1.6	3.0		Not Usually Employed	
2.0	1.4	1.4	1.6	1.2	1.8	3.7	1.4	1.6	2.9
1.5	1.4	1.5	1.7	1.2	1.9	4.6	1.4	1.8	3.3
1.0	1.4	1.5	1.8	1.3	2.1	5.3	1.4	2.2	4.0

the next-higher-wattage lamp, and recalculate the spacing.

(*For fluorescent lamps.*) Determine the lumens per meter (foot) of window produced by a single row of fluorescent lamps of the size, type and color selected:

$$\text{Lumens per Meter (Foot) for One Row} = \frac{\text{Total Lamp Lumens for One Row}}{\text{Length of Window in Meters (Feet)}}$$

$$\text{Number of Rows Needed} = \frac{\text{Lamp Lumens Needed per Meter (Foot) (from Step 1)}}{\text{Lumens per Meter (Foot) for One Row}}$$

B. For lamps with internal reflectors (Fig. 9-49). Design data for show window lighting with reflector lamps must take into account the various possible lamp aiming angles. The following procedure includes this element:

1. Determine window height in meters (feet) from floor to lamps (H), and window depth in meters (feet) from glass to background (D). [The ratio of these items (H/D) provides an index of window proportions for use in the first column of Fig. 9-49.]

2. Find the most suitable lamp type and aiming angle by referring to Fig. 9-49 and finding the illumination factors in bold type appearing opposite the proper value of H/D. (Suitable lamp types are indicated by the column in which the bold numbers appear, and suitable aiming angles are indicated by the angle on the same line in which the bold numbers appear.) If more than one suitable lamp type or aiming angle is found in this manner, then select either the one that produces the desired degree of emphasis in the most important zone, or provides greatest uniformity among zones. Relative illuminance in the three zones will be in proportion to the illumination factors (F_A, F_B and F_C) for the zones.

3. Find the center-to-center lamp spacing in millimeters (inches) that will produce an average of 2000 lux [200 footcandles] maintained in the zone considered most important (usually zone A or zone B):

$$\text{Spacing in Millimeters} = \frac{7.74 \times F}{H \text{ in Meters}}$$

Fig. 9-48 *Continued*

B. For Fluorescent Lamps

K Factors**

Cross Section	Zone A			Zone B			Zone C		
H/D	Wide	Semi-Conc.	Conc.	Wide	Semi-Conc.	Conc.	Wide	Semi-Conc.	Conc.
4.0	4.9	5.7	4.2	11.3	10.3	7.6	2.7	1.5	1.7
3.5	4.6	5.2	3.9	9.2	9.2	6.8	2.3	1.6	1.9
3.0	4.2	4.2	3.4	7.9	6.9	5.6	2.5	1.8	2.2
2.5	4.1	3.9	3.0	6.2	5.0	4.7	2.8	2.2	2.7
2.0	4.1	3.8	2.7	5.7	4.3	3.7	3.3	2.8	3.7
1.5	4.0	3.6	2.4	5.2	4.4	4.1	4.3	4.1	5.7
1.0	4.3	3.6	2.3	5.0	5.0	6.8	7.6	7.6	15.1

P (Length) Factors

Window Length Divided by Height (*L/H*)							
0.5		1.0		1.5		2.0	
1 Glass End	Solid Ends	1 Glass End	Solid Ends	1 Glass End	Solid Ends	1 Glass End	Solid Ends
1.55	1.45	1.20	1.10	1.00	0.95	0.95	0.90

Q (Shielding) Factors (Factor for Unlouvered Lamps = 1.00)

H/D Ratio	Shielding Factors (S): Cellular Louvers (Matte White) Shielding Lamps to 45 degree Crosswise and 25 degree Lengthwise		
	Zone A	Zone B	Zone C
4.0	1.11	1.20	1.14
3.5	1.11	1.20	1.14
3.0	1.11	1.18	1.14
2.5	1.11	1.18	1.14
2.0	1.11	1.18	1.16
1.5	1.09	1.15	1.19
1.0	1.09	1.10	1.27

* Based on typical commerical equipments. A light loss factor of 0.75 has been assumed.

** Based on equipments of typical distribution. Efficiencies: Wide 65%, Semi-Concentrating 85%, Concentrating 80%. A light loss factor of 0.75 has been assumed.

Note: This simplified table is based on four rows of fluorescent lamps in reflectors of typical widths, at typical angles of tilt. Some variation in results is to be expected depending on type of equipment selected, number of rows, and mounting angle. As the number of rows is increased, rear rows contribute progressively less light to the A and B zones, and more to the upper background.

$$\text{Spacing in Inches} = \frac{F}{H \text{ in Feet}}$$

(Where F is either F_A or F_B from Fig. 9-49 for the selected lamp type and aiming angle, and H is window height in meters or feet.)

4. Check to be sure that the lamp spacing just found does not exceed the maximum spacing in millimeters (inches) for generally satisfactory lateral uniformity (absence of streaks and scallops) in all zones.

Maximum Permissible Spacing (Millimeters)

$$= 83.3 \times U \times D$$

Maximum Permissible Spacing (Inches)

$$= U \times D$$

(Where U is the uniformity factor from Fig. 9-50 for the lamp type used, and D is window depth in meters or feet. Unless this maximum spacing is radically exceeded, the effect may not be objectionable, because the streaks or scallops would be at the top of the background in a closed-back window, and would probably be unnoticed in an open-back one.)

5. Illuminance in the other zones will be in accordance with these relationships:

Illuminance in Zone *A*

$$= \frac{F_A \times (\text{Illuminance in Zone } B)}{F_B} \quad \text{or} \quad \frac{F_A \times (\text{Illuminance in Zone } C)}{F_C}$$

Illuminance in Zone *B*

$$= \frac{F_B \times (\text{Illuminance in Zone } A)}{F_A} \quad \text{or} \quad \frac{F_B \times (\text{Illuminance in Zone } C)}{F_C}$$

Fig. 9-49. Illumination Factors* for Lamps with Internal Reflectors

H/D	Aiming Angle (degrees from vertical)	PAR-38 Spot			150-Watt R-40 Spot†			PAR-46 Medium Flood‡			PAR-38 Flood			150-Watt R-40 Flood†		
		F_A	F_B	F_C	F_A	F_B	F_C	F_A	F_B	F_C	F_A	F_B	F_C	F_A	F_B	F_C
1.0	0	55	10	5	55	15	5	80	10	5	75	10	5	50	20	5
	10	100	15	5	75	20	10	110	10	5	85	15	10	55	25	5
	20	125	20	5	85	25	10	115	15	5	90	30	10	60	35	10
	30	90	40	10	65	45	15	90	65	10	75	55	10	50	45	20
	35	70	70	15	55	60	20	70	85	10	65	70	15	**45**	**50**	**25**
	40	50	105	15	45	80	20	45	100	15	50	85	20	**40**	**55**	**30**
	45	30	125	20	30	90	25	30	115	20	35	90	30	**35**	**60**	**35**
	50	20	135	25	20	90	30	20	130	25	25	95	40	30	55	40
	55	15	130	35	10	90	40	15	125	45	20	90	50	30	50	45
1.5	0	75	15	5	80	20	10	120	15	15	115	20	10	70	35	10
	10	135	25	5	100	30	15	140	20	15	115	35	10	65	40	15
	20	130	50	10	90	45	20	105	55	20	100	60	15	60	50	25
	25	85	90	15	65	70	25	75	90	20	80	80	20	**55**	**55**	**30**
	30	50	130	20	50	90	30	50	115	25	60	90	30	**50**	**55**	**40**
	35	40	140	30	35	95	40	35	125	35	40	100	40	**40**	**55**	**50**
	40	30	125	45	30	90	50	25	115	50	30	95	55	35	55	60
	45	25	100	60	25	80	65	20	95	70	25	85	75	30	55	70
2.0	0	125	20	5	100	25	10	140	15	20	125	25	10	70	35	15
	10	150	40	10	110	40	15	140	40	20	115	45	15	65	40	25
	15	145	60	15	95	50	20	120	60	25	100	60	20	**60**	**45**	**30**
	20	100	90	25	80	70	30	85	75	30	85	75	30	**60**	**45**	**40**
	25	60	120	35	50	80	40	45	90	40	**65**	**80**	**40**	**50**	**50**	**50**
	30	40	125	45	40	85	55	25	100	55	45	80	55	45	50	65
	35	30	100	60	30	70	75	20	105	70	30	75	85	40	45	80
	40	25	60	85				20	90	95						
2.5	0	185	20	10	120	25	20	150	20	25	125	30	15	70	35	25
	5	175	30	10	115	35	20	145	30	25	120	40	15	70	40	30
	10	155	50	15	100	45	25	125	60	30	110	55	20	**65**	**40**	**35**
	15	120	80	25	**80**	**60**	**35**	**90**	**75**	**40**	95	65	30	**60**	**45**	**45**
	20	70	105	35	**55**	**80**	**45**	**50**	**80**	**50**	**70**	**75**	**45**	55	45	55
	25	40	110	50	40	75	65	30	80	65	50	70	65	45	40	70
	30	30	90	75	35	60	85	20	70	85	35	65	90	40	40	85
3.0	0	185	20	15	130	25	25	155	25	30	125	35	20	65	35	30
	5	170	30	20	115	35	30	140	40	30	115	45	25	**65**	**40**	**35**
	10	145	60	25	95	50	35	115	60	40	100	55	35	60	40	45
	15	100	85	35	**70**	**65**	**45**	**80**	**75**	**55**	**85**	**65**	**50**	55	40	55
	20	50	105	50	45	70	60	40	80	70	**60**	**70**	**65**	50	40	70
	25	30	95	75	35	60	85	25	65	90	40	65	90	45	35	85
3.5	0	180	25	20	130	25	25	160	40	30	125	40	20	65	30	35
	5	170	40	25	110	35	30	140	50	40	110	50	30	60	35	45
	10	**135**	**65**	**30**	**85**	**50**	**40**	**105**	**65**	**50**	95	55	40	60	35	55
	15	**75**	**85**	**40**	**60**	**65**	**55**	**65**	**75**	**70**	**75**	**60**	**60**	55	35	70
	20	45	85	65	40	60	75	35	70	95	55	60	85	45	35	85
4.0	0	175	25	20	125	25	30	155	35	30	120	40	25	60	30	40
	5	160	35	25	105	35	35	130	55	45	105	50	35	60	35	50
	10	**120**	**65**	**35**	**70**	**50**	**45**	**100**	**65**	**65**	**90**	**55**	**50**	55	35	65
	15	**65**	**80**	**50**	50	55	65	60	65	90	70	50	75	50	30	80
	20	40	70	75												

* These numbers are not illuminances, but must be divided by window height to find lamp spacing (see text). Based on assumed light loss factor of 0.80.

† Multiply illumination factors by 2.0 for 300-watt R-40, 3.1 for 500-watt R-40, 0.4 for 75-watt R-30.

‡ Wide section of beam perpendicular to glass. Multiply illumination factors by 1.7 for 300-watt PAR-56 medium flood.

Illuminance in Zone C

$$= \frac{F_C \times (\text{Illuminance in Zone } B)}{F_B}$$

$$\text{or } \frac{F_C \times (\text{Illuminance in Zone } A)}{F_A}$$

6. If an illuminance other than 2000 lux or 200 footcandles is desired in a particular zone, proceed as in steps 1 through 3 above. Then multiply the lamp spacing found in step 3 by 2000/desired lux or 200/desired footcandles. Illuminance in the other zones will be greater or less in the ratio

of desired lux/2000 or footcandles/200. Uniformity should be checked as in step 4.

Showcase Illumination. When lighting is supplied from within the showcase by lamps at the top front edge, illuminances in each of three typical trimplanes can be estimated. See Fig. 9-47. Plane *A* extends from the lower front edge to a line at ⅓ the height of the case. Plane *B–C* runs from the top rear edge to a line at ½ the case depth on the bottom. Zones *B* and *C* are equal. Any or all of these zones may be of prime importance, depending on the method of displaying merchandise.

In Fig. 9-51 typical glass showcases 500 millimeters (20 inches) deep are assumed. Constants *K* are given for each zone, for three typical case heights, and for four lighting methods. The values of *K* are based on an assumed light loss factor of 0.75.

When lamp and reflector have been selected, the illuminance in any zone can be estimated:

$$\text{Lux} = 3.28 \times K \times (\text{Lumens-per-Meter of Case Length})$$

$$\text{Footcandles} = K \times (\text{Lumens-per-Foot of Case Length})$$

The lumens-per-meter (foot) value for a continuous row of fluorescent lamps running the entire case length will be that of the lamps used. Otherwise, for either fluorescent or incandescent filament lamps it will be:

$$\text{Lumens-per-Meter (Foot) of Case Length} = \frac{\text{Lumens per Lamp} \times \text{Number of Lamps}}{\text{Length of Case in Meters (Feet)}}$$

Vertical Surface Illumination. Fig. 9-52 may be used to calculate the illuminance normal to a large vertical surface from a row of reflector lamps mounted close by. Care should be exercised to avoid a scalloping effect caused by having the lamps too far apart in relation to their distance from the lighted surface.

Fig. 9-50. Uniformity Factors for Lamps with Internal Reflectors

Lamp Type	Uniformity Factor
PAR-38 spot	2
R-30 spot	2
R-40 spot	2
PAR-46 med. flood	2
PAR-56 med. flood	2
PAR-38 flood	6
R-40 flood	12

FLOODLIGHTING CALCULATIONS

The two most commonly used systems for floodlight calculations are the point method and the beam-lumen method. The point method (page 9-10) permits the determination of illuminance at any point and orientation on a surface. This method is valuable since it permits a visualization of the degree of lighting uniformity realized for any given set of conditions. The beam-lumen method is quite similar to the lumen method for interior lighting except that it must take into consideration the fact that floodlights are not usually directly above the surface, but are aimed at various angles to the surface.

The Beam-Lumen Method for Floodlights

The basic formula is:

$$\text{Illuminance (Average Maintained)} = \frac{\text{Quantity of Floodlights} \times \text{Beam Lumens} \times \text{Coefficient of Beam Utilization} \times \text{Light Loss Factor}}{\text{Area}}$$

Fig. 9-51. Constants *K* for Showcases*
D = 500 millimeters (20 inches)

Lamp	Lighting Method	*H* = 250 millimeters (10 inches)			*H* = 500 millimeters (20 inches)			*H* = 690 millimeters (27 inches)		
		Zone			Zone			Zone		
		A	B	C	A	B	C	A	B	C
Fluorescent	White Diffusing Reflector	.199	.091	.038	.128	.103	.061	.097	.084	.071
	Concentrating Reflector	.241	.148	.070	.124	.144	.104	.092	.109	.113
Incandescent filament	Clear T-10 in Semidiffusing Reflector	.164	.103	.071	.107	.089	.091	.081	.070	.091
	T-10 Reflector Showcase	.267	.269	.096	.170	.250	.138	.134	.178	.155

* Assumed 0.75 light loss factor.

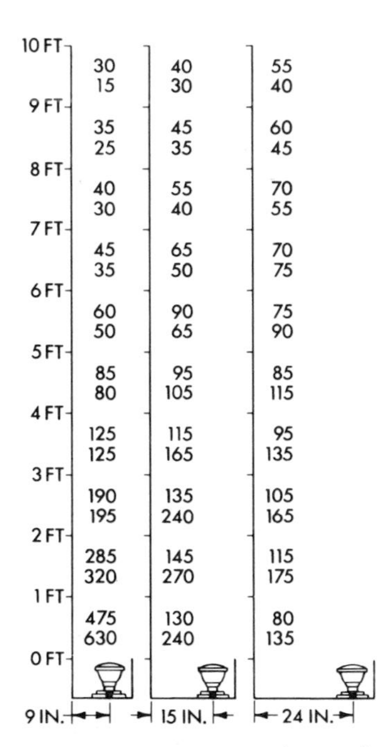

Fig. 9-52. Approximate illuminance in footcandles produced by incandescent reflector lamps on 8-inch centers. Upper figures are footcandles from PAR-38 flood lamps. Lower figures are from 150-watt R-40 flood lamps. Assumed 0.75 light loss factor.

Illuminance for Any Lamp Spacing

$$= \frac{\text{Footcandles from Chart} \times 8}{\text{Lamp Spacing in Inches}}$$

For estimating results from other reflector flood lamps, footcandle values from the R-40 can be multiplied by the following factors:

Lamp	*Factor*
500-watt R-40 flood	3.1
300-watt R-40 flood	2.0
75-watt R-40 flood	0.4

or

Quantity of Floodlights

$$= \frac{\text{Area} \times \text{Average Maintained Illuminance}}{\text{Beam Lumens} \times \text{Coefficient of Beam Utilization} \times \text{Light Loss Factor}}$$

Beam Lumens. Beam lumens are defined as the quantity of light that is contained within the beam limits as described in Section 1 under "beam spread."

Coefficient of Beam Utilization. This factor, CBU, written as a decimal fraction, expresses the following ratio:

Coefficient of Beam Utilization

$$= \frac{\text{Lumens Initially Reaching the Specified Area Directly from the Floodlight}}{\text{Total Beam Lumens}}$$

To determine the number of lumens from each floodlight beam that initially strikes the area, it is necessary to first locate and determine the aiming of each floodlight with respect to the area. For each floodlight location and direction the area is replotted in angular coordinates upon the isocandela curve of the floodlight, as in Fig. 9-53, and the lumens within the area boundaries summed. In this example the floodlight at F is directed at point O on building $ABCD$, and angle LFO is the angle whose tangent is 25/40 or 32 degrees. Angle EFL is the angle whose tangent is 50/40 or 51 degrees. Angle EFO must then be 19 degrees. By similar calculations all the angles in Fig. 9-53 can be determined and the building can then be plotted on the grid of the isocandela

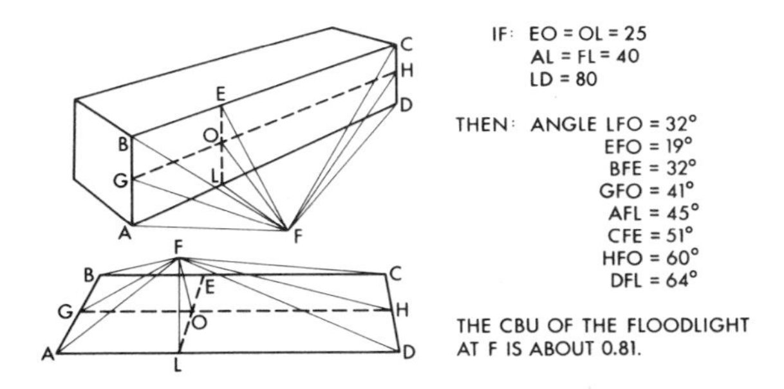

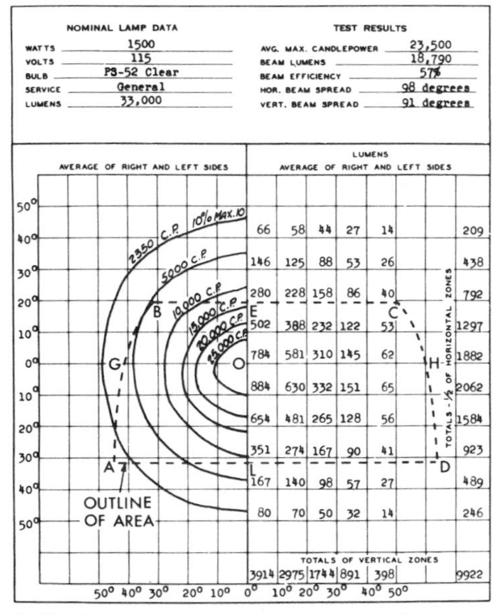

Fig. 9-53. Procedure for determining the coefficient of beam utilization for floodlighting equipment.

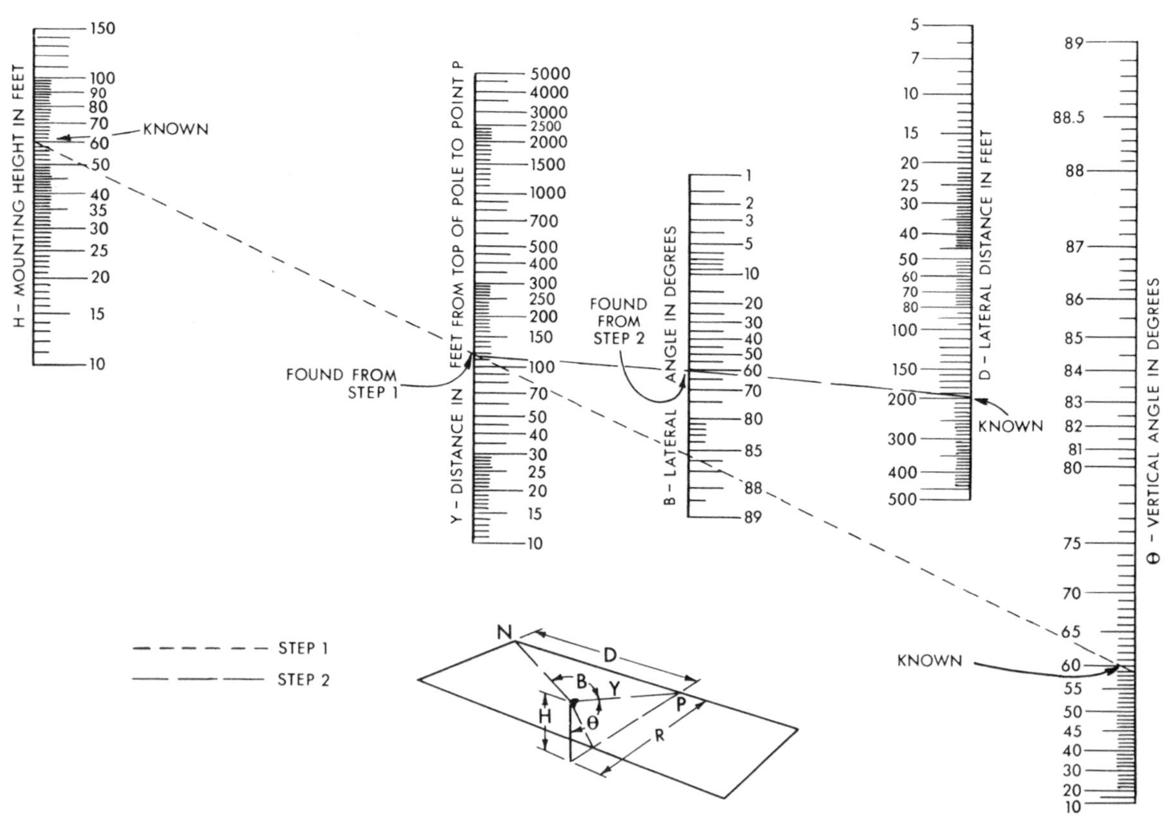

Fig. 9-54. Nomogram for determining lateral angle B in terms of vertical angle θ, mounting height H and distances Y and D.

curve of the floodlight. See Fig. 9-54 for nomograms of vertical and lateral angles. Because of the manner in which floodlights are photometered, all horizontal lines parallel to a line perpendicular to the beam axis appear as straight horizontal lines on the grid. All vertical lines except the one through the beam axis appear slightly curved.

Coverage.[31] It is recommended that sufficient point calculations be made for each job to check uniformity and coverage. A check for coverage alone can be made either by means of a scale drawing and protractor or by a coverage chart provided by the floodlight manufacturer.

CALCULATION OF VISUAL COMFORT PROBABILITY (VCP)

Discomfort glare is the sensation of discomfort caused by high luminances in an observer's field of view. Since there is no acceptable scale of discomfort, *visual comfort probability* (VCP) is the metric used to evaluate discomfort glare caused by luminaires directly in the field of view. VCP is an estimator of the probability that an observer will accept a lighting system as comfortable when viewed under defined conditions. It also may be interpreted as that percentage of a large observer population which would accept the lighting system as comfortable under the defined conditions. For use of VCP, see the Application Volume.

Equations for the calculation of VCP are based upon correlation with a large body of experimental data[32-39] rather than upon a model of the processes involved. Experiments in simulated rooms have been used to confirm the extension of laboratory experiments to actual lighting installations.[40,41]

The basic equations[42-48] determine the index of sensation M for each source and sums these individual source effects to produce a discomfort glare rating (DGR) for the full field of view. The DGR's are more or less arbitrary numbers which increase with discomfort; they are then converted to VCP's:

$$M = \frac{0.50 L_s Q}{P F^{0.44}}$$

$$\text{DGR} = [\sum_n M]^{n^{-0.0914}}$$

The factor 0.50 in the above equation for M is a correction from the deprecated original lambertian luminance unit of footlambert to the currently preferred candela per square meter (*i.e.*, $\text{lm} \cdot \text{m}^{-2} \cdot \text{sr}^{-1}$) for luminance. There is less chance of confusion if this is explicitly treated by a conversion factor than if the DGR scale were revised and the DGR to VCP conversion were modified.

Q is a function of the solid angle (ω_s) subtended by the glare source at the observer's eye.

$$Q = 20.4\omega_s + 1.52\omega_s^{0.2} - 0.075$$

F is the average luminance of the entire field of view.

$$F = \frac{L_w\omega_w + L_f\omega_f + L_c\omega_c + \sum L_s\omega_s}{5}$$

The factor of 5 in the denominator arises from the assumption that the total field of view is 5 steradians.[49]

P is the position index, an inverse measure of the relative sensitivity to glare throughout the field of view. Selected values or families of curves were published in early references. The data are best represented by the equation:[47]

$$\ln P = [35.2 - 0.31889\tau - 1.22\, e^{-2\tau/9}]10^{-3}\sigma + [21 + 0.26667\tau - 0.002963\tau^2]10^{-5}\sigma^2$$

where σ and τ are angles relating the source location to the line of sight (see Fig. 9-55).

The relation between DGR and VCP can be shown graphically as in Fig. 9-56, or it may be expressed analytically[50] by the relation

$$\text{VCP} = \frac{100}{\sqrt{2\pi}} \int_{-\infty}^{6.374-1.3227\ln(\text{DGR})} e^{-t^2/2}\, dt$$

The various terms used in the preceeding equations are defined as follows.

DGR—discomfort glare rating

F—average luminance of entire field of view, candelas per square meter ($\text{cd} \cdot \text{m}^{-2}$)

L_c—average luminance of ceiling, $\text{cd} \cdot \text{m}^{-2}$

L_f—average luminance of floor, $\text{cd} \cdot \text{m}^{-2}$

L_s—luminance of a source, $\text{cd} \cdot \text{m}^{-2}$

L_w—average luminance of walls, $\text{cd} \cdot \text{m}^{-2}$

M—index of sensation of a source

n—number of sources in the field of view

P—position index of a source

Q—function of solid angle of a source

σ—angle between line of sight and line from observer to source (see Fig. 9-55), degrees

τ—angle from vertical of plane containing source and line of sight (see Fig. 9-55), degrees

ω_c—solid angle subtended at observer by ceiling, steradians

ω_f—solid angle subtended at observer by floor, steradians

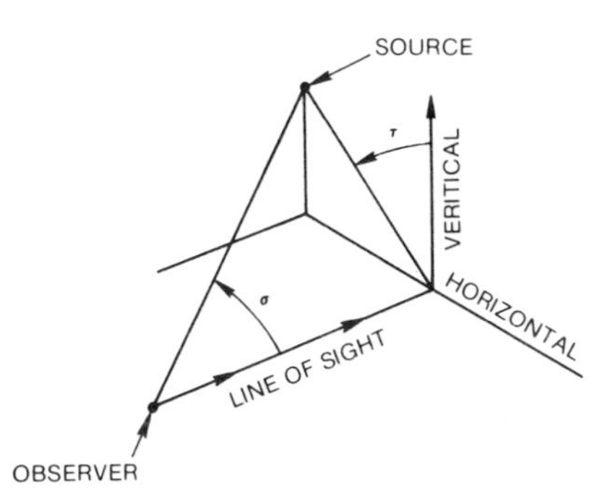

Fig. 9-55. Geometry defining position index as used in VCP calculations.

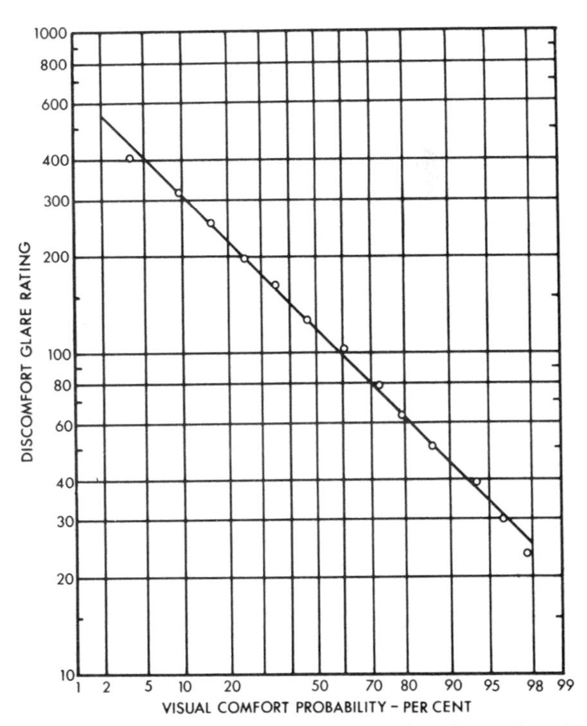

Fig. 9-56. A chart for converting discomfort glare ratings to visual comfort probabilities (VCP) (the per cent of observers who would be expected to judge a lighting condition to be at or more comfortable than the borderline between comfort and discomfort).

ω_s—solid angle subtended at observer by a source, steradians

ω_w—solid angle subtended at observer by walls, steradians

These basic relations may be applied through a variety of techniques. Generally, the room cavity concept is used in which the actual ceiling and floor luminances and solid angles are replaced by their equivalent cavity values. The computation can be performed by summing throughout the enclosure.[51]

VCP values are frequently associated with luminaires assuming standardized conditions of use. The specific luminaire arrangement in a room is not used; rather, the luminaires are fractionally apportioned over the ceiling according to a standard scheme. VCP values are determined for:

1. An initial average horizontal illuminance of 1000 lux [100 footcandles].
2. Room reflectances of $\rho_{CC} = 0.80$, $\rho_W = 0.50$, and $\rho_{FC} = 0.20$.
3. Luminaire mounting heights above the floor of 2.6, 3, 4 and 4.9 meters (8.5, 10, 13, and 16 feet).
4. A (given) range of room dimensions including square, long-narrow, and short-wide rooms.
5. An observation point 1.2 meters (4 feet) in front of the center of the rear wall and 1.2 meters (4 feet) above the floor.
6. A horizontal line of sight, directly forward.
7. An upward limit to the field of view corresponding to an angle of 53 degrees above and directly forward from the observer.

This standardized procedure is extensively treated in references 42, 43, 44, 47, and 48. Further modifications are under development, and the current literature should be reviewed. There are two objectives in applying the standardized procedure. First, it simplifies calculations by permitting organization of various procedural steps. Second, it allows comparisons between those luminaires for which the standardized values have been tabulated even before a specific lighting layout has been made. Fig. 9-57 illustrates a typical tabulation of visual comfort probabilities as developed by the standardized procedure.

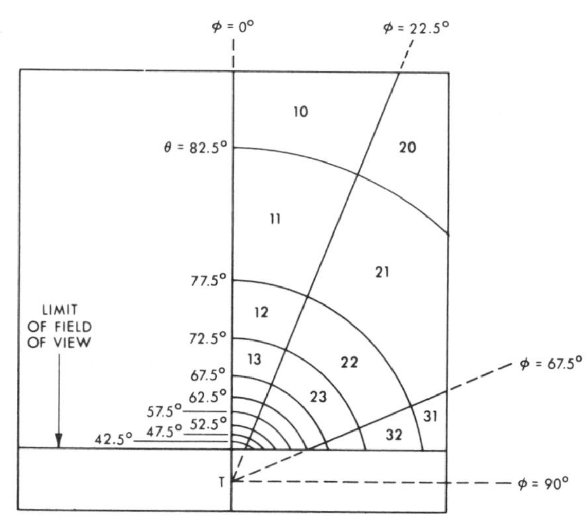

Fig. 9-58. Division of the ceiling into zones and sectors for the simplified procedure for assessment of visual comfort.

Fig. 9-57. Example of a Typical Tabulation of Visual Comfort Probability Values*

WALL REFL 50%, EFF CEILING CAV REFL 80%, EFF FLOOR CAV REFL 20%
LUMINAIRE NO. 000
WORK-PLANE ILLUMINATION 100 FOOTCANDLES

Room		Luminaires Lengthwise				Luminaires Crosswise			
W	L	8.5	10.0	13.0	16.0	8.5	10.0	13.0	16.0
20	20	78	82	90	94	77	81	89	93
20	30	73	76	82	88	72	75	81	86
20	40	71	73	78	82	70	72	76	80
20	60	69	71	74	78	68	70	73	76
30	20	78	82	88	92	77	81	87	92
30	30	73	75	80	85	72	74	79	84
30	40	70	72	75	78	69	71	74	77
30	60	68	69	71	74	67	69	70	73
30	80	67	69	69	72	67	68	68	71
40	20	79	82	87	92	79	82	87	91
40	30	74	76	79	84	73	75	78	83
40	40	71	72	74	77	70	71	73	76
40	60	68	69	70	72	68	69	69	71
40	80	67	68	68	70	67	68	67	69
40	100	67	68	67	69	67	67	66	68
60	30	75	76	79	83	74	76	78	82
60	40	71	72	74	76	71	72	73	76
60	60	69	69	69	71	68	69	68	70
60	80	68	68	67	69	67	68	66	68
60	100	67	67	66	67	67	67	65	66
100	40	74	75	75	78	74	74	75	77
100	60	71	71	71	72	71	71	70	72
100	80	70	70	68	69	70	69	67	69
100	100	69	68	66	67	69	68	66	67

* This example is for use when the unit of length is the foot. VCP values will be the same when the unit of length is the meter and the illuminance is in lux.

Simplified Procedure. An approximate assessment can be obtained by an alternate simplified procedure.[52-54] While this simplified procedure is applicable only to flat-bottomed, nonluminous sided equipment in a 18- by 18-meter (60- by 60-foot) room with a 3-meter (10-foot) mounting height, its extension to other types is

Fig. 9-59. Work Sheet for Computing $\bar{L}$ for Determining the Acceptability of a Luminaire from a VCP Standpoint for Use in Large Rooms

Luminaire		Viewing Direction			
	Angles Above Nadir (degrees)	L Average Luminance (candela per square meter)	T Multiplier	L × T	Totals
Viewing Direction	85		.0375		
	80		.1080		
	75		.0884		
	70		.0703		
	65		.0543		
	60		.0406		
	55		.0312		
	50		.0229		
	45		.0159		
	40		.0102		
Diagonal (45°)	85		.0203		
	80		.1065		
	75		.1022		
	70		.0841		
	65		.0681		
	60		.0507		
	55		.0333		
	50		.0214		
	45		.0109		
	40		.0021		
90° to Viewing Direction	80		.0046		
	75		.0096		
	70		.0052		
	65		.0017		

$\bar{L} = 0.292 \sum L \times T =$

To be acceptable $\bar{L}$ should be equal to or less than 320.*

* When the ratio B = Flux in 0–60° zone/Flux in 60–90° zone as calculated from available luminaire zonal flux data, is greater than 10 or less than 4 it becomes necessary to correct $\bar{L}$ by multiplying it by a correction factor $\phi = 2.34\ (B^{-0.48})$. Then the value of 320 for $\phi\bar{L}$ is used as the cut off level for regarding a luminaire as acceptable. If B falls between 4 and 10 no correction (ϕ) is necessary but may be used if desired.

A table relating values of ϕ for values of B is shown below.

B	ϕ	B	ϕ
1	2.34	22	.531
2	1.68	24	.509
3	1.38	26	.490
4	1.20	28	.473
5	1.08	30	.457
6	.990		
7	.920	35	.425
8	.862	40	.398
9	.815	45	.376
10	.775	50	.358
11	.740	55	.342
12	.710	60	.328
13	.683	65	.316
14	.659	70	.304
15	.638	75	.295
16	.613	80	.286
17	.601	85	.277
18	.584	90	.270
19	.569	95	.263
20	.556	100	.257

being studied. Although the method has been derived from the basic equation for M, it involves several assumptions:

1. Discomfort glare is independent of the number of luminaires used to provide 1000 lux [100 footcandles].
2. The average luminance of the field of view is proportional to the number of luminaires in the field of view.
3. VCP may be regarded as a function of $\bar{L}$ which is the summation of the product LT for 24 segments into which the ceiling is divided (see Fig. 9-58). L is equal to the average luminance of a luminaire in the lengthwise, diagonal and crosswise vertical planes at the midzone angles above nadir listed in Fig. 9-59; T is a factor which takes into account the position index and angular size of the segment.
4. The effect of the luminous area of the luminaires is included in the quantity $\bar{L}$.
5. If the value of $\bar{L}$ is as low as, or lower than 320, this may be taken as an indication that the VCP is 70 or higher.

A work sheet for computing $\bar{L}$ is shown in Fig. 9-59 on which are included the values of T for the different ceiling segments. When luminaires are viewed crosswise, crosswise values of luminance L are entered in the upper section and the lengthwise values in the lower section. For lengthwise viewing, lengthwise values are entered in the upper section and crosswise in the lower section. The sum of the LT values represents the rating. It is emphasized that if VCP ratings using the comprehensive and simplified procedures are different, the former takes precedence provided both are calculated correctly.

LUMINAIRE SPACING CRITERION

The Luminaire Spacing Criterion[55-57] (SC), a refinement of the earlier spacing-to-mounting height ratio[58], is a classification technique for interior luminaires relating to the spread or distribution provided by the direct component of illuminance from the luminaires. It tests the uniformity of horizontal illuminance at two pairs of selected points to estimate the probable extreme limit of acceptable luminaire spacing. It is not a specification of the spacing-to-mounting height ratio to be used in a lighting installation, and, in fact, installation of luminaires at this nominal value may produce a poor lighting system.

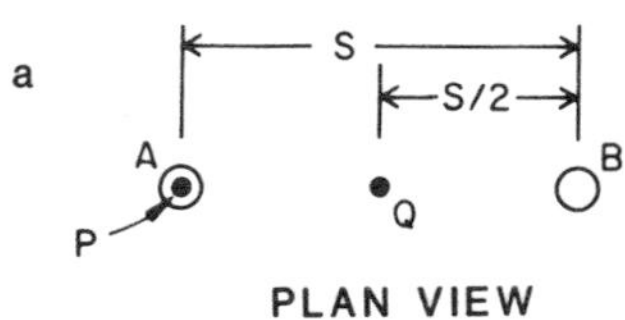

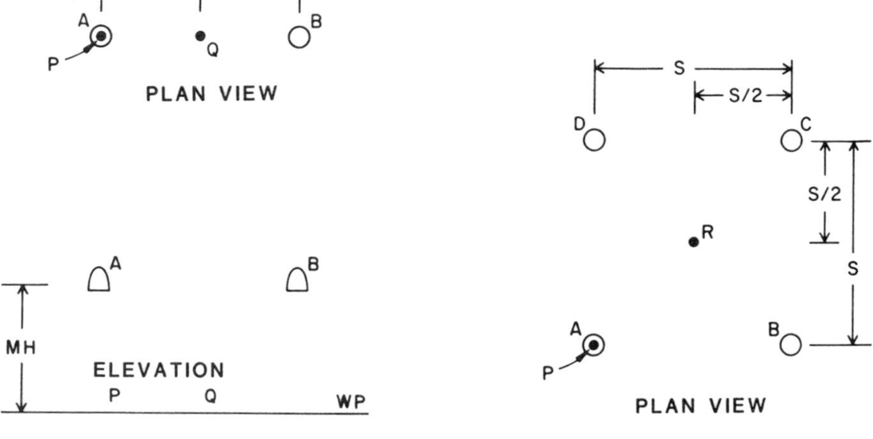

Fig. 9-60. Points Q and R represent points of most probable low illuminance for configuration of two luminaires (a) and square array of four luminaires (b).

The classification technique is to aid lighting designers in rapidly assessing one aspect of the potential of a luminaire with respect to its applications. It is analogous to schemes such as the NEMA Outdoor Floodlighting Luminaire Designation and the CIE Luminaire Classification (see Application Volume) in that it gives some idea about the distribution of flux from a luminaire and its subsequent effect on lighting system parameters using only a single number.

The basis of the luminaire spacing criterion is horizontal illuminance on the work-plane due to direct illuminance from nearby luminaires. As a first order approximation it is assumed that this represents a limiting case since the reflected component of illuminance and the illuminance due to more distant luminaires generally will tend to increase the uniformity of horizontal illuminance from point to point.

When two similar conventional luminaires are near their maximum spacing, the illuminance directly under a luminaire (P) is principally due to the overhead luminaire (A) (see Fig. 9-60a). Further, a very probable point of low illuminance will be one-half the way between two luminaires (Q). The maximum spacing at a given mounting height above the work-plane is chosen such that the illuminance one-half the way between the two luminaires (Q) due to both luminaires (A and B) equals the illuminance under one (P) due to that one luminaire (A) only.

Another likely point (R) for low illuminance is at the center of a square array of adjacent luminaires (see Fig. 9-60b). The maximum spacing at a given mounting height above the work-plane is chosen such that the illuminance at the center of the luminaires (R) due to all four luminaires (A, B, C and D) equals the illuminance under one (P) due to that one luminaire (A) only.

The maximum spacing (expressed as a spacing-to-mounting height ratio) that fulfills each of the above conditions is easily determined on a special graph using the intensity distribution of the luminaire. For the purpose of establishing this criterion, it is assumed that the inverse-square law (see page 9-10) is valid. This is the only assumption for the computations.

Procedure

A. For luminaires whose intensity distribution is nominally symmetric about the nadir:

1. Plot the relative intensity of the luminaire on the chart of Fig. 9-61.
2. Locate the point of one-half the intensity at 0 degrees on the ordinate, and draw a line through that point and parallel to the diagonal lines. If the intensity varies significantly in the vicinity of 0 degrees, use an average of the intensity over the 0 to 5 degrees polar angle.
3. Read scale A above the intersection of this line with the intensity curve.
4. Repeat step 2 using the point of one-quarter the intensity at 0 degrees.
5. Read scale B above the intersection of this line with the intensity curve.
6. The lower of the values found in steps 3 and 5 is the Luminaire Spacing Criterion. Round off the value to the nearest 0.1.

B. For luminaires with significantly asymmetric intensity distributions about the nadir:

1. Independently evaluate the longitudinal and transverse (parallel and perpendicular, 0 de-

grees and 90 degrees, etc.) intensity distributions.

2. Apply steps A.1, A.2, and A.3 for each of the intensity curves. Round off to the nearest 0.1.

Interpretation. The value from scale *A* corresponds to the criterion of point *Q* (Fig. 9–60a), and the value from scale *B* corresponds to the criterion of point *R* (Fig. 9–60b). Thus for symmetrical intensity distributions, the luminaire spacing criterion requires that the direct horizontal illuminance at neither test point be less than the illuminance directly below a single luminaire. For the non-symmetrical intensity distribution, it generally is found that independent testing at point *Q* for each orientation will be adequate.

A point directly under a luminaire will have relatively high illuminance and receive its principal contribution from the luminaire overhead when the spacing is large. If the illuminance at a probable low point due to the closest luminaires is no lower than that at a probable high point and due to the main component, then it is likely that reasonable uniformity will be achieved over the entire work-plane.

The luminaire spacing criterion only suggests a maximum spacing at which the horizontal illuminance will be reasonably uniform. When other criteria such as overlap between luminaires, vertical illuminance, shadowing, illuminance distribution above the work-plane, etc., are considered, it generally is found that luminaires must be installed at some spacing-to-mounting height ratio less than the nominal value of the criterion.

In cases where it is important, the lighting designer must calculate the illuminance at points throughout the room to assess uniformity of illuminance in detail. It is suggested that a maximum value of 1.5 be assigned as the luminaire spacing criterion for any luminaire since the use of larger values frequently does not produce acceptable lighting installations when all performance criteria are examined. Also, certain luminaires are designed to be installed only with specific spacing relations, and neither larger nor smaller spacings are desireable. In such cases, the concept of a luminaire spacing criterion is not an applicable concept. Specific limits or ranges for the spacing-to-mounting height ratios should be recommended by the manufacturer, and the basis for the recommendation should be stated, *e.g.*, a certain degree of horizontal illuminance uniformity (determined by point calculations).

The beam spread classification* of former systems will no longer be used due to potential errors when specifying luminaires.[58] One alternate method of specifying beam spread is to set a minimum value (*e.g.*, greater or equal to 1.3). However, it should be noted that the luminaire

* Highly concentrating, concentrating, medium spread, spread, and wide spread.

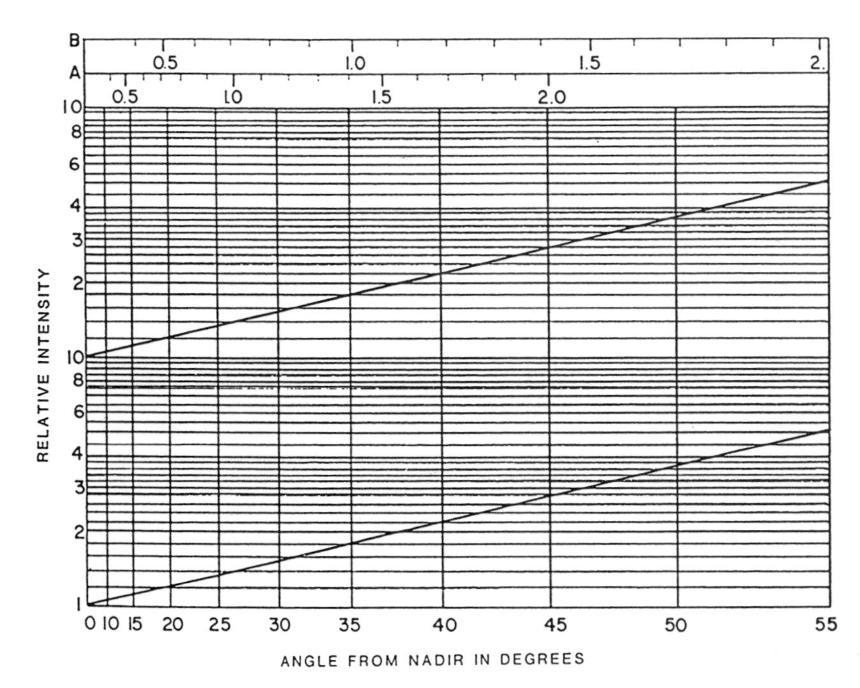

Fig. 9-61. Chart for determining the Luminaire Spacing Criterion.

spacing criterion is a design tool. It is not especially appropriate as a specification quantity.

In addition to the Luminaire Spacing Criterion, it is possible to determine illuminance uniformity for specific installation conditions and report spacing-to-mounting height limits for such conditions. If such is the case, the luminaire layout, the room conditions, the uniformity criteria, and any application restrictions should be explicitly described whenever such a ratio is reported.

COEFFICIENTS OF UTILIZATION[59-65]

Tables of coefficients of utilization, wall exitance coefficients, and ceiling cavity exitance coefficients can be prepared by systemized procedures.[85] It is desirable to standardize published tables of these values to prevent misunderstandings and to facilitate direct comparisons of the data for different luminaires.

These coefficients are derived from the equations described under radiative transfer theory (page 9-3). In the past the coefficients were hand-calculated and their generation standardized in terms of graphs, charts and calculation tables. However, in the mid-1970's hand-methods were superceded by computer generated coefficient tables. Consequently, only the basic form to permit computer programming is presented here. The exact form will be arranged to suit the needs of the users and their particular computers. Thus, only an outline plus the necessary equations will be described.

The basic assumptions used to develop the zonal-cavity coefficients are: (1) the room surfaces are lambertian reflectors; (2) incident flux on each surface is uniformly distributed over that surface; (3) the luminaires are uniformly distributed throughout the room, *i.e.*, uniformly dense but not necessarily in a uniform pattern; (4) the room is empty; and (5) the room surfaces are spectrally neutral.

Fig. 9-62 shows the recommended form for unabridged tables. It is recognized that space limitations often necessitate abridgements, and when necessary, only the columns for 80, 50, and 10 per cent ρ_{CC} are recommended for luminaires having 0 to 35 per cent of their output in the 90- to 180-degree zone; and 80, 70, and 50 per cent for luminaires having over 35 per cent of their output in the 90- to 180-degree zone. Also, 10 per cent ρ_W columns are not required for exitance coefficient abridged tables. It is recommended that coefficients of utilization be published to two decimal places, wall exitance coefficients to three decimal places, and ceiling cavity exitance coefficients to three decimal places. A wall direct radiation coefficient (WDRC) should be published to three decimal places for each room cavity ratio adjacent to the wall exitance coefficient table. The three significant figures are not justified in terms of coefficient accuracy, but are required since certain computational methods require the small differences between these coefficients.

Computation

1. Define 18 conic solid angle zones of 10-degree width from the nadir to the zenith about the luminaire as shown in Fig. 9-63 where:

$$N = \text{index of zones; integer, } 1 \leq N \leq 18$$

2. Determine the flux Φ_N (lumens) in the various zones:

a. The flux in a conic solid angle (see Fig. 9-64) is given by

$$\Phi = 2\pi I\,(\cos\theta_1 - \cos\theta_2)$$

where I is the midzone intensity in candelas.

b. If the intensity is not rotationally symmetric about the vertical axis, average the intensity about the vertical axis at each vertical angle θ. Note that the intensity must be sampled at equal angular intervals about the vertical axis. For example, if intensity is known for 3 vertical planes [$I_\perp$(perpendicular), $I_{45°}$, and I(parallel)],

$$I = \tfrac{1}{4}(I_\perp + 2I_{45°} + I_\parallel)$$

While 3 planes are sufficient for luminaires of nominal rotational symmetry, photometric data at 15- or 22½-degree increments about the vertical axis are preferred for luminaires without this symmetry.

c. If the intensity is taken at 10-degree vertical intervals ($\theta = 5°, 15°, 25°, \ldots$), then the flux Φ_N is determined by the application of the equation in (a) above to the full zone. It is preferred to have intensity values at 5-degree vertical angles ($\theta = 2½°, 7½°, 12½°, \ldots$). Then the zone N is divided into two parts, and the equation is applied to each part with the resulting flux being summed.

(*Continued on page 9-72.*)

Instructions and Notes for Use of Fig. 9–62.

(The following notes and instructions have been prepared to guide the user of Fig. 9–62.)

1. The luminaires in this table are organized by source type and luminaire form rather than by applications for convenience in locating luminaires.

   In some cases, luminaire data in this table are based on an actual typical luminaire; in other cases, the data represent a composite of generic luminaire types. Therefore, whenever possible, specific luminaire data should be used in preference to this table of typical luminaires.
2. The polar intensity sketch (candlepower distribution curve) and the corresponding luminaire spacing criterion are representative of many luminaires of each type shown. A specific luminaire may differ in perpendicular plane (crosswise) and parallel plane (lengthwise) intensity distributions and in spacing criterion from the values shown. However, the various coefficients depend only on the average intensity at each polar angle from nadir. The tabulated coefficients can be applied to any luminaire whose average intensity distribution matches the values used to generate the coefficients. The average intensity values used to generate the coefficients are given at the end of the table, normalized to a per thousand lamp lumen basis.
3. The various coefficients depend only on the average intensity distribution curve and are linearly related to the total luminaire efficiency. Consequently, the tabulated coefficients can be applied to luminaires with similarly shaped average intensity distributions using a correcting multiplier equal to the new luminaire total efficiency divided by the tabulated luminaire total efficiency. The use of polarizing diffusers or lenses on fluorescent luminaires has no effect on the coefficients given in this table except as they affect total luminaire efficiency.
4. Satisfactory installations depend on many factors including the environment, space utilization, luminous criteria, etc. as well as the luminaire itself. Consequently, a definitive spacing recommendation cannot be assigned independently to the luminaire. The spacing criterion (SC) values given are only general guides (see page 9–49). SC values are not assigned to semi-indirect and indirect luminaires since the basis of this technique does not apply to such situations. Also, SC values are not given for those bat-wing luminaires which must be located by criteria other than that of horizontal illuminance.
5. Key: ρ_{CC} = ceiling cavity reflectance (per cent)
   ρ_W = wall reflectance (per cent)
   ρ_{FC} = floor cavity reflectance (per cent)
   RCR = room cavity ratio
   WDRC = wall direct radiation coefficient
   SC = luminaire spacing criterion
   NA = not applicable
6. Many of the luminaires in this figure appeared in earlier editions of the *IES Lighting Handbook*. The identifying number may be different due to a reordering of the luminaires. In some cases, the data have been modified in terms of more recent or more extensive information. The user should specifically refer to this handbook when referencing luminaires from this handbook.
7. Efficiency, and consequently the coefficients, of fluorescent luminaires is a function of the number of lamps in relation to the size of the luminaire. This is due to temperature changes and due to changes in the blocking of light. In this figure, fluorescent luminaires have been chosen for typical luminaire sizes and numbers of lamps; these are identified under the typical luminaire drawings. Variations of the coefficients with size and number of lamps depend on the many details of luminaire construction. The following correction factors are average values:

   4 lamp, 610 mm (2 ft) wide:
   ×1.05 for 8 lamp, 1220 mm (4 ft) wide
   ×1.05 for 3 lamp, 610 mm (2 ft) wide
   ×1.1 for 2 lamp, 610 mm (2 ft) wide
   ×0.9 for 2 lamp, 300 mm (1 ft) wide

   2 lamp wraparound:
   ×0.95 for 4 lamp
8. Photometric data for fluorescent luminaires in this table are based upon tests using standard wattage fluorescent lamps. Reduced wattage fluorescent lamps cause lower lamp operating temperatures with some luminaires. Consequently, the efficiency and coefficients may be slightly increased. It is desirable to obtain specific correction factors from the manufacturers. Typical factors for reduced wattage fluorescent lamps (approximately 10 per cent below standard lamp watts) are:

2-lamp strip, surface mounted	× 1.03
4-lamp troffer, enclosed, non air handling	× 1.07
4-lamp wrap around, surface mounted	× 1.07
2-lamp industrial, vented	× 1.00

Electronic ballasts can be designed for any arbitrary operating condition. The manufacturer must be consulted for specific data.

Fig. 9–62 Coefficients of Utilization, Wall Exitance Coefficients, Ceiling Cavity Exitance
(See page 9-53 for

Typical Luminaire	Typical Intensity Distribution and Per Cent Lamp Lumens		ρ_{CC} →	80			70			50			30			10			0	WDRC	ρ_{CC} →
			ρ_W →	50	30	10	50	30	10	50	30	10	50	30	10	50	30	10	0		ρ_W →
	Maint. Cat.	SC	RCR ↓	Coefficients of Utilization for 20 Per Cent Effective Floor Cavity Reflectance (ρ_{FC} = 20)																	RCR ↓
1	V	1.5	0	.87	.87	.87	.81	.81	.81	.70	.70	.70	.59	.59	.59	.49	.49	.49	.45		
			1	.71	.66	.62	.65	.61	.58	.55	.52	.49	.46	.44	.42	.38	.36	.34	.30	.368	1
			2	.60	.53	.48	.55	.50	.45	.47	.42	.38	.39	.35	.32	.31	.29	.26	.23	.279	2
	35½% ↑		3	.52	.44	.38	.48	.41	.36	.40	.35	.31	.33	.29	.26	.27	.24	.21	.18	.227	3
			4	.45	.37	.32	.42	.35	.29	.35	.30	.25	.29	.25	.21	.23	.20	.17	.14	.192	4
			5	.40	.32	.27	.37	.30	.25	.31	.25	.21	.26	.21	.18	.21	.17	.14	.12	.166	5
	45% ↓		6	.35	.28	.23	.33	.26	.21	.28	.22	.18	.23	.19	.15	.19	.15	.12	.10	.146	6
			7	.32	.25	.19	.29	.23	.18	.25	.20	.16	.21	.16	.13	.17	.13	.11	.09	.130	7
			8	.29	.22	.17	.27	.20	.16	.23	.17	.14	.19	.15	.12	.15	.12	.09	.07	.117	8
Pendant diffusing sphere with incandes-			9	.26	.19	.15	.24	.18	.14	.21	.16	.12	.17	.13	.10	.14	.11	.08	.07	.107	9
cent lamp			10	.24	.17	.13	.22	.16	.12	.19	.14	.11	.16	.12	.09	.13	.10	.08	.06	.098	10
2	II	N.A.	0	.83	.83	.83	.72	.72	.72	.50	.50	.50	.30	.30	.30	.12	.12	.12	.03		
			1	.72	.69	.66	.62	.60	.57	.43	.42	.40	.26	.25	.25	.10	.10	.10	.03	.018	1
			2	.63	.58	.54	.54	.50	.47	.38	.35	.33	.23	.22	.20	.09	.09	.08	.02	.015	2
			3	.55	.49	.45	.47	.43	.39	.33	.30	.28	.20	.19	.17	.08	.07	.07	.02	.013	3
			4	.48	.42	.37	.42	.37	.33	.29	.26	.23	.18	.16	.15	.07	.06	.06	.02	.012	4
	83% ↑		5	.43	.36	.32	.37	.32	.28	.26	.23	.20	.16	.14	.12	.06	.06	.05	.01	.011	5
			6	.38	.32	.27	.33	.28	.24	.23	.20	.17	.14	.12	.11	.06	.05	.04	.01	.010	6
	3½% ↓		7	.34	.28	.23	.30	.24	.21	.21	.17	.15	.13	.11	.09	.05	.04	.04	.01	.009	7
			8	.31	.25	.20	.27	.21	.18	.19	.15	.13	.12	.10	.08	.05	.04	.03	.01	.008	8
Concentric ring unit with incandescent			9	.28	.22	.18	.24	.19	.16	.17	.14	.11	.10	.09	.07	.04	.03	.03	.01	.008	9
silvered-bowl lamp			10	.25	.20	.16	.22	.17	.14	.16	.12	.10	.10	.08	.06	.04	.03	.03	.01	.007	10
3	IV	1.3	0	.99	.99	.99	.97	.97	.97	.93	.93	.93	.89	.89	.89	.85	.85	.85	.83		
			1	.87	.84	.81	.85	.82	.79	.82	.79	.77	.79	.76	.74	.76	.74	.72	.71	.323	1
	0% ↑		2	.76	.70	.65	.74	.69	.65	.71	.67	.63	.69	.65	.62	.66	.63	.60	.59	.311	2
			3	.66	.59	.54	.65	.59	.53	.62	.57	.53	.60	.56	.52	.58	.54	.51	.49	.288	3
			4	.58	.51	.45	.57	.50	.45	.55	.49	.44	.53	.48	.44	.51	.47	.43	.41	.264	4
			5	.52	.44	.39	.51	.44	.38	.49	.43	.38	.47	.42	.37	.46	.41	.37	.35	.241	5
	83½% ↓		6	.46	.39	.33	.46	.38	.33	.44	.38	.33	.43	.37	.33	.41	.36	.32	.31	.221	6
			7	.42	.34	.29	.41	.34	.29	.40	.33	.29	.39	.33	.29	.38	.32	.28	.27	.203	7
			8	.38	.31	.26	.37	.31	.26	.36	.30	.26	.35	.30	.25	.34	.29	.25	.24	.187	8
Porcelain-enameled ventilated standard			9	.35	.28	.23	.34	.28	.23	.33	.27	.23	.32	.27	.23	.32	.26	.23	.21	.173	9
dome with incandescent lamp			10	.32	.25	.21	.32	.25	.21	.31	.25	.21	.30	.24	.21	.29	.24	.20	.19	.161	10
4	V	1.3	0	.89	.89	.89	.85	.85	.85	.77	.77	.77	.70	.70	.70	.63	.63	.63	.60		
			1	.77	.74	.71	.74	.71	.68	.67	.65	.63	.61	.59	.57	.55	.54	.53	.50	.264	1
	18½% ↑		2	.68	.63	.59	.65	.61	.57	.59	.56	.53	.54	.51	.49	.49	.47	.45	.42	.224	2
			3	.61	.55	.50	.58	.53	.48	.53	.49	.45	.49	.45	.42	.44	.42	.39	.37	.197	3
			4	.54	.48	.43	.52	.46	.42	.48	.43	.39	.44	.40	.37	.40	.37	.34	.32	.176	4
			5	.49	.42	.38	.47	.41	.37	.43	.38	.35	.40	.36	.33	.37	.33	.31	.29	.159	5
	60½% ↓		6	.44	.38	.33	.43	.37	.32	.39	.34	.31	.36	.32	.29	.34	.30	.27	.26	.145	6
			7	.40	.34	.30	.39	.33	.29	.36	.31	.27	.33	.29	.26	.31	.27	.25	.23	.133	7
			8	.37	.31	.27	.36	.30	.26	.33	.28	.25	.31	.27	.24	.29	.25	.22	.21	.124	8
			9	.34	.28	.24	.33	.27	.24	.31	.26	.22	.29	.24	.21	.27	.23	.20	.19	.115	9
Prismatic square surface drum			10	.32	.26	.22	.30	.25	.21	.28	.24	.21	.27	.23	.20	.25	.21	.19	.17	.108	10
5	IV	0.8	0	1.19	1.19	1.19	1.16	1.16	1.16	1.11	1.11	1.11	1.06	1.06	1.06	1.02	1.02	1.02	1.00		
			1	1.08	1.05	1.03	1.06	1.03	1.01	1.02	1.00	.98	.98	.97	.95	.95	.93	.92	.90	.241	1
	0% ↑		2	.99	.94	.89	.97	.92	.88	.93	.90	.86	.90	.87	.84	.88	.85	.83	.81	.238	2
			3	.90	.84	.79	.88	.83	.78	.86	.81	.77	.83	.79	.76	.81	.77	.74	.73	.227	3
			4	.82	.75	.70	.81	.75	.70	.79	.73	.69	.77	.72	.68	.75	.71	.67	.66	.215	4
			5	.76	.68	.63	.75	.68	.63	.73	.67	.62	.71	.66	.62	.69	.65	.61	.59	.202	5
	100% ↓		6	.70	.62	.57	.69	.62	.57	.67	.61	.57	.66	.60	.56	.64	.60	.56	.54	.191	6
			7	.65	.57	.52	.64	.57	.52	.62	.56	.52	.61	.56	.52	.60	.55	.51	.50	.180	7
			8	.60	.53	.48	.59	.53	.48	.58	.52	.48	.57	.52	.47	.56	.51	.47	.46	.169	8
			9	.56	.49	.44	.55	.49	.44	.54	.48	.44	.53	.48	.44	.52	.47	.44	.42	.160	9
R-40 flood without shielding			10	.52	.46	.41	.52	.45	.41	.51	.45	.41	.50	.45	.41	.49	.44	.41	.39	.152	10
6	IV	0.7	0	1.01	1.01	1.01	.99	.99	.99	.94	.94	.94	.90	.90	.90	.87	.87	.87	.85		
			1	.95	.93	.91	.93	.91	.89	.89	.88	.87	.86	.85	.84	.83	.82	.82	.80	.115	1
	0% ↑		2	.89	.86	.83	.87	.84	.82	.85	.82	.80	.82	.80	.79	.80	.78	.77	.76	.115	2
			3	.83	.80	.77	.82	.79	.76	.80	.77	.75	.78	.76	.74	.76	.74	.72	.71	.113	3
			4	.79	.74	.71	.78	.74	.71	.76	.73	.70	.74	.71	.69	.73	.70	.68	.67	.110	4
			5	.74	.70	.67	.74	.69	.66	.72	.68	.66	.71	.68	.65	.69	.67	.65	.63	.107	5
	85% ↓		6	.70	.66	.62	.70	.65	.62	.68	.65	.62	.67	.64	.61	.66	.63	.61	.60	.104	6
			7	.67	.62	.59	.66	.62	.59	.65	.61	.58	.64	.61	.58	.63	.60	.58	.57	.100	7
			8	.63	.59	.56	.63	.58	.55	.62	.58	.55	.61	.58	.55	.60	.57	.55	.54	.097	8
R-40 flood with specular anodized reflec-			9	.60	.56	.53	.60	.56	.53	.59	.55	.52	.58	.55	.52	.58	.54	.52	.51	.094	9
tor skirt; 45° cutoff			10	.57	.53	.50	.57	.53	.50	.56	.52	.50	.56	.52	.50	.55	.52	.49	.48	.091	10

Coefficients, Luminaire Spacing Criterion and Maintenance Categories of Typical Luminaires.
instructions and notes.)

80			70			50			30			10			80			70			50			30			10		
50	30	10	50	30	10	50	30	10	50	30	10	50	30	10	50	30	10	50	30	10	50	30	10	50	30	10	50	30	10
Wall Exitance Coefficients for 20 Per Cent Effective Floor Cavity Reflectance (ρ_{FC} = 20)															Ceiling Cavity Exitance Coefficients for 20 Per Cent Floor Cavity Reflectance (ρ_{FC} = 20)														
															.423	.423	.423	.361	.361	.361	.246	.246	.246	.142	.142	.142	.045	.045	.045
.328	.187	.059	.311	.178	.056	.280	.161	.051	.252	.145	.047	.226	.131	.042	.422	.396	.373	.361	.340	.321	.247	.234	.222	.142	.135	.129	.046	.044	.042
.275	.150	.046	.259	.143	.044	.231	.129	.040	.205	.115	.036	.181	.102	.032	.417	.379	.347	.357	.327	.300	.245	.226	.209	.141	.131	.123	.045	.043	.040
.240	.128	.038	.226	.121	.036	.200	.108	.033	.176	.097	.030	.154	.085	.026	.412	.367	.332	.353	.317	.287	.242	.220	.202	.140	.128	.119	.045	.042	.039
.214	.111	.033	.201	.105	.031	.177	.094	.028	.155	.083	.025	.135	.073	.022	.406	.358	.321	.348	.309	.279	.239	.215	.196	.138	.126	.116	.045	.041	.038
.193	.098	.028	.181	.093	.027	.160	.083	.024	.139	.073	.022	.120	.064	.019	.400	.350	.314	.343	.303	.273	.236	.212	.193	.137	.124	.114	.044	.041	.038
.176	.088	.025	.165	.084	.024	.145	.074	.022	.126	.066	.019	.109	.057	.017	.394	.344	.309	.338	.298	.269	.234	.209	.190	.135	.123	.113	.044	.040	.037
.162	.080	.023	.152	.076	.022	.133	.067	.019	.116	.059	.017	.100	.052	.015	.388	.339	.305	.334	.294	.266	.231	.206	.188	.134	.122	.112	.043	.040	.037
.150	.073	.021	.140	.069	.020	.123	.062	.018	.107	.054	.016	.092	.047	.014	.383	.335	.302	.330	.291	.264	.228	.204	.187	.133	.120	.111	.043	.039	.037
.139	.067	.019	.131	.064	.018	.115	.057	.016	.099	.050	.014	.085	.043	.013	.378	.332	.300	.326	.288	.262	.226	.202	.186	.131	.119	.111	.043	.039	.037
.130	.062	.017	.122	.059	.016	.107	.052	.015	.093	.046	.013	.080	.040	.011	.374	.328	.298	.322	.285	.260	.223	.201	.185	.130	.119	.110	.042	.039	.037
															.796	.796	.796	.680	.680	.680	.464	.464	.464	.267	.267	.267	.085	.085	.085
.226	.128	.041	.195	.111	.035	.137	.078	.025	.083	.048	.015	.034	.020	.006	.790	.772	.756	.676	.663	.651	.462	.456	.450	.266	.264	.262	.085	.085	.085
.207	.114	.035	.179	.099	.030	.126	.070	.022	.077	.043	.013	.031	.018	.006	.784	.755	.731	.671	.650	.632	.460	.450	.441	.265	.262	.258	.085	.085	.084
.191	.102	.030	.165	.088	.027	.116	.063	.019	.071	.039	.012	.029	.016	.005	.778	.743	.715	.667	.641	.620	.458	.445	.435	.265	.260	.256	.085	.084	.084
.177	.092	.027	.153	.080	.024	.108	.057	.017	.066	.035	.011	.027	.014	.004	.773	.734	.703	.664	.634	.611	.456	.442	.430	.264	.259	.255	.085	.084	.084
.164	.084	.024	.142	.073	.021	.100	.052	.015	.061	.032	.010	.025	.013	.004	.768	.726	.696	.660	.629	.605	.455	.439	.427	.263	.258	.253	.085	.084	.084
.153	.077	.022	.133	.067	.019	.094	.048	.014	.057	.030	.009	.023	.012	.004	.764	.721	.690	.656	.624	.601	.453	.437	.425	.263	.257	.253	.085	.084	.083
.143	.071	.020	.124	.062	.018	.088	.044	.013	.054	.027	.008	.022	.011	.003	.759	.716	.686	.653	.621	.598	.451	.435	.423	.262	.256	.252	.085	.084	.083
.134	.066	.018	.116	.057	.016	.082	.041	.012	.050	.026	.007	.020	.010	.003	.755	.712	.683	.650	.618	.595	.450	.434	.422	.262	.256	.252	.085	.084	.083
.126	.061	.017	.109	.053	.015	.077	.038	.011	.047	.024	.007	.019	.010	.003	.751	.709	.680	.647	.615	.593	.448	.432	.421	.261	.255	.251	.085	.084	.083
.119	.057	.016	.103	.050	.014	.073	.036	.010	.045	.022	.006	.018	.009	.003	.747	.706	.678	.644	.613	.592	.447	.431	.421	.261	.255	.251	.084	.084	.083
															.159	.159	.159	.136	.136	.136	.093	.093	.093	.053	.053	.053	.017	.017	.017
.248	.141	.045	.242	.138	.044	.231	.133	.042	.221	.128	.041	.212	.123	.040	.150	.130	.113	.128	.112	.097	.088	.077	.067	.050	.045	.039	.016	.014	.013
.240	.131	.040	.235	.129	.040	.225	.125	.039	.216	.121	.038	.208	.117	.037	.143	.110	.082	.123	.095	.071	.084	.066	.050	.048	.038	.029	.016	.012	.009
.225	.120	.036	.220	.118	.036	.212	.115	.035	.204	.112	.034	.196	.109	.034	.137	.095	.062	.118	.082	.054	.081	.057	.038	.047	.033	.022	.015	.011	.007
.209	.109	.032	.205	.107	.032	.197	.105	.031	.190	.102	.031	.184	.100	.030	.131	.084	.048	.113	.073	.042	.077	.051	.030	.045	.030	.018	.014	.010	.006
.194	.099	.029	.191	.098	.029	.184	.096	.028	.177	.094	.028	.171	.092	.028	.125	.076	.039	.108	.065	.034	.074	.046	.024	.043	.027	.014	.014	.009	.005
.181	.091	.026	.177	.090	.026	.171	.088	.026	.166	.086	.025	.160	.084	.025	.119	.069	.032	.103	.060	.028	.071	.042	.020	.041	.025	.012	.013	.008	.004
.168	.083	.024	.165	.082	.023	.160	.081	.023	.155	.079	.023	.150	.078	.023	.114	.063	.027	.098	.055	.024	.068	.038	.017	.039	.023	.010	.013	.007	.003
.157	.077	.022	.155	.076	.022	.150	.075	.021	.145	.074	.021	.141	.072	.021	.109	.058	.024	.093	.050	.021	.065	.035	.015	.038	.021	.009	.012	.007	.003
.147	.071	.020	.145	.071	.020	.141	.070	.020	.136	.068	.020	.133	.067	.019	.103	.054	.021	.089	.047	.018	.062	.033	.013	.036	.019	.008	.012	.006	.003
.138	.066	.018	.136	.066	.018	.132	.065	.018	.129	.064	.018	.125	.063	.018	.099	.051	.018	.085	.044	.016	.059	.031	.011	.034	.018	.007	.011	.006	.002
															.290	.290	.290	.248	.248	.248	.169	.169	.169	.097	.097	.097	.031	.031	.031
.243	.138	.044	.232	.132	.042	.211	.121	.039	.192	.111	.036	.175	.101	.033	.283	.264	.247	.242	.227	.213	.166	.156	.147	.095	.090	.085	.031	.029	.028
.216	.118	.036	.206	.114	.035	.187	.104	.032	.170	.095	.030	.154	.087	.027	.276	.246	.221	.236	.212	.191	.162	.147	.133	.093	.085	.078	.030	.028	.025
.196	.104	.031	.187	.100	.030	.170	.092	.028	.154	.085	.026	.140	.077	.024	.269	.233	.204	.231	.201	.177	.158	.139	.124	.092	.081	.073	.029	.026	.024
.180	.093	.027	.171	.090	.027	.156	.083	.025	.142	.076	.023	.128	.070	.021	.263	.223	.192	.226	.192	.167	.155	.134	.117	090	.079	.069	.029	.026	.023
.166	.084	.024	.158	.081	.024	.144	.075	.022	.131	.069	.021	.119	.064	.019	.257	.215	.183	.221	.186	.160	.152	.130	.113	.088	.076	.067	.028	.025	.022
.154	.077	.022	.147	.074	.021	.134	.069	.020	.122	.064	.019	.111	.058	.017	.252	.208	.177	.216	.180	.154	.149	.126	.109	.087	.074	.065	.028	.024	.021
.143	.071	.020	.137	.068	.019	.125	.063	.018	.114	.059	.017	.104	.054	.016	.246	.203	.173	.212	.176	.151	.146	.123	.107	.085	.073	.063	.027	.024	.021
.134	.066	.018	.129	.063	.018	.118	.059	.017	.108	.055	.016	.098	.050	.015	.242	.199	.169	.208	.172	.148	.144	.121	.105	.084	.071	.062	.027	.023	.021
.126	.061	.017	.121	.059	.017	.111	.055	.016	.102	.051	.015	.093	.047	.014	.237	.195	.166	.204	.169	.145	.142	.119	.103	.083	.070	.061	.027	.023	.020
.119	.057	.016	.114	.055	.015	.105	.051	.014	.096	.048	.014	.088	.044	.013	.233	.192	.164	.201	.167	.143	.139	.117	.102	.081	.069	.061	.026	.023	.020
															.190	.190	.190	.163	.163	.163	.111	.111	.111	.064	.064	.064	.020	.020	.020
.220	.125	.040	.213	.122	.039	.200	.115	.037	.189	.109	.035	.178	.103	.033	.174	.157	.141	.149	.135	.122	.102	.093	.084	.059	.054	.049	.019	.017	.016
.212	.116	.036	.206	.114	.035	.195	.109	.034	.185	.104	.033	.176	.099	.031	.161	.132	.107	.138	.114	.093	.095	.079	.065	.055	.046	.038	.018	.015	.012
.202	.107	.032	.197	.105	.032	.187	.101	.031	.178	.098	.030	.170	.094	.029	.151	.114	.084	.130	.098	.073	.089	.068	.051	.051	.040	.030	.017	.013	.010
.191	.099	.029	.186	.098	.029	.178	.094	.028	.170	.091	.028	.163	.089	.027	.142	.100	.067	.122	.086	.058	.084	.060	.041	.049	.035	.024	.016	.011	.008
.180	.092	.027	.176	.091	.026	.169	.088	.026	.162	.086	.025	.156	.083	.025	.135	.088	.055	.116	.077	.047	.080	.053	.033	.046	.031	.020	.015	.010	.007
.171	.086	.024	.167	.084	.024	.161	.082	.024	.155	.080	.024	.149	.078	.023	.128	.080	.045	.110	.069	.039	.076	.048	.028	.044	.028	.017	.014	.009	.005
.162	.080	.023	.158	.079	.022	.153	.077	.022	.147	.075	.022	.142	.074	.022	.121	.072	.038	.104	.063	.033	.072	.044	.024	.042	.026	.014	.014	.008	.005
.153	.075	.021	.150	.074	.021	.145	.073	.021	.140	.071	.021	.136	.070	.020	.115	.066	.033	.099	.058	.029	.069	.040	.020	.040	.024	.012	.013	.008	.004
.145	.070	.020	.143	.070	.020	.138	.068	.019	.134	.067	.019	.130	.066	.019	.110	.061	.029	.095	.053	.025	.066	.037	.018	.038	.022	.011	.012	.007	.003
.138	.066	.018	.136	.066	.018	.132	.065	.018	.128	.064	.018	.124	.062	.018	.105	.057	.025	.091	.050	.022	.063	.035	.016	.037	.021	.009	.012	.007	.003
															.162	.162	.162	.138	.138	.138	.094	.094	.094	.054	.054	.054	.017	.017	.017
.139	.079	.025	.133	.076	.024	.123	.070	.022	.113	.065	.021	.104	.060	.019	.144	.133	.124	.123	.115	.106	.084	.079	.074	.049	.046	.043	.016	.015	.014
.132	.072	.022	.127	.070	.022	.119	.066	.020	.110	.062	.019	.103	.058	.018	.131	.112	.097	.112	.097	.084	.077	.067	.059	.044	.039	.034	.014	.013	.011
.126	.067	.020	.122	.065	.020	.114	.062	.019	.107	.059	.018	.101	.056	.017	.120	.096	.078	.102	.083	.067	.070	.058	.047	.041	.034	.028	.013	.011	.009
.119	.062	.018	.116	.061	.018	.110	.058	.017	.104	.056	.017	.098	.053	.016	.110	.084	.063	.095	.072	.055	.065	.050	.039	.038	.029	.023	.012	.010	.008
.114	.058	.017	.111	.057	.017	.105	.055	.016	.100	.053	.016	.095	.051	.015	.103	.074	.052	.088	.064	.045	.061	.044	.032	.035	.026	.019	.011	.009	.006
.109	.055	.016	.106	.054	.015	.101	.052	.015	.097	.050	.015	.093	.049	.014	.096	.066	.044	.083	.057	.038	.057	.040	.027	.033	.023	.016	.011	.008	.005
.104	.052	.015	.102	.051	.014	.098	.049	.014	.094	.048	.014	.090	.047	.014	.091	.059	.037	.078	.051	.032	.054	.036	.023	.031	.021	.014	.010	.007	.004
.100	.049	.014	.098	.048	.014	.094	.047	.013	.090	.046	.013	.087	.045	.013	.086	.054	.032	.074	.047	.028	.051	.033	.020	.030	.019	.012	.010	.006	.004
.096	.046	.013	.094	.046	.013	.091	.045	.013	.087	.044	.013	.084	.043	.012	.081	.049	.027	.070	.043	.024	.049	.030	.017	.028	.018	.010	.009	.006	.003
.092	.044	.012	.091	.044	.012	.088	.043	.012	.085	.042	.012	.082	.041	.012	.077	.045	.024	.067	.039	.021	.046	.028	.015	.027	.016	.009	.009	.005	.003

Fig. 9-62. *Continued (see page 9-53 for instructions and notes)*

Typical Luminaire	Typical Intensity Distribution and Per Cent Lamp Lumens	Maint. Cat.	SC	ρ_{CC} →	80			70			50			30			10			0	WDRC	ρ_{CC} →
				ρ_W →	50	30	10	50	30	10	50	30	10	50	30	10	50	30	10	0		ρ_W →
				RCR ↓	Coefficients of Utilization for 20 Per Cent Effective Floor Cavity Reflectance (ρ_{FC} = 20)																	RCR ↓
7 — EAR-38 lamp above 51 mm (2'') diameter aperture (increase efficiency to 54½% for 76 mm (3'') diameter aperture)*	0% ↑ 43½% ↓	IV	0.7	0	.52	.52	.52	.51	.51	.51	.48	.48	.48	.46	.46	.46	.45	.45	.45	.44		
				1	.49	.48	.47	.48	.47	.46	.46	.45	.45	.44	.44	.43	.43	.43	.42	.41	.055	1
				2	.46	.44	.43	.45	.44	.43	.44	.43	.42	.43	.42	.41	.41	.41	.40	.39	.054	2
				3	.43	.41	.40	.43	.41	.40	.42	.40	.39	.41	.39	.38	.40	.39	.38	.37	.053	3
				4	.41	.39	.37	.41	.39	.37	.40	.38	.37	.39	.37	.36	.38	.37	.36	.35	.052	4
				5	.39	.37	.35	.39	.37	.35	.38	.36	.35	.37	.36	.34	.36	.35	.34	.34	.051	5
				6	.37	.35	.33	.37	.35	.33	.36	.34	.33	.35	.34	.33	.35	.34	.32	.32	.049	6
				7	.35	.33	.31	.35	.33	.31	.34	.33	.31	.34	.32	.31	.33	.32	.31	.30	.048	7
				8	.34	.31	.30	.33	.31	.30	.33	.31	.30	.32	.31	.29	.32	.31	.29	.29	.046	8
				9	.32	.30	.28	.32	.30	.28	.31	.30	.28	.31	.29	.28	.31	.29	.28	.28	.045	9
				10	.31	.28	.27	.31	.28	.27	.30	.28	.27	.30	.28	.27	.30	.28	.27	.26	.043	10
8 — Medium distribution unit with lens plate and inside frost lamp	0% ↑ 54½% ↓	V	1.0	0	.65	.65	.65	.63	.63	.63	.60	.60	.60	.58	.58	.58	.55	.55	.55	.54		
				1	.59	.57	.56	.58	.56	.55	.56	.54	.53	.53	.52	.52	.52	.51	.50	.49	.133	1
				2	.54	.51	.49	.53	.50	.48	.51	.49	.47	.49	.47	.46	.48	.46	.45	.44	.130	2
				3	.49	.46	.43	.48	.45	.43	.47	.44	.42	.45	.43	.41	.44	.42	.41	.40	.123	3
				4	.45	.41	.38	.44	.41	.38	.43	.40	.38	.42	.39	.37	.41	.39	.37	.36	.116	4
				5	.41	.37	.35	.41	.37	.34	.40	.36	.34	.39	.36	.34	.38	.35	.33	.32	.109	5
				6	.38	.34	.31	.38	.34	.31	.37	.33	.31	.36	.33	.31	.35	.33	.31	.30	.103	6
				7	.35	.31	.29	.35	.31	.29	.34	.31	.28	.33	.30	.28	.33	.30	.28	.27	.097	7
				8	.33	.29	.26	.32	.29	.26	.32	.28	.26	.31	.28	.26	.31	.28	.26	.25	.092	8
				9	.31	.27	.24	.30	.27	.24	.30	.26	.24	.29	.26	.24	.29	.26	.24	.23	.087	9
				10	.29	.25	.22	.28	.25	.22	.28	.25	.22	.27	.24	.22	.27	.24	.22	.21	.082	10
9 — Recessed baffled downlight, 140 mm (5 ½'') diameter aperture—150-PAR/FL lamp	0% ↑ 68½% ↓	IV	0.5	0	.82	.82	.82	.80	.80	.80	.76	.76	.76	.73	.73	.73	.70	.70	.70	.69		
				1	.78	.77	.75	.76	.75	.74	.74	.73	.72	.71	.70	.70	.69	.68	.68	.67	.051	1
				2	.74	.72	.71	.73	.71	.70	.71	.70	.68	.69	.68	.67	.67	.66	.66	.65	.050	2
				3	.71	.69	.67	.71	.68	.67	.69	.67	.66	.67	.66	.65	.66	.65	.64	.63	.049	3
				4	.69	.66	.64	.68	.66	.64	.67	.65	.63	.66	.64	.63	.64	.63	.62	.61	.048	4
				5	.67	.64	.62	.66	.63	.62	.65	.63	.61	.64	.62	.61	.63	.61	.60	.59	.047	5
				6	.64	.62	.60	.64	.61	.60	.63	.61	.59	.62	.60	.59	.61	.60	.59	.58	.045	6
				7	.63	.60	.58	.62	.60	.58	.61	.59	.57	.61	.59	.57	.60	.58	.57	.56	.044	7
				8	.61	.58	.56	.60	.58	.56	.60	.58	.56	.59	.57	.56	.59	.57	.56	.55	.043	8
				9	.59	.56	.55	.59	.56	.55	.58	.56	.54	.58	.56	.54	.57	.55	.54	.54	.042	9
				10	.58	.55	.53	.57	.55	.53	.57	.55	.53	.56	.54	.53	.56	.54	.53	.52	.041	10
10 — Recessed baffled downlight, 140 mm (5½'') diameter aperture—75ER30 lamp	0% ↑ 85% ↓	IV	0.5	0	1.01	1.01	1.01	.99	.99	.99	.95	.95	.95	.91	.91	.91	.87	.87	.87	.85		
				1	.96	.94	.93	.94	.93	.91	.91	.89	.88	.88	.87	.86	.85	.84	.83	.82	.085	1
				2	.91	.88	.86	.90	.87	.85	.87	.85	.83	.84	.83	.81	.82	.81	.80	.79	.084	2
				3	.87	.83	.81	.86	.83	.80	.83	.81	.79	.81	.79	.78	.80	.78	.77	.75	.082	3
				4	.83	.79	.76	.82	.79	.76	.80	.77	.75	.79	.76	.74	.77	.75	.73	.72	.080	4
				5	.79	.76	.73	.79	.75	.72	.77	.74	.72	.76	.73	.71	.75	.72	.71	.70	.078	5
				6	.76	.72	.70	.76	.72	.69	.74	.71	.69	.73	.71	.68	.72	.70	.68	.67	.076	6
				7	.73	.69	.67	.73	.69	.67	.72	.69	.66	.71	.68	.66	.70	.68	.66	.65	.073	7
				8	.71	.67	.64	.70	.67	.64	.69	.66	.64	.69	.66	.64	.68	.65	.63	.62	.071	8
				9	.68	.64	.62	.68	.64	.62	.67	.64	.62	.66	.63	.61	.66	.63	.61	.60	.069	9
				10	.66	.62	.60	.66	.62	.60	.65	.62	.59	.64	.61	.59	.64	.61	.59	.58	.067	10
11 — Wide distribution unit with lens plate and inside frost lamp	0% ↑ 53½% ↓	V	1.4	0	.63	.63	.63	.62	.62	.62	.59	.59	.59	.57	.57	.57	.54	.54	.54	.53		
				1	.57	.55	.54	.56	.54	.53	.54	.52	.51	.52	.51	.50	.50	.49	.48	.47	.153	1
				2	.51	.48	.46	.50	.48	.45	.48	.46	.44	.47	.45	.43	.45	.44	.42	.41	.150	2
				3	.46	.42	.40	.45	.42	.39	.44	.41	.39	.42	.40	.38	.41	.39	.37	.36	.142	3
				4	.42	.38	.35	.41	.37	.34	.40	.36	.34	.39	.36	.33	.37	.35	.33	.32	.133	4
				5	.38	.34	.30	.37	.33	.30	.36	.33	.30	.35	.32	.30	.34	.32	.29	.28	.124	5
				6	.34	.30	.27	.34	.30	.27	.33	.29	.27	.32	.29	.27	.31	.28	.26	.25	.117	6
				7	.31	.27	.24	.31	.27	.24	.30	.27	.24	.29	.26	.24	.29	.26	.24	.23	.109	7
				8	.29	.25	.22	.28	.24	.22	.28	.24	.22	.27	.24	.22	.27	.24	.21	.20	.103	8
				9	.26	.22	.20	.26	.22	.20	.26	.22	.20	.25	.22	.20	.25	.22	.19	.19	.097	9
				10	.24	.21	.18	.24	.20	.18	.24	.20	.18	.23	.20	.18	.23	.20	.18	.17	.091	10
12 — Recessed unit with dropped diffusing glass	1½% ↑ 50½% ↓	V	1.3	0	.62	.62	.62	.60	.60	.60	.57	.57	.57	.54	.54	.54	.52	.52	.52	.51		
				1	.52	.50	.48	.51	.49	.47	.49	.47	.45	.46	.45	.43	.44	.43	.42	.40	.256	1
				2	.45	.41	.38	.44	.40	.37	.42	.39	.36	.40	.37	.35	.38	.36	.34	.33	.222	2
				3	.39	.35	.31	.38	.34	.31	.37	.33	.30	.35	.32	.29	.33	.31	.28	.27	.195	3
				4	.35	.30	.26	.34	.29	.26	.32	.28	.25	.31	.27	.25	.29	.27	.24	.23	.173	4
				5	.31	.26	.22	.30	.25	.22	.29	.25	.22	.28	.24	.21	.26	.23	.21	.20	.154	5
				6	.28	.23	.19	.27	.22	.19	.26	.22	.19	.25	.21	.19	.24	.21	.18	.17	.139	6
				7	.25	.20	.17	.25	.20	.17	.24	.20	.17	.23	.19	.16	.22	.19	.16	.15	.127	7
				8	.23	.18	.15	.22	.18	.15	.22	.18	.15	.21	.17	.15	.20	.17	.14	.13	.116	8
				9	.21	.16	.14	.21	.16	.13	.20	.16	.13	.19	.16	.13	.18	.15	.13	.12	.107	9
				10	.19	.15	.12	.19	.15	.12	.18	.15	.12	.18	.14	.12	.17	.14	.12	.11	.099	10

* Also, reflector downlight with baffles and inside frosted lamp.

Wall Exitance Coefficients for 20 Per Cent Effective Floor Cavity Reflectance (ρ_{FC} = 20)

80			70			50			30			10		
50	30	10	50	30	10	50	30	10	50	30	10	50	30	10
.069	.039	.012	.066	.038	.012	.061	.035	.011	.056	.032	.010	.051	.030	.010
.065	.036	.011	.063	.035	.011	.058	.032	.010	.054	.030	.010	.050	.028	.009
.062	.033	.010	.060	.032	.010	.056	.030	.009	.052	.029	.009	.049	.027	.008
.059	.031	.009	.057	.030	.009	.054	.028	.009	.051	.027	.008	.048	.026	.008
.056	.029	.008	.054	.028	.008	.052	.027	.008	.049	.026	.008	.046	.025	.007
.053	.027	.008	.052	.026	.008	.050	.025	.007	.047	.025	.007	.045	.024	.007
.051	.025	.007	.050	.025	.007	.048	.024	.007	.046	.023	.007	.044	.023	.007
.049	.024	.007	.048	.024	.007	.046	.023	.007	.044	.022	.006	.043	.022	.006
.047	.023	.006	.046	.023	.006	.044	.022	.006	.043	.021	.006	.041	.021	.006
.045	.022	.006	.045	.022	.006	.043	.021	.006	.041	.021	.006	.040	.020	.006
.121	.069	.022	.117	.067	.021	.110	.063	.020	.104	.060	.019	.098	.057	.018
.116	.063	.019	.113	.062	.019	.107	.059	.018	.101	.057	.018	.096	.054	.017
.110	.058	.017	.107	.057	.017	.102	.055	.017	.097	.053	.016	.092	.051	.016
.103	.054	.016	.101	.053	.016	.097	.051	.015	.092	.050	.015	.088	.048	.015
.098	.050	.014	.096	.049	.014	.092	.048	.014	.088	.046	.014	.084	.045	.014
.092	.046	.013	.091	.046	.013	.087	.045	.013	.084	.043	.013	.081	.042	.013
.088	.043	.012	.086	.043	.012	.083	.042	.012	.080	.041	.012	.077	.040	.012
.083	.041	.011	.082	.040	.011	.079	.039	.011	.076	.039	.011	.074	.038	.011
.079	.038	.011	.078	.038	.011	.075	.037	.011	.073	.036	.010	.070	.036	.010
.075	.036	.010	.074	.036	.010	.072	.035	.010	.069	.035	.010	.067	.034	.010
.090	.051	.016	.086	.049	.016	.077	.044	.014	.069	.040	.013	.062	.036	.012
.083	.046	.014	.079	.044	.013	.072	.040	.013	.066	.037	.012	.060	.034	.011
.077	.041	.012	.074	.040	.012	.068	.037	.011	.063	.034	.011	.058	.032	.010
.072	.038	.011	.070	.036	.011	.065	.034	.010	.060	.032	.010	.056	.030	.009
.068	.035	.010	.066	.034	.010	.062	.032	.009	.058	.030	.009	.054	.029	.009
.064	.032	.009	.062	.031	.009	.059	.030	.009	.055	.029	.008	.052	.028	.008
.061	.030	.009	.059	.030	.008	.056	.028	.008	.053	.027	.008	.051	.026	.008
.058	.028	.008	.057	.028	.008	.054	.027	.008	.052	.026	.008	.049	.025	.007
.056	.027	.008	.054	.026	.007	.052	.026	.007	.050	.025	.007	.048	.024	.007
.053	.026	.007	.052	.025	.007	.050	.025	.007	.048	.024	.007	.046	.023	.007
.123	.070	.022	.117	.067	.021	.107	.061	.020	.097	.056	.018	.088	.051	.016
.115	.063	.019	.110	.061	.019	.102	.057	.018	.094	.053	.016	.086	.049	.015
.108	.058	.017	.104	.056	.017	.097	.052	.016	.090	.049	.015	.084	.046	.014
.102	.053	.016	.099	.052	.015	.092	.049	.015	.087	.046	.014	.081	.044	.013
.096	.049	.014	.094	.048	.014	.088	.046	.014	.083	.044	.013	.079	.042	.013
.092	.046	.013	.089	.045	.013	.085	.043	.013	.080	.042	.012	.076	.040	.012
.087	.043	.012	.085	.042	.012	.081	.041	.012	.077	.040	.012	.074	.038	.011
.084	.041	.011	.082	.040	.011	.078	.039	.011	.075	.038	.011	.072	.037	.011
.080	.039	.011	.078	.038	.011	.075	.037	.011	.072	.036	.010	.069	.035	.010
.077	.037	.010	.075	.036	.010	.072	.036	.010	.070	.035	.010	.067	.034	.010
.131	.074	.024	.127	.072	.023	.120	.069	.022	.114	.066	.021	.108	.062	.020
.126	.069	.021	.123	.068	.021	.117	.065	.020	.111	.062	.020	.106	.060	.019
.119	.064	.019	.117	.062	.019	.111	.060	.018	.107	.058	.018	.102	.057	.017
.113	.059	.017	.110	.058	.017	.106	.056	.017	.101	.054	.016	.097	.053	.016
.106	.054	.016	.104	.053	.016	.100	.052	.015	.096	.051	.015	.093	.050	.015
.100	.050	.014	.098	.050	.014	.095	.049	.014	.091	.047	.014	.088	.046	.014
.095	.047	.013	.093	.046	.013	.090	.045	.013	.087	.044	.013	.084	.044	.013
.090	.044	.012	.088	.043	.012	.085	.043	.012	.082	.042	.012	.080	.041	.012
.085	.041	.011	.083	.041	.011	.081	.040	.011	.078	.039	.011	.076	.039	.011
.081	.039	.011	.079	.038	.011	.077	.038	.011	.075	.037	.011	.073	.037	.010
.187	.106	.034	.182	.104	.033	.174	.100	.032	.167	.096	.031	.160	.093	.030
.168	.092	.028	.164	.090	.028	.157	.087	.027	.150	.084	.026	.144	.082	.026
.151	.081	.024	.148	.079	.024	.142	.077	.023	.136	.074	.023	.130	.072	.022
.138	.072	.021	.135	.070	.021	.129	.068	.020	.124	.066	.020	.119	.065	.020
.126	.064	.019	.123	.063	.018	.118	.062	.018	.113	.060	.018	.109	.058	.018
.116	.058	.017	.113	.057	.016	.109	.056	.016	.105	.054	.016	.101	.053	.016
.107	.053	.015	.105	.052	.015	.101	.051	.015	.097	.050	.014	.094	.049	.014
.100	.049	.014	.098	.048	.014	.094	.047	.013	.091	.046	.013	.087	.045	.013
.093	.045	.013	.091	.044	.012	.088	.043	.012	.085	.043	.012	.082	.042	.012
.087	.042	.012	.086	.041	.012	.083	.040	.011	.080	.040	.011	.077	.039	.011

Ceiling Cavity Exitance Coefficients for 20 Per Cent Floor Cavity Reflectance (ρ_{FC} = 20)

80			70			50			30			10		
50	30	10	50	30	10	50	30	10	50	30	10	50	30	10
.083	.083	.083	.071	.071	.071	.048	.048	.048	.028	.028	.028	.009	.009	.009
.074	.069	.064	.063	.059	.055	.043	.041	.038	.025	.023	.022	.008	.008	.007
.067	.058	.050	.057	.050	.043	.039	.034	.030	.023	.020	.018	.007	.006	.006
.061	.050	.040	.052	.043	.035	.036	.030	.025	.021	.017	.014	.007	.006	.005
.056	.043	.033	.048	.037	.029	.033	.026	.020	.019	.015	.012	.006	.005	.004
.052	.038	.027	.045	.033	.024	.031	.023	.017	.018	.013	.010	.006	.004	.003
.049	.034	.023	.042	.029	.020	.029	.020	.014	.017	.012	.008	.005	.004	.003
.046	.030	.019	.039	.026	.017	.027	.018	.012	.016	.011	.007	.005	.004	.002
.043	.027	.017	.037	.024	.014	.026	.017	.010	.015	.010	.006	.005	.003	.002
.041	.025	.014	.035	.022	.013	.024	.015	.009	.014	.009	.005	.005	.003	.002
.039	.023	.013	.034	.020	.011	.023	.014	.008	.014	.008	.005	.004	.003	.002
.104	.104	.104	.088	.088	.088	.060	.060	.060	.035	.035	.035	.011	.011	.011
.095	.085	.077	.081	.073	.066	.055	.050	.046	.032	.029	.027	.010	.009	.009
.088	.072	.058	.075	.062	.051	.052	.043	.035	.030	.025	.021	.010	.008	.007
.082	.062	.046	.070	.053	.040	.048	.037	.028	.028	.022	.016	.009	.007	.005
.077	.054	.036	.066	.047	.032	.046	.033	.022	.026	.019	.013	.008	.006	.004
.073	.048	.030	.063	.042	.026	.043	.029	.018	.025	.017	.011	.008	.006	.004
.069	.043	.025	.060	.038	.021	.041	.026	.015	.024	.015	.009	.008	.005	.003
.066	.039	.021	.057	.034	.018	.039	.024	.013	.023	.014	.008	.007	.005	.003
.063	.036	.018	.054	.031	.016	.037	.022	.011	.022	.013	.007	.007	.004	.002
.060	.033	.016	.052	.029	.014	.036	.020	.010	.021	.012	.006	.007	.004	.002
.057	.031	.014	.049	.027	.012	.034	.019	.008	.020	.011	.005	.006	.004	.002
.131	.131	.131	.112	.112	.112	.076	.076	.076	.044	.044	.044	.014	.014	.014
.115	.108	.102	.099	.093	.088	.068	.064	.061	.039	.037	.035	.012	.012	.011
.103	.091	.082	.088	.078	.070	.060	.054	.049	.035	.032	.029	.011	.010	.009
.092	.078	.066	.079	.067	.058	.054	.047	.040	.031	.027	.024	.010	.009	.008
.083	.067	.055	.072	.058	.048	.049	.040	.034	.028	.024	.020	.009	.008	.007
.076	.059	.046	.065	.051	.040	.045	.035	.028	.026	.021	.017	.008	.007	.006
.070	.052	.039	.060	.045	.034	.041	.031	.024	.024	.018	.014	.008	.006	.005
.064	.046	.033	.055	.040	.029	.038	.028	.020	.022	.016	.012	.007	.005	.004
.060	.041	.029	.052	.036	.025	.036	.025	.018	.021	.015	.011	.007	.005	.003
.056	.037	.025	.048	.032	.022	.033	.023	.015	.019	.013	.009	.006	.004	.003
.053	.034	.022	.045	.030	.019	.032	.021	.014	.018	.012	.008	.006	.004	.003
.162	.162	.162	.139	.139	.139	.095	.095	.095	.054	.054	.054	.017	.017	.017
.144	.134	.126	.123	.115	.108	.084	.079	.075	.048	.046	.043	.016	.015	.014
.129	.113	.100	.110	.097	.086	.076	.067	.060	.044	.039	.035	.014	.013	.011
.117	.097	.081	.100	.083	.070	.069	.058	.049	.040	.034	.029	.013	.011	.009
.106	.084	.066	.091	.072	.057	.063	.050	.040	.036	.029	.024	.012	.010	.008
.098	.073	.055	.084	.063	.048	.058	.044	.034	.034	.026	.020	.011	.008	.007
.091	.065	.046	.078	.056	.040	.054	.039	.029	.031	.023	.017	.010	.008	.006
.084	.058	.039	.073	.050	.034	.050	.035	.024	.029	.021	.015	.009	.007	.005
.079	.052	.034	.068	.045	.030	.047	.032	.021	.027	.019	.013	.009	.006	.004
.074	.048	.030	.064	.041	.026	.044	.029	.018	.026	.017	.011	.008	.006	.004
.070	.044	.026	.061	.038	.023	.042	.027	.016	.025	.016	.010	.008	.005	.003
.101	.101	.101	.087	.087	.087	.059	.059	.059	.034	.034	.034	.011	.011	.011
.094	.084	.074	.080	.072	.064	.055	.049	.044	.032	.029	.026	.010	.009	.008
.088	.070	.056	.075	.061	.048	.052	.042	.034	.030	.024	.020	.010	.008	.006
.083	.061	.043	.071	.052	.037	.049	.036	.026	.028	.021	.015	.009	.007	.005
.079	.053	.034	.067	.046	.030	.046	.032	.021	.027	.019	.012	.009	.006	.004
.075	.048	.028	.064	.041	.024	.044	.029	.017	.026	.017	.010	.008	.006	.003
.071	.043	.023	.061	.037	.020	.042	.026	.014	.024	.015	.008	.008	.005	.003
.068	.039	.019	.058	.034	.017	.040	.024	.012	.023	.014	.007	.008	.005	.002
.065	.036	.016	.056	.031	.014	.039	.022	.010	.022	.013	.006	.007	.004	.002
.062	.034	.014	.053	.029	.013	.037	.020	.009	.022	.012	.005	.007	.004	.002
.059	.031	.013	.051	.027	.011	.035	.019	.008	.021	.011	.005	.007	.004	.002
.112	.112	.112	.095	.095	.095	.065	.065	.065	.037	.037	.037	.012	.012	.012
.108	.094	.080	.092	.080	.069	.063	.055	.048	.036	.032	.028	.012	.010	.009
.104	.081	.062	.089	.070	.053	.061	.048	.037	.035	.028	.022	.011	.009	.007
.100	.072	.050	.086	.062	.043	.059	.043	.030	.034	.025	.018	.011	.008	.006
.096	.065	.042	.083	.057	.036	.057	.039	.026	.033	.023	.015	.011	.008	.005
.092	.060	.036	.079	.052	.032	.055	.036	.022	.032	.021	.013	.010	.007	.004
.088	.056	.032	.076	.048	.028	.052	.034	.020	.030	.020	.012	.010	.007	.004
.085	.052	.030	.073	.045	.026	.050	.032	.018	.029	.019	.011	.009	.006	.004
.081	.049	.027	.070	.043	.024	.048	.030	.017	.028	.018	.010	.009	.006	.003
.078	.047	.026	.067	.040	.022	.046	.028	.016	.027	.017	.009	.009	.006	.003
.075	.044	.024	.064	.039	.021	.045	.027	.015	.026	.016	.009	.008	.005	.003

Fig. 9–62. *Continued (see page 9-53 for instructions and notes)*

Typical Luminaire	Typical Intensity Distribution and Per Cent Lamp Lumens		ρ_{CC} →	80			70			50			30			10			0		ρ_{CC} →
			ρ_W →	50	30	10	50	30	10	50	30	10	50	30	10	50	30	10	0	WDRC	ρ_W →
	Maint. Cat.	SC	RCR ↓	Coefficients of Utilization for 20 Per Cent Effective Floor Cavity Reflectance (ρ_{FC} = 20)																	RCR ↓
13	V	N.A.	0	.87	.87	87	.85	.85	.85	.80	.80	.80	.76	.76	.76	.73	.73	.73	.71		
			1	.75	.72	.69	.73	.70	.68	.70	.67	.65	.66	.64	.63	.63	.62	.60	.59	.312	1
	2½%↑		2	.66	.60	.56	.64	.59	.55	.61	.57	.54	.58	.55	.52	.56	.53	.51	.49	.279	2
			3	.58	.51	.47	.56	.51	.46	.54	.49	.45	.51	.47	.44	.49	.46	.43	.41	.251	3
			4	.51	.44	.39	.50	.44	.39	.48	.42	.38	.46	.41	.37	.44	.40	.37	.35	.226	4
	71%↓		5	.45	.39	.34	.44	.38	.33	.42	.37	.33	.41	.36	.32	.39	.35	.32	.30	.206	5
			6	.41	.34	.29	.40	.33	.29	.38	.33	.28	.37	.32	.28	.35	.31	.28	.26	.188	6
			7	.37	.30	.26	.36	.30	.25	.35	.29	.25	.33	.28	.25	.32	.28	.24	.23	.173	7
			8	.33	.27	.23	.33	.27	.22	.31	.26	.22	.30	.25	.22	.29	.25	.22	.20	.159	8
Bilateral batwing distribution—clear HID			9	.30	.24	.20	.30	.24	.20	.29	.23	.20	.28	.23	.19	.27	.22	.19	.18	.148	9
with dropped prismatic lens			10	.28	.22	.18	.27	.22	.18	.26	.21	.18	.26	.21	.17	.25	.20	.17	.16	.138	10
14	V	1.3	0	.78	.78	.78	.77	.77	.77	.73	.73	.73	.70	.70	.70	.67	.67	.67	.66		
			1	.71	.69	.67	.69	.67	.65	.67	.65	.63	.64	.63	.61	.62	.61	.60	.58	.188	1
	0%↑		2	.64	.60	.57	.62	.59	.56	.60	.57	.55	.58	.56	.54	.56	.54	.53	.51	.183	2
			3	.57	.53	.49	.56	.52	.49	.54	.51	.48	.53	.50	.47	.51	.49	.46	.45	.173	3
			4	.52	.47	.43	.51	.46	.43	.49	.46	.42	.48	.45	.42	.47	.44	.41	.40	.161	4
			5	.47	.42	.38	.46	.42	.38	.45	.41	.38	.44	.40	.37	.43	.40	.37	.36	.151	5
	66%↓		6	.43	.38	.34	.42	.38	.34	.41	.37	.34	.40	.36	.34	.39	.36	.33	.32	.141	6
			7	.39	.34	.31	.39	.34	.31	.38	.34	.30	.37	.33	.30	.36	.33	.30	.29	.132	7
			8	.36	.31	.28	.36	.31	.28	.35	.31	.28	.34	.30	.27	.34	.30	.27	.26	.124	8
Clear HID lamp and glass refractor above			9	.34	.29	.25	.33	.28	.25	.32	.28	.25	.32	.28	.25	.31	.28	.25	.24	.117	9
plastic lens panel			10	.31	.26	.23	.31	.26	.23	.30	.26	.23	.30	.26	.23	.29	.25	.23	.22	.110	10
15	V	1.4	0	.85	.85	.85	.83	.83	.83	.80	.80	.80	.76	.76	.76	.73	.73	.73	.72		
			1	.77	.75	.73	.76	.74	.72	.73	.71	.69	.70	.69	.67	.67	.66	.65	.64	.189	1
	0%↑		2	.70	.66	.63	.68	.65	.62	.66	.63	.60	.64	.61	.59	.61	.60	.58	.56	.190	2
			3	.63	.58	.54	.62	.57	.54	.60	.56	.53	.58	.54	.52	.56	.53	.51	.50	.183	3
			4	.56	.51	.47	.56	.51	.47	.54	.50	.46	.52	.49	.46	.51	.48	.45	.44	.174	4
	71½%↓		5	.51	.46	.42	.50	.45	.41	.49	.44	.41	.48	.44	.40	.46	.43	.40	.39	.164	5
			6	.46	.41	.37	.46	.41	.37	.45	.40	.36	.43	.39	.36	.42	.39	.36	.34	.155	6
			7	.42	.37	.33	.42	.37	.33	.41	.36	.33	.40	.36	.32	.39	.35	.32	.31	.146	7
			8	.39	.33	.30	.38	.33	.29	.37	.33	.29	.37	.32	.29	.36	.32	.29	.28	.137	8
Enclosed reflector with an incandescent			9	.36	.30	.27	.35	.30	.27	.35	.30	.27	.34	.30	.26	.33	.29	.26	.25	.129	9
lamp			10	.33	.28	.24	.33	.28	.24	.32	.27	.24	.31	.27	.24	.31	.27	.24	.23	.122	10
16	III	0.7	0	.93	.93	.93	.90	.90	.90	.86	.86	.86	.82	.82	.82	.78	.78	.78	.77		
	1½%↑		1	.86	.84	.82	.84	.82	.80	.80	.79	.78	.77	.76	.75	.74	.74	.73	.71	.138	1
			2	.79	.76	.73	.78	.75	.72	.75	.73	.71	.73	.71	.69	.70	.69	.67	.66	.136	2
			3	.74	.70	.66	.73	.69	.66	.70	.67	.65	.68	.66	.63	.66	.64	.62	.61	.132	3
			4	.69	.64	.61	.68	.64	.60	.66	.62	.60	.64	.61	.59	.63	.60	.58	.57	.126	4
	77%↓		5	.64	.60	.56	.63	.59	.56	.62	.58	.55	.60	.57	.55	.59	.56	.54	.53	.120	5
			6	.60	.55	.52	.60	.55	.52	.58	.54	.51	.57	.54	.51	.56	.53	.50	.49	.115	6
			7	.57	.52	.49	.56	.52	.48	.55	.51	.48	.54	.50	.48	.53	.50	.47	.46	.109	7
			8	.53	.49	.45	.53	.48	.45	.52	.48	.45	.51	.47	.45	.50	.47	.44	.43	.104	8
"High bay" narrow distribution ventilated			9	.51	.46	.43	.50	.46	.43	.49	.45	.42	.48	.45	.42	.48	.44	.42	.41	.100	9
reflector with clear HID lamp			10	.48	.43	.40	.48	.43	.40	.47	.43	.40	.46	.42	.40	.45	.42	.40	.39	.095	10
17	III	1.0	0	.91	.91	.91	.89	.89	.89	.85	.85	.85	.81	.81	.81	.78	.78	.78	.76		
			1	.83	.81	.79	.81	.79	.77	.78	.76	.75	.75	.74	.72	.72	.71	.70	.68	.187	1
	1%↑		2	.75	.71	.68	.74	.70	.67	.71	.68	.65	.68	.66	.64	.66	.64	.62	.61	.189	2
			3	.68	.63	.59	.67	.62	.59	.65	.61	.58	.62	.59	.57	.61	.58	.56	.54	.183	3
			4	.62	.56	.52	.61	.56	.52	.59	.54	.51	.57	.53	.50	.55	.52	.50	.48	.174	4
	76%↓		5	.56	.50	.46	.55	.50	.46	.54	.49	.45	.52	.48	.45	.51	.47	.44	.43	.165	5
			6	.51	.46	.41	.51	.45	.41	.49	.44	.41	.48	.44	.40	.47	.43	.40	.39	.155	6
			7	.47	.41	.37	.47	.41	.37	.45	.40	.37	.44	.40	.37	.43	.39	.36	.35	.147	7
			8	.43	.38	.34	.43	.37	.34	.42	.37	.33	.41	.36	.33	.40	.36	.33	.32	.138	8
"High bay" intermediate distribution ven-			9	.40	.35	.31	.40	.34	.31	.39	.34	.31	.38	.34	.30	.37	.33	.30	.29	.131	9
tilated reflector with clear HID lamp			10	.37	.32	.28	.37	.32	.28	.36	.31	.28	.35	.31	.28	.35	.31	.28	.27	.124	10
18	III	1.5	0	.93	.93	.93	.91	.91	.91	.87	.87	.87	.83	.83	.83	.79	.79	.79	.78		
			1	.84	.81	.79	.82	.80	.78	.79	.77	.75	.76	.74	.73	.73	.72	.70	.69	.217	1
	½%↑		2	.75	.71	.67	.74	.70	.66	.71	.68	.65	.68	.66	.63	.66	.64	.62	.60	.219	2
			3	.67	.62	.57	.66	.61	.57	.64	.59	.56	.61	.58	.55	.59	.56	.54	.52	.211	3
			4	.60	.54	.50	.59	.54	.49	.57	.52	.48	.55	.51	.48	.54	.50	.47	.46	.200	4
	77½%↓		5	.54	.48	.43	.53	.47	.43	.52	.46	.42	.50	.45	.42	.49	.45	.41	.40	.189	5
			6	.49	.42	.38	.48	.42	.38	.47	.41	.37	.45	.41	.37	.44	.40	.37	.35	.177	6
			7	.44	.38	.34	.44	.38	.33	.42	.37	.33	.41	.36	.33	.40	.36	.33	.31	.166	7
			8	.40	.34	.30	.40	.34	.30	.39	.33	.30	.38	.33	.29	.37	.32	.29	.28	.156	8
"High bay" wide distribution ventilated			9	.37	.31	.27	.37	.31	.27	.36	.30	.27	.35	.30	.26	.34	.29	.26	.25	.146	9
reflector with clear HID lamp			10	.34	.28	.24	.34	.28	.24	.33	.28	.24	.32	.27	.24	.31	.27	.24	.22	.138	10

80			70			50			30			10			80			70			50			30			10		
50	30	10	50	30	10	50	30	10	50	30	10	50	30	10	50	30	10	50	30	10	50	30	10	50	30	10	50	30	10
Wall Exitance Coefficients for 20 Per Cent Effective Floor Cavity Reflectance (ρ_{FC} = 20)															Ceiling Cavity Exitance Coefficients for 20 Per Cent Floor Cavity Reflectance (ρ_{FC} = 20)														
															.159	.159	.159	.136	.136	.136	.093	.093	.093	.053	.053	.053	.017	.017	.017
.238	.135	.043	.232	.132	.042	.220	.126	.040	.210	.121	.039	.201	.116	.038	.152	.134	.117	.130	.115	.101	.089	.079	.070	.051	.046	.041	.016	.015	.013
.218	.119	.037	.212	.117	.036	.202	.113	.035	.193	.108	.034	.185	.105	.033	.146	.116	.091	.125	.100	.079	.086	.069	.055	.050	.040	.032	.016	.013	.010
.200	.106	.032	.195	.105	.031	.186	.101	.031	.178	.098	.030	.171	.095	.029	.141	.103	.074	.121	.089	.064	.083	.062	.045	.048	.036	.026	.015	.012	.009
.184	.096	.028	.180	.094	.028	.172	.091	.027	.165	.089	.027	.158	.086	.026	.135	.094	.062	.116	.081	.054	.080	.056	.038	.046	.033	.022	.015	.011	.007
.170	.087	.025	.167	.086	.025	.160	.083	.024	.153	.081	.024	.147	.079	.024	.130	.086	.054	.111	.075	.047	.077	.052	.033	.044	.031	.020	.014	.010	.006
.158	.079	.023	.155	.078	.023	.149	.076	.022	.143	.074	.022	.137	.072	.021	.125	.080	.048	.107	.069	.042	.074	.049	.030	.043	.029	.018	.014	.009	.006
.147	.073	.021	.144	.072	.021	.139	.070	.020	.133	.068	.020	.128	.067	.020	.120	.075	.044	.103	.065	.038	.071	.046	.027	.041	.027	.016	.013	.009	.005
.138	.067	.019	.135	.067	.019	.130	.065	.019	.125	.063	.018	.121	.062	.018	.115	.071	.041	.099	.061	.035	.068	.043	.025	.040	.025	.015	.013	.008	.005
.129	.063	.017	.127	.062	.017	.122	.060	.017	.118	.059	.017	.114	.058	.017	.111	.067	.038	.095	.058	.033	.066	.041	.024	.038	.024	.014	.012	.008	.005
.122	.058	.016	.119	.058	.016	.115	.056	.016	.111	.055	.016	.107	.054	.015	.106	.064	.036	.092	.056	.031	.064	.039	.022	.037	.023	.013	.012	.008	.004
															.126	.126	.126	.107	.107	.107	.073	.073	.073	.042	.042	.042	.013	.013	.013
.161	.091	.029	.156	.089	.028	.148	.085	.027	.140	.081	.026	.133	.077	.025	.116	.103	.092	.099	.089	.079	.068	.061	.055	.039	.035	.032	.013	.011	.010
.154	.085	.026	.151	.083	.026	.143	.080	.025	.136	.077	.024	.130	.074	.023	.109	.087	.069	.093	.075	.060	.064	.052	.042	.037	.030	.024	.012	.010	.008
.146	.078	.023	.143	.076	.023	.136	.074	.022	.130	.071	.022	.125	.069	.021	.102	.075	.054	.088	.065	.046	.060	.045	.033	.035	.026	.019	.011	.009	.006
.138	.072	.021	.135	.070	.021	.129	.068	.020	.124	.066	.020	.119	.065	.020	.097	.066	.042	.083	.057	.037	.057	.040	.026	.033	.023	.015	.011	.008	.005
.130	.066	.019	.127	.065	.019	.122	.063	.019	.117	.062	.018	.113	.060	.018	.092	.059	.034	.079	.051	.030	.054	.036	.021	.032	.021	.013	.010	.007	.004
.122	.061	.018	.120	.061	.017	.115	.059	.017	.111	.058	.017	.107	.056	.017	.087	.053	.028	.075	.046	.025	.052	.032	.018	.030	.019	.010	.010	.006	.003
.115	.057	.016	.113	.056	.016	.109	.055	.016	.106	.054	.016	.102	.053	.016	.083	.048	.024	.072	.042	.021	.050	.029	.015	.029	.017	.009	.009	.006	.003
.109	.053	.015	.107	.053	.015	.104	.052	.015	.100	.051	.015	.097	.050	.015	.080	.045	.021	.068	.039	.018	.047	.027	.013	.028	.016	.008	.009	.005	.003
.103	.050	.014	.102	.050	.014	.098	.049	.014	.095	.048	.014	.093	.047	.014	.076	.041	.018	.065	.036	.016	.045	.025	.011	.026	.015	.007	.009	.005	.002
.098	.047	.013	.097	.047	.013	.094	.046	.013	.091	.045	.013	.088	.044	.013	.073	.039	.016	.063	.033	.014	.043	.024	.010	.025	.014	.006	.008	.005	.002
															.137	.137	.137	.117	.117	.117	.080	.080	.080	.046	.046	.046	.015	.015	.015
.167	.095	.030	.162	.092	.029	.152	.087	.028	.144	.083	.027	.136	.079	.025	.126	.113	.102	.108	.097	.087	.074	.067	.060	.043	.039	.035	.014	.012	.011
.163	.089	.027	.159	.088	.027	.151	.084	.026	.143	.080	.025	.137	.077	.024	.118	.096	.077	.101	.082	.066	.069	.057	.046	.040	.033	.027	.013	.011	.009
.157	.083	.025	.153	.082	.025	.146	.079	.024	.139	.076	.023	.133	.074	.023	.112	.083	.059	.096	.071	.051	.066	.049	.036	.038	.029	.021	.012	.009	.007
.149	.077	.023	.146	.076	.023	.139	.074	.022	.134	.072	.022	.128	.070	.021	.106	.073	.047	.091	.063	.041	.063	.044	.029	.036	.026	.017	.012	.008	.006
.141	.072	.021	.138	.071	.021	.133	.069	.020	.128	.067	.020	.123	.066	.020	.101	.065	.038	.087	.056	.033	.060	.039	.023	.035	.023	.014	.011	.008	.005
.134	.067	.019	.131	.066	.019	.126	.065	.019	.122	.063	.019	.117	.062	.018	.096	.059	.032	.083	.051	.028	.057	.036	.020	.033	.021	.012	.011	.007	.004
.127	.063	.018	.124	.062	.018	.120	.061	.017	.116	.059	.017	.112	.058	.017	.092	.054	.027	.079	.047	.023	.055	.033	.017	.032	.019	.010	.010	.006	.003
.120	.059	.016	.118	.058	.016	.114	.057	.016	.110	.056	.016	.107	.055	.016	.088	.049	.023	.076	.043	.020	.052	.030	.014	.030	.018	.009	.010	.006	.003
.114	.055	.015	.112	.055	.015	.108	.054	.015	.105	.053	.015	.102	.052	.015	.084	.046	.020	.072	.040	.018	.050	.028	.013	.029	.017	.007	.009	.005	.002
.108	.052	.014	.106	.051	.014	.103	.051	.014	.100	.050	.014	.097	.049	.014	.081	.043	.018	.069	.037	.016	.048	.026	.011	.028	.016	.007	.009	.005	.002
															.158	.158	.158	.135	.135	.135	.092	.092	.092	.053	.053	.053	.017	.017	.017
.147	.084	.026	.141	.081	.026	.131	.075	.024	.121	.070	.022	.112	.065	.021	.144	.133	.122	.123	.114	.105	.084	.078	.073	.049	.045	.042	.016	.015	.014
.140	.077	.024	.136	.075	.023	.127	.070	.022	.118	.066	.021	.111	.063	.020	.133	.113	.097	.114	.097	.084	.078	.067	.058	.045	.039	.034	.014	.013	.011
.133	.071	.021	.129	.069	.021	.121	.066	.020	.114	.063	.019	.107	.059	.018	.123	.099	.079	.106	.085	.068	.073	.059	.048	.042	.035	.028	.013	.011	.009
.126	.066	.019	.123	.064	.019	.116	.061	.018	.110	.059	.018	.104	.056	.017	.116	.087	.066	.099	.075	.057	.068	.053	.040	.039	.031	.024	.013	.010	.008
.120	.061	.018	.117	.060	.017	.111	.058	.017	.105	.055	.016	.100	.053	.016	.109	.078	.056	.094	.068	.049	.064	.047	.034	.037	.028	.020	.012	.009	.007
.114	.057	.016	.111	.056	.016	.106	.054	.016	.101	.052	.015	.096	.051	.015	.103	.071	.048	.089	.062	.042	.061	.043	.030	.035	.025	.018	.011	.008	.006
.108	.053	.015	.106	.053	.015	.101	.051	.015	.097	.050	.014	.092	.048	.014	.098	.065	.042	.084	.057	.037	.058	.040	.026	.034	.023	.016	.011	.008	.005
.103	.050	.014	.101	.050	.014	.097	.048	.014	.093	.047	.014	.089	.046	.013	.093	.060	.038	.080	.052	.033	.056	.037	.023	.032	.022	.014	.010	.007	.005
.098	.048	.013	.096	.047	.013	.093	.046	.013	.089	.045	.013	.086	.043	.013	.089	.056	.034	.077	.049	.030	.053	.034	.021	.031	.020	.013	.010	.007	.004
.094	.045	.013	.092	.045	.012	.089	.043	.012	.086	.042	.012	.082	.041	.012	.086	.053	.031	.074	.046	.027	.051	.032	.019	.030	.019	.012	.010	.006	.004
															.153	.153	.153	.131	.131	.131	.089	.089	.089	.051	.051	.051	.016	.016	.016
.171	.098	.031	.166	.095	.030	.156	.089	.029	.146	.084	.027	.137	.080	.026	.140	.127	.115	.120	.109	.099	.082	.075	.068	.047	.043	.040	.015	.014	.013
.168	.092	.028	.163	.090	.028	.154	.086	.027	.146	.082	.026	.138	.078	.025	.131	.108	.089	.113	.093	.077	.077	.064	.053	.044	.037	.031	.014	.012	.010
.161	.086	.026	.157	.084	.025	.149	.081	.025	.142	.078	.024	.135	.075	.023	.124	.094	.070	.106	.081	.061	.073	.056	.043	.042	.033	.025	.014	.011	.008
.153	.080	.023	.149	.078	.023	.143	.076	.023	.136	.073	.022	.130	.071	.022	.118	.083	.057	.101	.072	.050	.069	.050	.035	.040	.029	.021	.013	.010	.007
.145	.074	.021	.142	.073	.021	.136	.071	.021	.130	.069	.020	.125	.067	.020	.112	.075	.048	.096	.065	.041	.066	.045	.029	.038	.027	.017	.012	.009	.006
.138	.069	.020	.135	.068	.020	.129	.066	.019	.124	.065	.019	.120	.063	.019	.107	.068	.041	.092	.059	.035	.063	.041	.025	.037	.024	.015	.012	.008	.005
.131	.065	.018	.128	.064	.018	.123	.062	.018	.119	.061	.018	.114	.059	.017	.102	.063	.035	.088	.055	.031	.061	.038	.022	.035	.023	.013	.011	.007	.004
.124	.061	.017	.122	.060	.017	.117	.059	.017	.113	.057	.017	.109	.056	.016	.098	.058	.031	.084	.051	.027	.058	.036	.019	.034	.021	.012	.011	.007	.004
.118	.057	.016	.116	.056	.016	.112	.055	.016	.108	.054	.015	.104	.053	.015	.094	.055	.028	.081	.047	.024	.056	.033	.017	.033	.020	.010	.011	.006	.003
.112	.054	.015	.110	.053	.015	.106	.052	.015	.103	.051	.015	.100	.050	.014	.090	.051	.026	.078	.045	.022	.054	.031	.016	.031	.019	.009	.010	.006	.003
															.154	.154	.154	.132	.132	.132	.090	.090	.090	.052	.052	.052	.017	.017	.017
.188	.107	.034	.183	.104	.033	.172	.099	.032	.162	.094	.030	.154	.089	.029	.143	.128	.115	.122	.110	.099	.084	.076	.068	.048	.044	.040	.015	.014	.013
.186	.102	.031	.181	.100	.031	.172	.095	.030	.163	.092	.029	.155	.088	.028	.135	.109	.087	.115	.094	.076	.079	.065	.053	.046	.038	.031	.015	.012	.010
.178	.095	.028	.174	.093	.028	.166	.090	.027	.158	.087	.027	.152	.084	.026	.128	.095	.068	.110	.082	.059	.075	.057	.042	.043	.033	.024	.014	.011	.008
.170	.088	.026	.166	.087	.026	.159	.084	.025	.152	.082	.025	.146	.079	.024	.122	.084	.055	.105	.073	.048	.072	.051	.034	.042	.030	.020	.013	.010	.007
.161	.082	.024	.157	.081	.024	.151	.079	.023	.145	.076	.023	.139	.074	.022	.117	.076	.045	.100	.065	.039	.069	.046	.028	.040	.027	.017	.013	.009	.005
.152	.076	.022	.149	.075	.022	.143	.073	.021	.138	.072	.021	.133	.070	.021	.112	.069	.038	.096	.060	.033	.066	.042	.024	.038	.025	.014	.012	.008	.005
.143	.071	.020	.141	.070	.020	.136	.069	.020	.131	.067	.020	.126	.066	.019	.107	.064	.033	.092	.055	.029	.064	.039	.020	.037	.023	.012	.012	.007	.004
.136	.066	.019	.133	.066	.019	.129	.064	.018	.124	.063	.018	.120	.062	.018	.102	.059	.029	.088	.051	.026	.061	.036	.018	.036	.021	.011	.011	.007	.004
.128	.062	.017	.126	.062	.017	.122	.060	.017	.118	.059	.017	.1'4	.058	.017	.098	.055	.026	.085	.048	.023	.059	.034	.016	.034	.020	.010	.011	.007	.003
.122	.058	.016	.120	.058	.016	.116	.057	.016	.112	.056	.016	.109	.055	.016	.094	.052	.024	.081	.045	.021	.056	.032	.015	.033	.019	.009	.011	.006	.003

Fig. 9-62. *Continued (see page 9-53 for instructions and notes)*

Typical Luminaire	Typical Intensity Distribution and Per Cent Lamp Lumens: Maint. Cat.	SC	ρCC → / ρW → / RCR ↓	80: 50	80: 30	80: 10	70: 50	70: 30	70: 10	50: 50	50: 30	50: 10	30: 50	30: 30	30: 10	10: 50	10: 30	10: 10	0: 0	WDRC	ρCC → / ρW → / RCR ↓
				Coefficients of Utilization for 20 Per Cent Effective Floor Cavity Reflectance (ρ_{FC} = 20)																	
19	III	1.0	0	.96	.96	.96	.93	.93	.93	.88	.88	.88	.83	.83	.83	.78	.78	.78	.76		
	6½% ↑		1	.88	.86	.83	.86	.83	.81	.81	.79	.78	.77	.75	.74	.73	.72	.71	.69	.167	1
			2	.80	.76	.73	.78	.74	.71	.74	.71	.69	.71	.68	.66	.68	.66	.64	.62	.168	2
			3	.73	.68	.64	.71	.67	.63	.68	.64	.61	.65	.62	.60	.63	.60	.58	.56	.162	3
			4	.67	.61	.57	.65	.60	.57	.63	.59	.55	.60	.57	.54	.58	.55	.52	.51	.155	4
			5	.61	.56	.52	.60	.55	.51	.58	.53	.50	.56	.52	.49	.54	.50	.48	.46	.147	5
	75½% ↓		6	.57	.51	.47	.56	.50	.46	.54	.49	.45	.52	.48	.45	.50	.46	.44	.42	.139	6
			7	.52	.47	.43	.51	.46	.42	.50	.45	.42	.48	.44	.41	.47	.43	.40	.39	.132	7
"High bay" intermediate distribution ventilated reflector with phosphor coated HID lamp			8	.49	.43	.39	.48	.42	.39	.46	.42	.38	.45	.41	.38	.44	.40	.37	.36	.125	8
			9	.45	.40	.36	.45	.39	.36	.43	.39	.35	.42	.38	.35	.41	.37	.34	.33	.118	9
			10	.42	.37	.33	.42	.37	.33	.41	.36	.33	.39	.35	.32	.38	.35	.32	.31	.112	10
20	III	1.5	0	.93	.93	.93	.90	.90	.90	.83	.83	.83	.77	.77	.77	.72	.72	.72	.69		
	12% ↑		1	.85	.82	.80	.82	.79	.77	.76	.74	.73	.71	.70	.69	.66	.65	.65	.62	.168	1
			2	.76	.72	.69	.74	.70	.67	.69	.66	.64	.65	.63	.61	.61	.59	.58	.56	.168	2
			3	.69	.64	.60	.67	.62	.59	.63	.59	.56	.59	.56	.54	.56	.54	.51	.49	.163	3
			4	.62	.57	.52	.61	.55	.51	.57	.53	.50	.54	.51	.48	.51	.48	.46	.44	.156	4
	69% ↓		5	.57	.51	.46	.55	.50	.46	.52	.48	.44	.49	.46	.43	.47	.44	.41	.39	.148	5
			6	.52	.45	.41	.50	.45	.40	.48	.43	.39	.45	.41	.38	.43	.40	.37	.35	.141	6
			7	.47	.41	.37	.46	.40	.36	.44	.39	.35	.42	.37	.34	.40	.36	.33	.32	.133	7
"High bay" wide distribution ventilated reflector with phosphor coated HID lamp			8	.43	.37	.33	.42	.36	.33	.40	.35	.32	.38	.34	.31	.37	.33	.30	.29	.126	8
			9	.40	.34	.30	.39	.33	.29	.37	.32	.29	.35	.31	.28	.34	.30	.27	.26	.120	9
			10	.37	.31	.27	.36	.30	.27	.34	.29	.26	.33	.28	.25	.31	.28	.25	.23	.114	10
21	V	1.8	0	.82	.82	.82	.80	.80	.80	.76	.76	.76	.73	.73	.73	.70	.70	.70	.68		
			1	.73	.70	.68	.71	.69	.67	.68	.66	.64	.65	.64	.62	.63	.62	.61	.59	.231	1
	0% ↑		2	.64	.60	.56	.63	.59	.55	.60	.57	.54	.58	.55	.53	.56	.54	.52	.50	.227	2
			3	.56	.51	.47	.55	.51	.47	.53	.49	.46	.52	.48	.45	.50	.47	.44	.43	.213	3
		45°	4	.50	.44	.40	.49	.44	.40	.48	.43	.39	.46	.42	.39	.44	.41	.38	.37	.199	4
			5	.45	.39	.34	.44	.38	.34	.42	.38	.34	.41	.37	.33	.40	.36	.33	.32	.184	5
	68½% ↓		6	.40	.34	.30	.39	.34	.30	.38	.33	.29	.37	.33	.29	.36	.32	.29	.28	.171	6
		∥, ⊥	7	.36	.30	.26	.36	.30	.26	.35	.29	.26	.34	.29	.26	.33	.29	.25	.24	.159	7
"Low bay" rectangular pattern, lensed bottom reflector unit with clear HID lamp			8	.33	.27	.23	.32	.27	.23	.31	.26	.23	.31	.26	.23	.30	.26	.23	.21	.148	8
			9	.30	.24	.20	.29	.24	.20	.29	.24	.20	.28	.23	.20	.27	.23	.20	.19	.138	9
			10	.27	.22	.18	.27	.22	.18	.26	.22	.18	.26	.21	.18	.25	.21	.18	.17	.129	10
22	V	1.9	0	.83	.83	.83	.81	.81	.81	.77	.77	.77	.73	.73	.73	.70	.70	.70	.68		
			1	.72	.69	.66	.70	.67	.65	.67	.64	.62	.63	.62	.60	.60	.59	.57	.56	.302	1
	3% ↑		2	.62	.57	.53	.61	.56	.52	.58	.54	.50	.55	.52	.49	.52	.50	.47	.46	.279	2
			3	.54	.48	.43	.53	.47	.43	.50	.45	.41	.48	.44	.40	.46	.42	.39	.38	.253	3
			4	.47	.41	.36	.46	.40	.35	.44	.39	.35	.42	.37	.34	.40	.36	.33	.31	.229	4
			5	.42	.35	.30	.41	.34	.30	.39	.33	.29	.37	.32	.29	.36	.31	.28	.26	.208	5
	68% ↓		6	.37	.30	.26	.36	.30	.25	.35	.29	.25	.33	.28	.25	.32	.27	.24	.23	.189	6
			7	.33	.27	.22	.33	.26	.22	.31	.26	.22	.30	.25	.21	.29	.24	.21	.19	.173	7
			8	.30	.24	.19	.29	.23	.19	.28	.23	.19	.27	.22	.19	.26	.22	.18	.17	.159	8
"Low bay" lensed bottom reflector unit with clear HID lamp			9	.27	.21	.17	.27	.21	.17	.26	.20	.17	.25	.20	.17	.24	.19	.16	.15	.147	9
			10	.25	.19	.15	.24	.19	.15	.24	.18	.15	.23	.18	.15	.22	.18	.15	.13	.137	10
23	IV	1.7	0	.67	.67	.67	.65	.65	.65	.62	.62	.62	.60	.60	.60	.57	.57	.57	.56		
			1	.60	.58	.56	.58	.57	.55	.56	.55	.53	.54	.53	.52	.52	.51	.50	.49	.177	1
			2	.53	.49	.46	.52	.48	.46	.50	.47	.45	.48	.46	.44	.46	.44	.43	.42	.179	2
	0% ↑		3	.46	.42	.39	.46	.42	.38	.44	.41	.38	.42	.40	.37	.41	.39	.37	.35	.172	3
			4	.41	.36	.33	.40	.36	.33	.39	.35	.32	.38	.34	.32	.37	.34	.31	.30	.161	4
			5	.37	.32	.28	.36	.31	.28	.35	.31	.28	.34	.30	.27	.33	.30	.27	.26	.150	5
	56% ↓		6	.33	.28	.24	.32	.28	.24	.31	.27	.24	.30	.27	.24	.30	.26	.24	.23	.139	6
			7	.30	.25	.21	.29	.25	.21	.28	.24	.21	.28	.24	.21	.27	.23	.21	.20	.129	7
Wide spread, recessed, small open bottom reflector with low wattage diffuse HID lamp			8	.27	.22	.19	.26	.22	.19	.26	.22	.19	.25	.21	.19	.24	.21	.19	.17	.120	8
			9	.25	.20	.17	.24	.20	.17	.24	.20	.17	.23	.19	.17	.22	.19	.17	.16	.112	9
			10	.22	.18	.15	.22	.18	.15	.22	.18	.15	.21	.18	.15	.21	.17	.15	.14	.105	10
24	VI	N.A.	0	.74	.74	.74	.63	.63	.63	.43	.43	.43	.25	.25	.25	.08	.08	.08	.00		
			1	.64	.62	.59	.55	.53	.51	.38	.36	.35	.22	.21	.20	.07	.07	.07	.00	.000	1
			2	.56	.52	.48	.48	.45	.42	.33	.31	.29	.19	.18	.17	.06	.06	.06	.00	.000	2
			3	.49	.44	.40	.42	.38	.35	.29	.26	.24	.17	.15	.14	.05	.05	.05	.00	.000	3
			4	.43	.38	.34	.37	.33	.29	.26	.23	.20	.15	.13	.12	.05	.04	.04	.00	.000	4
			5	.38	.33	.28	.33	.28	.25	.23	.20	.17	.13	.12	.10	.04	.04	.03	.00	.000	5
	78% ↑		6	.34	.28	.24	.29	.25	.21	.20	.17	.15	.12	.10	.09	.04	.03	.03	.00	.000	6
			7	.31	.25	.21	.26	.22	.18	.18	.15	.13	.11	.09	.08	.03	.03	.03	.00	.000	7
Open top, indirect, reflector type unit with HID lamp (mult. by 0.9 for lens top)	0% ↓		8	.28	.22	.18	.24	.19	.16	.16	.13	.11	.10	.08	.07	.03	.03	.02	.00	.000	8
			9	.25	.20	.16	.21	.17	.14	.15	.12	.10	.09	.07	.06	.03	.02	.02	.00	.000	9
			10	.23	.17	.14	.20	.15	.12	.14	.11	.09	.08	.06	.05	.03	.02	.02	.00	.000	10

80			70			50			30			10			80			70			50			30			10		
50	30	10	50	30	10	50	30	10	50	30	10	50	30	10	50	30	10	50	30	10	50	30	10	50	30	10	50	30	10
Wall Exitance Coefficients for 20 Per Cent Effective Floor Cavity Reflectance (ρ_{FC} = 20)															Ceiling Cavity Exitance Coefficients for 20 Per Cent Floor Cavity Reflectance (ρ_{FC} = 20)														
															.207	.207	.207	.177	.177	.177	.121	.121	.121	.069	.069	.069	.022	.022	.022
.176	.100	.032	.168	.096	.030	.154	.088	.028	.141	.081	.026	.128	.074	.024	.194	.180	.168	.166	.155	.144	.113	.106	.100	.065	.062	.058	.021	.020	.019
.170	.093	.029	.163	.090	.028	.150	.084	.026	.139	.078	.024	.128	.073	.023	.184	.160	.140	.157	.138	.121	.108	.095	.085	.062	.055	.050	.020	.018	.016
.162	.086	.026	.156	.083	.025	.145	.078	.024	.135	.074	.023	.125	.069	.021	.175	.145	.121	.150	.125	.105	.103	.087	.074	.060	.051	.043	.019	.017	.014
.153	.080	.023	.148	.078	.023	.138	.073	.022	.129	.069	.021	.121	.066	.020	.168	.134	.107	.144	.116	.093	.099	.081	.066	.057	.047	.039	.018	.015	.013
.145	.074	.021	.141	.072	.021	.132	.069	.020	.124	.065	.019	.116	.062	.019	.162	.125	.097	.139	.108	.085	.096	.075	.060	.055	.044	.035	.018	.014	.012
.138	.069	.020	.133	.067	.019	.126	.064	.019	.118	.061	.018	.111	.058	.017	.156	.118	.090	.134	.102	.078	.093	.071	.055	.054	.042	.033	.017	.014	.011
.131	.065	.018	.127	.063	.018	.120	.060	.017	.113	.058	.017	.106	.055	.016	.151	.112	.084	.130	.097	.073	.090	.068	.052	.052	.040	.031	.017	.013	.010
.124	.061	.017	.120	.059	.017	.114	.057	.016	.108	.055	.016	.102	.052	.015	.147	.107	.080	.126	.093	.069	.087	.065	.049	.051	.038	.029	.016	.013	.010
.118	.057	.016	.115	.056	.016	.109	.054	.015	.103	.052	.015	.098	.050	.014	.142	.103	.076	.123	.089	.066	.085	.063	.047	.049	.037	.028	.016	.012	.009
.112	.054	.015	.109	.053	.015	.104	.051	.014	.099	.049	.014	.094	.047	.014	.138	.099	.073	.119	.086	.064	.083	.061	.046	.048	.036	.027	.016	.012	.009
															.244	.244	.244	.209	.209	.209	.143	.143	.143	.082	.082	.082	.026	.026	.026
.183	.104	.033	.174	.099	.032	.157	.090	.029	.141	.081	.026	.127	.073	.024	.232	.218	.205	.199	.187	.177	.136	.129	.122	.078	.075	.071	.025	.024	.023
.177	.097	.030	.168	.093	.029	.153	.085	.026	.139	.078	.024	.126	.071	.023	.223	.199	.178	.191	.171	.154	.131	.118	.107	.076	.069	.063	.024	.022	.021
.168	.090	.027	.161	.086	.026	.148	.080	.024	.135	.074	.023	.123	.068	.021	.216	.184	.159	.185	.159	.138	.127	.111	.097	.073	.065	.057	.024	.021	.019
.160	.083	.024	.154	.080	.024	.141	.075	.022	.130	.070	.021	.119	.065	.020	.209	.173	.146	.180	.150	.127	.124	.104	.089	.071	.061	.053	.023	.020	.017
.152	.077	.022	.146	.075	.022	.135	.070	.021	.125	.066	.020	.115	.061	.018	.204	.165	.136	.175	.143	.119	.121	.100	.084	.070	.059	.050	.023	.019	.016
.144	.072	.021	.139	.070	.020	.129	.066	.019	.119	.062	.018	.110	.058	.017	.199	.158	.129	.171	.137	.112	.118	.096	.080	.068	.056	.047	.022	.018	.016
.137	.068	.019	.132	.066	.019	.123	.062	.018	.114	.058	.017	.106	.055	.016	.194	.153	.124	.167	.132	.108	.115	.093	.076	.067	.055	.045	.022	.018	.015
.130	.063	.018	.125	.062	.017	.117	.058	.017	.109	.055	.016	.101	.052	.015	.190	.148	.120	.163	.128	.104	.113	.090	.074	.066	.053	.044	.021	.017	.015
.124	.060	.017	.119	.058	.016	.112	.055	.016	.104	.052	.015	.097	.049	.014	.186	.144	.116	.160	.125	.101	.111	.088	.072	.065	.052	.043	.021	.017	.014
.118	.056	.016	.114	.055	.015	.106	.052	.015	.100	.049	.014	.093	.047	.013	.182	.141	.114	.157	.122	.099	.109	.086	.071	.063	.051	.042	.021	.017	.014
															.130	.130	.130	.112	.112	.112	.076	.076	.076	.044	.044	.044	.014	.014	.014
.186	.106	.033	.181	.103	.033	.172	.099	.032	.164	.094	.030	.156	.091	.029	.122	.107	.094	.104	.092	.081	.071	.063	.056	.041	.037	.033	.013	.012	.011
.181	.099	.030	.177	.097	.030	.169	.094	.029	.162	.091	.028	.155	.088	.028	.115	.090	.069	.099	.078	.060	.068	.054	.042	.039	.031	.025	.013	.010	.008
.171	.091	.027	.168	.090	.027	.161	.087	.026	.154	.085	.026	.148	.082	.025	.110	.078	.053	.094	.067	.046	.065	.047	.032	.037	.027	.019	.012	.009	.006
.161	.084	.025	.158	.083	.024	.152	.081	.024	.146	.078	.024	.141	.076	.023	.105	.069	.041	.090	.060	.036	.062	.042	.025	.036	.024	.015	.012	.008	.005
.151	.077	.022	.148	.076	.022	.143	.074	.022	.138	.073	.022	.133	.071	.021	.100	.062	.033	.086	.053	.029	.059	.037	.020	.034	.022	.012	.011	.007	.004
.142	.071	.020	.139	.070	.020	.134	.069	.020	.130	.067	.020	.126	.066	.020	.096	.056	.027	.082	.049	.024	.057	.034	.017	.033	.020	.010	.011	.007	.003
.133	.066	.019	.131	.065	.019	.127	.064	.018	.122	.063	.018	.119	.062	.018	.092	.051	.023	.079	.045	.020	.054	.031	.014	.032	.018	.009	.010	.006	.003
.125	.061	.017	.123	.061	.017	.119	.060	.017	.116	.059	.017	.112	.058	.017	.088	.047	.020	.075	.041	.017	.052	.029	.012	.030	.017	.007	.010	.006	.002
.118	.057	.016	.116	.057	.016	.113	.056	.016	.109	.055	.016	.106	.054	.016	.084	.044	.017	.072	.038	.015	.050	.027	.011	.029	.016	.006	.009	.005	.002
.112	.053	.015	.110	.053	.015	.107	.052	.015	.104	.051	.015	.101	.051	.015	.080	.041	.015	.069	.036	.013	.048	.025	.010	.028	.015	.006	.009	.005	.002
															.156	.156	.156	.133	.133	.133	.091	.091	.091	.052	.052	.052	.017	.017	.017
.230	.131	.041	.224	.128	.041	.213	.122	.039	.203	.117	.038	.194	.112	.036	.149	.131	.115	.128	.113	.099	.087	.078	.069	.050	.045	.040	.016	.014	.013
.216	.118	.036	.210	.116	.036	.201	.112	.035	.192	.108	.034	.183	.104	.033	.144	.114	.089	.124	.099	.077	.085	.068	.054	.049	.040	.032	.016	.013	.010
.200	.106	.032	.195	.104	.031	.186	.101	.031	.178	.098	.030	.171	.094	.029	.139	.102	.073	.119	.088	.063	.082	.061	.044	.047	.036	.026	.015	.012	.009
.184	.096	.028	.180	.094	.028	.172	.091	.027	.165	.089	.027	.158	.086	.026	.134	.093	.061	.115	.080	.053	.079	.056	.038	.046	.033	.022	.015	.011	.007
.170	.087	.025	.166	.085	.025	.159	.083	.024	.153	.081	.024	.147	.078	.024	.129	.086	.054	.111	.074	.047	.077	.052	.033	.044	.030	.020	.014	.010	.006
.158	.079	.023	.154	.078	.022	.148	.076	.022	.142	.074	.022	.137	.072	.021	.124	.080	.048	.107	.069	.042	.074	.049	.030	.043	.029	.018	.014	.009	.006
.147	.072	.021	.144	.072	.020	.138	.070	.020	.132	.068	.020	.127	.066	.019	.120	.075	.044	.103	.065	.039	.071	.046	.027	.041	.027	.016	.013	.009	.005
.137	.067	.019	.134	.066	.019	.129	.064	.018	.124	.063	.018	.119	.061	.018	.115	.071	.041	.099	.062	.036	.069	.043	.026	.040	.026	.015	.013	.008	.005
.128	.062	.017	.125	.061	.017	.121	.060	.017	.116	.058	.017	.112	.057	.016	.111	.068	.039	.095	.059	.034	.066	.041	.024	.039	.024	.014	.012	.008	.005
.120	.058	.016	.118	.057	.016	.113	.056	.016	.109	.054	.015	.106	.053	.015	.107	.065	.037	.092	.056	.032	.064	.040	.023	.037	.023	.014	.012	.008	.005
															.107	.107	.107	.091	.091	.091	.062	.062	.062	.036	.036	.036	.011	.011	.011
.146	.083	.026	.142	.081	.026	.135	.077	.025	.128	.074	.024	.122	.071	.023	.099	.088	.077	.085	.075	.067	.058	.052	.046	.033	.030	.027	.011	.010	.009
.145	.079	.024	.141	.078	.024	.135	.075	.023	.129	.072	.023	.123	.070	.022	.094	.074	.057	.080	.064	.049	.055	.044	.034	.032	.026	.020	.010	.008	.007
.138	.074	.022	.136	.073	.022	.130	.070	.021	.125	.068	.021	.120	.066	.021	.090	.064	.043	.077	.055	.038	.053	.038	.026	.030	.022	.016	.010	.007	.005
.131	.068	.020	.128	.067	.020	.123	.065	.020	.119	.064	.019	.114	.062	.019	.086	.056	.034	.074	.049	.029	.051	.034	.021	.029	.020	.012	.009	.006	.004
.123	.063	.018	.121	.062	.018	.116	.061	.018	.112	.059	.018	.108	.058	.017	.082	.050	.027	.070	.044	.024	.048	.031	.017	.028	.018	.010	.009	.006	.003
.116	.058	.017	.114	.057	.017	.110	.056	.016	.106	.055	.016	.102	.054	.016	.078	.046	.022	.067	.040	.020	.046	.028	.014	.027	016	.008	.009	.005	.003
.109	.054	.015	.107	.053	.015	.103	.052	.015	.100	.051	.015	.097	.050	.015	.075	.042	.019	.064	.036	.017	.044	.026	.012	.026	.015	.007	.008	.005	.002
.102	.050	.014	.101	.050	.014	.097	.049	.014	.094	.048	.014	.091	.047	.014	.071	.039	.016	.062	.034	.014	.043	.024	.010	.025	.014	.006	.008	.005	.002
.096	.047	.013	.095	.046	.013	.092	.045	.013	.089	.045	.013	.087	.044	.013	.068	.036	.014	.059	.031	.012	.041	.022	.009	.024	.013	.005	.008	.004	.002
.091	.044	.012	.090	.043	.012	.087	.043	.012	.084	.042	.012	.082	.041	.012	.065	.034	.013	.056	.029	.011	.039	.021	.008	.023	.012	.005	.007	.004	.002
															.743	.743	.743	.635	.635	.635	.433	.433	.433	.249	.249	.249	.080	.080	.080
.201	.114	.036	.172	.098	.031	.117	.067	.022	.068	.039	.013	.022	.013	.004	.737	.721	.707	.631	.619	.609	.431	.426	.421	.248	.247	.245	.080	.079	.079
.184	.101	.031	.158	.087	.027	.108	.060	.019	.062	.035	.011	.020	.011	.004	.732	.706	.685	.627	.608	.592	.430	.421	.413	.248	.245	.242	.079	.079	.079
.170	.091	.027	.146	.078	.024	.100	.054	.017	.058	.032	.010	.019	.010	.003	.727	.695	.670	.623	.600	.581	.428	.417	.408	.247	.243	.240	.079	.079	.079
.158	.082	.024	.135	.071	.021	.093	.049	.015	.054	.029	.009	.017	.009	.003	.723	.687	.660	.620	.594	.573	.426	.414	.404	.247	.242	.239	.079	.079	.079
.147	.075	.022	.126	.065	.019	.087	.045	.013	.050	.027	.008	.016	.009	.003	.718	.681	.653	.617	.589	.568	.425	.411	.401	.246	.242	.238	.079	.079	.078
.137	.069	.020	.117	.059	.017	.081	.042	.012	.047	.024	.007	.015	.008	.002	.714	.676	.648	.614	.585	.564	.423	.410	.399	.246	.241	.237	.079	.079	.078
.128	.063	.018	.110	.055	.016	.076	.038	.011	.044	.023	.007	.014	.007	.002	.710	.671	.644	.611	.582	.562	.422	.408	.398	.245	.240	.237	.079	.079	.078
.120	.059	.016	.103	.051	.014	.071	.036	.010	.042	.021	.006	.013	.007	.002	.706	.668	.642	.608	.579	.559	.421	.407	.397	.245	.240	.236	.079	.079	.078
.113	.054	.015	.097	.047	.013	.067	.033	.009	.039	.020	.006	.013	.006	.002	.703	.665	.639	.605	.577	.558	.419	.406	.396	.244	.240	.236	.079	.079	.078
.106	.051	.014	.091	.044	.012	.063	.031	.009	.037	.018	.005	.012	.006	.002	.699	.662	.638	.603	.575	.556	.418	.405	.395	.244	.239	.236	.079	.079	.078

Fig. 9-62. *Continued (see page 9-53 for instructions and notes)*

Typical Luminaire	Typical Intensity Distribution and Per Cent Lamp Lumens		ρcc →	80			70			50			30			10			0	WDRC	ρcc →
			ρw →	50	30	10	50	30	10	50	30	10	50	30	10	50	30	10	0		ρw →
	Maint. Cat.	SC	RCR ↓	Coefficients of Utilization for 20 Per Cent Effective Floor Cavity Reflectance (ρ_{FC} = 20)																	RCR ↓
25	II	1.3	0	.99	.99	.99	.94	.94	.94	.85	.85	.85	.77	.77	.77	.69	.69	.69	.65		
			1	.87	.84	.81	.83	.80	.77	.75	.73	.71	.68	.66	.65	.62	.60	.59	.56	.236	1
			2	.77	.71	.67	.73	.68	.64	.67	.63	.60	.60	.58	.55	.55	.53	.51	.48	.220	2
	22½% ↑		3	.68	.62	.56	.65	.59	.54	.59	.55	.51	.54	.50	.47	.49	.46	.44	.41	.203	3
			4	.61	.54	.48	.58	.52	.47	.53	.48	.44	.48	.44	.41	.44	.41	.38	.35	.186	4
			5	.54	.47	.42	.52	.46	.41	.48	.42	.38	.44	.39	.36	.40	.36	.33	.31	.170	5
	65% ↓		6	.49	.42	.37	.47	.40	.36	.43	.38	.34	.40	.35	.32	.36	.33	.30	.27	.157	6
			7	.45	.37	.32	.43	.36	.32	.39	.34	.30	.36	.32	.28	.33	.29	.26	.24	.145	7
			8	.41	.34	.29	.39	.33	.28	.36	.31	.27	.33	.29	.25	.31	.27	.24	.22	.135	8
Porcelain-enameled reflector with			9	.37	.31	.26	.36	.30	.25	.33	.28	.24	.31	.26	.23	.28	.24	.22	.20	.126	9
35°CW shielding			10	.34	.28	.24	.33	.27	.23	.31	.25	.22	.28	.24	.21	.26	.22	.20	.18	.118	10
26	II	1.5/1.3	0	.95	.95	.95	.91	.91	.91	.83	.83	.83	.76	.76	.76	.69	.69	.69	.66		
			1	.85	.82	.79	.81	.79	.76	.75	.73	.71	.69	.67	.66	.63	.62	.61	.58	.197	1
	17% ↑		2	.75	.71	.67	.72	.68	.65	.67	.63	.61	.62	.59	.57	.57	.55	.53	.51	.194	2
			3	.67	.61	.57	.65	.59	.55	.60	.56	.52	.55	.52	.49	.51	.49	.46	.44	.184	3
			4	.60	.54	.49	.58	.52	.48	.54	.49	.45	.50	.46	.43	.46	.43	.41	.39	.173	4
			5	.54	.47	.43	.52	.46	.42	.49	.43	.40	.45	.41	.38	.42	.39	.36	.34	.162	5
	66% ↓		6	.49	.42	.37	.47	.41	.37	.44	.39	.35	.41	.37	.33	.38	.35	.32	.30	.151	6
			7	.44	.38	.33	.43	.37	.32	.40	.35	.31	.38	.33	.30	.35	.31	.28	.27	.141	7
			8	.40	.34	.29	.39	.33	.29	.37	.31	.28	.34	.30	.27	.32	.28	.26	.24	.132	8
Diffuse aluminum reflector with			9	.37	.31	.26	.36	.30	.26	.34	.29	.25	.32	.27	.24	.30	.26	.23	.21	.124	9
35°CW shielding			10	.34	.28	.24	.33	.27	.23	.31	.26	.23	.29	.25	.22	.28	.24	.21	.19	.117	10
27	II	1.0	0	.91	.91	.91	.86	.86	.86	.77	.77	.77	.68	.68	.68	.61	.61	.61	.57		
			1	.80	.77	.75	.76	.74	.71	.69	.67	.65	.62	.60	.59	.55	.54	.53	.50	.182	1
			2	.71	.67	.63	.68	.64	.60	.61	.58	.55	.55	.53	.51	.50	.48	.46	.43	.174	2
	23½% ↑		3	.63	.58	.53	.60	.55	.51	.55	.51	.47	.50	.46	.44	.45	.42	.40	.38	.163	3
			4	.57	.51	.46	.54	.49	.44	.49	.45	.41	.45	.41	.38	.41	.38	.35	.33	.151	4
			5	.51	.45	.40	.49	.43	.39	.45	.40	.36	.41	.37	.34	.37	.34	.31	.29	.140	5
	57% ↓		6	.46	.40	.35	.44	.38	.34	.41	.36	.32	.37	.33	.30	.34	.30	.28	.26	.130	6
			7	.42	.36	.31	.40	.35	.30	.37	.32	.29	.34	.30	.27	.31	.28	.25	.23	.121	7
			8	.38	.32	.28	.37	.31	.27	.34	.29	.26	.31	.27	.24	.29	.25	.23	.21	.113	8
Porcelain-enameled reflector with			9	.35	.29	.25	.34	.28	.25	.31	.27	.23	.29	.25	.22	.27	.23	.21	.19	.106	9
30°CW × 30°LW shielding			10	.33	.27	.23	.31	.26	.22	.29	.24	.21	.27	.23	.20	.25	.21	.19	.17	.099	10
28	II	1.5/1.1	0	.83	.83	.83	.79	.79	.79	.72	.72	.72	.65	.65	.65	.59	.59	.59	.56		
			1	.74	.72	.70	.71	.69	.67	.65	.63	.62	.59	.58	.57	.54	.53	.52	.50	.160	1
	17% ↑		2	.66	.62	.59	.64	.60	.57	.58	.56	.53	.54	.51	.49	.49	.47	.46	.44	.158	2
			3	.59	.54	.50	.57	.53	.49	.53	.49	.46	.48	.46	.43	.45	.42	.40	.38	.150	3
			4	.53	.48	.44	.51	.46	.42	.47	.43	.40	.44	.41	.38	.40	.38	.36	.34	.141	4
			5	.48	.42	.38	.46	.41	.37	.43	.39	.35	.40	.36	.33	.37	.34	.32	.30	.132	5
	56½% ↓		6	.44	.38	.34	.42	.37	.33	.39	.35	.31	.36	.33	.30	.34	.31	.28	.27	.124	6
			7	.40	.34	.30	.38	.33	.29	.36	.31	.28	.33	.30	.27	.31	.28	.25	.24	.116	7
			8	.36	.31	.27	.35	.30	.26	.33	.28	.25	.31	.27	.24	.29	.25	.23	.21	.109	8
Diffuse aluminum reflector with 35°CW			9	.33	.28	.24	.32	.27	.24	.30	.26	.23	.28	.24	.22	.26	.23	.21	.19	.102	9
× 35°LW shielding			10	.31	.25	.22	.30	.25	.22	.28	.24	.21	.26	.22	.20	.25	.21	.19	.18	.096	10
29	II	1.1	0	.75	.75	.75	.69	.69	.69	.57	.57	.57	.46	.46	.46	.37	.37	.37	.32		
			1	.66	.64	.62	.61	.59	.57	.51	.50	.48	.42	.41	.40	.33	.33	.32	.28	.094	1
	39% ↑		2	.59	.55	.52	.54	.51	.48	.46	.43	.41	.38	.36	.34	.30	.29	.28	.25	.091	2
			3	.52	.48	.44	.48	.44	.41	.41	.38	.35	.34	.32	.30	.27	.26	.25	.22	.085	3
			4	.47	.42	.38	.43	.39	.35	.37	.33	.31	.31	.28	.26	.25	.23	.22	.19	.079	4
			5	.42	.37	.33	.39	.34	.31	.33	.30	.27	.28	.25	.23	.23	.21	.20	.17	.073	5
			6	.38	.33	.29	.35	.31	.27	.30	.27	.24	.25	.23	.21	.21	.19	.18	.16	.068	6
	32% ↓		7	.35	.29	.26	.32	.28	.24	.28	.24	.21	.23	.21	.19	.19	.17	.16	.14	.063	7
			8	.32	.26	.23	.29	.25	.22	.25	.22	.19	.22	.19	.17	.18	.16	.15	.13	.059	8
Metal or dense diffusing sides with			9	.29	.24	.21	.27	.23	.20	.23	.20	.17	.20	.17	.15	.17	.15	.13	.12	.056	9
45°CW × 45°LW shielding			10	.27	.22	.19	.25	.21	.18	.22	.18	.16	.19	.16	.14	.16	.14	.12	.11	.052	10
30	IV	1.0	0	.61	.61	.61	.58	.58	.58	.55	.55	.55	.51	.51	.51	.48	.48	.48	.46		
			1	.54	.52	.50	.52	.50	.49	.49	.47	.46	.46	.45	.43	.43	.42	.41	.40	.159	1
	6% ↑		2	.48	.45	.42	.46	.44	.41	.44	.41	.39	.41	.39	.38	.39	.37	.36	.34	.145	2
			3	.43	.39	.36	.42	.38	.35	.39	.36	.34	.37	.35	.33	.35	.33	.31	.30	.132	3
			4	.39	.35	.32	.38	.34	.31	.36	.32	.30	.34	.31	.29	.32	.30	.28	.27	.121	4
			5	.35	.31	.28	.34	.30	.27	.32	.29	.27	.31	.28	.26	.29	.27	.25	.24	.111	5
	46% ↓		6	.32	.28	.25	.31	.27	.25	.30	.26	.24	.28	.25	.23	.27	.25	.23	.22	.102	6
			7	.29	.25	.22	.29	.25	.22	.27	.24	.22	.26	.23	.21	.25	.23	.21	.20	.095	7
			8	.27	.23	.20	.27	.23	.20	.25	.22	.20	.24	.21	.19	.23	.21	.19	.18	.088	8
Same as unit #29 except with top reflec-			9	.25	.21	.19	.25	.21	.18	.24	.20	.18	.23	.20	.18	.22	.19	.17	.16	.083	9
tors			10	.23	.20	.17	.23	.19	.17	.22	.19	.17	.21	.18	.16	.20	.18	.16	.15	.077	10

80			70			50			30			10			80			70			50			30			10		
50	30	10	50	30	10	50	30	10	50	30	10	50	30	10	50	30	10	50	30	10	50	30	10	50	30	10	50	30	10
Wall Exitance Coefficients for 20 Per Cent Effective Floor Cavity Reflectance (ρ_{FC} = 20)															Ceiling Cavity Exitance Coefficients for 20 Per Cent Floor Cavity Reflectance (ρ_{FC} = 20)														
															.339	.339	.339	.290	.290	.290	.198	.198	.198	.114	.114	.114	.036	.036	.036
.243	.138	.044	.230	.131	.042	.206	.118	.038	.184	.106	.034	.163	.095	.031	.329	.311	.293	.282	.267	.253	.193	.183	.175	.111	.106	.102	.036	.034	.033
.228	.125	.038	.216	.119	.037	.195	.108	.034	.174	.098	.031	.156	.088	.028	.322	.290	.264	.276	.250	.228	.189	.173	.159	.109	.101	.093	.035	.033	.030
.212	.113	.034	.202	.108	.032	.182	.098	.030	.163	.090	.027	.146	.081	.025	.315	.275	.244	.270	.238	.212	.185	.165	.148	.107	.096	.087	.034	.031	.029
.197	.102	.030	.187	.098	.029	.169	.090	.027	.153	.082	.025	.137	.074	.023	.308	.264	.231	.265	.228	.200	.182	.159	.141	.105	.093	.083	.034	.030	.027
.183	.093	.027	.175	.090	.026	.158	.082	.024	.143	.075	.022	.129	.069	.021	.302	.255	.221	.260	.221	.192	.179	.154	.136	.104	.091	.081	.033	.030	.027
.171	.086	.025	.163	.082	.024	.148	.076	.022	.134	.070	.020	.121	.064	.019	.297	.248	.214	.255	.215	.186	.176	.151	.132	.102	.089	.078	.033	.029	.026
.160	.079	.022	.153	.076	.022	.139	.070	.020	.126	.065	.019	.114	.059	.017	.291	.243	.209	.250	.210	.182	.173	.148	.129	.101	.087	.077	.032	.028	.025
.150	.073	.021	.144	.071	.020	.131	.065	.019	.119	.060	.017	.107	.055	.016	.286	.238	.205	.246	.206	.179	.170	.145	.127	.099	.085	.076	.032	.028	.025
.141	.068	.019	.135	.066	.018	.123	.061	.017	.112	.056	.016	.101	.052	.015	.281	.234	.202	.242	.203	.176	.168	.143	.125	.098	.084	.075	.032	.028	.025
.133	.064	.018	.128	.062	.017	.117	.057	.016	.106	.053	.015	.096	.048	.014	.277	.230	.199	.239	.200	.174	.165	.141	.124	.097	.083	.074	.031	.027	.024
															.286	.286	.286	.244	.244	.244	.167	.167	.167	.096	.096	.096	.031	.031	.031
.209	.119	.038	.198	.113	.036	.178	.102	.033	.159	.092	.029	.142	.082	.027	.275	.259	.244	.235	.222	.210	.161	.153	.145	.093	.088	.084	.030	.028	.027
.200	.110	.034	.191	.105	.032	.173	.096	.030	.156	.087	.027	.140	.079	.025	.267	.239	.216	.229	.206	.187	.157	.143	.130	.090	.083	.076	.029	.027	.025
.190	.101	.030	.181	.097	.029	.164	.089	.027	.149	.082	.025	.135	.075	.023	.260	.225	.197	.223	.194	.171	.153	.135	.120	.088	.079	.071	.028	.026	.023
.178	.093	.027	.170	.089	.026	.156	.083	.025	.142	.076	.023	.129	.070	.021	.254	.214	.184	.218	.185	.159	.150	.129	.112	.087	.076	.066	.028	.025	.022
.168	.085	.025	.161	.082	.024	.147	.077	.023	.134	.071	.021	.123	.065	.020	.248	.206	.174	.213	.178	.151	.147	.124	.107	.085	.073	.063	.027	.024	.021
.158	.079	.023	.151	.076	.022	.139	.071	.021	.127	.066	.019	.116	.061	.018	.243	.199	.167	.209	.172	.145	.144	.121	.103	.084	.071	.061	.027	.023	.020
.148	.073	.021	.142	.071	.020	.131	.066	.019	.120	.062	.018	.110	.057	.017	.238	.193	.162	.205	.168	.141	.142	.118	.100	.082	.069	.060	.027	.023	.020
.140	.068	.019	.135	.066	.019	.124	.062	.018	.114	.058	.017	.105	.054	.016	.234	.189	.158	.201	.164	.138	.139	.115	.098	.081	.068	.058	.026	.022	.019
.132	.064	.018	.127	.062	.017	.118	.058	.016	.108	.054	.016	.100	.051	.015	.229	.185	.155	.197	.160	.135	.137	.113	.096	.080	.067	.057	.026	.022	.019
.125	.060	.017	.121	.058	.016	.112	.055	.015	.103	.051	.015	.095	.048	.014	.225	.182	.153	.194	.158	.133	.135	.111	.095	.079	.066	.056	.025	.022	.019
															.334	.334	.334	.286	.286	.286	.195	.195	.195	.112	.112	.112	.036	.036	.036
.210	.119	.038	.197	.113	.036	.173	.099	.032	.151	.087	.028	.131	.076	.025	.325	.308	.294	.278	.265	.253	.190	.182	.175	.109	.105	.102	.035	.034	.033
.199	.109	.033	.187	.103	.032	.166	.092	.029	.146	.082	.026	.127	.072	.023	.317	.290	.267	.272	.249	.231	.186	.173	.161	.107	.100	.094	.034	.032	.031
.186	.099	.030	.176	.094	.028	.156	.085	.026	.138	.076	.023	.121	.067	.021	.311	.276	.249	.266	.238	.215	.183	.165	.151	.106	.097	.089	.034	.031	.029
.174	.090	.027	.164	.086	.025	.146	.078	.023	.130	.070	.021	.115	.062	.019	.305	.266	.236	.261	.230	.205	.180	.160	.144	.104	.094	.085	.033	.031	.028
.162	.083	.024	.154	.079	.023	.137	.072	.021	.122	.065	.019	.108	.058	.017	.299	.258	.227	.257	.223	.198	.177	.156	.139	.103	.091	.083	.033	.030	.027
.152	.076	.022	.144	.073	.021	.129	.066	.019	.115	.060	.018	.102	.054	.016	.294	.251	.221	.253	.218	.192	.174	.152	.136	.101	.090	.081	.033	.029	.027
.143	.071	.020	.135	.067	.019	.122	.062	.018	.109	.056	.016	.097	.050	.015	.289	.246	.216	.249	.213	.188	.172	.149	.133	.100	.088	.079	.032	.029	.026
.134	.066	.018	.128	.063	.018	.115	.057	.016	.103	.052	.015	.091	.047	.014	.284	.241	.212	.245	.209	.185	.169	.147	.131	.099	.087	.078	.032	.028	.026
.127	.061	.017	.120	.059	.016	.109	.054	.015	.097	.049	.014	.087	.044	.013	.280	.238	.209	.241	.206	.182	.167	.145	.129	.097	.086	.077	.032	.028	.026
.120	.057	.016	.114	.055	.015	.103	.050	.014	.092	.046	.013	.082	.041	.012	.276	.234	.207	.238	.204	.180	.165	.143	.128	.096	.085	.077	.031	.028	.025
															.268	.268	.268	.229	.229	.229	.156	.156	.156	.090	.090	.090	.029	.029	.029
.180	.103	.032	.170	.097	.031	.151	.087	.028	.134	.077	.025	.118	.068	.022	.259	.245	.232	.221	.210	.200	.151	.144	.138	.087	.084	.080	.028	.027	.026
.173	.095	.029	.164	.090	.028	.146	.081	.025	.131	.073	.023	.116	.066	.021	.251	.227	.207	.215	.196	.179	.148	.135	.125	.085	.079	.073	.027	.025	.024
.163	.087	.026	.155	.083	.025	.139	.076	.023	.125	.069	.021	.112	.062	.019	.245	.215	.191	.210	.185	.165	.144	.129	.116	.083	.075	.068	.027	.024	.022
.153	.080	.023	.146	.076	.023	.132	.070	.021	.119	.064	.019	.107	.058	.018	.240	.205	.179	.206	.177	.156	.141	.124	.110	.082	.072	.065	.026	.024	.021
.144	.074	.021	.138	.071	.021	.125	.065	.019	.113	.060	.018	.102	.054	.016	.235	.198	.171	.202	.171	.148	.139	.120	.105	.080	.070	.062	.026	.023	.020
.136	.068	.019	.130	.065	.019	.118	.060	.018	.107	.056	.016	.097	.051	.015	.230	.192	.165	.198	.166	.143	.136	.116	.101	.079	.068	.060	.026	.022	.020
.128	.063	.018	.122	.061	.017	.111	.056	.016	.101	.052	.015	.092	.048	.014	.226	.187	.160	.194	.162	.139	.134	.114	.099	.078	.067	.059	.025	.022	.019
.121	.059	.017	.116	.057	.016	.106	.053	.015	.096	.049	.014	.087	.045	.013	.222	.183	.156	.191	.159	.136	.132	.111	.097	.077	.066	.058	.025	.022	.019
.114	.055	.015	.109	.053	.015	.100	.050	.014	.091	.046	.013	.083	.042	.012	.218	.180	.154	.188	.156	.134	.130	.110	.095	.076	.065	.057	.025	.021	.019
.108	.052	.014	.104	.050	.014	.095	.047	.013	.087	.043	.012	.079	.040	.011	.214	.177	.152	.185	.153	.132	.128	.108	.094	.075	.064	.056	.024	.021	.019
															.433	.433	.433	.370	.370	.370	.253	.253	.253	.145	.145	.145	.046	.046	.046
.180	.102	.032	.163	.093	.030	.132	.076	.024	.103	.060	.019	.077	.044	.014	.426	.411	.399	.364	.353	.343	.249	.243	.237	.143	.141	.138	.046	.045	.045
.168	.092	.028	.153	.084	.026	.125	.069	.022	.098	.055	.017	.074	.042	.013	.419	.396	.377	.359	.341	.326	.246	.236	.227	.142	.137	.133	.046	.044	.043
.157	.083	.025	.143	.077	.023	.117	.063	.019	.093	.051	.016	.070	.039	.012	.414	.385	.362	.355	.332	.314	.244	.231	.220	.141	.135	.130	.045	.044	.042
.146	.076	.022	.133	.070	.021	.109	.058	.017	.087	.047	.014	.066	.036	.011	.409	.376	.351	.351	.325	.305	.241	.227	.215	.140	.133	.127	.045	.043	.042
.136	.069	.020	.125	.064	.019	.102	.053	.016	.082	.043	.013	.063	.034	.010	.404	.370	.344	.347	.320	.299	.239	.223	.211	.139	.131	.125	.045	.043	.041
.127	.064	.018	.117	.059	.017	.096	.049	.014	.077	.040	.012	.059	.031	.009	.400	.364	.339	.344	.315	.295	.237	.221	.209	.138	.130	.124	.044	.042	.041
.119	.059	.017	.109	.055	.016	.091	.046	.013	.073	.037	.011	.056	.029	.009	.396	.360	.335	.341	.312	.292	.235	.219	.207	.137	.129	.123	.044	.042	.041
.112	.055	.015	.103	.051	.014	.085	.043	.012	.069	.035	.010	.053	.027	.008	.392	.356	.331	.338	.309	.289	.234	.217	.205	.136	.128	.122	.044	.042	.040
.106	.051	.014	.097	.047	.013	.081	.040	.011	.065	.033	.009	.051	.026	.007	.388	.353	.329	.335	.306	.287	.232	.215	.204	.135	.127	.122	.044	.042	.040
.100	.048	.013	.092	.044	.012	.077	.037	.011	.062	.031	.009	.048	.024	.007	.385	.350	.327	.332	.304	.286	.230	.214	.203	.134	.127	.121	.044	.042	.040
															.145	.145	.145	.124	.124	.124	.085	.085	.085	.049	.049	.049	.016	.016	.016
.142	.081	.026	.137	.078	.025	.127	.073	.023	.117	.068	.022	.108	.063	.020	.139	.128	.118	.119	.110	.102	.081	.076	.070	.047	.044	.041	.015	.014	.013
.132	.072	.022	.127	.070	.022	.118	.066	.020	.109	.061	.019	.102	.058	.018	.134	.116	.100	.115	.099	.087	.079	.069	.060	.045	.040	.035	.015	.013	.012
.122	.065	.019	.118	.063	.019	.109	.059	.018	.102	.056	.017	.095	.053	.016	.129	.106	.088	.111	.092	.077	.076	.064	.054	.044	.037	.032	.014	.012	.010
.113	.059	.017	.109	.057	.017	.102	.054	.016	.095	.051	.015	.089	.048	.015	.125	.099	.080	.107	.086	.070	.074	.060	.049	.043	.035	.029	.014	.011	.010
.105	.054	.016	.102	.052	.015	.095	.050	.015	.089	.047	.014	.083	.044	.013	.121	.094	.074	.104	.081	.065	.071	.057	.046	.041	.033	.027	.013	.011	.009
.098	.049	.014	.095	.048	.014	.089	.046	.013	.083	.043	.013	.078	.041	.012	.117	.090	.070	.101	.078	.061	.069	.054	.043	.040	.032	.026	.013	.010	.008
.092	.045	.013	.089	.044	.013	.084	.042	.012	.079	.040	.012	.074	.038	.011	.114	.086	.067	.098	.075	.058	.068	.052	.041	.039	.031	.024	.013	.010	.008
.086	.042	.012	.084	.041	.012	.079	.039	.011	.074	.038	.011	.070	.036	.010	.111	.083	.064	.095	.072	.056	.066	.051	.040	.038	.030	.024	.012	.010	.008
.081	.039	.011	.079	.039	.011	.074	.037	.010	.070	.035	.010	.066	.034	.010	.108	.080	.062	.093	.070	.054	.064	.049	.038	.037	.029	.023	.012	.010	.008
.077	.037	.010	.075	.036	.010	.071	.035	.010	.067	.033	.009	.063	.032	.009	.105	.078	.060	.091	.068	.053	.063	.048	.037	.037	.028	.022	.012	.009	.007

Fig. 9–62. *Continued* *(see page 9-53 for instructions and notes)*

Typical Luminaire	Typical Intensity Distribution and Per Cent Lamp Lumens		ρ_{CC} →	80			70			50			30			10			0	WDRC	ρ_{CC} →
			ρ_W →	50	30	10	50	30	10	50	30	10	50	30	10	50	30	10	0		ρ_W →
	Maint. Cat.	SC	RCR ↓	Coefficients of Utilization for 20 Per Cent Effective Floor Cavity Reflectance (ρ_{FC} = 20)																	RCR ↓
31	IV	1.5/1.2	0	.69	.69	.69	.67	.67	.67	.64	.64	.64	.62	.62	.62	.59	.59	.59	.58		
			1	.62	.61	.59	.61	.59	.58	.59	.57	.56	.57	.55	.54	.55	.54	.53	.52	.159	1
	0%↑		2	.56	.53	.50	.55	.52	.50	.53	.50	.48	.51	.49	.47	.49	.48	.46	.45	.160	2
			3	.50	.46	.43	.49	.46	.43	.48	.44	.42	.46	.43	.41	.45	.42	.41	.39	.155	3
			4	.45	.41	.37	.44	.40	.37	.43	.39	.36	.42	.38	.36	.40	.38	.36	.34	.147	4
	58%↓		5	.40	.36	.32	.40	.36	.32	.39	.35	.32	.38	.34	.32	.37	.34	.31	.30	.139	5
			6	.37	.32	.29	.36	.32	.28	.35	.31	.28	.34	.31	.28	.33	.30	.28	.27	.131	6
			7	.33	.29	.25	.33	.28	.25	.32	.28	.25	.31	.28	.25	.30	.27	.25	.24	.123	7
150 mm × 150 mm (6 × 6″) cell parabolic wedge louver—multiply by 1.1 for 250 × 250 mm (10 × 10″) cells			8	.30	.26	.23	.30	.26	.22	.29	.25	.22	.28	.25	.22	.28	.25	.22	.21	.115	8
			9	.28	.23	.20	.27	.23	.20	.27	.23	.20	.26	.23	.20	.26	.22	.20	.19	.109	9
			10	.26	.21	.18	.25	.21	.18	.25	.21	.18	.24	.21	.18	.24	.20	.18	.17	.102	10
32	I	1.3	0	1.02	1.02	1.02	.99	.99	.99	.92	.92	.92	.86	.86	.86	.81	.81	.81	.78		
			1	.85	.80	.76	.82	.78	.74	.76	.73	.70	.71	.68	.66	.67	.64	.62	.60	.467	1
	9½%↑		2	.72	.65	.59	.70	.63	.58	.65	.60	.55	.61	.56	.52	.57	.53	.50	.47	.387	2
			3	.63	.55	.48	.60	.53	.47	.56	.50	.45	.53	.47	.43	.49	.45	.41	.38	.331	3
			4	.55	.46	.40	.53	.45	.39	.50	.43	.37	.46	.41	.36	.43	.38	.34	.32	.289	4
	78%↓		5	.49	.40	.34	.47	.39	.33	.44	.37	.32	.41	.35	.31	.39	.34	.29	.27	.255	5
2-lamp, surface mounted, bare lamp unit—photometry with 460 mm (18″) wide panel above luminaire—lamps on 150 mm (6″) centers			6	.43	.35	.29	.42	.34	.29	.40	.33	.28	.37	.31	.27	.35	.30	.26	.23	.228	6
			7	.39	.31	.25	.38	.30	.25	.36	.29	.24	.34	.28	.23	.32	.26	.22	.20	.206	7
			8	.36	.28	.22	.35	.27	.22	.33	.26	.21	.31	.25	.21	.29	.24	.20	.18	.188	8
			9	.33	.25	.20	.32	.25	.20	.30	.24	.19	.28	.23	.18	.27	.22	.18	.16	.173	9
			10	.30	.23	.18	.29	.22	.18	.28	.21	.17	.26	.21	.17	.25	.20	.16	.14	.159	10
33	VI	N.A.	0	.77	.77	.77	.68	.68	.68	.50	.50	.50	.34	.34	.34	.19	.19	.19	.12		
			1	.67	.64	.61	.59	.56	.54	.43	.42	.41	.29	.29	.28	.17	.16	.16	.10	.048	1
			2	.58	.54	.50	.51	.48	.44	.38	.36	.34	.26	.24	.23	.14	.14	.13	.08	.045	2
			3	.51	.46	.42	.45	.41	.37	.33	.30	.28	.23	.21	.19	.13	.12	.11	.07	.041	3
	66%↑		4	.45	.39	.35	.40	.35	.31	.30	.26	.24	.20	.18	.17	.11	.10	.10	.06	.037	4
			5	.40	.34	.30	.35	.30	.26	.26	.23	.20	.18	.16	.14	.10	.09	.08	.05	.034	5
			6	.36	.30	.25	.31	.26	.23	.24	.20	.17	.16	.14	.12	.09	.08	.07	.04	.031	6
			7	.32	.26	.22	.28	.23	.20	.21	.18	.15	.15	.12	.11	.08	.07	.06	.04	.028	7
	12%↓		8	.29	.23	.19	.26	.21	.17	.19	.16	.13	.13	.11	.09	.08	.06	.06	.03	.026	8
Luminous bottom suspended unit with extra-high output lamp			9	.26	.21	.17	.23	.18	.15	.17	.14	.12	.12	.10	.08	.07	.06	.05	.03	.024	9
			10	.24	.19	.15	.21	.17	.13	.16	.13	.10	.11	.09	.07	.06	.05	.04	.03	.022	10
34	VI	1.4/1.2	0	.91	.91	.91	.85	.85	.85	.74	.74	.74	.64	.64	.64	.54	.54	.54	.50		
			1	.80	.77	.74	.75	.72	.70	.65	.63	.61	.57	.55	.54	.49	.47	.47	.43	.179	1
			2	.70	.65	.61	.66	.62	.58	.58	.54	.52	.50	.48	.46	.43	.42	.40	.37	.166	2
	33%↑		3	.62	.56	.51	.58	.53	.49	.51	.47	.44	.45	.42	.39	.39	.37	.35	.32	.153	3
			4	.55	.49	.44	.52	.46	.42	.46	.41	.38	.40	.37	.34	.35	.32	.30	.27	.140	4
			5	.50	.43	.38	.47	.41	.36	.41	.37	.33	.36	.33	.30	.32	.29	.26	.24	.129	5
	50%↓		6	.45	.38	.33	.42	.36	.32	.37	.33	.29	.33	.29	.26	.29	.26	.23	.21	.119	6
			7	.40	.34	.29	.38	.32	.28	.34	.29	.26	.30	.26	.23	.26	.23	.21	.19	.111	7
			8	.37	.30	.26	.35	.29	.25	.31	.26	.23	.28	.24	.21	.24	.21	.19	.17	.103	8
Prismatic bottom and sides, open top, 4-lamp suspended unit—see note 7			9	.34	.27	.23	.32	.26	.22	.29	.24	.21	.25	.22	.19	.22	.19	.17	.15	.096	9
			10	.31	.25	.21	.29	.24	.20	.26	.22	.19	.23	.20	.17	.21	.18	.15	.14	.090	10
35	V	1.5/1.2	0	.81	.81	.81	.78	.78	.78	.72	.72	.72	.66	.66	.66	.61	.61	.61	.59		
			1	.71	.68	.66	.68	.66	.63	.63	.61	.59	.58	.57	.56	.54	.53	.52	.50	.223	1
	11½%↑		2	.63	.58	.55	.60	.56	.53	.56	.53	.50	.52	.50	.47	.48	.46	.45	.43	.201	2
			3	.56	.50	.46	.54	.49	.45	.50	.46	.43	.47	.43	.41	.43	.41	.39	.37	.183	3
			4	.50	.44	.40	.48	.43	.39	.45	.40	.37	.42	.38	.35	.39	.36	.34	.32	.167	4
	58½%↓		5	.45	.39	.34	.43	.38	.34	.40	.36	.32	.38	.34	.31	.35	.32	.30	.28	.153	5
			6	.40	.34	.30	.39	.34	.30	.37	.32	.28	.34	.30	.27	.32	.29	.26	.25	.142	6
			7	.37	.31	.27	.35	.30	.26	.33	.29	.25	.31	.27	.24	.30	.26	.23	.22	.131	7
			8	.33	.28	.24	.32	.27	.23	.30	.26	.23	.29	.25	.22	.27	.24	.21	.20	.122	8
2-lamp prismatic wraparound—see note 7			9	.31	.25	.21	.30	.25	.21	.28	.24	.20	.26	.23	.20	.25	.22	.19	.18	.114	9
			10	.28	.23	.19	.27	.22	.19	.26	.21	.18	.24	.21	.18	.23	.20	.17	.16	.107	10
36	V	1.2	0	.82	.82	.82	.77	.77	.77	.69	.69	.69	.61	.61	.61	.53	.53	.53	.50		
			1	.71	.67	.65	.67	.64	.61	.59	.57	.55	.52	.51	.49	.46	.45	.44	.40	.234	1
			2	.62	.57	.53	.59	.54	.51	.52	.49	.46	.46	.44	.41	.41	.39	.37	.34	.194	2
	24%↑		3	.55	.49	.45	.52	.47	.43	.46	.42	.39	.41	.38	.36	.37	.34	.32	.30	.168	3
			4	.49	.43	.39	.47	.41	.37	.42	.37	.34	.37	.34	.31	.33	.30	.28	.26	.150	4
			5	.44	.38	.34	.42	.36	.32	.38	.33	.30	.34	.30	.27	.30	.27	.25	.23	.135	5
	50%↓		6	.40	.34	.29	.38	.32	.28	.34	.30	.26	.31	.27	.24	.28	.25	.22	.20	.123	6
			7	.36	.30	.26	.35	.29	.25	.31	.27	.23	.28	.25	.22	.25	.22	.20	.18	.112	7
			8	.33	.27	.23	.32	.26	.23	.29	.24	.21	.26	.22	.20	.23	.20	.18	.16	.104	8
2-lamp prismatic wraparound—see note 7			9	.30	.25	.21	.29	.24	.20	.26	.22	.19	.24	.20	.18	.22	.19	.16	.15	.097	9
			10	.28	.23	.19	.27	.22	.18	.25	.20	.17	.22	.19	.16	.20	.17	.15	.14	.090	10

80			70			50			30			10			80			70			50			30			10		
50	30	10	50	30	10	50	30	10	50	30	10	50	30	10	50	30	10	50	30	10	50	30	10	50	30	10	50	30	10
Wall Exitance Coefficients for 20 Per Cent Effective Floor Cavity Reflectance (ρ_{FC} = 20)															Ceiling Cavity Exitance Coefficients for 20 Per Cent Floor Cavity Reflectance (ρ_{FC} = 20)														
															.111	.111	.111	.094	.094	.094	.064	.064	.064	.037	.037	.037	.012	.012	.012
.138	.078	.025	.134	.076	.024	.126	.072	.023	.119	.069	.022	.113	.065	.021	.102	.091	.081	.087	.078	.070	.060	.054	.048	.034	.031	.028	.011	.010	.009
.136	.074	.023	.132	.073	.022	.126	.070	.022	.120	.067	.021	.114	.065	.020	.095	.077	.061	.082	.066	.053	.056	.046	.037	.032	.027	.022	.010	.009	.007
.130	.069	.021	.127	.068	.021	.122	.066	.020	.117	.064	.020	.112	.062	.019	.090	.066	.047	.077	.057	.041	.053	.040	.028	.031	.023	.017	.010	.008	.005
.124	.065	.019	.122	.064	.019	.117	.062	.018	.112	.060	.018	.108	.058	.018	.086	.058	.037	.074	.050	.032	.051	.035	.023	.029	.021	.013	.009	.007	.004
.118	.060	.017	.115	.059	.017	.111	.058	.017	.107	.056	.017	.103	.055	.017	.082	.052	.030	.070	.045	.026	.049	.031	.018	.028	.018	.011	.009	.006	.004
.112	.056	.016	.109	.055	.016	.105	.054	.016	.102	.053	.016	.098	.052	.015	.078	.047	.024	.067	.041	.021	.046	.028	.015	.027	.017	.009	.009	.005	.003
.106	.052	.015	.104	.052	.015	.100	.051	.015	.097	.050	.014	.094	.049	.014	.075	.043	.021	.064	.037	.018	.044	.026	.013	.026	.015	.008	.008	.005	.003
.100	.049	.014	.098	.048	.014	.095	.047	.014	.092	.047	.013	.089	.046	.013	.072	.040	.018	.062	.034	.015	.043	.024	.011	.025	.014	.007	.008	.005	.002
.095	.046	.013	.093	.045	.013	.090	.045	.013	.087	.044	.013	.085	.043	.012	.068	.037	.015	.059	.032	.013	.041	.022	.010	.024	.013	.006	.008	.004	.002
.090	.043	.012	.088	.043	.012	.086	.042	.012	.083	.041	.012	.081	.041	.012	.066	.034	.014	.057	.030	.012	.039	.021	.008	.023	.012	.005	.007	.004	.002
															.239	.239	.239	.205	.205	.205	.140	.140	.140	.080	.080	.080	.026	.026	.026
.345	.196	.062	.335	.191	.061	.318	.182	.058	.302	.174	.056	.287	.166	.054	.236	.209	.185	.202	.180	.159	.138	.123	.110	.080	.071	.064	.025	.023	.021
.300	.164	.050	.292	.161	.049	.276	.153	.048	.262	.147	.046	.248	.140	.044	.230	.189	.154	.197	.163	.133	.135	.112	.093	.078	.065	.054	.025	.021	.018
.267	.142	.043	.259	.139	.042	.245	.133	.040	.232	.127	.039	.220	.122	.038	.224	.174	.135	.192	.150	.117	.132	.104	.082	.076	.061	.048	.024	.020	.016
.240	.125	.037	.233	.122	.036	.220	.117	.035	.209	.112	.034	.198	.107	.033	.217	.163	.122	.186	.141	.106	.128	.098	.075	.074	.058	.044	.024	.019	.015
.218	.111	.032	.212	.109	.032	.200	.104	.031	.190	.100	.030	.180	.096	.029	.210	.154	.113	.180	.134	.099	.124	.093	.070	.072	.055	.041	.023	.018	.014
.199	.100	.029	.194	.098	.028	.183	.094	.027	.174	.090	.027	.165	.087	.026	.203	.147	.107	.175	.128	.093	.121	.089	.066	.070	.053	.039	.023	.017	.013
.184	.091	.026	.179	.089	.025	.169	.086	.025	.160	.082	.024	.152	.079	.023	.197	.141	.103	.169	.123	.089	.117	.086	.063	.068	.051	.038	.022	.017	.012
.170	.083	.023	.166	.082	.023	.157	.078	.022	.149	.075	.022	.141	.073	.021	.191	.137	.099	.164	.118	.086	.114	.083	.061	.066	.049	.037	.021	.016	.012
.158	.077	.021	.154	.075	.021	.146	.072	.021	.139	.070	.020	.132	.067	.019	.185	.132	.096	.160	.115	.084	.111	.081	.060	.064	.048	.036	.021	.016	.012
.148	.071	.020	.144	.070	.019	.137	.067	.019	.130	.065	.018	.124	.062	.018	.180	.129	.094	.155	.112	.082	.108	.079	.058	.063	.046	.035	.020	.015	.012
															.653	.653	.653	.558	.558	.558	.381	.381	.381	.219	.219	.219	.070	.070	.070
.206	.117	.037	.181	.103	.033	.133	.077	.024	.090	.052	.017	.049	.029	.009	.647	.631	.616	.553	.541	.530	.378	.372	.367	.218	.215	.213	.070	.069	.069
.191	.104	.032	.167	.092	.028	.124	.069	.021	.084	.047	.015	.047	.026	.008	.641	.615	.593	.549	.529	.512	.376	.366	.357	.217	.213	.209	.070	.069	.068
.176	.094	.028	.155	.083	.025	.115	.062	.019	.078	.043	.013	.043	.024	.007	.636	.603	.577	.545	.521	.501	.374	.362	.351	.216	.211	.207	.069	.069	.068
.163	.085	.025	.144	.075	.022	.107	.057	.017	.072	.039	.012	.040	.022	.007	.631	.595	.567	.542	.514	.493	.373	.358	.347	.216	.210	.205	.069	.068	.067
.152	.077	.022	.133	.069	.020	.099	.052	.015	.067	.036	.011	.038	.020	.006	.627	.588	.560	.538	.509	.487	.371	.356	.344	.215	.209	.204	.069	.068	.067
.141	.071	.020	.124	.063	.018	.093	.047	.014	.063	.033	.010	.035	.019	.006	.623	.583	.554	.535	.505	.483	.369	.353	.342	.214	.208	.203	.069	.068	.067
.132	.065	.018	.116	.058	.017	.087	.044	.013	.059	.030	.009	.033	.017	.005	.618	.578	.551	.532	.501	.480	.367	.352	.340	.214	.207	.202	.069	.068	.067
.124	.060	.017	.109	.054	.015	.081	.041	.012	.055	.028	.008	.031	.016	.005	.614	.575	.548	.529	.499	.478	.366	.350	.339	.213	.206	.202	.069	.068	.067
.116	.056	.016	.103	.050	.014	.077	.038	.011	.052	.026	.007	.029	.015	.004	.611	.572	.545	.526	.496	.476	.364	.349	.338	.212	.206	.201	.069	.068	.067
.109	.052	.015	.097	.047	.013	.072	.035	.010	.049	.025	.007	.028	.014	.004	.607	.569	.544	.523	.494	.474	.363	.348	.337	.212	.206	.201	.069	.068	.067
															.409	.409	.409	.350	.350	.350	.239	.239	.239	.137	.137	.137	.044	.044	.044
.226	.129	.041	.210	.120	.038	.181	.104	.033	.154	.089	.028	.129	.075	.024	.401	.383	.367	.343	.329	.316	.235	.226	.219	.135	.131	.127	.043	.042	.041
.210	.115	.035	.196	.108	.033	.169	.094	.029	.145	.081	.025	.122	.069	.022	.394	.365	.340	.337	.314	.294	.231	.217	.205	.133	.126	.120	.043	.041	.039
.195	.104	.031	.182	.098	.029	.158	.086	.026	.135	.074	.023	.115	.063	.020	.388	.351	.322	.332	.303	.279	.228	.211	.196	.132	.123	.116	.042	.040	.038
.182	.094	.028	.170	.089	.026	.147	.078	.023	.127	.068	.021	.107	.058	.018	.382	.341	.310	.328	.295	.269	.225	.205	.190	.130	.120	.112	.042	.039	.037
.169	.086	.025	.158	.081	.024	.138	.072	.021	.119	.063	.019	.101	.054	.016	.376	.333	.301	.323	.288	.262	.223	.201	.185	.129	.118	.110	.042	.039	.036
.158	.079	.023	.148	.075	.022	.129	.066	.019	.111	.058	.017	.095	.050	.015	.371	.327	.295	.319	.283	.257	.220	.198	.182	.128	.116	.108	.041	.038	.036
.148	.073	.021	.139	.069	.020	.121	.061	.018	.105	.054	.016	.089	.046	.014	.366	.321	.290	.315	.279	.253	.218	.195	.179	.126	.115	.107	.041	.038	.035
.139	.068	.019	.130	.064	.018	.114	.057	.016	.099	.050	.014	.084	.043	.013	.361	.317	.286	.311	.275	.250	.215	.193	.177	.125	.114	.106	.041	.037	.035
.131	.063	.018	.123	.060	.017	.108	.053	.015	.093	.047	.013	.080	.041	.012	.357	.313	.284	.308	.272	.247	.213	.191	.176	.124	.113	.105	.040	.037	.035
.123	.059	.016	.116	.056	.016	.102	.050	.014	.089	.044	.012	.076	.038	.011	.353	.310	.281	.304	.269	.246	.211	.189	.174	.123	.112	.104	.040	.037	.035
															.221	.221	.221	.189	.189	.189	.129	.129	.129	.074	.074	.074	.024	.024	.024
.202	.115	.036	.193	.110	.035	.178	.102	.033	.163	.094	.030	.150	.087	.028	.213	.198	.183	.183	.170	.158	.125	.117	.109	.072	.068	.064	.023	.022	.021
.186	.102	.031	.178	.098	.030	.164	.091	.028	.151	.085	.027	.139	.079	.025	.207	.181	.160	.177	.156	.138	.121	.108	.096	.070	.063	.056	.022	.020	.018
.172	.091	.027	.165	.088	.027	.153	.083	.025	.141	.077	.024	.130	.072	.022	.201	.169	.144	.172	.146	.125	.118	.101	.087	.068	.059	.051	.022	.019	.017
.159	.083	.024	.153	.080	.024	.142	.075	.022	.131	.071	.021	.121	.066	.020	.196	.160	.133	.168	.138	.115	.115	.096	.081	.067	.056	.048	.021	.018	.016
.148	.076	.022	.143	.073	.021	.133	.069	.020	.123	.065	.019	.114	.061	.018	.191	.153	.125	.164	.132	.109	.113	.092	.077	.065	.054	.045	.021	.018	.015
.139	.069	.020	.134	.068	.019	.124	.064	.019	.115	.060	.018	.107	.056	.017	.186	.147	.119	.160	.127	.104	.110	.089	.073	.064	.052	.044	.021	.017	.014
.130	.064	.018	.125	.062	.018	.117	.059	.017	.108	.056	.016	.101	.052	.015	.182	.142	.115	.156	.123	.100	.108	.087	.071	.063	.051	.042	.020	.017	.014
.122	.060	.017	.118	.058	.016	.110	.055	.016	.102	.052	.015	.095	.049	.014	.178	.138	.112	.153	.120	.097	.106	.084	.069	.062	.050	.041	.020	.016	.014
.115	.056	.016	.111	.054	.015	.104	.051	.015	.097	.048	.014	.090	.046	.013	.174	.135	.109	.150	.117	.095	.104	.082	.068	.060	.049	.040	.020	.016	.013
.108	.052	.014	.105	.051	.014	.098	.048	.014	.092	.046	.013	.086	.043	.012	.170	.132	.107	.147	.115	.094	.102	.081	.066	.059	.048	.040	.019	.016	.013
															.324	.324	.324	.277	.277	.277	.189	.189	.189	.108	.108	.108	.035	.035	.035
.232	.132	.042	.219	.125	.040	.196	.112	.036	.175	.101	.032	.155	.090	.029	.318	.300	.284	.272	.257	.244	.186	.177	.169	.107	.102	.098	.034	.033	.032
.204	.112	.034	.193	.106	.033	.172	.096	.030	.152	.086	.027	.134	.076	.024	.312	.283	.260	.267	.244	.224	.183	.169	.157	.105	.098	.092	.034	.032	.030
.185	.098	.029	.174	.093	.028	.155	.084	.026	.137	.075	.023	.121	.067	.021	.305	.271	.244	.262	.234	.211	.180	.163	.148	.104	.095	.087	.033	.031	.029
.169	.088	.026	.160	.083	.025	.142	.075	.022	.126	.067	.020	.111	.060	.018	.300	.262	.233	.257	.226	.202	.177	.158	.143	.102	.092	.084	.033	.030	.028
.156	.079	.023	.147	.076	.022	.131	.068	.020	.116	.061	.018	.102	.055	.016	.294	.255	.225	.253	.220	.196	.174	.154	.138	.101	.090	.082	.033	.029	.027
.144	.072	.021	.136	.069	.020	.122	.062	.018	.108	.056	.016	.095	.050	.015	.289	.249	.220	.249	.216	.191	.172	.151	.135	.100	.089	.080	.032	.029	.027
.134	.066	.019	.127	.063	.018	.114	.057	.017	.101	.052	.015	.089	.046	.014	.285	.244	.216	.245	.212	.188	.169	.148	.133	.098	.087	.079	.032	.029	.026
.126	.061	.017	.119	.059	.017	.107	.053	.015	.095	.048	.014	.084	.043	.012	.280	.240	.212	.241	.208	.185	.167	.146	.131	.097	.086	.078	.031	.028	.026
.118	.057	.016	.112	.055	.015	.100	.050	.014	.089	.045	.013	.079	.040	.012	.276	.237	.210	.238	.205	.183	.165	.144	.130	.096	.085	.078	.031	.028	.026
.111	.053	.015	.106	.051	.014	.095	.046	.013	.084	.042	.012	.075	.038	.011	.272	.234	.208	.235	.203	.181	.163	.143	.129	.095	.084	.077	.031	.028	.026

Fig. 9–62. *Continued (see page 9-53 for instructions and notes)*

Typical Luminaire	Typical Intensity Distribution and Per Cent Lamp Lumens		ρ_{CC} →	80			70			50			30			10			0	WDRC	ρ_{CC} →
			ρ_W →	50	30	10	50	30	10	50	30	10	50	30	10	50	30	10	0		ρ_W →
	Maint. Cat.	SC	RCR ↓	Coefficients of Utilization for 20 Per Cent Effective Floor Cavity Reflectance (ρ_{FC} = 20)																	RCR ↓
37	V	1.3	0	.52	.52	.52	.50	.50	.50	.46	.46	.46	.43	.43	.43	.39	.39	.39	.38		
			1	.44	.42	.40	.42	.40	.39	.39	.37	.36	.36	.35	.33	.33	.32	.31	.30	.201	1
	8%↑		2	.38	.35	.32	.37	.33	.31	.34	.31	.29	.31	.29	.27	.28	.27	.25	.24	.171	2
			3	.33	.29	.26	.32	.28	.25	.29	.26	.24	.27	.25	.22	.25	.23	.21	.20	.149	3
			4	.29	.25	.22	.28	.24	.21	.26	.23	.20	.24	.21	.19	.22	.20	.18	.17	.132	4
	37½%↓		5	.26	.22	.19	.25	.21	.18	.23	.20	.17	.21	.18	.16	.20	.17	.15	.14	.117	5
			6	.23	.19	.16	.22	.18	.16	.21	.17	.15	.19	.16	.14	.18	.15	.13	.12	.106	6
			7	.21	.17	.14	.20	.16	.14	.19	.15	.13	.17	.15	.12	.16	.14	.12	.11	.096	7
			8	.19	.15	.12	.18	.15	.12	.17	.14	.12	.16	.13	.11	.15	.12	.11	.10	.088	8
			9	.17	.14	.11	.17	.13	.11	.16	.13	.10	.15	.12	.10	.14	.11	.09	.09	.081	9
2-lamp diffuse wraparound—see note 7			10	.16	.12	.10	.15	.12	.10	.14	.11	.09	.14	.11	.09	.13	.10	.09	.08	.075	10
38	IV	1.0	0	.60	.60	.60	.58	.58	.58	.56	.56	.56	.53	.53	.53	.51	.51	.51	.50		
	0%↑		1	.53	.51	.49	.52	.50	.49	.50	.48	.47	.48	.47	.46	.46	.45	.44	.43	.168	1
			2	.47	.44	.42	.46	.43	.41	.44	.42	.40	.43	.41	.39	.41	.40	.38	.37	.159	2
			3	.42	.38	.36	.41	.38	.35	.40	.37	.35	.39	.36	.34	.37	.35	.34	.32	.146	3
			4	.38	.34	.31	.37	.34	.31	.36	.33	.30	.35	.32	.30	.34	.32	.30	.29	.135	4
	50%↓		5	.34	.30	.27	.34	.30	.27	.33	.29	.27	.32	.29	.27	.31	.28	.26	.25	.124	5
			6	.31	.27	.24	.31	.27	.24	.30	.27	.24	.29	.26	.24	.28	.26	.24	.23	.114	6
			7	.29	.25	.22	.28	.24	.22	.28	.24	.22	.27	.24	.21	.26	.23	.21	.20	.106	7
			8	.26	.22	.20	.26	.22	.20	.25	.22	.20	.25	.22	.20	.24	.21	.19	.19	.099	8
4-lamp, 610 mm (2′) wide troffer with			9	.24	.21	.18	.24	.21	.18	.24	.20	.18	.23	.20	.18	.23	.20	.18	.17	.092	9
45° plastic louver—see note 7			10	.23	.19	.17	.22	.19	.17	.22	.19	.16	.22	.19	.16	.21	.18	.16	.16	.086	10
39	IV	0.9	0	.55	.55	.55	.54	.54	.54	.51	.51	.51	.49	.49	.49	.47	.47	.47	.46		
	0%↑		1	.49	.48	.46	.48	.47	.46	.46	.45	.44	.45	.44	.43	.43	.42	.42	.41	.137	1
			2	.44	.42	.40	.43	.41	.39	.42	.40	.38	.40	.39	.37	.39	.38	.37	.36	.131	2
			3	.40	.37	.34	.39	.36	.34	.38	.36	.33	.37	.35	.33	.36	.34	.32	.32	.122	3
			4	.36	.33	.30	.36	.33	.30	.35	.32	.30	.34	.31	.29	.33	.31	.29	.28	.113	4
			5	.33	.30	.27	.33	.29	.27	.32	.29	.27	.31	.28	.26	.30	.28	.26	.25	.104	5
	46%↓		6	.30	.27	.24	.30	.27	.24	.29	.26	.24	.29	.26	.24	.28	.25	.24	.23	.097	6
			7	.28	.25	.22	.28	.24	.22	.27	.24	.22	.26	.24	.22	.26	.23	.22	.21	.090	7
			8	.26	.23	.20	.26	.22	.20	.25	.22	.20	.25	.22	.20	.24	.22	.20	.19	.085	8
4-lamp, 610 mm (2′) wide troffer with			9	.24	.21	.19	.24	.21	.19	.23	.20	.18	.23	.20	.18	.23	.20	.18	.18	.079	9
45° white metal louver—see note 7			10	.23	.19	.17	.22	.19	.17	.22	.19	.17	.22	.19	.17	.21	.19	.17	.16	.075	10
40	V	1.2	0	.73	.73	.73	.71	.71	.71	.68	.68	.68	.65	.65	.65	.62	.62	.62	.60		
			1	.63	.60	.58	.62	.59	.57	.59	.57	.55	.56	.55	.53	.54	.53	.51	.50	.259	1
	1%↑		2	.55	.51	.47	.54	.50	.46	.51	.48	.45	.49	.46	.44	.47	.45	.43	.42	.236	2
			3	.48	.43	.39	.47	.42	.39	.45	.41	.38	.43	.40	.37	.42	.39	.36	.35	.212	3
			4	.43	.37	.33	.42	.37	.33	.40	.36	.32	.39	.35	.32	.37	.34	.31	.30	.191	4
	60½%↓		5	.38	.33	.29	.37	.32	.28	.36	.31	.28	.35	.31	.28	.33	.30	.27	.26	.173	5
			6	.34	.29	.25	.34	.29	.25	.33	.28	.24	.31	.27	.24	.30	.27	.24	.23	.158	6
			7	.31	.26	.22	.31	.26	.22	.30	.25	.22	.29	.25	.21	.28	.24	.21	.20	.144	7
			8	.28	.23	.20	.28	.23	.20	.27	.23	.19	.26	.22	.19	.25	.22	.19	.18	.133	8
Fluorescent unit dropped diffuser, 4-			9	.26	.21	.18	.26	.21	.18	.25	.21	.17	.24	.20	.17	.24	.20	.17	.16	.123	9
lamp 610 mm (2′) wide—see note 7			10	.24	.19	.16	.24	.19	.16	.23	.19	.16	.22	.19	.16	.22	.18	.16	.15	.115	10
41	V	1.2	0	.69	.69	.69	.67	.67	.67	.64	.64	.64	.61	.61	.61	.59	.59	.59	.58		
			1	.60	.58	.55	.59	.57	.55	.56	.55	.53	.54	.53	.51	.52	.51	.50	.49	.227	1
	0%↑		2	.52	.49	.45	.51	.48	.45	.49	.46	.44	.47	.45	.43	.46	.44	.42	.40	.214	2
			3	.46	.41	.38	.45	.41	.37	.43	.40	.37	.42	.39	.36	.40	.38	.35	.34	.196	3
			4	.41	.36	.32	.40	.35	.32	.39	.34	.31	.37	.34	.31	.36	.33	.30	.29	.178	4
	57½%↓		5	.36	.31	.28	.36	.31	.27	.35	.30	.27	.33	.30	.27	.32	.29	.26	.25	.162	5
			6	.33	.28	.24	.32	.27	.24	.31	.27	.24	.30	.26	.23	.29	.26	.23	.22	.148	6
			7	.30	.25	.21	.29	.25	.21	.28	.24	.21	.28	.24	.21	.27	.23	.21	.20	.136	7
			8	.27	.22	.19	.27	.22	.19	.26	.22	.19	.25	.21	.19	.25	.21	.19	.17	.126	8
Fluorescent unit with flat bottom diffuser,			9	.25	.20	.17	.25	.20	.17	.24	.20	.17	.23	.20	.17	.23	.19	.17	.16	.116	9
4-lamp 610 mm (2′) wide—see note 7			10	.23	.18	.15	.23	.18	.15	.22	.18	.15	.22	.18	.15	.21	.18	.15	.14	.108	10
42	V	1.4/1.2	0	.75	.75	.75	.73	.73	.73	.70	.70	.70	.67	.67	.67	.64	.64	.64	.63		
			1	.67	.64	.62	.65	.63	.61	.63	.61	.59	.60	.59	.58	.58	.57	.56	.55	.208	1
			2	.59	.56	.52	.58	.55	.52	.56	.53	.51	.54	.52	.49	.52	.50	.48	.47	.199	2
	0%↑		3	.53	.48	.45	.52	.48	.44	.50	.46	.43	.48	.45	.43	.47	.44	.42	.41	.186	3
			4	.47	.42	.38	.46	.42	.38	.45	.41	.38	.44	.40	.37	.42	.39	.37	.35	.172	4
	63%↓	60°	5	.43	.37	.34	.42	.37	.33	.41	.36	.33	.39	.36	.33	.38	.35	.32	.31	.160	5
			6	.39	.33	.30	.38	.33	.29	.37	.32	.29	.36	.32	.29	.35	.31	.29	.27	.148	6
			7	.35	.30	.26	.35	.30	.26	.34	.29	.26	.33	.29	.26	.32	.28	.26	.24	.138	7
			8	.32	.27	.24	.32	.27	.23	.31	.26	.23	.30	.26	.23	.29	.26	.23	.22	.128	8
Fluorescent unit with flat prismatic lens,			9	.30	.25	.21	.29	.24	.21	.28	.24	.21	.28	.24	.21	.27	.24	.21	.20	.120	9
4-lamp 610 mm (2′) wide—see note 7			10	.27	.22	.19	.27	.22	.19	.26	.22	.19	.26	.22	.19	.25	.22	.19	.18	.113	10

80			70			50			30			10			80			70			50			30			10		
50	30	10	50	30	10	50	30	10	50	30	10	50	30	10	50	30	10	50	30	10	50	30	10	50	30	10	50	30	10
Wall Exitance Coefficients for 20 Per Cent Effective Floor Cavity Reflectance (ρ_{FC} = 20)															Ceiling Cavity Exitance Coefficients for 20 Per Cent Floor Cavity Reflectance (ρ_{FC} = 20)														
															.147	.147	.147	.125	.125	.125	.085	.085	.085	.049	.049	.049	.016	.016	.016
.162	.092	.029	.156	.089	.028	.145	.083	.027	.136	.078	.025	.127	.073	.024	.144	.131	.120	.123	.113	.103	.084	.077	.071	.048	.045	.041	.016	.014	.013
.144	.079	.024	.139	.076	.024	.129	.072	.022	.120	.068	.021	.112	.064	.020	.141	.121	.104	.120	.104	.090	.083	.072	.063	.048	.042	.037	.015	.014	.012
.130	.069	.021	.125	.067	.020	.116	.063	.019	.108	.059	.018	.101	.056	.017	.137	.113	.094	.118	.098	.081	.081	.068	.057	.047	.040	.034	.015	.013	.011
.118	.061	.018	.114	.059	.018	.106	.056	.017	.098	.053	.016	.092	.050	.015	.134	.107	.087	.115	.093	.076	.079	.065	.053	.046	.038	.032	.015	.012	.010
.108	.055	.016	.104	.053	.016	.097	.050	.015	.090	.048	.014	.084	.045	.013	.130	.103	.083	.112	.089	.072	.077	.062	.051	.045	.037	.030	.014	.012	.010
.099	.050	.014	.096	.048	.014	.089	.046	.013	.083	.043	.013	.077	.041	.012	.127	.099	.079	.109	.086	.069	.075	.060	.049	.044	.035	.029	.014	.012	.010
.092	.045	.013	.088	.044	.013	.083	.042	.012	.077	.039	.011	.072	.037	.011	.124	.096	.077	.107	.083	.067	.074	.059	.047	.043	.034	.028	.014	.011	.009
.085	.042	.012	.082	.041	.011	.077	.038	.011	.072	.036	.010	.067	.034	.010	.121	.094	.075	.104	.081	.065	.072	.057	.046	.042	.034	.028	.014	.011	.009
.080	.038	.011	.077	.037	.011	.072	.036	.010	.067	.034	.010	.063	.032	.009	.118	.092	.074	.102	.079	.064	.071	.056	.046	.041	.033	.027	.013	.011	.009
.075	.036	.010	.072	.035	.010	.067	.033	.009	.063	.031	.009	.059	.030	.009	.116	.090	.072	.100	.078	.063	.069	.055	.045	.040	.032	.027	.013	.011	.009
															.095	.095	.095	.082	.082	.082	.056	.056	.056	.032	.032	.032	.010	.010	.010
.135	.077	.024	.132	.075	.024	.125	.072	.023	.119	.069	.022	.114	.066	.021	.089	.078	.069	.076	.067	.059	.052	.046	.041	.030	.027	.024	.010	.009	.008
.128	.070	.022	.125	.069	.021	.120	.066	.021	.114	.064	.020	.109	.062	.020	.084	.066	.051	.072	.057	.044	.049	.039	.031	.028	.023	.018	.009	.007	.006
.120	.064	.019	.117	.063	.019	.112	.061	.018	.108	.059	.018	.103	.057	.018	.079	.057	.039	.068	.049	.034	.047	.034	.024	.027	.020	.014	.009	.006	.005
.112	.058	.017	.109	.057	.017	.105	.056	.017	.101	.054	.016	.097	.053	.016	.075	.050	.031	.065	.043	.027	.044	.030	.019	.026	.018	.011	.008	.006	.004
.104	.053	.015	.102	.052	.015	.098	.051	.015	.094	.050	.015	.091	.049	.015	.071	.045	.025	.061	.039	.022	.042	.027	.016	.024	.016	.009	.008	.005	.003
.097	.049	.014	.095	.048	.014	.092	.047	.014	.089	.046	.014	.086	.045	.013	.068	.041	.021	.058	.035	.018	.040	.025	.013	.023	.014	.008	.008	.005	.003
.091	.045	.013	.089	.045	.013	.086	.044	.013	.083	.043	.012	.081	.042	.012	.065	.037	.018	.056	.032	.015	.038	.023	.011	.022	.013	.007	.007	.004	.002
.086	.042	.012	.084	.041	.012	.081	.041	.012	.079	.040	.012	.076	.039	.011	.062	.034	.015	.053	.030	.013	.037	.021	.009	.021	.012	.006	.007	.004	.002
.081	.039	.011	.079	.039	.011	.077	.038	.011	.075	.037	.011	.072	.037	.011	.059	.032	.013	.051	.027	.012	.035	.019	.008	.020	.011	.005	.007	.004	.002
.076	.037	.010	.075	.036	.010	.073	.036	.010	.071	.035	.010	.069	.035	.010	.056	.029	.012	.048	.026	.010	.033	.018	.007	.020	.011	.004	.006	.003	.001
															.088	.088	.088	.075	.075	.075	.051	.051	.051	.029	.029	.029	.009	.009	.009
.115	.065	.021	.112	.064	.020	.106	.061	.019	.100	.058	.019	.095	.055	.018	.081	.072	.064	.069	.062	.055	.048	.043	.038	.027	.025	.022	.009	.008	.007
.109	.060	.018	.107	.059	.018	.102	.056	.018	.097	.054	.017	.092	.052	.016	.076	.061	.048	.065	.052	.042	.045	.036	.029	.026	.021	.017	.008	.007	.006
.103	.055	.016	.100	.054	.016	.096	.052	.016	.092	.050	.015	.088	.049	.015	.072	.053	.037	.061	.045	.032	.042	.031	.023	.024	.018	.013	.008	.006	.004
.096	.050	.015	.094	.049	.015	.090	.048	.014	.086	.046	.014	.083	.045	.014	.068	.046	.030	.058	.040	.026	.040	.028	.018	.023	.016	.011	.007	.005	.004
.090	.046	.013	.088	.045	.013	.085	.044	.013	.081	.043	.013	.078	.042	.013	.064	.041	.024	.055	.036	.021	.038	.025	.015	.022	.015	.009	.007	.005	.003
.084	.042	.012	.083	.042	.012	.080	.041	.012	.077	.040	.012	.074	.039	.012	.061	.037	.020	.052	.032	.017	.036	.022	.012	.021	.013	.007	.007	.004	.002
.079	.039	.011	.078	.039	.011	.075	.038	.011	.073	.037	.011	.070	.036	.011	.058	.034	.017	.050	.029	.015	.034	.021	.010	.020	.012	.006	.006	.004	.002
.075	.037	.010	.074	.036	.010	.071	.035	.010	.069	.035	.010	.067	.034	.010	.055	.031	.015	.047	.027	.013	.033	.019	.009	.019	.011	.005	.006	.004	.002
.071	.034	.010	.070	.034	.010	.067	.033	.009	.065	.033	.009	.063	.032	.009	.052	.029	.013	.045	.025	.011	.031	.018	.008	.018	.010	.005	.006	.003	.002
.067	.032	.009	.066	.032	.009	.064	.031	.009	.062	.031	.009	.060	.030	.009	.050	.027	.011	.043	.023	.010	.030	.016	.007	.017	.010	.004	.006	.003	.001
															.123	.123	.123	.105	.105	.105	.072	.072	.072	.041	.041	.041	.013	.013	.013
.196	.111	.035	.191	.109	.035	.182	.105	.033	.174	.101	.032	.167	.097	.031	.118	.102	.089	.101	.088	.076	.069	.060	.053	.040	.035	.031	.013	.011	.010
.181	.099	.030	.177	.098	.030	.170	.094	.029	.163	.091	.029	.156	.088	.028	.113	.087	.066	.096	.075	.057	.066	.052	.040	.038	.030	.023	.012	.010	.008
.167	.089	.027	.163	.087	.026	.156	.085	.026	.150	.082	.025	.144	.080	.025	.108	.077	.052	.092	.066	.045	.063	.046	.032	.037	.027	.019	.012	.009	.006
.153	.080	.023	.150	.079	.023	.144	.076	.023	.139	.074	.022	.133	.072	.022	.103	.069	.042	.088	.059	.037	.061	.041	.026	.035	.024	.015	.011	.008	.005
.141	.072	.021	.139	.071	.021	.133	.069	.020	.128	.068	.020	.124	.066	.020	.098	.062	.036	.085	.054	.031	.058	.038	.022	.034	.022	.013	.011	.007	.004
.131	.066	.019	.128	.065	.019	.124	.063	.018	.119	.062	.018	.115	.061	.018	.094	.057	.031	.081	.050	.027	.056	.035	.019	.032	.020	.011	.010	.007	.004
.122	.060	.017	.119	.060	.017	.115	.058	.017	.111	.057	.017	.107	.056	.016	.090	.053	.027	.077	.046	.024	.053	.032	.017	.031	.019	.010	.010	.006	.003
.114	.056	.016	.112	.055	.016	.108	.054	.015	.104	.053	.015	.101	.052	.015	.086	.049	.024	.074	.043	.021	.051	.030	.015	.030	.018	.009	.010	.006	.003
.106	.051	.014	.105	.051	.014	.101	.050	.014	.098	.049	.014	.095	.048	.014	.082	.046	.022	.071	.040	.019	.049	.028	.014	.029	.017	.008	.009	.005	.003
.100	.048	.013	.098	.048	.013	.095	.047	.013	.092	.046	.013	.089	.045	.013	.079	.044	.021	.068	.038	.018	.047	.027	.013	.027	.016	.008	.009	.005	.003
															.110	.110	.110	.094	.094	.094	.064	.064	.064	.037	.037	.037	.012	.012	.012
.174	.099	.031	.170	.097	.031	.162	.093	.030	.155	.089	.029	.149	.086	.028	.104	.090	.078	.089	.077	.067	.061	.053	.046	.035	.031	.027	.011	.010	.009
.165	.090	.028	.161	.089	.027	.155	.086	.027	.149	.083	.026	.143	.081	.025	.099	.076	.057	.085	.065	.049	.058	.045	.034	.033	.026	.020	.011	.009	.007
.153	.082	.024	.150	.080	.024	.144	.078	.024	.139	.076	.023	.134	.074	.023	.094	.066	.043	.081	.057	.037	.056	.039	.026	.032	.023	.015	.010	.007	.005
.142	.074	.022	.139	.073	.022	.134	.071	.021	.129	.069	.021	.124	.068	.021	.090	.058	.034	.077	.050	.029	.053	.035	.021	.031	.021	.012	.010	.007	.004
.131	.067	.019	.129	.066	.019	.124	.065	.019	.120	.063	.019	.116	.062	.019	.086	.052	.027	.074	.045	.024	.051	.032	.017	.029	.018	.010	.009	.006	.003
.122	.061	.018	.120	.061	.017	.116	.059	.017	.112	.058	.017	.108	.057	.017	.082	.047	.023	.070	.041	.020	.048	.029	.014	.028	.017	.008	.009	.006	.003
.114	.056	.016	.112	.056	.016	.108	.055	.016	.104	.054	.016	.101	.053	.015	.078	.043	.019	.067	.038	.017	.046	.026	.012	.027	.016	.007	.009	.005	.002
.106	.052	.015	.104	.051	.015	.101	.051	.014	.098	.050	.014	.095	.049	.014	.074	.040	.017	.064	.035	.014	.044	.024	.010	.026	.014	.006	.008	.005	.002
.100	.048	.013	.098	.048	.013	.095	.047	.013	.092	.046	.013	.090	.046	.013	.070	.037	.015	.061	.032	.013	.042	.023	.009	.025	.013	.005	.008	.004	.002
.094	.045	.012	.092	.045	.012	.090	.044	.012	.087	.043	.012	.085	.043	.012	.067	.035	.013	.058	.030	.011	.040	.021	.008	.023	.013	.005	.008	.004	.002
															.120	.120	.120	.103	.103	.103	.070	.070	.070	.040	.040	.040	.013	.013	.013
.168	.096	.030	.164	.093	.030	.156	.089	.029	.148	.085	.027	.141	.082	.026	.112	.099	.087	.096	.085	.075	.065	.058	.052	.038	.034	.030	.012	.011	.010
.161	.088	.027	.157	.087	.027	.150	.083	.026	.143	.080	.025	.137	.078	.024	.105	.083	.064	.090	.072	.056	.062	.050	.039	.036	.029	.023	.011	.009	.007
.152	.081	.024	.148	.079	.024	.142	.077	.023	.136	.075	.023	.131	.072	.022	.100	.072	.049	.086	.062	.043	.059	.043	.030	.034	.025	.018	.011	.008	.006
.142	.074	.022	.139	.073	.022	.134	.071	.021	.128	.069	.021	.124	.067	.020	.095	.063	.039	.082	.055	.034	.056	.038	.024	.032	.022	.014	.010	.007	.005
.133	.068	.020	.131	.067	.020	.126	.065	.019	.121	.064	.019	.117	.062	.019	.091	.057	.031	.078	.049	.027	.054	.034	.019	.031	.020	.011	.010	.007	.004
.125	.063	.018	.123	.062	.018	.118	.061	.018	.114	.059	.017	.110	.058	.017	.086	.051	.026	.074	.044	.023	.051	.031	.016	.030	.018	.010	.010	.006	.003
.117	.058	.016	.115	.057	.016	.111	.056	.016	.108	.055	.016	.104	.054	.016	.082	.047	.022	.071	.041	.019	.049	.028	.014	.028	.017	.008	.009	.005	.003
.110	.054	.015	.109	.053	.015	.105	.052	.015	.102	.052	.015	.099	.051	.015	.079	.043	.019	.068	.037	.016	.047	.026	.012	.027	.016	.007	.009	.005	.002
.104	.050	.014	.102	.050	.014	.099	.049	.014	.096	.048	.014	.093	.047	.014	.075	.040	.017	.065	.035	.014	.045	.024	.010	.026	.014	.006	.008	.005	.002
.098	.047	.013	.097	.047	.013	.094	.046	.013	.091	.045	.013	.089	.045	.013	.072	.037	.015	.062	.032	.013	.043	.023	.009	.025	.014	.005	.008	.004	.002

Fig. 9-62. *Continued (see page 9-53 for instructions and notes)*

Typical Luminaire	Typical Intensity Distribution and Per Cent Lamp Lumens: Maint. Cat.	SC	ρCC →	80			70			50			30			10			0	WDRC	ρCC →
			ρW →	50	30	10	50	30	10	50	30	10	50	30	10	50	30	10	0		ρW →
			RCR ↓	Coefficients of Utilization for 20 Per Cent Effective Floor Cavity Reflectance (ρ_{FC} = 20)																	RCR ↓
43	V	1.4/1.3	0	.78	.78	.78	.76	.76	.76	.73	.73	.73	.70	.70	.70	.67	.67	.67	.66		
4-lamp, 610 mm (2′) wide unit with sharp cutoff (high angle—low luminance) flat prismatic lens—see note 7			1	.71	.68	.66	.69	.67	.65	.66	.65	.63	.64	.63	.61	.62	.61	.60	.58	.181	1
			2	.63	.60	.57	.62	.59	.56	.60	.57	.55	.58	.56	.54	.56	.54	.52	.51	.180	2
			3	.57	.52	.49	.56	.52	.48	.54	.51	.48	.52	.49	.47	.51	.48	.46	.45	.173	3
			4	.51	.46	.43	.50	.46	.42	.49	.45	.42	.47	.44	.41	.46	.43	.41	.39	.164	4
			5	.46	.41	.37	.46	.41	.37	.44	.40	.37	.43	.39	.36	.42	.39	.36	.35	.154	5
			6	.42	.37	.33	.41	.37	.33	.40	.36	.33	.39	.35	.32	.38	.35	.32	.31	.145	6
			7	.38	.33	.29	.38	.33	.29	.37	.32	.29	.36	.32	.29	.35	.32	.29	.28	.136	7
			8	.35	.30	.26	.35	.30	.26	.34	.29	.26	.33	.29	.26	.32	.29	.26	.25	.127	8
			9	.32	.27	.24	.32	.27	.24	.31	.27	.24	.31	.27	.24	.30	.26	.24	.22	.120	9
			10	.30	.25	.22	.30	.25	.22	.29	.25	.22	.28	.24	.22	.28	.24	.21	.20	.113	10
44	IV	N.A.	0	.71	.71	.71	.70	.70	.70	.66	.66	.66	.64	.64	.64	.61	.61	.61	.60		
Bilateral batwing distribution—louvered fluorescent unit			1	.64	.62	.60	.63	.61	.60	.60	.59	.58	.58	.57	.56	.56	.55	.54	.53	.167	1
			2	.57	.54	.51	.56	.53	.51	.54	.52	.50	.52	.50	.48	.51	.49	.47	.46	.170	2
			3	.51	.47	.44	.50	.46	.43	.49	.45	.43	.47	.44	.42	.46	.43	.41	.40	.165	3
			4	.46	.41	.38	.45	.41	.37	.44	.40	.37	.42	.39	.36	.41	.38	.36	.35	.157	4
			5	.41	.36	.33	.40	.36	.32	.39	.35	.32	.38	.35	.32	.37	.34	.31	.30	.148	5
			6	.37	.32	.28	.36	.32	.28	.35	.31	.28	.34	.31	.28	.34	.30	.28	.27	.139	6
			7	.33	.29	.25	.33	.28	.25	.32	.28	.25	.31	.27	.25	.30	.27	.24	.23	.130	7
			8	.30	.26	.22	.30	.25	.22	.29	.25	.22	.28	.25	.22	.28	.24	.22	.21	.122	8
			9	.28	.23	.20	.27	.23	.20	.27	.23	.20	.26	.22	.20	.25	.22	.19	.18	.115	9
			10	.25	.21	.18	.25	.21	.18	.25	.20	.18	.24	.20	.18	.23	.20	.18	.17	.108	10
45	V	N.A.	0	.57	.57	.57	.56	.56	.56	.53	.53	.53	.51	.51	.51	.49	.49	.49	.48		
Bilateral batwing distribution—4-lamp, 610 mm (2′) wide fluorescent unit with flat prismatic lens and overlay—see note 7			1	.50	.48	.46	.49	.47	.45	.47	.45	.44	.45	.43	.42	.43	.42	.41	.40	.204	1
			2	.43	.40	.37	.42	.39	.36	.40	.38	.35	.39	.37	.35	.37	.36	.34	.33	.192	2
			3	.37	.33	.30	.37	.33	.30	.35	.32	.29	.34	.31	.29	.33	.30	.28	.27	.175	3
			4	.33	.28	.25	.32	.28	.25	.31	.27	.24	.30	.27	.24	.29	.26	.24	.23	.159	4
			5	.29	.24	.21	.28	.24	.21	.27	.24	.21	.26	.23	.20	.25	.23	.20	.19	.145	5
			6	.26	.21	.18	.25	.21	.18	.24	.21	.18	.24	.20	.18	.23	.20	.17	.16	.132	6
			7	.23	.19	.16	.23	.18	.15	.22	.18	.15	.21	.18	.15	.21	.17	.15	.14	.122	7
			8	.21	.17	.14	.21	.16	.14	.20	.16	.13	.19	.16	.13	.19	.16	.13	.12	.112	8
			9	.19	.15	.12	.19	.15	.12	.18	.14	.12	.18	.14	.12	.17	.14	.12	.11	.104	9
			10	.17	.13	.11	.17	.13	.11	.17	.13	.11	.16	.13	.11	.16	.13	.10	.10	.096	10
46	V	N.A.	0	.87	.87	.87	.84	.84	.84	.77	.77	.77	.72	.72	.72	.66	.66	.66	.64		
Bilateral batwing distribution—one-lamp, surface mounted fluorescent with prismatic wraparound lens			1	.75	.72	.69	.72	.69	.66	.67	.64	.62	.62	.60	.58	.57	.56	.54	.52	.296	1
			2	.65	.60	.56	.63	.58	.54	.58	.54	.51	.54	.51	.48	.50	.47	.45	.43	.261	2
			3	.57	.51	.46	.55	.49	.45	.51	.46	.42	.47	.43	.40	.44	.41	.38	.36	.232	3
			4	.50	.44	.39	.48	.42	.38	.45	.40	.36	.42	.38	.34	.39	.35	.32	.30	.209	4
			5	.45	.38	.33	.43	.37	.32	.40	.35	.31	.37	.33	.29	.35	.31	.28	.26	.189	5
			6	.40	.33	.28	.39	.32	.28	.36	.31	.26	.34	.29	.25	.31	.27	.24	.22	.172	6
			7	.36	.29	.25	.35	.29	.24	.32	.27	.23	.30	.26	.22	.28	.24	.21	.19	.158	7
			8	.33	.26	.22	.31	.25	.21	.29	.24	.20	.28	.23	.20	.26	.22	.19	.17	.146	8
			9	.30	.23	.19	.29	.23	.19	.27	.22	.18	.25	.21	.17	.24	.20	.17	.15	.135	9
			10	.27	.21	.17	.26	.21	.17	.25	.20	.16	.23	.19	.16	.22	.18	.15	.13	.126	10
47	V	1.7	0	.71	.71	.71	.69	.69	.69	.66	.66	.66	.63	.63	.63	.61	.61	.61	.60		
Radial batwing distribution—4-lamp, 610 mm (2′) wide fluorescent unit with flat prismatic lens—see note 7			1	.62	.59	.57	.60	.58	.56	.58	.56	.54	.55	.54	.52	.53	.52	.51	.50	.251	1
			2	.53	.49	.46	.52	.48	.45	.50	.47	.44	.48	.45	.43	.46	.44	.42	.41	.237	2
			3	.46	.41	.37	.45	.41	.37	.44	.40	.36	.42	.39	.36	.40	.38	.35	.34	.216	3
			4	.41	.35	.31	.40	.35	.31	.38	.34	.30	.37	.33	.30	.36	.32	.30	.28	.196	4
			5	.36	.30	.26	.35	.30	.26	.34	.29	.26	.33	.29	.26	.32	.28	.25	.24	.178	5
			6	.32	.27	.23	.32	.26	.23	.31	.26	.22	.29	.25	.22	.29	.25	.22	.21	.162	6
			7	.29	.24	.20	.28	.23	.20	.28	.23	.19	.27	.22	.19	.26	.22	.19	.18	.149	7
			8	.26	.21	.17	.26	.21	.17	.25	.20	.17	.24	.20	.17	.24	.20	.17	.16	.137	8
			9	.24	.19	.15	.24	.19	.15	.23	.18	.15	.22	.18	.15	.22	.18	.15	.14	.127	9
			10	.22	.17	.14	.22	.17	.14	.21	.17	.14	.20	.16	.14	.20	.16	.14	.12	.118	10
48	I	1.6/1.2	0	1.01	1.01	1.01	.96	.96	.96	.87	.87	.87	.79	.79	.79	.72	.72	.72	.68		
2-lamp fluorescent strip unit			1	.84	.79	.75	.80	.76	.72	.72	.69	.66	.65	.63	.60	.59	.57	.55	.52	.414	1
			2	.72	.65	.59	.68	.62	.57	.62	.57	.52	.56	.52	.48	.50	.47	.44	.41	.343	2
			3	.62	.54	.48	.59	.52	.46	.53	.47	.42	.48	.43	.39	.43	.39	.36	.33	.293	3
			4	.54	.46	.39	.52	.44	.38	.47	.40	.35	.42	.37	.33	.38	.34	.30	.27	.255	4
			5	.48	.40	.33	.46	.38	.32	.41	.35	.30	.38	.32	.28	.34	.29	.26	.23	.225	5
			6	.43	.35	.29	.41	.33	.28	.37	.31	.26	.34	.28	.24	.30	.26	.22	.20	.202	6
			7	.38	.30	.25	.37	.29	.24	.34	.27	.22	.31	.25	.21	.28	.23	.19	.17	.182	7
			8	.35	.27	.22	.33	.26	.21	.31	.24	.20	.28	.22	.18	.25	.21	.17	.15	.166	8
			9	.32	.24	.19	.30	.24	.19	.28	.22	.18	.26	.20	.16	.23	.19	.15	.13	.152	9
			10	.29	.22	.17	.28	.21	.17	.26	.20	.16	.24	.18	.15	.22	.17	.14	.12	.140	10

80			70			50			30			10			80			70			50			30			10		
50	30	10	50	30	10	50	30	10	50	30	10	50	30	10	50	30	10	50	30	10	50	30	10	50	30	10	50	30	10
Wall Exitance Coefficients for 20 Per Cent Effective Floor Cavity Reflectance (ρ_{FC} = 20)															Ceiling Cavity Exitance Coefficients for 20 Per Cent Floor Cavity Reflectance (ρ_{FC} = 20)														
															.125	.125	.125	.107	.107	.107	.073	.073	.073	.042	.042	.042	.013	.013	.013
.156	.089	.028	.152	.087	.028	.143	.082	.026	.136	.078	.025	.128	.074	.024	.115	.103	.092	.098	.088	.079	.067	.061	.055	.039	.035	.032	.012	.011	.010
.153	.084	.026	.149	.082	.025	.142	.079	.024	.135	.076	.024	.129	.073	.023	.108	.087	.069	.092	.075	.060	.063	.052	.042	.037	.030	.024	.012	.010	.008
.146	.078	.023	.143	.076	.023	.136	.074	.022	.130	.071	.022	.125	.069	.021	.102	.075	.053	.087	.065	.046	.060	.045	.032	.035	.026	.019	.011	.008	.006
.139	.072	.021	.136	.071	.021	.130	.069	.021	.125	.067	.020	.120	.065	.020	.097	.066	.042	.083	.057	.036	.057	.040	.026	.033	.023	.015	.011	.008	.005
.131	.067	.019	.129	.066	.019	.124	.064	.019	.119	.063	.019	.115	.061	.018	.092	.059	.034	.079	.051	.029	.055	.035	.021	.032	.021	.012	.010	.007	.004
.124	.062	.018	.122	.061	.018	.117	.060	.017	.113	.059	.017	.109	.057	.017	.088	.053	.028	.076	.046	.024	.052	.032	.017	.030	.019	.010	.010	.006	.003
.117	.058	.016	.115	.057	.016	.111	.056	.016	.107	.055	.016	.104	.054	.016	.084	.048	.024	.072	.042	.021	.050	.029	.015	.029	.017	.009	.009	.006	.003
.111	.054	.015	.109	.054	.015	.105	.053	.015	.102	.052	.015	.099	.051	.015	.080	.045	.020	.069	.039	.018	.048	.027	.013	.028	.016	.007	.009	.005	.002
.105	.051	.014	.103	.050	.014	.100	.050	.014	.097	.049	.014	.094	.048	.014	.077	.041	.018	.066	.036	.015	.046	.025	.011	.027	.015	.006	.009	.005	.002
.100	.048	.013	.098	.047	.013	.095	.047	.013	.092	.046	.013	.090	.045	.013	.073	.039	.016	.063	.034	.014	.044	.024	.010	.026	.014	.006	.008	.005	.002
															.114	.114	.114	.097	.097	.097	.066	.066	.066	.038	.038	.038	.012	.012	.012
.144	.082	.026	.140	.080	.025	.132	.076	.024	.125	.072	.023	.118	.069	.022	.105	.094	.084	.090	.080	.072	.061	.055	.050	.035	.032	.029	.011	.010	.009
.142	.078	.024	.139	.076	.024	.132	.073	.023	.126	.071	.022	.120	.068	.021	.099	.079	.063	.085	.068	.054	.058	.047	.038	.033	.027	.022	.011	.009	.007
.137	.073	.022	.134	.072	.022	.128	.069	.021	.123	.067	.021	.118	.065	.020	.094	.068	.048	.080	.059	.042	.055	.041	.029	.032	.024	.017	.010	.008	.006
.131	.068	.020	.128	.067	.020	.123	.065	.019	.118	.063	.019	.113	.062	.019	.089	.060	.038	.077	.052	.033	.053	.036	.023	.030	.021	.014	.010	.007	.004
.124	.063	.018	.122	.062	.018	.117	.061	.018	.113	.059	.018	.109	.058	.017	.085	.054	.030	.073	.046	.026	.050	.032	.019	.029	.019	.011	.009	.006	.004
.117	.059	.017	.115	.058	.017	.111	.057	.017	.107	.056	.016	.104	.055	.016	.082	.049	.025	.070	.042	.022	.048	.029	.015	.028	.017	.009	.009	.006	.003
.111	.055	.016	.109	.054	.015	.105	.053	.015	.102	.052	.015	.099	.051	.015	.078	.044	.021	.067	.039	.018	.046	.027	.013	.027	.016	.008	.009	.005	.003
.105	.051	.014	.103	.051	.014	.100	.050	.014	.097	.049	.014	.094	.048	.014	.075	.041	.018	.064	.036	.016	.044	.025	.011	.026	.015	.007	.008	.005	.002
.100	.048	.013	.098	.048	.013	.095	.047	.013	.092	.046	.013	.089	.045	.013	.071	.038	.016	.062	.033	.014	.043	.023	.010	.025	.014	.006	.008	.005	.002
.094	.045	.013	.093	.045	.013	.090	.044	.012	.088	.044	.012	.085	.043	.012	.068	.036	.014	.059	.031	.012	.041	.022	.009	.024	.013	.005	.008	.004	.002
															.092	.092	.092	.078	.078	.078	.053	.053	.053	.031	.031	.031	.010	.010	.010
.153	.087	.027	.149	.085	.027	.143	.082	.026	.137	.079	.025	.131	.076	.025	.087	.075	.064	.074	.064	.055	.051	.044	.038	.029	.026	.022	.009	.008	.007
.145	.079	.024	.142	.078	.024	.136	.076	.024	.131	.074	.023	.126	.071	.023	.084	.063	.047	.072	.055	.040	.049	.038	.028	.028	.022	.016	.009	.007	.005
.135	.072	.022	.132	.071	.021	.127	.069	.021	.123	.067	.021	.118	.065	.020	.080	.055	.035	.069	.048	.030	.047	.033	.021	.027	.019	.013	.009	.006	.004
.125	.065	.019	.123	.064	.019	.118	.063	.019	.114	.061	.018	.110	.060	.018	.077	.049	.027	.066	.042	.024	.045	.029	.017	.026	.017	.010	.008	.006	.003
.116	.059	.017	.114	.058	.017	.110	.057	.017	.106	.056	.017	.102	.055	.016	.073	.044	.022	.063	.038	.019	.043	.026	.014	.025	.016	.008	.008	.005	.003
.107	.054	.015	.106	.053	.015	.102	.052	.015	.099	.051	.015	.096	.050	.015	.070	.040	.018	.060	.035	.016	.042	.024	.011	.024	.014	.007	.008	.005	.002
.100	.049	.014	.098	.049	.014	.095	.048	.014	.092	.047	.014	.089	.046	.014	.067	.037	.015	.057	.032	.013	.040	.022	.010	.023	.013	.006	.007	.004	.002
.093	.046	.013	.092	.045	.013	.089	.044	.013	.086	.044	.013	.084	.043	.013	.064	.034	.013	.055	.029	.012	.038	.021	.008	.022	.012	.005	.007	.004	.002
.087	.042	.012	.086	.042	.012	.083	.041	.012	.081	.041	.012	.079	.040	.012	.061	.031	.012	.052	.027	.010	.036	.019	.007	.021	.011	.004	.007	.004	.001
.082	.039	.011	.081	.039	.011	.079	.038	.011	.076	.038	.011	.074	.037	.011	.058	.029	.010	.050	.026	.009	.035	.018	.006	.020	.011	.004	.007	.003	.001
															.236	.236	.236	.201	.201	.201	.138	.138	.138	.079	.079	.079	.025	.025	.025
.247	.140	.044	.238	.136	.043	.221	.127	.040	.205	.118	.038	.191	.111	.036	.230	.210	.193	.196	.181	.166	.134	.124	.115	.077	.072	.067	.025	.023	.022
.224	.123	.038	.216	.119	.037	.201	.112	.035	.187	.105	.033	.174	.098	.031	.224	.193	.167	.192	.166	.144	.131	.115	.101	.076	.067	.059	.024	.022	.019
.205	.109	.033	.198	.106	.032	.184	.100	.030	.171	.094	.029	.160	.088	.027	.218	.180	.150	.187	.155	.130	.128	.108	.091	.074	.063	.054	.024	.020	.018
.188	.098	.029	.182	.095	.028	.169	.090	.027	.158	.085	.026	.147	.080	.024	.213	.170	.138	.182	.147	.120	.125	.103	.084	.073	.060	.050	.023	.020	.016
.174	.089	.026	.168	.086	.025	.157	.082	.024	.146	.077	.023	.136	.073	.022	.207	.163	.130	.178	.141	.113	.123	.098	.080	.071	.058	.047	.023	.019	.016
.161	.081	.023	.156	.079	.023	.146	.075	.022	.136	.071	.021	.127	.067	.020	.202	.157	.124	.174	.136	.108	.120	.095	.076	.069	.056	.045	.022	.018	.015
.150	.074	.021	.145	.072	.021	.136	.069	.020	.127	.065	.019	.119	.062	.018	.197	.152	.120	.169	.131	.104	.117	.092	.074	.068	.054	.044	.022	.018	.015
.140	.069	.019	.136	.067	.019	.127	.063	.018	.119	.060	.017	.111	.057	.017	.192	.147	.117	.166	.128	.102	.115	.090	.072	.067	.053	.043	.022	.017	.014
.132	.064	.018	.127	.062	.017	.119	.059	.017	.112	.056	.016	.105	.053	.015	.188	.144	.114	.162	.125	.100	.112	.088	.071	.065	.052	.042	.021	.017	.014
.124	.059	.016	.120	.058	.016	.112	.055	.016	.106	.052	.015	.099	.050	.014	.184	.141	.112	.158	.122	.098	.110	.086	.070	.064	.051	.041	.021	.017	.014
															.114	.114	.114	.097	.097	.097	.066	.066	.066	.038	.038	.038	.012	.012	.012
.188	.107	.034	.184	.105	.033	.176	.101	.032	.169	.097	.031	.162	.094	.030	.108	.093	.080	.092	.080	.069	.063	.055	.047	.036	.032	.028	.012	.010	.009
.179	.098	.030	.176	.097	.030	.169	.094	.029	.162	.091	.028	.156	.088	.028	.103	.079	.058	.089	.068	.050	.061	.047	.035	.035	.027	.020	.011	.009	.007
.167	.089	.027	.163	.087	.026	.157	.085	.026	.151	.083	.025	.146	.081	.025	.099	.068	.043	.085	.059	.038	.058	.041	.026	.034	.024	.016	.011	.008	.005
.154	.080	.024	.151	.079	.023	.145	.077	.023	.140	.075	.023	.136	.074	.022	.095	.060	.034	.081	.052	.029	.056	.036	.021	.032	.021	.012	.010	.007	.004
.142	.073	.021	.140	.072	.021	.135	.070	.021	.130	.069	.020	.126	.067	.020	.091	.054	.027	.078	.047	.024	.054	.033	.017	.031	.019	.010	.010	.006	.003
.132	.066	.019	.130	.065	.019	.125	.064	.019	.121	.063	.018	.117	.062	.018	.086	.049	.023	.074	.043	.020	.051	.030	.014	.030	.018	.008	.010	.006	.003
.123	.061	.017	.121	.060	.017	.117	.059	.017	.113	.058	.017	.110	.057	.017	.082	.045	.019	.071	.039	.017	.049	.027	.012	.028	.016	.007	.009	.005	.002
.115	.056	.016	.113	.055	.016	.109	.055	.016	.106	.054	.015	.103	.053	.015	.078	.042	.017	.068	.036	.014	.047	.025	.010	.027	.015	.006	.009	.005	.002
.107	.052	.014	.106	.051	.014	.102	.051	.014	.099	.050	.014	.097	.049	.014	.075	.039	.015	.064	.034	.013	.045	.024	.009	.026	.014	.005	.008	.005	.002
.101	.048	.013	.099	.048	.013	.096	.047	.013	.094	.046	.013	.091	.046	.013	.071	.036	.013	.061	.032	.011	.043	.022	.008	.025	.013	.005	.008	.004	.002
															.325	.325	.325	.278	.278	.278	.189	.189	.189	.109	.109	.109	.035	.035	.035
.335	.191	.060	.323	.184	.058	.299	.172	.055	.277	.160	.051	.257	.149	.048	.321	.295	.272	.275	.253	.234	.188	.174	.162	.108	.101	.094	.035	.032	.030
.293	.161	.049	.282	.155	.048	.260	.145	.045	.241	.135	.042	.222	.126	.040	.316	.275	.241	.270	.237	.208	.185	.164	.145	.107	.095	.085	.034	.031	.028
.262	.139	.042	.251	.135	.040	.232	.126	.038	.214	.117	.036	.197	.109	.034	.309	.261	.222	.265	.225	.192	.182	.156	.135	.105	.091	.080	.034	.030	.026
.236	.123	.036	.226	.119	.035	.209	.111	.033	.192	.103	.031	.177	.096	.029	.303	.250	.209	.260	.216	.182	.179	.150	.128	.103	.088	.076	.033	.029	.025
.215	.109	.032	.206	.106	.031	.190	.099	.029	.175	.092	.027	.161	.086	.026	.296	.241	.201	.254	.209	.175	.175	.146	.123	.101	.086	.073	.033	.028	.024
.197	.099	.028	.189	.095	.027	.174	.089	.026	.160	.083	.024	.147	.078	.023	.289	.234	.195	.249	.203	.169	.172	.142	.120	.100	.084	.071	.032	.027	.024
.181	.090	.025	.174	.087	.025	.161	.081	.023	.148	.076	.022	.136	.071	.021	.283	.228	.190	.244	.198	.166	.168	.139	.117	.098	.082	.070	.032	.027	.023
.168	.082	.023	.162	.080	.022	.149	.075	.021	.137	.070	.020	.126	.065	.019	.277	.224	.187	.239	.194	.163	.165	.136	.115	.096	.080	.069	.031	.026	.023
.157	.076	.021	.151	.073	.021	.139	.069	.020	.128	.064	.018	.118	.060	.017	.272	.219	.184	.234	.190	.160	.162	.134	.114	.095	.079	.068	.031	.026	.023
.147	.070	.020	.141	.068	.019	.130	.064	.018	.120	.060	.017	.111	.056	.016	.267	.216	.182	.230	.187	.159	.160	.132	.113	.093	.078	.067	.030	.026	.022

Fig. 9-62. *Continued (see page 9-53 for instructions and notes)*

Typical Luminaire	Typical Intensity Distribution and Per Cent Lamp Lumens		ρ_{CC} →	80			70			50			30			10			0	WDRC	ρ_{CC} →
			ρ_W →	50	30	10	50	30	10	50	30	10	50	30	10	50	30	10	0		ρ_W →
	Maint. Cat.	SC	RCR ↓	Coefficients of Utilization for 20 Per Cent Effective Floor Cavity Reflectance (ρ_{FC} = 20)																	RCR ↓
49	I	1.4/1.2	0	1.13	1.13	1.13	1.09	1.09	1.09	1.01	1.01	1.01	.94	.94	.94	.88	.88	.88	.85		
			1	.95	.90	.86	.92	.87	.83	.85	.82	.78	.79	.76	.74	.74	.72	.69	.66	.464	1
			2	.82	.74	.68	.79	.72	.66	.73	.68	.63	.68	.64	.60	.63	.60	.56	.53	.394	2
	12½%↑		3	.71	.62	.55	.69	.61	.54	.64	.57	.52	.59	.54	.49	.55	.51	.47	.44	.342	3
			4	.62	.53	.46	.60	.52	.45	.56	.49	.43	.52	.46	.41	.49	.44	.40	.37	.300	4
			5	.55	.46	.39	.54	.45	.39	.50	.43	.37	.47	.40	.36	.44	.38	.34	.32	.267	5
	85%↓		6	.50	.41	.34	.48	.40	.33	.45	.38	.32	.42	.36	.31	.39	.34	.30	.27	.240	6
			7	.45	.36	.30	.43	.35	.29	.41	.34	.28	.38	.32	.27	.36	.30	.26	.24	.218	7
			8	.41	.32	.26	.40	.32	.26	.37	.30	.25	.35	.29	.24	.33	.27	.23	.21	.199	8
2-lamp fluorescent strip unit with 235°			9	.37	.29	.24	.36	.28	.23	.34	.27	.22	.32	.26	.22	.30	.25	.21	.19	.183	9
reflector fluorescent lamps			10	.34	.26	.21	.33	.26	.21	.32	.25	.20	.30	.24	.20	.28	.23	.19	.17	.170	10

Typical Luminaires	ρ_{CC} →	80			70			50			30			10			0
	ρ_W →	50	30	10	50	30	10	50	30	10	50	30	10	50	30	10	0
	RCR ↓	Coefficients of utilization for 20 Per Cent Effective Floor Cavity Reflectance, ρ_{FC}															
50	1	.42	.40	.39	.36	.35	.33	.25	.24	.23	Coves are not recommended for lighting areas having low reflectances.						
	2	.37	.34	.32	.32	.29	.27	.22	.20	.19							
	3	.32	.29	.26	.28	.25	.23	.19	.17	.16							
	4	.29	.25	.22	.25	.22	.19	.17	.15	.13							
	5	.25	.21	.18	.22	.19	.16	.15	.13	.11							
	6	.23	.19	.16	.20	.16	.14	.14	.12	.10							
	7	.20	.17	.14	.17	.14	.12	.12	.10	.09							
	8	.18	.15	.12	.16	.13	.10	.11	.09	.08							
Single row fluorescent lamp cove without reflector, mult. by 0.93	9	.17	.13	.10	.15	.11	.09	.10	.08	.07							
for 2 rows and by 0.85 for 3 rows.	10	.15	.12	.09	.13	.10	.08	.09	.07	.06							
51 ρ_{CC} from below ~65%	1				.60	.58	.56	.58	.56	.54							
	2				.53	.49	.45	.51	.47	.43							
	3				.47	.42	.37	.45	.41	.36							
	4				.41	.36	.32	.39	.35	.31							
	5				.37	.31	.27	.35	.30	.26							
	6				.33	.27	.23	.31	.26	.23							
Diffusing plastic or glass	7				.29	.24	.20	.28	.23	.20							
1) Ceiling efficiency ~60%; diffuser transmittance ~50%;	8				.26	.21	.18	.25	.20	.17							
diffuser reflectance ~40%. Cavity with minimum obstructions	9				.23	.19	.15	.23	.18	.15							
and painted with 80% reflectance paint—use ρ_c = 70.	10				.21	.17	.13	.21	.16	.13							
2) For lower reflectance paint or obstructions—use ρ_c = 50.																	
52 ρ_{CC} from below ~60%	1				.71	.68	.66	.67	.66	.65	.65	.64	.62				
	2				.63	.60	.57	.61	.58	.55	.59	.56	.54				
	3				.57	.53	.49	.55	.52	.48	.54	.50	.47				
	4				.52	.47	.43	.50	.45	.42	.48	.44	.42				
	5				.46	.41	.37	.44	.40	.37	.43	.40	.36				
	6				.42	.37	.33	.41	.36	.32	.40	.35	.32				
Prismatic plastic or glass.	7				.38	.32	.29	.37	.31	.28	.36	.31	.28				
1) Ceiling efficiency ~67%; prismatic transmittance ~72%;	8				.34	.28	.25	.33	.28	.25	.32	.28	.25				
prismatic reflectance ~18%. Cavity with minimum obstructions	9				.30	.25	.22	.30	.25	.21	.29	.25	.21				
and painted with 80% reflectance paint—use ρ_c = 70.	10				.27	.23	.19	.27	.22	.19	.26	.22	.19				
2) For lower reflectance paint or obstructions—use ρ_c = 50.																	
53 ρ_{CC} from below ~45%	1							.51	.49	.48				.47	.46	.45	
	2							.46	.44	.42				.43	.42	.40	
	3							.42	.39	.37				.39	.38	.36	
	4							.38	.35	.33				.36	.34	.32	
	5							.35	.32	.29				.33	.31	.29	
Louvered ceiling.	6							.32	.29	.26				.30	.28	.26	
1) Ceiling efficiency ~50%; 45° shielding opaque louvers of	7							.29	.26	.23				.28	.25	.23	
80% reflectance. Cavity with minimum obstructions and	8							.27	.23	.21				.26	.23	.21	
painted with 80% reflectance paint—use ρ_c = 50.	9							.24	.21	.19				.24	.21	.19	
2) For other conditions refer to Fig. 6–18.	10							.22	.19	.17				.22	.19	.17	

80			70			50			30			10			80			70			50			30			10		
50	30	10	50	30	10	50	30	10	50	30	10	50	30	10	50	30	10	50	30	10	50	30	10	50	30	10	50	30	10
Wall Exitance Coefficients for 20 Per Cent Effective Floor Cavity Reflectance (ρ_{FC} = 20)															Ceiling Cavity Exitance Coefficients for 20 Per Cent Floor Cavity Reflectance (ρ_{FC} = 20)														
															.280	.280	.280	.239	.239	.239	.163	.163	.163	.094	.094	.094	.030	.030	.030
.357	.203	.064	.346	.197	.063	.326	.187	.060	.307	.177	.057	.290	.168	.054	.275	.247	.222	.235	.212	.191	.161	.146	.132	.093	.084	.077	.030	.027	.025
.316	.173	.053	.306	.169	.052	.288	.160	.050	.271	.152	.048	.256	.145	.046	.268	.224	.188	.230	.193	.162	.157	.134	.113	.091	.078	.066	.029	.025	.022
.284	.151	.045	.275	.147	.044	.259	.140	.043	.243	.133	.041	.229	.127	.039	.261	.208	.166	.224	.180	.144	.154	.125	.101	.089	.073	.060	.028	.024	.019
.257	.134	.039	.249	.130	.039	.234	.124	.037	.221	.118	.036	.208	.113	.034	.253	.196	.152	.217	.169	.132	.150	.118	.093	.086	.069	.055	.028	.022	.018
.235	.120	.035	.227	.117	.034	.214	.111	.033	.202	.106	.032	.190	.101	.030	.246	.186	.142	.211	.161	.123	.145	.112	.087	.084	.066	.052	.027	.022	.017
.215	.108	.031	.209	.106	.030	.197	.101	.029	.185	.096	.028	.175	.092	.027	.239	.178	.135	.205	.154	.117	.142	.108	.083	.082	.064	.049	.027	.021	.016
.199	.098	.028	.193	.096	.027	.182	.092	.027	.172	.088	.026	.162	.084	.025	.232	.172	.130	.199	.149	.113	.138	.104	.080	.080	.061	.048	.026	.020	.016
.185	.090	.025	.180	.088	.025	.169	.085	.024	.160	.081	.023	.151	.077	.023	.225	.166	.126	.194	.144	.109	.134	.101	.078	.078	.060	.046	.025	.020	.015
.172	.083	.023	.168	.082	.023	.158	.078	.022	.150	.075	.021	.141	.072	.021	.219	.161	.122	.189	.140	.107	.131	.099	.076	.076	.058	.045	.025	.019	.015
.162	.077	.022	.157	.076	.021	.149	.073	.020	.140	.070	.020	.133	.067	.019	.214	.157	.120	.184	.137	.105	.128	.096	.074	.074	.057	.044	.024	0.19	.015

54

910 mm x 910 mm (3′ x 3′) fluorescent troffer with 1220 mm (48″) lamps mounted along diagonals—use units 40, 41 or 42 as appropriate

55

610 mm x 610 mm (2′ x 2′) fluorescent troffer with two "U" lamps—use units 40, 41 or 42 as appropriate

Tabulation of Luminous Intensities Used to Compute Above Coefficients Normalized Average Intensity (Candelas per 1000 lumens)

Angle ↓	Luminaire No.													
	1	2	3	4	5	6	7	8	9	10	11	12	13	14
5	72.5	6.5	256.0	238.0	808.0	1320.0	695.0	374.0	2680.0	2610.0	208.0	152.0	190.0	316.0
15	72.5	8.0	246.0	264.0	671.0	1010.0	630.0	357.0	1150.0	1200.0	220.0	148.0	196.0	311.0
25	72.5	9.5	238.0	248.0	494.0	584.0	286.0	305.0	209.0	411.0	254.0	141.0	199.0	301.0
35	72.5	10.0	238.0	191.0	340.0	236.0	88.0	212.0	13.5	97.0	220.0	125.0	212.0	271.0
45	72.5	8.0	203.0	122.0	203.0	22.0	5.0	81.0	0	15.0	130.0	106.0	206.0	156.0
55	72.0	6.5	168.0	62.5	91.0	0	0	40.5	0	0	59.0	87.5	125.0	63.0
65	71.5	4.5	130.0	45.5	33.0	0	0	20.5	0	0	26.0	69.5	68.5	31.5
75	70.5	2.5	34.0	38.0	12.5	0	0	9.5	0	0	11.0	47.0	41.5	17.5
85	70.0	2.0	7.0	32.0	4.0	0	0	2.5	0	0	3.5	23.5	26.0	4.0
95	67.0	15.0	0	28.0	0	0	0	0	0	0	0	9.5	12.5	0
105	62.5	147.0	0	28.0	0	0	0	0	0	0	0	4.5	6.0	0
115	58.0	170.0	0	41.0	0	0	0	0	0	0	0	1.0	3.5	0
125	54.5	168.0	0	42.5	0	0	0	0	0	0	0	0	1.5	0
135	51.0	183.0	0	33.0	0	0	0	0	0	0	0	0	0	0
145	48.0	159.0	0	22.5	0	0	0	0	0	0	0	0	0	0
155	46.5	139.0	0	9.0	0	0	0	0	0	0	0	0	0	0
165	45.0	94.5	0	3.0	0	0	0	0	0	0	0	0	0	0
175	44.0	50.5	0	1.0	0	0	0	0	0	0	0	0	0	0

Angle ↓	Luminaire No.													
	15	16	17	18	19	20	21	22	23	24	25	26	27	28
5	288.0	999.0	470.0	294.0	576.0	274.0	203.0	136.0	155.0	0	263.0	246.0	284.0	244.0
15	321.0	775.0	384.0	282.0	519.0	302.0	192.0	151.0	169.0	0	258.0	260.0	262.0	248.0
25	331.0	475.0	344.0	294.0	426.0	344.0	194.0	171.0	185.0	0	236.0	264.0	226.0	242.0
35	260.0	188.0	290.0	294.0	274.0	321.0	252.0	175.0	188.0	0	210.0	248.0	187.0	218.0
45	202.0	90.5	210.0	246.0	127.0	209.0	230.0	182.0	162.0	0	163.0	192.0	145.0	152.0
55	114.0	32.0	86.5	137.0	69.5	45.5	119.0	158.0	119.0	0	98.0	98.0	83.0	70.0
65	13.5	8.5	18.0	26.0	20.0	8.0	52.5	90.0	57.0	0	55.5	32.5	36.5	26.0

Fig. 9–62. *Continued* *(see page 9-53 for instructions and notes)*

Angle ↓	Luminaire No.													
	15	16	17	18	19	20	21	22	23	24	25	26	27	28
75	6.0	6.0	5.0	6.5	2.5	3.0	21.0	41.0	4.5	0	29.5	12.5	18.5	10.0
85	2.0	1.0	1.0	1.0	1.5	2.5	3.5	17.0	0	0	11.0	4.0	5.5	3.0
95	1.0	0.5	0.5	0.5	0.5	3.5	0	8.0	0	19.0	8.0	3.5	3.5	2.5
105	0	0.5	0.5	0.5	0.5	8.0	0	7.0	0	64.0	14.5	6.5	11.0	6.0
115	0	0.5	0.5	0.5	4.5	15.5	0	7.0	0	212.0	21.5	12.0	21.0	13.0
125	0	1.0	0.5	0.5	10.5	22.5	0	5.0	0	205.0	31.0	21.5	34.5	24.0
135	0	1.5	1.0	0.5	16.5	29.0	0	0	0	160.0	47.0	33.5	51.5	36.0
145	0	8.0	3.0	1.5	20.5	33.5	0	0	0	128.0	59.5	50.0	71.5	49.5
155	0	8.5	8.0	7.5	32.0	42.0	0	0	0	115.0	82.5	70.5	92.0	70.0
165	0	0.5	0.5	0.5	33.0	27.5	0	0	0	106.0	105.0	92.0	109.0	88.5
175	0	0.5	0.5	0.5	16.5	2.5	0	0	0	102.0	111.0	102.0	115.0	95.5

Angle ↓	Luminaire No.																				
	29	30	31	32	33	34	35	36	37	38	39	40	41	42	43	44	45	46	47	48	49
5	189.0	270.0	218.0	199.0	41.5	194.0	210.0	206.0	107.0	272.0	312.0	218.0	206.0	253.0	288.0	197.0	90.0	132.0	135.0	157.0	238.0
15	176.0	249.0	220.0	194.0	38.5	192.0	211.0	199.0	104.0	244.0	268.0	207.0	202.0	249.0	284.0	196.0	104.0	144.0	142.0	156.0	232.0
25	147.0	200.0	224.0	184.0	35.5	187.0	212.0	185.0	98.5	202.0	213.0	187.0	183.0	236.0	271.0	199.0	125.0	181.0	167.0	153.0	218.0
35	110.0	144.0	222.0	170.0	32.5	169.0	204.0	158.0	90.0	156.0	148.0	164.0	162.0	214.0	246.0	235.0	140.0	202.0	171.0	147.0	200.0
45	64.0	86.5	187.0	154.0	29.0	123.0	164.0	108.0	79.5	106.0	87.0	135.0	133.0	172.0	190.0	223.0	131.0	173.0	151.0	137.0	176.0
55	34.5	53.5	99.0	137.0	22.0	77.5	78.5	51.5	66.5	68.0	51.0	106.0	104.0	95.5	97.0	99.5	104.0	113.0	120.0	122.0	149.0
65	20.5	34.0	15.5	117.0	14.5	37.5	36.5	35.5	52.0	42.0	30.0	74.0	70.5	45.0	25.0	18.5	65.5	63.0	82.0	104.0	119.0
75	10.0	20.5	3.5	88.5	7.0	18.5	26.0	34.5	36.0	21.5	15.5	42.5	36.5	19.0	6.0	3.0	27.5	42.5	41.5	79.0	86.5
85	2.5	10.0	1.0	59.0	2.0	10.5	17.5	32.0	21.5	6.0	4.0	15.5	7.0	7.0	2.5	0.5	8.0	27.5	7.5	52.5	50.5
95	4.0	7.0	0	49.5	11.0	14.5	15.5	32.5	14.5	0	0	5.5	0	0	0	0	0	23.0	0	45.0	32.5
105	19.0	8.5	0	32.5	49.5	40.0	22.0	49.0	14.5	0	0	2.5	0	0	0	0	0	31.0	0	43.5	27.5
115	40.5	9.5	0	6.5	96.0	57.0	27.0	49.0	14.0	0	0	0	0	0	0	0	0	30.0	0	38.5	22.0
125	67.0	10.0	0	0	130.0	68.5	23.0	44.5	13.0	0	0	0	0	0	0	0	0	19.5	0	33.0	17.5
135	93.0	11.0	0	0	155.0	71.5	18.5	36.0	12.0	0	0	0	0	0	0	0	0	10.0	0	27.0	13.5
145	117.0	11.0	0	0	172.0	67.5	12.0	28.5	10.0	0	0	0	0	0	0	0	0	7.5	0	20.0	10.5
155	136.0	11.5	0	0	183.0	65.0	7.5	24.0	8.5	0	0	0	0	0	0	0	0	4.5	0	13.0	7.5
165	151.0	12.0	0	0	189.0	67.5	4.5	21.0	6.5	0	0	0	0	0	0	0	0	1.5	0	7.0	5.0
175	155.0	13.0	0	0	201.0	73.5	4.0	18.0	5.5	0	0	0	0	0	0	0	0	0	0	2.5	2.5

3. Determine the additional flux functions:

$$\Phi_D = \frac{1}{\Phi_T} \sum_{N=1}^{9} \Phi_N$$

$$\Phi_U = \frac{1}{\Phi_T} \sum_{N=10}^{N=18} \Phi_N$$

where Φ_T is the total flux (lumens) of the lamps in the luminaire.

4. Determine the direct ratio, D_G, the fraction of luminaire flux below the horizontal which is directly incident on the work-plane.

$$D_G = \frac{1}{\Phi_D \Phi_T} \sum_{N=1}^{9} (K_{GN} \Phi_N)$$

where

G = room cavity ratio (RCR);*
integer, $1 \leq G \leq 10$

K_{GN} = zonal multipliers

* G is an integer for generating the standard zonal-cavity coefficient tables. The equations remain valid for nonintegral values of G.

Zonal multipliers are the fraction of downward directed flux directly incident on the work-plane (lower surface of room cavity) for each zone N. These are functions of RCR.

$$K_{GN} = \exp(-AG^B)$$

The values of the constants A and B are given in Fig. 9–65.

5. Determine the parameters C_1, C_2, C_3, and C_0 as an intermediate step. ρ_1 is the wall reflectance, ρ_2 is the ceiling cavity reflectance, and ρ_3 is the floor cavity reflectance which is taken as 0.2 for standard coefficient tables. $f_{2\to3}$ is the form factor described above under flux transfer theory.

$$C_1 = \frac{(1 - \rho_1)(1 - f_{2\to3}^2)G}{2.5\rho_1(1 - f_{2\to3}^2) + Gf_{2\to3}(1 - \rho_1)}$$

$$C_2 = \frac{(1 - \rho_2)(1 + f_{2\to3})}{1 + \rho_2 f_{2\to3}}$$

$$C_3 = \frac{(1 - \rho_3)(1 + f_{2\to3})}{1 + \rho_3 f_{2\to3}}$$

$$C_0 = C_1 + C_2 + C_3$$

6. Determine the coefficient of utilization (CU), the wall exitance coefficient (WEC), and the ceiling cavity exitance coefficient (CCEC) for each applicable combination of reflectances and RCR.

$$\mathrm{CU} = \frac{2.5\rho_1 C_1 C_3 (1 - D_G)\Phi_D}{G(1-\rho_1)(1-\rho_3)C_0} + \frac{\rho_2 C_2 C_3 \Phi_U}{(1-\rho_2)(1-\rho_3)C_0} + \left[1 - \frac{\rho_3 C_3 (C_1 + C_2)}{(1-\rho_3)C_0}\right]\frac{D_G \Phi_D}{(1-\rho_3)}$$

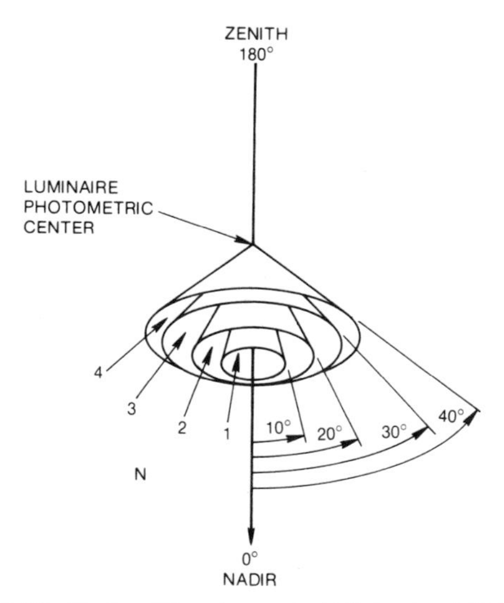

Fig. 9-63. Conic solid angle zones of 10-degree width for use in calculating zonal flux.

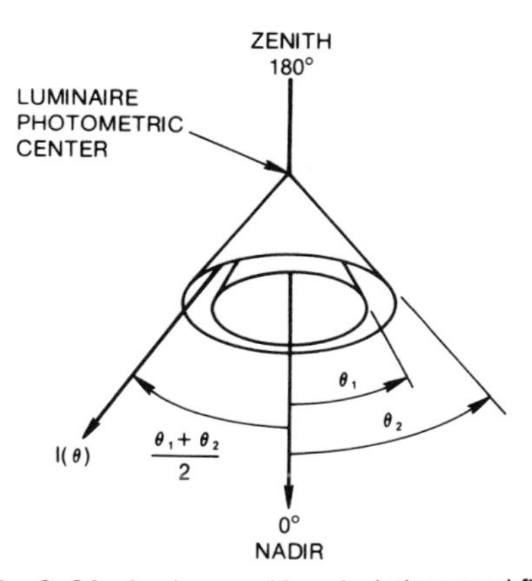

Fig. 9-64. Angles used in calculating zonal flux.

Fig. 9-65. Constants for the Zonal Multiplier Equation

Zone (N)	A	B
1	0.	0.
2	0.041	0.98
3	0.070	1.05
4	0.100	1.12
5	0.136	1.16
6	0.190	1.25
7	0.315	1.25
8	0.640	1.25
9	2.10	0.80

$$\mathrm{WEC} = \frac{2.5}{G}\left\{\frac{\rho_1(1-D_G)\Phi_D}{(1-\rho_1)} \times \left[1 - \frac{2.5\rho_1 C_1 (C_2 + C_3)}{G(1-\rho_1)C_0}\right] + \frac{\rho_1\rho_2 C_1 C_2 \Phi_U}{(1-\rho_1)(1-\rho_2)C_0} + \frac{\rho_1\rho_3 \mathrm{C}_1 \mathrm{C}_3 \mathrm{D}_G \Phi_D}{(1-\rho_1)(1-\rho_3)\mathrm{C}_0}\right\}$$

$$\mathrm{CCEC} = \frac{2.5\rho_1\rho_2 C_1 C_2 (1-D_G)\Phi_D}{G(1-\rho_1)(1-\rho_2)C_0} + \frac{\rho_2\Phi_U}{(1-\rho_2)}\left[1 - \frac{\rho_2 C_2 (C_1 + C_3)}{(1-\rho_2)C_0}\right] + \frac{\rho_2\rho_3 C_2 C_3 D_G \Phi_D}{(1-\rho_2)(1-\rho_3)C_0}$$

7. Determine the wall direct radiation coefficient (WDRC) for each RCR.

$$\mathrm{WDRC} = \frac{2.5\Phi_D(1-D_G)}{G}$$

8. The above equations can be used to calculate the coefficient of utilization and the ceiling cavity exitance coefficient for RCR = 0, but the forms of the equations must be arranged to avoid division by zero. It is simplest to use the following relationships:

$$\mathrm{CU}(G = 0) = \frac{\Phi_D + \rho_2\Phi_U}{1 - \rho_2\rho_3}$$

$$\mathrm{CCEC}(G = 0) = \frac{\rho_2(\Phi_U + \rho_3\Phi_D)}{1 - \rho_2\rho_3}$$

REFERENCES

1. Siegel, R. and Howel, J. R.: *Thermal Radiative Heat Transfer*, 2nd Edition, New York, McGraw-Hill Book Company, 1980.
2. Levin, R. E.: "Cavities, Coefficients, and Direct Ratios," *J. Illum. Eng. Soc.*, Vol. 11, p. 178, April, 1982.
3. Lighting Design Practice Committee of the IES: "General Procedure for Calculating Maintained Illumination," *Illum. Eng.*, Vol. 65, p. 602, October, 1970.
4. Clark, F.: "Accurate Maintenance Factors," *Illum. Eng.*, Vol. LVIII, p. 124, March, 1963. Part Two, Vol. LXI, p. 37, January, 1966.
5. Clark, F.: "Light Loss Factors in the Design Process," *Illum. Eng.*, Vol. 63, p. 515, November, 1968.
6. *American National Standard Methods of Measurement of Fluorescent Lamp Ballasts*, ANSI C82.2-1977, American National Standards Institute, New York, NY, 1977.
7. Testing Procedures Committee of the IES: "IES Approved Method for Determining Luminaire—Lamp—Ballast Combination Operating Factors for High Intensity Discharge Luminaires," *Illum. Eng.*, Vol. 65, p. 718, December, 1970.
8. McNamara, A. C. *et al*: "High Wattage HID Lamp Fixture Coordination—Vertical Versus Horizontal Versus Somewhere in Between," *IEEE Trans. Ind. Appl.*, Vol. IA-10, No. 5, September/October, 1974.
9. Levin, R. E. and Lemons, T. M.: "High-Intensity Discharge Lamps and Their Environment," *IEEE Trans. Ind. Gen. Appl.*, Vol. IGA-7, No. 2, March/April, 1971.
10. Lighting Design Practice Committee of the IES: "The Determination of Illumination at a Point in Interior Spaces," *J. Illum. Eng. Soc.*, Vol. 3, p. 170, January, 1974.
11. Goodbar, I.: "New Methods for Point by Point Calculations," Illum. Eng., Vol. XLI, p. 39, January, 1946.
12. Jones, J. R., LeVere, R. C., Ivanicki, N., and Chesebrough, P.: "Angular Coordinate System for Computing Illumination at a Point," *Illum. Eng.*, Vol. LXIV, Section I, p. 296, April, 1969.
13. "The Calculation of Direct Illumination from Linear Sources," IES (London) Technical Report No. 11, The Illuminating Engineering Society. York House, Westminster Bridge Road, London S.E. 1, England.
14. Spencer, D. E.: "Exact and Approximate Formulae for Illumination from Troffers," *Illum. Eng.*, Vol. XXXVII, p. 596, November, 1942. Wakefield, E. H., and McCord, C.: "Discussion of Illumination Distribution from Linear Strip and Surface Sources," *Illum. Eng.*, Vol. XXXVI, p. 1330, December, 1941. Wakefield, E. H.: "A Simple Graphical Method of Finding Illumination Values from Tubular, Ribbon, and Surface Sources," *Illum. Eng.*, Vol. XXXV, p. 142, February, 1940.
15. Woblauer, A. A.: "The Flux from Lines of Light," *Trans. Illum. Eng. Soc.*, Vol. XXXI, p. 694, July, 1936.
16. Whipple, R. R.: "Rapid Computation of Illumination from Certain Line Sources," *Trans. Illum. Eng. Soc.*, Vol. XXX, p. 492, June, 1935.
17. Burnham, R. D.: "The Illumination at a Point from an Industrial Fluorescent Luminaire," *Illum. Eng.*, Vol. XLV, p. 753, December, 1950.
18. O'Brien, P. F. and Balogh, E.: "Configuration Factors for Computing Illumination Within Interiors," *Illum. Eng.*, Vol. 62, p. 169, April, 1967.
19. Murdoch, J. B.: "Extension of the Configuration Factor Method," *J. Illum. Eng. Soc.*, Vol. 13, January 1984.
20. DiLaura, D. L.: "On the Computation of Equivalent Sphere Illumination," *J. Illum. Eng. Soc.*, Vol. 4, p. 129, January, 1975.
21. DiLaura, D. L.: "On the Simplification of Radiative Transfer Calculations," *J. Illum. Eng. Soc.*, Vol. 12, p. 12, October, 1982.
22. Lighting Design Practice Committee of the IES: "Zonal-Cavity Method of Calculating and Using Coefficients of Utilization," *Illum. Eng.*, Vol. LIX, p. 309, May, 1964.
23. Jones, J. R. and Jones, B. F.: "Using the Zonal-Cavity System in Lighting Calculations," *Illum. Eng.*, Vol. LIX; Part I, p. 413, May, 1964; Part II, p. 448, June, 1964; Part III, p. 501, July, 1964; Part IV, p. 556, August, 1964.
24. Jones, B. F.: "Zonal-Cavity—A Three-Level Approach," *Illum. Eng.*, Vol. 64, p. 149, March, 1969.
25. Levin, R. E.: "The Photometric Connection," *Light. Des. Appl.*, Vol. 12, Part I, p. 28, September, 1982; Part II, p. 60, October, 1982; Part III, p. 42, November, 1982; Part IV, p. 16, December, 1982.
26. Levin, R. E.: "On the Nonrectangular Cavity Approximation," *J. Illum. Eng. Soc.*, Vol. 13, p. 107, October, 1983.
27. O'Brien, P. F.: "Luminous Transfer in Discrete Spaces," *Appl. Opt.*, No. 6, p. 1469, 1967.
28. Toupes, K. A., *Confac II*, North American Aviation, Inc.
29. Levin, R. E.: "Luminance—A Tutorial Paper," *J. of the SMPTE*, Vol. 77, p. 1005, October, 1968.
30. Committee on Lighting Design Practice of the IES: "Zonal-Cavity Method of Calculating and Using Coefficients of Utilization," *Illum. Eng.*, Vol. 59, No. 5, p. 309, May, 1964.
31. Hallman, E. D.: "Floodlighting Design Procedure as Applied to Modern Setback Construction," *Trans. Illum. Eng. Soc.*, Vol. XXIX, p. 287, April, 1934. Dearborn, R. L.: "Floodlighting Design by Graphical Method," *Illum. Eng.*, Vol. XL, p. 514, September, 1945.
32. Luckiesh, M. and Guth, S. K.: "Brightnesses in Visual Field at Borderline Between Comfort and Discomfort (BCD)," *Illum. Eng.*, Vol. XLIV, p. 650, November, 1949.
33. Hopkinson, R. G.: "Evaluation of Glare," *Illum. Eng.*, Vol. 52, p. 305, June, 1957.
34. Guth, S. K. and McNelis, J. F.: "A Discomfort Glare Evaluator," *Illum. Eng.*, Vol. LIV, p. 398, June, 1959.
35. Guth, S. K. and McNelis, J. F.: "Further Data on Discomfort Glare From Multiple Sources," *Illum. Eng.*, Vol. LVI, p. 46, January, 1961.
36. Bradley, R. D. and Logan, H. L.: "A Uniform Method for Computing the Probability of Comfort Responses in a Visual Field," *Illum. Eng.*, Vol. LIX, p. 189, March, 1964.
37. Guth, S. K.: "A Method for the Evaluation of Discomfort Glare," *Illum. Eng.*, Vol. LVIII, p. 351, May, 1963.
38. Allphin, W.: "Influence of Sight Line on BCD Judgments of Direct Discomfort Glare," *Illum. Eng.*, Vol. LXI, p. 629, October, 1966.
39. Allphin, W.: "Further Studies of Sight Line and Direct Discomfort Glare," *Illum. Eng.*, Vol. 63, p. 26, January, 1968.
40. Allphin, W.: "BCD Appraisals of Luminaire Brightness in a Simulated Office," *Illum. Eng.*, Vol. 56, p. 31, January, 1961.
41. Allphin, W.: "Further Appraisals of Luminaire Brightness," *Illum. Eng.*, Vol. 56, p. 700, December, 1961.
42. Committee on Recommendations of Quality and Quantity of Illumination of the IES: "Outline of a Standard Procedure for Computing Visual Comfort Ratings for Interior Lighting—Report No. 2," *Illum. Eng.*, Vol. LXI, p. 643, October, 1966.
43. Guth, S. K.: "Computing Visual Comfort Ratings for a Specific Interior Lighting System," *Illum. Eng.*, Vol. LXI, p. 634, October, 1966.
44. McGowan, T. K. and Guth, S. K.: "Extending and Applying the IES Visual Comfort Rating Procedure," *Illum. Eng.*, Vol. 64, p. 253, April, 1969.
45. Committee on Recommendations of Quality and Quantity of Illumination of the IES: "A Statement Concerning Visual Comfort Probability (VCP)—Naive vs Experienced Observers," *Illum. Eng.*, Vol. 64, p. 604, September, 1969.
46. Committee on Testing Procedures of the IES: "Determination of Average Luminance of Luminaires," *J. Illum. Eng. Soc.*, Vol. 1, p. 181, January, 1972.
47. Committee on Recommendations of Quality and Quantity of Illumination of the IES: "Outline of a Standard Procedure for Computing Visual Comfort Ratings for Interior Lighting—Report No. 2 (1972)," *J. Illum. Eng. Soc.*, Vol. 2, p. 328, April, 1973.
48. Committee on Recommendations of Quality and Quantity of Illumination of the IES: "Determination of Effective Candlepower of Modular and Linear Regressed Systems—Appendix to Report No. 2 (1972)," *J. Illum. Eng. Soc.*, Vol. 2, p. 504, July, 1973.
49. Levin, R. E.: "Position Index in VCP Calculations," *J. Illum. Eng. Soc.*, Vol. 4, p. 99, January, 1975.
50. Levin, R. E.: "An Evaluation of VCP Calculations," *J. Illum. Eng. Soc.*, Vol. 2, p. 355, July, 1973.
51. DiLaura, D. L.: "On the Computation of Visual Comfort Probability," *J. Illum. Eng. Soc.*, Vol. 5, p. 207, July, 1976.
52. Committee on Recommendations of Quality and Quantity of Illumination of the IES: "An Alternate Simplified Method for Determining the Acceptability of a Luminaire, from the VCP

Standpoint for Use in Large Rooms—Report No. 3," *J. Illum. Eng. Soc.*, Vol. 1, p. 256, April, 1972.

53. Fry, G. A.: "A Simplified Formula for Discomfort Glare," *J. Illum. Eng. Soc.*, Vol. 6, p. 10, October, 1976.

54. Goodbar, I.: "A Simplified Method for Determining the Acceptability of a Luminaire from the VCP Standpoint," *J. Illum. Eng. Soc.*, Vol. 6, p. 21, October, 1976.

55. Design Practice Committee of the IES: "Recommended Practice for Classification of Interior Luminaires by Distribution: Luminaire Spacing Criterion," *Light. Des. Appl.*, Vol. 7, p. 20, August, 1977.

56. Levin, R. E.: "Revision of the S/MH Concept," *Light. Des. Appl.*, Vol. 7, p. 22, August, 1977.

57. *IES Lighting Handbook*, 4th edition, Fig. A-20, p. A-18, Illuminating Engineering Society, New York, 1966.

58. LeVere, R. C., Levin, R. E., and Primrose, W.: "Spacing Criteria for Interior Luminaires—the Practice and Pitfalls," *J. Illum. Eng. Soc.*, Vol. 3, p. 41, October, 1973.

59. O'Brien, P. F.: "Lighting Calculations for Thirty-Five Thousand Rooms," *Illum. Eng.*, Vol. LV, p. 215, April, 1960.

60. O'Brien, P. F.: "Numerical Analysis for Lighting Design," *Illum. Eng.*, Vol. LX, p. 169, April, 1965.

61. Design Practice Committee of the IES: "Zonal-Cavity Method of Calculating and Using Coefficients of Utilization," *Illum. Eng.*, Vol. LIX, p. 309, May, 1964.

62. Design Practice Committee of the IES: "Calculation of Luminance Coefficients Based Upon the Zonal-Cavity Methods," *Illum. Eng.*, Vol. 63, p. 423, August, 1968.

63. CIE Committee TC-1.5: *Calculations for Interior Lighting—Basic Method*, CIE Publication No. 40, 1978.

64. CIE Committee TC-1.5: *Calculations for Interior Lighting—Applied Method*, CIE Publication, in preparation.

65. Calculation Procedures Committee of the IES: "Recommended Procedure for Calculating Coefficients of Utilization, Wall Exitance Coefficients, and Ceiling Cavity Exitance Coefficients," *J. Illum. Eng. Soc.*, Vol. 12, p. 3, October, 1982.

Credits for Illustrations and Tables

The Illuminating Engineering Society is indebted to the many individuals, committees and organizations which contributed the multitude of illustrations and tables published in this Handbook. Many of the illustrations and tables omitted from the following listing appeared in previous publications issued by the Society or were supplied by committees of the Society especially for use in this volume.

Contributors

1. The American Institute of Physics, Woodbury, NY, *Journal of the Optical Society of America*
2. American Medical Association, Chicago, IL, *Journal of the American Medical Association*
3. Bausch & Lomb Optical Company, Rochester NY
4. Blackwell, H. R., Ohio State University, Columbus, OH
5. Boynton, R. M., LaJolla, CA
6. Canadian Standards Association, Rexdale, Ont., Canada
7. Chapman & Hall Ltd., London, *Vision and the Eye*, Pirenne
8. Corning Glass Works, Corning, NY
9. Crouse-Hinds Company, Syracuse, NY
10. Eastman Kodak Company, Rochester, NY
11. General Electric Company, Nela Park, Cleveland, OH
12. GTE Products Corporation, Danvers, MA
13. Inter-Society Color Council, Washington, DC
14. Johns-Manville, Holophane Division, Columbus, OH
15. Kelly, K. L., Washington, DC
16. McGraw-Hill Book Co., Inc., New York, NY,
    - **a.** *The Principle of Optics*, Hardy & Perrin
    - **b.** *Scientific Basis of Illuminating Engineering*, Moon
17. Middleton, W. E. K., Ottawa, Ont., Canada
18. Minolta Corporation, Industrial Meters, Ramsey, NJ
19. Monsanto Chemical Co., St. Louis, MO
20. C. V. Mosby Co., St. Louis, MO, *Investigative Opthalmology*
21. Munsell Color Company, Baltimore, MD
22. National Bureau of Standards, U. S. Dept. of Commerce, Washington, DC
23. National Carbon Company, Cleveland, OH
24. Optical Society of America, Washington, DC
25. Owens-Illinois Glass Co., Toledo, OH, *Daylight in School Classrooms*, Paul
26. Pergamon Press, Inc., Elmsford, NY, *Vision Research*
27. Philadelphia Electric Company, Philadelphia, PA
28. Photo Research, Div. Kollmorgen Corp., Burbank, CA
29. Physiological Society, Cambridge, England, *Journal of Physiology*
30. Polaroid Corporation, Cambridge, MA
31. The Rockefeller University Press, New York, NY, *The Journal of General Physiology*
32. Smith, Hinchman & Grylls Associates, Inc., Detroit, MI
33. Society of Automotive Engineers, Inc., Warrendale, PA
34. Tektronix Inc., Beaverton, OR
35. D. Van Nostrand Co., Inc., New York, NY, *The Science of Seeing*, Luckiesh & Moss
36. Westinghouse Electric Corporation, Bloomfield, NJ
37. Weston Electrical Instrument Corp., Newark, NJ
38. John Wiley & Sons, New York, NY,
    - **a.** *Color Science*, Wyszecki & Stiles
    - **b.** *Color Research & Application*

Credits

Section 2.

2-4: **16a.** 2-5: **16b.** 2-6: **19.** 2-7: **36.** 2-9: **36.** 2-20: **37.** 2-24: **16a.** 2-32a: **14.** 2-32c: **8.** 2-32d: **3.** 2-34: **3.** 2-35: **36.** 2-36: **11.** 2-39: **31.**

Section 3.

3-2: **20.** 3-3: **25.** 3-5: **5, 26.** 3-6: **7.** 3-8: **1.** 3-9: **2.** 3-10: **1.** 3-13: **35.** 3-14: **35.** 3-15: **24.** 3-16: **11.** 3-20: **24.** 3-28: **4.** 3-30: **29.** 3-33: **1.** 3-34: **31.** 3-35: **1.** 3-36: **31.** 3-37: **1.** 3-38: **1.** 3-39: **4.** 3-41: **22.** 3-43: **17.** 3-45: **1.**

Section 4.

4-3a: **12.** 4-3b: **29.** 4-3c: **28.** 4-3d: **28.** 4-3e: **28.** 4-3f: **34.** 4-3g: **24.** 4-6a: **14.** 4-6b: **32.** 4-6c: **14.** 4-9: **9.** 4-13: **6.** 4-20: **22.**

Section 5.

5-1a: **38a.** 5-1b: **38a.** 5-2: **38a.** 5-3: **38a.** 5-5: **38b.** 5-8: **21.** 5-9: **21.** 5-13: **15.** 5-15: **38a.** 5-23a: **27.** 5-23c: **27.** 5-24: **10.** 5-27: **1.** 5-28: **13.**

Section 7.

7-4: **25.**

Section 8.

8-6: **11.** 8-7: **11.** 8-9: **11.** 8-16: **11.** 8-18: **11.** 8-23: **12.** 8-44: **11.** 8-73: **23.** 8-101: **33.**

Section 9.

9-16: **11.** 9-18: **36.** 9-19: **36.** 9-20: **36.** 9-47: **11.**

Index*

Pages are numbered consecutively within each section

* Pages or groups of page numbers preceded by an R (*e.g.,* R1-36, 1-37) denote pages in this Handbook. Those preceded by an A (*e.g.,* A10-15, 10-16) denote pages in the *IES Lighting Handbook, 1981 Application Volume.*

* Pages or groups of page numbers preceded by an R (*e.g.*, R1–36, 1–37) denote pages in this Handbook. Those preceded by an A (*e.g.*, A10–15, 10–16) denote pages in the *IES Lighting Handbook, 1981 Application Volume.*

* Pages or groups of page numbers preceded by an R (*e.g.,* R1–36, 1–37) denote pages in this Handbook. Those preceded by an A (*e.g.,* A10–15, 10–16) denote pages in the *IES Lighting Handbook, 1981 Application Volume.*

* Pages or groups of page numbers preceded by an R (*e.g.,* R1–36, 1–37) denote pages in this Handbook. Those preceded by an A (*e.g.,* A10–15, 10–16) denote pages in the *IES Lighting Handbook, 1981 Application Volume.*

* Pages or groups of page numbers preceded by an R (*e.g.,* R1-36, 1-37) denote pages in this Handbook. Those preceded by an A (*e.g.,* A10-15, 10-16) denote pages in the *IES Lighting Handbook, 1981 Application Volume.*

* Pages or groups of page numbers preceded by an R (*e.g.,* R1-36, 1-37) denote pages in this Handbook. Those preceded by an A (*e.g.,* A10-15, 10-16) denote pages in the *IES Lighting Handbook, 1981 Application Volume.*

---

* Pages or groups of page numbers preceded by an R (*e.g.,* R1-36, 1-37) denote pages in this Handbook. Those preceded by an A (*e.g.,* A10-15, 10-16) denote pages in the *IES Lighting Handbook, 1981 Application Volume.*

* Pages or groups of page numbers preceded by an R (*e.g.,* R1–36, 1–37) denote pages in this Handbook. Those preceded by an A (*e.g.,* A10–15, 10–16) denote pages in the *IES Lighting Handbook, 1981 Application Volume.*

* Pages or groups of page numbers preceded by an R (*e.g.,* R1–36, 1–37) denote pages in this Handbook. Those preceded by an A (*e.g.,* A10–15, 10–16) denote pages in the *IES Lighting Handbook, 1981 Application Volume.*

* Pages or groups of page numbers preceded by an R (*e.g.*, R1-36, 1-37) denote pages in this Handbook. Those preceded by an A (*e.g.*, A10-15, 10-16) denote pages in the *IES Lighting Handbook, 1981 Application Volume.*

* Pages or groups of page numbers preceded by an R (*e.g.,* R1-36, 1-37) denote pages in this Handbook. Those preceded by an A (*e.g.,* A10-15, 10-16) denote pages in the *IES Lighting Handbook, 1981 Application Volume.*

* Pages or groups of page numbers preceded by an R (*e.g.,* R1–36, 1–37) denote pages in this Handbook. Those preceded by an A (*e.g.,* A10–15, 10–16) denote pages in the *IES Lighting Handbook, 1981 Application Volume.*

---

* Pages or groups of page numbers preceded by an R (*e.g.,* R1–36, 1–37) denote pages in this Handbook. Those preceded by an A (*e.g.,* A10–15, 10–16) denote pages in the *IES Lighting Handbook, 1981 Application Volume.*

* Pages or groups of page numbers preceded by an R (*e.g.,* R1-36, 1-37) denote pages in this Handbook. Those preceded by an A (*e.g.,* A10-15, 10-16) denote pages in the *IES Lighting Handbook, 1981 Application Volume.*

---

* Pages or groups of page numbers preceded by an R (*e.g.,* R1-36, 1-37) denote pages in this Handbook. Those preceded by an A (*e.g.,* A10-15, 10-16) denote pages in the *IES Lighting Handbook, 1981 Application Volume.*

* Pages or groups of page numbers preceded by an R (*e.g.,* R1–36, 1–37) denote pages in this Handbook. Those preceded by an A (*e.g.,* A10–15, 10–16) denote pages in the *IES Lighting Handbook, 1981 Application Volume.*

---

* Pages or groups of page numbers preceded by an R (*e.g.,* R1-36, 1-37) denote pages in this Handbook. Those preceded by an A (*e.g.,* A10-15, 10-16) denote pages in the *IES Lighting Handbook, 1981 Application Volume.*

* Pages or groups of page numbers preceded by an R (*e.g.,* R1-36, 1-37) denote pages in this Handbook. Those preceded by an A (*e.g.,* A10-15, 10-16) denote pages in the *IES Lighting Handbook, 1981 Application Volume.*

* Pages or groups of page numbers preceded by an R (*e.g.,* R1–36, 1–37) denote pages in this Handbook. Those preceded by an A (*e.g.,* A10–15, 10–16) denote pages in the *IES Lighting Handbook, 1981 Application Volume.*

* Pages or groups of page numbers preceded by an R (*e.g.,* R1–36, 1–37) denote pages in this Handbook. Those preceded by an A (*e.g.,* A10–15, 10–16) denote pages in the *IES Lighting Handbook, 1981 Application Volume.*

* Pages or groups of page numbers preceded by an R (*e.g.,* R1–36, 1–37) denote pages in this Handbook. Those preceded by an A (*e.g.,* A10–15, 10–16) denote pages in the *IES Lighting Handbook, 1981 Application Volume.*

* Pages or groups of page numbers preceded by an R (*e.g.,* R1–36, 1–37) denote pages in this Handbook. Those preceded by an A (*e.g.,* A10–15 10–16) denote pages in the *IES Lighting Handbook, 1981 Application Volume.*

* Pages or groups of page numbers preceded by an R (*e.g.,* R1–36, 1–37) denote pages in this Handbook. Those preceded by an A (*e.g.,* A10–15, 10–16) denote pages in the *IES Lighting Handbook, 1981 Application Volume.*

* Pages or groups of page numbers preceded by an R (*e.g.,* R1–36, 1–37) denote pages in this Handbook. Those preceded by an A (*e.g.,* A10–15, 10–16) denote pages in the *IES Lighting Handbook, 1981 Application Volume.*

Illuminating Engineering Society of North America

IES of North America is a recognized technical authority for the illumination field. For over 75 years its objective has been to communicate information about all aspects of good lighting practice to individual members, the lighting industry and consumers through a variety of programs, publications and services. The strength of IES is in its diversified membership: engineers, architects, designers, educators, students, contractors, distributors, utility personnel, manufacturers and scientists, all *contributing to* and *benefiting from* the Society.

PROGRAMS

IES local, regional and transnational meetings, and conferences, symposiums, seminars, workshops and lighting exhibitions provide an access to the latest developments in the field through audio-visual presentations and expert speakers. Basic and advanced IES lighting courses are offered by local IES Sections and in cooperation with other organizations. Other Society programs include liaison with school and colleges and career information for students and counselors.

PUBLICATIONS

Lighting Design & Application (LD&A), and the *Journal of the Illuminating Engineering Society* are the official magazines of the Society. *LD&A* is a popular application-oriented monthly magazine. Every issue contains special feature articles and news of practical and innovative lighting layouts, systems, equipment and economics, and news of the industry and its people. The *Journal,* a technical quarterly, contains official transactions: American National Standards, IES recommended practices, technical committee reports, conference papers and research reports; and other technically-oriented materials.

In addition to Handbooks, the Society also publishes the *IES Times,* a timely newsletter, and offers nearly 100 varied publications including: education courses; IES technical committee reports covering many specific lighting applications; forms and guides used for measuring and reporting lighting values, lighting calculations, performance of light sources and luminaires, energy management, etc. Also, the *IES Lighting Library* provides a complete reference package in an oversized loose-leaf binder. The Library, encompassing all essential IES documents, is available with a yearly maintenance service to keep it up-to-date.

Complete lists of current and available IES publications are published periodically in its official magazines or may be obtained by writing to the Publications Office of the Illuminating Engineering Society.

SERVICES

IES provides professional staff assistance with technical problems, reference help and interprofessional liaison with AIA, AID, IEEE, NAED, NECA, NEMA, NSID and other groups. IES is a forum for exchange, professional development and recognition. It correlates the vast amount of research, investigation and discussion through hundreds of qualified members of its technical committees to guide lighting experts and laymen on research-based lighting recommendations.

The Society has two types of membership: individual and sustaining. Applications and current dues schedules are available upon request from the Membership Department of the Illuminating Engineering Society.

RESEARCH AND EDUCATION

The Lighting Research and Education Fund (LREF) is a major fund-raising campaign initiated to further lighting research and education. To fully utilize the funds raised by LREF in meeting its research objectives, the Lighting Research Institute (LRI), a not-for-profit corporation, was established in 1982. LRI promotes and sponsors basic and applied research applicable to all aspects of lighting phenomena: the fundamental life and behavioral sciences; bases for practices of lighting design; and needs of consumers and users of lighting.

Hand-in-hand with the research program of the LRI, an equally comprehensive educational program has been set in motion by expanding the role of the IES in education. As recommended by its established Educational Advisory Council, a two-fold approach to education is being implemented—academic (teaching materials, faculty development, continuing education of teachers, scholarships and fellowships, etc.) and public (career development for the lighting community, continuing education for the opinion makers and the general public). The LREF campaign is international in its scope as it seeks the support for building an endowment from the lighting industry, its members and the users located all across North America and Europe.